Next Generation Materials for Sustainable Engineering

Shrikaant Kulkarni
Padmashri Dr. V.B. Kolte College of Engineering, India

Boris Kharisov
Autonomous University of Nuevo Leon, Mexico

A.K. Haghi
University of Guilan, UK

Vipul Srivastava
Lovely Professional University, India

IGI Global
PUBLISHER of TIMELY KNOWLEDGE

A volume in the Advances in Chemical and Materials Engineering (ACME) Book Series

Published in the United States of America by
IGI Global
Engineering Science Reference (an imprint of IGI Global)
701 E. Chocolate Avenue
Hershey PA, USA 17033
Tel: 717-533-8845
Fax: 717-533-8661
E-mail: cust@igi-global.com
Web site: http://www.igi-global.com

Library of Congress Cataloging-in-Publication Data

Names: Kulkarni, Shrikaant, editor. | Kharisov, Boris I., editor. | Haghi, A. K., editor. | Srivastava, Vipul, 1973- editor.
Title: Next generation materials for sustainable engineering / edited by: Shrikaant Kulkarni, Boris Ildusovich Kharisov, A.K. Haghi, Vipul Srivastava.
Description: Hershey PA : Engineering Science Reference, [2024] | Includes bibliographical references. | Summary: "This book focuses on various synthesis protocols for the development of various kinds of next-generation novel materials, their characterization and applications"-- Provided by publisher.
Identifiers: LCCN 2023050900 (print) | LCCN 2023050901 (ebook) | ISBN 9798369313060 (hardcover) | ISBN 9798369313077 (ebook)
Subjects: LCSH: Sustainable engineering. | Materials.
Classification: LCC TA163 .N49 2024 (print) | LCC TA163 (ebook) | DDC 620.1/1--dc23/eng/20240424
LC record available at https://lccn.loc.gov/2023050900
LC ebook record available at https://lccn.loc.gov/2023050901

This book is published in the IGI Global book series Advances in Chemical and Materials Engineering (ACME) (ISSN: 2327-5448; eISSN: 2327-5456)

British Cataloguing in Publication Data
A Cataloguing in Publication record for this book is available from the British Library.

Advances in Chemical and Materials Engineering (ACME) Book Series

J. Paulo Davim
University of Aveiro, Portugal

ISSN:2327-5448
EISSN:2327-5456

Mission

The cross disciplinary approach of chemical and materials engineering is rapidly growing as it applies to the study of educational, scientific and industrial research activities by solving complex chemical problems using computational techniques and statistical methods.

The **Advances in Chemical and Materials Engineering (ACME) Book Series** provides research on the recent advances throughout computational and statistical methods of analysis and modeling. This series brings together collaboration between chemists, engineers, statisticians, and computer scientists and offers a wealth of knowledge and useful tools to academics, practitioners, and professionals through high quality publications.

Coverage

- Multifuncional and Smart Materials
- Sustainable Materials
- Wear of Materials
- Ceramics and glasses
- Toxicity of Materials
- Materials to Renewable Energies
- Electrochemical and Corrosion
- Artificial Intelligence Methods
- Polymers Engineering
- Numerical Techniques

IGI Global is currently accepting manuscripts for publication within this series. To submit a proposal for a volume in this series, please contact our Acquisition Editors at Acquisitions@igi-global.com or visit: http://www.igi-global.com/publish/.

The Advances in Chemical and Materials Engineering (ACME) Book Series (ISSN 2327-5448) is published by IGI Global, 701 E. Chocolate Avenue, Hershey, PA 17033-1240, USA, www.igi-global.com. This series is composed of titles available for purchase individually; each title is edited to be contextually exclusive from any other title within the series. For pricing and ordering information please visit http://www.igi-global.com/book-series/advances-chemical-materials-engineering/73687. Postmaster: Send all address changes to above address.

Titles in this Series

For a list of additional titles in this series, please visit:
www.igi-global.com/book-series/advances-chemical-materials-engineering/73687

Machining Polymer Matrix Composites Tools, Techniques, and Sustainability
Francisco Mata Cabrera (Universidad de Castilla-La Mancha, Spain) and Issam Hanafi (Université Abdelmalek Essaâdi, Morocco)
Engineering Science Reference • copyright 2024 • 320pp • H/C (ISBN: 9781668499276) • US $300.00 (our price)

Building a Low-Carbon Society Through Applied Environmental Materials Science
Takaomi Kobayashi (Nagaoka University of Technology, Japan)
Engineering Science Reference • copyright 2024 • 300pp • H/C (ISBN: 9798369300039) • US $255.00 (our price)

FlexPDE and Finite Element Method Applications in Thermal Energy Storage and Cavities
Khashayar Hosseinzadeh (Sharif University of Technology, Iran) Mehdi Mahboobtosi (Babol Noshirvani University of Technology, Iran) and Davood Domairry Ganji (Babol Noshirvani University of Technology, Iran)
Engineering Science Reference • copyright 2024 • 300pp • H/C (ISBN: 9798369309322) • US $225.00 (our price)

Advancements in Powder Metallurgy Processing, Applications, and Properties
Shashanka Rajendrachari (Bartin University, Turkey) and Rayappa Shrinivas Mahale (Jain College of Engineering and Research, India)
Engineering Science Reference • copyright 2024 • 300pp • H/C (ISBN: 9781668493854) • US $275.00 (our price)

Innovative Multifunctional Nanomaterial for Photocatalysis, Sensing, and Imaging
Azad Kumar (Siddharth University, India)
Engineering Science Reference • copyright 2023 • 283pp • H/C (ISBN: 9781668487433) • US $270.00 (our price)

Innovations in Materials Chemistry, Physics, and Engineering Research
Eugene de Silva (Virginia Research Institute, USA) and Pramudi Abeydeera (Polymeric Ltd., USA)
Engineering Science Reference • copyright 2023 • 446pp • H/C (ISBN: 9781668468302) • US $270.00 (our price)

Handbook of Research on Advanced Functional Materials for Orthopedic Applications
R. Ranjith (SNS College of Technology, India) and J. Paulo Davim (University of Aveiro, Portugal)
Engineering Science Reference • copyright 2023 • 371pp • H/C (ISBN: 9781668474129) • US $325.00 (our price)

IGI Global
PUBLISHER of TIMELY KNOWLEDGE

701 East Chocolate Avenue, Hershey, PA 17033, USA
Tel: 717-533-8845 x100 • Fax: 717-533-8661
E-Mail: cust@igi-global.com • www.igi-global.com

List of Reviewers

Pedro Aguilar-Zárate, *Technological Institute of Ciudad Valles, Mexico*
Pezhman Ahmadi, *Heriot-watt University, UK*
H. Hariri Asli, *Lamar University, USA*
Devrim Balkose, *Izmir Institute of Technology, Turkey*
Chin Hua Chia, *National University of Malaysia, Malaysia*
Theresa Obiajulu Egbuchunam, *Federal University of Petroleum Resources, Nigeria*
Ana Cristina Faria Ribeiro, *University of Coimbra, Portugal*
Ricardo Gómez-García, *Universidade Católica Portuguesa, Portugal*
Ali Pourhashemi, *Christian Brothers University, USA*
Tamara Tatrishvili, *Ivane Javakhishvili Tbilisi State University (TSU), Georgia*

Table of Contents

Detailed Table of Contents

Carbon dots (CDs) represent a next-generation class of materials with immense potential for various biomedical applications. Their unique combination of optical properties, biocompatibility, and tunable surface chemistry positions them as versatile tools for bioimaging, drug delivery, biosensing, and photothermal therapy. As researchers continue to explore and refine the properties of carbon dots, these nanomaterials are poised to play a pivotal role in shaping the future of biomedical technologies. The future of carbon dots in biomedical applications appears promising with ongoing research focusing on enhancing their properties and expanding their functionalities. Further advancements in surface engineering and functionalization strategies will likely lead to the development of CDs with improved stability, specificity, and tailored functionalities for targeted applications.

This review delves into the application of bio-based phase change materials (PCMs) within geopolymer matrices, with a specific focus on their environmental advantages. Geopolymers, renowned for their eco-friendly characteristics as substitutes for conventional cement-based materials, have attracted considerable attention. The review covers topics such as geopolymer technology, bio-based PCMs, recent advancements, emerging trends, and the existing challenges in upscaling, durability, and standardization when integrating bio-based PCMs into geopolymer matrices. In conclusion, this review represents a valuable resource for individuals dedicated to promoting environmental sustainability in the realms of construction and materials. Bio-based geopolymer composites incorporating PCMs are poised to play a pivotal role in the pursuit of more environmentally friendly and energy-efficient building materials.

In order for humans to meet their fundamental needs, energy is required. The main sources of natural energy are carbon-based fuels such as biomass and fossil fuels. These fuels made of carbon are scarce and harmful to the environment. Innovative clean and environmentally friendly energy solutions are desperately needed. Particularly in light of the global issues posed by population expansion, the environment, and economic growth, solar energy may be a workable alternative for sustainable development. The Scheffler is a unique kind of solar concentrator that can quickly attain a high temperature. By adding a heat storage unit, Scheffler's capacity to retain heat may be greatly boosted. This study compares the thermal and physical properties, such as density, thermal conductivity, and melting point, of various thermal storage materials that are used to store heat energy for a longer amount of time, such as bricks, concrete, water, phase change material (PCM), latent heat storage material (LHS), aluminum, iron, molten salts, silicon, etc.

Nanomaterials have become a viable way to improve the performance of batteries and supercapacitors in the quest for effective and sustainable energy storage solutions. This chapter provides an in-depth look at current advances in synthesis processes and applications of VS2-based nanomaterials in energy storage devices. The discussion carefully assesses electrochemical characteristics of VS2-based nanomaterials while critically analyzing alternative synthesis techniques and composite approaches. Furthermore, the chapter highlights current research discoveries concerning the performance of VS2-based nanomaterials in various types of batteries. Additionally, potential uses of VS2-based nanomaterials in energy storage devices, such as lithium-ion batteries, sodium-ion batteries, potassium-ion batteries, zinc-ion batteries, magnesium-ion batteries, aluminum-ion batteries, and ammonium-ion batteries, are investigated. This review emphasizes the importance of VS2-based nanomaterials in energy storage applications and offers possible future research directions in this rapidly expanding sector.

Piezoceramics, stemming from PZT and BaTiO3 discoveries, revolutionize technology with exceptional piezoelectricity. This exploration traces their evolution, highlighting PZT and BaTiO3's foundational roles. Diverse piezoceramics like BiFeO3, BiNaTiO3, and high-d33 KNN are featured. Enhancing

The internet of things (IoT) concept has long been regarded as a crucial driver of the Fourth Industrial Revolution, having the potential to change the way we live. However, the incorporation of nanotechnology promises to increase its impact significantly more. The proliferation of intelligent settings implies application interconnection and the use of the internet. As a result, the concept of the internet of things (IoT) has emerged. The internet of nano things (IoNT) is made possible by the development of the IoT idea, a new communication networks paradigm based on nanotechnology and IoT, that is, a paradigm capable of interconnecting nano-scale devices via existing networks. The concept of the internet of nano things (IoNT) arose from the connecting of these nano devices with the internet. The internet of nano-things (IoNT) is a network of nanoconnected devices, objects, or organisms with unique IDs that wirelessly transport data to the cloud over a computer or cellular network. Data delivery, caching, and energy usage are three of the most important topics in IoNT today.

This chapter examines the importance of materials in capturing carbon dioxide (CO2) and the criteria for selecting these materials using modeling tools. The discussion starts by analyzing the adsorption process of CO2 capture as a phenomenon and selecting suitable adsorbent materials under different considerations. After that, an exploration of important modeling tools that can be used to simulate the CO2 process and to evaluate the performance of these materials is conducted. In addition, a brief overview of the relationship between theoretical mathematical models and experimental methods to validate and improve both is addressed. Additionally, it discusses the integration of mathematical programming and optimization approaches in the selection and performance of materials in the CO2 capture process and related technologies. Finally, it examines the economic and sustainability considerations when selecting materials, highlighting the need for technology to be economically feasible and environmentally sustainable.

The integration of artificial intelligence (AI) in the renewable energy sector has emerged as a disruptive force that has the potential to fundamentally alter the ways in which clean energy is produced, distributed, and used. Artificial intelligence (AI) technologies, such as machine learning, predictive analytics, optimization algorithms, and smart control systems, have enabled unprecedented improvements in renewable energy systems. These advancements include optimizing energy storage systems to increase grid dependability and stability as well as optimizing solar and wind turbine efficiency via the use of adaptive control and real-time data analysis. Artificial intelligence-driven predictive maintenance has decreased operational

Preface

In the realm of contemporary engineering, the significance of materials cannot be overstated. Our endeavor in this edited reference book, *Next Generation Materials for Sustainable Engineering*, stems from the profound understanding that materials are the bedrock of innovation, progress, and sustainability.

The multifaceted demands of modern living continually drive a revolution in material needs. The dynamic landscape of these requirements propels relentless research pursuits to address ever-evolving needs. Polymers, metals, composites, ceramics, alloys, organometallics, biomaterials, and their distinctive attributes have woven themselves into the fabric of our existence, finding application across diverse domains.

Adapting materials to align with shifting needs and preferences presents a challenge that demands an in-depth comprehension of their structures and an adeptness in manipulating these structures to elicit specific properties. Research initiatives unfurl a path toward material evolution, harmonizing their advancements with frontier areas such as energy, environment, health, agriculture, industry, and construction.

The urgency for sustainable practices looms large in the face of the heedless depletion of natural resources, exploited not merely to meet needs but to satiate human greed. Harnessing available resources judiciously and ushering in novel, intelligent materials form the crux of securing our future. Research and development initiatives poised to anticipate challenges and mitigate them through sustainable action are the bedrock upon which the sustenance of our planet and future generations rests.

This comprehensive book delves into diverse synthesis protocols for forging novel materials, their characterization, and applications. It illuminates the path toward a sustainable future by discussing the challenges in material fabrication, their alignment with sustainable development goals, and their integration into the broader scope of engineering.

Our intended audience comprises academics, researchers, material scientists, and engineers seeking enlightenment, enrichment, and a roadmap for research promotion in this domain. This book aims to bridge the gap between theoretical knowledge and practical applications, guiding the pursuit of excellence in material science and engineering. By highlighting how research findings can fulfill sustainable development goals, it charts a course for the integration of research and development efforts with practical, real-world applications.

The chapters cover an extensive array of topics, from materials for renewable energy and spintronics to nanocomposites, bio-sensing applications, energy storage, and beyond. Each chapter provides a window into cutting-edge developments, encouraging a deeper conceptual understanding and practical relevance of research initiatives.

We believe this collection will serve as a cornerstone for those aiming to grasp the intricacies of materials science, advance research, and contribute to the principles of sustainable engineering for a better and enduring future.

CHAPTER OVERVIEW

Chapter 1: The Emerging Next-Generation Materials for Biomedical Applications

Ranjit Barua and Sudipto Datta's chapter explores carbon dots' potential in biomedical applications. It highlights their optical properties, biocompatibility, and application areas like bioimaging, drug delivery, biosensing, and photothermal therapy, emphasizing ongoing research for enhanced functionalities.

Chapter 2: Unveiling the Promise of Phase Change Materials in Geopolymer Composites: A Comprehensive Review

Nidhya Rathinavel and Abdul Aleem discuss bio-based phase change materials in geopolymer matrices. The chapter covers geopolymer technology, recent advancements, challenges, and their role in fostering environmental sustainability.

Chapter 3: Materials for Phase Change Heat Storage in Scheffler Reflectors: A Sustainable and Perceptive Analysis of Renewable Energy

Sabyasachi Pramanik examines thermal storage materials for Scheffler reflectors, emphasizing the importance of sustainable energy alternatives and comparing various materials' thermal and physical properties.

Chapter 4: Advancements in Vanadium Disulphide (VS_2) Nanomaterials for Enhanced Energy Storage Synthesis Techniques, Electrochemical Properties, and Prospective

Harikrishnan S, Abhijith A R, and team delve into VS_2-based nanomaterials for energy storage devices. The chapter reviews synthesis, electrochemical properties, and potential applications in various batteries, outlining research directions in this rapidly expanding sector.

Chapter 5: Exploring Lead-Free Piezoelectric Ceramics for Enhanced Energy Harvesting Applications

Chiranjit Chaliha and Haobam Singh delve into lead-free piezoelectric ceramics, emphasizing their structural properties, phase transitions, and the quest for materials that can replace traditional piezoceramics like PZT.

Chapter 6: Halide Perovskite Materials for Photovoltaics and Lighting: Halide Perovskite Materials for Photovoltaics and Lighting

Veena Thakur and Sadhna Singh provide an in-depth exploration of halide perovskite materials for high-efficiency solar cells and light-emitting devices. Their chapter encompasses synthesis, properties, applications, and future prospects, emphasizing the promising scalability and advancements in this field.

Chapter 7: Piezoelectric Nanogenerators

Devaraj E elucidates the potential of piezoelectric nanoparticles in energy harvesting systems. Focusing on ZnO's diverse applications, the chapter discusses nanogenerator technologies and their utility in generating electric energy from mechanical stimulation.

Chapter 8: Status and Prospects of GdIG Garnet Ferrites for Energy Storage Devices: A Review

Anjori Sharma and Dipesh Dalal review GdIG garnet ferrites for energy storage devices. The chapter explores their structural characteristics, usage in energy storage systems, challenges, and advancements, providing insights for future research.

Chapter 9: Supercapacitors - Revolutionizing Energy Storage: A Sustainable Energy Solution

Hamnesh Mahajan and Anjori Sharma provide an overview of supercapacitors as efficient energy storage solutions. The chapter covers construction, working principles, applications, and quantitative modeling, highlighting their potential in sustainable energy landscapes.

Chapter 10: Nanowires for Bio-sensing Applications

Nipun Jain and co-authors discuss nanomaterials' significance in biomedical sensors. The chapter touches upon fabrication, applications in biomarker detection, RNA/virus detection, drug discovery, and highlights advancements aiming for improved stability and sensitivity.

Chapter 11: Lead-Free Alkali Niobate Piezoelectric Materials: Synthesis and Characterisations

Haobam Singh and colleagues explore lead-free piezoelectric ceramics' importance for energy harvesters. Highlighting the significance of KNN, the chapter examines its properties, phase transitions, and challenges in achieving high-density ceramics.

Chapter 12: Improving Water Quality through Mathematical Modeling Adsorption With Carbonaceous Materials

Luis Cavazos-Cuello and team explore the use of carbonaceous materials in water treatment through adsorption. The chapter discusses mathematical modeling, batch vs. continuous regime, and their importance in large-scale pollutant removal.

Chapter 13: Exploring X-Ray Techniques for Comprehensive Material Characterization and Analysis: Exploring X-Ray Techniques for Comprehensive Material Characterization and Analysis

Sangeeta Prasher and Mukesh Kumar delve into various X-ray techniques for material analysis. The chapter discusses XRD, XRF, XPS, AES, and SIMS, their principles, operations, and applications, aiding readers in material characterization and technological advancement.

Chapter 14: Analysis of Various Aspects of the Internet of Nano-Things (IoNT) Its Integration in Machine Learning and Its Diverse Applications

Preeti Singh Bahadur presents IoNT's integration with IoT and nanotechnology. Focusing on communication networks, the chapter explores IoNT's potential applications, data transmission, caching, and energy management.

Chapter 15: Modeling Tools for Evaluating Materials in CO_2 Capture: Performance and Sustainability

Ilse Hernández-Romero, Michael Pérez-Rodríguez, and team shed light on selecting materials for CO_2 capture using modeling tools. They analyze adsorption processes, modeling techniques, experimental validation, and the crucial economic and sustainability aspects in material selection for CO_2 capture.

Chapter 16: AI's Function in Sustainable Development's Renewable Energy Planning

Sabyasachi Pramanik's chapter explores the disruptive integration of artificial intelligence (AI) in renewable energy. It delineates AI's impact on optimizing energy storage systems, improving solar and wind turbine efficiency, predictive maintenance, and encouraging energy-saving habits among consumers.

Chapter 17: Recent Trends in Photonic Crystal

Kannadhasan Suriyan and Nagarajan R offer insights into photonic crystals, outlining their potential in various fields like acoustics, bio-photonics, quantum engineering, and optoelectronics, emphasizing their role in integrated optics.

These chapters encompass a wide range of materials, technologies, and applications, aiming to enlighten researchers, engineers, and enthusiasts about the ongoing advancements and potential directions in sustainable engineering with a focus on next-generation materials.

CONCLUSION

In the vast landscape of modern engineering, the centrality of materials stands undeniably profound. *Next Generation Materials for Sustainable Engineering*, the edited reference book compiled by our esteemed editors and authors, aims to underscore this fundamental facet of innovation, progression, and sustainability.

The relentless evolution of modern living standards continues to fuel the demand for materials that can adapt and cater to ever-changing needs. Materials ranging from polymers, metals, composites, and ceramics, to biomaterials have woven themselves into the very fabric of our existence, finding applications across diverse domains.

The pivotal challenge lies in manipulating the structures of these materials to evoke specific properties, aligning them with shifting preferences and needs. Through tireless research initiatives, we chart a trajectory toward material evolution that resonates with advancements in energy, environment, health, agriculture, industry, and construction.

The pressing need for sustainable practices looms large amidst the reckless depletion of natural resources driven not solely by necessity but by human greed. Our future rests on harnessing available resources judiciously and ushering in intelligent materials that contribute to a sustainable world. Research and development initiatives aimed at anticipating and mitigating challenges through sustainable actions form the cornerstone of preserving our planet for future generations.

This comprehensive book journeys through diverse synthesis protocols for forging novel materials, exploring their characterization, and delineating their applications. It illuminates the path toward a sustainable future by discussing challenges in material fabrication, their alignment with sustainable development goals, and their integration into the broader scope of engineering.

Our intended audience, spanning academics, researchers, material scientists, and engineers, seeks enlightenment, enrichment, and a roadmap for promoting research in this domain. Bridging the chasm between theoretical knowledge and practical applications, this book serves as a guiding beacon, steering the pursuit of excellence in material science and engineering. Showcasing how research findings can fulfill sustainable development goals, it paves the way for the integration of research and development efforts into practical, real-world applications.

The chapters encompass an expansive array of topics, from materials for renewable energy and biosensing applications to advanced energy storage solutions and material characterization techniques. Each chapter opens a window into cutting-edge developments, fostering a deeper conceptual understanding and demonstrating the practical relevance of ongoing research initiatives.

We believe this collection will serve as a cornerstone for those delving into the intricacies of materials science, fostering research advancement, and contributing to the principles of sustainable engineering for a better and enduring future.

Shrikaant Kulkarni
Padmashri Dr. V.B. Kolte College of Engineering, India

Boris Kharisov
Autonomous University of Nuevo Leon, Mexico

A. K. Haghi
University of Guilan, UK

Vipul Srivastava
Lovely Professional University, India

Chapter 1
The Emerging Next-Generation Materials for Biomedical Applications

Ranjit Barua
https://orcid.org/0000-0003-2236-3876
Indian Institute of Engineering Science and Technology, Shibpur, India

Sudipto Datta
https://orcid.org/0000-0003-2161-6878
Indian Institute of Science, Bangalore, India

ABSTRACT

Carbon dots (CDs) represent a next-generation class of materials with immense potential for various biomedical applications. Their unique combination of optical properties, biocompatibility, and tunable surface chemistry positions them as versatile tools for bioimaging, drug delivery, biosensing, and photothermal therapy. As researchers continue to explore and refine the properties of carbon dots, these nanomaterials are poised to play a pivotal role in shaping the future of biomedical technologies. The future of carbon dots in biomedical applications appears promising with ongoing research focusing on enhancing their properties and expanding their functionalities. Further advancements in surface engineering and functionalization strategies will likely lead to the development of CDs with improved stability, specificity, and tailored functionalities for targeted applications.

1. INTRODUCTION

Carbon dots (CDs) have emerged as a promising class of nanomaterials with unique properties, making them highly attractive for various biomedical applications. These nanoscale carbon-based structures exhibit exceptional optical, electronic, and biocompatible properties, rendering them suitable for applications ranging from bioimaging and drug delivery to biosensing and photothermal therapy. Since the groundbreaking discovery of nanoscience and nanotechnology, it has emerged as a new frontier in the

DOI: 10.4018/979-8-3693-1306-0.ch001

global scientific community. Nanomaterials' unique optoelectronic and physicochemical properties, as opposed to their bulk structures, have been the driving force behind this development which is reported by Mondal et al., 2014. Composite materials consisting of Carbon Dots (CDs) and polymers have garnered significant attention for applications in biochemistry, biology, and biomedicine due to their easy preparation, cost-effective processing, and biocompatibility reported by Ganguly et al., 2019. CDs can be easily synthesized from a range of natural and synthetic sources. The production of CD/polymer composites involves the straightforward mixing of polymers and CDs, with or without additional treatments (Waidi et al., 2023). While blending polymers and nanomaterials is not a new synthesis process, the ongoing research focuses on developing improved CD/polymer composite materials that demonstrate favorable biocompatibility and optoelectrical properties. Various polymer architectures, including homo- and co-polymers, hyperbranched polymers, and diverse polymeric chains, have been utilized to anchor and coat CDs, resulting in the fabrication of biocompatible CD/polymer hybrid composite materials. Recent studies indicate that several methods, such as ligand exchange between polymers and CDs, grafting polymers to CDs, grafting polymers from CDs, capping polymers onto CDs, and growing CDs within the polymer template, are commonly employed for the production of biocompatible CD/polymer composite materials (Foubert et al., 2016) (Zhou et al., 2017). The straightforward preparation and outstanding properties of CDs, along with the cross-linked polymeric attributes, have sparked interest in potential applications, such as drug delivery systems, photodynamic therapy, bioimaging, etc., within the field of biomedical sciences reported by Sharker et al., 2021. Carbon dots, characterized by small carbon nanoparticles with a size <10 nm and surface passivation, have become particularly noteworthy (Datta et al., 2023). Their composition, derived from carbon—an abundant and non-toxic element, a fundamental building block of the human body—renders them attractive for various applications. This is especially true in fields where concerns about toxicity are paramount, such as biological imaging, drug/gene delivery, solar cell fabrication, and photonic device fabrication. Unlike other nanomaterials and quantum dots, carbon dots exhibit biocompatibility, eliminating the need to worry about environmental effects, toxicity, or production costs. The unique structural and electronic properties arising from the carbon composition set carbon dots apart from other nanomaterials.

Recent years have witnessed extensive research on the synthesis, properties, and applications of carbon dots by scholars like Lee et al., 2012, Baker et al., 2010, Hong et al., 2015, Lim et al., 2015, and Wang et al., 2014, owing to their intriguing properties. In comparison to semiconductor quantum dots and organic dyes, carbon dots surpass in aqueous solubility, chemical inertness, biocompatibility, easy modification, and resistance to photobleaching and photoblinking. The outstanding electronic properties of carbon dots enable them to function as both electron donors and acceptors, resulting in photoluminescence, chemiluminescence, and electrochemical luminescence. This versatility positions carbon dots at the forefront of optoelectronics, catalysis, and biomedical applications. Their low toxicity and excellent biocompatibility make them particularly promising for biomedical applications such as drug/gene delivery, bioimaging, and photothermal and photodynamic therapy. This review comprehensively outlines the recent progress in the biomedical applications of carbon dots, covering synthetic methods, properties, biocompatibility, various biomedical applications, and future prospects.

Figure 1. (a) Schematic diagram of the synthesis of CDs from starting materials (first) and fabrication of CD/polymer composite materials (second). (b) CD/polymer composite materials are being proposed for biomedical applications (Sharker et al., 2021).

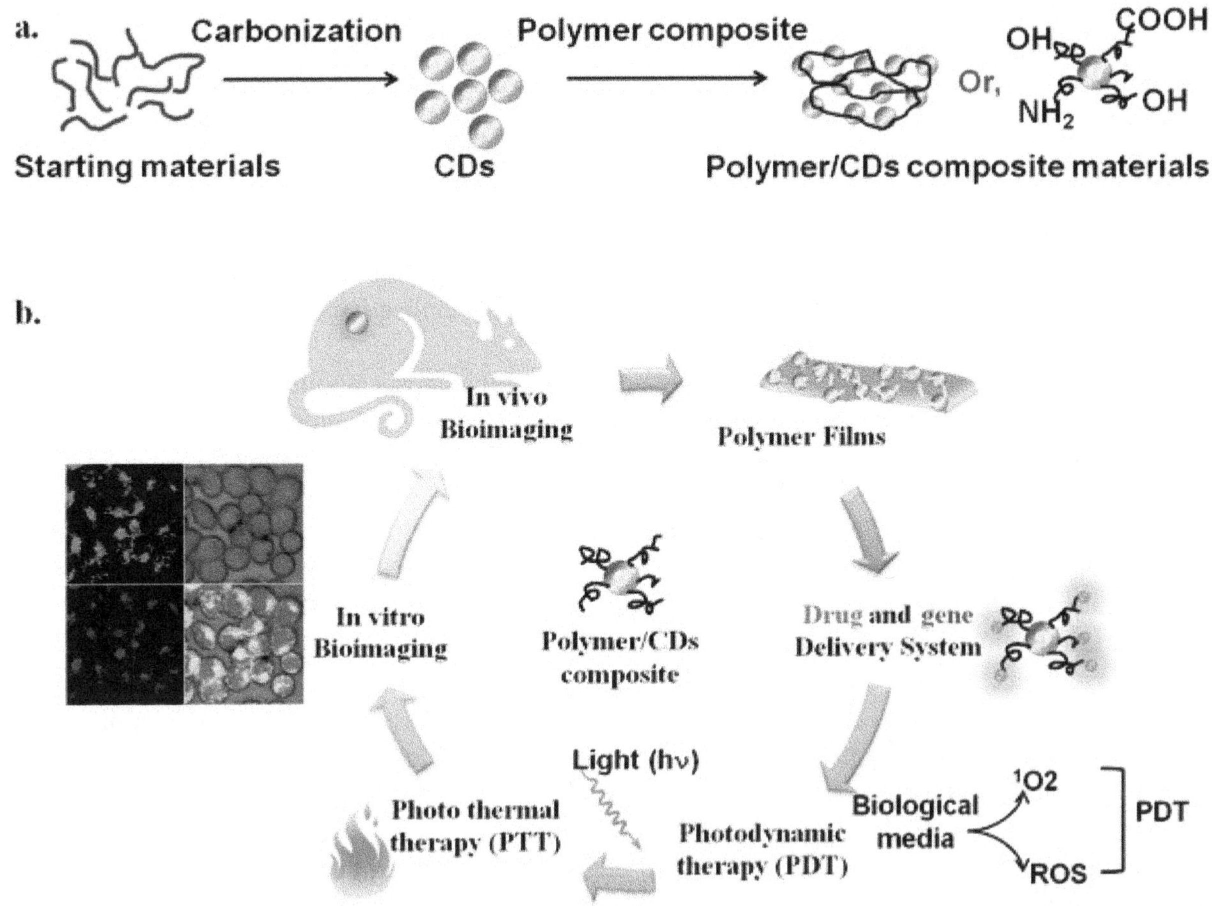

2. APPLICATION OF CARBON DOTS (CDS) IN BURGEONING FIELD

The burgeoning field of biomedical applications is witnessing a surge in the utilization of Carbon Dots (CDs) and their composite materials. These nanomaterials are finding increasing relevance and demonstrating promising potential in various biomedical applications. Here, we explore the emerging applications of CDs and CD-based composite materials in the realm of biomedicine.

a. **Drug Delivery Systems:** CDs, with their unique properties, are being actively explored for drug delivery applications. The facile surface modification of CDs allows for the efficient loading and controlled release of therapeutic agents. Composite materials, combining CDs with polymers or other drug carriers, enhance the overall drug delivery system, providing improved biocompatibility and targeted delivery.
b. **Photodynamic Therapy (PDT):** The photoluminescent properties of CDs make them excellent candidates for photodynamic therapy. CDs can efficiently generate reactive oxygen species upon

light irradiation, contributing to the destruction of targeted cells. Composite materials incorporating CDs enhance the effectiveness of PDT, offering a versatile approach for cancer treatment and other therapeutic interventions.

c. **Bioimaging:** CDs exhibit remarkable optical properties, including strong fluorescence and excellent photostability. These characteristics make them ideal for bioimaging applications. Composite materials, particularly those involving CDs and polymers, are designed to optimize imaging contrast, enabling high-resolution imaging of biological structures. This has implications for diagnostics and real-time monitoring of physiological processes.
d. **Biosensing:** The superior optoelectronic properties of CDs are leveraged in the development of biosensors. CDs and their composites serve as sensitive probes for detecting various biomolecules and pathogens. The integration of CDs with different sensing platforms enhances the specificity and sensitivity of biosensing applications, with potential implications for disease diagnosis and monitoring.
e. **Theranostics:** CDs are increasingly being explored for theranostic applications, where diagnosis and therapy are integrated into a single platform. The combination of diagnostic imaging capabilities and therapeutic functionalities in CD-based composite materials holds promise for personalized medicine and targeted treatment strategies.
f. **Wound Healing and Tissue Engineering:** The biocompatibility of CDs and their composite materials makes them suitable for applications in wound healing and tissue engineering. CDs can be incorporated into scaffolds or wound dressings to facilitate cell growth and tissue regeneration, contributing to enhanced healing processes.
g. **Neuroscience Applications:** CDs have shown potential in neuroscience research, including imaging and drug delivery to the central nervous system. Composite materials tailored for neuroapplications aim to address specific challenges related to crossing the blood-brain barrier and targeted drug delivery to neural tissues.

In summary, the emerging applications of Carbon Dots and their composite materials in biomedicine encompass a wide range of areas, from diagnostics and imaging to therapeutic interventions and regenerative medicine. The unique properties of CDs, combined with the synergistic effects achieved through composite materials, are paving the way for innovative solutions in the field of biomedical research and healthcare.

3. BIOCOMPATIBILITY OF CARBON DOTS (CDS)

Ensuring biocompatibility and assessing toxicity are pivotal considerations for materials intended for use in biological applications (Barua et al., 2020). While numerous studies have delved into the in vitro cytotoxicity of Carbon Dots (CDs) conducted by various research groups of Kong et al., 2014, and Shang et al., 2014, the scope of in vivo investigations remains limited. To further advance the applications of CDs, there is a critical need for extensive exploration of in vivo biocompatibility reported by Kumar et al., 2013, Wang et al., 2011, and, Huang et al., 2011. For instance, Datta et al., 2023 conducted groundbreaking research, investigating the toxicity of CDs both in vitro and in vivo, with their surfaces passivated by oligomeric PEG molecules. Incubating CDs with the human colorectal adenocarcinoma HT-29 cell line and human breast cancer MCF7 cell line, they examined the impact on cell proliferation, mortality, and

Figure 2. Carbon dots (CDs) have been used on agricultural systems to prevent water loss, increase agricultural produce storage, and enhance photosynthesis (Adam et al., 2022)

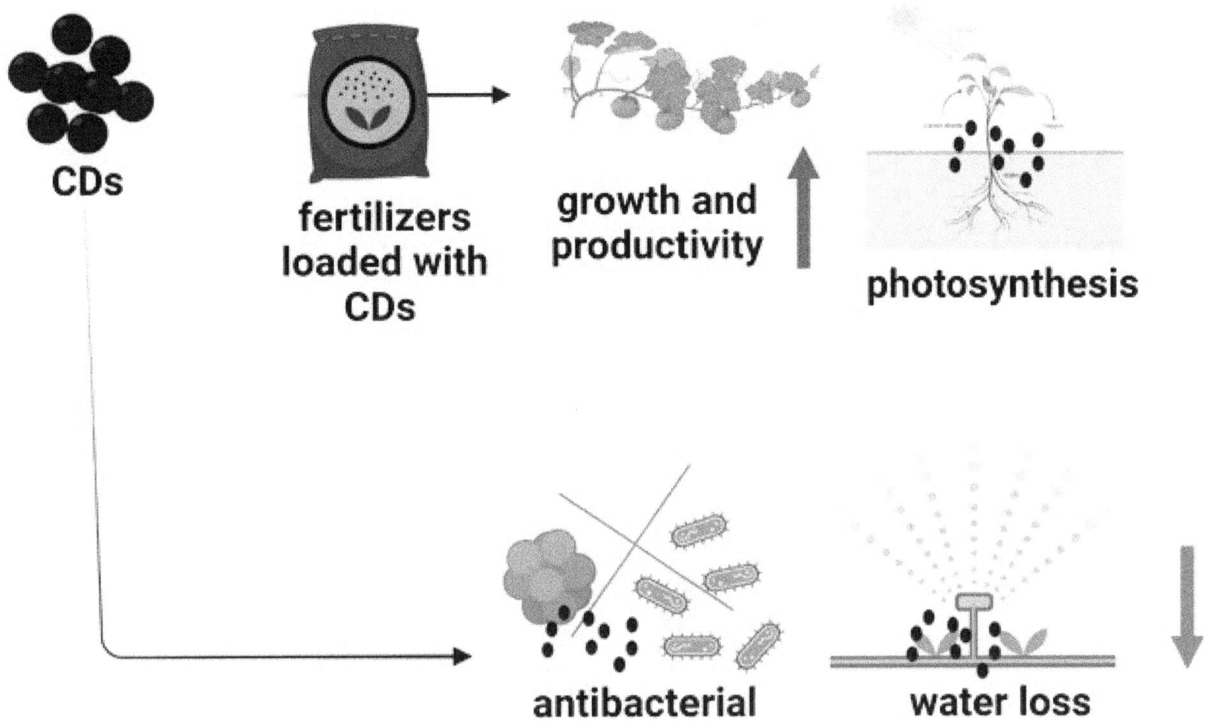

viability. Results indicated that CDs exhibited no significant toxic effects on cells. Furthermore, they conducted in vivo toxicity studies in mice following intravenous injection of varying CD dosages. Even after prolonged exposure, no notable toxicity was observed at concentrations necessary for bioimaging as reported by Kumar et al., 2013.

Additionally, researchers assessed the cytotoxicity of CDs functionalized with PPEI-EI (poly-(propionylethylenimine-co-ethylenimine)) as reported by Cao et al., 2007, BPEI (branched poly(ethylenimine)) by Dong et al., 2012, PEI (polyethylenimine) and PAA (poly(acrylic acid)) by Wang et al., 2018. Their findings revealed that PPEI-EI-functionalized CDs demonstrated non-toxicity when compared to an equivalent amount of bare PPEI-EI. Conversely, PEI-functionalized CDs exhibited higher cytotoxicity on HT29 cells than CDs functionalized with PPEI-EI, potentially attributed to the higher concentration of ethylenimine units within PEI molecules compared to PPEI-EI. Free PAA nonaqueous solution was found to be toxic to cells even at minute concentrations. When CDs were functionalized with PAA, the cytotoxicity mirrored that of free PAA reported by Wang et al., 2018. Consequently, it can be concluded that CDs functionalized with molecules exhibiting little to no cytotoxicity are well-suited for biological applications, even at higher concentrations.

Figure 3. Carbon dots (CDs) have a theranostic effect on coronavirus infection (Adam et al., 2022)

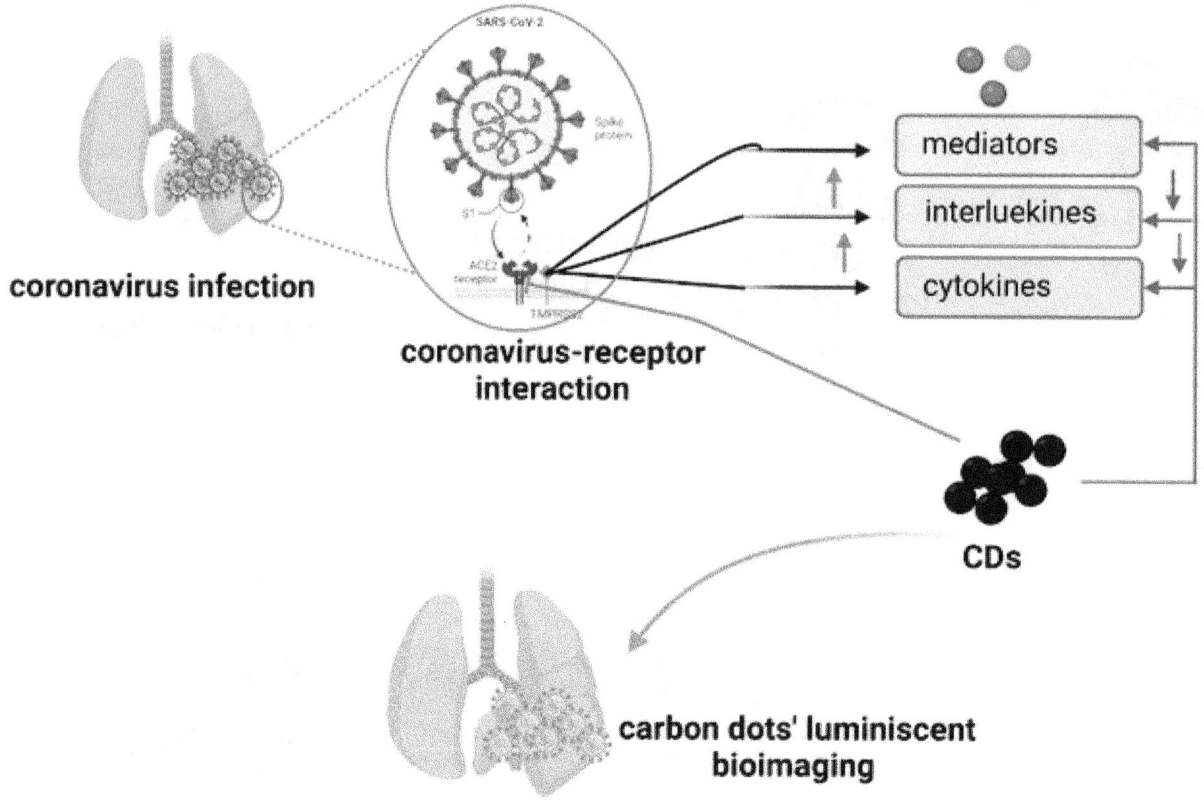

4. PERSPECTIVES OF CARBON DOTS (CDS)

The unprecedented advancements in nanoscience and nanotechnology have paved the way for innovative materials with remarkable properties, and among them, Carbon Dots (CDs) have emerged as a versatile and promising player in the biomedical field (Datta et al., 2023). As we delve into the conclusions and perspectives of CDs in upcoming biomedical applications, it becomes evident that their unique characteristics offer a multitude of opportunities for transformative developments in diagnostics, therapeutics, and beyond.

a. **Multimodal Imaging and Theranostics:** The integration of CDs into multimodal imaging platforms holds tremendous potential. By combining the unique optical properties of CDs with other imaging modalities, such as magnetic resonance imaging (MRI) or computed tomography (CT), researchers can develop comprehensive imaging approaches. Additionally, the theranostic potential of CDs, where diagnosis and therapy are seamlessly integrated, could usher in a new era of personalized and precise medicine.
b. **Targeted Drug Delivery and Precision Medicine:** CDs' ability to functionalize and target specific tissues opens avenues for targeted drug delivery. The development of CDs tailored for specific cell types or tissues could enhance the precision of drug delivery, minimizing off-target effects and

maximizing therapeutic efficacy. This aligns with the broader goals of precision medicine, where treatments are tailored to individual patients based on their unique characteristics.

c. **Biosensing and Point-of-Care Diagnostics:** Exploiting the excellent sensing capabilities of CDs, there is potential for the development of rapid and sensitive biosensors. CDs could be employed for point-of-care diagnostics, enabling quick and reliable detection of biomarkers associated with various diseases. This could revolutionize disease diagnosis and monitoring, particularly in resource-limited settings.

d. **Regenerative Medicine and Tissue Engineering:** The biocompatibility and unique properties of CDs position them as valuable components in regenerative medicine and tissue engineering. CDs could be integrated into scaffolds or biomaterials to enhance cellular growth and tissue regeneration. This opens avenues for the development of advanced materials for repairing and replacing damaged tissues.

e. **Environmental Sensing and Monitoring:** Beyond biomedical applications, CDs could find utility in environmental sensing and monitoring. Their sensitivity to various environmental factors makes them potential candidates for detecting pollutants or monitoring changes in environmental conditions. This interdisciplinary application highlights the versatility of CDs beyond the biomedical realm.

5. CONCLUSION

The journey of Carbon Dots in the realm of biomedicine is marked by significant strides and immense potential. From their origins in nanoscience laboratories to the forefront of biomedical research, CDs have demonstrated their worth as multifaceted tools. The convergence of optical, chemical, and biological properties in CDs propels them into a pivotal role in the future of diagnostics, therapeutics, and beyond. As researchers continue to unravel the intricacies of CDs and refine their applications, the horizon of possibilities expands, promising a future where these tiny dots leave a colossal impact on the landscape of biomedical innovation. A critical factor for any biomedical application is the material's biocompatibility. CDs have exhibited notable biocompatibility in numerous in vitro studies, making them conducive for interaction with biological systems. The absence of significant toxic effects observed in these studies, even at higher concentrations required for various applications, positions CDs as a safe and viable option for biomedical use. While in vitro studies have provided valuable insights into the cytotoxicity of CDs, the need for comprehensive in vivo evaluations is paramount. CDs offer a versatile platform for functionalization, allowing researchers to tailor their properties for specific applications. Functionalization with polymers, such as PPEI-EI and PAA, has demonstrated varied effects on cytotoxicity, emphasizing the importance of strategic molecular design. This versatility enables the customization of CDs to meet the specific requirements of diverse biomedical applications. The exceptional optical properties of CDs, including strong fluorescence and photostability, position them as valuable tools for bioimaging. Their potential in this regard has been demonstrated in various studies, showcasing the feasibility of CDs for high-resolution imaging of biological structures. This capability holds immense promise for advancing diagnostic techniques. Beyond imaging, CDs show great promise in therapeutic applications. Their use in drug delivery systems, photodynamic therapy, and theranostics underscores their potential to revolutionize treatment modalities. CDs can serve as efficient carriers for therapeutic agents, allowing for targeted and controlled drug delivery, while their photodynamic properties make them effective agents for light-mediated therapies.

CONFLICTS OF INTEREST

The authors declare that there are no conflicts of interest regarding the publication of this chapter.

ETHICAL APPROVAL

Not required.

ACKNOWLEDGMENT

The authors would like to thank IIEST-Shibpur, Centre for Healthcare Science and Technology lab, and thanks to Mrs. Nibedita Bardhan for language proof reading.

REFERENCES

Adam, G. O., Sharker, S. M., & Ryu, G. (2022). Emerging biomedical applications of carbon dot and polymer composite materials. *Applied Sciences (Basel, Switzerland)*, *12*(20), 10565. doi:10.3390/app122010565

Baker, S. N., & Baker, G. A. (2010). Luminescent carbon nanodots: Emergent nanolights. *Angewandte Chemie International Edition*, *49*(38), 6726–6744. doi:10.1002/anie.200906623 PMID:20687055

Barua, R., Datta, S., Datta, P., & Chowdhury, A. R. (2020, September). Experimental analysis the tissue deformation of needle insertion process in tissue engineering. In *2020 IEEE 1st International conference for convergence in engineering (ICCE)* (pp. 83-85). IEEE. 10.1109/ICCE50343.2020.9290598

Cao, L., Wang, X., Meziani, M. J., Lu, F., Wang, H., Luo, P. G., Lin, Y., Harruff, B. A., Veca, L. M., Murray, D., Xie, S.-Y., & Sun, Y. P. (2007). Carbon dots for multiphoton bioimaging. *Journal of the American Chemical Society*, *129*(37), 11318–11319. doi:10.1021/ja073527l PMID:17722926

Datta, S., & Barua, R. (2023). Advanced Materials for Surgical Tools and Biomedical Implants. *Advanced Materials and Manufacturing Techniques for Biomedical Applications*, 25-35.

Datta, S., Barua, R., & Prasad, A. (2023). Additive Manufacturing for the Development of Artificial Organs. *Advanced Materials and Manufacturing Techniques for Biomedical Applications*, 411-427.

Dong, Y., Wang, R., Li, G., Chen, C., Chi, Y., & Chen, G. (2012). Polyamine-functionalized carbon quantum dots as fluorescent probes for selective and sensitive detection of copper ions. *Analytical Chemistry*, *84*(14), 6220–6224. doi:10.1021/ac3012126 PMID:22686413

Foubert, A., Beloglazova, N. V., Rajkovic, A., Sas, B., Madder, A., Goryacheva, I. Y., & De Saeger, S. (2016). Bioconjugation of quantum dots: Review & impact on future application. *Trends in Analytical Chemistry*, *83*, 31–48. doi:10.1016/j.trac.2016.07.008

Ganguly, S., Das, P., Banerjee, S., & Das, N. C. (2019). Advancement in science and technology of carbon dot-polymer hybrid composites: A review. *Functional Composites and Structures*, *1*(2), 022001. doi:10.1088/2631-6331/ab0c80

Hong, G., Diao, S., Antaris, A. L., & Dai, H. (2015). Carbon nanomaterials for biological imaging and nanomedicinal therapy. *Chemical Reviews*, *115*(19), 10816–10906. doi:10.1021/acs.chemrev.5b00008 PMID:25997028

Huang, X., Zhang, F., Zhu, L., Choi, K. Y., Guo, N., Guo, J., Tackett, K., Anilkumar, P., Liu, G., Quan, Q., Choi, H. S., Niu, G., Sun, Y.-P., Lee, S., & Chen, X. (2013). Effect of injection routes on the biodistribution, clearance, and tumor uptake of carbon dots. *ACS Nano*, *7*(7), 5684–5693. doi:10.1021/nn401911k PMID:23731122

Kong, W., Liu, J., Liu, R., Li, H., Liu, Y., Huang, H., Li, K., Liu, J., Lee, S.-T., & Kang, Z. (2014). Quantitative and real-time effects of carbon quantum dots on single living HeLa cell membrane permeability. *Nanoscale*, *6*(10), 5116–5120. doi:10.1039/c3nr06590a PMID:24722927

Kumar, V., Toffoli, G., & Rizzolio, F. (2013). Fluorescent carbon nanoparticles in medicine for cancer therapy. *ACS Medicinal Chemistry Letters*, *4*(11), 1012–1013. doi:10.1021/ml400394a PMID:24936239

Li, H., Kang, Z., Liu, Y., & Lee, S. T. (2012). Carbon nanodots: Synthesis, properties and applications. *Journal of Materials Chemistry*, *22*(46), 24230–24253. doi:10.1039/c2jm34690g

Lim, S. Y., Shen, W., & Gao, Z. (2015). Carbon quantum dots and their applications. *Chemical Society Reviews*, *44*(1), 362–381. doi:10.1039/C4CS00269E PMID:25316556

Mondal, P., Ghosal, K., Bhattacharyya, S. K., Das, M., Bera, A., Ganguly, D., Kumar, P., Dwivedi, J., Gupta, R. K., Martí, A. A., Gupta, B. K., & Maiti, S. (2014). Formation of a gold–carbon dot nanocomposite with superior catalytic ability for the reduction of aromatic nitro groups in water. *RSC Advances*, *4*(49), 25863–25866. doi:10.1039/C4RA02837F

Shang, W., Zhang, X., Zhang, M., Fan, Z., Sun, Y., Han, M., & Fan, L. (2014). The uptake mechanism and biocompatibility of graphene quantum dots with human neural stem cells. *Nanoscale*, *6*(11), 5799–5806. doi:10.1039/C3NR06433F PMID:24740121

Sharker, S. M., & Do, M. (2021). Nanoscale carbon-polymer dots for theranostics and biomedical exploration. *Journal of Nanotheranostics*, *2*(3), 118–130. doi:10.3390/jnt2030008

Waidi, Y. O., Barua, R., & Datta, S. (2023). Metals, Polymers, Ceramics, Composites Biomaterials Used in Additive Manufacturing for Biomedical Applications. In A. Kumar, P. Kumar, A. Srivastava, & V. Goyat (Eds.), *Modeling, Characterization, and Processing of Smart Materials* (pp. 165–184). IGI Global. doi:10.4018/978-1-6684-9224-6.ch008

Wang, Y., Anilkumar, P., Cao, L., Liu, J. H., Luo, P. G., Tackett, K. N. II, Sahu, S., Wang, P., Wang, X., & Sun, Y. P. (2011). Carbon dots of different composition and surface functionalization: Cytotoxicity issues relevant to fluorescence cell imaging. *Experimental Biology and Medicine*, *236*(11), 1231–1238. doi:10.1258/ebm.2011.011132 PMID:22036734

Wang, Y., Bao, L., Liu, Z., & Pang, D. W. (2011). Aptamer biosensor based on fluorescence resonance energy transfer from upconverting phosphors to carbon nanoparticles for thrombin detection in human plasma. *Analytical Chemistry*, *83*(21), 8130–8137. doi:10.1021/ac201631b PMID:21923110

Wang, Y., & Hu, A. (2014). Carbon quantum dots: Synthesis, properties and applications. *Journal of Materials Chemistry. C, Materials for Optical and Electronic Devices*, *2*(34), 6921–6939. doi:10.1039/C4TC00988F

Zhou, Y., Sharma, S. K., Peng, Z., & Leblanc, R. M. (2017). Polymers in carbon dots: A review. *Polymers*, *9*(2), 67. doi:10.3390/polym9020067 PMID:30970747

KEY TERMS AND DEFINITIONS

Biocompatibility: Biocompatibility is a critical factor in the development of materials for biomedical applications. Carbon dots, often derived from carbon-rich precursors like citric acid or glucose, exhibit inherent biocompatibility. Studies have shown that CDs are generally non-toxic and do not induce significant cytotoxicity, making them suitable candidates for in vivo applications. This biocompatibility is crucial for their use in drug delivery systems and other biomedical platforms, where the safety of the material is paramount.

Biomaterial: Biomaterials are substances engineered for interaction with biological systems, commonly used in medical applications. Designed to be biocompatible, they facilitate therapeutic or diagnostic functions, ranging from artificial implants to drug delivery systems. Their diverse compositions cater to specific medical needs, enhancing patient outcomes and medical advancements.

Carbon Dots: Carbon dots (CDs) are a class of nanomaterials that have gained significant attention in recent years due to their unique properties and versatile applications. These nanoscale carbon-based structures typically have sizes ranging from a few to tens of nanometers and are composed of carbon, hydrogen, and oxygen atoms. The carbon framework can be amorphous or crystalline, and the surface of CDs is rich in functional groups, such as hydroxyl, carboxyl, and amino groups.

Drug Delivery: Drug delivery systems are designed mechanisms for transporting pharmaceutical compounds into the body to achieve therapeutic effects. These systems enhance drug efficacy, improve bioavailability, and control release rates. From nanoparticles to implants, they play a crucial role in optimizing treatment outcomes while minimizing side effects in various medical conditions.

Optical Properties: One of the key features of carbon dots is their exceptional optical properties. CDs possess strong fluorescence, allowing for efficient bioimaging. The tunable fluorescence of CDs makes them versatile imaging agents, with applications in both in vitro and in vivo imaging. Moreover, their excellent photostability ensures prolonged imaging sessions without signal degradation. This property is particularly advantageous for tracking biological processes over extended periods, providing valuable insights into cellular behavior and interactions.

Chapter 2
Unveiling the Promise of Phase Change Materials in Geopolymer Composites:
A Comprehensive Review

Nidhya Rathinavel
https://orcid.org/0000-0002-9962-5478
PSG Institute of Technology and Applied Research, India

Abdul Aleem
https://orcid.org/0000-0002-6334-5773
PSG Institute of Technology and Applied Research, India

Arun Murugesan
https://orcid.org/0000-0002-2662-8483
PSG Institute of Technology and Applied Research, India

ABSTRACT

This review delves into the application of bio-based phase change materials (PCMs) within geopolymer matrices, with a specific focus on their environmental advantages. Geopolymers, renowned for their eco-friendly characteristics as substitutes for conventional cement-based materials, have attracted considerable attention. The review covers topics such as geopolymer technology, bio-based PCMs, recent advancements, emerging trends, and the existing challenges in upscaling, durability, and standardization when integrating bio-based PCMs into geopolymer matrices. In conclusion, this review represents a valuable resource for individuals dedicated to promoting environmental sustainability in the realms of construction and materials. Bio-based geopolymer composites incorporating PCMs are poised to play a pivotal role in the pursuit of more environmentally friendly and energy-efficient building materials.

DOI: 10.4018/979-8-3693-1306-0.ch002

Figure 1. Bio PCM

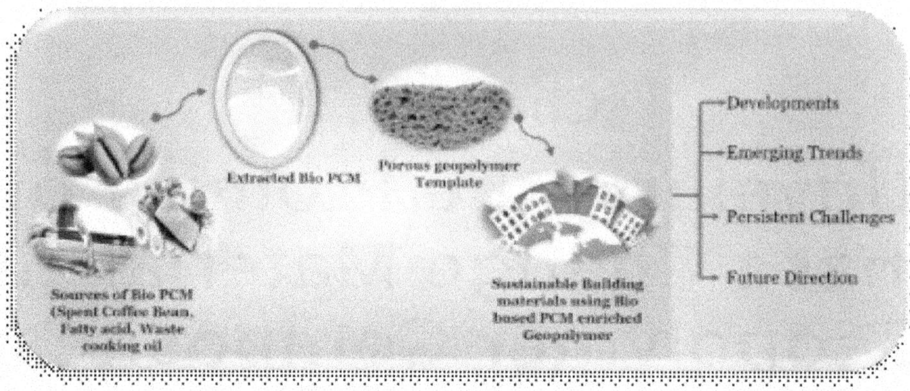

1. INTRODUCTION

The continuous release of carbon dioxide (CO_2) and the high energy demands associated with the production of Ordinary Portland Cement (OPC) are contributing to ongoing ozone layer depletion and global warming. However, the adoption of geopolymer concrete (GPC) technology in the construction sector is promoting sustainable development and a cleaner environment by reducing environmental pollution (Alhassan et al., 2023; Kanagaraj et al., 2023). Geopolymers are advanced materials known for their eco-friendly properties and versatile applications. They are inorganic, amorphous, three-dimensional aluminosilicate networks formed through a process called geopolymerization. This process involves dissolving silicon (Si) and aluminium (Al) atoms from source materials and reorganizing precursor elements, resulting in a newly synthesized solid structure (Alhassan et al., 2023; Imtiaz et al., 2020; Kanagaraj et al., 2023; Mehrizi et al., 2023). Common aluminosilicate source materials for geopolymers include clays and clay minerals, such as kaolin, kaolinite, metakaolinite, calcined clay, and zeolite. In the context of sustainability as a substitute for Ordinary Portland Cement (OPC) composites, Geopolymer Concrete (GPC) has demonstrated remarkable strength and durability (Amran et al., 2021; L. Assi et al., 2018; L. N. Assi et al., 2020; Festus et al., 2022; G. Xu & Shi, 2018; Yu et al., 2014).

Energy and its economic implications are crucial for a nation's economic well-being. Achieving self-sufficiency in energy supply and demand can undoubtedly boost a country's economy. Among the sectors with significant energy consumption, buildings rank third, following the transport and industrial sectors (Alva et al., 2018). Buildings allocate a considerable portion of their energy use to HVAC (Heating, Ventilation, and Air Conditioning) systems, primarily for ensuring occupant and equipment comfort. The utilization of energy storage systems provides an effective way to bridge the gap between energy supply and demand (Abdallah et al., 2016; Ahmed, 2014).

One such energy storage system is Latent Heat Thermal Energy Storage (LHTES) employing Phase Change Materials (PCMs). These systems are widely used in various domains, including solar thermal power plants, nuclear power facilities, and building thermal management, thanks to their isothermal properties, substantial thermal energy storage (TES) capacity, and phase change occurring within a narrow temperature range. Unlike power generation facilities, buildings using Phase Change Materials (PCMs) can store surplus solar energy during the day, reducing heat ingress into the building, and releasing it at night when temperatures drop below the PCM's melting point, making PCMs promising for thermal management in buildings (Arivazhagan et al., 2020; Sharshir et al., 2023).

Phase Change Materials (PCMs) are substances that can absorb and release substantial amounts of energy during phase transitions, depending on their specific temperature range and threshold values. To fully harness PCM capabilities, certain criteria related to thermal, physical, chemical, and kinetic attributes must be met, alongside economic feasibility and material availability. Essential chemical requirements for PCMs include non-toxicity, non-flammability, and low risk of explosion (Navarro et al., 2019; A. Sharma et al., 2009).

Common materials used in Latent Heat Thermal Energy Storage (LHTES) include paraffin waxes, salt hydrates, and eutectic compositions, which can be organic-organic, inorganic-inorganic, or organic-inorganic. Fatty acids are also frequently employed. The challenge now is to identify the optimal bio-based Phase Change Material (PCM) or bio-composite PCM that is cost-effective, practical, and environmentally friendly, particularly for building applications (Saxena et al., 2019).

Recent research has shown increasing interest in using bio-based Phase Change Materials (PCMs) in Latent Heat Thermal Energy Storage (LHTES) within buildings due to their cost-effectiveness, renewability, non-toxicity, wide availability, and accessibility (de Albuquerque Landi et al., 2020).

The main objective of this research is to provide an overview of the key elements to be discussed in subsequent sections. It begins by addressing the environmental imperatives driving the need for sustainable construction materials and then delves into the principles of geopolymer technology with a focus on its eco-friendly features. It introduces the concept of biobased PCMs and their unique qualities that enhance the environmental sustainability of geopolymer matrices. This field is interdisciplinary, where materials science, engineering, and environmental sustainability intersect, offering a compelling case for bridging sustainability and technological innovation to reshape the construction and materials industry.

In the following sections of this review, we will explore the latest developments, innovative material synthesis approaches, PCM encapsulation techniques, and manufacturing methods. Emerging trends, such as the use of waste materials, nanotechnology applications, and sustainable supply chain management, will be discussed in the context of biobased geopolymer composites. Further, subsequent paragraphs, will also addresses the persistent challenges faced by researchers and industries when adopting biobased PCMs in geopolymer matrices, including scalability, durability, cost-effectiveness, and standardization. This review aims to be a valuable resource for researchers, practitioners, and policymakers with a common goal of advancing environmental sustainability in the construction and materials industry. The pursuit of greener, more energy-efficient building materials remains essential, and biobased geopolymer composites infused with phase change materials are poised to play a crucial role in this endeavour.

2. EXPLORING POROUS GEOPOLYMER AS POTENTIAL TEMPLATE FOR PHASE CHANGE MATERIALS

In the quest for sustainable and energy-efficient building materials, the incorporation of phase change materials (PCMs) has emerged as a promising strategy. PCMs have the ability to store and release thermal energy during phase transitions, thereby contributing to improved energy efficiency in various applications, including building and construction. To maximize the effectiveness of PCMs, their encapsulation within suitable matrices is essential. One such matrix with considerable potential is porous geopolymer. This study aims to explore the feasibility of porous geopolymer as a template for phase change materials.

Geopolymers, known for their low carbon footprint and remarkable durability, present an environmentally friendly alternative to traditional cementitious materials (Ji et al., 2022; Laasri et al., 2022).

The introduction of porosity into geopolymer matrices opens up new possibilities, particularly when it comes to housing PCMs, making them a viable candidate for thermal energy storage in buildings (Gencel et al., 2022).

2.1 Properties and Advantages of Geopolymers as a Matrix Material for PCM Storage

Geopolymers have exceptional thermal stability, making them an ideal choice for encapsulating phase change materials (PCMs) (Asefi et al., 2022). This property ensures that the geopolymer matrix can withstand the repeated thermal cycles associated with PCM phase changes, which is crucial for the long-term durability and reliability of PCM storage systems (Tyagi et al., 2022). Geopolymers can be used in various applications such as residential heating or cooling, industrial processes, and renewable energy systems (Feng et al., 2022). The use of geopolymer matrices provides a stable environment for PCMs, allowing them to efficiently store and release thermal energy (Gao et al., 2021). This combination of geopolymer and PCM offers a promising solution for energy storage and management, contributing to the development of sustainable and efficient building materials and systems (Degefu et al., 2021).

One of the significant advantages of geopolymers is their flexibility to be engineered with tailored thermal conductivities. The ability to customize the thermal conductivity of the geopolymer matrix allows for precise control over heat transfer within the PCM storage system. Effective heat transfer between the PCM and its surroundings is essential for efficient energy storage and release, ensuring optimal system performance (Aziz et al., 2019; Fiala et al., 2019). Geopolymers have been found to exhibit high thermal stability and low thermal conductivity, making them suitable for thermal energy storage applications. Additionally, the addition of electrically conductive carbon black admixture to geopolymers has been shown to enhance their electrical properties, which can further improve heat evolution and distribution within the material (Parhizi & Jain, 2019; Sabbatini et al., 2017). Experimental studies have also been conducted to analyse the thermal resistance and dilatometry of geopolymer formulations, providing insights into the critical parameters that influence their properties (Wołoszyn et al., 2021).

Geopolymers exhibit remarkable resistance to various chemical agents. Chemical property is indeed crucial for PCM storage systems, as it ensures compatibility with the PCM material and other components within the system, contributing to long-term stability and functionality. The chemical resistance of PCMs is an important factor in their thermal stability and reliability. The review of different PCMs for heat storage applications discusses their thermal stability and reliability, including organic, inorganic, eutectic, and composite materials (Madeswaran et al., 2021; Munir & Hussain, 2022; Tyagi et al., 2022). Additionally, the life cycle assessment (LCA) and life cycle inventory (LCI) analysis of PCMs provide information on their material and energy requirements, as well as their environmental impact. The selection of suitable PCMs for thermal energy storage is based on their physical and chemical properties, as well as their state and melting temperatures. Therefore, considering the chemical properties of PCMs is essential for ensuring the long-term stability and performance of PCM storage systems (Sawadogo et al., 2021; Yousefi et al., 2021).

Geopolymers possess excellent adhesion properties, allowing them to securely hold the PCM in place and enhance the overall performance and reliability of the system. The geopolymer matrix's ability to prevent phase separation or leakage is a key advantage when integrating geopolymers into various applications (Abulencia et al., 2021; Chen et al., 2022; Chu & Kong, 2021; Díaz & Barrios, 2022).

Geopolymers can be formulated with different compositions, allowing for the customization of their properties to match specific PCM requirements. This includes tailoring the melting and freezing temperatures of the PCM to align with the desired application (Tyagi et al., 2022; Wan et al., 2021). Geopolymers typically have low densities, which can be advantageous for PCM storage applications. Lightweight PCM storage systems can be particularly valuable in construction or transportation applications (Sawadogo et al., 2021).

Geopolymers are crucial in minimizing phase separation issues between the PCM and the matrix material during the phase change process, ensuring consistent and reliable thermal performance. They can be engineered to have good heat transfer properties, promoting efficient heat exchange between the PCM and its surroundings. Geopolymer-based PCM storage systems facilitate rapid charging and discharging of thermal energy, making them highly effective in managing temperature variations. Geopolymers are known for their long-term durability, contributing to the extended operational life of PCM storage systems. This durability ensures that the PCM storage system remains effective and reliable over time, leading to energy savings and sustainability (Degefu et al., 2021; Jiang et al., 2021; Q. Jin et al., 2022; Regin et al., 2008; Tyagi et al., 2022).

Geopolymer-based PCM storage systems can be designed and produced at various scales, from small residential applications to large district-level heating and cooling systems. This scalability makes geopolymer matrices suitable for a wide range of applications, enhancing energy efficiency and sustainability across different sectors (Abulencia et al., 2021; Shao et al., 2021). Geopolymers offer a compelling combination of properties and advantages, including thermal stability, tailored thermal conductivity, chemical resistance, adhesion, customizability, low density, phase separation minimization, enhanced heat transfer, durability, low environmental impact, and scalability. These attributes make geopolymer-based PCM storage systems a promising solution for a variety of applications, contributing to energy efficiency and sustainability goals.

2.2 Methods of Geopolymer Synthesis and Characterization

The processes of synthesizing and characterizing geopolymers are pivotal steps in harnessing their full potential for various applications, including construction, infrastructure, and advanced materials. In this discussion, we will explore the common methods employed for geopolymer synthesis and the techniques used for characterizing these materials.

2.2.1. Geopolymer Synthesis

Geopolymers can be synthesized using different methods. The most widely used method is alkaline activation, which involves mixing a reactive aluminosilicate precursor with an alkaline activator solution. This process forms a three-dimensional geopolymer network through a polycondensation reaction. Another method is the sol-gel process, where a geopolymer precursor solution is prepared by hydrolysing and condensing alkoxysilanes or organo-silanes. This solution is allowed to gel and harden, resulting in the formation of a geopolymer matrix. Geopolymers can also be synthesized by blending different precursor materials, such as metakaolin, fly ash, and slag. This blending approach allows for the optimization of geopolymer characteristics for specific purposes (Güngör & Özen, 2021; Krishna et al., 2021; Kushwah et al., 2021).

2.2.2. Characterization

X-ray Diffraction (XRD) is a fundamental technique used to analyse the crystalline structure of geopolymers, identify mineral phases, and determine the degree of amorphousness, which influences material properties. Fourier Transform Infrared Spectroscopy (FTIR) is employed to investigate the chemical bonding and functional groups within geopolymers, providing insights into their structure and composition. Scanning Electron Microscopy (SEM) allows for the observation of the microstructure and morphology of geopolymers, providing information on particle distribution, porosity, and interfacial characteristics. Thermal Analysis (TG-DTA) techniques, such as thermogravimetry (TG) and differential thermal analysis (DTA), assess the thermal stability and decomposition behavior of geopolymers, especially for high-temperature applications. Mechanical testing, including compressive strength, flexural strength, and modulus of elasticity, is essential for assessing the suitability of geopolymers for structural applications. Porosity analysis is crucial for understanding the permeability and durability of geopolymers, particularly in construction materials, by evaluating pore size distribution and specific surface area. Nuclear Magnetic Resonance (NMR) techniques, such as solid-state NMR, provide insights into the atomic-level structure and dynamics of geopolymers, contributing to a comprehensive understanding of their properties (Azam et al., 2022; Meng et al., 2021; Richard et al., 2022).

The synthesis and characterization of geopolymers are complex processes that necessitate a combination of techniques to tailor their properties for specific applications. These methods, in conjunction with ongoing research and development, continue to enhance the role of geopolymers as sustainable and high-performance alternatives in the construction and materials industry.

2.3 Role of Geopolymers in Enhancing the Thermal Performance of PCM Composites

Geopolymers play a significant role in enhancing the thermal performance of Phase Change Material (PCM) composites, which are materials designed to store and release thermal energy during phase transitions. This combination offers several advantages for various applications, particularly in the field of energy-efficient building materials and thermal energy storage systems.

Geopolymers can be tailored to have low thermal conductivity, which is crucial for maintaining the efficiency of PCM composites by minimizing heat loss and gain (Alves et al., 2023; Jia et al., 2020; Raut et al., 2023). Geopolymers have been found to have lower thermal conductivity compared to traditional bricks, making them suitable for thermally efficient building materials. In addition, the use of light geopolymers in cladding panels for facades has shown an increase in thermal resistance, indicating their effectiveness as thermal control materials. Geopolymers also have excellent heat resistance and can be converted into ceramic materials with controllable thermal properties, making them promising precursors for high-temperature applications (Matsimbe et al., 2022).

Geopolymer-based PCM composites have been shown to effectively store and release thermal energy over a range of temperatures, making them ideal for applications such as solar energy storage. These composites have been developed using different materials and techniques. For example, one study used hierarchically porous kaolinite-based geopolymer (PKG) embedded with polyethylene glycol 4000

(PEG4000) to create shape-stabilized composite PCMs with high thermal energy storage properties (H. Zhang et al., 2023). Another study employed artificial geopolymer aggregate (GPA) as a novel PCM carrier, achieving excellent mechanical strength and thermal conductivity. Additionally, a PCM composite was designed using a combination of metal-organic frameworks (MOFs) and expanded graphite (EG), resulting in enhanced thermal conductivity and shape stability. Furthermore, geopolymer-based concrete (GEO) has been found to withstand high temperatures and exhibit higher thermal storage capacity compared to ordinary Portland cement (OPC)-based materials. Overall, these studies demonstrate the potential of geopolymer-based PCM composites for efficient thermal energy storage in various applications, including solar energy storage (Fang et al., 2023; Rahjoo et al., 2022; Xiao et al., 2023).

3. BIOBASED PHASE CHANGE MATERIALS

Biobased Phase Change Materials (PCMs) represent an innovative class of materials with significant potential in various applications, particularly in the fields of energy efficiency and thermal management. These materials undergo phase transitions, absorbing or releasing latent heat, which makes them crucial for storing and managing thermal energy. Biobased PCMs, as the name suggests, are derived from renewable, biological sources, setting them apart from traditional PCMs that often use non-renewable resources. This environmentally conscious approach aligns with the growing emphasis on sustainability and reducing the carbon footprint of various industries.

3.1 Classification and Selection Criteria for Biobased PCMs

When considering biobased PCMs, it is essential to classify them and establish selection criteria to identify the most suitable materials for specific applications. The classification of biobased PCMs can be based on various factors, including their source, phase transition temperature, and chemical composition.

Biobased PCMs can be derived from plant-based, animal-based, or microbial sources. Plant-based PCMs are derived from crops such as soybeans, palm, or corn starch, while animal-based PCMs come from sources like beeswax or tallow. Microbial PCMs are produced through microbial fermentation processes using microorganisms like bacteria or yeast. The phase transition temperature of biobased PCMs determines their suitability for different applications. Low-temperature biobased PCMs are suitable for applications in the range of -20°C to 0°C, while medium-temperature PCMs are ideal for applications between 0°C to 50°C. High-temperature biobased PCMs are designed for applications exceeding 50°C. The chemical composition of biobased PCMs includes fatty acids, such as triglycerides found in vegetable oils, alcohols like erythritol or xylitol, and sugar alcohols like mannitol and sorbitol derived from plant sources (Kang et al., 2015; Ma et al., 2012; Rasta & Suamir, 2019; Yu et al., 2014).

3.2 Environmental Benefits and Sustainability Aspects of Biobased PCMs

Phase change materials are substances that can absorb and release a significant amount of thermal energy when they change from one phase (usually solid) to another (usually liquid) at a specific temperature. This property makes them useful for thermal energy storage applications, such as in buildings, solar energy systems, and electronics cooling. Fatty acids, which are organic compounds composed of long hydrocarbon chains with a carboxyl group (-COOH) at one end, can exhibit phase change behavior under specific conditions (Cellat et al., 2015). In prior scientific literature, fatty acids with melting points falling within the thermal comfort range were identified as capric acid (CA), lauric acid (LA), myristic acid (MA), and palmitic acid (PA). Table 1 was compiled for the purpose of customizing the phase change material (PCM) to meet specific property requirements. The objective of evaluating various binary combinations is to pinpoint a composition that exhibits a narrow and desirable melting temperature range (de Jong et al., 2012). Fatty acids, owing to their notably reduced flammability and origins derived from non-fossil fuel sources, render them better suited for sustainable building applications when compared to paraffins. The thermogravimetric analysis (TGA) data from prior research investigations demonstrates the capability of these mixtures to endure temperatures as high as 120 °C without experiencing decomposition. This observation underscores the potential utility of the phase change materials (PCMs) for passive solar energy storage applications within building structures (Cellat et al., 2015).

Bio-based PCMs represent a sustainable evolution of traditional PCMs. They are derived from renewable sources, including plant oils and fatty acids, making them environmentally friendly alternatives. The utilization of bio-based PCMs reduces the reliance on fossil fuels, decreases carbon emissions associated with PCM production, and supports the transition to a more sustainable energy landscape. Multiple research investigations have illustrated the favourable suitability of employing bio-based Phase Change Materials (PCMs) for Latent Heat Thermal Energy Storage (LHTES) in building applications. This preference is primarily attributed to the economic viability, renewable nature, non-toxic properties, and ready availability and accessibility of these substances (de Albuquerque Landi et al., 2020; Mehrizi et al., 2023).

Untreated fats and vegetable oils exhibit beneficial phase change characteristics that render them suitable for Latent Heat Thermal Energy Storage in building applications, presenting a potential biomaterial replacement for paraffin and salt hydrates. In a previous research publication, an investigation was conducted into the thermal and physico-chemical properties of composite PCMs exclusively composed of bio-based constituents, each with distinct melting points. Specifically, unprocessed coconut fat and natural Palm Kernel vegetable fat were employed as the bio-based PCMs in their native states (Boussaba et al., 2019; Suppes et al., 2003).

Previous research has examined the dynamic thermal behavior of insulation materials enhanced by incorporating bio-based Phase Change Materials (PCMs). Additionally, investigations have assessed the thermal performance of a composite material comprising Boron Nitride and a bio-based PCM with a melting point of 29.38°C. Furthermore, research has explored the thermal and mechanical properties of composites featuring bio-based PCMs (Hu & Yu, 2014; Jeong et al., 2014).

Earlier studies have also demonstrated that unmodified natural fats and vegetable oils possess valuable phase change characteristics suitable for Latent Heat Thermal Energy Storage (LHTES) in buildings. This suggests their potential as viable biomaterial alternatives to conventional options like paraffin and salt hydrates. Moreover, prior scholarly investigations have scrutinized the thermal and physicochemical properties of bio-based Phase Change Materials (PCMs) with melting points of 26.53°C and 22.63°C, respectively. These composites were exclusively composed of bio-based PCMs, with coconut fat and Palm Kernel vegetable fat in their natural states serving as the PCM constituents(Boussaba et al., 2018; Jeong et al., 2014).

3.3 Methods for Incorporating Biobased PCMs Into Geopolymers

Incorporating biobased Phase Change Materials (PCMs) into geopolymers involves a carefully planned process to create composite materials with enhanced thermal properties. Geopolymers, known for their versatility and environmental benefits, can be an excellent matrix for the integration of biobased PCMs. When incorporating biobased PCMs into geopolymers, it's essential to consider factors such as PCM loading, compatibility, and the intended application. Proper mixing, dispersion, and curing conditions are critical to achieving the desired thermal performance and long-term stability of the resulting composite materials. Each method has its advantages and may be more suitable for specific applications based on the required properties of the final product.

3.3.1 Encapsulation Technique

Phase Change Materials (PCMs) and bio-based Phase Change Materials (bio-PCMs) encapsulation both involve the process of encapsulating materials that undergo phase transitions to enhance their performance in various applications, but they differ primarily in the source and nature of the PCM used. The scientific literature presents a wide array of methodologies for the fabrication of Microencapsulated Phase Change Materials (MPCM). The selection of a particular approach is contingent upon the intended application. Nevertheless, the overarching goal remains unchanged: to prevent the escape of molten PCM while concurrently augmenting the thermal and mechanical properties of the MPCM. These methodologies can be broadly categorized into three principal production techniques: physico-mechanical, chemical, and physio-chemical methods.

Encapsulation serves as a conventional strategy employed for the purposes of PCM containment, ameliorating heat transfer efficiency, and enhancing mechanical durability. These encapsulation methods can be categorized into macroencapsulation and microencapsulation, as well as matrix encapsulation, each characterized by notable disparities in cost and intended applications. Emulsion-templated materials or encapsulation matrices are composed of polymeric structures derived from emulsions (Silverstein, 2017). Porous matrices offer versatile opportunities for the inclusion of diverse materials, including Phase Change Materials (PCMs). In contrast to the conventional spherical encapsulation method frequently applied to non-renewable source-based PCMs, an innovative approach involved encapsulating soybean wax PCM within electro-spun fibers that were shielded by a protective polyurethane layer. This encapsulation process yielded PCMs with commendable thermal stability and robust thermal resistance, thereby warranting consideration for applications in thermal storage and thermal shielding (Hu & Yu, 2012).

Microencapsulation of a Phase Change Material derived from coconut oil, possessing a melting enthalpy of 119 J/g, was accomplished using the method of suspension polymerization, employing stearyl methacrylate and hydroxyethyl methacrylate as key components.

Numerous experimental and computational investigations have been conducted to augment the thermal conductivity, diminish thermal resistances, and mitigate supercooling within biobased Phase Change Materials (PCMs) by introducing highly conductive elements, such as diverse nanomaterials. These investigations encompass techniques such as the dispersion of carbon nanotubes and graphite nanoplatelets within biobased PCMs, leading to a remarkable 375% enhancement in thermal conductivity. Furthermore, research has explored the utilization of graphite nanoplatelets in conjunction with fumed silica for lightweight applications within architectural structures (Jeong et al., 2013; Kang et al., 2015).

3.3.2 Shape Stabilization Technique

To date, several techniques have been documented for the fabrication of shape-stabilized composite Phase Change Materials (PCMs). These methods encompass microcapsule technology, porous inorganic carrier utilization, the sol-gel approach, and the melt-blending method[58c]. Conventional porous inorganic carriers, such as expanded graphite, diatomite, and expanded vermiculite, exhibit exceptional thermal and physical characteristics. Nevertheless, there remains potential for further enhancement in the specific capacity of the aforementioned carriers (Novais et al., 2020). Hierarchically porous materials, characterized by the presence of pores with varying sizes and exceptional interconnectivity, have demonstrated exceptional absorption capabilities. Hierarchically porous materials, distinguished by their diverse pore size distribution and remarkable interconnectivity, have exhibited outstanding absorption capabilities. Furthermore, in similar applications, several studies have documented the development of hierarchically porous geopolymers (Gao et al., 2020). Organic Phase Change Materials (PCMs) are known to have inherent limitations, including low thermal conductivity and the potential for leakage during phase transitions. To address these drawbacks, supplementary materials are introduced into the PCM matrix. These additional materials fall into two categories: porous materials, which serve to prevent PCM leakage, and nanomaterials, which enhance the thermal properties of the PCM. When the PCM is combined with both of these supporting materials, it forms a composite known as Shape-Stabilized Composite PCM (ss-CPCM) or Shape-Stabilized PCM (Rathore & Shukla, 2021).

Porous materials facilitate PCM support owing to their augmented surface area, elevated thermal conductivity, and superior chemical compatibility (Huang et al., 2019).

4. ENVIRONMENTAL SUSTAINABILITY

Bio-based PCMs represent a sustainable evolution of traditional PCMs. They are derived from renewable sources, including plant oils and fatty acids, making them environmentally friendly alternatives. The utilization of bio-based PCMs reduces the reliance on fossil fuels, decreases carbon emissions associated with PCM production, and supports the transition to a more sustainable energy landscape[30c]. Multiple research investigations have illustrated the favourable suitability of employing bio-based Phase Change Materials (PCMs) for Latent Heat Thermal Energy Storage (LHTES) in building applications. This preference is primarily attributed to the economic viability, renewable nature, non-toxic properties, and ready availability and accessibility of these substances (de Albuquerque Landi et al., 2020).

Earlier studies have addressed the assessment of the embodied carbon footprint associated with the production of artificial geopolymer aggregates (GPAs) using varying mix proportions. Additionally, the literature includes data on the embodied carbon content of alternative lightweight aggregates previously reported as carriers for Phase Change Materials (PCMs). These aggregates encompass expanded perlite, synthetic zeolite, expanded foam glass, ceramic foam, and expanded clay (Hammond & Jones, 2011; Tobarameekul & Worathanakul, 2021; Zimele et al., 2019). The objective is to facilitate comparative analysis.

Furthermore, considering an emission factor of 0.231 kgCO2e/kWh for energy consumption [6e], the carbon emissions resulting from the curing process were quantified, yielding a value of 0.164 kgCO2e/kg. While the carbon emissions associated with the curing process alone exceeded the embodied carbon content of the raw materials, it's noteworthy that this energy demand can potentially be offset by utilizing biomethane, biofuel, and waste heat steam sourced from wet waste plants and waste incineration facilities. Furthermore, this accelerated curing methodology holds the potential to enhance aggregate strength and quality while expediting production efficiency, thereby delivering additional economic and environmental advantages.

The utilization of artificial aggregates is driven by a multifaceted objective, aiming to simultaneously address three critical challenges: (1) the repurposing of industrial solid wastes, such as fly ash, red mud, glass waste, and calcium carbide residue, (2) the containment of hazardous waste materials, such as incineration bottom ash (IBA), and (3) the mitigation of natural aggregate depletion. This strategic deployment of various waste materials within artificial geopolymer aggregates (GPAs) has the potential to yield cost savings in waste treatment processes. Beyond the environmental advantages inherent to GPAs, it is anticipated that ongoing advancements in alternative alkali materials will contribute to narrowing the cost differential between GPAs and natural aggregates in the future (L.-Y. Xu, Qian, et al., 2021).

5. RECENT DEVELOPMENTS

The development of Phase Change Materials (PCMs) enriched geopolymer composites is an exciting field with significant advancements. These composites combine the unique properties of geopolymer materials with the thermal energy storage capabilities of PCMs.

This research aims to evaluate the potential of utilizing artificial geopolymer aggregates (GPAs) as a versatile and porous substrate for creating form-stable Phase Change Material (PCM) composites. GPA, a recently developed porous aggregate variant, is produced through the alkali-activation process using aluminosilicate-rich source materials (Qian et al., 2022; L.-Y. Xu et al., 2022; L.-Y. Xu, Huang, et al., 2021; L.-Y. Xu, Qian, et al., 2021). The GPA-PCM composite is prepared using a vacuum impregnation technique.

Drawing from previous studies, the addition of incinerated bottom ash to the GPA mixture has proven effective in increasing its porosity. This results in the creation of lightweight GPA with a reduced bulk density ranging from 1.52 to 1.62 and an increased water absorption capacity between 18.7% and 21.0%. Remarkably, despite its reduced density, this lightweight GPA maintains a robust compressive strength, ranging from 36.4 to 60.2 MPa.

The GPA-PCM composite is generated using a vacuum impregnation method. Building on earlier research, the inclusion of incinerated bottom ash into the GPA blend has shown to be a successful approach for enhancing its porosity. As a result, lightweight GPA with reduced bulk density (1.52–1.62) and improved water absorption capacity (18.7% to 21.0%) is achieved, while maintaining a substantial compressive strength (36.4 to 60.2 MPa).

Impregnating PCM into GPA leads to a partial reduction in the crushing strength of GPA-PCM, with reductions ranging from 4.13% to 25.46% when compared to GPA alone. The initial strength of GPA contributes to a more significant strength decline in GPA-PCM, primarily due to expansion stress during the phase transformation process. However, the GPA-PCM composites demonstrate remarkable thermal stability and exhibit significant melting enthalpy ranging from approximately 4.29 to 24.74 J/g. Additionally, the thermal conductivity of GPA-PCM ranges from 0.510 to 0.589 W/mK, surpassing that of GPA. Importantly, all GPA-PCM composites meet the established criteria for classification as insulation materials.

Geopolymers, known for their environmental benefits and versatility, have emerged as a sustainable alternative to conventional construction materials. Integrating biobased PCMs into geopolymer matrices presents an exciting avenue for enhancing the thermal performance and sustainability of these materials.

In a previous study, researchers conducted a comparative investigation involving three distinct form-stable composite Phase Change Materials (PCMs) designed for building applications. This research aimed to introduce the selected PCMs into pumice, a naturally occurring pozzolan known for its non-crystalline structure and porous nature. Pumice possesses exceptionally low density, substantial porosity, an extensive surface area, and limited chemical reactivity. These unique characteristics make pumice a promising candidate for efficiently absorbing PCM through surface tension and capillarity forces while maintaining a cost-effective preparation process (Karaipekli & Sarı, 2016).

Numerous researchers have dedicated their efforts to examining these bio-based PCMs that differ from paraffin-based ones. They have assessed these PCMs' appropriateness by evaluating their thermophysical properties, heat storage capacity, and parameters like phase-change enthalpy, specific heat capacity, and melting temperature. Despite these research efforts, it's essential to note that the existing body of literature provides only limited thermal data on these bio-based PCMs. Additionally, there are currently no established norms or standards for characterizing bio-based PCMs (Baetens et al., 2010).

The PCMs under scrutiny in this research investigation include a eutectic blend of capric acid and palmitic acid, dodecanol, and heptadecane. The study entails a comprehensive characterization of these composite materials, with a specific focus on evaluating their thermal properties, chemical stability, thermal stability, thermal cycling stability, and conducting microscopic examinations both before and after the impregnation process. The vacuum impregnation technique is employed for infusing the PCMs into the materials.

The study findings indicate that the Phase Change Material (PCM) exhibits no significant chemical reactivity with the carrier material, although some interactions between the PCM and the carrier were observed. Morphological analysis is conducted using scanning electron microscope (SEM) images. To assess the thermal stability of the PCM composite, thermogravimetric analysis is performed. Furthermore, the composite PCM undergoes 3000 thermal cycles involving phase transitions from solid to liquid and vice versa. The results from these cycling experiments demonstrate that the thermal performance of the composite PCM remains consistent throughout the entire spectrum of cycles examined.

Based on differential scanning calorimetry (DSC) tests, it is determined that the composite Phase Change Materials (PCM) are well-suited for applications within the temperature range of 21–23 °C. The research findings indicate that the pumice-heptadecane composite Phase Change Material (PCM) exhibits relatively superior thermal performance [46c].

Prior research has extensively investigated and documented the feasibility of utilizing bio-based fatty acids as thermal energy storage materials. These bio-based fatty acids are sourced from animal and plant oils, setting them apart from paraffins, which originate from finite, non-renewable fossil fuel reservoirs (Cellat et al., 2015).

Bio-based fatty acids, specifically sourced from animal and plant oils and distinct from paraffins extracted from non-renewable fossil fuel deposits, are explored. The research encompasses an investigation into organic compounds such as palmitic acid (PA), lauric acid (LA), myristic acid (MA), and capric acid.

A total of twelve binary mixtures are meticulously formulated, considering the desired indoor temperature range of 22–27 °C. These mixtures undergo thorough systematic evaluation of their thermal properties, with a particular emphasis on the CA-LA and CA-MA combinations. These combinations are singled out for potential integration into concrete to augment its thermal inertia. The selection criteria for these combinations are based on latent heat and phase transition temperatures obtained from measurements conducted using differential scanning calorimetry (DSC).

The analysis considers PCM concentrations of 1% and 2% by weight. Additionally, the incorporation of these PCMs into concrete leads to a substantial reduction in the concrete's hardening time.

This comprehensive review delves into the existing and prospective applications of biobased Phase Change Material (PCM)-enriched geopolymer matrices in the realms of building materials and energy-efficient construction.

Recent developments in Bio-Based PCM with geopolymer for Environmental Sustainability reflect the growing interest in eco-friendly construction materials and their potential to enhance energy efficiency. Notable trends include researchers exploring new and diverse sources for bio-based PCMs, such as agricultural waste, algae-derived lipids, and other sustainable feedstocks. These alternative sources increase the availability of bio-based PCMs and align with the goal of reducing waste and promoting circular economy principles (Bai & Colombo, 2018; Boussaba et al., 2019; de Albuquerque Landi et al., 2020; Fang et al., 2023; Mehrizi et al., 2023).

Advancements in microencapsulation and shape stabilization technologies enable the efficient incorporation of bio-based PCMs into geopolymer matrices. The development of hybrid materials that combine bio-based PCM-enriched geopolymers with other sustainable additives, such as aerogels or nanomaterials, is gaining traction. These combinations leverage the strengths of multiple materials to achieve superior thermal properties and overall performance (Fang et al., 2023).

Simulation and modeling tools are employed to assess the energy-saving potential of buildings incorporating Bio-Based PCM with geopolymer. This trend aids in optimizing the design and placement of these materials to maximize energy efficiency (Berardi & Gallardo, 2019). There is a growing emphasis on life cycle assessments, evaluating the environmental impact of these materials throughout their entire lifecycle. This approach provides a holistic view of their sustainability, from raw material extraction to disposal or recycling (Kumar et al., 2023).

Furthermore, emerging and evolving techniques include additive manufacturing, immersion and impregnation methods, reactive emulsion templating, particle compaction methods, and combined approaches (X. Zhang et al., 2021).

PCMs have been recognized as a favorable choice for advanced building insulation since the mid-1970s. Their application in buildings is aimed at enhancing energy efficiency by leveraging the specific characteristics of the surrounding environmental conditions (Rathore et al., 2020). PCMs primarily enhance the thermal attributes of structural components within which they are incorporated. They become active at distinct temperature thresholds, capturing thermal energy, which reduces the need for indoor heating and cooling, thereby advancing energy efficiency. PCMs can be integrated into building components via various methods, including direct admixture, immersion in liquid PCM, or encapsulation within other materials in the form of micro and macro capsules (Vicente & Silva, 2014).

Research focuses on optimizing the use of PCMs to reduce energy consumption in buildings and enhance energy efficiency. Investigations examine the integration of PCMs into traditional building materials. For instance, PCM can be incorporated into concrete to improve its thermal energy storage capacity (Cabeza et al., 2007; Ling & Poon, 2013; Ryms & Klugmann-Radziemska, 2019). In this context, a study investigated clay geopolymer mortar integrated with PCM, revealing a significant reduction in heat transfer compared to alternative mortar materials (Wang et al., 2016). Proper orientation is crucial for achieving optimal thermal comfort in indoor spaces, aligning construction materials with specific climatic conditions. Selection of PCM properties depends on climatic conditions, impacting the material's ideal melting temperature (Lagou et al., 2019; Park et al., 2021).

In addition to optimizing the thermal performance of PCMs, the focus has shifted toward minimizing the environmental impact of these materials. Life cycle assessments reveal variations in the environmental metrics associated with PCM production and efficacy, influenced by differences in assessment scope and boundary conditions. Embodied energy plays a pivotal role in PCM environmental characteristics, typically included in LCA analyses (Carbonaro et al., 2015; De Gracia et al., 2010; Kylili & Fokaides, 2016; Zhuk, 2019) [25e].

This comprehensive review explores the latest developments in PCM-enriched geopolymer matrices, emphasizing their applications, properties, and environmental benefits. By encapsulating PCMs within geopolymer structures, these innovative composites enhance thermal energy storage, improve temperature regulation in buildings, and reduce energy consumption. Through a thorough synthesis of research findings, this review underscores the potential of PCM-enriched geopolymer matrices as a sustainable solution for addressing environmental challenges in the construction sector.

In a previous research investigation, the primary focus was on utilizing artificial geopolymer aggregates (GPAs) as a medium for housing Phase Change Materials (PCMs), specifically designed for energy storage applications. The physical, mechanical, and thermal properties of GPA-PCM composites were thoroughly evaluated, showing that these properties can be adjusted by manipulating raw material constituents, such as slag content, water-to-binder ratio, and the inclusion of incinerated bottom ash (IBA). Increasing the IBA content was found to enhance the porosity of GPAs, thus increasing their PCM containment capacity.

Through a vacuum suction process, the absorption capacity reached a maximum of 16 wt% for Phase Change Material (PCM) within the artificial geopolymer aggregate (GPA), resulting in a substantial melting enthalpy of 24.74 J/g. Additionally, the GPA-PCM composite exhibited outstanding mechanical strength, surpassing 53.2 MPa, and demonstrated notable thermal conductivity ranging between 0.510 to 0.589 W/mK. Analysis of time-temperature history curves for GPA underscored that PCM impregnation resulted in impressive thermal regulation capabilities, achieving temperature differentials of up to 10.5 °C (Fang et al., 2023).

In a previous investigation, the focus was on integrating an innovative Phase Change Material (PCM) comprised of fatty acids, enclosed within a polyurethane foam matrix, into a geopolymer mortar matrix. The research involved a comprehensive analysis of the structural and thermal attributes of the developed geopolymer mortar (GP) and the form-stable PCM composite (PU@PCM). The inquiry also delved into the impact of PU@PCM inclusion on the mechanical properties of the geopolymer mortar, determined through compressive testing. Thermal behavior was systematically explored through thermogravimetric analysis (TGA), differential scanning calorimetry (DSC), and thermal conductivity measurements using the laser flash analysis (LFA) technique.

The findings from this study revealed that the synthesized geopolymer mortar forms a continuous and robust three-dimensional network structure. Simultaneously, the PU@PCM composite exhibited an operational temperature range from 10 to 22 °C, ideal for energy-efficient building design. The PU@PCM composite demonstrated an impressive energy storage capacity of $\Delta Hf = 164$ J/g. Despite a reduction in compressive strength due to PU@PCM incorporation, the composites continue to meet mechanical requirements for concrete applications, with the PU@PCM-rich composite achieving a notable strength of 29.5 MPa.

Moreover, the evaluation of thermal performance highlighted the substantial benefits of PU@PCM integration, particularly in terms of increased heat capacity and only a slight reduction in thermal conductivity within the composites. Specifically, the PU@PCM-rich composite exhibited heat capacity and thermal conductivity values of 1.1 J/g.°C and 0.84 W/m.°C, respectively.

Cyclic differential scanning calorimetry (DSC) tests confirmed the sustained thermal stability of the composites, even after enduring 20 heating and cooling cycles. In summary, the geopolymer mortar enriched with the highest PU@PCM content demonstrates robust potential as a promising candidate for the development of sustainable and energy-efficient buildings. This material combines commendable thermal performance with mechanical strength, making it suitable for practical applications in the construction industry (Haily et al., 2023).

In a prior study, a thermal energy storing cementitious composite was developed. This composite included microencapsulated Phase Change Material (PCM) encased in a bio-based polymer shell, referred to as m-PCM. The study also assessed the biodegradability of the PCM encapsulation through Nuclear Magnetic Resonance (NMR), confirming the high biodegradability of the m-PCM shell under specific environmental conditions.

The thermophysical properties of the m-PCM were characterized using Differential Scanning Calorimetry (DSC) and Thermogravimetric analysis (TGA). DSC analysis revealed exothermic enthalpies of 84.7 J/g and 140.2 J/g, along with associated peak temperatures of -3.2 °C and -22.9 °C, respectively. Subsequently, the m-PCM was incorporated into cement-based mortar at varying proportions of 5%, 10%, and 15% by weight of the binder.

The findings demonstrate that, in comparison to other sets of microcapsules, LP-4 exhibits a favorable level of biodegradability, characterized by an average particle size of 6.6 µm. Furthermore, LP-4 boasts a notably higher core material active content, surpassing other samples by 42.6%. Analysis using Nuclear Magnetic Resonance (NMR) revealed that LP-4 capsules exhibit the most significant biodegradability when exposed to environmental conditions, showing a remarkable 56.39% reduction in peak intensity.

Furthermore, the thermal assessment of the LP-4 PCM microcapsules, as developed in this study, exhibited commendable thermal performance and can be effectively integrated into construction materials to enhance their thermal inertia. Differential Scanning Calorimetry (DSC) results indicated that the m-PCM released substantial heat energy, measuring at 84.76 J/g and 140.2 J/g, with peak temperatures of -3.2 °C and -22.9 °C, respectively.

The nanomodified thermal energy storage cementitious composite developed in this study possesses adequate mechanical strength for application as a construction material in low-temperature scenarios. Thermal cycling tests corroborate that the inclusion of m-PCM in the mortar substantially enhances their thermal inertia. However, it is important to note that due to the supercooling effect, P5 and CSP 5 demonstrated lower efficiency compared to P10 and P15. However, the introduction of highly thermally conductive Multi-Walled Carbon Nanotubes (MWCNTs) improved the thermal efficiency of the specimens (X. Jin et al., 2023).

In summary, the development of PCM-enriched geopolymer composites is an evolving field with ongoing research aimed at enhancing their thermal properties, reducing costs, and expanding their applications. These developments have the potential to revolutionize multiple industries by providing efficient and sustainable solutions for thermal energy storage and management.

6. APPLICATIONS AND EMERGING TRENDS

Phase Change Materials (PCMs) enriched geopolymer composites are a fascinating and innovative area of research in materials science and construction engineering. Geopolymers are a type of environmentally friendly, cementitious material that can be used as an alternative to traditional concrete, and when combined with PCMs, they offer a range of applications and emerging trends.

The emergence of 3D printing has ushered in a paradigm shift in conventional construction practices, offering a cost-effective and expeditious means of fabricating intricate architectural structures. Nonetheless, there persist unresolved challenges concerning the efficacious incorporation of functional additives into 3D printing construction materials. A preceding research article has introduced a straightforward and environmentally conscientious approach designed to advance the cause of sustainable building construction while concurrently curbing energy consumption.

This approach involves the amalgamation of Macroencapsulated Phase Change Materials (MEPCM) into a 3D printable geopolymer paste (GPP), which is synthesized from fly ashes. The research was executed meticulously, encompassing a comprehensive evaluation of the immediate and long-term properties of geopolymer pastes with varying proportions of MEPCM. Additionally, the study involved an in-depth analysis of the thermal properties of these formulations and an assessment of their thermal performance, accomplished by the fabrication of miniature structures using 3D printing technology. This assessment compared structures produced without MEPCM to those containing 20% MEPCM.

Remarkably, MEPCM displayed a dual functionality, serving as a pivotal component for the management of thermal energy within the structures and as a modifier of the viscosity of the 3D printable geopolymer paste. In summary, this study introduces an innovative trajectory towards sustainable construction, accentuating the paramount importance of enhancing energy efficiency and curtailing wasteful practices within the construction sector (Rahemipoor et al., 2023).

Previous research article outlines an experimental investigation involving the characterization of a wall assembly constructed from geopolymer concrete and phase change materials (PCMs) in the context of the climate of southern Italy. Data collected during the period from May to July 2021 were analyzed, focusing on surface temperatures and heat fluxes. This analysis introduced several novel indices, including the thermal stress reduction amplitude, the percentage of time within the melting range, and extreme surface stresses.

The results of the study demonstrate that the incorporation of PCMs effectively mitigates thermal stress and safeguards the integrity of the geopolymer blocks. Notably, during periods of peak solar exposure, the indoor surface temperature experiences a reduction of up to 3°C. This reduction contributes significantly to alleviating localized thermal discomfort. On a broader scale, the heat flux measured within the geopolymer layer located behind the PCM exhibits attenuation, resulting in a more consistent trend. This reduction in heat flux translates to decreased heat ingress through the opaque building envelope, ultimately leading to a reduction in energy demand (De Masi et al., 2023).

This study investigates the energy and environmental performance of Phase Change Material (PCM) coatings as a retrofit solution for walls in existing buildings. A validated model was utilized to calculate the total heat flux over the course of one year, considering three distinct geographical locations and four different orientations of the building. Additionally, the study establishes the energy payback period for the proposed PCM solutions, which is defined as the ratio between the embodied energy of the materials and the reduction in heat flux through the building element.

The findings indicate that the placement of the PCM layer has a significant impact on the energy performance of the building element. Notably, positioning the PCM layer on the external side of the wall is identified as the optimal solution, irrespective of the building's orientation. Furthermore, the study discerns variations in the performance of PCM coatings for different orientations of the building.

The effectiveness of the PCM coatings is observed to be more pronounced for structures with eastern and western orientations in warm-dominant climates (e.g., Athens), and for those with northern orientations in cold-dominant climates (e.g., Copenhagen). The research also reveals specific instances where PCM coatings yield substantial energy savings, such as 4.8 kWh/m2 per annum with an energy payback period of 1.5 years for western orientation in Athens, and 29.84 kWh/m2 per annum with an energy payback period of 0.8 years for northern orientation in Copenhagen.

The energy payback period for PCM coatings is shown to vary significantly, ranging from 0.8 to 30 years, depending on the installation location and the building's configuration. This underscores that there are situations where the installation of PCM coatings may not be environmentally and economically feasible (Georgiou et al., 2023).

7. PERSISTENT CHALLENGES

Persistent challenges in Bio-Based PCM-with geopolymer for environmental Sustainability are essential considerations for researchers and industry professionals aiming to maximize the potential of these materials.

7.1 Phase Transition Temperatures

Selecting PCMs with appropriate phase transition temperatures that align with the desired thermal performance of the geopolymer composite can be complex. The PCM's transition temperature should ideally match the building's temperature fluctuations, which vary by climate and location (Rehman et al., 2022; Sun et al., 2022).

7.2 PCM Dispersion

Achieving uniform dispersion of PCMs within the geopolymer matrix without agglomeration or settling is crucial for maintaining consistent thermal properties. Further, achieving compatibility between bio-based PCMs and geopolymer matrices can be challenging. Ensuring that these components interact seamlessly without causing separation, degradation, or phase separation is crucial for the materials' effectiveness and long-term stability (Marani & Nehdi, 2019).

7.3 PCM Compatibility and Leakage

Ensuring that the PCM is chemically compatible with the geopolymer binder and does not compromise the mechanical strength or durability of the composite (Messenger et al., 2023). Preventing PCM leakage, especially in high-temperature environments, is essential to avoid structural and environmental issues. The quality assurance of composite matrices necessitates addressing challenging research areas, including the complete prevention of Phase Change Material (PCM) leakage into the matrix, ensuring robust long-term cyclic stability, and minimizing any adverse impact on compressive strength (R. Sharma et al., 2022).

7.4 Production Costs and Long-Term Performance

PCMs can be relatively expensive, so cost-effectiveness must be considered when incorporating them into geopolymer composites for widespread use. Assessing the long-term durability and stability of PCM-geopolymer composites, including their ability to withstand repeated phase transitions and cyclic thermal stresses (Łach et al., 2021).

7.5 Impact on Workability; Scale-Up Challenges; Environmental Impact

Evaluating how the addition of PCMs affects the workability and setting characteristics of the geopolymer mix, which may require adjustments to the manufacturing process. Moving from laboratory-scale experiments to large-scale production presents logistical and economic challenges that need to be addressed. Conducting a comprehensive Life Cycle Assessment (LCA) to determine the overall environmental impact of PCM-geopolymer composites and ensure that they align with sustainability goals. Further researchers must assess factors such as material aging, resistance to moisture, thermal cycling, and structural integrity over time and compatibility (Regin et al., 2008; Reyez-Araiza et al., 2021).

Overcoming these challenges is essential for the successful integration of PCMs into geopolymer composites, ultimately contributing to enhanced thermal performance and energy efficiency in buildings and construction applications.

8. FUTURE PROSPECTS

Predictions for the future of biobased PCM-enriched geopolymers include their potential use as thermoregulatory materials (Korniejenko et al., 2022). Bio-based PCMs have large latent heat, low vapor pressure, and good chemical stability, making them promising candidates for various applications. Geopolymers and their composites, including those enriched with biobased PCMs, have advantages such as low-temperature preparation, low cost, and heat resistance. These materials can be used in conventional construction, metallurgy, and aerospace industries. The development of bio-based polymers from renewable resources, including those used in geopolymers, is gaining attention due to environmental concerns and the finite nature of petroleum resources (Jeong et al., 2014; Jia et al., 2020). The combination of biobased PCMs and geopolymers offers the potential for enhanced thermal properties and durability, making them suitable for future applications in fields such as civil engineering and energy storage.

Future directions in Bio-Based PCM-enriched geopolymer for environmental sustainability involve innovative strategies and research priorities to further advance these materials' contributions to greener and more energy-efficient construction practices. Continued exploration of sustainable and diverse sources for bio-based PCMs is crucial. Research should focus on identifying novel feedstocks and extraction methods to ensure a stable and environmentally friendly supply of bio-based PCMs (Nazari Sam et al., 2020). Enhancing microencapsulation and shape stabilisation methods to improve the stability, dispersion, and thermal performance of bio-based PCMs within geopolymer matrices. Innovations in shape stabilisation technologies can lead to materials with superior thermal properties (Marani &

Nehdi, 2019). Developing Bio-Based PCM Enriched Geopolymer materials with tailored properties for specific climate zones and building applications. This customization ensures optimal performance and energy savings in diverse environmental conditions (Boussaba et al., 2019). Conducting comprehensive life cycle assessments to evaluate the environmental impact of these materials throughout their entire lifecycle. This approach provides a holistic view of their sustainability, guiding further improvements. Continued development and refinement of energy simulation and modelling tools to accurately predict the energy-saving potential of buildings incorporating these materials. This aids architects and engineers in optimizing designs for maximum efficiency (Kumar et al., 2023). Investigating opportunities for recycling and repurposing of Bio-Based PCM Enriched Geopolymer materials at the end of their lifecycle to minimize waste and promote circular economy principles (Ong et al., 2023).

The amalgamation of PCM performance data with Life Cycle Assessment (LCA) indicators also creates opportunities for future research endeavors aimed at extracting and establishing indicators linked to the overall energy performance of materials over their entire lifecycle.

The scope of this study is amenable to future expansion, with the potential inclusion of alternative PCM variants and consideration for a broader spectrum of climatic conditions and weather patterns (Georgiou et al., 2023).

Lastly, delving into the potential utilization of alternative polymers for the encapsulation of the investigated energy storage material holds promise for fostering greater innovation and sustainability within the realm of construction materials. These research domains have the capacity to fine-tune the developed geopolymer mortars and chart the course towards the emergence of ecologically innovative building materials, which are of paramount importance in the pursuit of a sustainable future and in mitigating the adverse consequences of climate change (Walubita et al., 2022). In summary the future direction of Phase Change Materials (PCM)-based geopolymer composites holds significant potential for advancing sustainable construction and energy-efficient building practices. Several key directions and trends are likely to shape this field are tailored PCM, improved PCM dispersion, multifunctional composites, nanomaterials, large scale applications, life cycle assessment and smart building integration. As these directions unfold, PCM-based geopolymer composites are expected to play an increasingly important role in enhancing the energy efficiency, thermal comfort, and sustainability of buildings, contributing to global efforts to mitigate climate change (Georgiou et al., 2023; Walubita et al., 2022).

9. CONCLUSION

The prime objective of this research was to conduct a comprehensive analysis of the Developments, Trends, Challenges and Future direction associated with bio-based Phase Change Materials (PCMs) in geopolymer matrices. However, the continued research and innovation in this field hold the potential to revolutionize the construction industry, making buildings more energy-efficient, environmentally friendly, and sustainable. These materials are poised to play a vital role in shaping the future of green construction practices. These recent trends collectively highlight the growing momentum and potential of Bio-Based PCM Enriched Geopolymer materials in advancing environmental sustainability in the construction industry.

However, additional research endeavours are essential to explore the processing of these PCMs and to comprehensively assess their thermophysical and mechanical characteristics, both at the micro and macro scales. This evaluation is critical in determining the viability of substituting conventional petroleum-based PCMs with environmentally sustainable bio-based alternatives. Overcoming obstacles is crucial for unlocking the full potential of Bio-Based PCM in Geopolymer materials and advancing environmental sustainability in the construction sector. The future direction of Bio PCM-Geopolymer is characterized by a commitment to sustainability, innovation, and practical application. As these materials continue to evolve, they hold great potential to transform the construction industry and contribute significantly to energy-efficient, environmentally responsible building practices. By addressing these future directions, the construction industry can harness the full potential of Bio-Based PCM in Geopolymer materials to significantly reduce energy consumption, lower carbon emissions, and promote more sustainable and environmentally responsible building practices.

REFERENCES

Abdallah, M., El-Rayes, K., & Liu, L. (2016). Economic and GHG emission analysis of implementing sustainable measures in existing public buildings. *Journal of Performance of Constructed Facilities, 30*(6), 4016055. doi:10.1061/(ASCE)CF.1943-5509.0000911

Abulencia, A. B., Villoria, M. B. D., Libre, R. G. D. Jr, Quiatchon, P. R. J., Dollente, I. J. R., Guades, E. J., Promentilla, M. A. B., Garciano, L. E. O., & Ongpeng, J. M. C. (2021). Geopolymers as sustainable material for strengthening and restoring unreinforced masonry structures: A review. *Buildings, 11*(11), 532. doi:10.3390/buildings11110532

Ahmed, A. (2014). Energy smart buildings: Potential for conservation and efficiency of energy. *Pakistan Development Review, 53*(4II), 371–380. doi:10.30541/v53i4IIpp.371-381

Alhassan, I., Park, J., & Tang, X. (2023). Characterizing a Local Sewage Sludge Ash for Developing Alkali-Activated Geopolymer Mortar. *Transportation Research Record: Journal of the Transportation Research Board, 2677*(9). doi:10.1177/03611981231159112

Alva, G., Lin, Y., & Fang, G. (2018). An overview of thermal energy storage systems. *Energy, 144*, 341–378. doi:10.1016/j.energy.2017.12.037

Alves, C., Pelisser, F., Labrincha, J., & Novais, R. (2023). Effect of Hydrogen Peroxide on the Thermal and Mechanical Properties of Lightweight Geopolymer Mortar Panels. *Minerals (Basel), 13*(4), 542. doi:10.3390/min13040542

Amran, M., Debbarma, S., & Ozbakkaloglu, T. (2021). Fly ash-based eco-friendly geopolymer concrete: A critical review of the long-term durability properties. *Construction & Building Materials, 270*, 121857. doi:10.1016/j.conbuildmat.2020.121857

Arivazhagan, R., Geetha, N. B., Sivasamy, P., Kumaran, P., Gnanamithra, M. K., Sankar, S., Loganathan, G. B., & Arivarasan, A. (2020). Review on performance assessment of phase change materials in buildings for thermal management through passive approach. *Materials Today: Proceedings, 22*, 419–431. doi:10.1016/j.matpr.2019.07.616

Asefi, G., Ma, T., & Wang, R. (2022). Parametric investigation of photovoltaic-thermal systems integrated with porous phase change material. *Applied Thermal Engineering, 201*, 117727. doi:10.1016/j.applthermaleng.2021.117727

Assi, L., Carter, K., Deaver, E. E., Anay, R., & Ziehl, P. (2018). Sustainable concrete: Building a greener future. *Journal of Cleaner Production*, *198*, 1641–1651. doi:10.1016/j.jclepro.2018.07.123

Assi, L. N., Carter, K., Deaver, E., & Ziehl, P. (2020). Review of availability of source materials for geopolymer/sustainable concrete. *Journal of Cleaner Production*, *263*, 121477. doi:10.1016/j.jclepro.2020.121477

Azam, M. A., Safie, N. E., & Abdul Aziz, M. F. (2022). Characterization of Hierarchical Porous Materials. *Advanced Functional Porous Materials: From Macro to Nano Scale Lengths*, 407–429.

Aziz, I. H., Al Bakri Abdullah, M. M., Yong, H. C., Ming, L. Y., Hussin, K., Surleva, A., & Azimi, E. A. (2019). Manufacturing parameters influencing fire resistance of geopolymers: A review. *Proceedings of the Institution of Mechanical Engineers, Part L: Journal of Materials: Design and Applications, 233*(4), 721–733. 10.1177/1464420716668203

Baetens, R., Jelle, B. P., & Gustavsen, A. (2010). Phase change materials for building applications: A state-of-the-art review. *Energy and Building*, *42*(9), 1361–1368. doi:10.1016/j.enbuild.2010.03.026

Bai, C., & Colombo, P. (2018). Processing, properties and applications of highly porous geopolymers: A review. *Ceramics International*, *44*(14), 16103–16118. doi:10.1016/j.ceramint.2018.05.219

Berardi, U., & Gallardo, A. A. (2019). Properties of concretes enhanced with phase change materials for building applications. *Energy and Building*, *199*, 402–414. doi:10.1016/j.enbuild.2019.07.014

Boussaba, L., Foufa, A., Makhlouf, S., Lefebvre, G., & Royon, L. (2018). Elaboration and properties of a composite bio-based PCM for an application in building envelopes. *Construction & Building Materials*, *185*, 156–165. doi:10.1016/j.conbuildmat.2018.07.098

Boussaba, L., Makhlouf, S., Foufa, A., Lefebvre, G., & Royon, L. (2019). vegetable fat: A low-cost bio-based phase change material for thermal energy storage in buildings. *Journal of Building Engineering*, *21*, 222–229. doi:10.1016/j.jobe.2018.10.022

Cabeza, L. F., Castellon, C., Nogues, M., Medrano, M., Leppers, R., & Zubillaga, O. (2007). Use of microencapsulated PCM in concrete walls for energy savings. *Energy and Building*, *39*(2), 113–119. doi:10.1016/j.enbuild.2006.03.030

Carbonaro, C., Cascone, Y., Fantucci, S., Serra, V., Perino, M., & Dutto, M. (2015). Energy assessment of a PCM–embedded plaster: Embodied energy versus operational energy. *Energy Procedia*, *78*, 3210–3215. doi:10.1016/j.egypro.2015.11.782

Cellat, K., Beyhan, B., Güngör, C., Konuklu, Y., Karahan, O., Dündar, C., & Paksoy, H. (2015). Thermal enhancement of concrete by adding bio-based fatty acids as phase change materials. *Energy and Building*, *106*, 156–163. doi:10.1016/j.enbuild.2015.05.035

Chen, H., Dong, S., Zhang, Y., & He, P. (2022). Robust structure regulation of geopolymer as novel efficient amine support to prepare high-efficiency CO2 capture solid sorbent. *Chemical Engineering Journal*, *427*, 131577. doi:10.1016/j.cej.2021.131577

Chu, S. H., & Kong, Y. K. (2021). Mathematical model for strength of alkali-activated materials. *Journal of Building Engineering*, *44*, 103189. doi:10.1016/j.jobe.2021.103189 de Albuquerque Landi, F. F., Fabiani, C., & Pisello, A. L. (2020). Palm oil for seasonal thermal energy storage applications in buildings: The potential of multiple melting ranges in blends of bio-based fatty acids. *Journal of Energy Storage*, *29*, 101431. doi:10.1016/j.est.2020.101431

De Gracia, A., Rincón, L., Castell, A., Jiménez, M., Boer, D., Medrano, M., & Cabeza, L. F. (2010). Life Cycle Assessment of the inclusion of phase change materials (PCM) in experimental buildings. *Energy and Building*, *42*(9), 1517–1523. doi:10.1016/j.enbuild.2010.03.022

de Jong, E., Higson, A., Walsh, P., & Wellisch, M. (2012). Bio-based chemicals value added products from biorefineries. *IEA Bioenergy, Task42 Biorefinery, 34*, 1–33.

De Masi, R. F., Del Regno, N., Festa, V., Gigante, A., Ruggiero, S., & Vanoli, G. P. (2023). Innovative wall package made of macro-encapsulated phase change material and geopolymer concrete: In-field thermal analysis for a Mediterranean climate. *Architectural Science Review*, 1–21. doi:10.1080/00038628.2023.2243245

Degefu, D. M., Liao, Z., Berardi, U., & Doan, H. (2021). Salient parameters affecting the performance of foamed geopolymers as sustainable insulating materials. *Construction & Building Materials*, *313*, 125400. doi:10.1016/j.conbuildmat.2021.125400

Díaz, E. E. S., & Barrios, V. A. E. (2022). Development and use of geopolymers for energy conversion: An overview. *Construction & Building Materials*, *315*, 125774. doi:10.1016/j.conbuildmat.2021.125774

Fang, Y., Ahmad, M. R., Lao, J.-C., Qian, L.-P., & Dai, J.-G. (2023). Development of artificial geopolymer aggregates with thermal energy storage capacity. *Cement and Concrete Composites*, *135*, 104834. doi:10.1016/j.cemconcomp.2022.104834

Feng, L., Zhang, Y., Zhou, H., Kang, Y., Zhang, S., Bao, L., Lei, J., Bian, L., & Wang, J. (2022). Quasi-monodispersed nanocapsules with form stability at high temperature and under shear force for thermal energy storage. *Chemical Engineering Journal*, *428*, 131088. doi:10.1016/j.cej.2021.131088

Festus, N. (2022). A review on selected durability parameters on performance of geopolymers containing industrial by-products, agro-wastes and natural pozzolan. *Journal of Sustainable Construction Materials and Technologies*, *7*(4), 375–400. doi:10.47481/jscmt.1190244

Fiala, L., Petříková, M., & Černý, R. (2019). Basic physical, electrical and thermal properties of geopolymers with enhanced electrical properties. *AIP Conference Proceedings*, *2133*(1), 020013. doi:10.1063/1.5120143

Gao, H., Liao, L., Liang, Y., Tang, X., Liu, H., Mei, L., Lv, G., & Wang, L. (2021). Improvement of durability of porous perlite geopolymer-based thermal insulation material under hot and humid environment. *Construction & Building Materials*, *313*, 125417. doi:10.1016/j.conbuildmat.2021.125417

Gao, H., Liu, H., Liao, L., Mei, L., Zhang, F., Zhang, L., Li, S., & Lv, G. (2020). A bifunctional hierarchical porous kaolinite geopolymer with good performance in thermal and sound insulation. *Construction & Building Materials*, *251*, 118888. doi:10.1016/j.conbuildmat.2020.118888

Gencel, O., Nodehi, M., Hekimoğlu, G., Ustaoğlu, A., Sarı, A., Kaplan, G., Bayraktar, O. Y., Sutcu, M., & Ozbakkaloglu, T. (2022). Foam concrete produced with recycled concrete powder and phase change materials. *Sustainability (Basel)*, *14*(12), 7458. doi:10.3390/su14127458

Georgiou, L., Konatzii, P., Morsink-Georgali, P.-Z., Klumbyte, E., Christou, P., & Fokaides, P. A. (2023). Numerical and environmental analysis of post constructive application of PCM coatings for the improvement of the energy performance of building structures. *Construction & Building Materials*, *364*, 129984. doi:10.1016/j.conbuildmat.2022.129984

Güngör, D., & Özen, S. (2021). Development and characterization of clinoptilolite-, mordenite-, and analcime-based geopolymers: A comparative study. *Case Studies in Construction Materials*, *15*, e00576. doi:10.1016/j.cscm.2021.e00576

Haily, E., Ousaleh, H. A., Zari, N., Faik, A., Bouhfid, R., & Qaiss, A. (2023). Use of a form-stable phase change material to improve thermal properties of phosphate sludge-based geopolymer mortar. *Construction & Building Materials*, *386*, 131570. doi:10.1016/j.conbuildmat.2023.131570

Hammond, G., & Jones, C. (2011). *Embodied Carbon: The Inventory of Carbon and Energy (ICE), BSRIA Guide BG*. Services Research and Information Association. doi:10/2011. Building

Hu, W., & Yu, X. (2012). Encapsulation of bio-based PCM with coaxial electrospun ultrafine fibers. *RSC Advances*, *2*(13), 5580–5584. doi:10.1039/c2ra20532g

Hu, W., & Yu, X. (2014). Thermal and mechanical properties of bio-based PCMs encapsulated with nanofibrous structure. *Renewable Energy*, *62*, 454–458. doi:10.1016/j.renene.2013.07.047

Huang, X., Chen, X., Li, A., Atinafu, D., Gao, H., Dong, W., & Wang, G. (2019). Shape-stabilized phase change materials based on porous supports for thermal energy storage applications. *Chemical Engineering Journal*, *356*, 641–661. doi:10.1016/j.cej.2018.09.013

Imtiaz, L., Rehman, S. K. U., Ali Memon, S., Khizar Khan, M., & Faisal Javed, M. (2020). A review of recent developments and advances in eco-friendly geopolymer concrete. *Applied Sciences (Basel, Switzerland)*, *10*(21), 7838. doi:10.3390/app10217838

Jeong, S.-G., Chung, O., Yu, S., Kim, S., & Kim, S. (2013). Improvement of the thermal properties of Bio-based PCM using exfoliated graphite nanoplatelets. *Solar Energy Materials and Solar Cells*, *117*, 87–92. doi:10.1016/j.solmat.2013.05.038

Jeong, S.-G., Lee, J.-H., Seo, J., & Kim, S. (2014). Thermal performance evaluation of Bio-based shape stabilized PCM with boron nitride for energy saving. *International Journal of Heat and Mass Transfer*, *71*, 245–250. doi:10.1016/j.ijheatmasstransfer.2013.12.017

Ji, R., Li, X., Lv, C., & Dong, Y. (2022). Energy performance optimization of phase change materials considering the building micro-environment. *International Journal of Energy Research*, *46*(13), 18067–18078. doi:10.1002/er.8299

Jia, D., He, P., Wang, M., Yan, S., Jia, D., He, P., Wang, M., & Yan, S. (2020). Geopolymers and Their Matrix Composites: A State-of-the-Art Review. *Geopolymer and Geopolymer Matrix Composites*, 7–34.

Jiang, Z., Rivero, M. E. N., Liu, X., She, X., Xuan, Y., & Ding, Y. (2021). A novel composite phase change material for medium temperature thermal energy storage manufactured with a scalable continuous hot-melt extrusion method. *Applied Energy*, *303*, 117591. doi:10.1016/j.apenergy.2021.117591

Jin, Q., Long, X., & Liang, R. (2022). Numerical analysis on the thermal performance of PCM-integrated thermochromic glazing systems. *Energy and Building*, *257*, 111734. doi:10.1016/j.enbuild.2021.111734

Jin, X., Haider, M. Z., Ahn, J., Fang, G., & Hu, J. W. (2023). Development of nanomodified eco-friendly thermal energy storing cementitious composite using PCM microencapsulated in biosourced encapsulation shell. *Case Studies in Construction Materials*, *19*, e02447. doi:10.1016/j.cscm.2023.e02447

Kanagaraj, B., Anand, N., Lukose, J., Andrushia, D., & Lubloy, E. (2023). Influence of elevated temperature exposure on the interfacial shear strength capacity of binary blended high strength self-compacting geopolymer concrete. *Case Studies in Construction Materials*, *18*, e01974. doi:10.1016/j.cscm.2023.e01974

Kang, Y., Jeong, S.-G., Wi, S., & Kim, S. (2015). Energy efficient Bio-based PCM with silica fume composites to apply in concrete for energy saving in buildings. *Solar Energy Materials and Solar Cells*, *143*, 430–434. doi:10.1016/j.solmat.2015.07.026

Karaipekli, A., & Sarı, A. (2016). Development and thermal performance of pumice/organic PCM/gypsum composite plasters for thermal energy storage in buildings. *Solar Energy Materials and Solar Cells*, *149*, 19–28. doi:10.1016/j.solmat.2015.12.034

Korniejenko, K., Figiela, B., Kozub, B., Azzopardi, B., & Łach, M. (2022). Environmental degradation of foamed geopolymers. *Continuum Mechanics and Thermodynamics*, 1–15.

Krishna, R. S., Mishra, J., Zribi, M., Adeniyi, F., Saha, S., Baklouti, S., Shaikh, F. U. A., & Gökçe, H. S. (2021). A review on developments of environmentally friendly geopolymer technology. *Materialia*, *20*, 101212. doi:10.1016/j.mtla.2021.101212

Kumar, D., Alam, M., & Sanjayan, J. (2023). Experimental and numerical investigation of novel light weight concrete panels made with aerogel and phase change materials. *Energy and Building*, *283*, 112836. doi:10.1016/j.enbuild.2023.112836

Kushwah, S., Mudgal, M., & Chouhan, R. K. (2021). The Process, Characterization and Mechanical properties of fly Solid form geopolymer via mechanical activation. *South African Journal of Chemical Engineering*, *38*(1), 104–114. doi:10.1016/j.sajce.2021.09.002

Kylili, A., & Fokaides, P. A. (2016). Life cycle assessment (LCA) of phase change materials (PCMs) for building applications: A review. *Journal of Building Engineering*, *6*, 133–143. doi:10.1016/j.jobe.2016.02.008

Laasri, I. A., Es-sakali, N., Outzourhit, A., & Mghazli, M. O. (2022). Numerical building energy simulation with phase change materials including hysteresis effect for different square building cases in a semi-arid climate. *2022 13th International Renewable Energy Congress (IREC)*, 1–4.

Łach, M., Pławecka, K., Bąk, A., Adamczyk, M., Bazan, P., Kozub, B., Korniejenko, K., & Lin, W.-T. (2021). Review of solutions for the use of phase change materials in geopolymers. *Materials (Basel), 14*(20), 6044. doi:10.3390/ma14206044 PMID:34683631

Lagou, A., Kylili, A., Šadauskienė, J., & Fokaides, P. A. (2019). Numerical investigation of phase change materials (PCM) optimal melting properties and position in building elements under diverse conditions. *Construction & Building Materials, 225*, 452–464. doi:10.1016/j.conbuildmat.2019.07.199

Ling, T.-C., & Poon, C.-S. (2013). Use of phase change materials for thermal energy storage in concrete: An overview. *Construction & Building Materials, 46*, 55–62. doi:10.1016/j.conbuildmat.2013.04.031

Ma, Y., Chu, X., Li, W., & Tang, G. (2012). Preparation and characterization of poly (methyl methacrylate-co-divinylbenzene) microcapsules containing phase change temperature adjustable binary core materials. *Solar Energy, 86*(7), 2056–2066. doi:10.1016/j.solener.2012.04.008

Madeswaran, N., Desai, F. J., & Asmatulu, E. (2021). Life cycle inventory and performance analysis of phase change materials for thermal energy storages. *Emergent Materials, 4*(6), 1697–1709. doi:10.1007/s42247-021-00235-0

Marani, A., & Nehdi, M. L. (2019). Integrating phase change materials in construction materials: Critical review. *Construction & Building Materials, 217*, 36–49. doi:10.1016/j.conbuildmat.2019.05.064

Matsimbe, J., Dinka, M., Olukanni, D., & Musonda, I. (2022). Geopolymer: A systematic review of methodologies. *Materials (Basel), 15*(19), 6852. doi:10.3390/ma15196852 PMID:36234194

Mehrizi, A. A., Karimi-Maleh, H., Naddafi, M., & Karimi, F. (2023). Application of bio-based phase change materials for effective heat management. *Journal of Energy Storage, 61*, 106859. doi:10.1016/j.est.2023.106859

Meng, T., Yang, X., Yu, Y., Yu, H., & Huang, M. (2021). Hybrid effect of pva fibre and carbon nanotube on the mechanical properties and microstructure of geopolymers. *Frontiers in Materials, 8*, 773103. doi:10.3389/fmats.2021.773103

Messenger, M. A., Troxler, C. J., Melendez, I., Freeman, T. B., Reed, N., Rodriguez, R. M., & Boetcher, S. K. S. (2023). Mechanical and Thermal Characterization of Phase-Change Material and High-Density Polyethylene Functional Composites for Thermal Energy Storage. *Journal of Solar Energy Engineering, 145*(6), 61006. doi:10.1115/1.4062197

Munir, M. U., & Hussain, A. (2022). Thermal Stability Analysis of a PCM-Based Energy Storage System. *Engineering Proceedings, 23*(1), 20.

Navarro, L., Solé, A., Martín, M., Barreneche, C., Olivieri, L., Tenorio, J. A., & Cabeza, L. F. (2019). Benchmarking of useful phase change materials for a building application. *Energy and Building, 182*, 45–50. doi:10.1016/j.enbuild.2018.10.005

Nazari Sam, M., Caggiano, A., Mankel, C., & Koenders, E. (2020). A comparative study on the thermal energy storage performance of bio-based and paraffin-based PCMs using DSC procedures. *Materials (Basel), 13*(7), 1705. doi:10.3390/ma13071705 PMID:32260573

Novais, R. M., Pullar, R. C., & Labrincha, J. A. (2020). Geopolymer foams: An overview of recent advancements. *Progress in Materials Science*, *109*, 100621. doi:10.1016/j.pmatsci.2019.100621

Ong, P. J., Leow, Y., Soo, X. Y. D., Chua, M. H., Ni, X., Suwardi, A., Tan, C. K. I., Zheng, R., Wei, F., & Xu, J. (2023). Valorization of Spent coffee Grounds: A sustainable resource for Bio-based phase change materials for thermal energy storage. *Waste Management (New York, N.Y.)*, *157*, 339–347. doi:10.1016/j.wasman.2022.12.039 PMID:36603448

Parhizi, M., & Jain, A. (2019). The impact of thermal properties on performance of phase change based energy storage systems. *Applied Thermal Engineering*, *162*, 114154. doi:10.1016/j.applthermaleng.2019.114154

Park, J. H., Berardi, U., Chang, S. J., Wi, S., Kang, Y., & Kim, S. (2021). Energy retrofit of PCM-applied apartment buildings considering building orientation and height. *Energy*, *222*, 119877. doi:10.1016/j.energy.2021.119877

Qian, L.-P., Xu, L.-Y., Alrefaei, Y., Wang, T., Ishida, T., & Dai, J.-G. (2022). Artificial alkali-activated aggregates developed from wastes and by-products: A state-of-the-art review. *Resources, Conservation and Recycling*, *177*, 105971. doi:10.1016/j.resconrec.2021.105971

Rahemipoor, S., Hasany, M., Mehrali, M., Almdal, K., Ranjbar, N., & Mehrali, M. (2023). Phase change materials incorporation into 3D printed geopolymer cement: A sustainable approach to enhance the comfort and energy efficiency of buildings. *Journal of Cleaner Production*, *417*, 138005. doi:10.1016/j.jclepro.2023.138005

Rahjoo, M., Goracci, G., Gaitero, J. J., Martauz, P., Rojas, E., & Dolado, J. S. (2022). Thermal Energy Storage (TES) Prototype Based on Geopolymer Concrete for High-Temperature Applications. *Materials (Basel)*, *15*(20), 7086. doi:10.3390/ma15207086 PMID:36295154

Rasta, I. M., & Suamir, I. N. (2019). Study on thermal properties of bio-PCM candidates in comparison with propylene glycol and salt based PCM for sub-zero energy storage applications. *IOP Conference Series. Materials Science and Engineering*, *494*(1), 12024. doi:10.1088/1757-899X/494/1/012024

Rathore, P. K. S., & Shukla, S. K. (2021). Enhanced thermophysical properties of organic PCM through shape stabilization for thermal energy storage in buildings: A state of the art review. *Energy and Building*, *236*, 110799. doi:10.1016/j.enbuild.2021.110799

Rathore, P. K. S., Shukla, S. K., & Gupta, N. K. (2020). Potential of microencapsulated PCM for energy savings in buildings: A critical review. *Sustainable Cities and Society*, *53*, 101884. doi:10.1016/j.scs.2019.101884

Raut, A. N., Murmu, A. L., & Alomayri, T. (2023). Physico-Mechanical and thermal behavior of prolong heat Cured geopolymer blocks. *Construction & Building Materials*, *370*, 130309. doi:10.1016/j.conbuildmat.2023.130309

Regin, A. F., Solanki, S. C., & Saini, J. S. (2008). Heat transfer characteristics of thermal energy storage system using PCM capsules: A review. *Renewable & Sustainable Energy Reviews*, *12*(9), 2438–2458. doi:10.1016/j.rser.2007.06.009

Rehman, A. U., Sheikh, S. R., Kausar, Z., Grimes, M., & McCormack, S. J. (2022). Experimental Thermal Response Study of Multilayered, Encapsulated, PCM-Integrated Building Construction Materials. *Energies*, *15*(17), 6356. doi:10.3390/en15176356

Reyez-Araiza, J. L., Pineda-Piñón, J., López-Romero, J. M., Gasca-Tirado, J. R., Arroyo Contreras, M., Jáuregui Correa, J. C., Apátiga-Castro, L. M., Rivera-Muñoz, E. M., Velazquez-Castillo, R. R., Pérez Bueno, J., & Manzano-Ramirez, A. (2021). Thermal energy storage by the encapsulation of phase change materials in building elements—A review. *Materials (Basel)*, *14*(6), 1420. doi:10.3390/ma14061420 PMID:33804151

Richard, D., Martinez, J. M., Mizrahi, M., Andrini, L., & Rendtorff, N. M. (2022). Assessment of structural order indices in kaolinites: A multi-technique study including EXAFS. *Journal of Electron Spectroscopy and Related Phenomena*, *254*, 147128. doi:10.1016/j.elspec.2021.147128

Ryms, M., & Klugmann-Radziemska, E. (2019). Possibilities and benefits of a new method of modifying conventional building materials with phase-change materials (PCMs). *Construction & Building Materials*, *211*, 1013–1024. doi:10.1016/j.conbuildmat.2019.03.277

Sabbatini, A., Vidal, L., Pettinari, C., Sobrados, I., & Rossignol, S. (2017). Control of shaping and thermal resistance of metakaolin-based geopolymers. *Materials & Design*, *116*, 374–385. doi:10.1016/j.matdes.2016.12.039

Sawadogo, M., Duquesne, M., Belarbi, R., Hamami, A. E. A., & Godin, A. (2021). Review on the integration of phase change materials in building envelopes for passive latent heat storage. *Applied Sciences (Basel, Switzerland)*, *11*(19), 9305. doi:10.3390/app11199305

Saxena, R., Rakshit, D., & Kaushik, S. C. (2019). Phase change material (PCM) incorporated bricks for energy conservation in composite climate: A sustainable building solution. *Solar Energy*, *183*, 276–284. doi:10.1016/j.solener.2019.03.035

Shao, Z., Wang, J., Jiang, Y., Zang, J., Wu, T., Ma, F., Qian, B., Wang, L., Hu, Y., & Ma, B. (2021). The performance of micropore-foamed geopolymers produced from industrial wastes. *Construction & Building Materials*, *304*, 124636. doi:10.1016/j.conbuildmat.2021.124636

Sharma, A., Tyagi, V. V., Chen, C. R., & Buddhi, D. (2009). Review on thermal energy storage with phase change materials and applications. *Renewable & Sustainable Energy Reviews*, *13*(2), 318–345. doi:10.1016/j.rser.2007.10.005

Sharma, R., Jang, J.-G., & Hu, J.-W. (2022). Phase-change materials in concrete: Opportunities and challenges for sustainable construction and building materials. *Materials (Basel)*, *15*(1), 335. doi:10.3390/ma15010335 PMID:35009481

Sharshir, S. W., Joseph, A., Elsharkawy, M., Hamad, M. A., Kandeal, A. W., Elkadeem, M. R., Thakur, A. K., Ma, Y., Moustapha, M. E., & Rashad, M. (2023). Thermal energy storage using phase change materials in building applications: A review of the recent development. *Energy and Building*, *285*, 112908. doi:10.1016/j.enbuild.2023.112908

Silverstein, M. S. (2017). Emulsion-templated polymers: Contemporary contemplations. *Polymer*, *126*, 261–282. doi:10.1016/j.polymer.2017.07.046

Sun, X., Zhang, Y., Xie, K., & Medina, M. A. (2022). A parametric study on the thermal response of a building wall with a phase change material (PCM) layer for passive space cooling. *Journal of Energy Storage*, *47*, 103548. doi:10.1016/j.est.2021.103548

Suppes, G. J., Goff, M. J., & Lopes, S. (2003). Latent heat characteristics of fatty acid derivatives pursuant phase change material applications. *Chemical Engineering Science*, *58*(9), 1751–1763. doi:10.1016/S0009-2509(03)00006-X

Tobarameekul, P., & Worathanakul, P. (2021). Heat management options to reduce carbon footprint of green zeolite faujasite synthesis from rice husk ash. *Thai Environmental Engineering Journal*, *35*(3), 59–68.

Tyagi, V. V., Chopra, K., Sharma, R. K., Pandey, A. K., Tyagi, S. K., Ahmad, M. S., Sarı, A., & Kothari, R. (2022). A comprehensive review on phase change materials for heat storage applications: Development, characterization, thermal and chemical stability. *Solar Energy Materials and Solar Cells*, *234*, 111392. doi:10.1016/j.solmat.2021.111392

Vicente, R., & Silva, T. (2014). Brick masonry walls with PCM macrocapsules: An experimental approach. *Applied Thermal Engineering*, *67*(1–2), 24–34. doi:10.1016/j.applthermaleng.2014.02.069

Walubita, L. F., Martinez-Arguelles, G., Polo-Mendoza, R., Ick-Lee, S., & Fuentes, L. (2022). Comparative environmental assessment of rigid, flexible, and perpetual pavements: A case study of Texas. *Sustainability (Basel)*, *14*(16), 9983. doi:10.3390/su14169983

Wan, Q., Zhang, Y., & Zhang, R. (2021). The effect of pore behavior and gel structure on the mechanical property at different initial water content. *Construction & Building Materials*, *309*, 125146. doi:10.1016/j.conbuildmat.2021.125146

Wang, Z., Su, H., Zhao, S., & Zhao, N. (2016). Influence of phase change material on mechanical and thermal properties of clay geopolymer mortar. *Construction & Building Materials*, *120*, 329–334. doi:10.1016/j.conbuildmat.2016.05.091

Wołoszyn, J., Szopa, K., & Czerwiński, G. (2021). Enhanced heat transfer in a PCM shell-and-tube thermal energy storage system. *Applied Thermal Engineering*, *196*, 117332. doi:10.1016/j.applthermaleng.2021.117332

Xiao, T., Shi, X., Gen, L., Dai, Y., Zhao, J., & Liu, C. (2023). *Synergistic Enhancement of Phase Change Materials through Three-Dimensional Macropore Lamellar Structured MOF/EG Composite for Solar Energy Storage and Beyond*. Eg Composite for Solar Energy Storage and Beyond. doi:10.1016/j.applthermaleng.2023.121378

Xu, G., & Shi, X. (2018). Characteristics and applications of fly ash as a sustainable construction material: A state-of-the-art review. *Resources, Conservation and Recycling*, *136*, 95–109. doi:10.1016/j.resconrec.2018.04.010

Xu, L.-Y., Huang, B.-T., & Dai, J.-G. (2021). Development of engineered cementitious composites (ECC) using artificial fine aggregates. *Construction & Building Materials*, *305*, 124742. doi:10.1016/j.conbuildmat.2021.124742

Xu, L.-Y., Huang, B.-T., Li, V. C., & Dai, J.-G. (2022). High-strength high-ductility Engineered/Strain-Hardening Cementitious Composites (ECC/SHCC) incorporating geopolymer fine aggregates. *Cement and Concrete Composites*, *125*, 104296. doi:10.1016/j.cemconcomp.2021.104296

Xu, L.-Y., Qian, L.-P., Huang, B.-T., & Dai, J.-G. (2021). Development of artificial one-part geopolymer lightweight aggregates by crushing technique. *Journal of Cleaner Production*, *315*, 128200. doi:10.1016/j.jclepro.2021.128200

Yousefi, A., Tang, W., Khavarian, M., & Fang, C. (2021). Development of novel form-stable phase change material (PCM) composite using recycled expanded glass for thermal energy storage in cementitious composite. *Renewable Energy*, *175*, 14–28. doi:10.1016/j.renene.2021.04.123

Yu, S., Jeong, S.-G., Chung, O., & Kim, S. (2014). Bio-based PCM/carbon nanomaterials composites with enhanced thermal conductivity. *Solar Energy Materials and Solar Cells*, *120*, 549–554. doi:10.1016/j.solmat.2013.09.037

Zhang, H., Gao, H., Bernardo, E., Lei, S., & Wang, L. (2023). Thermal energy storage performance of hierarchical porous kaolinite geopolymer based shape-stabilized composite phase change materials. *Ceramics International*, *49*(18), 29808–29819. doi:10.1016/j.ceramint.2023.06.238

Zhang, X., Bai, C., Qiao, Y., Wang, X., Jia, D., Li, H., & Colombo, P. (2021). Porous geopolymer composites: A review. *Composites. Part A, Applied Science and Manufacturing*, *150*, 106629. doi:10.1016/j.compositesa.2021.106629

Zhuk, P. (2019). Lifecycle analysis of finishing products enhanced with phase changing materials. *IOP Conference Series. Earth and Environmental Science*, *323*(1), 12154. doi:10.1088/1755-1315/323/1/012154

Zimele, Z., Sinka, M., Korjakins, A., Bajare, D., & Sahmenko, G. (2019). Life Cycle Assessment of Foam Concrete Production in Latvia. *Environmental & Climate Technologies, 23*(3).

Chapter 3
Materials for Phase Change Heat Storage in Scheffler Reflectors:
A Sustainable and Perceptive Analysis of Renewable Energy

Sabyasachi Pramanik
https://orcid.org/0000-0002-9431-8751
Haldia Institute of Technology, India

ABSTRACT

In order for humans to meet their fundamental needs, energy is required. The main sources of natural energy are carbon-based fuels such as biomass and fossil fuels. These fuels made of carbon are scarce and harmful to the environment. Innovative clean and environmentally friendly energy solutions are desperately needed. Particularly in light of the global issues posed by population expansion, the environment, and economic growth, solar energy may be a workable alternative for sustainable development. The Scheffler is a unique kind of solar concentrator that can quickly attain a high temperature. By adding a heat storage unit, Scheffler's capacity to retain heat may be greatly boosted. This study compares the thermal and physical properties, such as density, thermal conductivity, and melting point, of various thermal storage materials that are used to store heat energy for a longer amount of time, such as bricks, concrete, water, phase change material (PCM), latent heat storage material (LHS), aluminum, iron, molten salts, silicon, etc.

INTRODUCTION

Fossil fuel supplies across the planet are finite and will run out by 2088. In many nations, a significant section of the population still relies on carbon fuels including wood, biomass, and other hydrocarbons (Aniket & Akhade, n.d.). When these carbon-based fuels burn, a significant number of hazardous

DOI: 10.4018/979-8-3693-1306-0.ch003

chemicals, including greenhouse gases, are released (Patel & Patel, 2022). Fossil fuels take thousands of years to produce, and cutting them down quickly might cause them to run out. To address the people's fundamental energy needs, alternatives to these sources of energy should be investigated (Kumar et al., 2017). Clean and environmentally friendly, renewable energy sources, particularly solar energy, may be excellent choices (Jayasimha, n.d.). Solar energy may be used in a variety of ways and with a variety of equipment, including solar heaters, panels, concentrators, and their combinations. Solar concentrators are utilised in a wide range of applications, including cooking, power generating, heating and cooling, and regions that need high temperatures. An effective solar device for concentrating sunrays at a fixed focus is a Scheffler reflector (see Figure 1). It must follow the Sun's daily and seasonal movements in order to keep a constant focus (Kim & Schwartz, 2013). Scheffler's tracking system, which tracks the movement of the Sun, may be either automated or manual. Figure 1 illustrates the steps involved in the functioning of a Scheffler. The sun's rays strike the reflector and are focused on a fixed focus, or receiver, which converts the incoming light into heat that may be utilised for a variety of tasks. It is possible to cover the receiver with a substance that retains heat well. Cooking is its main use, although it may also be used for desalination, power production, cremation, and other processes. (Reddy et al., 2018; Santosh Junare et al., 2017) as well, which call for a lot of heat. It is used in communal kitchens to prepare meals for a large crowd. The world's biggest solar kitchen, constructed by Brahma Kumari, has the capacity to prepare 38,500 meals daily. Scheffler depends on solar energy for operation, hence it will not function well on cloudy or gloomy days. The use of heat-storage materials, which retain heat for a long time and allow Scheffler to operate for longer periods of time, may help to reduce this issue.

Figure 1. Scheffler reflector used for desalination of water

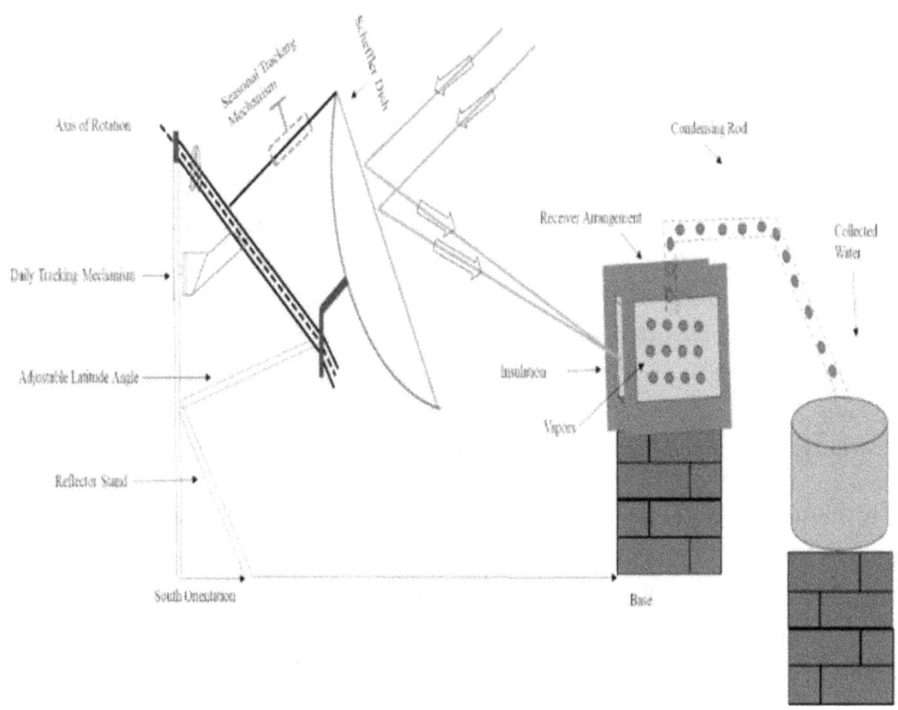

The present study will be conducted as follows: The varieties of heat storage materials and their features that are appropriate for use in Scheffler reflectors are covered in Part II, along with a comparison of the desirable attributes in each material. Part III covered the difficulties encountered with generating heat using phase-change materials. The last section of this research discusses the future and conversation in this subject.

There are two ways to store energy as heat: one way is to change the phase, and another is to do it without altering it (Hassan et al., 2022). Since phase change heat storage can store more heat for a given unit volume and maintains a constant temperature during the transfer process, it seems more appealing than more practical heat storage methods. A substance's temperature rises when heat is applied, and thermal energy is just stored there (Ahmed et al., 2022; Cárdenas & León, 2013). As the temperature dropped, this energy may be collected from the designated medium. They also support a broad range of applications and can function in a variety of temperatures. The following characteristics are present in a phase-changing material:

- As phases change, there shouldn't be a significant difference in volume.
- A large latent heat capacity is required of the material (Chandel & Agarwal, 2017).
- It must not be corrosive or react with the container it is stored in.
- Its thermos-physical characteristics shouldn't be lost too quickly during phase transitions (Salunkhe & Jaya Krishna, 2017).
- There should be no damage to the ecosystem.
- The material should be inexpensive (Lingayat & Suple, n.d.).
- It must be easy to access.
- When functioning at a greater temperature, it shouldn't catch fire.

Table 1 displays the thermophysical properties and heat storage of the LHS and non-latent materials, respectively. Density, melting and boiling temperatures, thermal conductivity, and heat capacity are some of these properties.

As shown in Fig. 2, non-LHS materials have larger densities than LHS materials, such as silicon (Kg/m3), iron (7874 Kg/m3), aluminium, pebbles (1900 Kg/m3), and aluminium (2710 Kg/m3), but their heat capacities are much lower (0.45 J/gK, 0.89 J/gK, 0.88 J/gK, and 0.71 J/gK, respectively) (Aydin, Casey, & Riffat, 2015; Reddy et al., 2018). Thus, across all materials, density is not a function of heat capacity; rather, as was previously said, a high density material would take less space than a low density one. As shown by Table 2, a good material (LHS) for a Scheffler Reflector can be one that has a good heat capacity and a compromised density, such as acetamide (263 J/g-K and 1159 Kg/m3), erythritol (339.8 J/g-K and 1480 Kg/m3), sodium dihydrogen phosphate dodecahydrate (279.6 J/g-K and 1520 Kg/m3), etc.

Figure 3 illustrates that materials with low melting points often have large heat capacities. Brick and iron have low heat capacities of 0.84 J/g-K and 0.45 J/g-K, respectively, but high melting points of 1811 K and 2000 K. With the exception of water, PCM materials such as acetamide and lauric acid have low melting points of 355 K and 315 K, respectively, but high heat capacities of 263 J/g-K and 200 J/g-K, making them excellent candidates for use as storage materials in Scheffler reactors.

As can be seen in Fig. 4, PCM material has a very low heat conductivity when compared to other common materials such as silicon (230 W/m-K), iron (79.5 W/m-K), aluminium (237 W/m-K), etc. Thus,

Table 1. Thermo-physical properties of various materials (Chandel & Agarwal, 2017)

MATERIALS *(PROPERTIES)*	VALUES
BRICKS	
Density	1700 kg/m^3
Melting Point	2000 K
Thermal Conductivity	1.31 W/m-K
Heat Capacity	0.84 J/g-K
CONCRETE	
Density	2400 kg/m^3
Melting Point	1800 K
Thermal Conductivity	2.25 W/m-K
Heat Capacity	0.88J/g-K
SAND	
Density	1442 kg/m^3
Melting Point	1973 K
Thermal Conductivity	2.05 W/m-K
Heat Capacity	0.830 J/g-K
PEBBLES	
Density	(1846 & 2399.8) kg/m^3
Melting Point	1753 K
Thermal Conductivity	2.5 W/m-K
Heat Capacity	0.88050 J/g-K
ALUMINUM	
Density	2710 Kg/m^3
Melting Point	933.47 K
Thermal Conductivity	237 W/m-K
Heat Capacity	0.89 J/g-K
IRON	
Density	7874 Kg/m^3
Melting Point335	1811 K
Thermal Conductivity	79.5 W/m-K
Heat Capacity	0.450 J/g-K
MOLTEN SALTS	
Density	1549 kg/m^3
Melting Point	773 K
Thermal Conductivity	0.652–0.927 W/m-K
Heat Capacity	1.53 J/g-K
SILICON	
Density	2330 Kg/m^3
Melting Point	1687 K
Thermal Conductivity	230 W/m-K
Heat Capacity	0.71 J/g-K

Table 2. Thermo-physical properties of PCM (Chandel & Agarwal, 2017)

PARAFFIN WAX	VALUES
Density	**900 Kg/m3**
Melting Point	**643 K**
Thermal Conductivity	**7.1 W/m-k**
Heat Capacity	**(2.14–2.9) J/g-K**
ACETAMIDE	
Density	**1159 Kg/m3**
Melting Point	**355 K**
Thermal Conductivity	**0.5 W/m-K**
Heat Capacity	**263 J/g-K**
STEARIC ACID	
Density	**940 Kg/m3**
Melting Point	**337.5 K**
Thermal Conductivity	**1.60 W/m-K**
Heat Capacity	**196 J/g-K**
MALIC ACID	
Density	**860 Kg/m3**
Melting Point	**325 K**
Thermal Conductivity	**0.17 W/m-K**
Heat Capacity	**192 J/g-K**
Calcium Chloride Hexahydrate	
Density	**1710 Kg/m3**
Melting Point	**303 K**
Thermal Conductivity	**1.08 W/m-K**
Heat Capacity	**178 J/g-K**
Magnesium Chloride Hexahydrate	
Density	**1450 Kg/m3**
Melting Point	**390 K**
Thermal Conductivity	**0.570 W/m-K**
Heat Capacity	**168.6 J/g-K**
PALMITIC ACID	
Density	**942 Kg/m3**
Melting Point	**335 K**
Thermal Conductivity	**0.162 W/m-K**
Heat Capacity	**205 J/g-K**
LAURIC ACID	
Density	**862 Kg/m3**
Melting Point	**315 K**
Thermal Conductivity	**0.16 W/m-K**
Heat Capacity	**200 J/g-K**
ERYTHRITOL	
Density	**1480 Kg/m3**
Melting Point	**391 K**
Thermal Conductivity	**0.326 W/m-K**
Heat Capacity	**339.8 J/g-K**
Sodium Dihydrogen Phosphate Dodecahydrate	
Density	**1520 Kg/m3**
Melting Point	**313 K**
Thermal Conductivity	**0.476 W/m-K**
Heat Capacity	**279.6 J/g-K**
WATER	
Density	**998 kg/m3**
Melting Point	**373 K**
Thermal Conductivity	**0.598 W/m-K**
Heat Capacity	**4.182 J/g-K**

Figure 2. Graph between heat capacity and density

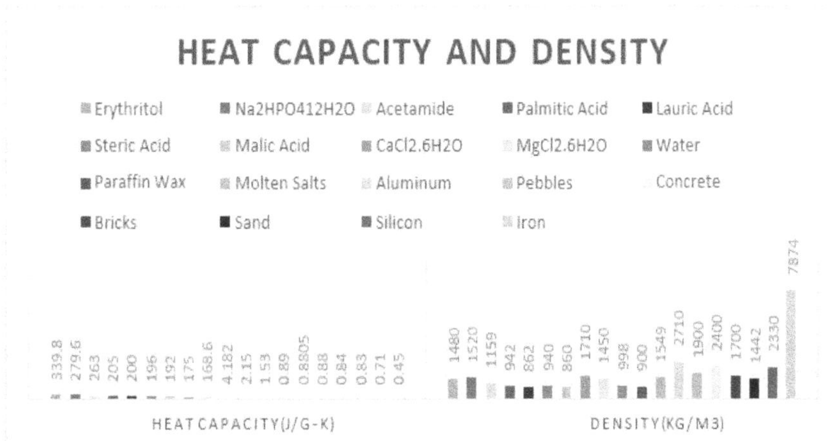

Figure 3. Graph between heat capacity and melting point

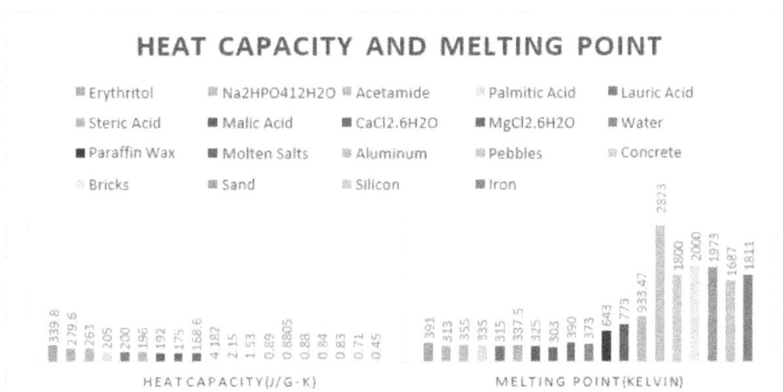

Figure 4. Graph between heat capacity and thermal conductivity

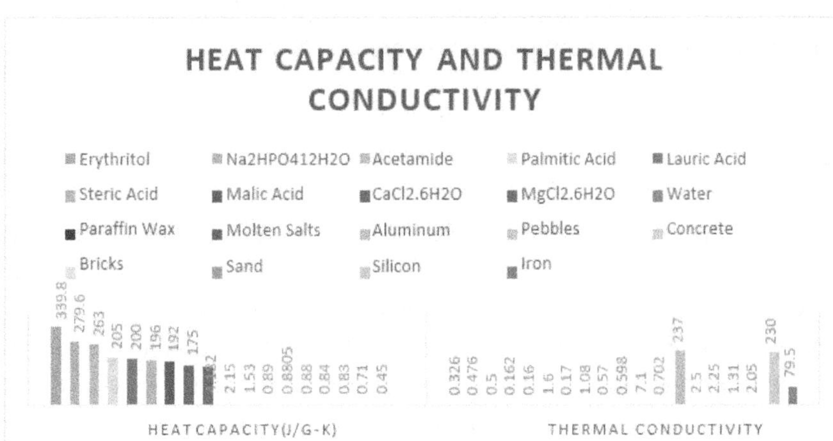

it has been discovered that low heat storage (THS) materials with high heat storage qualities, such as erythritol (0.326 W/m-K) and acetamide (0.5 W/m-K), will have low thermal conductivity.

ACADEMIES

A. Safety

It makes logical to utilise a heat storage material with higher LHS characteristics and improved safety since safety is an important consideration when selecting one for a unit (Salunkhe & Jaya Krishna, 2017). Despite having a low ignition temperature, certain organic materials, such paraffin and non-paraffin, are appropriate LHS compounds (Lingayat & Suple, n.d.). Their use is restricted to temperatures in the middle range.

B. Expensive For High Density Of Energy

Thermochemical storage is one of the possibilities because of its high energy density of storage, despite the fact that both the apparatus and the procedure are relatively costly (Aydin, Casey, & Riffat, 2015). Thermal absorption is chosen over it because it is less costly, especially in applications needing solar heat (Abedin, 2011).

C. PCM Selection Standard

It is possible to create a phase-change material having both thermal and physical properties (Leong et al., 2019). Certain compounds, such as paraffin, are better suited for usage in low-temperature applications, whilst fatty acid is used in high-temperature applications (Ferrer et al., 2015).

D. Material Reactivity for Storage

Despite having excellent heat storage capacity and low associated costs, some substances, such as salt compounds (Ozeryanaya, 1985) and metal alloys (mainly used in heat-based applications in the solar energy area) (Ding et al., 2018), are not suitable for use as heat-absorbing substances for Scheffler usage due to their corrosive nature and rapid temperature decrease (Nene et al., 2019).

CONCLUSION

The generation of thermal energy and its use as a main source of renewable energy have been shown to be significantly aided by the Scheffler reflector's functioning. The dish's size may be adjusted to provide heat for applications that need a broad temperature range. The kind of materials utilised to store heat as a unit affects how heat is managed and controlled in Scheffler. In order to manage heat over an extended length of time, a variety of heat storage materials were compared and contrasted based on thermos-physical properties such density, melting point, heat capacity, and thermal conductivity.

A phase-change material (PCM) is a material that, when its phase changes, emits or stores energy to produce heating or cooling. PCMs are known for having a large heat capacity and can retain heat for a long period. Due to the fact that a substance's latent heat is much greater than its specific heat, latent heating may store the same amount of heat in a smaller space than specific heat storage. The results show that LHS materials, such as erythritol, acetamide, and sodium dihydrogen phosphate dodecahydrate, have higher heat capacities than other heat storage materials. The findings also demonstrate that specific heat does not immediately correlate with density, but a high-density material would take up less space in a thermal storage device. Therefore, the best materials for heat storage in a Scheffler reflector are compromised materials like acetamide (263 J/g-K and 1159 Kg/m3), erythritol (339.8 J/g-K and 1480 Kg/m3), and sodium dihydrogen phosphate dodecahydrate (279.6 J/g-K and 1520 Kg/m3), among others. Research indicates that substances with low melting or boiling points will also have high heat capacities. For instance, acetamide and lauric acid have low melting points of 355 K and 315 K, respectively, but high heat capacities of 263 J/g-K and 200 J/g-K. Results showed that materials with greater specific heat values had lower thermal conductivities. As a result, LHS materials are the ideal option when it comes to heat storage units for Scheffler reflectors since they have a high heat-storage capacity.

There are several possible uses for LHS material in the Scheffler reflector industry. To increase the practical material's uses and societal exposure, the industrial and inventive domains should develop dedicated awareness initiatives. Despite the excellent outcomes that thermochemical energy storage systems have yielded, the construction of Scheffler instruments using this technology is now quite expensive.

REFERENCES

Abedin, A. H. (2011, August). A Critical Review of Thermochemical Energy Storage Systems. *The Open Renewable Energy Journal*, *4*(1), 42–46. doi:10.2174/1876387101004010042

Ahmed, S. F., Rafa, N., Mehnaz, T., Ahmed, B., Islam, N., Mofijur, M., Hoang, A. T., & Shafiullah, G. M. (2022, September). Integration of phase change materials in improving the performance of heating, cooling, and clean energy storage systems: An overview. *Journal of Cleaner Production*, *364*, 132639. Advance online publication. doi:10.1016/j.jclepro.2022.132639

Aniket & Akhade. (n.d.). *Review of Scheffler reflector*. Academic Press.

Aydin, D., Casey, S. P., & Riffat, S. (2015, January). The latest advancements on thermochemical heat storage systems. *Renewable & Sustainable Energy Reviews*, *41*, 356–367. doi:10.1016/j.rser.2014.08.054

Aydin, D., Casey, S. P., & Riffat, S. (2015, January). The latest advancements on thermochemical heat storage systems. *Renewable & Sustainable Energy Reviews*, *41*, 356–367. doi:10.1016/j.rser.2014.08.054

Cárdenas, B., & León, N. (2013). High temperature latent heat thermal energy storage: Phase change materials, design considerations and performance enhancement techniques. *Renewable & Sustainable Energy Reviews*, *27*, 724–737. doi:10.1016/j.rser.2013.07.028

Chandel & Agarwal. (2017). Review of current state of research on energy storage, toxicity, health hazards and commercialization of phase changing materials. *Renewable and Sustainable Energy Reviews, 67*, 581–596. . doi:10.1016/j.rser.2016.09.070

Ding, Bonk, & Bauer. (2018). Corrosion behavior of metallic alloys in molten chloride salts for thermal energy storage in concentrated solar power plants: A review. *Frontiers of Chemical Science and Engineering, 12*(3), 564–576. . doi:10.1007/s11705-018-1720-0

Ferrer, G., Solé, A., Barreneche, C., Martorell, I., & Cabeza, L. F. (2015, April). Corrosion of metal containers for use in PCM energy storage. *Renewable Energy*, *76*, 465–469. doi:10.1016/j.renene.2014.11.036

Hassan, F., Jamil, F., Hussain, A., Ali, H. M., Janjua, M. M., Khushnood, S., Farhan, M., Altaf, K., Said, Z., & Li, C. (2022, February). Recent advancements in latent heat phase change materials and their applications for thermal energy storage and buildings: A state of the art review. *Sustainable Energy Technologies and Assessments*, *49*, 101646. Advance online publication. doi:10.1016/j.seta.2021.101646

Jayasimha. (n.d.). *Application of Scheffler Reflectors for Process Industry*. Academic Press.

Kim & Schwartz. (2013). *An Inexpensive Scheffler Reflector for Developing Nations*. Academic Press.

Kumar, A., Prakash, O., & Kaviti, A. K. (2017, September). A comprehensive review of Scheffler solar collector. *Renewable & Sustainable Energy Reviews*, *77*, 890–898. doi:10.1016/j.rser.2017.03.044

Leong, K. Y., Abdul Rahman, M. R., & Gurunathan, B. A. (2019, February). Nano-enhanced phase change materials: A review of thermo-physical properties, applications and challenges. *Journal of Energy Storage*, *21*, 18–31. doi:10.1016/j.est.2018.11.008

Lingayat, A. B., & Suple, Y. R. (n.d.). *Review on Phase Change Material as Thermal Energy Storage Medium: Materials, Application*. Available: www.ijera.com

Lingayat, A. B., & Suple, Y. R. (n.d.). *Review On Phase Change Material As Thermal Energy Storage Medium: Materials, Application*. Available: www.ijera.com

Nene, A. A., Ramachandran, S., & Suyambazhahan, S. (2019). Design and analysis of solar thermal energy storage system for Scheffler solar concentrator. *Comptes Rendus de l'Academie Bulgare des Sciences, 72*(10), 1321–1325. Available:https://inis.iaea.org/Search/search.aspx?orig_q=RN:51004309

Ozeryanaya, I. N. (1985, March). Corrosion of metals by molten salts in heat-treatment processes. *Metal Science and Heat Treatment*, *27*(3), 184–188. doi:10.1007/BF00699649

Patel & Patel. (2022). Scheffler reflector for cooking application: a review. *International Journal of Ambient Energy, 43*(1), 2606–2618. . doi:10.1080/01430750.2020.1758778

Reddy, D. S., Khan, M. K., Alam, M. Z., & Rashid, H. (2018, March). Design charts for Scheffler reflector. *Solar Energy*, *163*, 104–112. doi:10.1016/j.solener.2018.01.081

Salunkhe, P. B., & Jaya Krishna, D. (2017, August). Investigations on latent heat storage materials for solar water and space heating applications. *Journal of Energy Storage*, *12*, 243–260. doi:10.1016/j.est.2017.05.008

Santosh Junare, S., Vilas Zamre, S., & Madhukar Aware, M. (2017). *Scheffler Dish and its Applications*. Available: http://www.ijritcc.org

Chapter 4
Advancements in Vanadium Disulphide (VS_2) Nanomaterials for Enhanced Energy Storage Synthesis Techniques, Electrochemical Properties, and Prospective Outcomes

S. Harikrishnan
https://orcid.org/0009-0003-2738-1042
Lovely Professional University, India

Deepak Sharma
https://orcid.org/0000-0002-6358-2453
Lovely Professional University, India

A. R. Abhijith
https://orcid.org/0009-0002-5175-020X
Lovely Professional University, India

Amar Srivastava
Lovely Professional University, India

Jyoti Rajput
Lovely Professional University, India

ABSTRACT

Nanomaterials have become a viable way to improve the performance of batteries and supercapacitors in the quest for effective and sustainable energy storage solutions. This chapter provides an in-depth look at current advances in synthesis processes and applications of VS_2-based nanomaterials in energy storage devices. The discussion carefully assesses electrochemical characteristics of VS_2-based nanomaterials while critically analyzing alternative synthesis techniques and composite approaches. Furthermore, the chapter highlights current research discoveries concerning the performance of VS_2-based nanomaterials in various types of batteries. Additionally, potential uses of VS_2-based nanomaterials in energy storage devices, such as lithium-ion batteries, sodium-ion batteries, potassium-ion batteries, zinc-ion batteries, magnesium-ion batteries, aluminum-ion batteries, and ammonium-ion batteries, are investigated. This review emphasizes the importance of VS_2-based nanomaterials in energy storage applications and offers possible future research directions in this rapidly expanding sector.

DOI: 10.4018/979-8-3693-1306-0.ch004

INTRODUCTION

Batteries and supercapacitors have become essential energy storage devices in modern society. With the widespread adoption of renewable energy sources, electric vehicles, and portable electronic devices, there is a rising in the demand for energy storage systems that offer superior performance, long cycle life, and fast charging/discharging rates (Speirs et al., 2014). Energy storage systems play a vital role in ensuring a reliable, efficient, and sustainable energy supply. They help to balance the electricity grid, reduce energy consumption during peak demand, and provide backup power during emergencies (Das et al., 2018). Furthermore, energy storage systems can enable the integration of renewable energy sources into the grid by addressing issues such as intermittency and variability (Worighi et al., 2019).

When compared to other energy storage devices, batteries offer several advantages. One of the most significant benefits is their high energy density, which allows for the storage of large amounts of energy in a small volume (Dunn et al., 2011; Tan et al., 2013). This makes them ideal for use in electric vehicles, where a high energy-density battery can provide the necessary range for long journeys. Additionally, batteries can store energy for long periods without the need for maintenance or frequent recharging, making them suitable for backup power systems (Fichtner, 2022; Sharma et al., 2020). Another advantage of batteries is their portability, which makes them perfect for use in portable electronic devices such as smartphones, laptops, and cameras. Finally, batteries can charge and discharge quickly, which is crucial for applications that require rapid energy delivery, for example, batteries in electric vehicles or energy storage systems at the grid-scale level (Assefi et al., 2020; Hu, 2019; Liang et al., 2019).

Batteries come in various types and are classified based on their electrochemical reaction. Primary and secondary batteries are the two primary categories of batteries (Goel et al., 2020). Primary batteries, also known as disposable batteries, are intended for single use and cannot be recharged. On the other hand, secondary batteries, commonly referred to as rechargeable batteries, can be recharged and reused multiple times (Gayathri et al., 2021). Examples of secondary batteries include Lead-acid batteries, Lithium-ion batteries, and Nickel-Cadmium batteries, which are used in a wide range of applications, from powering electric vehicles to storing renewable energy generated from solar or wind sources (Fan et al., 2020; Jeyaseelan, 2020; Wang, Tan, Bai et al, 2021). Other types of batteries include flow batteries, which are rechargeable batteries that store energy in an electrolyte solution outside the battery, and sodium-ion batteries, which are gaining attention for their potential as an economical and environmentally friendly substitute for lithium-ion batteries (Hirsh, 2020; Lourenssen et al., 2019; Yao et al., 2021).

Lead-acid batteries, recognized as one of the oldest and most prevalent rechargeable battery types, are extensively employed in too many applications including vehicles, backup power systems, and other scenarios that necessitate a dependable and economically viable energy storage solution (Jiang & Song, 2022). Nickel-cadmium batteries were once the most commonly used rechargeable batteries because of their long cycle life and durability, but they have been largely replaced by lithium-ion batteries (Banerjee & Kaur, 2022). Lithium-ion batteries provide notable advantages, including high energy density, extended cycle life and rapid charging capabilities. These attributes make them well-suited for a range of applications such as electric vehicles, portable electronic devices, and grid-scale energy storage systems (Wong, 2021). Sodium-ion batteries, an emerging battery technology, are garnering attention for their promising potential as a cost-effective and environmentally friendly substitute for lithium-ion batteries. These batteries exhibit high energy density and find applicability in diverse domains, including electric vehicles and grid-scale energy storage systems (Hijazi et al., 2021). While each of these batteries has

Figure 1. Vanadium disulfide

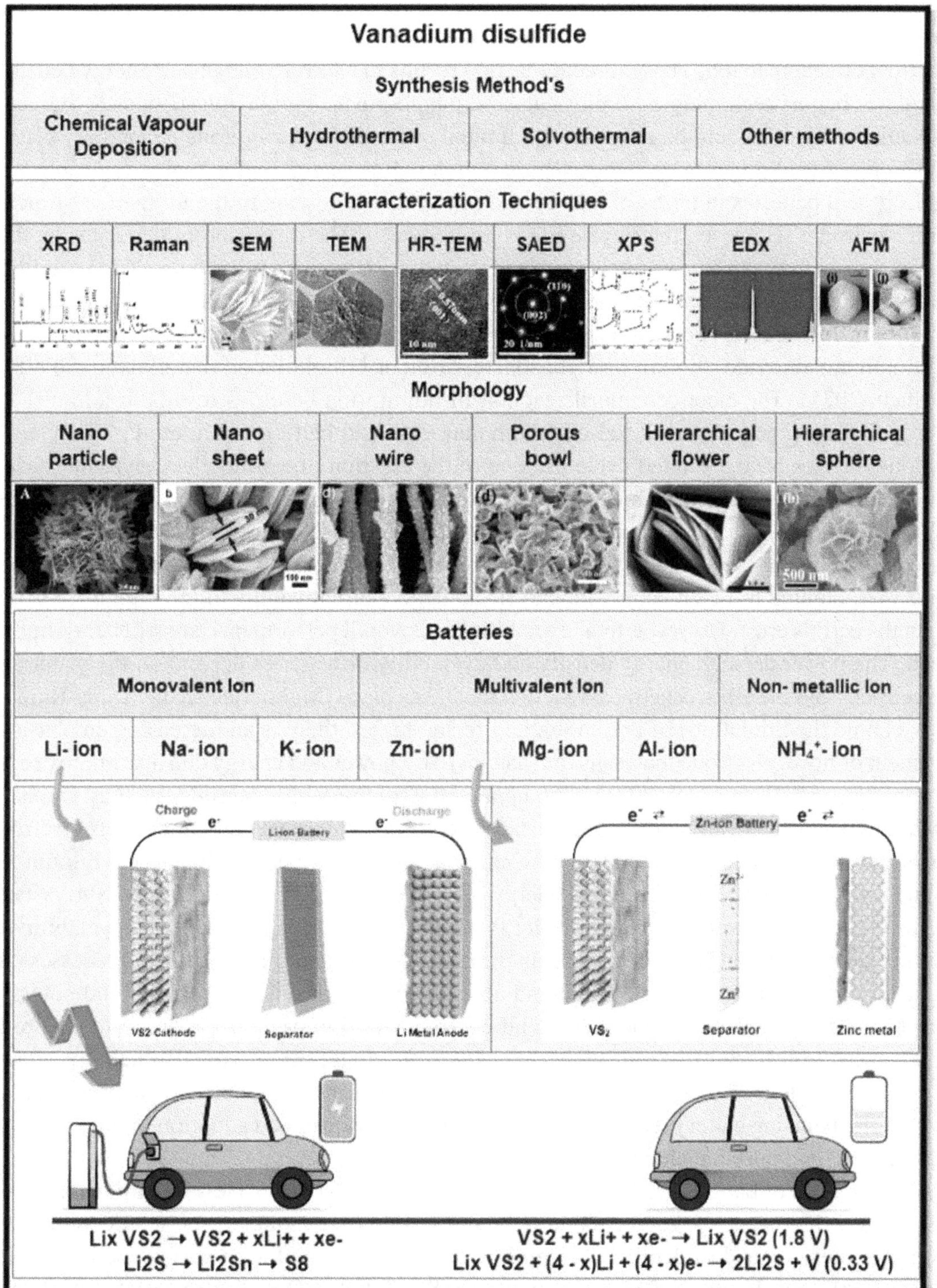

its advantages and disadvantages, they all play an important role in meeting the growing demand for reliable and sustainable energy storage solutions.

Lead-acid batteries have a relatively low-power and energy density due to the limitations of the lead-acid electrochemical reaction. The lead-acid reaction results in a lower voltage and energy density compared to newer battery technologies, limiting its use in high-power applications (Lopes & Stamenkovic, 2020). Additionally, lead-acid batteries have a limited cycle life and are prone to memory effects due to the build-up of lead sulphate crystals on the electrodes (Bais et al., 2022). Nickel-cadmium batteries surpass lead-acid batteries in terms of energy and power density owing to the utilization of nickel and cadmium electrodes. However, their cycle life is limited due to the formation of dendrites on the electrodes over time, which can cause a short circuit in the battery (Petrovic & Petrovic, 2021). Additionally, cadmium is a toxic material that can be harmful to the environment if not disposed of properly (Genchi, 2020). Lithium-ion batteries possess a higher power density as well as higher energy density due to the use of lithium cobalt oxide, lithium iron phosphate, or other lithium-based materials in the electrodes (Xu & Shen, 2021). The electrochemical reaction in lithium-ion batteries results in a higher voltage and energy density compared to nickel-cadmium and lead-acid batteries (Liu et al., 2019). However, lithium-ion batteries have a limited cycle life due to the creation of a solid electrolyte interface (SEI) layer on the electrodes over time, which can reduce the battery's performance and capacity (Banerjee & Kaur, 2022). Sodium-ion batteries, a relatively new technology, use sodium ions instead of lithium ions in the electrodes. While they have lower power and energy densities compared to lithium-ion batteries, they have the potential to be an economic and sustainable alternative due to the higher availability of sodium in the earth's crust. However, their cyclability and overall performance are still being studied and improved. The power density, energy density and cyclability of batteries depend on the materials used in the electrodes and the electrochemical reaction that takes place (Mukherjee, 2019; Tapia-Ruiz, 2021).

To overcome the limitations of current battery technologies, there is an increasing emphasis on the advancement of novel electrode materials that can increase power and energy density, improve cyclability, and reduce costs (Han et al., 2021). One approach is to use nanomaterials, such as graphene and carbon nanotubes, which can increase the surface area of the electrode and enhance its electrochemical properties. Another approach is to use new materials, such as silicon, sulphur, and lithium-sulphur compounds, which have higher theoretical energy densities than traditional materials (Shrivastav et al., 2019). However, these materials can also suffer from rapid capacity fading and poor cyclability, which needs to be addressed through the development of new synthesis and manufacturing processes (Ansari, 2022). Furthermore, the use of sustainable and abundant materials, such as sodium and potassium, is being investigated as a way to reduce costs and environmental impacts. In addition to electrode materials, advances in battery management systems and control algorithms can also help to improve the performance and safety of batteries. Overall, the development of new electrode materials is a critical component of advancing battery technology and enabling the widespread adoption of renewable energy, electric vehicles, and portable electronics (Ali & Allam, 2019).

The performance of batteries relies on the materials used in the electrodes. Electrode materials having a higher specific capacity, good conductivity and stability are essential for achieving high energy density, power density, and cyclability. Currently, researchers are exploring various new materials that can be used for the development of advanced batteries (Shen, 2021). One such material is sulphur, which possesses a substantial theoretical specific capacity of 1672 mAh/g and its very cheap (Alzahrani et al., 2021). However, the high solubility and low electrical conductivity of polysulfides in sulphur-based electrodes present significant challenges. To address these challenges, researchers have developed strategies such as using conductive additives, coating sulphur particles with conductive materials, designing new sulphur-based compounds with improved electronic conductivity, and trapping polysulfides within the electrode structure using porous carbon or designing new sulphur-based materials with stable and reversible electrochemical behaviour (Huang, 2020).

Dichalcogenide metal-sulphur-based nanomaterials have received extensive attention in research due to their exceptional physical and chemical properties. These materials typically have a layered structure consisting of two-dimensional sheets that are stacked together through weak van der Waals forces. This structure allows for easy intercalation of ions and promotes rapid electron transfer, resulting in high electrical conductivity. In addition to their excellent electrochemical properties, dichalcogenide metal-sulphur-based nanomaterials have excellent mechanical and thermal stability, making them attractive candidates for high-performance battery applications (Liu, Liang, Ren et al, 2022; Van Noorden, 2013). They also have a high theoretical capacity, which is a measure of the maximum amount of charge that can be stored in a material. High theoretical capacity is critical for achieving high energy density in a battery, which is the amount of energy that can be stored per unit mass or volume. Furthermore, dichalcogenide metal-sulphur-based nanomaterials are abundant and cost-effective, which is crucial for their practical application in large-scale battery production. With the rising demand for cost-effective and high-performing energy storage devices, these materials have garnered substantial attention as highly promising electrode materials for advanced batteries (Hart et al., 2015; Qie & Manthiram, 2016).

One such material that has been extensively studied is Vanadium disulphide (VS_2). Vanadium disulphide has a layered structure and exhibits unique electrochemical properties that make it an attractive candidate for electrode materials in batteries. It has a high theoretical capacity and can undergo reversible redox reactions, providing excellent cyclability. Additionally, Vanadium disulphide-based electrodes have shown high-rate capability due to their exceptional electrical conductivity and distinctive structure (Xue, 2019). Vanadium disulphide is a layered transition metal dichalcogenide (TMD) that belongs to the same family as molybdenum disulphide (MoS2) and tungsten disulphide (WS_2) (Hussain, 2023). The layered structure of VS_2 consists of vanadium atoms sandwiched between two sulphur layers, bound together by weak Van der Waals forces, which allow for the intercalation of various cations and anions. This intercalation ability, combined with the high electrical conductivity and excellent electrochemical properties of Vanadium disulphide-based nanomaterials, makes them promising candidates for application in energy storage devices (Feng et al., 2011).

The electrochemical properties of Vanadium disulphide-based nanomaterials depend on their synthesis method, crystal structure, and morphology. Latest advances in the synthesis and characterization of Vanadium disulphide-based nanomaterials have enabled the production of materials with controlled crystal structure and morphology, resulting in improved electrochemical properties. Furthermore, the development of scalable synthesis methods has made it possible to produce Vanadium disulphide-based nanomaterials in large quantities, which is essential for commercialization. Vanadium disulphide-based nanomaterials have shown excellent electrochemical performance in various types of batteries, such as Lithium-ion batteries (LIBs), Sodium-ion batteries (SIBs), Potassium-ion batteries (KIBs), Zinc-ion batteries (ZIBs), Magnesium-ion batteries (MIBs), Aluminium-ion batteries (AIBs), and Ammonium-ion batteries. For example, Vanadium disulphide-based nanomaterials have shown high specific capacity, good charge/discharge rate, and long cycle life performance in LIBs, which are the most widely used type of battery (Li, 2019).

Vanadium disulphide (VS_2) is a promising electrode material for advanced batteries because of its exceptional theoretical capacity, excellent electrical conductivity, and good mechanical stability. There are several methods of synthesizing Vanadium disulphide, including chemical vapour deposition (CVD), solvothermal synthesis, hydrothermal synthesis, and solid-state reaction. Chemicalvapour deposition (CVD) is a popular method for synthesizing Vanadium disulphide because it allows for precise control over the size of the resulting Vanadium disulphide crystals and their surface morphology. In CVD method, a precursor gas containing vanadium and sulphur is introduced into a high-temperature reactor, where the gas reacts to form Vanadium disulphide crystals on a substrate. The substrate can be varied to tailor the properties of the Vanadium disulphide crystals, such as their thickness and crystal orientation (Su, 2020). Solvothermal synthesis is another method used to synthesize Vanadium disulphide, in which a vanadium precursor and sulphur source are dissolved in a solvent and heated under pressure to promote crystal growth. This method is advantageous because it can produce high-quality Vanadium disulphide crystals with controlled size and morphology (Yin et al., 2020). Hydrothermal synthesis is similar to solvothermal synthesis, except that water is used as the solvent instead of an organic solvent. This method is preferred for large-scale production of Vanadium disulphide due to its low cost and simplicity (Zhu et al., 2020). The solid-state reaction is a method of synthesizing VS_2 by heating a mixture of vanadium oxide and sulphur under high temperatures in a sealed container. This method is relatively simple and cost-effective, but the resulting Vanadium disulphide crystals are often of lower quality compared to other methods (Guo, Yang, Wang, Cao, & Dong, 2018).

In light of the significant potential of Vanadium disulphide-based nanomaterials in energy storage applications, It is crucial to conduct a comprehensive review and conduct a critical assessment of the latest advancements in this field. In this paper, we provide a comprehensive review of the properties of vanadium disulphide, structural and morphological studies, various types of synthesis methods, vanadium disulphide as an electrode material in batteries and their mechanism, vanadium disulphidecomposites and at last the challenges that are facing in the field and some solutions.

PROPERTIES AND SYNTHESIS METHODS

Vanadium Disulphide

Vanadium disulphide (VS_2) belongs to the transition metal dichalcogenide (TMD) family, characterized by its layered structure comprising transition metal atoms positioned between two layers of chalcogen atoms (Feng et al., 2011). Vanadium disulphide was first synthesized and characterized in 1970. They prepared Vanadium disulphide using a high-temperature reaction between vanadium oxide and sulphur. Since then, Vanadium disulphide has been extensively studied due to its unique properties and potential applications (Li, Hirbod, & Li, 2020).

Vanadium disulphide has a hexagonal crystal structure, with the vanadium atoms forming distorted octahedral coordination geometry. Each vanadium atom is bonded covalently with four sulphur atoms in the equatorial plane and two sulphur atoms above and below the plane, forming a puckered honeycomb lattice. The layers of sulphur atoms are bound together by weak van der Waals forces, which allow for easy exfoliation into single-layer nanosheets (Yu et al., 2018). The thickness of Vanadium disulphide can range from bulk to a single layer, which can be easily tuned by mechanical exfoliation or chemical vapour deposition. The properties of Vanadium disulphide depend on its thickness, with the bandgap increasing from 1.5 eV in bulk to 2.5 eV in monolayer. Vanadium disulphide is a semiconductor, with valence and conduction bands separated by a bandgap. The properties of Vanadium disulphide can be modified by doping with other elements or by modifying its surface with functional groups (Wang et al., 2016).

Vanadium disulphide has shown promising potential in various fields, including electronic and optoelectronic devices, energy storage devices, catalysis, and sensors. In electronics, Vanadium disulphide$_2$ has been investigated as a channel material for field-effect transistors (FETs) (Hossain, 2018), due to its high carrier mobility and tuneable bandgap. In energy storage, Vanadium disulphide has shown high stabilityandcapacity as alithium-ion battery electrode material. In catalysis, Vanadium disulphide has shown excellent activity for Hydrogen evolution reaction (HER) (Yuan, Wu, Hardy, Loya, Lou, Yang, Najmaei, Jiang, Qin, Keyshar, Ji, Gao, Bao, Kono, Natelson, Ajayan, & Lou, 2015), O2 reduction reaction (ORR) (Moradbeigi et al., 2023), and CO2 reduction reaction (CO2RR), due to its high surface area and unique electronic structure (Parsons & Alotaibi, 2023; Qin et al., 2022).

Vanadium disulphide is a solid material at room temperature and pressure, with a molecular weight of 154.2 g/mol. It's insoluble in water and most organic solvents but can be dispersed in polar solvents including dimethyl sulfoxide (DMSO) and N-methyl-2-pyrrolidone (NMP) (Shafaatian & Baharian, 2017). The chemical reactivity of VS_2 is dependent on its surface chemistry and thickness, with thin-layer VS_2 exhibiting higher reactivity due to its larger surface area.

Compared to other transition metal disulphides, such as MoS_2 and WS_2, VS_2 has a smaller bandgap, which makes it more suitable for certain electronic and optoelectronic applications (Pang, 2020; Sharma et al., 2022). However, the weak interlayer interactions in Vanadium disulphide make it mechanically less stable than other TMDs, which may limit its practical applications in certain areas (Meng et al., 2023). In terms of isotopes, there are two naturally occurring isotopes of vanadium, ^50V and ^51V, with ^51V being the more abundant isotope. VS_2 can be synthesized using either of these isotopes for isotopic labelling experiments (Prytulak, 2016).

Structural and Morphological Properties

The structural and morphological properties of a material are key factors that determine its chemical and physical properties, as well as its potential applications. In the case of Vanadium disulphide (VS_2), these properties have been extensively studied nowadays due to its promising applications in various fields.

Structural properties refer to the arrangement of atoms in a material and their interactions with each other. Vanadium disulphide has a layered structure consisting of covalently bonded Sulphur atoms (chalcogen atom) and Vanadium atoms (transition metal atom). The crystal structure of VS_2 consists of layers comprising edge-sharing VS6 octahedra. These layers are stacked together through weak van der Waals interactions, resulting in a layered structure with a 5.76 Å interlayer distance. Each VS6 octahedron is coordinated with six S atoms in a distorted octahedral arrangement (Pu et al., 2019; Wiegers et al., 1974). The interlayer spacing in VS_2 is larger than that of graphite, but has a similar honeycomb lattice structure as graphene, with strong covalent bonding within the plane (Allen et al., 2010). The material exhibits two forms, known as the hexagonal (H) and trigonal (T) phases, which are determined by the coordination of vanadium with sulphur. In the H-Vanadium disulphide and T-Vanadium disulphide layers, vanadium atoms are positioned at the centre of a trigonal prism and an octahedron of sulphur atoms, respectively.

Both structures exhibit a two-dimensional hexagonal lattice with regularity along the a/b directions, running parallel to the plane of Vanadium disulphide (Kan et al., 2015; Wang et al., 2016). The material exhibits metallic behaviour characterized by a substantial local density of states (DOS) crossing the Fermi level, as well as a zero band gap, signifying high microscopic two-dimensional conductivity. The metallic behaviour of VS_2 arises from the V-3d and S-3p states. The Vanadium disulphide H phase is the ground state structure with a lower total energy of 43 meV than the T phase (Li, Hirbod, & Li, 2020; Pan, 2016; Zhang, Wei, Sun et al, 2018).

Various techniques can be used to study the structural properties of Vanadium disulphide, such as X-ray diffraction (XRD, Scanning tunnelling microscopy (STM)), and transmission electron microscopy (TEM). XRD provides information about the crystal structure, lattice parameters, and crystal orientation. STM and TEM allow direct visualization of the atomic structure of materials (Lee et al., 2022).

Morphological properties, on the other hand, refer to the shape, size, and surface features of a material. The morphology of Vanadium disulphide can be controlled by various synthesis methods, such as Chemical vapour deposition (CVD) (Yu, Tang, Shao, Dai, Wang, Yan, Pan, & Chua, 2019), Hydrothermal synthesis, and Solvothermal synthesis. The morphology of Vanadium disulphide can vary from 2D nanosheets to 3D hierarchical structures. 2D nanosheets have a high surface area and excellent electrical conductivity, making them ideal candidates for various applications such as energy storage, catalysis, and electronic devices. The thickness of Vanadium disulphide nanosheets can vary from a few atomic layers to tens of nanometres, and the lateral size can be several micrometres. By adjusting the synthesis conditions such as temperature, pressure, and precursor concentration, it is possible to manipulate the morphology of Vanadium disulphide nanosheets (Zhang, Chang, Zhang et al, 2022). Three-dimensional hierarchical structures, such as nanoflowers, nanorods, and nanoplates, possess a substantial surface area, allowing for a greater number of active sites for catalytic reactions. The morphology of VS_2 hierarchical structures can be regulated by changing synthesis conditions, such as the choice of precursor, solvent, and reaction duration (Liang et al., 2016).

Various Synthesis Method

Vanadium disulphide (VS_2) can be synthesized through various methods depending on the specific properties and applications required. One of the methods is the Chemical Vapor Deposition (CVD) Method (Shen et al., 2018), where Vanadium chloride and Sulphur powder are heated together in a sealed tube under a vacuum or an inert gas atmosphere (Su, 2020). Another method is the Hydrothermal Method, where Vanadium pentoxide and thioacetamide are reacted in a hydrothermal reactor at high temperature and pressure (Huang et al., 2014). The Solvothermal Method involves reacting Vanadium compounds and Sulphur in a solvent at high temperature and pressure (Xie et al., 2020). The Calcination technique is commonly used for solid-state preparation of inorganic nanostructured electrode materials (Hu, 2023). Each method results in the formation of VS_2 (Maeyoshi et al., 2021; Mohan et al., 2015), but the specific method chosen will depend on the desired properties and applications of the material.

Chemical Vapour Deposition

Chemical vapour deposition (CVD) is a well-established synthesis technique for producing 2D materials with exceptional crystalline quality and minimal contamination (Shen et al., 2018). VS_2 nanosheets with high crystallinity have been successfully synthesized using VCl_3 and Sulphur as sources. In 2015, Yuan et al. pioneered an ambient-pressure CVD method to grow 1T-VS_2 single crystal nanosheets on a SiO2/Si substrate (Yuan, Wu, Hardy, Loya, Lou, Yang, Najmaei, Jiang, Qin, Keyshar, Ji, Gao, Bao, Kono, Natelson, Ajayan, & Lou, 2015), Ji et al. (2017) demonstrated adept control over the domain sizes and thicknesses of VS_2 nanosheets through precise adjustment of the VCl_3 evaporation temperature

Figure 2. Chemical vapour deposition (CVD) for the VS_2 growth

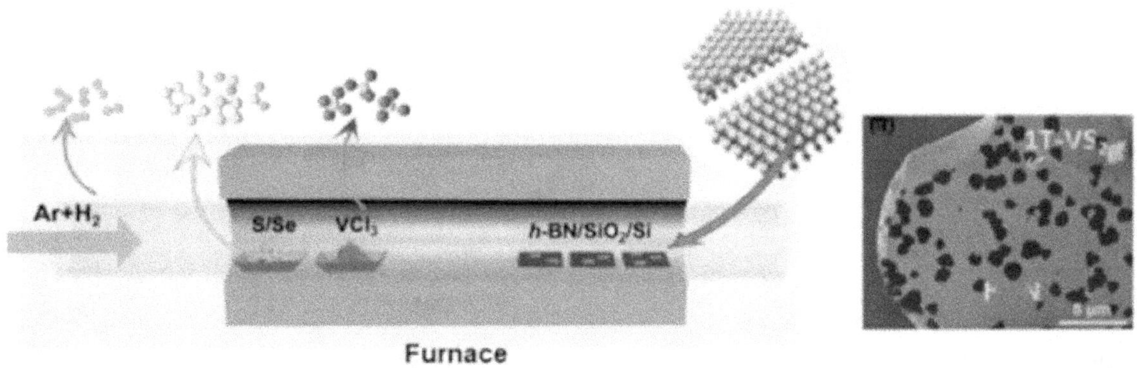

(Ji et al., 2017). Hossain et al.(2021) achieved excellent results by depositing 2D 1T-VS_2 on hexagonal BN/SiO2/Si and mica substrate (Hossain et al., 2021). Additionally, Li et al.(2019) devised a dual-zone atmospheric pressure CVD synthesis method, this enabled the creation of densely packed forests of Vanadium disulphide sheets incorporated into a carbon nanotube substrate to make a current collector. Furthermore, they demonstrated the floating catalyst CVD approach, which facilitated the growth of Vanadium disulphide nanosheets on a single-walledthin film made of carbon nanotubes (CNTs), forming a core-branch structure (Littlejohn et al., 2019). Yu et al. employed a multi-zone chemical vapour deposition (CVD) technique to synthesize and integrate two complementary materials, resulting in the formation of lateral MoS2-VS_2 heterostructures. Notably, Dong et al. achieved the successful fabrication of smooth VS_2 films, as well as VS_2 nanowire-like structuresandVS_2 nanorod-like structures, on a double-polished sapphire (c-plane) substrate by precisely adjusting the growth pressure. This study demonstrated the versatility and effectiveness of chemical vapour deposition (CVD) in producing high-quality two-dimensional (2D) materials (Dong et al., 2019).

Hydrothermal Synthesis

The hydrothermal method has emerged as a prominent liquid-phase synthesis technique for fabricating inorganic nanostructured electrode materials. It has garnered significant attention owing to its ability to control the reaction process, achieve rapid reaction kinetics, offer cost-effectiveness, and facilitate the regulation of nanostructures (Gan et al., 2020). In the realm of VS_2-based nanostructures, it is the most widely used synthetic strategy. The success of the ammonia-assisted hydrothermal synthesis method in the production of ultrathin Vanadium disulphide nanosheets marked a turning point for the practical use of hydrothermal synthesis in the fabrication of VS_2-based nanostructures (Feng et al., 2011). Subsequently, various Vanadium disulphide nanostructures such as flower-like (Lia, 2015), petal-like (Wang, Zhang, Zheng et al, 2019), layer-by-layer-like Vanadium disulphide nanosheets with different thicknesses (Huang et al., 2018), and Vanadium disulphide quantum dots have been directly produced by hydrothermal synthesis method (Du, Shang, Shang et al, 2018) with different precursors Ssuch as Na3VO4·12H2O/Na3VO4 (sodium orthovanadate) (Du, Liang, Shang et al, 2018) or NH4VO3 (ammonium vanadate) (Ma et al., 2018) as Vanadiumsource and CH3CSNH2 (thioacetamide, TAA) (Joseph &

Figure 3. Solvothermal method

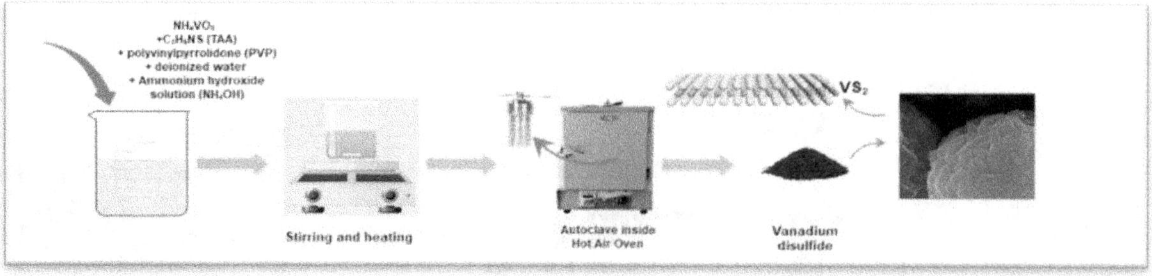

Aneesh, 2022) or C5H9NO3S (N-acetyl-L-cysteine) (Yang, Chen, Wang et al, 2021) as Sulphur source (reductant). In addition, hydrothermal synthesis has been employed to prepare a range of VS_2-based composites such as VS_2/graphene (Lia, 2015), Na2Ti2O5/VS_2 (Liao & Manthiram, 2015), VS_2/carbon paper (Qu et al., 2017), TiO2-B@VS_2 (Cao et al., 2017), rGO/VS_2 (Mohan et al., 2015), VOOH/VS_2 (Li et al., 2017), VS_2/MWCNTs (Zhang, Chen, Zhang et al, 2018), VS_2/CNT (Meyer et al., 2020), Ni3S2/VS_2 (Zhong et al., 2018), VS_2/SS (stainless steel mesh) (Patil et al., 2020), and NiCo2S4@VS_2 (Zhang et al., 2019) nanocomposites with the aid of pure, CTAB (cetyltrimethyl ammonium bromide) or Sodium hydroxide-assisted hydrothermal reactions (Li, Hirbod, & Li, 2020).

Solvothermal Method

Over the last decade, solvothermal methods have emerged as a highly promising liquid-phase preparation method for Vanadium disulphide-based nanostructures, building upon the hydrothermal method by offering several advantages, such as the ability to adjust solvent polarity, density, viscosity, and dispersion. Furthermore, solvothermal methods allow for high sintering activity, complete crystal growth, and low agglomeration (Liu et al., 2020). Chen et al. were the first to fabricate randomly distributed VS_2 nanosheets by reacting VCl4 and sulphur in a mixed solution of oleylamine and 1-octadecene at 300°C under N2 atmosphere. Subsequently, the researchers transformed these sheets into ultrasmall VS_2 nanodots with a size of approximately 2 nm. This conversion was accomplished by utilizing the coordination bond created between the phosphate group in C39H73O8P(1,2-dioleoyl-sn-glycero-3-phosphate) and V4+ (Chen, Tang, Pan et al, 2020). Cai et al. developed a CTAB-assisted solvothermal process to prepare flower-like VS_2 nanosheets with different crystallinities, utilizing a mixed solution of deionized water and diethylene glycol as the solvent (Cai et al., 2018). Sarkar et al. introduced a simple method for producing VS_2 nanoparticles. This method involved the solvothermal decomposition of the [VO(acda)2] complex in triethylenetetramine (TETA) at a temperature of 220°C (Sarkar et al., 2018), while Zhang et al. employed octylamine as the solvent in a one-pot solvothermal process to fabricate a VS_2 nanoflower structure. This approach resulted in expanded interlayers and a high concentration of defects within the nanoflower (Zhang, Li, Ma et al, 2022). Finally, Li et al. achieved the successful fabrication of VS_2 microrods composed of assembled nanosheets. They regulated the crystal structure of the nanosheets using an ethanol-assisted in-situ etching synthetic approach (Li, Hirbod, & Li, 2020).

Other Methods

The first synthetic method reported for the production of the Vanadium disulphide phase is the calcination technique, commonly used for the solid-state synthesis of inorganic nanostructured electrode materials. Two methods are typically used to prepare Vanadium disulphide nanosheets: gas-solid reaction of Li2CO3 and V2O5/V2O3 under H2S atmosphere, and solid-solid reactions of Vanadium metal and quartz ampoules or Sulphur powder containing stoichiometric V/S quartz ampoules under high pressure. VS_2/rGO hybrids can be obtained by directly calcining VS4/rGO hybrids in vacuum. However, the calcination method is limited by its unmanageable morphology, high reaction temperature, and large agglomeration, thereby restricting its extensive application in nanostructured VS_2 synthesis (Li, Hirbod, & Li, 2020).

In addition to these methods, several alternative strategies have been introduced to produce Vanadium disulphide with diverse structures. For example, Pandit et al. utilized the successive ionic layer adsorption and reaction (SILAR) technique to deposit Vanadium disulphide hexagons onto the hexagonal matrix of a multi-walled carbon nanotubes (MWCNTs) nanonetwork. They also developed a two-dimensional nanostructured Vanadium disulphide thin film on a stainless steel (SS) substrate (Pandit et al., 2022). Guo et al. employed a colloidal chemical technique to produce Vanadium disulphide nanoplates with an ultrathin two-dimensional (2D) structure. They achieved the formation of nanoplates with abundant in-plane and out-of-plane defects by adjusting the molar ratio of the precursor's vanadium (V) to sulphur (S) (Guo, Yang, Wang, Cao, & Dong, 2018). Li et al. utilized a combination of the hydrothermal method and ultrasonic stripping technique to fabricate coin-like Vanadium disulphide structures with varying degrees of stacking. They employed a mixed solution of waterand polyvinylpyrrolidone (PVP-K30) as the solvent during the synthesis process (Li, 2018).

VANADIUM DISULPHIDE AS AN ELECTRODE MATERIAL IN BATTERIES

Vanadium disulphide (VS_2) has attracted substantial interest as a highly promising electrode material for batteries due to its distinctive structure and exceptional electrochemical properties. VS_2 possesses a layered crystal structure with vanadium atoms sandwiched between two sulphur atoms, providing an excellent platform for ion intercalation and deintercalation (Feng et al., 2011). The process of intercalating ions into the layers of Vanadium disulphide enables a reversible redox reaction, which plays a crucial role in facilitating the charging and discharging process of the battery. Additionally, the high electronic conductivity and large interlayer spacing of Vanadium disulphide allow for fast ion diffusion and efficient charge transport, resulting in high power and energy densities (Meng et al., 2018). Several studies have demonstrated the effectiveness of VS_2 as electrode material in different types of batteries, including lithium-ion batteries (Li, 2019), sodium-ion batteries (Wang et al., 2016) and magnesium-ion batteries (Xue, 2019). Enhancing the electrochemical performance of VS_2 can be achieved through surface modification or the integration of other materials, resulting in composite structures that exhibit improved properties (Liu, Zhao, Zhang et al, 2022). However, there are still challenges in the large-scale production and commercialization of VS_2-based batteries, and further research is needed to address these issues.

Monovalent Ion Batteries

Li-Ion Battery

Lithium-ion batteries offer several advantages, including their high energy density and the absence of a memory effect (Nitta et al., 2015). Consequently, they are extensively researched and utilized in our day-to-day life. Nonetheless, The practical implementation of commercial lithium-ion batteries (LIBs) is limited by the theoretical specific capacity of graphite, which is 372 mAh g−1 (Moyassari, 2022). However, a VS_2 monolayer presents a promising alternative. The restriction stems from the relatively low energy barrier of 0.22 eV (Agyapong-Fordjour, 2021) that is necessary to overcome the diffusion barrier of Li+ ions. This barrier is equivalent to that of graphite. With three layers of adsorption, a single layer of Vanadium disulphide has a theoretical capacity of 1397 mAh g−1 (Xu et al., 2022). Additionally, it possesses excellent lithium-ion mobility, lithiation stability,and electrical conductivity. Consequently, it is widely recognized as one of the most promising electrode materials for lithium-ion batteries (LIBs).

The energy storage mechanism in MX_2 is primarily determined by the interaction between MX and $LiMX_2$ (Chianelli, 1976). Lithium sulphide and Vanadium metal nanoparticles are produced as a consequence of conversion events that occur during electrochemical processes when VS_2 exhibits a smaller MX interaction and a greater Li binding energy (Wang, 2023). This results in the structural breakdown of VS_2, leading to the deposition of Sulphur on its surface, which ultimately affects the electrochemical stability and performance. A more thorough investigation of the storage mechanism is made possible by in situ characterization technology, which provides real-time monitoring of the development of material during (dis)charge as well as information on the micromorphology,crystal structure, electrical structure, and etc. (Chen, Chua, & Lee, 2020).

For instance, Li and colleagues(2019) successfully generated extremely crystalline Vanadium disulphide nanosheets anchored carbon nanotubes during an early (de)lithiation phase using in situ TEM and HRTEM. The intercalation reaction, which results in the formation of amorphous-phase $LixVS_2$, corresponds to the first lithiation. $LixVS_2$ is changed into a crystalline Li2S matrix and V nanoparticles as a result of the ongoing lithiation, and triangular Li2S nanosheet progressively forms on the edge surface of the lithiatedVanadium disulphide nanosheet. The dark-contrast nanoparticles were found to be immersed in a crystalline matrix and to be growing in size, as seen by the TEM pictures. The crystalline matrix was identified as Li2S by analysing the lattice gap, and the black particles were identified as crystalline Vanadium nanoparticles (eqn3 and 4). With the removal of lithium ions during charging, $LixVS_2$ progressively changes back to the original VS_2 state (eqn1). On the surface of lithiatedVanadium disulphidenanosheet, the triangular Li2S NS vanishes, and Li2S is permanently oxidised to S8 (eq2). V experiences a minor oxidation process as well. The S component is lost, active materials lose some of their activity, and capacity gradually degrades as a result of the partly irreversible conversion process. In another investigation, it was found that the conversion reaction exhibited partial reversibility, which reduced with an increasing number of cycles.

The electrochemical mechanism is significantly influenced by the cut-off voltage. It is interesting that altering the cut-off voltage may only cause the intercalation process to occur in VS_2. In the study conducted by Sun et al. (2021), where the voltage range was controlled between 1.5 and 3.5 V, only intercalation reactions were observed within this range. However, partly irreversible events were observed during the first cycle.In situ XRD measurements provided insights into the first discharge process, which involved a two-step Li+ insertion leading to the formation of hexagonal-phase $Li0.33VS_2$ and $LiVS_2$. There were

Figure 4. Electrochemical mechanism

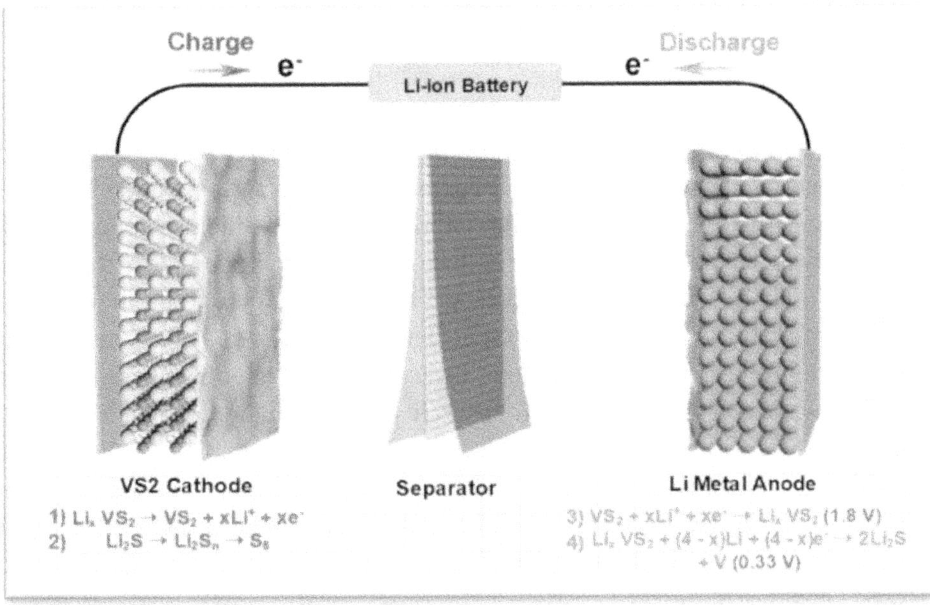

two processes in the charging procedure for the lithium deintercalation. Mesophase hexagonal-phase Li0.33VS_2 was discovered after Li+ was removed. Just a portion of the hexagonal-phase Li0.33VS_2 after full charge changed into VS_2, with the remaining portion changing into the more stable monoclinic-phase Li0.33VS_2. The fundamental cause of the loss of electrochemical performance is this insufficient reversibility. With an impressive capacity retention rate of 97% after 50 cycles at a current density of 0.1 A g^-1, the specific capacity of the material may reach 172 mAh g^-1 (Sun et al., 2021). In another study, Fang and colleagues (2015) successfully synthesized hierarchical Vanadium disulphide/graphene nanosheetsusing a one-step hydrothermal process aided by cetyl trimethylammonium bromide (CTAB). The structural alterations of Vanadium disulphide during Li storage were discovered by ex situ XRD in the 1.5–3.0 V range. The observed changes during the lithium-ion insertion process include an increase in the layer spacing and a leftward shift of the peak location at 15.45°. The peak position restored to its previous position after the full removal of lithium ions, demonstrating the excellent reversibility of Vanadium disulphide during lithium storage. After undergoing 200 cycles at a current density of 0.2 C, the material can still exhibit a specific capacity of 180.1 mAh g^-1. This performance can be attributed to the synergistic effects of heterostructures that enhance electron transfer and graphene nanosheets that mitigate the volume expansion of Vanadium disulphide during cycling\.;'] (Lia, 2015). Thin 1T-VS_2 nanosheets were created by Liang et al. and successfully prevent the phase change that occurs during the electrochemical reaction. According to DFT calculations, 1T-VS_2 can tolerate significant lithiation without structural change (Liang et al., 2020).

During the process of deep discharge, a partly reversible conversion reaction occurs in which LixVS_2 decomposes to create lithium sulphide (Li2S) and metal vanadium (V). Moreover, Li2S may go through an oxidation-reduction cycle to create Sulphur during charging, and some oxidation of Vanadium may also take place. The instability in the material's structure and partial deactivation of the active species are caused by the imperfect reversibility of these reactions during cycling. Unsatisfactory electrochemical performance is as a result shown. More research is necessary to ascertain if both Sulphur and Vanadium participate in the redox process as active centres as well as the order in which they do so.

Na-Ion Battery

A replacement for LIBs is envisaged for sodium-ion batteries because of their availability and affordability. The development of this technology is hampered by the high radius of sodium ions (1.02 A), which results in a number of problems, including delayed electrochemical reaction kinetics,slow ion diffusion,separation of active materials from the collector, and the volume expansion (Abraham, 2020; Nayak et al., 2018). To get beyond these restrictions, Vanadium disulphide provides numerous active sites for sodium ions, effectively reducing the distance that sodium ions and electrons need to travel. Additionally, it exhibits a high specific surface area and layer spacing, further enhancing its electrochemical performance (Qi, Wang, Zuo et al, 2020). Due to its ability to control the incidence of intercalation and conversion processes, the cut-off voltage is an important SIB parameter. When only the intercalation process occurs within a voltage window range of 1.5 V to 2.8 V while discharging, the phase change from $NaVS_2$ (R2m) to $NaVS_2$ (R3m) occurs. Unfortunately, the battery's rate performance and cycle stability are compromised by the challenging phase transition process and the high diffusion barrier for sodium ions (Na+) (Zhang, 2020).

Phase changes often happen when a conversion reaction is present. Although having a large capacity, poor stability is often the consequence of volume expansion and active ingredient fragmentation. Wang et al (2018) noticed charge transfer in both cations and anions. The extent of the reaction and the resulting products were observed to vary as the voltage ranges were changed, as confirmed by high-resolution transmission electron microscopy (HRTEM) and selected area electron diffraction (SAED) patterns. These techniques provided valuable insights into the structural changes and composition of the materials at different voltage conditions.During the redox reaction, only sulfur ions were involved in the electrochemical process for Na+ injection quantities ranging from 0 to 1.0. However, for Na+ injection quantities ranging from 1.0 to 2.0, both vanadium and sulfur ions participated in the reaction. This resulted in the formation of $NaxVS_2$, while the layered structure of the host material was maintained.The two stages of the redox reaction exhibited stability and reversibility, indicating their potential for high cycle stability.However, when the Na+ insertion level reached 2.0 to 3.0, a conversion reaction took place, resulting in the formation of Na2S and V. This conversion event caused a significant change in the structure of VS_2, as evidenced by the sodium process. Despite its large theoretical capacity of 3,699 mAh g1, the battery faced challenges such as significant volume expansion, poor cycle stability, and a shortened cycle life.

Sun et al. (2021) conducted a study using in situ XRD, SAED and ex-situ HRTEM techniques to investigate the effect of cut-off voltage on the electrochemical reaction during the charging and discharging processes. The researchers observed a microstructural evolution in the sodium intercalation and deintercalation processes. Specifically, during the first stage of discharge within a voltage range of 0.4-2.2 V, the (001) and (011) planes of VS_2 shifted to the left, and the peak corresponding to VS_2 disappeared. Instead, a new phase called $NaxVS_2$ was formed.. It completely reversed after being fully charged, returning to its former high of VS_2. Nevertheless, an irreversible conversion event took place and produced Na2S when the voltage window was increased from 0.01 V to 2.2 V. When fully charged, it could be returned to its initial state of VS_2, but the host structure was harmed and could not be repaired

Figure 5. Potassium-ion battery

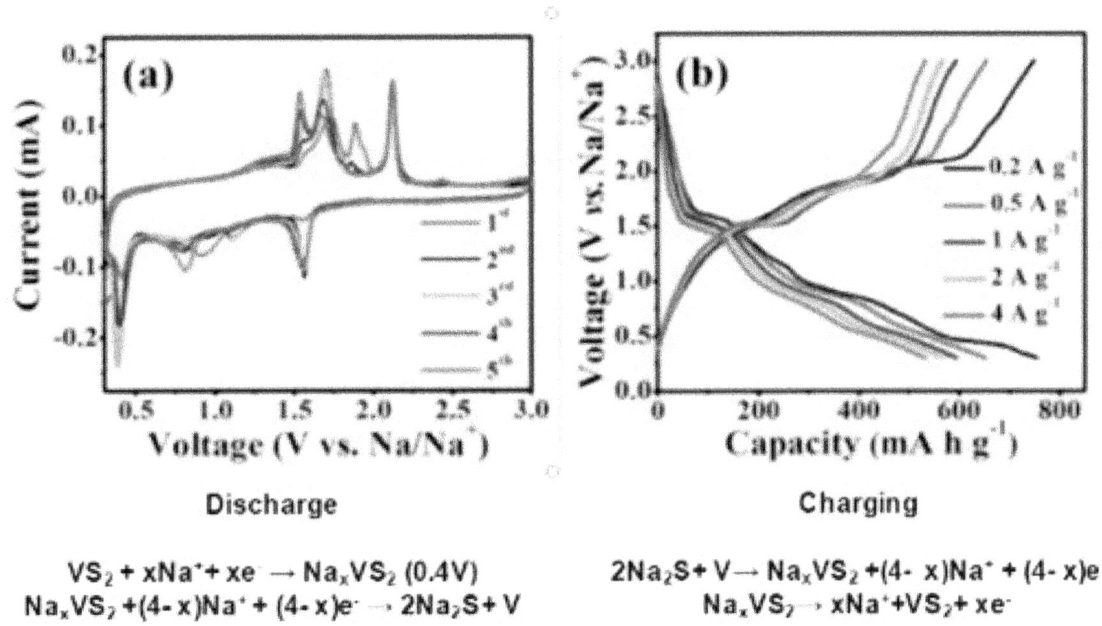

completely. The host structure's integrity could be successfully protected, the conversion reaction could be prevented, and the cycle stability could be improved by managing the voltage window (Sun et al., 2021).

Despite the conversion reaction increasing the initial discharge capacity to 700 mAh g1, Yu et al. discovered that raising the working potential to 0.01 V resulted in significant volume changes and shrinkage between the active material and the current collector. This led to a rapid decay in capacity, with the battery barely retaining any capacity after 15 cycles (Yu et al., 2018). Putungan et al. (2016) conducted calculations and found that monolayer VS_2 exhibited superior structural stability compared to other transition metal dichalcogenides (TMDs). It was less susceptible to phase transitions from the hexagonal (H) to the trigonal (T) phase during the sodiation process. This was attributed to the increased instability of the T-phase and the elevated energy barrier between the H- and T-phases induced by sodiation (Putungan et al., 2016).

Consequently, one of the difficulties to take into account is finding a balance between capacity and stability. Additionally, by enhancing the electrode material and regulating the voltage range, the issue of a short cycle life brought on by volume growth may be further addressed.

Potassium-Ion Battery

Potassium-ion batteries (PIBs) have attracted interest as prospective electrochemical energy storage devices because to their availability of resources, low redox potential (2.936V versus SHE) and inexpensive (Deng, 2018; Rajagopalan, 2020). Nevertheless, their future development has been hampered by the sluggish electrochemical reaction kinetics and diffusion rate brought on by the higher K+ ion radiusas compared to that of Li+ (1.38 vs 0.76) (Wu et al., 2019). Yet, the robust design of Vanadium disulphide may provide a high number of K+ active sites and has a broad layer gap, resulting in a minimal diffusion barrier (Sun et al., 2022).

The storage method of potassium-ion batteries (PIBs) is entirely reversible and requires intercalation processes, which only take place within an appropriate voltage window, in contrast to LIBs and SIBs (Wang, Ang, Yang et al, 2021). Zhang et al. used in situ XRD to examine the first cycle phase transition of PIBs. The $K0.6VS_2$ phase appeared as a new set of peaks, demonstrating that the intercalation reaction occurred incrementally with K+ insertion. The system went back to its initial Vanadium disulphide state after being fully charged to a voltage value of 3.5 V, demonstrating a highly reversible phase transition process (Zhang, 2020).

An easy and efficient way to enhance the (de)insertion of working ions is to widen the interlayer gap. With an extremely wide interlayer spacing of 10.2, carbon intercalated VS_2 nanosheets have been successfully manufactured by Xie et al.(2020). According to the cyclic voltammetry (CV) curves, the reduction peaks at 1.37 V and 0.68 V represent, respectively, the first step of K+ intercalation ($KxVS_2$, where x is the quantity of inserted K+), and the following conversion process. The extraction of K+ is indicated by the presence of an oxidation peak at 2.4 V. The great reversibility is shown by the high CV curve coincidence. The expansion in volume during the solvothermal reaction and carbonization process can be effectively minimized by the 10.2 interlayer spacing of VS_2/NC (VS_2/N-doped porous carbon). Moreover, comparable processes have been seen in other TMDs, including MoS2 and VSe2, which are transition metal dichalcogenides (Xie et al., 2020).

Discharging: $VS_2 + xK^+ + xe^- \rightarrow K_xVS_2$

$K_xVS_2 + (4 - x)K^+ + (4 - x)e^- \rightarrow V + 2K_2S$

Charging: $2K_2S + V \rightarrow K_xVS_2 + (4 - x)K^+ + (4 - x)e^-$

$KVS_2 \rightarrow VS_2 + xNa^+ + xe^-$

Multivalent Ion Batteries

Zinc-Ion Battery

Zinc-ion batteries (ZIBs) have received a lot of research attention because of their affordability,environmental friendly, and The redox potential of VS_2 is relatively low, with a value of -0.76 V vs SHE (Standard Hydrogen Electrode) (Shin et al., 2020; Xu et al., 2012). And it gained significant interest due to its multielectronic properties, offering the potential for high theoretical specific capacity of 820 mAh g-1. These characteristics make ZIBs a promising candidate for various applications, including portable and flexible electronic devices (Ling et al., 2020). It is crucial to identify optimal cathode materials to enhance the performance of zinc-ion batteries (ZIBs). One promising candidate is a material with a distinctive layered structure and relatively wide layer separation of 5.7 Å (Wu, 2023), VS_2 makes a strong case for being used as the cathode material. This property allows Zn ions, which have a 0.74 A diameter (Kang et al., 2020), to diffuse quickly.

Owing to the high charge density of divalent Zn ions, VS_2 electrode materials display simultaneous intercalation and conversion processes, similar to monovalent ion batteries. Both reactions are significantly influenced by the cut-off voltage, which is a key element. It's also important to take into account the sulphur shuttle phenomenon, which causes unfavourable side effects and poor material utilisation rates (Jiao et al., 2019).

As an example, He's group demonstrated the insertion process of Zn2+, which is a highly reversible intercalation reaction occurring in two steps. Initially, 0.09 Zn2+ intercalates to form Zn0.09VS_2, followed by further insertion and transformation of Zn2+ into Zn0.23VS_2, providing a substantial capacity contribution. The degree of an electrochemical reaction in electrode materials, such as VS_2, can be effectively controlled by regulating the cut-off voltage. By carefully selecting the appropriate cut-off voltage, it is possible to inhibit the occurrence of irreversible reactions, thereby improving the reversible capacity and cycle stability of the battery system (He, 2017). Zhu et al. analysed the energy storage mechanism of VS_2 using X-ray photoelectron spectroscopy (XPS) and Raman spectroscopy. They found that the electrochemical reaction at the voltage range of 0.4-0.8 V exhibited high reversibility. By analysing the V 2p XPS peak position, they were able to deconvolute it into four subpeaks (516.8, 517.7, 524.1, and 525.2 eV), which closely resembled the discharge state of the VS_2 electrode.At a voltage of 1 V, a conversion reaction takes place in the VS_2 electrode material. X-ray photoelectron spectroscopy (XPS) analysis of the S 2p peak reveals that it can be deconvoluted into four small peaks, with the peak at 164.1 eV corresponding to the S element. Additionally, a small peak at 481 cm^−1 is observed in the Raman spectrum. This indicates the involvement of sulphur in the conversion reaction. However, after the second discharge and throughout 50 cycles, the presence of Sulphur is barely detected in the electrode sheet during Raman measurements. This suggests that sulphur has minimal participation in the reaction processes occurring during the cycling of the VS_2 electrode (Zhu et al., 2021).

In a more recent study, 1T-VS_2 colloidal nanoflowers were successfully synthesized using a one-step hydrothermal method. The synthesis process involved the use of ZnSO4 (2.5M) as the electrolyte. The energy storage mechanism of VS_2 in zinc ion storage was investigated by controlling the cut-off voltage and conducting ex-situ tests. The study revealed that by controlling the cut-off voltage at 0.4-0.85 V, some zinc ions remained in the interlayer of VS_2 during the discharge process, acting as stability-enhancing pillars. The material exhibited specific capacities of 212.9 mAh g−1 at 0.1 A g−1, 102.1 mAh g−1 at 5 A g−1, and 203.4 mAh g−1 at 0.1 A g−1, indicating long cycle life and maintaining 86.7% of the initial specific capacity after 200 cycles at 2 A g−1. This study highlights the potential of 1T-VS_2 colloidal nanoflowers as a promising electrode material for zinc ion batteries, offering improved stability and performance (Yun, 2022).

Jiao et al. conducted a study on the energy storage mechanism of VS_2 by growing it on a stainless-steel substrate and investigating the microstructure and valence changes using XRD and XPS methods. They have identified a unique phase transition mechanism in which the intercalation reaction produces ZnxVS_2 (0 x 0.12, the Zn2+ insertion quantity is x) before Zn2+ insertion takes place. The XRD peak moves to the left as a result. New phases of V3S4 are produced when additional zinc ions are injected into the interlayer. Moreover, a new phase of $Zn_{0.42}V_3S_4$ with high reversibility will arise with an increase in the insertion of zinc ions. Throughout the charging process, the peak of this new phase steadily diminishes until it vanishes and reverts to the old peak (Jiao et al., 2019). In a different work, Liu et al. produced $Zn_3V_2O_7$ $(OH)_{22}H_2O$ for ZIBs by electrochemically activating VS_2. The (001) peak location of $Zn_3V_2O_7$ $(OH)_{22}H_2O$ moved to the left after the introduction of Zn ions. Yet, after being completely charged, the (001) peak reverted to its initial location, suggesting high cycle stability and reversibility (Liu et al., 2023).

Figure 6. Magnesium-ion battery

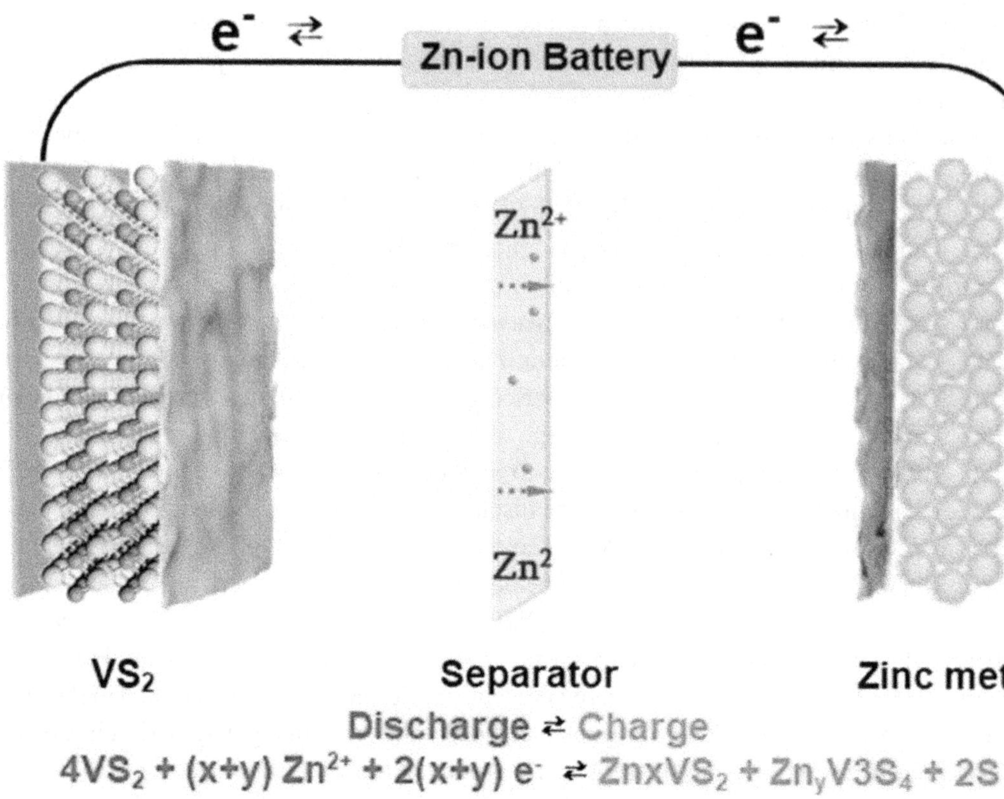

The reversibility of Vanadium disulphide in aqueous Zinc-Ion batteries has not yet been fully established. Improved characterisation techniques are required to conduct additional research on the products at various charge and discharge phases.

Magnesium-Ion Battery

Magnesium-ion batteries (MIBs) have gained significant interest due to several advantageous properties. One of the key factors is the abundance of magnesium in nature, making it a readily accessible and cost-effective material for battery applications. Another advantage is the absence of dendrite formation, which is a common issue in lithium-ion batteries. Additionally, MIBs offer a low reduction potential of -2.4 V versus the standard hydrogen electrode (SHE), enabling high energy density and efficient electrochemical reactions. The theoretical capacity of magnesium-ion batteries is also high, reaching 2205 mAh g-1, indicating its potential for high energy storage. Moreover, MIBs exhibit a promising volumetric capacity of 3833 mAh cm3, making them attractive for applications where compact and efficient energy storage is desired (Bucur et al., 2015; Muldoon et al., 2017; Walter et al., 2015). Unfortunately, finding a suitable cathode material is very difficult due to the Mg2+'s interfacial charge transfer and slow diffusion kinetics (Wang et al., 2017). To maintain its structural integrity while facilitating fast ion diffusion, the VS_2

material adopts a loose accumulation structure that features numerous open channels between its layers. This unique structure allows for efficient ion transport within the material. Density functional theory (DFT) calculations have indicated that magnesium ions can readily and stably adsorb onto single-layer, double-layer, and bulk VS_2. All three states all exhibit the same adsorption capacity, which results in a uniform theoretical specific capacity of 233 mAh g1. The adsorption of magnesium ions onto the VS_2 material leads to a reduction in the binding energy of the Vanadium-Sulphur bond. This weakening of the bond results in structural instability, which can negatively impact the cycle stability of magnesium-ion batteries using Vanadium disulphide as an electrode material (Yang, Wang, Dong et al, 2021).

Stabilising the host structure and increasing the interlayer gap and ion channelare the two main ways to improve the electrochemical characteristics of MIBs. The VS_2 follows an intercalation and conversion storage technique when it comes to MIB electrode materials. Additionally, the relatively large ionic radius of magnesium ions enables the formation of a significant complex of cation and anion ions, which acts as the active species in magnesium-ion batteries.For instance, the direct insertion of 2-ethylhexamine molecules into the interlayer of Vanadium disulphide as interlayer pillars may be made easier by the solvothermal approach. The voltage range that results is between 0.2 and 2.2 V. Over numerous intercalation processes, magnesium ions exhibit high reversibility. First, when discharged to 0.6 V, the intercalation process takes place, causing a substantial rise in the interlayer gap of VS_2. High-resolution transmission electron microscopy (HRTEM) imaging has revealed that the lattice spacing of the (001) plane in VS_2 increases from 9.93 to 10.70 during certain processes or conditions. This expansion in lattice spacing indicates structural changes or intercalation of species within the material. Additionally, X-ray diffraction (XRD) patterns of the (001) and (002) peaks show weakening and a shift towards the left.The conversion process occurs when the discharge is extended to 0.2 V, and the X-Ray diffraction peak of Vanadium disulphide totally vanishes. When the lattice spacing of Magnesium sulphide(111) and Vanadium(110) is measured, the weak diffraction peak of magnesium sulphide and the metal vanadium peak concurrently emerge. The structure eventually reverts back to the original extended VS_2 configuration after a full charge, exhibiting distinct lattice spacings of 2.77 and 10.10, which correspond to the (100) and (001) crystal planes, respectively. Anions (A-) may generate MgA+ using Mg2+ as carriers to reduce the issue of a high diffusion barrier. Hyun et al. employed MgCl+ as a carrier to facilitate the (de)intercalation process between the layers of TiS2. The presence of negatively charged Cl- ions bonded with Mg2+ ions results in a low polarization intensity, effectively reducing the diffusion barrier. This is advantageous because only desolvation is needed for the insertion process, with an activation energyof 0.8 eV. Another potential strategy involves the development of magnesium electrolytes that can effectively operate on MoS2 by utilizing boron cluster anions (Xue, 2019).

Aluminium-Ion Battery

Aluminium-ion batteries (AIBs) are an appealing option for energy storage systems becauseof their many advantages, including their economic viability, high abundance, high specific capacity (2980 mAh g1), excellent volume energy density (8046 mAh cm3), and suitably low redox potential (Abu Nayem et al., 2022; Elia et al., 2021). Finding the right cathode materials is essential for attaining the best performance since larger charge densities are hindered by the host material's slow diffusion of Al3+. The electrode materials that are now being researched the most include V2O5, TiO2, and Prussian blue analogues (De et al., 2023; Koketsu et al., 2017). The favourable crystal and electrical structures make VS_2 a strong candidate as the cathode material for AIBs.

Using in-situ X-ray diffraction, Wu and colleagues have discovered that VS_2 is completely reversible throughout the (dis)charge process. Although deposition and dissolution can be seen at the anode, the crystallinity and crystal phase of VS_2 remain unaltered. VS_2 is a good contender for use as electrode material in AIBs (Wu et al., 2018), as shown by this research. Despite continuous study, there is still a great need to learn more about how aluminium ions are stored in VS_2.

$$VS_2 + xAl^{3+} + 3xe^- \rightleftarrows Al_xVS_2$$

Non-Metallic Ion Storage

Ammonium-Ion Battery

The interaction between different types of charge carriers and a particular electrode material differs from the interaction between identical types of charge carriers and diverse electrode materials. As a result, metallic and non-metalliccarrier batteries exhibit different electrochemical characteristics and performances (Liang et al., 2021). Benefits of non-metallic ion energy storage include affordability, high levels of safety, and the possibility for extensive power grid energy storage (Janoschka et al., 2015). Ammonium ion batteries, which use NH4+ ions with a lower hydrated ion radius and a lighter molar mass of 18 g mol, are one possible non-metallic ion battery type (Chen et al., 2022). Yu et al.(2021) studied the electrochemical performance of heterostructural VS_2/VOx material as a cathode for aqueous-ion batteries. His study shows that this material has an enhanced working potential of 0.25 V and a specific capacity of 156 mA h g−1 at 10 A g−1, making it superior in rate capability to other materials. And it possessesa high reversible capacity of over 150 mA h g−1 and a long lifespan of over 1000 cycles as a cathode for aqueous NH4+ ion storage (Yu, 2021).

CHALLENGES AND SOLUTIONS

Challenges

Volume Expansion

Volume expansion is a common problem in battery electrodes, particularly those that use high-capacity materials. This expansion and contraction during charge and discharge cycles can lead to mechanical stress, cracking, and even electrode failure. This can result in a decrease in electrical conductivity, loss of contact between the active material and current collector, and a reduction in battery capacity, efficiency, and cycle life (Cheng & Verbrugge, 2010; Jerliu et al., 2014). When Vanadium disulphide is used as a material for energy storage, it expands during operation, which can cause cracks in the material. These cracks can lead to the destruction of contact with the current collector, which is responsible for collecting the current and transporting it out of the battery. When the contact is lost, the battery's capacity drops rapidly.

Furthermore, the expansion of Vanadium disulphide can also create a great deal of stress within the battery, which can cause physical damage, such as extruding the pole piece or even breaking the battery.

The pole piece is a critical component of a battery that separates the anode and cathode, and damage to it can result in a loss of battery function (Li et al., 2019).

Conversion Reaction

Conversion reactions in electrode materials refer to a type of chemical reaction that occurs during the charging and discharging of batteries or other electrochemical devices. These reactions involve a change in the chemical composition or oxidation state of the electrode material, often through the insertion or extraction of ions. Conversion reactions can offer high energy density, but they can also lead to structural changes and degradation of the electrode material over time, which can limit the cycle life and performance of the battery. Understanding and controlling conversion reactions is important for the development of high-performance and durable electrochemical devices (Wang et al., 2011). During the discharge phase of a zinc-ion battery using Vanadium disulphide as an electrode material, a conversion reaction takes place. Due to the weak interaction between Vanadium and Sulphur (V-S) and the high binding energy of the working ions, the conversion process occurs easily. This leads to the formation of MS2 and metallic Vanadium (V) particles. The conversion reaction has a negative impact on the long-term cycling performance of the battery. For instance, when the voltage window ranges from 0.4 to 1.0 V, the battery capacity using VS_2 as the electrode material decreases to only 27.1 mAh g1 after 50 cycles. However, by adjusting the voltage window to 0.4-0.8 V, a higher capacity of 104.4 mAh g1 can be achieved even after 350 cycles. This suggests that controlling the voltage range can help mitigate the degradation associated with the conversion reaction and improve the cycling stability of the battery (Zhu et al., 2021).

Poor Irreversibility

Poor irreversibility in electrode materials refers to the inability of the material to undergo reversible reactions during charging and discharging cycles. This can occur due to structural changes, side reactions, or the loss of active material. Poor irreversibility can lead to a loss of capacity and poor cycling performance over time. Researchers are exploring various approaches, such as the use of stable electrolytes and the design of electrode structures that minimize structural changes, to improve the reversibility of electrode materials (Aravindan et al., 2017). By enhancing the reversibility of electrode materials, it may be possible to improve the performance and longevity of electrochemical devices.Another challenge is the low reversibility of by-products produced, such as the oxidation or dissolutionof polysulfides to form Sulphur, which results in structural damage of Vanadium disulphide. Dissolution and shuttle phenomena have an effect on the discharge processes and reversibility of charge, resulting in decreased cycle stability. In simpler terms, the byproducts formed during the charge-discharge process cause damage to the structure of VS_2, leading to poor cycle stability. The limited reversibility of these byproducts has an impact on the charge and discharge process, making it less efficient and incurring further material degradation (Zheng et al., 2021).

Solutions

VS_2 has received a lot of interest because of its potential in energy storage applications. One notable issue with the VS_2 is the battery's high volume expansion during use, which may cause fractures and physical damage. Another difficulty is the poor performance caused by the conversion reaction during the discharge process, as well as the inability of the by-products created to be reversed.

To address these concerns, many methods to modify the structure and features of VS_2 have been pursued. Intercalating molecules between the layers of VS_2, for example, may increase the layer gap and stabilise the structure. In situ polymerization of EDOT and VS_2 may introduce PEDOT into the VS_2 interlayer, producing strong hydrogen bonds and increasing the layer gap. These changes may assist to relieve the volume expansion issue while also improving the electrochemical performance of VS_2 (Murugan, 2005; Yang, 2021). Another way to overcome these issues is to make a VS_2 composite with other materials. Composite materials can provide structural support to prevent the volume expansion of VS_2 and improve its cycling stability. They can also enhance the kinetics of the conversion reaction and reduce the irreversible capacity loss (Liu, Xu, Cai et al, 2022). For example, researchers have investigated the use of carbon-based materials, such as carbon nanotubes and reduced graphene oxide etc., as composites with VS_2. These materials can improve the electrical conductivity of the composite, enhance the electrochemical performance, and increase the reversible capacity of the electrode (Ghorui, 2022; Wang, n.d.). Other studies have focused on using metal oxides, such as TiO2 and MoS2, as composites with VS_2. These metal oxides can provide structural support, improve the stability of the composite, and enhance the electrochemical performance (Cao et al., 2017; Dong et al., 2020).

Moreover, the size and shape of the VS_2 material may have an impact on its electrochemical performance. A one-step synthesis technique for VS_2 microspheres made of multiple curved ultrathin nanosheets has been devised, which successfully eliminates nanosheet agglomeration and reaggregation, reduces ion diffusion distance, and relieves volume expansion. This structure has a high specific capacity and strong cycle stability for various current densities (Wang, Luo, Wu et al, 2019).

Composites

Carbon-Based Composites

Vanadium disulphide (VS_2) and carbon are combined to create a vanadium disulphide-carbon composite. The complementary nature of the two components gives this composite material its distinctive characteristics. Graphene, carbon nanotubes, and activated carbon are only a few of the several types of carbon, which is a versatile substance. It is renowned for its great surface area, thermal stability, and electrical conductivity.

A composite material with superior electrochemical performance, such as increased capacity, greater cycle stability, and quicker charging and discharging rates, is produced by combining VS_2 with carbon. This is because each component has distinct qualities. Although the VS_2 offers a high capacity for lithium-ion storage, the carbon contributes electrical conductivity and a large surface area for the electrochemical process. The vanadium disulphide-carbon composite's shape has a significant impact on how well it functions as an electrode material for energy storage. The form, size, and structure of the particles that make up the composite material are referred to as its morphology (An et al., 2015; Wang, n.d.).

Vanadium disulphide-carbon composites may be made using a variety of synthesis processes, such as chemical vapour deposition, hydrothermal synthesis, and solvothermal synthesis. By changing the synthesis parameters, such as temperature, pressure, and precursor concentration, the composite's shape may be managed. In general, the composite's morphology may have a variety of effects on how well it performs electrochemically. The diffusion of lithium ions into and out of the material may be influenced, first, by the size and form of the particles. Greater ion diffusion may be facilitated by smaller particles' greater surface areas, which improves electrochemical performance. Second, the material's conductivity may be impacted by the shape. Better conductivity and lower resistance may be found in materials with a more networked and porous topology, which improves electrochemical performance (Wang, n.d.).

Recent research has examined the impact of morphology on the electrochemical performance of composites made of vanadium disulphide and carbon as electrode materials. For instance, one research found that compared to vanadium disulphide-carbon composites with a porous bowl shape (Wu et al., 2020), those with heteroarchitecture (Huang et al., 2020) morphology had greater capacity and improved cycle stability. Vanadium disulphide-carbon composite's electrochemical performance as electrode materials may also be influenced by the amount of carbon present, the crystallinity present, and the surface area present. The conductivity of the composite may be influenced by the carbon content, with a larger carbon concentration often resulting in better conductivity and electrochemical performance. The degree of order in the atomic arrangement, or crystallinity, may have an impact on a material's capacity and rate of lithium-ion diffusion. Lithium-ion diffusion may occur more quickly and at larger capacities with increased crystallinity. In addition, the composite's surface area may influence electrochemical performance by increasing the number of active sites for lithium-ion adsorption and promoting quicker ion diffusion. More capacity and improved rate performance may be achieved by increasing the surface area by the management of particle size or the addition of porous features (Renault et al., 2013; Zhu et al., 2013). Vanadium disulphide-carbon composites must have their carbon content, crystallinity, and surface area optimised in order to have excellent electrochemical performance as electrode materials.

Reduced Graphene Oxide

Vanadium disulphide-reduced graphene oxide (VS_2-rGO) composite is a hybrid material that combines the unique properties of both vanadium disulphide and reduced graphene oxide. The electrochemical performance of VS_2-rGO composites as an electrode material in energy storage devices can be significantly affected by their morphology, reduced graphene oxide content, crystallinity, and surface area.

Reduced graphene oxide (rGO) is a graphene derivative formed via the reduction of graphene oxide (Alam et al., 2017). A highly oxidised version of graphene called graphene oxide has functional groups including hydroxyl, epoxy, and carboxyl on its surface (Zaaba et al., 2017). The elimination of oxygen-containing groups during the reduction of graphene oxide to rGO allows for the restoration of sp2 hybridization in the carbon lattice and the creation of a highly conductive material. There are many ways to carry out the reduction process, including thermal reduction, chemical reduction, and electrochemical reduction (Zhang et al., 2015).

A battery composed of vanadium disulphide has an insufficient cycle life and poor rate capability due to an inherent problem of high-volume fluctuation and low active material usage (Yu et al., 2018). Combining carbon-based materials to create a conductive matrix is a viable solution to this problem. Because of its outstanding electrical conductivity, improved mechanical strength and substantial specific surface area, reduced graphene oxide (rGO) is an attractive contender. Moreover, the functional groups on the surface of rGO provide strong interfacial interaction between the active components and the conductive skeleton, facilitating quick charge transfer and good structural stability during repeated electrochemical cycling (Li, 2021). For instance, Chen et al. showed that the new VS_2@rGO structure had an outstanding electrochemical performance for Zn storage. A high capacity of 238 mAh per gramme was shown, and even after 1000 cycles, capacity retention was above 93% (Chen, Zhu, Chen et al, 2020). Fang et al.

developed Vanadium disulphide anchored Graphene nanosheets, by combining VS_2 flakes with graphene nanosheets for testing as a LIB anode and cathode. The nanocomposites exhibited promising performance as a cathode material, maintaining a specific capacity of 180.1 mAh/g and demonstrating good characteristics at various charging rates. As an anode material, VS_2/GNS sustained a high capacity of 528 mAh/g even after 100 cycles at 200 mA/g (Fang et al., 2015). Superior electrochemical capabilities were shown as a consequence of the interactions between graphene and VS_2, which were found to enable efficient electron transport and mechanical strengthening. To completely comprehend the mechanics of lithium penetration and look into lithiation processes at the atomic level, further study is necessary. The VS_2 layer thickness is a constraint, and less stacked layers could enhance the composite's sorption capabilities. Metal dichalcogenides often age with structural degeneration, raising issues regarding material deterioration as well. An inert outer layer like graphene or h-BN combined with the dichalcogenide monolayer in a heterostructure might slow ageing and maintain the special features of the substance. The surface and interlayer space sorption energies and diffusion barriers of the heterostructure are computed and compared to bulk systems (Dimitropoulos, 2022).

Carbon Nanotubes (CNTs)

Carbon atoms are organised in a hexagonal lattice to form the cylinder-shaped carbon nanotubes (CNTs). CNTs are very favoured for a variety of applications because to their remarkable mechanical, thermal, and electrical capabilities that result from their distinctive structure (Zihan, 2022). Single-walled carbon nanotubes (SWCNTs) and Multi-walled carbon nanotubes are the two varieties of CNTs (MWCNTs). Whereas MWCNTs are made of many layers of carbon atoms coiled concentrically around a central axis, SWCNTs are made of a single layer of carbon atoms rolled up into a smooth cylinder (Okpalugo et al., 2005).

Li et al. effectively manufactured high crystallinity vanadium disulphide nanosheets (VS_2 NS) on the surface of a thin film made of carbon nanotubes (CNTs), resulting in a core/branch architecture (VS_2 NS@CNTs). It was carefully investigated how this material behaved electrochemically when used as an electrode in lithium-ion batteries (LIBs). A notable feature of the VS_2 NS@CNTs electrode was its outstanding coulombic efficiency of about 98%, high reversible discharge capacity of over 850 milliampere-hours per gramme (mA h g1) at a current density of 200 milliampere per gramme (mA g1), great cycle stability, and better rate capability (Li et al., 2019).

Carbon Nanofibers (CNFs)

Similar to carbon nanotubes, carbon nanofibers (CNFs) are cylinder-shaped nanostructures made of graphitic carbon. They do, however, have a more intricate, branching structure than carbon nanotubes. CNFs have lengths that may reach millimetres and widths that range from a few tens of nanometers to several micrometres. Depending on the synthesis technique used, CNFs may have a variety of morphologies. With a carbon core and an outer layer of metal catalyst particles, CNFs may have a core-shell structure. They might also have a bamboo-like structure, with amorphous portions between sections of crystalline structure resembling graphite (Qi, Xu, Wang et al, 2020).

Xu et al. successfully synthesised ultrathin bowl-like vanadium disulphide (VS_2) nanosheet arrays on highly conductive carbon nanofibers (CNF@VS_2) using a one-step solvothermal process. The resultant VS_2 array had a wide surface area, great homogeneity, and a vertical distribution that made it simple for sodium ions to quickly and easily reach the surface of VS_2, which led to good rate capability. Volume expansion from the conversion process may be effectively absorbed by VS_2 and carbon nanofibers. The unusual freestanding state was made possible by the two-dimensional nanosheet structure's generation of reversible surface redox pseudocapacitance. The CNF@VS_2 composite demonstrated exceptional mechanical flexibility together with a high reversible capacity of 659 milliampere-hours per gramme (mAh g-1) at a current density of 0.1 Ampere per gramme (Ag-1) and long-term cycle stability up to 6000 cycles at 2 Ag-1. The findings point to the possible use of the created CNF@VS_2 composite in high-performance sodium-ion batteries (Xu et al., 2020).

Carbon Cloths

A potential electrode material for energy storage applications is the vanadium disulphide-Carbon Cloth (CC) composite because tofits high specific capacity, extended cycle life and great rate capability. The electrochemical performance of the VS_2-CC composite is significantly influenced by its shape. Improved ion diffusion kinetics and an enhanced charge transfer reaction might result from the uniform distribution of VS_2 particles over the CC surface with a wide interfacial area.

The resulting material, known as carbon cloth, is made by weaving together carbon strands to produce a textile-like structure. The kind of fibre, weave pattern, and finishing procedure are only a few aspects of the variables that might affect the structure and morphology of the carbon fabric throughout the weaving process. Most carbon fabric is very porous, having a sizable surface area and linked pore structure. Due to this, it may be used in processes like gas separation and catalysis, as well as electrode materials for batteries and supercapacitors (Jost et al., 2013; Pandya et al., 2011).

Using the hydrothermal synthesis of vanadium disulphide (VS_2) nanosheets on the surface of carbon cloth (CC), Zhang et al. created a highly open network that efficiently encourages charge transfer and raises

Figure 7. Nitrogen-doped carbon

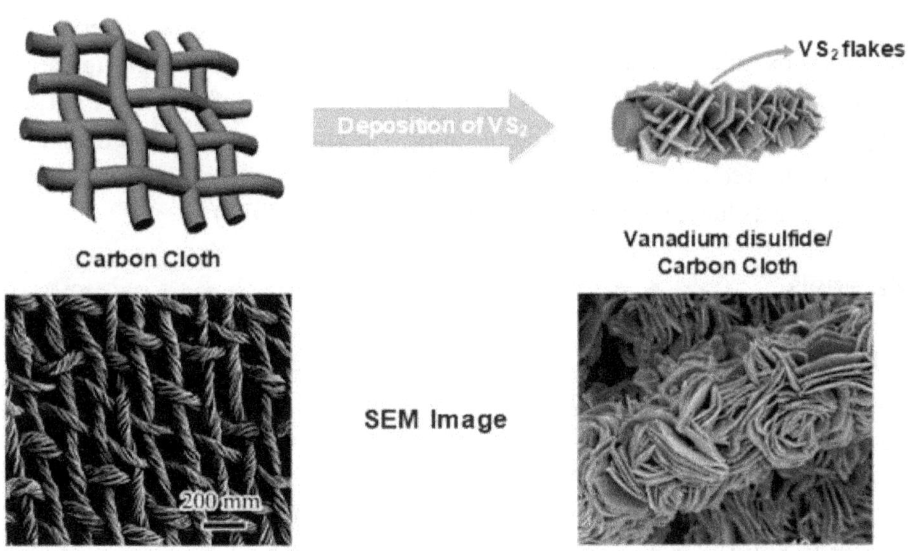

energy density. The electrochemical performance of the system is significantly improved by harnessing the synergistic effect between carbon cloth (CC) with double-layer capacitance properties and VS_2 with pseudocapacitance properties. At 1 mA cm2, the specific capacitance of Vanadium disulphide-Carbon cloth is 972.5 mF cm2, and after 10,000 charge and discharge cycles at 10 mA cm2, the capacitance retention is 77.5%. Moreover, a Vanadium disulphide-Carbon cloth-12 and activated carbon-based asymmetric supercapacitor was put together. At 4.24 mW cm2, the asymmetric supercapacitor has a higher energy density of 0.22 mW h cm2. Energy density remains at 0.16 mW h cm2 even while power density rises to 8.26 mW cm2. This shows that flexible supercapacitor electrode materials made on Vanadium disulphide-Carbon cloth composites have a wide range of possible uses (Zhang, Miao, Yan et al, 2022).

Nitrogen-Doped Carbon

Nitrogen-doped carbon is made by adding nitrogen atoms into the carbon lattice structure. This may be accomplished using a variety of techniques, including chemical vapour deposition, pyrolysis, and doping with molecules containing nitrogen.Depending on the synthesis procedure and environmental factors, the structure of carbon that has been doped with nitrogen may change (Naik et al., 2022; Podyacheva & Ismagilov, 2015). While hexagonal rings of carbon atoms are stacked in a layered pattern, nitrogen-doped carbon compounds often have a graphitic structure. Nitrogen atoms are introduced into the carbon lattice, which causes defects to occur and the material's electrical characteristics to alter. A potential material for uses in sensing, energy storage, and catalysis is carbon that has been doped with nitrogen. It is a flexible material with the potential for a wide range of applications since its characteristics may be tailored by adjusting the synthesis conditions and the nitrogen doping amount (Ismagilov et al., 2009).

Nitrogen doping has emerged as a promising method to enhance carbon reactivity by introducing external defects. Additionally, the strong interaction between vanadium-based compounds and N-doped carbon is facilitated by the high affinity of nitrogen dopants for transition metal species. This significantly improves interfacial electron transport and serves as a reliable buffer to mitigate stress and volume changes (Shuai, 2023). Vanadium disulphide (VS_2) nanocrystals that resemble spindles have been prepared using an effective in situ hybridization technique on a layer of carbon doped with nitrogen, and its electrochemical performance as a cathode material for rechargeable aqueous Zn-ion batteries have been thoroughly examined. The resultant cathode material had great long-term cycling stability with a capacity retention of 97% after 600 cycles at a current density of 1000 mA g-1, as well as good specific capacity of 203 milliampere-hours per gramme (mAh g-1) at a current density of 50 mA g-1. As a cathode material for rechargeable aqueous Zn-ion batteries, the findings point to the development of VS_2@N-C composite as having enormous potential (Liu et al., 2021).

Inorganic Materials

Vanadium disulphide (VS_2) can also be combined with various inorganic materials to form composites for use as electrode materials. The morphology of the vanadium disulphide-inorganic materials composite can vary depending on the type of inorganic material used. For example, when combined with molybdenum disulphide, it forms a sandwich-like structure (Liu, 2023). The content of the inorganic material in the composite can also affect its electrochemical performance, with higher content generally resulting in improved performance.

The crystallinity of the vanadium disulphide-inorganic materials composite can also affect its electrochemical performance. For example, when combined with zinc oxide, the composite with a higher degree of crystallinity exhibits better electrochemical performance (Karthik et al., 2018). Similarly, the surface area of the composite can also play a role in its performance, with higher surface area generally leading to improved performance. For example, when combined with titanium dioxide nanotubes, the composite with a higher surface area showed improved electrochemical performance compared to the one with lower surface area (Cao et al., 2017).

A series of compounds known as vanadium oxide (VOx) is made up of different vanadium and oxygen atoms. Depending on the precise composition, which might range from V2O3 to V2O5, the structure and morphology of VOx are affected. VOx compounds often display a variety of crystal morphologies and topologies, which may have an effect on their physical and chemical characteristics (Piccirillo et al., 2007). Yu et al. presented an in situ electrochemically driven production of heterostructuralVS_2/VOx material by taking advantage of the instability of TMS in water. The VS_2/VOx cathode demonstrated an improved Zn2+ storage performance, including a high reversible capacity of 310 mA h g1 and longer stability of 3000 cycles at 1A g1. These improvements were made possible by the special interface properties caused by the internal electric field at the heterointerfaces, high chemical stability of oxides, and high conductivity of sulphide. The high conductivity of sulphide and the great chemical stability of oxides, together with the peculiar interface features brought on by the internal electric field at the heterointerfaces, were credited with this achievement. In compared to pure VS_2, the average working potential was elevated by 0.25 V, which helped to raise the energy density. In addition, the heterostructuredVS_2/VOx were further assessed as the cathode for NH4+ ion storage, inspired by the performance in AZIBs. High reversible capacity (over 150 mA h g1 at 0.1 A g1) and higher rate capabilities (100 mA h g1 at 5 A g1) were provided by VS_2/VOx (Yu, 2021).

Na2Ti2O5-VS_2 is another high performing composite material for electrode making. The structural stability of the Na2Ti2O5 nanowire arrays and the increased conductivity and capacityof the two-dimensional VS_2 nanosheets are responsible for the larger capacity, enhanced rate capability, and superior stability of the composite Na2Ti2O5-VS_2 nanowire arrays. Liao et al. synthesized Na2Ti2O5-VS_2 composite nanowire array electrodes using a two-step hydrothermal method. The charge capacities of Na2Ti2O5-VS_2 remained at 298 and 203 mA h g-1 after 50 (C/10) and 100 (1C) cycles, respectively. These values significantly outperformed the identical cycled Na2Ti2O5 electrode, which achieved capacities of 157 and 93 mA h g-1 (Liao & Manthiram, 2015).

VS_2-TiO2-B composite is also a promising material for the production of high-performing battery electrodes. Due to its minimal volume expansion (3e4%) during lithium-ion insertion/desertion, which bestows it with exceptional structural stability, TiO2 demonstrates good cycle performance. Due to its open framework structure that permits lithium-ion mobility and its pseudocapacitive process for lithium storage, the bronze phase of TiO2 (TiO2-B) is the most alluring polymorph. TiO2-B@VS_2 heterogeneous nanowire arrays (TVNAs) were successfully created by Cao et al. using a two-step hydrothermal procedure and used as additive-free anodes for LIBs. TiO2-B vertical nanowire arrays were covered in an even distribution of VS_2 nanoparticles with a layered structure (TNAs). With a reversible capacity of 365.4 mA h g-1 over 500 cycles at a current density of 1 C and a high-rate capability of 171.2 mA h g-1 at 10 C rate, the TVNAs electrode successfully merged the good electrochemical capabilities of the constituent components (Cao et al., 2017).

Li et al. coated vanadium disulphide flakes on a titanium disulphide layer that was about 2.5 nm thick. The titanium disulphide coating was shown to be much less prone to Peier distortion during the lithiation-delithiation process, allowing it to stabilise the underlying vanadium disulphide material, according to simulations using density functional theory. The vanadium disulphide cathode covered with titanium disulphide had an operating voltage of 2 V, high specific capacity (180 mAh g1 @200 mA g1 current density), and rate capability (70 mAh g1 @1000 mA g1), while attaining capacity retention close to 100% after 400 charge-discharge stages (Li, 2019).

Dong et al. grew VS_2 nanosheets vertically on the MoS2 surface, forming abundant Mott-Schottky heterojunctions. They observed that vanadate encourages the creation of very thin hierarchical nanostructures with tunable 1T/2H ratios from irregular bulks by monitoring the evolution of the heterostructure. According to calculations using density functional theory (DFT), the VS_2/MoS2 hybrids enable energy band rearrangement towards the fermi level, which results in self-improving electrical conductivity. At the heterointerface, the built-in electrical field significantly increases Li adsorption energy (-3.60 eV), making it considerably stronger than the MoS2 surface (-1.57 eV). According to the charge/discharge process, Dong et al. also found that the Li-S bond breakage triggered by VS_2, and the sulphur anchored by a high Mo6+/Mo4+ ratio may work together to promote redox reversibility. The result was a high specific capacity of 1273 mAh g-1 at 0.1 A g-1 and 61% capacity retention even at 5 A g-1 for the VS_2/MoS2 hybrid. At 100 cycles at 0.2 A g-1 and 500 cycles at 1 A g-1, the hybrid also demonstrated excellent stability (Dong et al., 2020).

Pu et al. used a one-pot hydrothermal method to create hydrophilic VOOH on rose-like VS_2 composed of nanosheets (VS_2@VOOH). Improved electrochemical performance, particularly in terms ofextended cycle life and rate capability, results from the addition of O-H to VOOH, which also improves the infiltration of the electrolyte and electrode. As an example, this composite material has a superior capacity of 107.5 mA h g-1 after 350 cycles at 1.5 A g-1 when employed as an aqueous ZIB cathode. Even after 400 cycles of operation at a high current density of 2.5 A g-1, it still has a reversible capacity of 91.4 mA h g-1. As a result, the researchers think that VS_2@VOOH has a lot of promise as a cathode material for ZIBs (Pu et al., 2019).

CONCLUSION

Vanadium disulfide (VS_2) plays a crucial role as an electrode material in energy storage devices due to its remarkable properties. One key factor contributing to its importance is the unique crystal structure of VS_2, which consists of layers comprising edge-sharing VS_6 octahedra. This layered structure provides several advantages for energy storage applications. Firstly, it results in VS_2 nanosheets having a high surface area, allowing for a greater number of active sites for electrochemical reactions. This increased surface area facilitates the efficient storage and release of energy within the material. Secondly, VS_2 exhibits excellent electrical conductivity, enabling the smooth flow of electrons during the charging and discharging processes. The combination of high surface area and electrical conductivity makes vanadium disulfide a highly desirable electrode material, offering enhanced performance and contributing to the overall efficiency and effectiveness of energy storage devices.

The synthesis of vanadium disulfide offers various methods that can be chosen based on specific needs and availability. Two commonly used methods for preparing vanadium disulfide are chemical vapor deposition (CVD) and calcination. CVD allows for the production of high-quality, two-dimensional materials with excellent crystalline properties and minimal contamination. However, both CVD and calcination methods lack control over the morphology and size of the synthesized vanadium disulfide.

To overcome this limitation, alternative synthesis methods such as hydrothermal and solvothermal methods can be employed. Hydrothermal and solvothermal methods offer the advantage of allowing the synthesis of vanadium disulfide with different morphologies and sizes. Moreover, the solvothermal method provides additional benefits by allowing control over the solvent's polarity, density, viscosity, and dispersion. This level of control contributes to achieving complete crystal growth and minimizing agglomeration, leading to the production of vanadium disulfide with desired characteristics for energy storage applications.

Vanadium disulphide (VS_2) exhibits exceptional characteristics that make it a promising material for electrode applications in various energy storage devices. Its layered crystal structure, with vanadium atoms sandwiched between two sulphur atoms, provides an ideal platform for ion intercalation and deintercalation processes. This enables a reversible redox reaction, which is crucial for facilitating the charging and discharging processes in batteries. Additionally, the high electronic conductivity and large interlayer spacing of vanadium disulphide allow for rapid ion diffusion and efficient charge transport, resulting in high power and energy densities. In the realm of lithium-ion batteries, where the specific capacity of graphite is limited to 372 mAh g−1, a monolayer of VS_2 presents a promising alternative. With its low energy barrier and theoretical capacity of 1397 mAh g−1, it is widely recognized as a highly attractive electrode material, boasting excellent lithium-ion mobility, lithiation stability, and electrical conductivity.Vanadium disulphide also holds great potential for overcoming limitations in sodium-ion battery technology. The large radius of sodium ions poses challenges such as delayed electrochemical reaction kinetics, slow ion diffusion, separation of active materials, and volume expansion. However, the numerous active sites provided by vanadium disulphide for sodium ions effectively reduce the distance they need to travel, aided by its high specific surface area and layer spacing. These properties enhance electrochemical performance by controlling the incidence of intercalation and conversion processes. In the case of potassium-ion batteries, the higher radius of K+ ions impedes electrochemical reaction kinetics and diffusion rates. Nonetheless, the robust design of vanadium disulphide, offering a high number of K+ active sites and a broad layer gap, mitigates diffusion barriers, making it a viable option. Similarly, in zinc-ion batteries, the distinctive layered structure and wide layer separation of VS_2 (5.7 Å) make it an attractive cathode material, facilitating quick diffusion of Zn ions due to their smaller diameter. The high charge density of divalent Zn ions allows for simultaneous intercalation and conversion processes, significantly influenced by the cut-off voltage.When considering magnesium-ion and aluminum-ion batteries, finding suitable cathode materials poses challenges due to interfacial charge transfer, slow diffusion kinetics, and hindered larger charge densities. Vanadium disulphide addresses these issues by adopting a loose accumulation structure with open channels between its layers, ensuring both structural integrity and efficient ion transport within the material. In the realm of aluminum-ion batteries, the most researched electrode materials include V2O5, TiO2, and Prussian blue analogues, but the favorable crystal and electrical structures of VS_2 position it as a strong candidate for cathode materials in AIBs. While vanadium disulphide has shown good electrochemical performance in ammonium batteries, further improvements are required to reach optimal levels.

Vanadium disulphide (VS_2) exhibits several limitations that need to be addressed for its effective use as an electrode material in energy storage devices. One major limitation is the volume expansion during operation, leading to cracks and loss of contact with the current collector, resulting in a rapid decrease in battery capacity. Moreover, the expansion of VS_2 generates internal stress, potentially causing physical damage. Another limitation is the occurrence of conversion reactions, particularly in zinc-ion batteries, which can result in poor cycling stability and irreversible capacity loss. Additionally, vanadium disulphide shows limited reversibility, hindering long-term performance.

One effective solution to overcome the limitations of vanadium disulphide (VS_2) as an electrode material is the development of composite materials. Composites offer several advantages, such as providing structural support to prevent volume expansion and improving cycling stability. They can also enhance the kinetics of conversion reactions and reduce irreversible capacity loss. One popular type of composite material is carbon-based composites. By combining carbon with vanadium disulphide, a vanadium disulphide-carbon composite can be created. The combination of these two components results in a composite material with distinctive characteristics that complement each other. Additionally, inorganic composites have also shown promise in addressing these limitations. In this review paper, various types of carbon-based vanadium composites have been discussed, highlighting their applications and the importance of considering the existing morphology of vanadium disulphide and the specific properties that need to be enhanced or modified based on the intended application.

Vanadium disulphide (VS_2) holds significant promise as an electrode material for batteries due to its layered crystal structure, allowing for efficient ion intercalation and deintercalation processes. Its high electronic conductivity and large interlayer spacing enable fast ion diffusion and efficient charge transport, resulting in high power and energy densities. By addressing limitations such as volume expansion and conversion reactions through the use of composite materials, vanadium disulphide can further enhance its performance as a versatile and promising electrode material for battery applications.

REFERENCES

Abraham, K. M. (2020). How comparable are sodium-ion batteries to lithium-ion counterparts? *ACS Energy Letters*, *5*(11), 3544–3547. doi:10.1021/acsenergylett.0c02181

Abu Nayem, S. M., Ahmad, A., Shaheen Shah, S., Saeed Alzahrani, A., Saleh Ahammad, A. J., & Aziz, M. A. (2022). High Performance and Long-cycle Life Rechargeable Aluminum Ion Battery: Recent Progress, Perspectives and Challenges. *Chemical Record (New York, N.Y.)*, *22*(12), e202200181. doi:10.1002/tcr.202200181 PMID:36094785

Agyapong-Fordjour, F. (2021). Substitutional Vanadium SulfideNanodispersed in MoS2 Film for Pt-Scalable Catalyst. *Advanced Science*, *8*(16).

Alam, S. N., Sharma, N., & Kumar, L. (2017). Synthesis of graphene oxide (GO) by modified hummers method and its thermal reduction to obtain reduced graphene oxide (rGO). *Graphene (Irvine, Calif.)*, *6*(1), 1–18. doi:10.4236/graphene.2017.61001

Ali, B. A., & Allam, N. K. (2019). A first-principles roadmap and limits to design efficient supercapacitor electrode materials. *Physical Chemistry Chemical Physics*, *21*(32), 17494–17511. doi:10.1039/C9CP02614B PMID:31372620

Allen, M. J., Tung, V. C., & Kaner, R. B. (2010). Honeycomb carbon: A review of graphene. *Chemical Reviews*, *110*(1), 132–145. doi:10.1021/cr900070d PMID:19610631

Alzahrani, A. S., Otaki, M., Wang, D., Gao, Y., Arthur, T. S., Liu, S., & Wang, D. (2021). Confining sulfur in porous carbon by vapor deposition to achieve high-performance cathode for all-solid-state lithium–sulfur batteries. *ACS Energy Letters*, *6*(2), 413–418. doi:10.1021/acsenergylett.0c01956

An, G.-H., Ahn, H.-J., & Hong, W.-K. (2015). Electrochemical properties for high surface area and improved electrical conductivity of platinum-embedded porous carbon nanofibers. *Journal of Power Sources*, *274*, 536–541. doi:10.1016/j.jpowsour.2014.10.086

Ansari. (2022). Critical aspects of various techniques for synthesizing metal oxides and fabricating their composite-based supercapacitor electrodes: a review. *Nanomaterials, 12*(11).

Aravindan, Lee, & Madhavi. (2017). Best Practices for Mitigating Irreversible Capacity Loss of Negative Electrodes in Li-Ion Batteries. *Advanced Energy Materials, 7*(17).

Assefi, M., Maroufi, S., Yamauchi, Y., & Sahajwalla, V. (2020). Pyrometallurgical recycling of Li-ion, Ni–Cd and Ni–MH batteries: A minireview. *Current Opinion in Green and Sustainable Chemistry*, *24*, 26–31. doi:10.1016/j.cogsc.2020.01.005

Bais, A. R., Subhedhar, D. G., Joshi, N. C., & Panchal, S. (2022). Numerical investigation on thermal management system for lithium ion battery using phase change material. *Materials Today: Proceedings*, *66*, 1726–1733. doi:10.1016/j.matpr.2022.05.269

Banerjee, M., & Kaur, H. (2022). A Comparison Among Lithium-Ion, Nickel-Cadmium & Nickel-Metal-Hydride Batteries for Charging and Discharging in Electric Vehicle by Bidirectional Dc-Dc Converter. In *2022 IEEE IAS Global Conference on Emerging Technologies (GlobConET)*. IEEE. 10.1109/GlobConET53749.2022.9872335

Bucur, C. B., Gregory, T., Oliver, A. G., & Muldoon, J. (2015). Confession of a magnesium battery. *The Journal of Physical Chemistry Letters*, *6*(18), 3578–3591. doi:10.1021/acs.jpclett.5b01219 PMID:26722727

Cai, L., Zhang, Q., Mwizerwa, J. P., Wan, H., Yang, X., Xu, X., & Yao, X. (2018). Highly crystalline layered VS2 nanosheets for all-solid-state lithium batteries with enhanced electrochemical performances. *ACS Applied Materials & Interfaces*, *10*(12), 10053–10063. doi:10.1021/acsami.7b18798 PMID:29498503

Cao, M., Gao, L., Lv, X., & Shen, Y. (2017). TiO2-B@ VS2 heterogeneous nanowire arrays as superior anodes for lithium-ion batteries. *Journal of Power Sources*, *350*, 87–93. doi:10.1016/j.jpowsour.2017.03.070

Chen, Chua, & Lee. (2020). The advances of metal sulfides and in situ characterization methods beyond Li ion batteries: sodium, potassium, and aluminum ion batteries. *Small Methods, 4*(1).

Chen, J., Tang, Z., Pan, Z., Shi, W., Wang, Y., Tian, Z. Q., & Shen, P. K. (2020). Template-free growth of spherical vanadium disulfide nanoflowers as efficient anodes for sodium/potassium ion batteries. *Materials & Design*, *192*, 108780. doi:10.1016/j.matdes.2020.108780

Chen, Q., Song, M., Zhang, X., Zhang, J., Hou, G., & Tang, Y. (2022). Ammonium ion pre-intercalation stabilized tunnel h-WO 3 for fast NH 4+ storage. *Journal of Materials Chemistry. A, Materials for Energy and Sustainability*, *10*(29), 15614–15622. doi:10.1039/D2TA03643F

Chen, T., Zhu, X., Chen, X., Zhang, Q., Li, Y., Peng, W., Zhang, F., & Fan, X. (2020). VS2 nanosheets vertically grown on graphene as high-performance cathodes for aqueous zinc-ion batteries. *Journal of Power Sources*, *477*, 228652. doi:10.1016/j.jpowsour.2020.228652

Cheng & Verbrugge. (2010). Diffusion-induced stress, interfacial charge transfer, and criteria for avoiding crack initiation of electrode particles. *Journal of the Electrochemical Society, 157*(4).

Chianelli, R. R. (1976). Microscopic studies of transition metal chalcogenides. *Journal of Crystal Growth*, *34*(2), 239–244. doi:10.1016/0022-0248(76)90135-4

Das, C. K., Bass, O., Kothapalli, G., Mahmoud, T. S., & Habibi, D. (2018). Overview of energy storage systems in distribution networks: Placement, sizing, operation, and power quality. *Renewable & Sustainable Energy Reviews*, *91*, 1205–1230. doi:10.1016/j.rser.2018.03.068

De, P., Halder, J., Priya, S., Srivastava, A. K., & Chandra, A. (2023). Two-Dimensional V2O5 Nanosheets as an Advanced Cathode Material for Realizing Low-Cost Aqueous Aluminum-Ion Batteries. *ACS Applied Energy Materials*, *6*(2), 753–762. doi:10.1021/acsaem.2c02979

Deng. (2018). Layered P2-type K0. 65Fe0. 5Mn0. 5O2 microspheres as superior cathode for high-energy potassium-ion batteries. *Advanced Functional Materials, 28*(28).

Dimitropoulos. (2022). Wrinkle-mediated CVD synthesis of wafer scale Graphene/h-BN heterostructures. *Nanotechnology, 34*(2).

Dong, H., Duan, A., Zhong, S., Zhang, Y., & Zhou, X. (2019). Chemical vapor deposition growth of broadband tunable light absorption and anti-reflection properties of VS2 nanostructure films. *Materials Letters*, *252*, 227–230. doi:10.1016/j.matlet.2019.06.002

Dong, Y., Liu, Y., Hu, Y., Ma, K., Jiang, H., & Li, C. (2020). Boosting reaction kinetics and reversibility in Mott-Schottky VS2/MoS2 heterojunctions for enhanced lithium storage. *Science Bulletin*, *65*(17), 1470–1478. doi:10.1016/j.scib.2020.05.007 PMID:36747404

Du, C., Liang, D., Shang, M., Zhang, J., Mao, J., Liu, P., & Song, W. (2018). In situ engineering MoS2 NDs/VS2 lamellar heterostructure for enhanced electrocatalytic hydrogen evolution. *ACS Sustainable Chemistry & Engineering*, *6*(11), 15471–15479. doi:10.1021/acssuschemeng.8b03929

Du, C., Shang, A., Shang, M., Ma, X., & Song, W. (2018). Water-soluble VS2 quantum dots with unusual fluorescence for biosensing. *Sensors and Actuators. B, Chemical*, *255*, 926–934. doi:10.1016/j.snb.2017.08.070

Dunn, B., Kamath, H., & Tarascon, J.-M. (2011). Electrical energy storage for the grid: A battery of choices. *Science*, *334*(6058), 928–935. doi:10.1126/science.1212741 PMID:22096188

Elia, G. A., Kravchyk, K. V., Kovalenko, M. V., Chacón, J., Holland, A., & Wills, R. G. A. (2021). An overview and prospective on Al and Al-ion battery technologies. *Journal of Power Sources*, *481*, 228870. doi:10.1016/j.jpowsour.2020.228870

Fan, X., Liu, B., Liu, J., Ding, J., Han, X., Deng, Y., Lv, X., Xie, Y., Chen, B., Hu, W., & Zhong, C. (2020). Battery technologies for grid-level large-scale electrical energy storage. *Transactions of Tianjin University*, *26*(2), 92–103. doi:10.1007/s12209-019-00231-w

Fang, W., Zhao, H., Xie, Y., Fang, J., Xu, J., & Chen, Z. (2015). Facile hydrothermal synthesis of VS2/graphene nanocomposites with superior high-rate capability as lithium-ion battery cathodes. *ACS Applied Materials & Interfaces*, *7*(23), 13044–13052. doi:10.1021/acsami.5b03124 PMID:26016687

Feng, J., Sun, X., Wu, C., Peng, L., Lin, C., Hu, S., Yang, J., & Xie, Y. (2011). Metallic few-layered VS2 ultrathin nanosheets: High two-dimensional conductivity for in-plane supercapacitors. *Journal of the American Chemical Society*, *133*(44), 17832–17838. doi:10.1021/ja207176c PMID:21951158

Fichtner, M. (2022). Recent research and progress in batteries for electric vehicles. *Batteries & Supercaps, 5*(2).

Gan, Y. X., Jayatissa, A. H., Yu, Z., Chen, X., & Li, M. (2020). Hydrothermal synthesis of nanomaterials. *Journal of Nanomaterials*, *2020*, 1–3. doi:10.1155/2020/8917013

Gayathri, Manimegalai, & Krishnakumar. (2021). Challenging Issues and Solutions on Battery Thermal Management for Electric Vehicles. *Electrical and Electronic Devices, Circuits, and Materials: Technological Challenges and Solutions,* 535-553.

Genchi, G. (2020). The effects of cadmium toxicity. *International Journal of Environmental Research and Public Health, 17*(11).

Ghorui. (2022). Newly Designed One-Pot In-Situ Synthesis of VS2/rGO Nanocomposite to Explore Its Electrochemical Behavior towards Oxygen Electrode Reactions. *ChemElectroChem, 9*(15).

Goel, P., Dobhal, D., & Sharma, R. C. (2020). Aluminum–air batteries: A viability review. *Journal of Energy Storage*, *28*, 101287. doi:10.1016/j.est.2020.101287

Guo, Z., Yang, L., Wang, W., Cao, L., & Dong, B. (2018). Ultrathin VS 2 nanoplate with in-plane and out-of-plane defects for an electrochemical supercapacitor with ultrahigh specific capacitance. *Journal of Materials Chemistry. A, Materials for Energy and Sustainability*, *6*(30), 14681–14688. doi:10.1039/C8TA03812K

Guo, Z., Yang, L., Wang, W., Cao, L., & Dong, B. (2018). Ultrathin VS 2 nanoplate with in-plane and out-of-plane defects for an electrochemical supercapacitor with ultrahigh specific capacitance. *Journal of Materials Chemistry. A, Materials for Energy and Sustainability*, *6*(30), 14681–14688. doi:10.1039/C8TA03812K

Han, Z., Li, S., Wu, Y., Yu, C., Cheng, S., & Xie, J. (2021). Challenges and key parameters in exploring the cyclability limitation of practical lithium–sulfur batteries. *Journal of Materials Chemistry. A, Materials for Energy and Sustainability*, *9*(43), 24215–24240. doi:10.1039/D1TA06499A

Hart, C. J., Cuisinier, M., Liang, X., Kundu, D., Garsuch, A., & Nazar, L. F. (2015). Rational design of sulphur host materials for Li–S batteries: Correlating lithium polysulphide adsorptivity and self-discharge capacity loss. *Chemical Communications*, *51*(12), 2308–2311. doi:10.1039/C4CC08980D PMID:25562067

He. (2017). Layered VS2 nanosheet-based aqueous Zn ion battery cathode. *Advanced Energy Materials, 7*(11).

Hijazi, H., Desai, P., & Mariyappan, S. (2021). Non-Aqueous Electrolytes for Sodium-Ion Batteries: Challenges and Prospects Towards Commercialization. *Batteries & Supercaps*, *4*(6), 881–896. doi:10.1002/batt.202000277

Hirsh. (2020). Sodium-ion batteries paving the way for grid energy storage. *Advanced Energy Materials, 10*(32).

Hossain. (2018). Chemical vapor deposition of 2D vanadium disulfide and diselenide and Raman characterization of the phase transitions. *Advanced Materials Interfaces, 5*(16).

Hossain, Qin, & Sen. (2021). Chemical Synthesis and Substrate Temperature Effect on Morphology of 2D Vanadium Disulfide. *Crystal Research and Technology, 56*(3).

Hu. (2023). Recent Progresses on Vanadium Sulfide Cathodes for Aqueous Zinc-Ion Batteries. *Energies, 16*(2).

Hu. (2019). A portable and efficient solar-rechargeable battery with ultrafast photo-charge/discharge rate. *Advanced Energy Materials, 9*(28).

Huang, K.-J., Liu, Y.-J., Shi, G.-W., Yang, X.-R., & Liu, Y.-M. (2014). Label-free aptamer sensor for 17β-estradiol based on vanadium disulfide nanoflowers and Au nanoparticles. *Sensors and Actuators. B, Chemical*, *201*, 579–585. doi:10.1016/j.snb.2014.05.055

Huang. (2020). Electrode design for lithium–sulfur batteries: problems and solutions. *Advanced Functional Materials, 30*(22).

Huang, L., Zhu, W., Zhang, W., Chen, K., Wang, J., Wang, R., Yang, Q., Hu, N., Suo, Y., & Wang, J. (2018). Layered vanadium (IV) disulfide nanosheets as a peroxidase-like nanozyme for colorimetric detection of glucose. *Mikrochimica Acta*, *185*(1), 1–8. doi:10.1007/s00604-017-2552-1 PMID:29594417

Huang, Z., Han, X., Cui, X., He, C., Zhang, J., Wang, X., Lin, Z., & Yang, Y. (2020). Vertically aligned VS 2 on graphene as a 3D heteroarchitectured anode material with capacitance-dominated lithium storage. *Journal of Materials Chemistry. A, Materials for Energy and Sustainability*, *8*(12), 5882–5889. doi:10.1039/C9TA13835H

Hussain. (2023). Design of XS2 (X= W or Mo)-Decorated VS2 Hybrid Nano-Architectures with Abundant Active Edge Sites for High-Rate Asymmetric Supercapacitors and Hydrogen Evolution Reactions. *Small, 19*(8).

Ismagilov, Z. R., Shalagina, A. E., Podyacheva, O. Y., Ischenko, A. V., Kibis, L. S., Boronin, A. I., Chesalov, Y. A., Kochubey, D. I., Romanenko, A. I., Anikeeva, O. B., Buryakov, T. I., & Tkachev, E. N. (2009). Structure and electrical conductivity of nitrogen-doped carbon nanofibers. *Carbon*, *47*(8), 1922–1929. doi:10.1016/j.carbon.2009.02.034

Janoschka, T., Martin, N., Martin, U., Friebe, C., Morgenstern, S., Hiller, H., Hager, M. D., & Schubert, U. S. (2015). An aqueous, polymer-based redox-flow battery using non-corrosive, safe, and low-cost materials. *Nature*, *527*(7576), 78–81. doi:10.1038/nature15746 PMID:26503039

Jerliu, B., Hüger, E., Dörrer, L., Seidlhofer, B.-K., Steitz, R., Oberst, V., Geckle, U., Bruns, M., & Schmidt, H. (2014). Volume expansion during lithiation of amorphous silicon thin film electrodes studied by in-operando neutron reflectometry. *The Journal of Physical Chemistry. C, Nanomaterials and Interfaces*, *118*(18), 9395–9399. doi:10.1021/jp502261t

Jeyaseelan, C. (2020). *Ni-Cd Batteries. Rechargeable Batteries: History*, Progress, and Applications.

Ji, Q., Li, C., Wang, J., Niu, J., Gong, Y., Zhang, Z., Fang, Q., Zhang, Y., Shi, J., Liao, L., Wu, X., Gu, L., Liu, Z., & Zhang, Y. (2017). Metallic vanadium disulfide nanosheets as a platform material for multifunctional electrode applications. *Nano Letters*, *17*(8), 4908–4916. doi:10.1021/acs.nanolett.7b01914 PMID:28749686

Jiang, S., & Song, Z. (2022). A review on the state of health estimation methods of lead-acid batteries. *Journal of Power Sources*, *517*, 230710. doi:10.1016/j.jpowsour.2021.230710

Jiao, T., Yang, Q., Wu, S., Wang, Z., Chen, D., Shen, D., Liu, B., Cheng, J., Li, H., Ma, L., Zhi, C., & Zhang, W. (2019). Binder-free hierarchical VS 2 electrodes for high-performance aqueous Zn ion batteries towards commercial level mass loading. *Journal of Materials Chemistry. A, Materials for Energy and Sustainability*, *7*(27), 16330–16338. doi:10.1039/C9TA04798K

Joseph, A., & Aneesh, P. M. (2022). Efficient degradation of methylene blue: A comparative study using hydrothermally synthesised SnS2, WS2 and VS2 nanostructures. *Materials Research Bulletin*, *146*, 111623. doi:10.1016/j.materresbull.2021.111623

Jost, K., Stenger, D., Perez, C. R., McDonough, J. K., Lian, K., Gogotsi, Y., & Dion, G. (2013). Knitted and screen printed carbon-fiber supercapacitors for applications in wearable electronics. *Energy & Environmental Science*, *6*(9), 2698–2705. doi:10.1039/c3ee40515j

Kan, M., Wang, B., Lee, Y. H., & Sun, Q. (2015). A density functional theory study of the tunable structure, magnetism and metal-insulator phase transition in VS 2 monolayers induced by in-plane biaxial strain. *Nano Research*, *8*(4), 1348–1356. doi:10.1007/s12274-014-0626-5

Kang, L., Cui, M., Zhang, Z., & Jiang, F. (2020). Rechargeable Aqueous Zinc-Ion Batteries with Mild Electrolytes: A Comprehensive Review. *Batteries & Supercaps*, *3*(10), 966–1005. doi:10.1002/batt.202000060

Karthik, R., Vinoth Kumar, J., Chen, S.-M., Sundaresan, P., Mutharani, B., Chi Chen, Y., & Muthuraj, V. (2018). Simple sonochemical synthesis of novel grass-like vanadium disulfide: A viable non-enzymatic electrochemical sensor for the detection of hydrogen peroxide. *Ultrasonics Sonochemistry*, *48*, 473–481. doi:10.1016/j.ultsonch.2018.07.008 PMID:30080574

Koketsu, T., Ma, J., Morgan, B. J., Body, M., Legein, C., Dachraoui, W., Giannini, M., Demortière, A., Salanne, M., Dardoize, F., Groult, H., Borkiewicz, O. J., Chapman, K. W., Strasser, P., & Dambournet, D. (2017). Reversible magnesium and aluminium ions insertion in cation-deficient anatase TiO2. *Nature Materials*, *16*(11), 1142–1148. doi:10.1038/nmat4976 PMID:28920941

Lee, S.-H., Park, Y. C., Chae, J., Kim, G., Kim, H. J., Choi, B. K., Lee, I. H., Chang, Y. J., Chun, S.-H., Jung, M., Seo, J., & Lee, S. (2022). Strong electron–phonon coupling driven charge density wave states in stoichiometric 1T-VS 2 crystals. *Journal of Materials Chemistry. C, Materials for Optical and Electronic Devices*, *10*(44), 16657–16665. doi:10.1039/D2TC04186C

Li. (2019). Vanadium disulfide flakes with nanolayered titanium disulfide coating as cathode materials in lithium-ion batteries. *Nature Communications, 10*(1).

Li. (2018). Synthesis of coin-like vanadium disulfide and its sodium storage performance. *Acta ChimicaSinica, 76*(4).

Li, W., Hirbod, M. K. S., & Li, X. (2020). Emerging layered metallic vanadium disulfide for rechargeable metal-ion batteries: Progress and opportunities. *ChemSusChem*, *13*(6), 1172–1202. doi:10.1002/cssc.201903081 PMID:31777162

Li, W., Huang, J., Feng, L., Cao, L., Feng, Y., Wang, H., Li, J., & Yao, C. (2017). Facile in situ synthesis of crystalline VOOH-coated VS 2 microflowers with superior sodium storage performance. *Journal of Materials Chemistry. A, Materials for Energy and Sustainability*, *5*(38), 20217–20227. doi:10.1039/C7TA05205G

Li. (2021). Engineering textile electrode and bacterial cellulose nanofiber reinforced hydrogel electrolyte to enable high-performance flexible all-solid-state supercapacitors. *Advanced Energy Materials, 11*(12).

Li, X., Fu, J., Sun, Y., Sun, M., Cheng, S., Chen, K., Yang, X., Lou, Q., Xu, T., Shang, Y., Xu, J., Chen, Q., & Shan, C. (2019). Design and understanding of core/branch-structured VS 2 nanosheets@CNTs as high-performance anode materials for lithium-ion batteries. *Nanoscale*, *11*(28), 13343–13353. doi:10.1039/C9NR03581H PMID:31271407

Lia, W. (2015). Facile hydrothermal synthesis of VS2/graphene nanocomposites with superior high-rate capability as lithium-ion battery cathodes. *ACS Applied Materials & Interfaces*, *7*(23), 13044–13052. doi:10.1021/acsami.5b03124 PMID:26016687

Liang, G., Mo, F., Ji, X., & Zhi, C. (2021). Non-metallic charge carriers for aqueous batteries. *Nature Reviews. Materials*, *6*(2), 109–123. doi:10.1038/s41578-020-00241-4

Liang, H., Shi, H., Zhang, D., Ming, F., Wang, R., Zhuo, J., & Wang, Z. (2016). Solution growth of vertical VS2 nanoplate arrays for electrocatalytic hydrogen evolution. *Chemistry of Materials*, *28*(16), 5587–5591. doi:10.1021/acs.chemmater.6b01963

Liang, X., Chen, M., Zhu, H., Zhu, H., Cui, X., Yan, J., Chen, Q., Xia, X., & Liu, Q. (2020). Unveiling the solid-solution charge storage mechanism in 1T vanadium disulfide nanoarray cathodes. *Journal of Materials Chemistry. A, Materials for Energy and Sustainability*, *8*(18), 9068–9076. doi:10.1039/D0TA02922J

Liang, Y., Zhao, C.-Z., Yuan, H., Chen, Y., Zhang, W., Huang, J.-Q., Yu, D., Liu, Y., Titirici, M.-M., Chueh, Y.-L., Yu, H., & Zhang, Q. (2019). A review of rechargeable batteries for portable electronic devices. *InfoMat*, *1*(1), 6–32. doi:10.1002/inf2.12000

Liao, J.-Y., & Manthiram, A. (2015). High-performance Na2Ti2O5 nanowire arrays coated with VS2 nanosheets for sodium-ion storage. *Nano Energy*, *18*, 20–27. doi:10.1016/j.nanoen.2015.09.014

Ling, W., Wang, P., Chen, Z., Wang, H., Wang, J., Ji, Z., Fei, J., Ma, Z., He, N., & Huang, Y. (2020). Nanostructure design strategies for aqueous zinc-ion batteries. *ChemElectroChem*, *7*(14), 2957–2978. doi:10.1002/celc.202000372

Littlejohn, A. J., Li, Z., Lu, Z., Sun, X., Nawarat, P., Wang, Y., Li, Y., Wang, T., Chen, Y., Zhang, L., Li, H., Kisslinger, K., Shi, S., Shi, J., Raeliarijaona, A., Shi, W., Terrones, H., Lewis, K. M., Washington, M., ... Wang, G.-C. (2019). Large metallic vanadium disulfide ultrathin flakes for spintronic circuits and quantum computing devices. *ACS Applied Nano Materials*, *2*(6), 3684–3694. doi:10.1021/acsanm.9b00608

Liu, A., Liang, X., Ren, X., Guan, W., & Ma, T. (2022). Recent progress in MXene-based materials for metal-sulfur and metal-air batteries: Potential high-performance electrodes. *Electrochemical Energy Reviews*, *5*(1), 1–33. doi:10.1007/s41918-021-00110-w

Liu, D., Xu, S., Cai, Y., Wang, Y., Guo, J., & Li, Y. (2022). A coupling technology of capacitive deionization and carbon-supported petal-like VS2 composite for effective and selective adsorption of lead (II) ions. *Journal of Electroanalytical Chemistry (Lausanne, Switzerland)*, *910*, 116152. doi:10.1016/j.jelechem.2022.116152

Liu, F., Peng, C., Porvali, A., Wang, Z., Wilson, B. P., & Lundström, M. (2019). Synergistic recovery of valuable metals from spent nickel–metal hydride batteries and lithium-ion batteries. *ACS Sustainable Chemistry & Engineering*, *7*(19), 16103–16111. doi:10.1021/acssuschemeng.9b02863

Liu, H. (2023). Local Strain engineering on Janus MoSSe Nanoribbons Induced Tunable Electronic Structures and Remarkable Magnetic Moments. *Journal of Physics. D, Applied Physics.*

Liu, J., Peng, W., Li, Y., Zhang, F., & Fan, X. (2021). A VS 2@ N-doped carbon hybrid with strong interfacial interaction for high-performance rechargeable aqueous Zn-ion batteries. *Journal of Materials Chemistry. C, Materials for Optical and Electronic Devices*, *9*(19), 6308–6315. doi:10.1039/D1TC00531F

Liu, M., Zhao, Z., Zhang, W., & Zheng, W. (2022). Perspective of Vanadium Disulfide: A Rising Star Finds Plenty of Room in Single and Multielectron Energy Storage. *Energy & Fuels*, *36*(23), 13931–13955. doi:10.1021/acs.energyfuels.2c02642

Liu, Liu, & Wu. (2023). Pre-intercalation Strategies to Layered Vanadium-based Electrode Materials for Aqueous Zinc Batteries. *Batteries & Supercaps, 6*(2).

Liu, Y.-Y., Xu, L., Guo, X.-T., Lv, T.-T., & Pang, H. (2020). Vanadium sulfide based materials: Synthesis, energy storage and conversion. *Journal of Materials Chemistry. A, Materials for Energy and Sustainability*, *8*(40), 20781–20802. doi:10.1039/D0TA07436E

Lopes, P. P., & Stamenkovic, V. R. (2020). Past, present, and future of lead–acid batteries. *Science*, *369*(6506), 923–924. doi:10.1126/science.abd3352 PMID:32820114

Lourenssen, K., Williams, J., Ahmadpour, F., Clemmer, R., & Tasnim, S. (2019). Vanadium redox flow batteries: A comprehensive review. *Journal of Energy Storage*, *25*, 100844. doi:10.1016/j.est.2019.100844

Ma, X., Du, C., Shang, M., & Song, W. (2018). VS 2 quantum dot label-free fluorescent probe for sensitive and selective detection of ALP. *Analytical and Bioanalytical Chemistry*, *410*(5), 1417–1426. doi:10.1007/s00216-017-0778-8 PMID:29256077

Maeyoshi, Y., Yoshii, K., Shikano, M., & Sakaebe, H. (2021). Improving Cycling Stability of Vanadium Sulfide (VS4) as a Li Battery Cathode Material Using a Localized High-Concentration Carbonate-Based Electrolyte. *ACS Applied Energy Materials*, *4*(12), 13627–13635. doi:10.1021/acsaem.1c02312

Meng, P., Wang, W., Shang, J., Liu, P., Xu, H., Wang, Q., Wang, S., Wang, F., & Wang, X. (2023). 2D VS2@ MXene Based Zinc Ion Batteries with SPANI-Contained Electrolyte Enables Dendrite-Free Anode for Stable Cycling. *Small Methods*, *7*(4), 2201471. doi:10.1002/smtd.202201471 PMID:36720008

Meng, Y., Zhao, Y., Wang, D., Yang, D., Gao, Y., Lian, R., Chen, G., & Wei, Y. (2018). Fast Li+ diffusion in interlayer-expanded vanadium disulfide nanosheets for Li+/Mg 2+ hybrid-ion batteries. *Journal of Materials Chemistry. A, Materials for Energy and Sustainability*, *6*(14), 5782–5788. doi:10.1039/C8TA00418H

Meyer, E., Bede, A., Mutukwa, D., Taziwa, R., & Zingwe, N. (2020). Optimization, and analysis of carbon supported VS2 nanocomposites as potential electrodes in supercapacitors. *Journal of Energy Storage*, *27*, 101074. doi:10.1016/j.est.2019.101074

Mohan, P., Yang, J., Jena, A., & Suk Shin, H. (2015). VS2/rGO hybrid nanosheets prepared by annealing of VS4/rGO. *Journal of Solid State Chemistry*, *224*, 82–87. doi:10.1016/j.jssc.2014.06.031

Moradbeigi, N., Bahari, A., & Ghasemi, S. (2023). Effect of molybdenum doping on the catalytic activity of VS 2/CNT for the oxygen reduction reaction in alkaline media. *New Journal of Chemistry*, *47*(3), 1291–1298. doi:10.1039/D2NJ04721G

Moyassari. (2022). The role of silicon in silicon-graphite composite electrodes regarding specific capacity, cycle stability, and expansion. *Journal of The Electrochemical Society, 169*(1).

Mukherjee. (2019). Electrode materials for high-performance sodium-ion batteries. *Materials, 12*(12).

Muldoon, J., Bucur, C. B., & Gregory, T. (2017). Fervent hype behind magnesium batteries: An open call to synthetic chemists—electrolytes and cathodes needed. *Angewandte Chemie International Edition*, *56*(40), 12064–12084. doi:10.1002/anie.201700673 PMID:28295967

Murugan, A. V. (2005). Entrapment of poly (3, 4-ethylenedioxythiophene) between VS 2 layers to form a new organic–inorganic intercalative nanocomposite. *Journal of Materials Chemistry*, *15*(8), 902–909. doi:10.1039/B414922J

Naik, V. M., Bhosale, S. V., & Kolekar, G. B. (2022). A brief review on the synthesis, characterisation and analytical applications of nitrogen doped carbon dots. *Analytical Methods*, *14*(9), 877–891. doi:10.1039/D1AY02105B PMID:35174374

Nayak, P. K., Yang, L., Brehm, W., & Adelhelm, P. (2018). From lithium-ion to sodium-ion batteries: Advantages, challenges, and surprises. *Angewandte Chemie International Edition*, *57*(1), 102–120. doi:10.1002/anie.201703772 PMID:28627780

Nitta, N., Wu, F., Lee, J. T., & Yushin, G. (2015). Li-ion battery materials: Present and future. *Materials Today, 18*(5), 252–264. doi:10.1016/j.mattod.2014.10.040

Okpalugo, T. I. T., Papakonstantinou, P., Murphy, H., McLaughlin, J., & Brown, N. M. D. (2005). High resolution XPS characterization of chemical functionalised MWCNTs and SWCNTs. *Carbon, 43*(1), 153–161. doi:10.1016/j.carbon.2004.08.033

Pan, H. (2016). Tension-enhanced hydrogen evolution reaction on vanadium disulfide monolayer. *Nanoscale Research Letters, 11*(1).

Pandit, B., Rondiya, S. R., Cross, R. W., Dzade, N. Y., & Sankapal, B. R. (2022). Vanadium telluride nanoparticles on MWCNTs prepared by successive ionic layer adsorption and reaction for solid-state supercapacitor. *Chemical Engineering Journal, 429*, 132505. doi:10.1016/j.cej.2021.132505

Pandya, K. S., Veerraju, C., & Naik, N. K. (2011). Hybrid composites made of carbon and glass woven fabrics under quasi-static loading. *Materials & Design, 32*(7), 4094–4099. doi:10.1016/j.matdes.2011.03.003

Pang. (2020). Vanadium disulfide for ultrafast photonic application. *Nanotechnology, 32*(1).

Parsons & Alotaibi. (2023). The Application of Transition Metal Sulfide Nanomaterials and Their Composite Nanomaterials in the Electrocatalytic Reduction of CO2: A Review. *Applied Sciences, 13*(5).

Patil, S. J., Pujari, R. B., Hou, T., Park, J., & Lee, D.-W. (2020). Supercapacitive performance of vanadium sulfide deposited on stainless steel mesh: Effect of etching. *Micro and Nano Systems Letters, 8*(1), 1–7. doi:10.1186/s40486-020-00110-3

Petrovic & Petrovic. (2021). Nickel–cadmium batteries. *Battery Technology Crash Course: A Concise Introduction,* 73-88.

Piccirillo, C., Binions, R., & Parkin, I. P. (2007). Synthesis and functional properties of vanadium oxides: V2O3, VO2, and V2O5 deposited on glass by aerosol-assisted CVD. *Chemical Vapor Deposition, 13*(4), 145–151. doi:10.1002/cvde.200606540

Podyacheva, O. Yu., & Ismagilov, Z. R. (2015). Nitrogen-doped carbon nanomaterials: To the mechanism of growth, electrical conductivity and application in catalysis. *Catalysis Today, 249*, 12–22. doi:10.1016/j.cattod.2014.10.033

Prytulak, J. (2016). Stable vanadium isotopes as a redox proxy in magmatic systems. *Geochemical Perspectives Letters.*

Pu, X., Song, T., Tang, L., Tao, Y., Cao, T., Xu, Q., Liu, H., Wang, Y. G., & Xia, Y. (2019). Rose-like vanadium disulfide coated by hydrophilic hydroxyvanadium oxide with improved electrochemical performance as cathode material for aqueous zinc-ion batteries. *Journal of Power Sources, 437*, 226917. doi:10.1016/j.jpowsour.2019.226917

Putungan, D. B., Lin, S.-H., & Kuo, J.-L. (2016). Metallic VS2 monolayer polytypes as potential sodium-ion battery anode via ab initio random structure searching. *ACS Applied Materials & Interfaces, 8*(29), 18754–18762. doi:10.1021/acsami.6b03499 PMID:27373121

Qi, H., Wang, L., Zuo, T., Deng, S., Li, Q., Liu, Z.-H., Hu, P., & He, X. (2020). Hollow structure VS2@ reduced graphene oxide (RGO) architecture for enhanced sodium-ion battery performance. *ChemElectroChem*, *7*(1), 78–85. doi:10.1002/celc.201901626

Qi, X., Xu, H., Wang, X.-Q., Ma, W.-G., Qiu, C., An, M., Zhang, G., Wang, F., Zhang, X., & Bermak, A. (2020). Effective surface emissivity and heat dissipation among integrated bamboo-like super-black vertical carbon nanotube array electrodes in silicon via holes. *Carbon*, *158*, 846–856. doi:10.1016/j.carbon.2019.11.068

Qie, L., & Manthiram, A. (2016). High-energy-density lithium–sulfur batteries based on blade-cast pure sulfur electrodes. *ACS Energy Letters*, *1*(1), 46–51. doi:10.1021/acsenergylett.6b00033

Qin, Z., Wang, Z., & Zhao, J. (2022). Computational screening of single-atom catalysts supported by VS 2 monolayers for electrocatalytic oxygen reduction/evolution reactions. *Nanoscale*, *14*(18), 6902–6911. doi:10.1039/D2NR01671K PMID:35446333

Qu, Y., Shao, M., Shao, Y., Yang, M., Xu, J., Kwok, C. T., Shi, X., Lu, Z., & Pan, H. (2017). Ultra-high electrocatalytic activity of VS 2 nanoflowers for efficient hydrogen evolution reaction. *Journal of Materials Chemistry. A, Materials for Energy and Sustainability*, *5*(29), 15080–15086. doi:10.1039/C7TA03172F

Rajagopalan. (2020). Advancements and challenges in potassium ion batteries: a comprehensive review. *Advanced Functional Materials, 30*(12).

Renault, S., Brandell, D., Gustafsson, T., & Edström, K. (2013). Improving the electrochemical performance of organic Li-ion battery electrodes. *Chemical Communications*, *49*(19), 1945–1947. doi:10.1039/c3cc39065a PMID:23370005

Sarkar, A., Ghosh, A. B., Saha, N., Bhadu, G. R., & Adhikary, B. (2018). Newly designed amperometric biosensor for hydrogen peroxide and glucose based on vanadium sulfide nanoparticles. *ACS Applied Nano Materials*, *1*(3), 1339–1347. doi:10.1021/acsanm.8b00076

Shafaatian, B., & Baharian, A. (2017). New zinc (II), copper (II), nickel (II), and vanadium (IV) complexes containing disulfide ligand: Synthesis, spectroscopic studies and calf thymus DNA interaction investigation. *Phosphorus, Sulfur, and Silicon and the Related Elements*, *192*(10), 1102–1109. doi:10.1080/10426507.2017.1331232

Sharma, A., Kapse, S., Verma, A., Bisoyi, S., Pradhan, G. K., Thapa, R., & Rout, C. S. (2022). All-Solid-State Supercapacitor Based on Advanced 2D Vanadium Disulfide/Black Phosphorus Hybrids for Wearable Electronics. *ACS Applied Energy Materials*, *5*(8), 10315–10327. doi:10.1021/acsaem.2c02116

Sharma, S., Panwar, A. K., & Tripathi, M. M. (2020). Storage technologies for electric vehicles. *Journal of Traffic and Transportation Engineering*, *7*(3), 340–361. doi:10.1016/j.jtte.2020.04.004

Shen, P.-C., Lin, Y., Wang, H., Park, J.-H., Leong, W. S., Lu, A.-Y., Palacios, T., & Kong, J. (2018). CVD technology for 2-D materials. *IEEE Transactions on Electron Devices*, *65*(10), 4040–4052. doi:10.1109/TED.2018.2866390

Shen. (2021). Advanced electrode materials in lithium batteries: retrospect and prospect. *Energy Material Advances*.

Shin, J., Lee, J., Park, Y., & Choi, J. W. (2020). Aqueous zinc ion batteries: Focus on zinc metal anodes. *Chemical Science (Cambridge), 11*(8), 2028–2044. doi:10.1039/D0SC00022A PMID:32180925

Shrivastav, V., Sundriyal, S., Goel, P., Kaur, H., Tuteja, S. K., Vikrant, K., Kim, K.-H., Tiwari, U. K., & Deep, A. (2019). Metal-organic frameworks (MOFs) and their composites as electrodes for lithium battery applications: Novel means for alternative energy storage. *Coordination Chemistry Reviews, 393*, 48–78. doi:10.1016/j.ccr.2019.05.006

Shuai. (2023). Recent Advances of Transition Metal Sulfides/Selenides Cathodes for Aqueous Zinc-Ion Batteries. *Advanced Energy Materials, 13*(4).

Speirs, J., Contestabile, M., Houari, Y., & Gross, R. (2014). The future of lithium availability for electric vehicle batteries. *Renewable & Sustainable Energy Reviews, 35*, 183–193. doi:10.1016/j.rser.2014.04.018

Stukowski, A. (2012). Structure identification methods for atomistic simulations of crystalline materials. *Modelling and Simulation in Materials Science and Engineering, 20*(4).

Su. (2020). Sub-millimeter-scale monolayer p-type H-phase VS2. *Advanced Functional Materials, 30*(17).

Sun, J., Lian, G., Jing, L., Wu, D., Cui, D., Wang, Q., Yu, H., Zhang, H., & Wong, C.-P. (2022). Assembly of flower-like VS2/N-doped porous carbon with expanded (001) plane on rGO for superior Na-ion and K-ion storage. *Nano Research, 15*(5), 4108–4116. doi:10.1007/s12274-021-4060-1

Sun, Z., Zhang, Y., Liu, Y., Hou, L., & Yuan, C. (2021). Recent Progress on In Situ/Operando Characterization of Rechargeable Alkali Ion Batteries. *ChemPlusChem, 86*(10), 1487–1496. doi:10.1002/cplu.202100393 PMID:34674379

Tan, X., Li, Q., & Wang, H. (2013). Advances and trends of energy storage technology in microgrid. *International Journal of Electrical Power & Energy Systems, 44*(1), 179–191. doi:10.1016/j.ijepes.2012.07.015

Tapia-Ruiz. (2021). 2021 roadmap for sodium-ion batteries. *Journal of Physics: Energy, 3*(3).

Van Noorden, R. (2013). Sulphur back in vogue for batteries. *Nature, 498*(7455), 416–417. doi:10.1038/498416a PMID:23803819

Walter, M., Kravchyk, K. V., Ibáñez, M., & Kovalenko, M. V. (2015). Efficient and inexpensive sodium–magnesium hybrid battery. *Chemistry of Materials, 27*(21), 7452–7458. doi:10.1021/acs.chemmater.5b03531

Wang, B., Ang, E. H., Yang, Y., Zhang, Y., Ye, M., Liu, Q., & Li, C. C. (2021). Post-lithium-ion battery era: Recent advances in rechargeable potassium-ion batteries. *Chemistry (Weinheim an der Bergstrasse, Germany), 27*(2), 512–536. doi:10.1002/chem.202001811 PMID:32510710

Wang, D., Zhao, Y., Lian, R., Yang, D., Zhang, D., Meng, X., Liu, Y., Wei, Y., & Chen, G. (2018). Atomic insight into the structural transformation and anionic/cationic redox reactions of VS 2 nanosheets in sodium-ion batteries. *Journal of Materials Chemistry. A, Materials for Energy and Sustainability, 6*(33), 15985–15992. doi:10.1039/C8TA05396K

Wang, F., Robert, R., Chernova, N. A., Pereira, N., Omenya, F., Badway, F., Hua, X., Ruotolo, M., Zhang, R., Wu, L., Volkov, V., Su, D., Key, B., Whittingham, M. S., Grey, C. P., Amatucci, G. G., Zhu, Y., & Graetz, J. (2011). Conversion reaction mechanisms in lithium ion batteries: Study of the binary metal fluoride electrodes. *Journal of the American Chemical Society*, *133*(46), 18828–18836. doi:10.1021/ja206268a PMID:21894971

Wang, J., Luo, N., Wu, J., Huang, S., Yu, L., & Wei, M. (2019). Hierarchical spheres constructed by ultrathin VS 2 nanosheets for sodium-ion batteries. *Journal of Materials Chemistry. A, Materials for Energy and Sustainability*, *7*(8), 3691–3696. doi:10.1039/C8TA11950C

Wang, R., Chung, C.-C., Liu, Y., Jones, J. L., & Augustyn, V. (2017). Electrochemical intercalation of Mg2+ into anhydrous and hydrated crystalline tungsten oxides. *Langmuir*, *33*(37), 9314–9323. doi:10.1021/acs.langmuir.7b00705 PMID:28732164

WangS. (n.d.). Three-Dimensional Amorphous Carbon Rivet Vanadium Sulfide Enhanced Lithium Storage. *Available at* SSRN 4105060. doi:10.2139/ssrn.4105060

Wang, W., Sun, Z., Zhang, W., Fan, Q., Sun, Q., Cui, X., & Xiang, B. (2016). First-principles investigations of vanadium disulfide for lithium and sodium ion battery applications. *RSC Advances*, *6*(60), 54874–54879. doi:10.1039/C6RA07586J

Wang, X., Tan, G., Bai, Y., Wu, F., & Wu, C. (2021). Multi-electron reaction materials for high-energy-density secondary batteries: Current status and prospective. *Electrochemical Energy Reviews*, *4*(1), 35–66. doi:10.1007/s41918-020-00073-4

Wang, X., Zhang, Y., Zheng, J., Liu, X., & Meng, C. (2019). Hydrothermal synthesis of VS4/CNTs composite with petal-shape structures performing a high specific capacity in a large potential range for high-performance symmetric supercapacitors. *Journal of Colloid and Interface Science*, *554*, 191–201. doi:10.1016/j.jcis.2019.06.105 PMID:31299547

Wang. (2023). A customized strategy to design intercalation-type Li-free cathodes for all-solid-state batteries. *National Science Review*.

Wiegers, G. A., van der Meer, R., van Heiningen, H., Kloosterboer, H. J., & Alberink, A. J. A. (1974). The sodium intercalates of vanadium disulfide and their hydrolysis products. *Materials Research Bulletin*, *9*(9), 1261–1265. doi:10.1016/0025-5408(74)90045-2

Wong, K. L. (2021). Li-ion batteries state-of-charge estimation using deep lstm at various battery specifications and discharge cycles. *Proceedings of the Conference on Information Technology for Social Good*. 10.1145/3462203.3475878

Worighi, I., Maach, A., Hafid, A., Hegazy, O., & Van Mierlo, J. (2019). Integrating renewable energy in smart grid system: Architecture, virtualization and analysis. *Sustainable Energy, Grids and Networks*, *18*, 100226. doi:10.1016/j.segan.2019.100226

Wu, D., Wang, C., Wu, M., Chao, Y., He, P., & Ma, J. (2020). Porous bowl-shaped VS2 nanosheets/graphene composite for high-rate lithium-ion storage. *Journal of Energy Chemistry*, *43*, 24–32. doi:10.1016/j.jechem.2019.08.003

WuL. (2023). Interlayer Expanded Vs2 Nanosheet for Efficient Sodium Storage. *Available at* SSRN 4391411. doi:10.1016/j.matlet.2023.134551

Wu, L., Sun, R., Xiong, F., Pei, C., Han, K., Peng, C., Fan, Y., Yang, W., An, Q., & Mai, L. (2018). A rechargeable aluminum-ion battery based on a VS 2 nanosheet cathode. *Physical Chemistry Chemical Physics*, *20*(35), 22563–22568. doi:10.1039/C8CP04772C PMID:30159553

Wu, Z., Wang, L., Huang, J., Zou, J., Chen, S., Cheng, H., Jiang, C., Gao, P., & Niu, X. (2019). Loofah-derived carbon as an anode material for potassium ion and lithium ion batteries. *Electrochimica Acta*, *306*, 446–453. doi:10.1016/j.electacta.2019.03.165

Xie, X.-C., Shuai, H.-L., Wu, X., Huang, K.-J., Wang, L.-N., Wang, R.-M., & Chen, Y. (2020). Engineering ultra-enlarged interlayer carbon-containing vanadium disulfide composite for high-performance sodium and potassium ion storage. *Journal of Alloys and Compounds*, *847*, 156288. doi:10.1016/j.jallcom.2020.156288

Xu, C., Li, B., Du, H., & Kang, F. (2012). Energetic zinc ion chemistry: The rechargeable zinc ion battery. *Angewandte Chemie*, *124*(4), 957–959. doi:10.1002/ange.201106307 PMID:22170816

Xu, D., Wang, H., Qiu, R., Wang, Q., Mao, Z., Jiang, Y., Wang, R., He, B., Gong, Y., Li, D., & Hu, X. (2020). Coupling of bowl-like VS2 nanosheet arrays and carbon nanofiber enables ultrafast Na+-Storage and robust flexibility for sodium-ion hybrid capacitors. *Energy Storage Materials*, *28*, 91–100. doi:10.1016/j.ensm.2020.03.002

Xu, H., & Shen, M. (2021). The control of lithium-ion batteries and supercapacitors in hybrid energy storage systems for electric vehicles: A review. *International Journal of Energy Research*, *45*(15), 20524–20544. doi:10.1002/er.7150

Xu, J., Wang, S., Huang, T., Chen, J., & Li, H. (2022). Three-dimensional amorphous carbon rivet vanadium sulfide enhanced lithium storage. *Journal of Alloys and Compounds*, *929*, 166966. doi:10.1016/j.jallcom.2022.166966

Xue. (2019). One-step synthesis of 2-ethylhexylamine pillared vanadium disulfide nanoflowers with ultralarge interlayer spacing for high-performance magnesium storage. *Advanced Energy Materials, 9*(22).

Yang. (2021). Regulating intercalation of layered compounds for electrochemical energy storage and electrocatalysis. *Advanced Functional Materials, 31*(52).

Yang, D., Chen, L., Wang, L., Cui, Z., Wen, Z., Yang, B., He, H., Mei, S., Zhang, W., Xie, F., Ma, J., & Guo, R. (2021). One-step synthesis of high-quality vanadium disulfide quantum dots for long-term lysosome-targetable imaging. *Sensors and Actuators. B, Chemical*, *346*, 130544. doi:10.1016/j.snb.2021.130544

Yang, J., Wang, J., Dong, X., Zhu, L., Hou, D., Zeng, W., & Wang, J. (2021). The potential application of VS2 as an electrode material for Mg ion battery: A DFT study. *Applied Surface Science*, *544*, 148775. doi:10.1016/j.apsusc.2020.148775

Yao, Y., Lei, J., Shi, Y., Ai, F., & Lu, Y.-C. (2021). Assessment methods and performance metrics for redox flow batteries. *Nature Energy*, *6*(6), 582–588. doi:10.1038/s41560-020-00772-8

Yin, B.-S., Zhang, S.-W., Xiong, T., Shi, W., Ke, K., Lee, W. S. V., Xue, J., & Wang, Z.-B. (2020). Engineering sulphur vacancy in VS 2 as high performing zinc-ion batteries with high cyclic stability. *New Journal of Chemistry*, *44*(37), 15951–15957. doi:10.1039/D0NJ02917C

Yu. (2021). Boosting Zn2+ and NH4+ Storage in Aqueous Media via In-Situ Electrochemical Induced VS2/VOx Heterostructures. *Advanced Functional Materials, 31*(11).

Yu, D., Pang, Q., Gao, Y., Wei, Y., Wang, C., Chen, G., & Du, F. (2018). Hierarchical flower-like VS2 nanosheets–A high rate-capacity and stable anode material for sodium-ion battery. *Energy Storage Materials*, *11*, 1–7. doi:10.1016/j.ensm.2017.09.002

Yu, S. H., Tang, Z., Shao, Y., Dai, H., Wang, H. Y., Yan, J., Pan, H., & Chua, D. H. C. (2019). In situ hybridizing MoS2 microflowers on VS2 microflakes in a one-pot CVD process for electrolytic hydrogen evolution reaction. *ACS Applied Energy Materials*, *2*(8), 5799–5808. doi:10.1021/acsaem.9b00928

Yu, S. H., Tang, Z., Shao, Y., Dai, H., Wang, H. Y., Yan, J., Pan, H., & Chua, D. H. C. (2019). In situ hybridizing MoS2 microflowers on VS2 microflakes in a one-pot CVD process for electrolytic hydrogen evolution reaction. *ACS Applied Energy Materials*, *2*(8), 5799–5808. doi:10.1021/acsaem.9b00928

Yuan, J., Wu, J., Hardy, W. J., Loya, P., Lou, M., Yang, Y., Najmaei, S., Jiang, M., Qin, F., Keyshar, K., Ji, H., Gao, W., Bao, J., Kono, J., Natelson, D., Ajayan, P. M., & Lou, J. (2015). Facile synthesis of single crystal vanadium disulfide nanosheets by chemical vapor deposition for efficient hydrogen evolution reaction. *Advanced Materials*, *27*(37), 5605–5609. doi:10.1002/adma.201502075 PMID:26293810

Yun. (2022). Review on Oxygen-Free Vanadium-Based Cathodes for Aqueous Zinc-Ion Batteries. *Journal of Electrochemistry, 28*(11).

Zaaba, N. I., Foo, K. L., Hashim, U., Tan, S. J., Liu, W.-W., & Voon, C. H. (2017). Synthesis of graphene oxide using modified hummers method: Solvent influence. *Procedia Engineering*, *184*, 469–477. doi:10.1016/j.proeng.2017.04.118

Zhang, J., Xu, Y., Liu, Z., Yang, W., & Liu, J. (2015). A highly conductive porous graphene electrode prepared via in situ reduction of graphene oxide using Cu nanoparticles for the fabrication of high performance supercapacitors. *RSC Advances*, *5*(67), 54275–54282. doi:10.1039/C5RA07857A

Zhang, L., Wei, Q., Sun, D., Li, N., Ju, H., Feng, J., Zhu, J., Mai, L., Cairns, E. J., & Guo, J. (2018). Conversion reaction of vanadium sulfide electrode in the lithium-ion cell: Reversible or not reversible? *Nano Energy*, *51*, 391–399. doi:10.1016/j.nanoen.2018.06.076

Zhang, M. Y., Miao, J. Y., Yan, X. H., Zhu, Y. H., Li, Y. L., Zhang, W. J., Zhu, W., Pan, J. M., Javed, M. S., & Hussain, S. (2022). Vanadium disulfide nanosheets loaded on carbon cloth as electrode for flexible quasi-solid-state asymmetric supercapacitors: Energy storage mechanism and electrochemical performance. *Journal of Materials Chemistry. C, Materials for Optical and Electronic Devices*, *10*(2), 640–648. doi:10.1039/D1TC03903B

Zhang, S., Chang, P., Zhang, Y., Xu, X., Guan, L., & Tao, J. (2022). Effect of morphology and stacking on atomic interaction and magnetic characteristics in two-dimensional H-phase VS2 few layers. *Journal of Materials Science*, *57*(10), 5873–5884. doi:10.1007/s10853-022-06904-7

Zhang, W., Chen, X., Zhang, J., Tuo, C., Ji, L., Li, H., Zhang, X., & Yang, F. (2018). Exposure of active edge structure for electrochemical H2 evolution from VS2/MWCNTs hybrid catalysts. *International Journal of Hydrogen Energy*, *43*(51), 22949–22954. doi:10.1016/j.ijhydene.2018.10.168

Zhang. (2020). Insights into the Storage Mechanism of Layered VS2 Cathode in Alkali Metal-Ion Batteries. *Advanced Energy Materials, 10*(22).

Zhang, Y., Li, J., Ma, L., Li, H., Xu, X., Liu, X., Lu, T., & Pan, L. (2022). Insights into the storage mechanism of 3D nanoflower-like V3S4 anode in sodium-ion batteries. *Chemical Engineering Journal*, *427*, 130936. doi:10.1016/j.cej.2021.130936

Zhang, Z., Huang, X., Wang, H., Teo, S. H., & Ma, T. (2019). Free-standing NiCo2S4@ VS2 nanoneedle array composite electrode for high performance asymmetric supercapacitor application. *Journal of Alloys and Compounds*, *771*, 274–280. doi:10.1016/j.jallcom.2018.08.325

Zheng, X., Ahmad, T., & Chen, W. (2021). Challenges and strategies on Zn electrodeposition for stable Zn-ion batteries. *Energy Storage Materials*, *39*, 365–394. doi:10.1016/j.ensm.2021.04.027

Zhong, X., Tang, J., Wang, J., Shao, M., Chai, J., Wang, S., Yang, M., Yang, Y., Wang, N., Wang, S., Xu, B., & Pan, H. (2018). 3D heterostructured pure and N-Doped Ni3S2/VS2 nanosheets for high efficient overall water splitting. *Electrochimica Acta*, *269*, 55–61. doi:10.1016/j.electacta.2018.02.131

Zhu, J., Cai, L., Yin, X., Wang, Z., Zhang, L., Ma, H,, Ke, Y., Du, Y., Xi, S., Wee, A. T. S., Chai, Y., & Zhang, W. (2020). Enhanced electrocatalytic hydrogen evolution activity in single-atom Pt-decorated VS2 nanosheets. *ACS Nano*, *14*(5), 5600–5608. doi:10.1021/acsnano.9b10048 PMID:32320212

Zhu, J., Jian, T., Wu, Y., Ma, W., Lu, Y., Sun, L., Meng, F., Wang, B., Cai, F., Gao, J., Li, G., Yang, L., Yan, X., & Xu, C. (2021). A highly stable aqueous Zn/VS2 battery based on an intercalation reaction. *Applied Surface Science*, *544*, 148882. doi:10.1016/j.apsusc.2020.148882

Zhu, Z., Cheng, F., & Chen, J. (2013). Investigation of effects of carbon coating on the electrochemical performance of Li 4 Ti 5 O 12/C nanocomposites. *Journal of Materials Chemistry. A, Materials for Energy and Sustainability*, *1*(33), 9484–9490. doi:10.1039/c3ta00114h

Zihan, Y. (2022). A Review of the Extremely Structural Properties of Carbon Nanotubes, Synthesis and Purification Methods. *American Journal of Materials Synthesis and Processing*, *7*(2), 29–35.

Chapter 5
Exploring Lead-Free Piezoelectric Ceramics for Enhanced Energy Harvesting Applications

Chiranjit Chaliha
National Institute of Technology, Manipur, India

Haobam Samananda Singh
https://orcid.org/0000-0002-9931-096X
National Institute of Technology, Manipur, India

Irom Monika Aniz
National Institute of Technology, Manipur, India

Mamata Maisnam
National Institute of Technology, Manipur, India

ABSTRACT

Piezoceramics, stemming from PZT and $BaTiO_3$ discoveries, revolutionize technology with exceptional piezoelectricity. This exploration traces their evolution, highlighting PZT and $BaTiO_3$'s foundational roles. Diverse piezoceramics like $BiFeO_3$, $BiNaTiO_3$, and high-d33 KNN are featured. Enhancing KNN's piezoelectric performance via morphotropic phase boundaries, similar to PZT's MPB, is pivotal. The rhombohedral-tetragonal phase boundary in KNN is crucial for high d33 values, emphasizing phase transitions. Controlled sintering and poling, aligning domains, and reducing defects are vital for optimized piezoelectricity. In summary, this narrative explores historical origins, essential properties, KNN's remarkable d33 and Tc, the quest for MPBs, and the role of phase boundaries and meticulous processing in advancing piezoelectric materials.

DOI: 10.4018/979-8-3693-1306-0.ch005

Figure 1. (a)-(b) Schematic diagram showing direct and indirect piezoelectric effect respectively

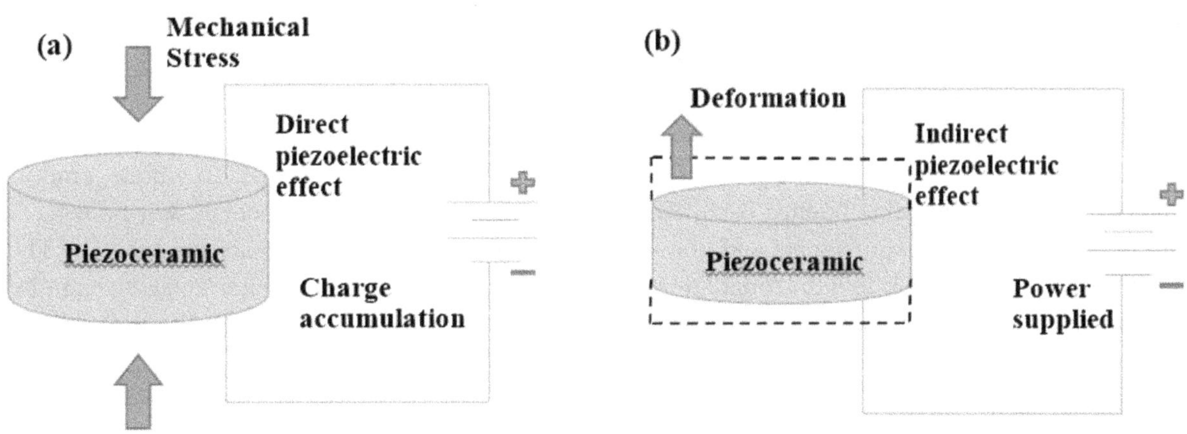

1. INTRODUCTION

1.1 Background

Piezoelectric materials are an enrapturing class with a unique characteristic feature called piezoelectricity. This characteristic enables them to develop electric charges which when subjected to mechanical stress or deformation properly called direct piezoelectricity Fig. 1(a), and conversely, to induce mechanical stress or deformation within the material when an electric field is applied popularly known as the indirect piezoelectric effect Fig. 1(b) given by equations 1 and 2 respectively.

$$D_x = d_{xyz} T_{yz} \tag{1}$$

$$S_{yz} = d^*_{xyz} E_x \tag{2}$$

Here d_{xyz} and d^*_{xyz} are piezoelectric constants, they are third-rank tensors coupling a first-rank (electric field/displacement) and second-rank tensor (stress/strain), D_x (electric displacement (C/m^2)), E_x (electric field component (V/m)), and S_{yz} (stress component (N/m^2)) (ANSI/IEEE Std 176-1987, 1987). Piezoelectric materials have a rich history and play a pivotal role in many technological applications because of their outstanding capability to efficiently convert mechanical energy into electrical and vice versa. This phenomenon was discovered in the mid-18th century (1880) by two brothers Pierre Curie and Jacques Curie. Their findings revealed that specific crystals, including quartz and tourmaline, can accumulate charges when subjected to mechanical stress, like bending or squeezing. They coined this property as "piezoelectricity" derived from the Greek word "piezein," meaning to squeeze or press, and subsequently the converse by Gabriel Lipmann in 1881 (Coursey & Brand, 1946).

During World War 1, Quartz and Rochelle salt (also known as sodium potassium tartrate [$NaKC_4H_4O_6 \cdot 4H_2O$]) single crystals were the only available piezoelectric materials used for numerous military applications making them very crucial for wartime technologies. Quartz single crystals were mainly used during that era in the development of oscillators and resonators because of their ability to give stability and high-precision results in maintaining a constant frequency when subjected to an electric current, making them extremely reliable for tuning and stabilizing the frequencies of radio equipment. Thus, improving the accuracy and reliability of the communication system during World War I led to better coordination in military operations and intelligence gathering. On the other hand, Rochelle salt was mainly used to develop early sonar technology. These single crystals yield piezoelectric properties when mechanically stressed. By applying an electric field to Rochelle salt crystals, they could be made to vibrate, emitting sound waves underwater that bounce back off objects in the water and, the Rochelle salt transducers would collect those acoustic signals and convert them back into electrical signals for analysis. Thus both of these single crystals laid the foundation for developing more advanced materials, which marked a significant step in the development of underwater detection systems (Moore, 1919; Wooster, 1953).

1.1.1 Piezoelectricity Prior and Post World War II

World War 2 brought about significant development in electronics and the need for more precise and reliable communication systems, radar technology, and sonar technology. Post World War 2, there was significant research to look for alternatives that could give more fruitful results than Quartz or Rochelle salt, as both these crystals were limited. This led to the discovery of Barium Titanate ($BaTiO_3$) in the mid-20th century with an attempt to create synthetic materials with enhanced piezoelectric properties (Coursey & Brand, 1946; Roberts, 1947). However, Barium Titanate was simultaneously developed independently by countries namely, the United States, Japan, and Russia, during World War II: Wainer and Salomon in 1942, Ogawa in 1944, and Vul in 1944, correspondingly (Uchino, 2015; Vul, 1946; Wainer, 1946). Researchers studied its crystallographic structure and properties to optimize its behaviour further. The reason for choosing Barium Titanate was that it has better electromechanical properties making it very useful in a wider range of applications (Coursey & Brand, 1946). Unlike Rochelle salt it was more competent to humidity and temperature, also with its versatility to be tailored through doping and synthesis methods its performances could be optimized for specific applications. Moreover, its legacy is continued today with researchers making an effort to enhance its properties, reduce cost, and, explore novel applications.

1.1.2 Origin and Discovery of PZT

Following the footsteps of Barium Titanate, studies were carried out on perovskite isomorphic oxides namely $PbTiO_3$, $PbZrO_3$, and, $SrTiO_3$. A group of researchers from Japan, E. Sawaguchi, G. Shirane, and Y. Takagi are worth mentioning because of their proposed $Pb(Zr, Ti)O_3$ system phase diagram Fig. 2, which laid the foundation for the evolution of PZT. Although this lead zirconate system was discovered by them, its outstanding capabilities were put forward in 1954 by a US researcher, Bernard Jaffe. He already knew the series of studies done by them, so he concentrated on the studies of piezoelectric measurements around the morphotropic phase boundary (MPB) separating rhombohedral and tetragonal structural phases. To his surprise, this MPB led to a unique feature in PZT. It was associated with some of its exceptional properties, including high piezoelectric coefficients, large electromechanical coupling factors, and enhanced strain capabilities (Bernard, 1955; Sawaguchi, 1953; Sawaguchi et al., 1951). Besides, we must keep in mind that this outstanding piezoelectric property is potentially due to lead (Pb) and MPB.

Figure 2. PZT phase diagram as proposed by Sawaguchi

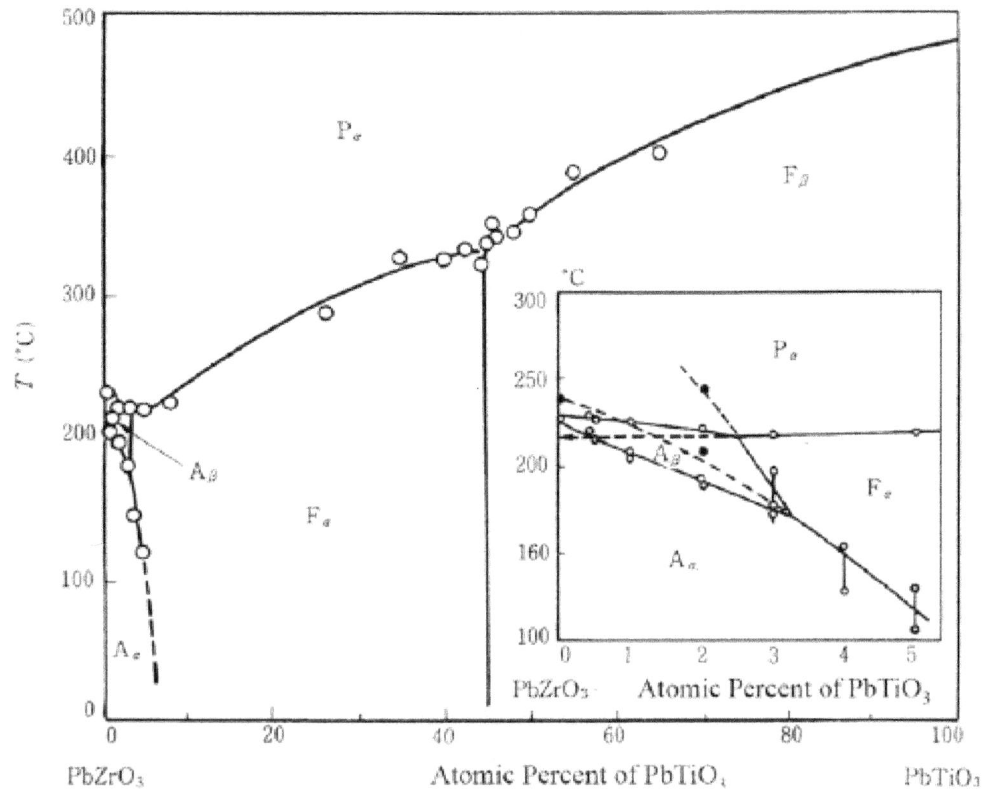

1.2 Important Parameters in Piezoelectricity

1.2.1 Piezoelectric Constants

Piezoelectric coefficient (d): Piezoelectric coefficient (d) measures the ability of a material to generate an electric field or charge (E) when it is externally subjected to mechanical stress or strain (x), or conversely to change its dimensions under an applied electric field which is typically expressed as charge per unit force (C/N) (important parameter in actuator applications):

$$d = \frac{x}{E} \tag{3}$$

The d constant is related to other important material properties through the following relationship:

$$d = k\sqrt{\varepsilon_0 k^T S^E} C / N \tag{4}$$

where:

k: electromechanical coupling coefficient
k^T: relative dielectric constant
S^E: elastic compliance (10 m/N) at a constant electric field

One of the most important d constants is:

$$d_{33} = k\sqrt{\varepsilon_0 k_3^T S_{33}^E} C / N \tag{5}$$

This suggests that the force acts along all three directions, in parallel with the polarization axis, and is exerted on the identical surface from which the charge is gathered.

Piezoelectric voltage coefficient (g): The piezoelectric voltage coefficient is denoted by 'g', and it represents the amount of voltage developed when subjected to mechanical deformation and its unit is in voltage-meter/Newton (Vm/N). The g constant can be calculated from the d constant and relative permittivity using the following relationship:

$$g = \frac{d}{\varepsilon} Vm / N \tag{6}$$

Corresponding to d_{33}, g_{33} is also present depending on the type of relative directions.

1.2.2 Electromechanical Coupling Coefficient (k)

The electromechanical coupling coefficient is given by the ratio of the amount of mechanical stored to the amount of electrical energy applied and vice versa:

$$k^2 = \frac{mechanical\ or\ electric\ energy\ stored}{electrical\ or\ mechanical\ energy\ applied} \tag{7}$$

Usually, k_p and k_t are generally used which is used to measure the electromechanical coupling coefficient along the radial and thickness axis of a circular-shaped disc but for an arbitrary shape k_{eff} is used to denote the electromechanical coupling factor and is given by:

$$k_{eff}^{\ 2} = 1 - \left(\frac{f_r}{f_a}\right)^2 \tag{8}$$

where f_r and f_a correspond to resonance and anti-resonance frequency.

1.2.3 Mechanical Quality Factor (Q_m)

The mechanical quality factor characterizes the efficiency of energy transfer between mechanical and electrical domains in a piezoelectric device. It provides information about the sharpness and selectivity of resonance peaks in a piezoelectric device frequency response. It can be found using the following relationship:

$$Q_m = \frac{f_r}{f_2 - f_1} \tag{9}$$

where the resonance frequency is f_r and the frequencies at 3 dB of the maximum admittance are f_1 and f_2.

Besides dielectric constant, Curie temperature, depolarization temperature, remanent polarization, spontaneous polarization (indicating the inherent polarization within each electric domain), and dielectric dissipation factor are also important parameters to be considered. Usually, a piezoelectric material will have high performance if it has high,, and and low and (Jaffe et al., 1971; Noheda, 2002; Uchino & Giniewicz, 2003; Wei, Wang, Xia, Cui, Shi, Dong, Liu, Ding, Zhang, Ma, Wang, Wang, Sun, Wei, & Guo, 2018).

2. LEAD-BASED PIEZOELECTRIC MATERIALS

Lead zirconate titanate (PZT) represents a widely recognized and extensively studied piezoelectric/ferroelectric material that has garnered attention due to its excellent electrical properties, including high dielectric constant, high piezoelectric constant, and high remnant polarization. PZT comprises a solid solution blending lead titanate ($PbTiO_3$) and lead zirconate ($PbZrO_3$), typically represented as $(PbZr_{1-x}Ti_x)O_3$, where x is the mole fraction of titanium. The structure of PZT depends on the composition of x and resembles perovskite (ABO_3)-like unit cells. The perovskite structure consists of a corner-sharing network of oxygen atoms enclosing cations in the octahedral coordination. The flexibility of the A-site cations (Pb^{2+}) and the B-site cations (Zr^{4+} and Ti^{4+}) allows for the solid solution of Zr and Ti ions into the lattice, resulting in a range of properties depending on the composition (Shirane et al., 1952; Takenaka & Nagata, 2005). The PZT undergoes several phase transitions with variations in composition. Besides one can achieve intensified piezoelectric capabilities around the phase boundary of rhombohedral and tetragonal. This phase boundary is known as the morphotropic phase boundary as shown in Fig. 3. This MPB enables the improvement of randomly aligned ferroelectric domains which will help in achieving maximum polarization in the presence of applied electric field (Jaffe, Roth, & Marzullo, 1954; Sawaguchi, 1953). Also in MPB, the electromechanical factors are maximum as a result of enhancement in polarizability which is achieved as of result of the coupling of equivalent energy states. This results in high piezoelectric coupling coefficients in ferroelectrics such as PZT due to its ability to polarize by switching electric field making it very efficient in energy harvesting applications, as it can show large deformation under application of very low voltages. Thus PZT piezoceramics are extensively used in various high-performance applications with a relatively low cost of making, and these in turn can be accountable for calling them as "king of piezoelectric materials'' (Coondoo et al., 2013; Jaffe, Roth, & Marzullo, 1954; Shirane et al., 1952; Takenaka & Nagata, 2005).

Figure 3. A simple PZT phase diagram (Coondoo et al., 2013)

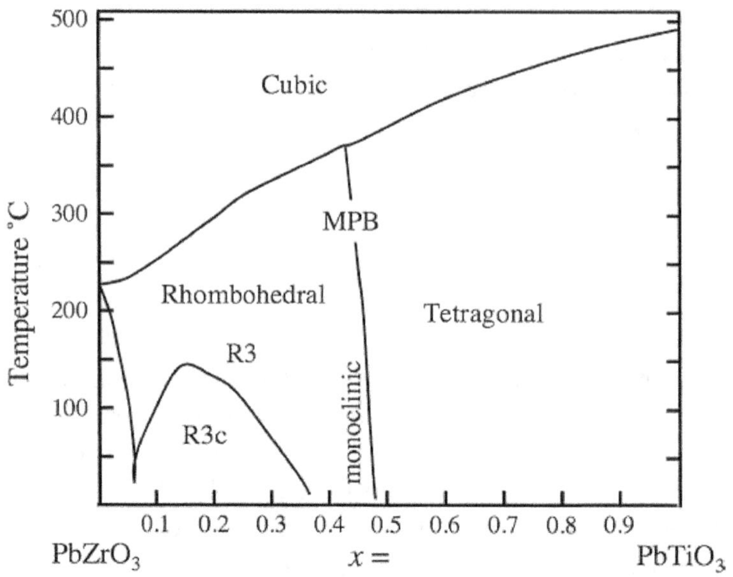

After the discovery of PZT, lead meta-niobate, lead-titanate, lead-barium-lithium-niobate, and modified lead-titanate were studied (Goodman, 1953; Jaffe, Roth, & Marzullo, 1954; Jaffe, 1958; Wei, Wang, Xia, Cui, Shi, Dong, Liu, Ding, Zhang, Ma, Wang, Wang, Sun, Wei, & Guo, 2018; Yokosuka, 1977). However, $PbZr_{1-x}Ti_xO_3$ (PZT) and (1-x) $PbMg_{0.33}Nb_{0.67}O_3$-$xPbTiO_3$ are the most commonly used ferroelectric ceramics to be used in MPB engineering and its applications. For instance in Fig. 2, with 52% lead zirconate and 48% lead titanate concentration MPB is achieved, with a realization of high remnant polarization of 35 μC/cm^2 and high coercive field of 1 kV/mm. Also, a high piezoelectric coefficient of ~779 pm/V was reported by calculating the slope of the electric field vs. strain curve (Cross, 2004; Donnelly et al., 2007). Earlier it was believed for PZT that MPB with rhombohedral and tetragonal phase boundary is a characteristic feature for obtaining excellent piezoelectric properties but some researchers found out there was an additional intermediate phase (monoclinic), particularly within the composition of x = 0.45-0.5 in between the rhombohedral and tetragonal phase and that the space group for this new monoclinic phase was Cm. Moreover, other researchers as such Noheda and colleagues, as well as Cox and their team, also documented the existence of an additional phase in the vicinity of MPB. However, there are still concerns over MPB in PZT (Cox et al., 2001; Noheda et al., 1999; Noheda et al., 2000). Another example with a similar phase illustration as PZT was $(1 - x)Pb(Zn_{0.33}Nb_{0.67})O_3$–$xPbTiO_3$ (PZN–xPT) single crystals, which exhibited substantial d_{33} values ranging from 1500 to 2500 pC/N, as well as high electromechanical coupling factor (k_{33}˃0.9) in the PZN-xPT system for the x values in between 0.28 and 0.33 and finally it can be ascribed that the MPB was responsible for the enhancement in piezoelectricity (Choi et al., 1989; Guo et al., 2003; Kuwata et al., 1981; Kuwata et al., 2014; Noheda et al., 2002; Park & Shrout, 1997; Singh & Pandey, 2001; Ye & Dong, 2000). Besides the enhancement in piezoelectricity, it was also necessary that we achieve a high Curie temperature T_c. This led to the invention of $(1 - x)BiMeO_3$–$xPbTiO_3$ solid solutions, with Me representing various elements like Sc, In, Yb, and Fe, however $(1 - x)BiScO_3$–$xPbTiO_3$(BS-xPT) solid solution displayed significant piezoelectric constant d_{33}, measuring 450 pC/N, along with high Curie temperature (T_c) of 450°C (Eitel et al., 2014).

All in all, it can be confirmed that the notion of the MPB shows potential for enhancing electrical traits. Also designing phase boundaries in piezoceramics through varying compositions could be of extensive help in developing Pb-free piezoelectric materials.

3. LEAD-FREE PIEZOELECTRIC MATERIALS

3.1 Why the Need?

The need for developing lead-free piezoceramics is driven by multifaceted concerns, including environmental hazards, health risks, regulatory compliance, consumer demand for eco-friendly products, energy harvesting applications, enhanced performance characteristics, durability requirements, market opportunities, and global collaborative efforts. Researchers and industries are actively engaged in creating safe, sustainable, and high-performance materials to replace lead-based piezoceramics in various applications, aligning with environmental regulations and consumer preferences while capitalizing on market potential and international cooperation. Moreover constructing MPBs in lead-free ones is also essential to promote piezoelectric capabilities. However, it won't be similar PZT phase as boundaries because of its sensitivity to composition and temperature. For example, in $BaTiO_3$ (BT) piezoceramics one can achieve enhancement in d_{33} value by introducing the R-T phase boundary but at a cost of a reduction in Curie temperature (T_c) less than 100°C (Gao et al., 2017). Besides real phase boundaries such as in PZT can also be achieved in $Bi_{0.5}Na_{0.5}TiO_3$ but its low d_{33} doesn't make it feasible to be used in practical applications (Lidjici et al., 2015). In one of our research, we too have created a phase boundary (orthorhombic and tetragonal) in $K_{0.47}Na_{0.47}Li_{0.06}NbO_3$ composite lead-free piezoceramics. However, its d_{33} value and ferroelectric properties are yet to be evaluated (Chaliha & Maisnam, 2023). Thus it is a necessary step to introduce phase boundary in lead-free piezoelectric ceramics for its optimum electrical properties.

3.2 Types of Piezoelectric Materials

3.2.1 Single Crystal

Single crystals are monocrystalline materials in which atoms or molecules form an ordered and uniform continuity. The constituent atoms, ions, or molecules form a repeated three-dimensional arrangement that extends unbroken straight through the entire chunk or crystal. Such uniformity in structure gives rise to unique material properties, like improved mechanical, electrical, and optical properties; consequently, single crystals are beneficially applicable in numerous scientific, industrial, and technological areas. Unlike polycrystalline materials, which consist of multiple small crystalline areas called grains that have separate orientations, single crystals are made up of a single large crystalline region. The same applies too for single-crystal piezoelectric materials, which have very ordered crystal structures that ensure higher efficiency in converting mechanical energy into electricity or vice versa compared with polycrystalline substances. As an example, PZT single crystals d_{33} values range is 450-550 pC/N along with remnant polarization P_r~20-45 μC/cm^2 depending upon the specific composition and processing conditions (Fang & Yang, 2014; Shinekumar & Dutta, 2015). Also, KNN possesses a d_{33} value in the range of 100-200 pC/N, along with remnant polarization P_r~5-15 μC/cm^2 (Lin et al., 2009). Besides chemical modifications followed by different preparation techniques with Mn and Li, an addition has resulted in electrical modification as well as a substantial d_{33} value of 405 pC/N (Bridgman Method) in KNN-LN and d_{33} of 305 pC/N (slow-cooling method) in KNN-Mn single crystals (Lin et al., 2010; Wu, Xiao, & Zhu, 2015).

3.2.2 Thin Films

Piezoelectric Materials in Thin Film Forms are a special type of materials that exhibit piezoelectric properties as thin films or very thin layers. Such materials play a vital role in such devices as MEMS, sensors, actuators, and others with limited space capabilities. Piezoelectricity can be applied to smaller-scale applications with these generators. The thin film piezoelectric materials are usually deposited on a substrate through methods including CVD, sputtering, or sol-gel processes. They are typically only a few micrometers in thickness but are engineered with unique piezoelectric characteristics for use in the appropriate application purpose. Depending upon specific composition and deposition process d_{33} value of PZT thin films is in the range of 100-150 pC/N with a remnant polarization P_r of 10-40 mC/cm^2 and for BT thin films it's in the range of 10-30 pC/N with P_r~1-10 μC/cm^2 and for KNN thin films d_{33}*~60 pm/V followed by P_r of 21 μC/cm^2. Usually, these thin films are synthesized using physical (RF magnetron Sputtering, Pulsed Layer Deposition) or chemical (Sol-Gel, Metal-Organic Deposition) methods are generally used (Guo et al., 2013; Pinho et al., 2023; Thery et al., 2015). However, due to low cost and easy processing, chemical methods are highly used to cope with the volatilization of elements such as K, Na, and Bi in KNN, BNT, and BFO thin films but do they have flaws such as poor roughness, and microstructure. Besides, a high P_r of 90.61 μC/cm^2 has been also reported through co-doping of Sr and Mn in BFO-based thin films (Fan et al., 2016). Thus it can be that said for maximum ferroelectric and piezoelectric capabilities the preparation parameter and chemical modifications are highly important.

3.2.3 Polymers

Piezoelectric polymers are a versatile grade of materials manifesting piezoelectric characteristics while being lightweight, flexible, and easy to process. They are highly valued in various applications particularly those that require conformability and low weight. Some examples of piezoelectric polymers are poly (vinylidene fluoride) PVDF, polyamides, and polylactic acid (PLLA). Despite having lower piezoelectric coefficients as compared to their ceramic counterparts, piezoelectric polymers offer advantages such as flexibility and biocompatibility. For enhancing the piezoelectric response of polymer techniques such as co-polymerization, the addition of additives or fillers and nanostructuring techniques have been used, thus modifying the structure and properties of polymers to increase piezoelectric properties. The use of poled PVDF films in tactile sensors for electronic devices, where pressure-sensitive touch panels can detect force through changes in charges or voltage. PVDF is also used in energy harvesting devices, converting mechanical energy to electrical energy because of its compliance with and storing mechanical energy as compared to piezoceramics. Piezoelectric polymers are also used in cell biology as these materials can influence and stimulate cellular function and are often as used substrates for culturing cells. Biocompatible polymers such as PVDF and PLLA with matching mechanical properties with the biological materials employed in this field. Nanostructuring of these polymers also allows for the generation of piezoelectric charge by cells themselves eliminating the need for external transducer (Jain, KJ, Sharma, Jain, & PN, 2015; Lu et al., 2020).

3.2.4 Composites

Composites refer to materials that are composed of two or more different constituents with different properties, combined in a way that results in a new material possessing a unique set of characteristics. Composites are designed to harness the strengths of each other constituents while mitigating their weaknesses. These materials can have a wide variety of applications in industries like aerospace, automobiles, construction, and electronics. Here are some of the piezoelectric composites such as PZT-PVDF which was reported to have $d_{33} \leq 100$ pC/N accompanied by a remnant polarization $P_r \sim 5\text{-}10\ \mu C/cm^2$. Similarly, KNN-$BaTiO_3$, BFO-PVDF, etc. are also studied yielding $d_{33}\sim40\text{-}200$ pC/N, 20-120 pC/N and remnant polarization $P_r\sim2\text{-}12\ \mu C/cm^2$, 1-8 $\mu C/cm^2$ respectively. These values are approximate and can vary widely on the specific composite formulation, processing techniques, and other factors (Jain, KJ, Sharma, Jain, & PN, 2015; Men et al., 2016).

3.2.5 Ceramics

Various methods can be adopted for the synthesis of polycrystalline piezo ceramics such as the solid-state reaction method, sol-gel method, hydrothermal method, etc. However, in every process, the basic step of stoichiometrically weighing the powder followed by compacting the powder and forming it into disc-like green bodies and ultimately sintering at the required temperature are the fundamental steps involved. Each of these methods has its strengths and weaknesses and the choice of methods depends upon the specific requirements of applications, the desired property of the piezoelectric material, and the available resources. Piezoceramics are mostly synthesized using the solid-state reaction method because it provide flexibility in providing suitable ceramic properties. Additionally, the method has a long history of use and is cost-effective, readily available, scalable for mass production as well as adaptable. Here we have discussed some of the piezoelectric ceramics below with an emphasis on potassium-sodium niobate $K_{0.5}Na_{0.5}NbO_3$ piezoelectric ceramics and methods to help in realizing MPB, for achieving outstanding piezoelectric capabilities (Li, Wang, Zhu, Cheng, & Yao, 2013; Malic et al., 2005; Rubio-Marcos et al., 2010; Thong et al., 2019).

3.2.5.1 Types of Piezoelectric Ceramics

$BaTiO_3$ (BT): Barium Titanate is the first discovered oxide that thrives in the perovskite family with an ABO_3 structure such that Ba^{2+} and Ti^{4+} occupy the A and B sites of the corner of the cubic and center of the unit cell respectively, while O^{2-} occupies the face center constituting the BO_6 octahedra. Thus due to the shift of A and O ions with relative B ions in the center develops a dipole moment coined as spontaneous polarization P_s. At an ambient temperature, BT displays adequate piezoelectricity, and low dielectric but its d_{33} and T_c are very low of the order of ≤ 190 pC/N and $\leq 120°C$ respectively, synthesized via traditional solid-state process (Wang et al., 2017). As a result, the focus was given to enhancing the piezoelectric properties by introducing different synthesizing techniques to manipulate the microstructure. This was achieved by adding sintering aids [4 mol% of LiF ($d_{33}\sim270$ pC/N, $k_p\sim0.45$)], different sintering techniques such as spark plasma [$d_{33}\sim410$ pC/N], and hot pressing [$d_{33}\sim485$ pC/N], different preparation methods [two step sintering method $d_{33}\sim$

460 pC/N] which led to the conclusion that better densification and smaller grain size ~2 µm can enhance piezoelectricity (Karaki et al., 2007; Shen & Li, 2010; Yang et al., 2012; Zhang et al., 2014).

One remarkable strategy is to introduce doping in the BT matrix by adding suitable donors to acceptors in the ceramic to enhance piezoelectric capabilities, consequently altering the phase boundary to MPB. This MPB possesses a vivid polarization orientation and facilitates the polar domains to easily orient themselves in the direction of the applied electric field (Habib et al., 2020; Jo et al., 2011). As Zhu et al. in (1-x)$Bi_{1+y}(Fe_{0.96}Mg_{0.02}Ti_{0.02})O_3$-$x$$BaTiO_3$ reported a rhombohedral (R) and tetragonal (T) coexistence phase boundary at x in between 0.26 and 0.32, with high quality piezoelectric constant d_{33} of 198 pC/N, k_p of 0.9%, and as well large curie temperature (T_c) of 497 °C (Zhu et al., 2017), Zhao, et al. synthesized a piezoceramic (1-x-y) $BaTiO_3$–$y$$CaTiO_3$–$x$$(BaZr_{1-z}Hf_z)O_3$ having O-T phase boundary yielding high (d_{33}=500 pC/N, k_p=0.52, P_r=14.0 µC/cm^2), however it was followed by a very low T_c less than 100 °C (Zhao et al., 2018). Moreover, Zhu et al. in their work $BaTiO_3$–$BiFeO_3$ doped by $Bi(Mg_{0.5}Ti_{0.5})O_3$ reported a d_{33} of 154 pC/N and T_c of 482 °C, by constructing a coexistence of pseudocubic and rhombohedral morphotropic phase boundary (MPB) (Zhu et al., 2016).

Further the enhancement of the piezoelectric capabilities, elevation of Curie temperature, optimization of temperature stability, and reduction of sintering temperature represent the subsequent steps essential for utilizing BT-based ceramics in piezoelectric applications. Achieving a delicate balance between piezoelectric properties, Curie temperature, and sintering temperature remains a challenging job for simultaneous improvement in all three aspects. Advancement in practical BT-based piezoceramics could be facilitated through the adoption of an updated phase boundary approach, the exploration of innovative material combinations, and the refinement of ceramic preparation methods (Wu, 2020).

$BiNaTiO_3$ (BNT): Discovered in 1960, BNT ceramics are well known for their large P_r ~ 38 µC/cm^2 as well as T_c ~ 320 °C. Despite that, the issue is with a low d_{33} ≤ 95 pC/N attributed to difficulty in poling the ceramic, as a result of a high coercive field and high conductivity. Thus to cope with this, compositional design and chemical modifications were employed to enhance piezoelectricity (Jo et al., 2012; Naderer et al., 2014; Smolenskii et al., 1960). For instance, Mn-doped BNT (d_{33} = 105 pC/N and T_c = 391 °C followed by depolarization temperature T_d = 197 °C) was achieved through ion substitution (Zuo et al., 2009). This is however not so satisfactory since there is no involvement of phase boundaries, hence ion substitution can't be considered for the improvement of piezoelectricity alone. So the necessity lies in the construction of MPB (R and T) in BNT, which is achieved by the addition of single or multiple perovskite structure ABO_3, and out of this BT and $B_{0.5}K_{0.5}TiO_3$ (BKT) addition have garnered attention. Lin et al. reported a d_{33} value 196 pC/N in $(Bi_{1-x-y}Nd_xNa_{1-y})_{0.5}Ba_yTiO3$ through the incorporation of the R-T phase boundary (Lin & Kwok, 2010). Also in BNT-BKT Pham et al. reported that the modification A and B-sites have simultaneously led to better piezoelectric constant (d_{33}*= 641 pm/V) and strain properties (44.8%) respectively as a result of the involvement of phase transition from polar R-T to non-polar domains Pseudocubic (P_c) (Pham et al., 2010). However, the concern is that although the R-T phase boundary involvement is present it is not as fruitful as PZT, besides the behavior of depolarization temperature T_d during the phase transition is still disputable, and this T_d presence also limits the practical applicability in piezoelectric devices.

$BiFeO_3$ (BFO): Bismuth ferrite ($BiFeO_3$) is the only single-phase perovskite oxide compound known for its multifunctional properties, including ferroelectricity, antiferromagnetism, and piezoelectricity.

It exhibits a rhombohedral crystal structure with a distorted crystal lattice (Catalan & Scott, 2009; Wang et al., 2003; Wu, Fan, Xiao et al, 2016). In the past chemical modifications have been adopted to get better physical properties by tuning the crystal structure, which resulted in decreasing high leakage current (easy in poling, much difficult earlier), oxygen vacancies, and mitigating the formation of secondary phases produced as a result of depletion of Bi^{3+}, reduction of Fe^{3+} to Fe^{2+} (Achenbach et al., 1967; Rojac et al., 2014; Valant et al., 2007; Wang et al., 2004). Substitution of ions at A-site, B-site, and simultaneous substitution at A-site and B-site have been seen to be remarkable when rare earth ions such as (La^{3+}, Sm^{3+}, Y^{3+}, Dy^{3+}, Pr^{3+}, $Nd^{3+,}$ etc.), particularly Sm and La co-doped BFO ceramics which yielded d_{33} = 50 pC/N and P_r = 40 μC/cm^2 by decreasing the leakage current by breaking down the space attenuated spin structure and also eradicating the secondary phases thus resulted in enhanced ferroelectricity (Chen et al., 2012; Ding et al., 2011; Haumont et al., 2006; Wang et al., 2011; Yu et al., 2008). However, it should be kept in mind that no deviation from the orthodox rhombohedral structure could be seen as a result of these ion substitutions. Since there was no phase boundary involved in our previous discussion, the priority has now shifted to the role of the ABO_3 modifiers to construct the phase boundary and attenuate the piezoelectric properties. As an example, Lee et al. after introducing BT got enhanced $d_{33,}$ measuring 402 pC/N, as well as a remnant polarization (P_r) of 22 μC/cm^2, coercive electric field (E_c) of 24 kV/cm, and Curie temperature (T_c) of 454 °C obtained via R-T phase boundary construction (Lee et al., 2015). Besides introducing ternary systems such as BFO-BTO-BZT a very high d_{33} of 324 pC/N, P_r of 24 μC/cm^2, E_c of 26 kV/m, and T_c = 466 °C was also reported (Lee et al., 2015).

$K_{0.5}Na_{0.5}NbO_3$ (KNN): Discovered in the 1960s, KNN is a leading contender to replace PZT because of its ability to manifest large piezoelectricity and high Curie temperature. KNN is a mixture of ferroelectric $KNbO_3$ and ferroelectric-antiferroelectric $NaNbO_3$ with an MPB coexistence (Fig. 4) such that the K/Na ratio belongs to a unity, moreover, this MPB is followed by a polymorphic phase transition (PPT) because of its lineage to change phase boundary due to temperature variation. Besides KNN possesses three types of phase transition as shown in Fig. 5, namely R-O, O-T, and

Figure 4. Schematic of the phase diagram of KNN (Li, Wang, Zhu, Cheng, & Yao, 2013)

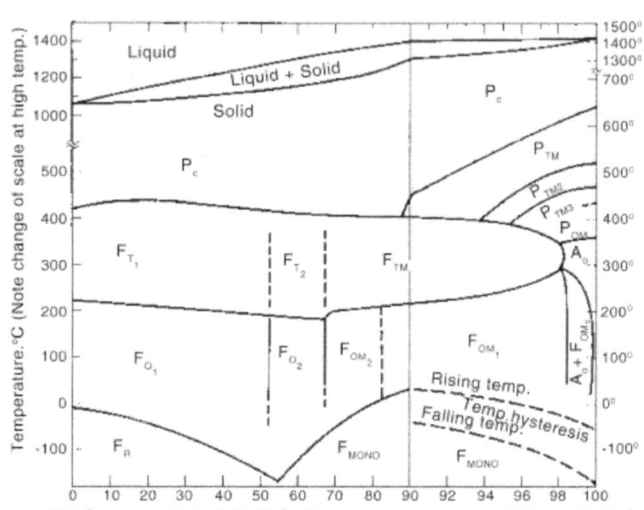

Figure 5. Behaviour of phase boundary along with temperature realized from Dielectric permittivity vs. temperature graph (Li, Wang, Zhu, Cheng, & Yao, 2013)

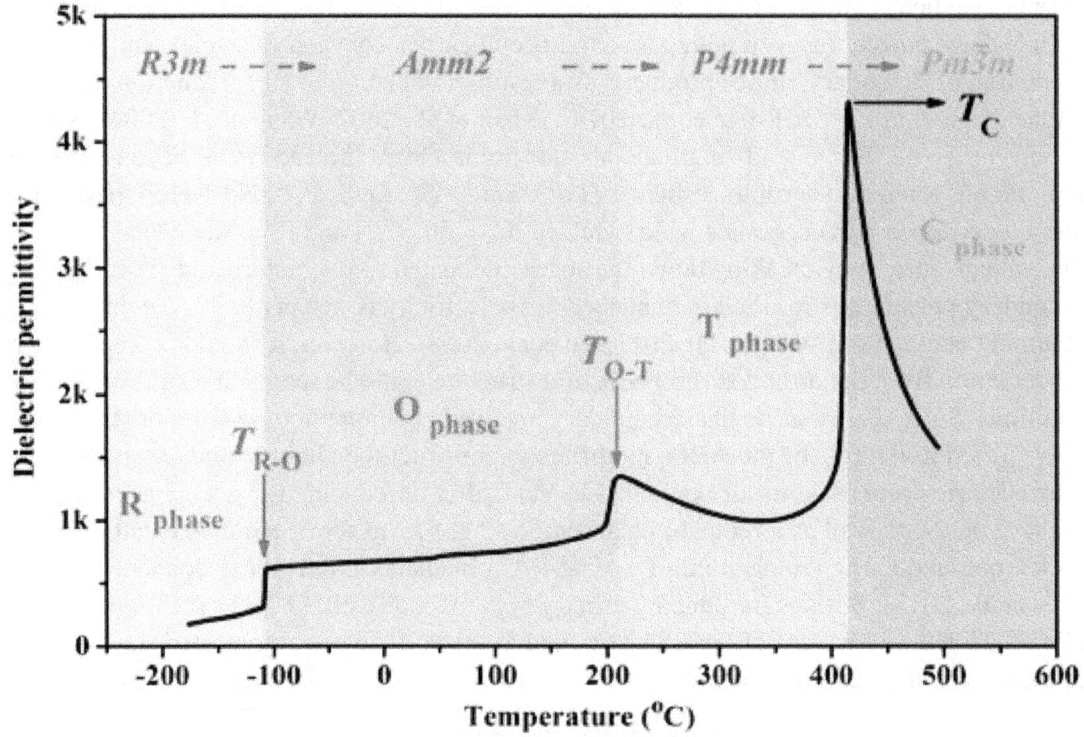

T-C (at T_c) as we plot the variation of dielectric constant along with temperature (Ge & Jouaneh, 1996; Li, Wang, Zhu, Cheng, & Yao, 2013).

Synthesizing KNN is a hard task, as it is difficult to maintain a dense structure which is a basic property for gaining enhanced piezoelectricity because of its narrow range of sintering temperature which results in the volatilization of alkalis (Cheng, Wang, Yao et al, 2013; Monika Aniz & Maisnam, 2022). Thus comes the necessity to tackle this problem without obtaining any undesirable properties to acquire excellent piezoelectric properties in KNN and to keep them in check, successful attempts have been made by introducing ion substitution, sintering aids, preparation techniques, and the most recent popular one is constructing phase boundaries at ambient temperature resulting in property enhancement due to increase amount of polarization states (Li, Wang, Zhu, Cheng, & Yao, 2013; Wu, Xiao, & Zhu, 2015). In 2004 Saito et al. made a breakthrough by adding ions Li^+, Ta^{5+}, and Sb^{5+} in KNN piezoceramics synthesized by reactive template grain growth method (RTGG) method resulting in large $d_{33} = 416$ pC/N, which credits go to forming O-T phase boundary (Saito, Takao, Tani, Nonoyama, Takatori, Homma, & Nakamura, 2004). Guo et al. reported a d_{33} value of 200 and 235 pC/N by adding $LiNbO_3$ or $LiTaO_3$ synthesized by traditional solid-state technique (Guo et al., 2004; Guo et al., 2005). However, no significant development was seen until 2014 which led to the generation of substantial piezoelectricity d_{33} of 490-570 pC/N as a result of R-T phase boundary coexistence. Subsequently, in 2019 ultrahigh piezoelectricity of $d_{33} = 650$ pC/N was reported due to the introduction of a new phase boundary (relaxor R-O-T). Also, it's not too

Figure 6. No. of publications for KNN lead-free piezoceramic in the year 2004-2023 (October) with keywords KNN and piezoceramic, data source from Google Scholar

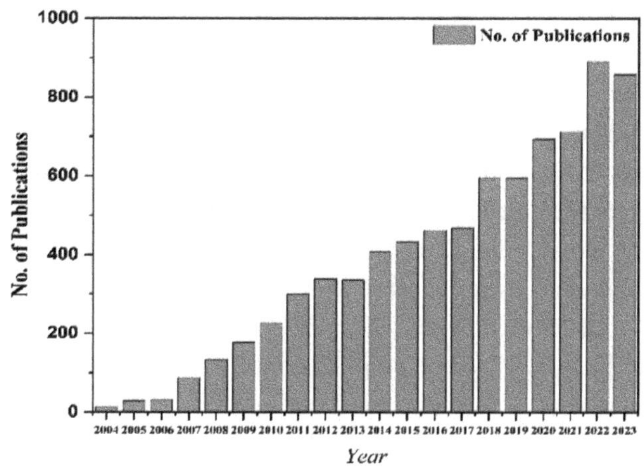

difficult to form an R-T or R-O-T phase boundary with the adequate and right amount of additives but it is hard to get enhanced piezoelectricity along with it (Guo et al., 2005; Wang, Wu, Xiao, Zhu, Cheng, Zheng, & Wang, 2014; Wu, Wu, Wu, Xiao, Zhu, & Pennycook, 2016; Xu, Li, Lv, Wu, Zhang, Xiao, & Zhu, 2016; Zheng, Wu, Yuan, Lv, Li, Men, Zhao, Xiao, Wu, Wang, Li, Gu, Zhu, & Pennycook, 2017). It is still controversial as to what might be the reason behind this enhanced piezoelectricity thus holding a barrier for further development. Fig. 5 shows the statistical data of publication on KNN lead-free piezoceramics showing an increment in publication every year, thereby providing an outlook to develop this material for future sustainable applications.

In the section below we will discuss the outcomes of various author's groups upon KNN piezoceramics, which were mainly concentrated on compositional designs and constructing phase boundaries. Starting with the addition of ions and compounds, followed by the addition of perovskite structure helping to realize phase boundary construction, and finally adopting further techniques in enhancing piezoelectricity. Fig. 6 displays the attention garnered in the publication of KNN piezoceramics with an increasing trend each year.

3.2.5.2 Construction and Importance of Morphotropic Phase Boundary (MPB)

The incorporation of chemical modification to develop phase boundaries at near ambient temperature is one of the important tools to acquire enhanced piezoelectricity in KNN piezoceramics and this was achieved after designing a binary or ternary ceramic system by a group of authors through relentless efforts in the past decade leading to a new phase boundary system. This new phase boundary enables the dipole polarization to rotate itself in the direction of the applied field or stress resulting in high piezoelectric properties. The key approach behind this was to increase and decrease phase transition temperature $T_{R\text{-}O}$ and $T_{O\text{-}T}$ to near ambient temperature respectively. This was achieved by the addition of compounds like $LiXO_3$ (X = Ta, Sb) and $YTiO_3$ ($Bi_{0.5}K_{0.5}$, $Bi_{0.5}Na_{0.5}$, Ba, Ca) to reduce $T_{O\text{-}T}$, simultaneously $T_{R\text{-}O}$ was increased by using $MZrO_3$ (M = Ba, $Bi_{0.5}K_{0.5}$, Mg, $Bi_{0.5}Na_{0.5}$) and $BiNO_3$ (N = Co, Fe). This pursuit has become a focal point for enhancing KNN-based piezoceramics (Jiang, Li, Xie, Wu, Chen, Xiao, & Zhu, 2017; Wu, Xiao, & Zhu, 2015).

O-T phase boundary

This phase boundary is achieved through the substitution of ions, the addition of single or multiple ABO_3 perovskite structures, or by manipulating the K: Na ratio, which drops the T_{O-T} phase transition temperature ~200 °C to near ambient temperature thus creating an O-T phase boundary with enhanced piezoelectric traits. The impact of Li, Ta and Sb doping upon microstructure, sintering temperature, and phase transition temperature was investigated vigorously and it was found that they were all responsible for decreasing T_{O-T} and slightly increasing T_{R-O} simultaneously and also increasing T_c (except Ta) thus creating O-T phase boundary (Wu, Xiao, & Zhu, 2015). Additionally, it was noticed that Ta restricted abnormal grain growth (AGG) accompanied by a decrease in T_c. In light of this Chang et al. reported a large d_{33} = 298 pC/N, k_p = 0.52, P_r = 28.68 μC/cm², and T_c = 366 °C in Ta-modified $(K_{0.458}Na_{0.542})0.96Li_{0.04}NbO_3$ ceramic system. They found that with increasing Ta concentration T_{O-T} and T_c values decreased and at x = 0.15 they got promising results as mentioned earlier (Chang et al., 2008). Also in Ag, modified ceramic system $(K_{0.50}Na_{0.50})_{0.95-x}Li_{0.05}Ag_x(Nb_{0.95}Ta_{0.05})O_3$ ceramic system Jiagang et al. reported that with an increase in Ag content T_c slightly got increased, followed by a decrease in T_{O-T}, and for x = 0.03 the optimum values of d_{33} = 252 pC/N, T_c = 438 °C, k_p = 45.4%, P_r = 30.1 μC/cm², and E_c = 13.8 kV/cm was obtained (Wang, Wu, Xiao et al, 2008).

The incorporation of single or multiple ABO_3 was also reported by various groups to realize the O-T phase boundary in the KNN ceramic system. This ABO_3 perovskite structure helps in decreasing the T_{O-T} value and hence constructing an O-T phase boundary at near ambient temperature which helps in realizing amplified piezoelectricity in KNN piezoceramics.

Authors discussed various ABO_3 perovskite compounds primarily consisting of $LiLO_3$ (L = Sb, Ta, Nb), $MTiO_3$ (M = $Bi_{0.5}Na_{0.5}$, $Bi_{0.5}K_{0.5}$, Ba, Sr), $BiNO_3$ (N = Sc, Al, Co), and $Bi_{0.5}Na_{0.5}XO_3$ (X = Zr, Hf,) which help them in shifting T_{O-T} decreasing to room temperature realizing O-T phase boundary and thus promoting piezoelectricity Table 1. Reports suggested that the addition of Bi and Zr helps in tuning phase boundary at room temperature thus promoting piezoelectricity. Our group also found an O-T phase boundary with BNZ addition in the KNLN ceramic system where Zr served as an entity to raise T_{R-O} near ambient temperature and Li drops slightly T_{O-T} to room temperature (Chaliha & Maisnam, 2023). Wang et al. also reported the O-T phase boundary in BNZ-modified KNLN with d_{33} of 255 pC/N, k_p of 0.41, and T_c of 360 °C (Wang, Wu, Zheng et al, 2014).

The improved d_{33} is always followed by a decrease in temperature, so it is very much essential for simultaneous enhancement in piezoelectricity as well as high-temperature stability for high-temperature applications. This is achieved by modifying KNN by adding multiple ABO_3 perovskite structures with the same additives as mentioned in Table 2 and ultimately developing an O-T phase boundary. Along with this, it was also reported by various authors that just changing the K/Na ratio can substantially modify the phase boundaries [(O-T) Fig. 7(a)] and hence enhance piezoelectricity (Fig. 7(b)). For example in $(K_{0.42}Na_{0.52}Li_{0.04}Ag_{0.02})(Nb_{0.91}Ta_{0.05}Sb_{0.04})O_3$, O-T phase boundary was achieved by a mere change in K:Na ratio, then good piezoelectricity was achieved d_{33} = 263, k_p = 45%, and T_c = 353 °C. Moreover, this K:Na ratio is very sensitive to the original KNN composition, hence an enhancement in piezoelectricity could be attained by phase evolution through commuting the K/Na ratio (Wu, Wang, Xiao et al, 2007; Wu, Xiao, Wang et al, 2007).

R-O phase boundary

Table 1. Piezoelectric properties of single ABO_3 modified KNN piezoceramics while attaining the O-T phase boundary

Ceramic System	d_{33} (pC/N)	k_p	T_c (°C)	Reference
KNN-LS	262	46%	373	(Wu, 2020)
KNN-LT	230	50.5%	----	(Saito & Takao, 2006)
KNN-LN	215	41%	450	(Du et al., 2007)
KNN-BNT	195	43%	375	(Zuo et al., 2007)
KNN-BKZT	260	38%	323	(Xiao et al., 2014)
KNLNS-BT	269	50%	349	(Wu et al., 2012)
KNN-ST	195	37%	289	(Wang, Xie, Hanada et al, 2008)
KNN-BS	165	40%	370	(Wu, Xiao, Wu, Liang et al, 2011)
KNN-BAl	202	46%	372	(Du et al., 2008)
KNN-BC	165	40%	390	(Wu et al., 2014)
KNN-BNZ	255	41%	360	(Wang, Wu, Zheng et al, 2014)
KNN-BNH	385	----	315	(Tao, Wu, & Wang, 2016)

Table 2. Piezoelectric properties of multiple ABO_3 modified KNN piezoceramics while attaining the O-T phase boundary

Ceramic System	d_{33} (pC/N)	k_p	T_c (°C)	Reference
KNN-BNZ-LA	354	46%	325	(Jiang, Li, Xie, Wu, Chen, Xiao, & Zhu, 2017)
KNN-BNZ-BA	355	49%	335	(Jiang et al., 2016)
KNN-LS-BS	305	54%	315	(Minhong et al., 2011)
KNN-LN-BFC	276	48%	340	(Zhao et al., 2013)

Figure7.(a)-(b)Phasediagramandvariationofd$_{33}$withdifferentvaluesofxin$K_xNa0._{96-x}Li_{0.04}Nb_{0.91}Ta_{0.05}Sb_{0.04}O_3$ ceramic system (Du et al., 2007)

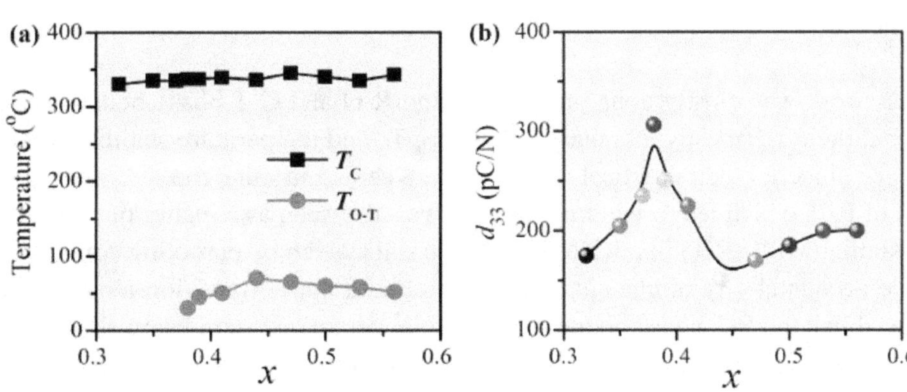

Figure 8. (a)-(b) XRD (44°-47°) and dielectric data of (1 − x − y)KNN-xBNKLZ-yBS ceramic system as a function x and y (Lee et al., 2019)

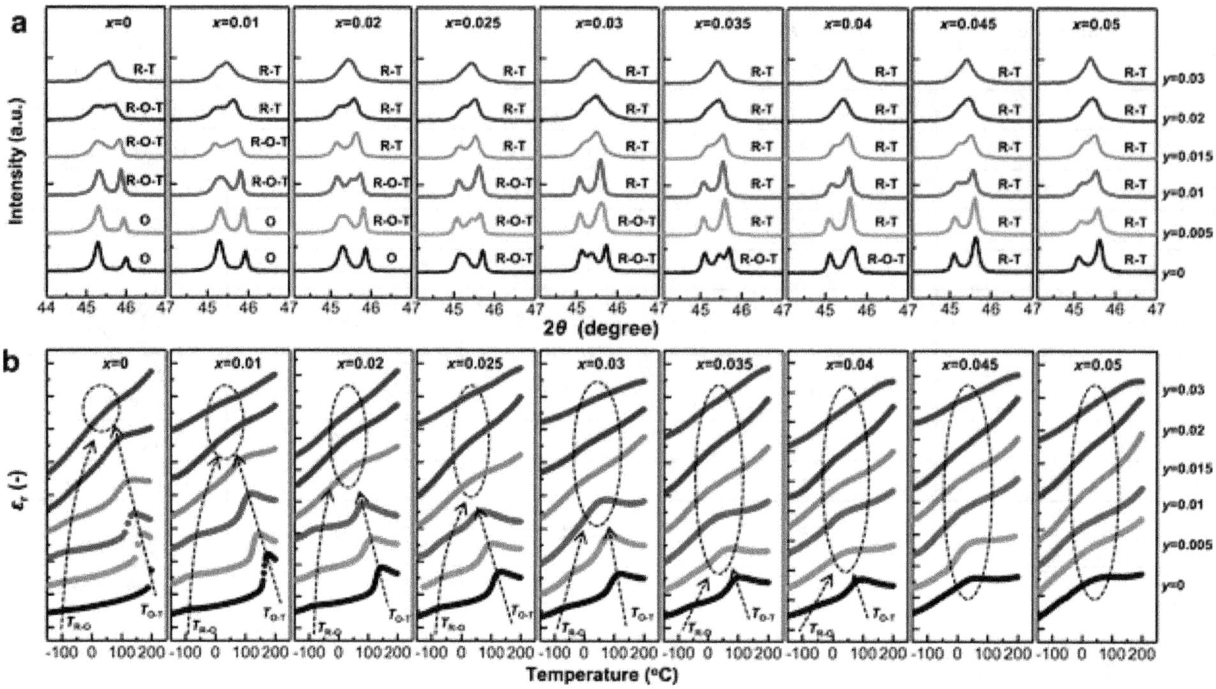

As mentioned earlier in Fig. 5 three-phase transition is possible and one of them is T_{R-O}. This phase transition is often kept aside owing to its unsatisfactory piezoelectric performance together with low T_c. As an example in $Na_{0.52}K_{0.48}Nb_{1-y}Sb_yO_3$ ceramic system, a d_{33} = 230 pC/N and a low T_c = 190 °C were reported, and if we try to enhance T_c, it is again at a cost of a reduction in d_{33}. This T_{R-O} has an important role in constructing a new phase transition T_{R-T} forming a new R-T phase boundary (Zuo et al., 2010). Various authors have reported that the addition of Sb, Ta, $CaZrO_3$, $SrZrO_3$, and $BaZrO_3$ has a significant influence on increasing T_{R-O} at low temperatures to room temperature. Out of this Sb has a remarkable impact not only on attaining T_{R-O} but also on the enhancement of its counterpart namely d_{33} and T_c, which is seen in $Na_{0.48}K_{0.52}Nb_{1-y}Sb_yO_3$ ceramic with d_{33} and T_c equal to 210 pC/N and 230 °C respectively (Wu, Tao, Yuan et al, 2015; Xie et al., 2022).

R-T phase boundary

Before 2013, work was mostly done on constructing R-O and O-T phase boundaries by chemical modifications but their d_{33} (mostly less than 300 pC/N), T_c, and temperature stability were not so satisfactory and hence hindered their practical applications. Keeping in mind the

MPB (R-T) of PZT which led to yielding high d_{33} was also seen as a matter of concern for research to construct a similar MPB (R-T) in a KNN-based material system by enrooting compositional design. Thus seeing the accountability of chemical composition and phase transition temperature, it became quite easy to establish the R-T phase boundary in a KNN ceramic system using similar techniques involved in the process of constructing O-T and R-O phase boundaries. Here the additives were doped in such a way that they simultaneously decreased and increased the phase transition temperatures $T_{O\text{-}T}$ and $T_{R\text{-}O}$ respectively near ambient temperatures such that the two dielectric peaks corresponding to phase transition temperatures (O-T) and (R-O) become one and hence finally forming the R-T phase boundary. For example, in (1-x-y) KNN-xBNKLZ-yBS ceramic system R-T phase boundary Fig. 8(a) was attained through the changing of the concentration of x and y. From Fig. 8(b) for various combinations of x and y, including x = 0 and y = 0.03, x = 0.01 and y = 0.02, x = 0.02 and y = 0.015, x = 0.03 and y = 0.01, x = 0.04 and y = 0.005, and x = 0.05 and y = 0, that the dielectric peak corresponds to only one peak conforming to the R-T phase boundary (Zhang et al., 2013).

In 2013, Zhang et al. achieved a high d_{33} = 425 pC/N in $KNLNS_x$–BZ by constructing the R-T phase boundary at near ambient temperature hence paving the way to construct R-T phase boundaries near ambient temperature with remarkable achievements achieved in the past few years (Wu et al., n.d.). Inspired by their previous work, they synthesized (1− x)KNNS–xBNKZ ceramic showing an R-T phase boundary which led to a high d_{33} of 490 pC/N and T_c of 454 °C by traditional technique (non-textured) and the result was much higher than to textured ceramic as was reported by Saito et al (Saito, Takao, Tani, Nonoyama, Takatori, Homma, Nagaya et al, 2004; Wu et al., n.d.). Thus continued massive research on establishing R-T phase boundary with reported high piezoelectricity confirms that R-T phase boundary construction is indeed essential in attaining optimum results in KNN-based piezoceramics.

The idea behind the construction of the R-T phase was formed by the addition of a single ABO_3 perovskite structure $Bi_{0.5}X_{0.5}ZrO_3$ (where X = Li, Na, K, and Ag), and Sb into the KNN ceramic system can abruptly tune phase transition temperatures (R-O and O-T) to just about ambient temperature, acquiring a remarkable feat in d_{33} value by fabrication of the R-T phase boundary. Also, it was found that with the addition of Li d_{33} value drops, and its lowest (335 pC/N) when Li completely takes up the X place. This was because the addition of Li content resulted in a deviation of the R-T phase boundary (≤ 10 °C) from room temperature and optimum d33 values were obtained R-T phase boundary was constructed at room temperature (Zheng, Wu, Xiao, Zhu, Wang, & Lou, 2015). Wu et al. in their work studied the addition of $Bi_{0.5}Na_{0.5}YO_3$ (where Y = Hf, Zr, Sn, and Ti) modified KNNS ceramic system and they found R-T phase boundary when the compositions BNH and BNZ lie in the range of 0.035-0.040 and 0.035-0.045 respectively. For BNT addition, an O-T phase boundary was noticed in the range of 0.005-0.02, and for BNS irrespective of the content it showed an orthorhombic phase (Zheng, Wu, Xiao, Zhu, Wang, & Lou, 2015). Table 3 displays the addition of a single ABO_3 perovskite structure-modified KNN-based piezoceramics to construct an R-T phase boundary.

Similarly, ABO_3 multi-components are also used as an additive to construct the R-T phase boundary. The most popular additives as discussed above are mainly $Bi_{0.5}X_{0.5}ZrO_3$ (where X = Li, Na, K, Ag, and also their amalgamation), $Bi_{0.5}Na_{0.5}YO_3$ (where Y = Hf, Zr, Sn, and Ti), and also Ba and Bi additives, to realize high d_{33} and T_c as formulated in Table 4.

3.2.5.3 Other Techniques for Enhancement of Piezoelectricity

Sintering aids: Sintering aids are essential in the fabrication of KNN piezoceramics, as they facilitate densification, enhance sinterability, control grain growth, stabilize desired phases, improve dielectric and piezoelectric properties, ensure material homogeneity, and reduce sintering temperatures (which will help loss of K/Na ratio volatilization at high sintering temperature). These additives, often oxides namely ZnO, CuO, MnO_2, LiF, Ga_2O_3, Co_2O_3, etc., play a crucial role in achieving the

Table 3. Piezoelectric properties of single ABO_3 modified KNN piezoceramics while attaining the R-T phase boundary

Ceramic System	d_{33} (pC/N)	k_p	T_c (°C)	Reference
KNLNS–BS	325	48%	358	(Cheng, Wu, Wang et al, 2013)
KNNS-BNKZ	490	46%	227	(Wang et al., 2013)
KNNS-BLZ	405	----	280	(Wang, Wu, Xiao, Zhu, Cheng, Zheng, Zhang et al, 2014)
KNANS-BNLZ	345	38.5%	260	(Zheng et al., 2014)
KNNS-BNH	419	45%	242	(Mazhao et al., 2015)
KNLNS-BNZ	390	45%	250	(Tao et al., 2015)
KNNS-BNH	418	----	242	(Yuan et al., 2015)
KNNS-BNLCZ	485	48%	227	(Tao, Wu, & Wu, 2016)
KNNS-BNSLZ	461	46%	228	(Xing et al., 2016)
KNNS-BNKH	525	----	210	143

Table 4. Piezoelectric properties of single ABO_3 modified KNN piezoceramics while attaining the R-T phase boundary

Ceramic System	d_{33} (pC/N)	k_p	T_c (°C)	Reference
KNNS-BF-BNZ	550	53.3%	----	(Zheng, Wu, Yuan, Lv, Li, Men, Zhao, Xiao, Wu, Wang, Li, Gu, Zhu, & Pennycook, 2017)
KNNS-BZ-BKF	570±10	----	190	(Wu, Wu, Wu, Xiao, Zhu, & Pennycook, 2016)
KNN-BS-BNKLZ	366	----	335	(Wu, Wu, Wu, Xiao, Zhu, & Pennycook, 2016)
KNN-BF-BNZ	428	51.6%	318	(Xu, Li, Lv, Wu, Zhang, Xiao, & Zhu, 2016)
KNN-BNT-BNZ	318	43%	326	(Wu, Wu, Xiao et al, 2015)

desired piezoelectric performance while minimizing defects and enhancing the material's overall quality, making them an integral component of the production process for KNN piezoceramics. Table 5 displays the properties of various studies done by authors who have used sintering aids in KNN-based piezoceramic systems. All the authors have concluded that the addition of sintering aids with an adequate amount can significantly contribute to obtaining a dense structure by employing a tiny amount of liquid phase (for example Zn, Cu) while sintering thus promoting the diffusion of ions within the ceramic, which is essential in the rearrangement of particles and closure of pores, thus promoting the densification process and also it can result in the enhancement of piezoelectric properties as well as temperature stability.

Table 5. Piezoelectric properties of sintering aids incorporated KNN-based piezoelectric ceramics

Ceramic System	d_{33} (pC/N)	k_p	T_c (°C)	Reference
KNN-BLT-BZ-BNZ-ZnO	350	0.50	250	(Gou et al., 2015)
KNN-BT-KNN-CuO	190	----	280	(Li et al., 2020)
KLNLT-AS-MnO_2	363	0.49	210	(Zhao et al., 2019)
LKNNTAS-LF	----	0.36	235	(Liu et al., 2016)
KNN-LT-Ga_2O_3	147	0.40	424	(Zhou et al., 2014)
KNLNS-Co_2O_3	262	0.40	352	(Tan et al., 2017)

Poling conditions: Enhancements in piezoelectricity in KNN piezoceramics are obtained through meticulous manipulation of poling conditions. By carefully selecting the electric field strength and temperature, controlling the poling time, employing voltage cycling, optimizing the crystal orientation and grain alignment, and even utilizing grain techniques, it is possible to improve the alignment of charge domains in the lattice structure. This accurate domain alignment enhances piezoelectric traits of KNN Pb-free ceramics, resulting in increased piezoelectric coefficients, charge sensitivity, and electromechanical coupling coefficients, all of which are crucial for applications for in sensors, actuators, transducers, and energy harvesting applications. Furthermore, the use of controlled atmospheres during poling and the choice of appropriate crystallographic orientation play vital roles in this process, ultimately contributing to the overall electromechanical performance of KNN piezoceramics. For example, by applying a DC poling field of 3-4 kV/mm Wu et al. in their work (1-x) $K_{0.45}Na_{0.55}Nb_{0.96}Sb_{0.04}O_3$-$xBi_{0.5}Na_{0.5}$ HfO_3 reported a high d_{33} = 419 pC/N, k_p = 0.45 and T_c= 242 °C (Wu, Xiao, Wu, Li et al, 2011).

Synthesizing Technique: To achieve high-density structures in KNN piezoceramics is very much an essential criterion to achieve superior piezoelectric properties and so several methods and techniques are employed. The traditional solid-state technique used in the synthesis is however not a viable method to achieve satisfactory densification because of the inevitable volatilization of K:Na at sintering temperatures. So to cope with this sintering temperature dependency other advanced methods have gained prominence, such as spark plasma sintering (SPS), hot pressing, high-energy ball milling, chemical precursor methods, pressure-less sintering, and the combination of processes have been used. These methods play vital roles in enhancing densification, controlling grain growth, and optimizing microstructures, ultimately leading to high-density KNN ceramics with improved properties, particularly in piezoelectric applications. Researchers continually explore and refine these methods to meet specific requirements of various applications and to keep up with the latest advancements in materials science. For example, Wu et al. in their work $0.955(K_{0.42}Na_{0.58})(Nb_{0.96}Sb_{0.04})O_3$-$0.045(Bi_{0.5}K_{0.5})_{0.90}Zn_{0.10}ZrO_3$ employed a two-step sintering which helped in obtaining a dense microstructure with a good d_{33} = 323-416 pC/N (Wu & Wang, 2014).

4. CONCLUSION

In this comprehensive book chapter, we have traversed the intriguing landscape of piezoelectric materials, embarking from the historical roots of this phenomenon and the breakthrough discovery of lead-based

PZT materials. We have elucidated the foundational parameters underpinning the piezoelectric effect, encompassing piezoelectric constants, electromechanical coupling coefficients, and the crucial mechanical quality factor. As environmental and health concern has taken center stage, we have delved into the realm of lead-free piezoelectric materials, underscoring the pressing need for eco-friendly alternatives. Our exploration has encompassed a wide spectrum of material categories, spanning single crystals, thin film polymers, composites, and ceramics. In particular, we have highlighted the promise laid by potassium sodium niobate (KNN) ceramics, as environmentally responsible and high-performance contenders for piezoelectric landscapes. Moreover, we've spotlighted techniques and methods taking part in a key role in enhancing KNN's piezoelectric traits, including ions, compound substitution, methods and the importance of phase boundary construction at near ambient temperature, use of sintering aids, manipulation of poling conditions, innovative synthesis approaches. The dynamic field of piezoelectric materials continues to inspire innovation, with potential applications ranging from advanced sensors to efficient energy harvesting, ultimately contributing to a more responsive, sustainable, and technologically advanced world.

REFERENCES

Achenbach, G. D., James, W. J., & Gerson, R. (1967). Preparation of single-phase polycrystalline BiFeO3. *Journal of the American Ceramic Society*, *50*(8), 437–437. doi:10.1111/j.1151-2916.1967.tb15153.x

ANSI/IEEE Std 176-1987. (1987). IEEE standard on piezoelectricity.

Bernard, J. (1955). *U.S. Patent No. 2,708,244*. Washington, DC: U.S. Patent and Trademark Office.

Catalan, G., & Scott, J. F. (2009). Physics and applications of bismuth ferrite. *Advanced Materials*, *21*(24), 2463–2485. doi:10.1002/adma.200802849

Chaliha, C., & Maisnam, M. (2023). Observation of the phase evolution and associated dielectric properties of K0. 47Na0. 47Li0. 06NbO3+(x) Bi0. 5Na0. 5ZrO3 lead free piezoceramic under varying frequency. *Materials Today: Proceedings*. Advance online publication. doi:10.1016/j.matpr.2023.02.233

Chang, Y., Yang, P., Ma, D., Liu, Z., & Wang, Z. (2008). Phase transitional behavior, microstructure, and electrical properties in Ta-modified [(K0.458Na0.542)0.96Li0.04]NbO3 lead-free piezoelectric ceramics. *Journal of Applied Physics*, *104*(2), 024109. doi:10.1063/1.2957591

Chen, X., Wang, J., Yuan, G., Wu, D., Liu, J., Yin, J., & Liu, Z. (2012). Structure, ferroelectric and piezoelectric properties of multiferroic Bi0. 875Sm0. 125FeO3 ceramics. *Journal of Alloys and Compounds*, *541*, 173–176. doi:10.1016/j.jallcom.2012.06.094

Cheng, L. Q., Wang, K., Yao, F. Z., Zhu, F., & Li, J. F. (2013). Composition Inhomogeneity due to Alkaline Volatilization in Li-Modified (K, Na) NbO 3 Lead-Free Piezoceramics. *Journal of the American Ceramic Society*, *96*(9), 2693–2695. doi:10.1111/jace.12497

Cheng, X., Wu, J., Wang, X., Zhang, B., Lou, X., Wang, X., Xiao, D., & Zhu, J. (2013). Mediating the contradiction of d33 and TC in potassium– sodium niobate lead-free piezoceramics. *ACS Applied Materials & Interfaces*, *5*(21), 10409–10417. doi:10.1021/am403448r PMID:24219128

Choi, S. W., Shrout, T. R., Jang, S. J., & Bhalla, A. S. (1989). Morphotropic phase boundary in Pb(MgNb) O3-PbTiO3 system. *Materials Letters*, *8*(6-7), 253–255. doi:10.1016/0167-577X(89)90115-8

Coondoo, I., Panwar, N., & Kholkin, A. (2013). Lead-free piezoelectrics: Current status and perspectives. *Journal of Advanced Dielectrics*, *3*(02), 1330002. doi:10.1142/S2010135X13300028

Coursey, P. R., & Brand, K. G. (1946). *Dielectric constants of some titanates*. Academic Press.

Cox, D. E., Noheda, B., Shirane, G., Uesu, Y., Fujishiro, K., & Yamada, Y. (2001). Universal phase diagram for high-piezoelectric perovskite systems. *Applied Physics Letters*, *79*(3), 400–402. doi:10.1063/1.1384475

Cross, E. (2004). Lead-free at last. *Nature*, *432*(7013), 24–25. doi:10.1038/nature03142 PMID:15516920

Ding, Y., Wang, T. H., Yang, W. C., Lin, T. C., Tu, C. S., Yao, Y. D., & Wu, K. T. (2011). Magnetization, Magnetoelectric Effect, and Structure Transition in $\hbox {BiFeO} _ {3} $ and $(\hbox {Bi} _ {0.95}\ hbox {La} _ {0.05})\hbox {FeO} _ {3} $ Multiferroic Ceramics. *IEEE Transactions on Magnetics*, *47*(3), 513–516. doi:10.1109/TMAG.2010.2103553

Donnelly, N. J., Shrout, T. R., & Randall, C. A. (2007). Addition of a Sr, K, Nb (SKN) combination to PZT (53/47) for high strain applications. *Journal of the American Ceramic Society*, *90*(2), 490–495. doi:10.1111/j.1551-2916.2006.01450.x

Du, H., Tang, F., Liu, D., Zhu, D., Zhou, W., & Qu, S. (2007). The microstructure and ferroelectric properties of (K0. 5Na0. 5) NbO3–LiNbO3 lead-free piezoelectric ceramics. *Materials Science and Engineering B*, *136*(2-3), 165–169. doi:10.1016/j.mseb.2006.09.031

Du, H., Zhou, W., Luo, F., Zhu, D., Qu, S., Li, Y., & Pei, Z. (2008). Design and electrical properties' investigation of (K0.5Na0.5)NbO3-BiMeO3 lead-free piezoelectric ceramics. *Journal of Applied Physics*, *104*(3), 034104. doi:10.1063/1.2964100

Eitel, R. E., Randall, C. A., Shrout, T. R., & Park, S.-E. (2014). Preparation and characterization of high temperature perovskite ferroelectrics, in the solid-solution (1-x)BiScO3–xPbTiO3. *Japanese Journal of Applied Physics*, *41*(4A), 2099–2104. doi:10.1143/JJAP.41.2099

Fan, S., Xie, X., Zhang, F., Guo, X., Yang, S., & Zhang, L. (2016). Improved leakage and ferroelectric properties of Sr doped BiFe 0.95 Mn 0.05 O 3 thin films. *Journal of Materials Science Materials in Electronics*, *27*(7), 6854–6858. doi:10.1007/s10854-016-4638-7

Fang, F., & Yang, W. (2014). Phase fragility and mechatronic reliability for Pb (Mg 1/3 Nb 2/3) O 3–PbTiO 3 ferroelectric single crystals—A review. *Journal of Advanced Dielectrics*, *4*(01), 1430001. doi:10.1142/S2010135X14300011

Gao, J., Xue, D., Liu, W., Zhou, C., & Ren, X. (2017, July). Recent progress on BaTiO3-based piezoelectric ceramics for actuator applications. *Actuators*, *6*(3), 24. doi:10.3390/act6030024

Ge, P., & Jouaneh, M. (1996). Tracking control of a piezoceramic actuator. *IEEE Transactions on Control Systems Technology*, *4*(3), 209–216. doi:10.1109/87.491195

Goodman, G. (1953). Ferroelectric properties of lead metaniobate. *Journal of the American Ceramic Society*, *36*(11), 368–372. doi:10.1111/j.1151-2916.1953.tb12820.x

Gou, Q., Xiao, D. Q., Wu, B., Xiao, M., Feng, S. S., Ma Zhao, D. D., Wu, J. G., & Zhu, J. G. (2015). New (1 _ xÞK0:5Na0:5NbO3–x (0.15Bi0:5Na0:5TiO3–0.85Bi0:5Na0:5ZrO3Þ ternary lead-free ceramics: Microstructure and electrical properties. *RSC Advances*, *5*, 30660. doi:10.1039/C4RA16780E

Guo, Q., Cao, G. Z., & Shen, I. Y. (2013). Measurements of piezoelectric coefficient d 33 of lead zirconate titanate thin films using a mini force hammer. *Journal of Vibration and Acoustics*, *135*(1), 011003. doi:10.1115/1.4006881

Guo, Y., Kakimoto, I., & Ohsato, H. (2004). Phase transitional behavior and piezoelectric properties of (Na0.5K0.5)NbO3-LiNbO3 ceramics. *Applied Physics Letters*, *85*(18), 4121–4123. doi:10.1063/1.1813636

Guo, Y., Kakimoto, I., & Ohsato, H. (2005). (Na0.5K0.5)NbO3-LiTaO3 lead-free piezoelectric ceramics. *Materials Letters*, *59*(2-3), 241–244. doi:10.1016/j.matlet.2004.07.057

Guo, Y., Luo, H., & Ling, D. (2003). The phase transition sequence and the location of the morphotropic phase boundary region in (1-x)[Pb(Mg1/3Nb2/3)O3]-xPbTiO3 single crystal. *J Phys: Conden Matter*, *15*(2), L77–L82.

Habib, M., Lee, M. H., Choi, H. I., Kim, M. H., Kim, W. J., & Song, T. K. (2020). Phase evolution and origin of the high piezoelectric properties in lead-free BiFeO3–BaTiO3 ceramics. *Ceramics International*, *46*(14), 22239–22252. doi:10.1016/j.ceramint.2020.05.301

Haumont, R., Kreisel, J., Bouvier, P., & Hippert, F. (2006). Phonon anomalies and the ferroelectric phase transition in multiferroic BiFeO 3. *Physical Review B: Condensed Matter and Materials Physics*, *73*(13), 132101. doi:10.1103/PhysRevB.73.132101

Jaffe, B., Cook, W. R. Jr, & Jaffe, H. (1971). *Piezoelectric Ceramics*. Academic.

Jaffe, B., Roth, R. S., & Marzullo, S. (1954). Piezoelectric properties of lead zirconate-lead titanate solid-solution ceramics. *Journal of Applied Physics*, *25*(6), 809–810. doi:10.1063/1.1721741

Jaffe, B., Roth, R. S., & Marzullo, S. (1954). Piezoelectric properties of lead zirconate-lead titanate solid-solution ceramics. *Journal of Applied Physics*, *25*(6), 809–810. doi:10.1063/1.1721741

Jaffe, H. (1958). Piezoelectric ceramics. *Journal of the American Ceramic Society*, *41*(11), 494–498. doi:10.1111/j.1151-2916.1958.tb12903.x

Jain, A., & KJ, PSharma, A. KJain, APN, R. (2015). Dielectric and piezoelectric properties of PVDF/PZT composites: A review. *Polymer Engineering and Science*, *55*(7), 1589–1616. doi:10.1002/pen.24088

Jain, A., & KJ, PSharma, A. KJain, APN, R. (2015). Dielectric and piezoelectric properties of PVDF/PZT composites: A review. *Polymer Engineering and Science*, *55*(7), 1589–1616. doi:10.1002/pen.24088

Jiang, L., Li, Y., Xie, L., Wu, J., Chen, Q., Xiao, D., & Zhu, J. (2017). Investigation of phase structure, microstructure, and electrical properties of LaAlO 3-modified alkali niobate lead-free perovskite. *Journal of Materials Science Materials in Electronics*, *28*(23), 17761–17771. doi:10.1007/s10854-017-7715-7

Jiang, L., Li, Y., Xie, L., Wu, J., Chen, Q., Xiao, D., & Zhu, J. (2017). Investigation of phase structure, microstructure, and electrical properties of LaAlO 3-modified alkali niobate lead-free perovskite. *Journal of Materials Science Materials in Electronics*, *28*(23), 17761–17771. doi:10.1007/s10854-017-7715-7

Jiang, L., Li, Y., Xing, J., Wu, J., Chen, Q., Liu, H., Xiao, D., & Zhu, J. (2017). Phase structure and enhanced piezoelectric properties in (1 _xÞ(K0:48Na0:52Þ(Nb0:95Sb0:05ÞO3–x(Bi0:5Na0:42Li0:08Þ0:9-Sr0:1ZrO3 lead-free piezoelectric ceramics. *Ceramics International*, *43*, 2100. doi:10.1016/j.ceramint.2016.10.189

Jiang, L., Tan, Z., Xing, J., Wu, J., Chen, Q., Zhang, W., Xiao, D., & Zhu, J. (2016). New potassium–sodium niobate ternary system with large piezoelectric coefficient and high Curie temperature. *Journal of Materials Science Materials in Electronics*, *27*(9), 9812–9820. doi:10.1007/s10854-016-5047-7

Jo, W., Daniels, J. E., Jones, J. L., Tan, X., Thomas, P. A., Damjanovic, D., & Rödel, J. (2011). Evolving morphotropic phase boundary in lead-free (Bi1/2Na1/2) TiO3–BaTiO3 piezoceramics. *Journal of Applied Physics*, *109*(1), 014110. doi:10.1063/1.3530737

Jo, W., Dittmer, R., Acosta, M., Zang, J., Groh, C., Sapper, E., Wang, K., & Rödel, J. (2012). Giant electric-field-induced strains in lead-free ceramics for actuator applications–status and perspective. *Journal of Electroceramics*, *29*(1), 71–93. doi:10.1007/s10832-012-9742-3

Karaki, T., Yan, K., & Adachi, M. (2007). Barium titanate piezoelectric ceramics manufactured by two-step sintering. *Japanese Journal of Applied Physics*, *46*(10S), 7035. doi:10.1143/JJAP.46.7035

Kuwata, J., Uchino, K., & Nomura, S. (1981). Phase transitions in the Pb(Zn1/3Nb2/3)O3-PbTiO3 system. *Ferroelectrics*, *37*(1), 579–582. doi:10.1080/00150198108223490

Kuwata, J., Uchino, K., & Nomura, S. (2014). Dielectric and piezoelectric properties of 0.91Pb (Zn1/3Nb2/3)O3-0.09PbTiO3 single crystals. *Japanese Journal of Applied Physics*, *21*(21), 1298–1302.

Lee, M. H., Kim, D. J., Park, J. S., Kim, S. W., Song, T. K., Kim, M. H., Kim, W.-J., Do, D., & Jeong, I. K. (2015). High-performance lead-free piezoceramics with high curie temperatures. *Advanced Materials*, *27*(43), 6976–6982. doi:10.1002/adma.201502424 PMID:26444562

Lee, M. K., Yang, S. A., Park, J. J., & Lee, G. J. (2019). Proposal of a rhombohedral-tetragonal phase composition for maximizing piezoelectricity of (K, Na) NbO3 ceramics. *Scientific Reports*, *9*(1), 4195. doi:10.1038/s41598-019-40943-6 PMID:30862966

Li, J. F., Wang, K., Zhu, F. Y., Cheng, L. Q., & Yao, F. Z. (2013). (K, Na) NbO 3-Based Lead-Free Piezoceramics: Fundamental Aspects, Processing Technologies, and Remaining Challenges. *Journal of the American Ceramic Society*, *96*(12), 3677–3696. doi:10.1111/jace.12715

Li, J. W., Liu, Y. X., Thong, H. C., Du, Z., Li, Z., Zhu, Z. X., Nie, J.-K., Geng, J.-F., Gong, W., & Wang, K. (2020). Effect of ZnO doping on (K, Na) NbO3-based lead-free piezoceramics: Enhanced ferroelectric and piezoelectric performance. *Journal of Alloys and Compounds*, *847*, 155936. doi:10.1016/j.jallcom.2020.155936

Lidjici, H., Lagoun, B., Berrahal, M., Rguitti, M., Hentatti, M. A., & Khemakhem, H. (2015). XRD, Raman and electrical studies on the (1− x)(Na0. 5Bi0. 5) TiO3− xBaTiO3 lead free ceramics. *Journal of Alloys and Compounds*, *618*, 643–648. doi:10.1016/j.jallcom.2014.08.161

Lin, D., & Kwok, K. W. (2010). Dielectric and piezoelectric properties of (Bi1− x− yNdxNa1− y) 0.5 BayTiO3 lead-free ceramics. *Current Applied Physics*, *10*(2), 422–427. doi:10.1016/j.cap.2009.06.044

Lin, D., Li, Z., Zhang, S., Xu, Z., & Yao, X. (2009). Dielectric/piezoelectric properties and temperature dependence of domain structure evolution in lead free (K0. 5Na0. 5) NbO3 single crystal. *Solid State Communications*, *149*(39-40), 1646–1649. doi:10.1016/j.ssc.2009.06.029

Lin, D., Li, Z., Zhang, S., Xu, Z., & Yao, X. (2010). Influence of MnO2 doping on the dielectric and piezoelectric properties and the domain structure in (K0. 5Na0. 5) NbO3 single crystals. *Journal of the American Ceramic Society*, *93*(4), 941–944. doi:10.1111/j.1551-2916.2009.03501.x

Liu, Q., Zhu, F. Y., Zhao, L., Wang, K., Li, L., & Li, J. F. (2016). Further enhancing piezoelectric properties by adding MnO2 in AgSbO3-modified (Li, K, Na)(Nb, Ta) O3 lead-free piezoceramics. *Journal of the American Ceramic Society*, *99*(11), 3670–3676. doi:10.1111/jace.14412

Lu, L., Ding, W., Liu, J., & Yang, B. (2020). Flexible PVDF based piezoelectric nanogenerators. *Nano Energy*, *78*, 105251. doi:10.1016/j.nanoen.2020.105251

Malic, B., Bernard, J., Holc, J., Jenko, D., & Kosec, M. (2005). Alkaline-earth doping in (K, Na) NbO3 based piezoceramics. *Journal of the European Ceramic Society*, *25*(12), 2707–2711. doi:10.1016/j.jeurceramsoc.2005.03.127

Mazhao, D., Xiao, D., Wu, J., & Zhu, J. (2015). Phase structure and electrical properties of 0.965(K0:4 5Na0:55Þ0:95Ag0:05(Nb1_xSbxÞ- O3–0.035Bi0:5(Na0:5Li0:5Þ0:5ZrO3 lead-free ceramics. *Journal of Materials Science Materials in Electronics*, *26*, 7309. doi:10.1007/s10854-015-3359-7

Men, T. L., Yao, F. Z., Zhu, Z. X., Wang, K., & Li, J. F. (2016). Piezoelectric properties of (K 0. 5 Na 0. 5) NbO3-BaTiO3 lead-free ceramics prepared by spark plasma sintering. *Journal of Advanced Dielectrics*, *6*(02), 1650013. doi:10.1142/S2010135X16500132

Messing, G. L., Trolier-McKinstry, S., Sabolsky, E. M., Duran, C., Kwon, S., Brahmaroutu, B., Park, P., Yilmaz, H., Rehrig, P. W., Eitel, K. B., Suvaci, E., Seabaugh, M., & Oh, K. S. (2004). Templated grain growth of textured piezoelectric ceramics. *Critical Reviews in Solid State and Material Sciences*, *29*(2), 45–96. doi:10.1080/10408430490490905

Minhong, J., Manjiao, D., Huaxin, L., Shi, W., & Xinyu, L. (2011). Piezoelectric and dielectric properties of K0.5Na0.5NbO3-LiSbO3-BiScO3 lead-free piezoceramics. *Materials Science and Engineering B*, *176*(2), 167–170. doi:10.1016/j.mseb.2010.10.007

Monika Aniz, I., & Maisnam, M. (2022). Microwave sintering of MnO2 added potassium sodium niobate ceramics and studies of their electrical properties. *Ferroelectrics*, *588*(1), 78–87. doi:10.1080/00150193.2022.2034441

Moore, R. W. (1919). A method of growing large perfect crystals from solution. *Journal of the American Chemical Society*, *41*(7), 1060–1066. doi:10.1021/ja02228a003

Naderer, M., Kainz, T., Schütz, D., & Reichmann, K. (2014). The influence of Ti-nonstoichiometry in Bi0. 5Na0. 5TiO3. *Journal of the European Ceramic Society*, *34*(3), 663–667. doi:10.1016/j.jeurceramsoc.2013.10.010

Noheda, B. (2002). Structure and high-piezoelectricity in lead oxide solid solutions. *Current Opinion in Solid State and Materials Science*, *6*(1), 27–34. doi:10.1016/S1359-0286(02)00015-3

Noheda, B., Cox, D. E., Shirane, G., Gao, J., & Ye, Z.-G. (2002). Phase diagram of the ferroelectric relaxor (1-x) PbMg1/3Nb2/3O3-xPbTiO3. *Physical Review B: Condensed Matter*, *66*(5), 340–351. doi:10.1103/PhysRevB.66.054104

Noheda, B., Cox, D. E., Shirane, G., Gonzalo, J. A., Cross, L. E., & Park, S. E. (1999). A monoclinic ferroelectric phase in the Pb (Zr 1− x Ti x) O 3 solid solution. *Applied Physics Letters*, *74*(14), 2059–2061. doi:10.1063/1.123756

Noheda, B., Cox, D. E., Shirane, G., Guo, R., Jones, B., & Cross, L. E. (2000). Stability of the monoclinic phase in the ferroelectric perovskite PbZr 1− x Ti x O 3. *Physical Review B: Condensed Matter*, *63*(1), 014103. doi:10.1103/PhysRevB.63.014103

Park, S. E., & Shrout, T. R. (1997). Ultrahigh strain and piezoelectric behavior in relaxor based ferroelectric single crystals. *Journal of Applied Physics*, *82*(4), 1804–1811. doi:10.1063/1.365983

Pham, K. N., Hussain, A., Ahn, C. W., Ill, W. K., Jeong, S. J., & Lee, J. S. (2010). Giant strain in Nb-doped Bi0. 5 (Na0. 82K0. 18) 0.5 TiO3 lead-free electromechanical ceramics. *Materials Letters*, *64*(20), 2219–2222. doi:10.1016/j.matlet.2010.07.048

Pinho, R., Vilarinho, R., Moreira, J. A., Zorro, F., Ferreira, P., Ivanov, M., Tkach, A., Costa, M. E., & Vilarinho, P. M. (2023). Stress induced effects on piezoelectric polycrystalline potassium sodium niobate thin films. *Journal of Materials Chemistry. C, Materials for Optical and Electronic Devices*, *11*(23), 7758–7771. doi:10.1039/D2TC05538D

Roberts, S. (1947). Dielectric and piezoelectric properties of barium titanate. *Physical Review*, *71*(12), 890–895. doi:10.1103/PhysRev.71.890

Rojac, T., Bencan, A., Malic, B., Tutuncu, G., Jones, J. L., Daniels, J. E., & Damjanovic, D. (2014). BiFeO 3 ceramics: Processing, electrical, and electromechanical properties. *Journal of the American Ceramic Society*, *97*(7), 1993–2011. doi:10.1111/jace.12982

Rubio-Marcos, F., Marchet, P., Merle-Méjean, T., & Fernandez, J. F. (2010). Role of sintering time, crystalline phases and symmetry in the piezoelectric properties of lead-free KNN-modified ceramics. *Materials Chemistry and Physics*, *123*(1), 91–97. doi:10.1016/j.matchemphys.2010.03.065

Rybyanets, A. N. (2011). Porous piezoceramics: Theory, technology, and properties. *IEEE Transactions on Ultrasonics, Ferroelectrics, and Frequency Control*, *58*(7), 1492–1507. doi:10.1109/TUFFC.2011.1968 PMID:21768033

Saito, Y., & Takao, H. (2006). High performance lead-free piezoelectric ceramics in the (K, Na) NbO3-LiTaO3 solid solution system. *Ferroelectrics*, *338*(1), 17–32. doi:10.1080/00150190600732512

Saito, Y., Takao, H., Tani, T., Nonoyama, T., Takatori, K., Homma, T., Nagaya, T., & Nakamura, M. (2004). Lead-free piezoceramics. *Nature*, *432*(7013), 84–87. doi:10.1038/nature03028 PMID:15516921

Saito, Y., Takao, H., Tani, T., Nonoyama, T., Takatori, K., Homma, T., & Nakamura, M. (2004). Lead-free piezoceramics. *Nature*, *432*(7013), 84–87. doi:10.1038/nature03028 PMID:15516921

Sawaguchi, E. (1953). Ferroelectricity versus antiferroelectricity in the solid solutions of PbZrO3 and PbTiO3. *Journal of the Physical Society of Japan*, *8*(5), 615–629. doi:10.1143/JPSJ.8.615

Sawaguchi, E., Shirane, G., & Takagi, Y. (1951). Phase transition in lead zirconate. *Journal of the Physical Society of Japan*, *6*(5), 333–339. doi:10.1143/JPSJ.6.333

Shen, Z. Y., & Li, J. F. (2010). Enhancement of piezoelectric constant d33 in BaTiO3 ceramics due to nano-domain structure. *Journal of the Ceramic Society of Japan*, *118*(1382), 940–943. doi:10.2109/jcersj2.118.940

Shinekumar, K., & Dutta, S. (2015). High-temperature piezoelectrics with large piezoelectric coefficients. *Journal of Electronic Materials*, *44*(2), 613–622. doi:10.1007/s11664-014-3534-2

Shirane, G., Suzuki, K., & Takeda, A. (1952). Phase transitions in solid solutions of PbZrO3 and PbTiO3 (II) X-ray study. *Journal of the Physical Society of Japan*, *7*(1), 12–18. doi:10.1143/JPSJ.7.12

Singh, A. K., & Pandey, D. (2001). Structure and the location of the morphotropic phase boundary region in (1-x)[Pb(Mg1/3Nb2/3)O3]-xPbTiO3. *Journal of Physics Condensed Matter*, *13*(48), 931–936. doi:10.1088/0953-8984/13/48/102

Smolenskii, G. A., Isupov, V. A., Agranovskaya, A. I., & Krainik, N. N. (1960). New materials of AIIBIVOVI type. *Sov Phys Solid State*, *2*, 2982–2985.

Takenaka, T., & Nagata, H. (2005). Current status and prospects of lead-free piezoelectric ceramics. *Journal of the European Ceramic Society*, *25*(12), 2693–2700. doi:10.1016/j.jeurceramsoc.2005.03.125

Tan, Z., Xing, J., Jiang, L., Zhu, J., & Wu, B. (2017). Ga 2 O 3 doping and vacancy effect in KNN—LT lead-free piezoceramics. *Frontiers of Materials Science*, *11*(4), 344–352. doi:10.1007/s11706-017-0403-y

Tao, H., Wu, J., & Wang, H. (2016). Modification of strain and piezoelectricity in (K, Na) NbO3–(Bi, Na) HfO3 lead-free ceramics with high Curie temperature. *Journal of Alloys and Compounds*, *684*, 217–223. doi:10.1016/j.jallcom.2016.04.322

Tao, H., Wu, J., Zheng, T., Wang, X., & Lou, X. (2015). New (1 _ xÞ-K0:45Na0:55Nb0:96Sb0:04O3–xBi0:5Na0:5HfO3 lead-free ceramics: Phase boundary and their electrical properties. *Journal of Applied Physics*, *118*(4), 044102. doi:10.1063/1.4927281

Tao, H., Wu, W., & Wu, J. (2016). Electrical properties of holmium doped (K,Na)(Nb,Sb)O3–(Bi,Na) HfO3 ceramics with wide sintering and poling temperature range. *Journal of Alloys and Compounds*, *689*, 759–766. doi:10.1016/j.jallcom.2016.08.037

Thery, V., Bayart, A., Blach, J. F., Roussel, P., & Saitzek, S. (2015). Effective piezoelectric coefficient measurement of BaTiO3 thin films using the X-ray diffraction technique under electric field available in a standard laboratory. *Applied Surface Science*, *351*, 480–486. doi:10.1016/j.apsusc.2015.05.155

Thong, H. C., Zhao, C., Zhou, Z., Wu, C. F., Liu, Y. X., Du, Z. Z., Li, J.-F., Gong, W., & Wang, K. (2019). Technology transfer of lead-free (K, Na) NbO3-based piezoelectric ceramics. *Materials Today*, *29*, 37–48. doi:10.1016/j.mattod.2019.04.016

Uchino, K. (2015). Glory of piezoelectric perovskites. *Science and Technology of Advanced Materials*, *16*(4), 046001. doi:10.1088/1468-6996/16/4/046001 PMID:27877827

Uchino, K., & Giniewicz, J. R. (2003). *Micromechatronics*. CRC. doi:10.4324/9780203912645

Valant, M., Axelsson, A. K., & Alford, N. (2007). Peculiarities of a solid-state synthesis of multiferroic polycrystalline BiFeO3. *Chemistry of Materials*, *19*(22), 5431–5436. doi:10.1021/cm071730+

Vul, B. M. (1946). *High and ultrahigh dielectric constant materials*. Elektrichestvo. [inRussian]

Wainer, E. (1946). High titania dielectrics. *Transactions of the Electrochemical Society*, *89*(1), 331. doi:10.1149/1.3071718

Wang, D., Hussain, F., Khesro, A., Feteira, A., Tian, Y., Zhao, Q., & Reaney, I. M. (2017). Composition and temperature dependence of structure and piezoelectricity in (1−x)(K1−yNay)NbO3-x (Bi1/2Na1/2) ZrO3 lead-free ceramics. *Journal of the American Ceramic Society*, *100*(2), 627–637. doi:10.1111/jace.14589

Wang, J. B. N. J., Neaton, J. B., Zheng, H., Nagarajan, V., Ogale, S. B., Liu, B., ... Ramesh, R. (2003). Epitaxial BiFeO3 multiferroic thin film heterostructures. *Science, 299*(5613), 1719-1722.

Wang, R., Xie, J., Hanada, K., Matsusaki, K., Kawanaka, H., Bando, H., & Itoh, M. (2008). Enhanced piezoelectricity around the tetragonal/orthorhombic morphotropic phase boundary in (Na, K) NbO3-ATiO3 solid solutions. *Journal of Electroceramics*, *21*(1-4), 263–266. doi:10.1007/s10832-007-9136-0

Wang, S. Y., Qiu, X., Gao, J., Feng, Y., Su, W. N., Zheng, J. X., Yu, D. S., & Li, D. J. (2011). Electrical reliability and leakage mechanisms in highly resistive multiferroic La0. 1Bi0. 9FeO3 ceramics. *Applied Physics Letters*, *98*(15), 152902. doi:10.1063/1.3580604

Wang, X., Wu, J., Cheng, X., Zhang, B., Xiao, D., Zhu, J., Wang, X., & Lou, X. (2013). Rhombohedral–tetragonal phase boundary and electrical properties of new K0:48Na0:52Nb0:98Sb0:02O3–Bi0:5Na0:5ZrO3 lead-free piezoceramics. *Journal of Physics. D, Applied Physics*, *46*(49), 495305. doi:10.1088/0022-3727/46/49/495305

Wang, X., Wu, J., Xiao, D., Zhu, J., Cheng, X., Zheng, T., Zhang, B., Lou, X., & Wang, X. (2014). Giant piezoelectricity in potassium–sodium niobate lead-free ceramics. *Journal of the American Chemical Society*, *136*(7), 2905–2910. doi:10.1021/ja500076h PMID:24499419

Wang, X., Wu, J., Xiao, D., Zhu, J., Cheng, X., Zheng, T., & Wang, X. (2014). Giant piezoelectricity in potassium–sodium niobate lead-free ceramics. *Journal of the American Chemical Society, 136*(7), 2905-2910.

Wang, X., Wu, J., Zheng, T., Cheng, X., Zhang, B., Xiao, D., & Zhu, J. (2014). Phase structure, electrical properties, and stability of 0.96 (K0. 48Na0. 52) 1− xLixNbO3–0.04 Bi0. 5Na0. 5ZrO3 lead-free piezoceramics. *Current Applied Physics*, *14*(5), 809–813. doi:10.1016/j.cap.2014.03.016

Wang, Y., Wu, J., Xiao, D., Wu, W., Zhang, B., Wu, L., & Zhu, J. (2008). Microstructure and Electrical Properties of [(K0. 50Na0. 50) 0.95− xLi0. 05Agx](Nb0. 95Ta0. 05) O3 Lead-Free Ceramics. *Journal of the American Ceramic Society*, *91*(8), 2772–2775. doi:10.1111/j.1551-2916.2008.02529.x

Wang, Y. P., Zhou, L., Zhang, M. F., Chen, X. Y., Liu, J. M., & Liu, Z. G. (2004). Room-temperature saturated ferroelectric polarization in BiFeO 3 ceramics synthesized by rapid liquid phase sintering. *Applied Physics Letters*, *84*(10), 1731–1733. doi:10.1063/1.1667612

Wei, H., Wang, H., Xia, Y., Cui, D., Shi, Y., Dong, M., Liu, C., Ding, T., Zhang, J., Ma, Y., Wang, N., Wang, Z., Sun, Y., Wei, R., & Guo, Z. (2018). An overview of lead-free piezoelectric materials and devices. *Journal of Materials Chemistry. C, Materials for Optical and Electronic Devices*, *6*(46), 12446–12467. doi:10.1039/C8TC04515A

Wei, H., Wang, H., Xia, Y., Cui, D., Shi, Y., Dong, M., Liu, C., Ding, T., Zhang, J., Ma, Y., Wang, N., Wang, Z., Sun, Y., Wei, R., & Guo, Z. (2018). An overview of lead-free piezoelectric materials and devices. *Journal of Materials Chemistry. C, Materials for Optical and Electronic Devices*, *6*(46), 12446–12467. doi:10.1039/C8TC04515A

Wooster, W. A. (1953). Physical properties and atomic arrangements in crystals. *Reports on Progress in Physics*, *16*(1), 62–82. doi:10.1088/0034-4885/16/1/302

Wu, B., Wu, H., Wu, J., Xiao, D., Zhu, J., & Pennycook, J. (2016). Giant piezoelectricity and high curie temperature in nanostructured alkali niobate lead-free piezoceramics through phase coexistence. *Journal of the American Chemical Society*, *138*(47), 15459–15464. doi:10.1021/jacs.6b09024 PMID:27933925

Wu, B., Wu, H., Wu, J., Xiao, D., Zhu, J., & Pennycook, S. J. (2016). Giant piezoelectricity and high Curie temperature in nanostructured alkali niobate lead-free piezoceramics through phase coexistence. *Journal of the American Chemical Society*, *138*(47), 15459–15464. doi:10.1021/jacs.6b09024 PMID:27933925

Wu, B., Wu, J., Xiao, D., & Zhu, J. (2015). Modification of both d33 and TC in a potassium–sodium niobate ternary system. *Dalton Transactions (Cambridge, England)*, *44*(48), 21141–21152. doi:10.1039/C5DT03680A PMID:26598931

Wu, G., Wang, P., Cheng, J., Xiao, Q., & Zhu, G. (n.d.). *High-piezoelectricity (1-x)(K1-yNay)(Nb1-zSbz) O3 + xBi0.5(Na1-uKu)0.5ZrO3 lead-free ceramics and their preparation method*. Chinese Patent, CN103482977A

Wu, J. (2020). Perovskite lead-free piezoelectric ceramics. *Journal of Applied Physics*, *127*(19), 190901. doi:10.1063/5.0006261

Wu, J., Fan, Z., Xiao, D., Zhu, J., & Wang, J. (2016). Multiferroic bismuth ferrite-based materials for multifunctional applications: Ceramic bulks, thin films and nanostructures. *Progress in Materials Science*, *84*, 335–402. doi:10.1016/j.pmatsci.2016.09.001

Wu, J., Tao, H., Yuan, Y., Lv, X., Wang, X., & Lou, X. (2015). Role of antimony in the phase structure and electrical properties of potassium-sodium niobate lead-free ceramics. *RSC Advances*, *5*(19), 14575–14583. doi:10.1039/C4RA14271C

Wu, J., & Wang, Y. (2014). Two-step sintering of new potassium sodium niobate ceramics: A high d 33 and wide sintering temperature range. *Dalton Transactions (Cambridge, England)*, *43*(34), 12836–12841. doi:10.1039/C4DT01712A PMID:25051530

Wu, J., Wang, Y., Xiao, D., Zhu, J., & Pu, Z. (2007). Effects of Ag content on the phase structure and piezoelectric properties of (K0.44-xNa0.52Li0.04Agx)(Nb0.91Ta0.05Sb0.04)O3 lead-free ceramics. *Applied Physics Letters*, *91*(13), 132914. doi:10.1063/1.2793507

Wu, J., Xiao, D., Wang, Y., Zhu, J., Wu, L., & Jiang, Y. (2007). Effects of K/Na ratio on the phase structure and electrical properties of (KxNa0.96-xLi0.04)(Nb0.91Ta0.05Sb0.04)O3 lead-free ceramics. *Applied Physics Letters*, *91*(25), 252907. doi:10.1063/1.2827573

Wu, J., Xiao, D., & Zhu, J. (2015). Potassium–sodium niobate lead-free piezoelectric materials: Past, present, and future of phase boundaries. *Chemical Reviews*, *115*(7), 2559–2595. doi:10.1021/cr5006809 PMID:25792114

Wu, J., Xiao, D., & Zhu, J. (2015). Potassium–sodium niobate lead-free piezoelectric ceramics: Recent advances and perspectives. *Journal of Materials Science Materials in Electronics*, *26*(12), 9297–9308. doi:10.1007/s10854-015-3084-2

Wu, W., Xiao, D., Ma, J., Wu, J., Li, J., & Liang, W. (2014). Temperature stability, phase structure and electrical behavior of Li-modified 0.99 (K0. 48Na0. 52) NbO3–0.01 BiCoO3 piezoelectric ceramics. *Ceramics International*, *40*(1), 1133–1137. doi:10.1016/j.ceramint.2013.06.114

Wu, W., Xiao, D., Wu, J., Li, J., & Zhu, J. (2011). Piezoelectric properties of (K0. 474Na0. 474Li0. 052) (Nb0. 948Sb0. 052) O3–Co2O3 lead-free ceramics. *Journal of the Ceramic Society of Japan*, *119*(1392), 654–657. doi:10.2109/jcersj2.119.654

Wu, W., Xiao, D., Wu, J., Li, J., Zhu, J., & Zhang, B. (2012). Microstructure and electrical properties of relaxor (1− x)[(K0. 5Na0. 5) 0.95 Li0. 05](Nb0. 95Sb0. 05) O3–xBaTiO3 piezoelectric ceramics. *Ceramics International*, *38*(3), 2277–2282. doi:10.1016/j.ceramint.2011.10.079

Wu, W., Xiao, D., Wu, J., Liang, W., Li, J., & Zhu, J. (2011). Polymorphic phase transition-induced electrical behavior of BiCoO3-modified (K0. 48Na0. 52) NbO3 lead-free piezoelectric ceramics. *Journal of Alloys and Compounds*, *509*(29), L284–L288. doi:10.1016/j.jallcom.2011.05.004

Xiao, M., Xiao, D., Wang, Z., Huang, T., Wu, J., Li, F., Wu, B., & Zhu, J. (2014). Microstructure and piezoelectric properties of lead-free 0.95 (Na 0.5 K 0.5) NbO 3–0.05 (Bi 0.5 K 0.5) Zr 1− x Ti x O 3 ceramics. *Journal of Materials Science Materials in Electronics*, *25*(4), 1938–1941. doi:10.1007/s10854-014-1824-3

Xie, Y., Xing, J., Tan, Z., Xie, L., Cheng, Y., Wu, X., Han, R., Chen, Q., & Zhu, J. (2022). High mechanical quality factor and piezoelectricity in potassium sodium niobate ceramics. *Ceramics International*, *48*(5), 6565–6573. doi:10.1016/j.ceramint.2021.11.204

Xing, J., Tan, Z., Jiang, L., Chen, Q., Wu, J., Zhang, W., Xiao, D., & Zhu, J. (2016). Phase structure and piezoelectric properties of (1 _ xÞ- K0:48Na0:52Nb0:95Sb0:05O3–x(Bi0:5Na0:5Þ0:9(Li0:5Ce0:5Þ0:1 ZrO3 lead-free piezoelectric ceramics. *Journal of Applied Physics*, *119*, 034101. doi:10.1063/1.4940191

Xu, K., Li, J., Lv, X., Wu, J., Zhang, X., Xiao, D., & Zhu, J. (2016). Superior piezoelectric properties in potassium–sodium niobate lead-free ceramics. *Advanced Materials*, *28*(38), 8519–8523. doi:10.1002/adma.201601859 PMID:27441456

Yang, W. G., Zhang, B. P., Ma, N., & Zhao, L. (2012). High piezoelectric properties of BaTiO3–xLiF ceramics sintered at low temperatures. *Journal of the European Ceramic Society*, *32*(4), 899–904. doi:10.1016/j.jeurceramsoc.2011.10.054

Ye, Z. G., & Dong, M. (2000). Morphotropic domain structures and phase transitions in relaxor-based piezo-/ferroelectric (1−x)Pb(Mg1/3Nb2/3)O3−xPbTiO3 single crystals. *Journal of Applied Physics*, *87*(5), 2312–2319. doi:10.1063/1.372180

Yokosuka, M. (1977). A new transparent ferroelectric ceramic, lanthanum-modified lead-barium metaniobate (PBLN). *Japanese Journal of Applied Physics*, *16*(2), 379–380. doi:10.1143/JJAP.16.379

Yu, B., Li, M., Wang, J., Pei, L., Guo, D., & Zhao, X. (2008). Enhanced electrical properties in multiferroic BiFeO3 ceramics co-doped by La3+ and V5+. *Journal of Physics. D, Applied Physics*, *41*(18), 185401. doi:10.1088/0022-3727/41/18/185401

Yuan, Y., Wu, J., Tao, H., Lv, X., Wang, X., & Lou, X. (2015). Composition design and electrical properties in (1 _ yÞ(K0:40- Na0:60Þ0:985Li0:015(Nb1_xSbxÞO3–yBi0:5Na0:5ZrO3 lead-free ceramics. *Journal of Applied Physics*, *117*, 084103. doi:10.1063/1.4913284

Zhang, B., Wu, J., Cheng, X., Wang, X., Xiao, D., Zhu, J., & Lou, X. (2013). Lead-free piezoelectrics based on potassium-sodium niobate with giant d33. *ACS Applied Materials & Interfaces*, *5*(16), 7718–7725. doi:10.1021/am402548x PMID:23937157

Zhang, J. L., Ji, P. F., Wu, Y. Q., Zhao, X., Tan, Y. Q., & Wang, C. L. (2014). Strong piezoelectricity exhibited by large-grained BaTiO3 ceramics. *Applied Physics Letters*, *104*(22), 222909. doi:10.1063/1.4881597

Zhao, C., Wu, B., Thong, H. C., & Wu, J. (2018). Improved temperature stability and high piezoelectricity in lead-free barium titanate-based ceramics. *Journal of the European Ceramic Society*, *38*(16), 5411–5419. doi:10.1016/j.jeurceramsoc.2018.08.004

Zhao, X., Wang, H., Yuan, C., Xu, J., Cui, Y., & Ma, J. (2013). Effects of Co doping on microstructure and properties of (K0.5Na0.5)NbO3-LiSbO3-BiFe1-xCoxO3 lead-free piezoelectric ceramics. *Journal of Materials Science Materials in Electronics*, *24*(5), 1480–1484. doi:10.1007/s10854-012-0958-4

Zhao, Z. H., Ge, R. F., & Dai, Y. (2019). Large electro-strain signal of the BNT–BT–KNN lead-free piezoelectric ceramics with CuO doping. *Journal of Advanced Dielectrics*, *9*(03), 1950022. doi:10.1142/S2010135X1950022X

Zheng, T., Wu, H., Yuan, Y., Lv, X., Li, Q., Men, T., Zhao, C., Xiao, D., Wu, J., Wang, K., Li, J.-F., Gu, Y., Zhu, J., & Pennycook, S. J. (2017). The structural origin of enhanced piezoelectric performance and stability in lead free ceramics. *Energy & Environmental Science*, *10*(2), 528–537. doi:10.1039/C6EE03597C

Zheng, T., Wu, H., Yuan, Y., Lv, X., Li, Q., Men, T., Zhao, C., Xiao, D., Wu, J., Wang, K., Li, J. F., Gu, Y., Zhu, J., & Pennycook, S. J. (2017). The structural origin of enhanced piezoelectric performance and stability in lead free ceramics. *Energy & Environmental Science*, *10*(2), 528–537. doi:10.1039/C6EE03597C

Zheng, T., Wu, J., Cheng, X., Wang, X., Zhang, B., Xiao, D., Zhu, J., & Lou, X. (2014). Wide phase boundary zone, piezoelectric properties and stability in 0.97(K0:4Na0:6Þ(Nb1_xSbxÞO3–0.03Bi0:5Li0:5-ZrO3 lead-free ceramics. *Dalton Transactions (Cambridge, England)*, *43*(25), 9419. doi:10.1039/c4dt00768a PMID:24848676

Zheng, T., Wu, J., Xiao, D., Zhu, J., Wang, X., & Lou, X. (2015). Potassium-sodium niobate lead-free ceramics: Modified strain as well as piezoelectricity. *Journal of Materials Chemistry. A, Materials for Energy and Sustainability*, *3*(5), 1868–1874. doi:10.1039/C4TA05423G

Zheng, T., Wu, J., Xiao, D., Zhu, J., Wang, X., & Lou, X. (2015). Composition-driven phase boundary and piezoelectricity in potassium–sodium niobate-based ceramics. *ACS Applied Materials & Interfaces*, *7*(36), 20332–20341. doi:10.1021/acsami.5b06033 PMID:26302094

Zhou, J. J., Cheng, L. Q., Wang, K., Zhang, X. W., Li, J. F., Liu, H., & Fang, J. Z. (2014). Low-temperature sintering of (K, Na) NbO3-based lead-free piezoceramics with addition of LiF. *Journal of the European Ceramic Society*, *34*(5), 1161–1167. doi:10.1016/j.jeurceramsoc.2013.11.029

Zhu, L. F., Zhang, B. P., Li, S., & Zhao, G. L. (2017). Large piezoelectric responses of Bi (Fe, Mg, Ti) O3-BaTiO3 lead-free piezoceramics near the morphotropic phase boundary. *Journal of Alloys and Compounds*, *727*, 382–389. doi:10.1016/j.jallcom.2017.08.014

Zhu, L. F., Zhang, B. P., Li, S., Zhao, L., Wang, N., & Shi, X. C. (2016). Enhanced piezoelectric properties of Bi (Mg1/2Ti1/2) O3 modified BiFeO3–BaTiO3 ceramics near the morphotropic phase boundary. *Journal of Alloys and Compounds*, *664*, 602–608. doi:10.1016/j.jallcom.2016.01.003

Zuo, R., Fang, X., & Ye, C. (2007). Phase structures and electrical properties of new lead-free (Na0. 5K0. 5) NbO3–(Bi0. 5Na0. 5) TiO3 ceramics. *Applied Physics Letters*, *90*(9), 092904. doi:10.1063/1.2710768

Zuo, R., Fu, J., Lv, D., & Liu, Y. (2010). Antimony tuned rhombohedral-orthorhombic phase transition and enhanced piezoelectric properties in sodium potassium niobate. *Journal of the American Ceramic Society*, *93*(9), 2783–2787. doi:10.1111/j.1551-2916.2010.03804.x

Zuo, R., Wang, H., Ma, B., & Li, L. (2009). Effects of Nb 5+ doping on sintering and electrical properties of lead-free (Bi 0.5 Na 0.5) TiO 3 ceramics. *Journal of Materials Science Materials in Electronics*, *20*(11), 1140–1143. doi:10.1007/s10854-008-9840-9

Chapter 6
Halide Perovskite Materials for Photovoltaics and Lighting

Veena Thakur
Barkatullah University, Bhopal, India

Sadhna Singh
Barkatullah University, Bhopal, India

ABSTRACT

Halide perovskites have recently emerged as promising materials for low-cost, high-efficiency solar cells. The unique optoelectronic properties of perovskites include high optical absorption, high carrier mobility, long diffusion lengths, and unique ambipolar charge transport properties. Over the last two decades, halide perovskites (HPs) have been identified as one of the most promising materials in photovoltaic and light-emitting devices. Halide perovskites can be synthesized easily. Therefore, these materials are expected to be highly scalable for a variety of applications. This will require extensive research, from the basic properties to practical devices. This chapter makes an effort to highlight the advancements made in the perovskite materials for Photovoltaics cell and lighting in the recent years.

INTRODUCTION

The perovskite crystal family is a group of materials that have been attracting attention in recent years due to their exceptional properties and potential applications in nanotechnology especially nanostructured solar cells. 'Perovskite' is a term used to describe a group of materials that have a distinctive crystal structure of cuboid and diamond shapes. They have long been of interest for their superconducting, electronic and ferroelectric properties. They are also remarkably efficient at absorbing photons of light and that this can be converted into an electric current, research interest in perovskite solar cells has skyrocketed.

A perovskite is any material with the same type of crystal structure as calcium titanium oxide ($CaTiO_3$), known as the perovskite structure. Perovskite take their name from the mineral, which was first discovered in the Ural Mountains of Russia by Gustav Rose in 1839 and is named after Russian mineralogist L.A. Perovski (1792-1856).

DOI: 10.4018/979-8-3693-1306-0.ch006

Figure 1. Perovskite structure ABX_3

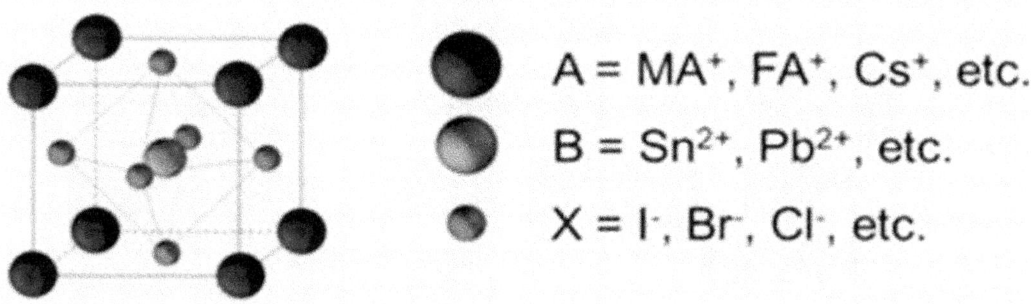

These materials have the potential to significantly improve the efficiency and cost-effectiveness of solar energy production, making them a highly promising area of research for scientists and engineers alike. With their unique physical and chemical characteristics, perovskites have the potential to revolutionize the way we harness and utilize solar energy.

Recently halide perovskites have attracted tremendous attention from many researchers, particularly for their excellent optoelectronic properties in applications such as photovoltaic solar cells and light emitting diodes. The family of hybrid organic-inorganic halide perovskites has the general composition of ABX_3, where the A-site cation is either methylammonium (MA), formamidinium (FA), or cesium, the B-site cation is lead or tin, and the anion (X) is iodide, bromide, or chloride as shown in Figure 1. The 'A' atoms are larger than the 'B' atoms. Dieter Weber first reported the three-dimensional (3D) perovskite structure of methylammonium lead iodide ($MAPBI_3$) in 1978 (Weber, 1978). There were a few other researchers who were investigating the optical properties of these materials and its feasibility for transistors and other electronics (Kagan et al., 1999; Mayshiyama & Kurihara, 1998; Kitazawa et al., 2002; Condeles et al., 2006). The material was further brought to light in 2009, when Miyasaka reported a 3.81% efficient dye-sensitized solar cell with $MAPBI_3$ coated on TiO2 surface (Teshima et al., 2009). This material was later heavily reported by various research groups across the globe due to its potential to transform photovoltaic technology and replace silicon (Takahashi et al., 2011; Snaith, 2010; Snaith & Gratzel, 2006; Im, Lee, Lee, Park & Park (2011). The hybrid organic–inorganic perovskites (HOIP) have strong intermolecular hydrogen bonds between the amino and halide group ions, whereas the weak vander waals exists among the organic ions.

The divalent metal ions function as the best metal cations for organic–inorganic framework. Among these combinations, those belonging to group 14 (including Sn^{2+} and Pb^{2+}) attracted more interest due to their good optoelectronic properties and potential for low-temperature device fabrication (Gao et al., 2014; Koutselas et al., 1996; Mitzi, Field, Harrison et al, 1994). These materials have been used in a wide variety of optoelectronic devices (Wang & Kim, 2017; Wang et al., 2018; Wei & Xing, 2019), but have had the most disruptive impact on photovoltaic technology (Huang, Yuan, Shao, & Yan, 2017; Jena et al., 2019; Stranks & Snaith, 2015). Power conversion efficiencies (PCEs) as high as 25% have now been reported for perovskite solar cells (PSCs) (Best, n.d.; Green et al., 2020), owing in large part to the excellent optoelectronic properties of lead halide perovskites. These include high absorption coefficients, direct band gaps, high charge carrier mobilities, long charge carrier diffusion lengths and a high degree of defect tolerance (Dong, Fang, Shao, Mulligan, Qiu, Cao, & Huang, 2015; Huang, Yuan,

Figure 2. Classification of perovskite materials

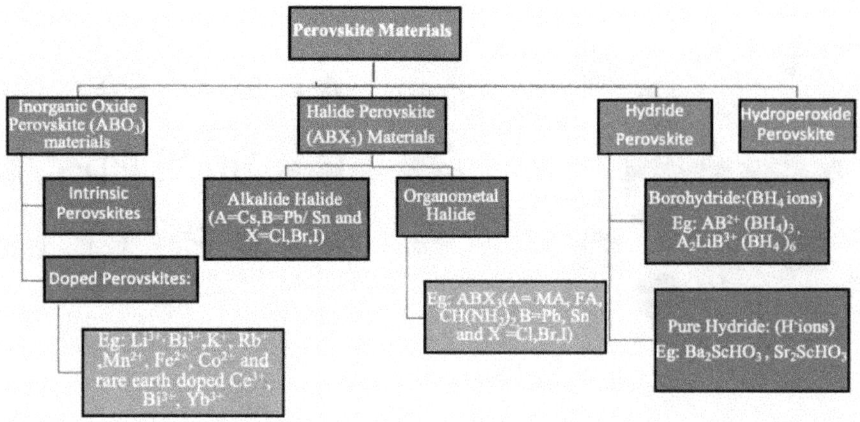

Shao, & Yan, 2017; Steirer et al., 2016; Stoumpos et al., 2013; Stranks et al., 2013; Walsh et al., 2015; Yin et al., 2014). However, in order for solar technology to be viable, three key requirements have to be met: high-performance, long lifetime, and low cost. Although PSCs have excellent performance and are projected to have low manufacturing costs (Song et al., 2017) and short energy payback times (Gong et al., 2015), long-term device stability remains a challenge (Rong et al., 2018). Perovskites face many challenges with respect to their stability (Boyd et al., 2019; Meng et al., 2018; Wang et al., 2019).

ABOUT THE MATERIAL

Halide perovskites (HPs), with an excellent photoactive nature, dielectric, piezoelectric, ferroelectric, and pyroelectric properties, have been potential candidates for obtaining flexible nanogenerator-based self-powered sensors including light, pressure, and temperature. Additionally, the photo-stimulated dielectric, piezoelectric, and triboelectric properties of HPs make them efficient entrants for developing bimodal and multimode sensors to sense multi-physical signals individually or simultaneously. Figure 2 depicts the classification of the perovskite system.

In the past decade, hybrid organic –inorganic perovskite (HOIP) have emerged as the exotic materials for the futuristic photovoltaics. The viability of low temperature, solution-processed manufacturing and a unique blend of electronic and optical properties, has further indicated its goals towards a potential commercialization. The Overview of layered and hybrid organic–inorganic metal halide perovskites is shown in Figure 3.

In the early 1990s, David et.al (Kagan et al., 1999; Mitzi, 1999; Mitzi, 2001; Mitzi et al., 2001; Mitzi et al., 1999) have extensively explored the structural dynamics of methylammonium lead iodide perovskite ($MAPBI_3$) materials engineered for niche applications such as electronics, thin film transistors, photovoltaics (PV) and light emitting diodes. In 2002, perovskite oxide $Pr_{0.7}Ca_{0.3}MnO_3$ was firstly utilized in a 64-bit RRAM array by a 500-nm complementary metal oxide semiconductor (CMOS) process (Hong et al., 2018). More than two decades passed, transition metal perovskite oxides are still one of the best materials for resistive random access memory (RRAM) active layers. Perovskite oxides possess stable

Figure 3. Overview of layered and hybrid organic–inorganic metal halides. (a) Hybrid organic–inorganic perovskite structure, (b) layered (2D) hybrid organic–inorganic metal halides, (c) 1D metal halides, (d) 0D hybrid organic–inorganic metal halides

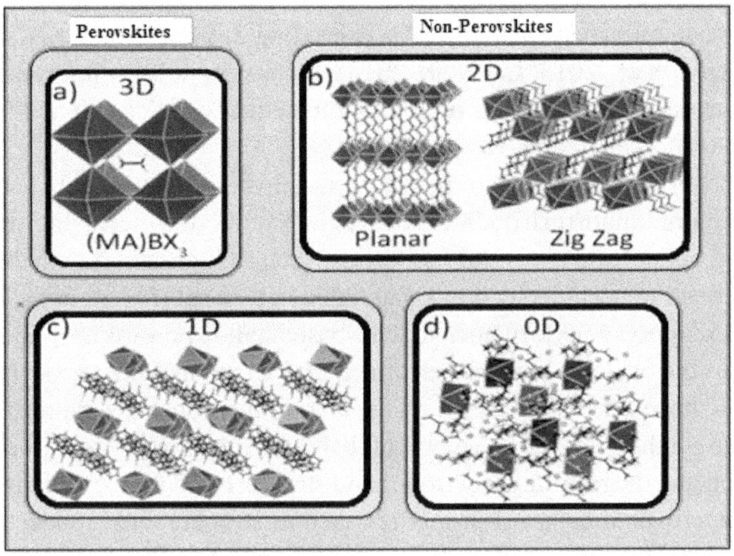

crystal structure with high defect tolerance and structure flexibility, which enables the accommodation of nonstoichiometric ions.

The nonstoichiometric ions contribute to the local ionic migration and thermochemical reaction therefore allowing for the resistance switching (RS) conversion. In addition, the strongly correlated electrons in perovskite oxides provide various electronic phases and bring out multifunctionality, e.g., colossal magnetoresistance/electroresistance, ferroelectricity, multiferroics, superconductivity, etc. (Jeong et al., 2012). Metal-insulator transition (MIT) effect has been found in many perovskite oxides, in which the electronic charges are injected into the insulating material to induct the current with an external bias (Asamitsu et al., 1997; Kim et al., 2006; Meijer et al., 2005). $Pr_{1-x}Ca_xMnO_3$ (PCMO), now one of the most developed memory materials for neuromorphic computing, is taken as an example, in which resistive switching behavior was first discovered in late 1990s (Asamitsu et al., 1997). In recent years, halide perovskites (HPs) have become a star material due to its excellent optical and charge transport properties. The rapid advance in power-conversion efficiency (PCE) of perovskite solar cells has exceeded by 20% (Green et al., 2018), and the simple and solution-based preparation enables low-cost production.

Halide perovskites received little attention until layered organometal halide perovskites were reported to exhibit a semiconductor-to-metal transition with increasing dimensionality (Mitzi, Feild, Harrison et al, 1994). In addition to changes in electrical properties, the band gap decreased with increased dimensionality from 2D to 3D. A narrow band gap is beneficial for solar cell applications. Miyasaka et al. (Kojima, Teshima, Shirai, & Miyasaka, 2009) applied 3D perovskite $CH_3NH_3PbX_3$ (X = Br, I) as an inorganic sensitizer in dye-sensitized solar cells in 2009. They demonstrated a power conversion efficiency (PCE) of 3.1% for X = Br and 3.8% for X = I. Park et al. (Im, Lee, Lee, Park, & Park, 2011) reported PCE of 6.5% using $CH_3NH_3PbI_3$. At the given TiO_2 film thickness of about 3.6 mm, $CH_3MH_3PbI_3$ perovskite showed an absorption coefficient that was 10 times greater than that of the conventional ruthenium-based molecular dye. Park et al. first reported long- term stable perovskite solar cells with PCE as high as 9.7% (Kim, Lee, Im, Lee, Moehl, Marchioro, Moon, Humphry-Baker, Yum, Moser, Grätzel, & Park, 2012). A non- sensitization type perovskite solar cell was proposed by Snaith et al. (Lee et al., 2012) in which the mixed halide perovskite $CH_3NH_3PbI_{3\_x}Cl_x$-coated Al_2O_3 film demonstrated PCE of 10.9%. Rapid progress was made over the year, and PCEs exceeding 15% were achieved using organolead halide perovskite (Burschka et al., 2013; Liu et al., 2013). Perovskite solar cell technology was selected as one of the biggest scientific breakthroughs of 2013 by the editors of Science and Nature (2013; Nature News Features, 2013).

Inorganic–organic hybrid perovskite materials such as methylammonium lead triiodide, formamidinium lead triiodide, and formamidinium tin triiodide are especially attracting increasing attention owing to their unique and interesting optical and electrical properties, such as high molar extinction a lone-pair state. Significant hybridization the occupied cation s and halogen p states (for example, overlap between Pb-6s and I-6p) endows exceptional performance to lead-based halide perovskites. The Fabini et al. (Fabini et al., 2020) introduces the $6s^2$ ($5s^2$) lone pair electrons on Pb (Sn) and the stereochemistry thereof. The lone pairs are shown to have a decisive impact on the electronic properties, directly yielding a favorable broad valence band and light hole states. They establish a connection between the electronic composition of the metal cation and the resulting electron- band structure and lattice dynamics of the compound.

HOIPs that inherit a diversified set of properties such as stronger and broader absorption, ultra-fast charge transportation, high dielectric, and swift charge recombination makes it the best suitable candidate material for PV (De Wolf et al., 2014; Kazim et al., 2014). Also, its ease of fabrication, stability, cost, efficiency, and performance rivals the best thin-film PV technologies.

In the recent years, the $MAPBI_3$ has attracted a great deal of interest for applications in high-performance optoelectronic devices (Yongping, 2015; Zhao & Zhu, 2016) due to their ability to tune optical band-gaps from the violet to near infrared (Knutson et al., 2005), as well as their higher absorption coefficients (Frost et al., 2014; Kojima, Teshima, Shirai, & Miyasaka, 2009) and potential for long-range electron and hole transport (Xing et al., 2013). Moreover, the optical and electrical properties of HOIP single crystals are far superior to their thin film counterparts (Dong, Fang, Shao, Mulligan, Qiu, Cao, & Huang, 2015; Shi et al., 2015). So far, researchers have been able to grow perovskite structured single crystals of methylammonium lead triiodide ($CH_3NH_3PbI_3$) and methylammonium lead tribromide ($CH_3NH_3PbBr_3$) with both size and shape controlled capabilities (Maculan et al., 2015; Saidaminov et al., 2015).

These materials potential has increased the power conversion efficiencies of solar cells from 1 to 24.2% (Heo, Im, Noh, Mandal, Lim, Chang, Lee, Kim, Sarkar, Nazeeruddin, Grätzel, & Seok, 2013; Yang, Noh, Jeon et al, 2015). Despite these materials were gaining efficiencies, they were losing on the stability and durability side due to the anomalous hysteretic effects that is typically exhibited by ferroelectric materials. So, it became critical to understand the significance of hysteresis in these materials.

The structural formability of the 3D-ABX_3 Halide Perovskites (HPs) and their stability were determined by the Goldschmidt tolerance factor (t) and the octahedral factors (μ) (Goldschmidt, 1927; Kieslich et al., 2014). The tolerance factor, $t = (r_A + r_X)/\sqrt{2}(r_B + r_X)$, where r_A, r_B, and r_X represent the effective ionic radii of A, B, and X in the HP, respectively. If the value of 't' is close to 1, they prefer an ideal cubic structure, while they distort into a low-symmetry structure, when 't' is smaller than 1.

The r_B/r_X is defined as an octahedral factor (μ) and is directly correlated with a BX_6 octahedron, playing an important role in the multifunctional properties of HP materials. The tolerance factor values for HP materials were found to be within the 0.8–1.11 range, while the octahedral factor values were in the range of 0.44–0.90 (Assadi et al., 2017; Park, 2015). The calculated t values for cubic, orthorhombic,

Figure 4. Metal halide perovskite (a) cubic perovskites, (b) anti perovskites, (c) distorted perovskites, (d) vacant perovskites, (e) ordered perovskites, (f) vacancy ordered perovskites

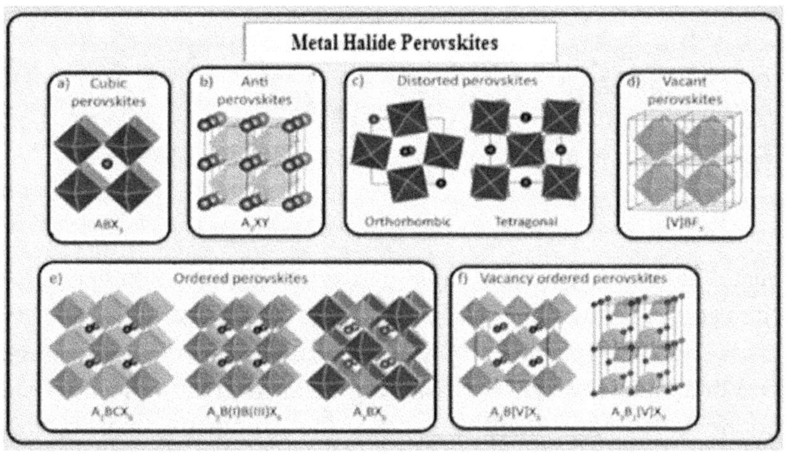

and hexagonal structured HPs were 0.8 < t < 1.0, t < 0.8, and t > 1, respectively (Goldschmidt, 1927; Li et al., 2016).

Metal halide perovskite films have exceptional optoelectronic features such as high absorption coefficient, tunable bandgap, long and balanced carrier diffusions, ambipolar transport of charge carriers, tolerance of defects (Xiao et al., 2016; Yamada et al., 2014), along with widely available raw materials and feasibility of film deposition via solution routes. The metal halide perovskites are shown in Figure 4.

In terms of grain size of metal-halide perovskite films, Huang et al. (Xiao et al., 2014) demonstrated that solvent annealing route can effectively increase the grain size and crystallinity of $MAPbI_3$ films, which resulted in much improved material electronic property and device performance. Similarly, Ohkita et al. (Kim, Ohkita, Benten et al, 2016) showed that the cells exhibited improved PV parameters with increase of grain size of $MAPbI_3$ films.

Figure 5. Overview of ternary metal halides not crystallizing in a perovskite structure

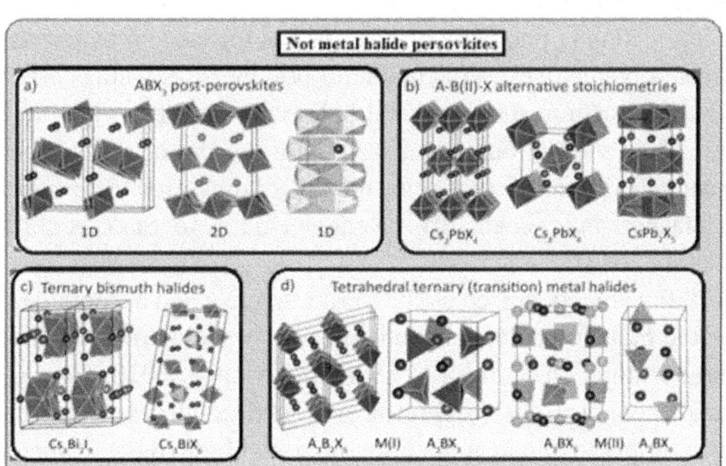

Figure 6. Antiperovskites of A_3BX_5 and AB_2X_5 phases

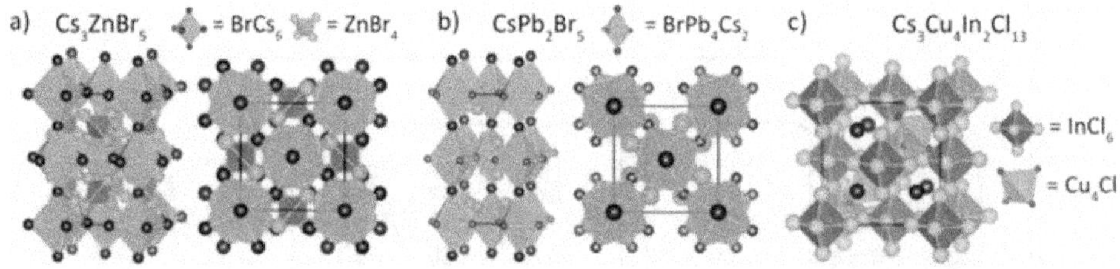

Not all ternary metal halides can form a stable perosvkite structure. In general, the HP framework collapses if the A-site cation is either too small or too large for the metal halide network (Stoumpos & Kanatzidis, 2015) As is shown in Figure 5, ABX_3 metal halides that cannot form geometrically stable perovskites crystallize in either 1D or 2D edge-sharing octahedral lattices (like $CsPbI_3$ or $RbPbBr_3$, in which the A-site cation would be too small to stabilize the hypothetical perovskite structure) or a 1D face-sharing crystal lattice (like $FAPbI_3$, in which the A-site cation would be too large for the perovskite structure). The structural stability of ternary metal halide perovskites depend not only on the size of the ions but also on temperature and pressure (Stoumpos & Kanatzidis, 2015). Antiperovskites of A_3BX_5 and AB_2X_5 phases shown in Figure 6.

APPLICATIONS

Over the last two decades, halide perovskites (HPs) have been identified as one of the most promising materials in photovoltaic and light-emitting devices (Correa-Baena et al., 2017; Lin, Xing, Quan, de Arquer et al, 2018). This has led to major breakthroughs in material science (Akkerman et al., 2018; Best Research-Cell Efficiency Chart, n.d.). The advances in perovskites for photovoltaic applications are shown in Figure 7.

Solar energy is one of the most promising alternatives to handle world energy demand, for which it is necessary to develop new materials and devices, such as solar cells, that allow it to be used more efficiently. The materials used for the manufacture of photovoltaic cells are those that convert photons into electrons through the photoelectric effect. This is why the search for materials that have the ability to absorbing light for application in photovoltaic devices has focused on materials ranging from metallic oxides, such as silicon, organic dyes, semiconductor polymers, as well as other chemical structures capable of performing this function, such as metal halide perovskites, which have a small band gap, charge mobility, low excitonic energy, long diffusion lengths, and high carrier mobility. Those materials are the basis for different generations of solar cells, the latter being used in perovskite solar cells (PSCs) (Chen et al., 2015; García-Fernandez et al., 2017; Nenon et al., 2016; Saski et al., 2019; Xu et al., 2019; Yang et al., 2017; Yu et al., 2017).

Halide perovskites (HPs) are recognized as the most attractive and acceptable materials in PV, thermoelectric, optoelectronic radiation detection, and emission optoelectronic applications owing to their remarkable optical and excitonic properties (Akkerman & Manna, 2020; Lin, Xing, Quan, García de Arquer et al, 2018; Orrea-Baena et al., 2017). Ge et al. evaluated recent developments in multidimensional metal halide perovskites for photovoltaic applications and highlighted the challenges (Ge et al., 2020).

Figure 7. Advances in perovskites for photovoltaic applications

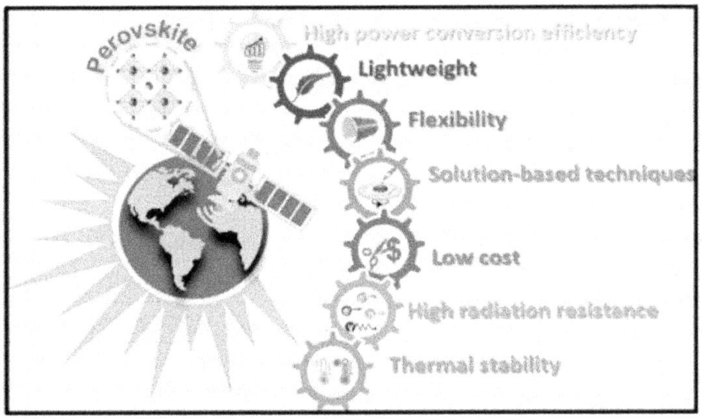

In particular, the dielectric and piezoelectric properties of HP-based materials are remarkable and nearly similar to those of ceramic materials (Ippili et al., 2018; Ippili et al., 2021; Jella et al., 2019; Jella et al., 2020; Kim, Dang, Choi et al, 2016; Yang, Gregori, Pellet et al, 2015). Many researchers have demonstrated the dielectric properties of HP-based materials and also systematically examined the changes in dielectric properties of compositionally tuned HP materials (Ippili et al., 2018; Jella et al., 2019; Jella et al., 2020; Yang, Gregori, Pellet et al, 2015). HP materials are the potential entrants to develop bi/multifunctional sensor devices in facile ways owing to the combined properties such as optoelectronic, piezo/ferroelectric, pyroelectric, and thermoelectric properties (Filip et al., 2014; Kim, Dang, Choi et al, 2016; Rakita et al., 2017; Röhm et al., 2017). Figure 8 shown the uses of perovskite materials.

Rapid advances in perovskite materials have increased the efficiency of solar cells from 3.8% to 22% in a very short period of time (Heo, Im, Noh, Mandal, Lim, Chang, Lee, Kim, Sarkar, Nazeeruddin, Grätzel, & Seok, 2013; Im, Lee, Lee, Park, & Park, 2011; Zhou et al., 2014). The advantages and optoelectronics applications of lead based perovskite materials are shown in Figure 9.The extensive de-

Figure 8. Uses of perovskite materials

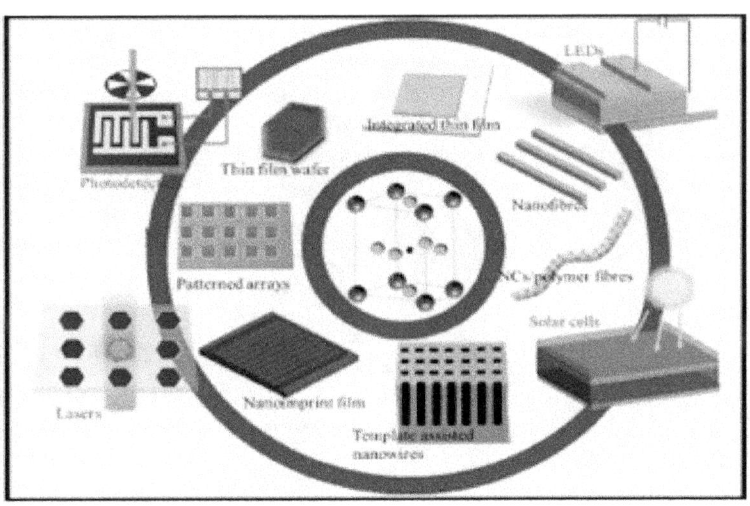

Table 1. Progress in hybrid perovskite solar cells and power conversion efficiency

S.No.	Hybrid Perovskite	HTM	Device Structure	Efficiency %
1	$CH_3NH_3PbI_3$	I^-A^{-3}	$CH_3NH_3PbI_3$ Quantum dots (QD)/TiO_2 substrate	6.5
2	$CH_3NH_3PbI_3$	CuI	$TiO_2/CH_3NH_3PbI_3$/spiroOMeTAD	6.0
3	$CH_3NH_3PbI_{3-x}CI_x$	NiO	FTO/NiO/CuSCN/$CH_3NH_3PbI_{3-x}CI_x$/ PCBM/Ag	7.3
4	$CH_3NH_3PbI_3$	CuSCN	$TiO_2/CH_3NH_3PbI_3$/CuSCN/Au	12.4
5	$CH_3NH_3PbI_{3-x}Br_x$	PTAA	$TiO_2/CH_3NH_3PbI_{3-x}Br_x$/DMSO/PTAA/Au	16.2
6	$CH_3NH_3PbI_3$	NiO_x	ITO/$NiO_x/CH_3NH_3PbI_3$/PCBM/Ag	16.47
7	$CH_3NH_3PbI_{3-x}CI_x$	spiroOMeTAD	ITO/PEIE/$TiO_2CH_3NH_3PbI_{3-x}CI_x$/ spiroOMeTAD/Au	19.32

velopments of methylammonium (MA) lead halide perovskites with respect to efficiency, as well as the testing of various HTMs, have resulted in significant advancements in thin-film optoelectronics. Table 1 describes the development of hybrid perovskite cells with PCE utilizing various HTMs (Christians et al., 2014; Im, Lee, Lee, Park, & Park, 2011; Jeon et al., 2014; Qin et al., 2014; Subbiah et al., 2014; Yin et al., 2016; Zhou et al., 2014).

Methylammonium lead iodide ($MAPbI_3$) perovskite has been most widely studied and used for solar cell applications. $CH_3NH_3PbI_3$ has higher absorption, which is an essential requirement for any solar material. Furthermore, the bandgap of this material is 1.55 eV. This feature, along with an extinction coefficient of 1.5 × 105 (mol/L)−1.cm−1 at about 550 nm, makes $CH_3NH_3PbI_3$ an excellent solar-radiation-absorbing material (Kim et al., 2013) and provides an exceptional quantum efficiency in the solar cell at 800 nm. Several studies have been conducted on the ambipolar behavior of $CH_3NH_3PbI_3$ perovskite. The optical and electrical characteristics of methylammonium lead iodide ($MAPbX_3$) perovskite

Figure 9. Advantages and optoelectronics applications of lead based perovskite materials

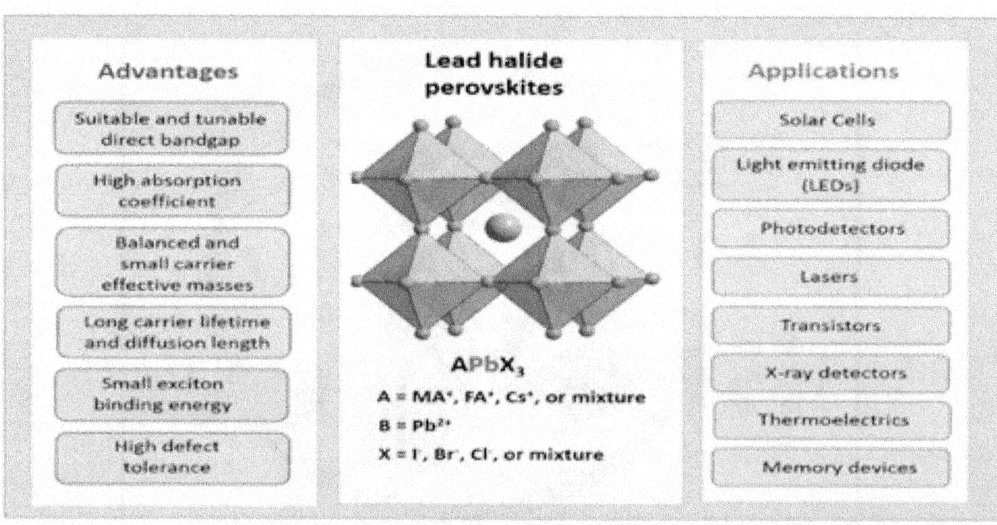

Table 2. Classification of perovskite materials based on their properties

Perovskite	Characterstics	Behavior Response
$CsPbI_3$	Bandgap	Spin-orbit coupling approximation 1.16 eV can be reduced to 0.39 eV
	Binding energy	169 eV
	Photoluminescence	λ_{em} = 680 nm, λ_{ex} = 365 nm
$CH_3NH_3PbBr_3$	Absorption	800 - 550 nm
	Bandgap	2.2eV
	Binding energy	150meV
	Photoluminescence	Emission at shorter wavelength owing to larger bandgap λ_{em} = 560 nm
$CH_3NH_3PbI_3$	Absorption	800 nm to complete visible spectrum
	The thickness of the film	600 nm
	Bandgap	1.55 eV comparable with the optical bandgap 1.4 eV
	Binding energy	50 meV
	Photoluminescence	Emission at longer wavelength owing to small bandgap (λ_{em} = 770nm for λ_{ex} = 546 nm)
Ambipolar behavior	Mesoporous TiO_2/ $CH_3NH_3PbI_3$	p-type charge transport. Solar efficiency of 5.5%
	Mesoporous ZnO_2/ $CH_3NH_3PbI_3$	n-type charge transport. Solar efficiency of 10.8%
	Dispersion range	130 nm and 100 nm: 1100 nm and 1200 nm in $CH_3NH_3PbI_{3-x}CI_x$

are remarkable (Baikie et al., 2013; Chen et al., 2014). Table 2 shows the classification of perovskite materials based on their properties.

Advancements in perovskite and an increase in conversion efficiency from 3.8% to 25.8% over the decade can be credited to the frantic research and mass production of perovskite materials. Recently, highly stabilized solar cells with negligible voltage loss have been explored (Yoo et al., 2019). A perovskite device with long-term stability has displayed a PCE of approximately 25.2%. According to numerous reports, this value is close to the C-Si efficiency of 26.2%, which has dominated the market.

In the last decade, perovskite solar cells based on formamidinium lead iodide ($FAPbI_3$) have attracted much attention. $FAPbI_3$ perovskites are believed to have higher thermal stability and a more ideal bandgap of 1.48 eV (Juarez-Perez et al., 2019). These materials have bandgap energies suitable for use in single-junction or multi-junction solar cells and show improved stability compared to their MA-based counterparts (Yadegarifard et al., 2023). This structure has a PCE of 25.7% (Li et al., 2023; National Renewable Energy Laboratory, n.d.). Although $FAPbI_3$ perovskites have a preferred bandgap below 1.55 eV for solar cell applications, they suffer from operational instability (Zhao et al., 2018).

Wang et al. developed novel $FA_{0.026}MA_{0.974}PbI_3$−yCly-Cu:NiO (formamidinium methylammonium (FAMA)-perovskite-Cu:NiO) and Al_2O_3/Cu:NiO composites and used them to fabricate highly stable and efficient PSCs through fully ambient air processes (Wang et al., 2020).

Lead based halide perovskites are preferred material for mass production of solar panels due to their stable structure and simple and economical manufacturing procedure. However, some of the urgent issues associated with these materials such as lead toxicity, poor stability and atmospheric impacts. Therefore, Lead-free perovskite seems to be a suitable alternative to mitigate environmental concerns.

The modified compounds are required to have optoelectronic properties that should be on par with Pb-based perovskites and fulfill the criteria for mass production and commercialization. Tin (Sn), bismuth (Bi), antimony (Sb), germanium (Ge), double-halide perovskite materials, and chalcogenide perovskites could become favorable compounds in perovskite solar cell applications.

FUTURE

In recent years, perovskite materials and technology have taken a step toward manufacturing and have explored the market for the commercialization of perovskite solar cell devices. Wang et al. opined that the mass production of PSCs is not yet feasible due to various factors (Wang et al., 2021). Challenges faced by perovskite solar cells and modules are (i) high-quality perovskite layer fabrication with low cost and large-scale reproducibility, (ii) the choice and deposition of charge extraction layers, (iii) the fabrication of bottom and top electrodes using low-cost processes, (iv) recombination, optical, and resistance losses, and (v) lower efficiency of large cell modules. These issues need to be addressed to build small- and large-area PSC devices. In a review report, Yang et al. presented the advancement of efficient large-area PVs with stability for commercialization (Yang et al., 2021). Recently, the use of computational methods and artificial intelligence (AI), referred to as machine learning (ML), has attracted worldwide attention for the prediction and assessment of perovskite materials (Wu et al., 2018). Future device performance enhancements are hampered by substantial charge recombination at the interface of the perovskite film and charge transport layers. Therefore, any further improvement in photovoltaic performance now requires a better understanding of the elementary physical processes governing energy conversion, such as charge carrier generation, transport, and recombination dynamics. To create high-efficiency perovskite cells, many strategies, such as additive engineering, defect passivation, interface engineering, and transmission material optimization, have been proposed. Chen and Park showed that trap-assisted nonradiative recombination, called Shockley–Read–Hall (SRH), in perovskite films and interface recombination are primarily responsible for the perovskite cell efficiency (Chen & Park, 2019).

CONCLUSION

The challenges encountered in the commercialization of perovskite devices, such as material and structural stability, device stability under high temperatures and humidity, lifetime, and manufacturing cost. Because of their advantages, such as low cost, simple fabrication techniques, and excellent crystallinity, perovskite materials are very practical and competitive for fabrication.

Overall, perovskite solar cells have great potential for producing efficient and low-cost solar energy. We believe that their instability and the presence of toxic materials, along with other issues, are major market restraints. The development of solar cells that are cost-effective, efficient, and stable is necessary so that they can be widespread and accessible to the general population. In conclusion, it is indicated that the successful replacement of Pb metal, environmental concerns, cost-effectiveness, and industrialization issues must be addressed in order for perovskite solar cells to be widely accepted.

REFERENCES

Akkerman, Q. A., & Manna, L. (2020). What defines a halide perovskite? *ACS Energy Letters*, *5*(2), 604–610. doi:10.1021/acsenergylett.0c00039 PMID:33344766

Akkerman, Q. A., Rainò, G., Kovalenko, M. V., & Manna, L. (2018). Genesis, Challenges and Opportunities for Colloidal Lead Halide Perovskite Nanocrystals. *Nature Materials*, *17*(5), 394–405. doi:10.1038/s41563-018-0018-4 PMID:29459748

Asamitsu, A., Tomioka, Y., Kuwahara, H., & Tokura, Y. (1997). Current switching of resistive states in magnetoresistive manganites. *Nature*, *388*(6637), 50–52. doi:10.1038/40363

Assadi, M. K., Bakhoda, S., Saidur, R., & Hanaei, H. (2017). Recent progress in perovskite solar cells. *Renewable & Sustainable Energy Reviews*, *81*, 2812–2822. doi:10.1016/j.rser.2017.06.088

Baikie, T., Fang, Y., Kadro, J. M., Schreyer, M., Wei, F., Mhaisalkar, S. G., Graetzel, M., & White, T. J. (2013). Synthesis and crystal chemistry of the hybrid perovskite (CH_3NH_3) PbI_3 for solid-state sensitised solar cell applications. *Journal of Materials Chemistry. A, Materials for Energy and Sustainability*, *1*(18), 5628–5641. doi:10.1039/c3ta10518k

Best, R.-C. E. C. (n.d.). *The National Renewable Energy Laboratory*. https://www.nrel.gov/pv/cell-efficiency.html

Best Research-Cell Efficiency Chart. (n.d.). https://www.nrel.gov/pv/cell-efficiency.html

Boyd, C. C., Cheacharoen, R., Leijtens, T., & McGehee, M. D. (2019). Understanding degradation mechanisms and improving stability of perovskite photovoltaics. *Chemical Reviews*, *119*(5), 119, 3418–3451. doi:10.1021/acs.chemrev.8b00336 PMID:30444609

Burschka, J., Pellet, N., Moon, S.-J., Humphry-Baker, R., Gao, P., Nazeeruddin, M. K., & Grätzel, M. (2013). Sequential deposition as a route to high-performance perovskite-sensitized solar cells. *Nature*, *499*(7458), 316–319. doi:10.1038/nature12340 PMID:23842493

Chen, J., & Park, N. G. (2019). Causes and Solutions of Recombination in Perovskite Solar Cells. *Advanced Materials*, *31*(47), 1803019. doi:10.1002/adma.201803019 PMID:30230045

Chen, Q., Zhou, H., Hong, Z., Luo, S., Duan, H. S., Wang, H. H., Liu, Y., Li, G., & Yang, Y. (2014). Planar heterojunction perovskite solar cells via vapor-assisted solution process. *Journal of the American Chemical Society*, *136*(2), 622–625. doi:10.1021/ja411509g PMID:24359486

Chen, W., Wu, Y., Yue, Y., Liu, J., Zhang, W., Yang, X., Chen, H., Bi, E., Ashraful, I., Grätzel, M., & Han, L. (2015). Efficient and stable large-area perovskite solar cells with inorganic charge extraction layers. *Science*, *350*(6263), 944–948. doi:10.1126/science.aad1015 PMID:26516198

Christians, J. A., Fung, R. C. M., & Kamat, P. V. (2014). An inorganic hole conductor for organo-lead halide perovskite solar cells. Improved hole conductivity with copper iodide. *Journal of the American Chemical Society*, *136*(2), 758–764. doi:10.1021/ja411014k PMID:24350620

Condeles, J. F., Lofrano, R. C. Z., Rosolen, J. M., & Mulato, M. (2006). Stoichiometry, surface and structural characterization of lead iodide thin films. *Brazilian Journal of Physics*, *36*(2a), 320–323. doi:10.1590/S0103-97332006000300023

Correa-Baena, J. P., Saliba, M., Buonassisi, T., Grätzel, M., Abate, A., Tress, W., & Hagfeldt, A. (2017). Promises and Challenges of Perovskite Solar Cells. *Science*, *358*(6364), 739–744. doi:10.1126/science.aam6323 PMID:29123060

De Wolf, S., Holovsky, J., Moon, S.-J., Löper, P., Niesen, B., Ledinsky, M., Haug, F.-J., Yum, J.-H., & Ballif, C. (2014). Organometallic halide Perovskites: Sharp optical absorption edge and its relation to photovoltaic performance. *The Journal of Physical Chemistry Letters*, *5*(6), 1035–1039. doi:10.1021/jz500279b PMID:26270984

Dong, Q., Fang, Y., Shao, Y., Mulligan, P., Qiu, J., Cao, L., & Huang, J. (2015). Electron-hole diffusion lengths > 175 μm in solution-grown $CH_3NH_3PbI_3$ single crystals. *Science*, *347*(6225), 967–970. doi:10.1126/science.aaa5760 PMID:25636799

Fabini, D. H., Seshadri, R., & Kanatzidis, M. G. (2020). The underappreciated lone pair in halide perovskites underpins their unusual properties. *MRS Bulletin*, *45*(6), 467–477. doi:10.1557/mrs.2020.142

Filip, M. R., Eperon, G. E., Snaith, H. J., & Giustino, F. (2014). Steric engineering of metal-halide perovskites with tunable optical band gaps. *Nature Communications*, *5*(1), 5757. doi:10.1038/ncomms6757 PMID:25502506

Frost, J. M., Butler, K. T., Brivio, F., Hendon, C. H., van Schilfgaarde, M., & Walsh, A. (2014). Atomistic origins of high-performance in hybrid halide perovskite solar cells. *Nano Letters*, *14*(5), 2584–2590. doi:10.1021/nl500390f PMID:24684284

Gao, P., Grätzel, M., & Nazeeruddin, M. K. (2014). Organohalide lead perovskites for photovoltaic applications. *Energy & Environmental Science*, *7*(8), 2448–2463. doi:10.1039/C4EE00942H

García-Fernandez, A., Bermúdez-García, J. M., Castro-García, S., Llamas-Saiz, A. L., Artiaga, R., López-Beceiro, J., Hu, S., Ren, W., Stroppa, A., Sánchez-Andújar, M., & Señarís-Rodríguez, M. A. (2017). Phase transition, dielectric properties, and ionic transport in the $[(CH_3)_2NH_2]PbI_3$ organic-inorganic hybrid with 2H-hexagonal perovskite structure. *Inorganic Chemistry*, *56*(9), 4918–4927. doi:10.1021/acs.inorgchem.6b03095 PMID:28375611

Ge, C., Xue, Y. Z. B., Li, L., Tang, B., & Hu, H. (2020). Recent Progress in 2D/3D Multidimensional Metal Halide Perovskites Solar Cells. *Frontiers in Materials*, *7*, 601179. doi:10.3389/fmats.2020.601179

Goldschmidt, V. M. (1927). Krystallbau und chemische Zusammensetzung. Ber. Dtsch. *Chem. Ges*, *60*, 1263–1296.

Gong, J., Darling, S. B., & You, F. (2015). Perovskite photovoltaics: Life-cycle assessment of energy and environmental impacts. *Energy & Environmental Science*, *8*(7), 1953–1968. doi:10.1039/C5EE00615E

Green, M. A., Dunlop, E. D., Hohl-Ebinger, J., Yoshita, M., Kopidakis, N., & Ho-Baillie, A. W. Y. (2020). Solar cell efficiency tables (version 55). *Progress in Photovoltaics: Research and Applications*, *28*(1), 3–15. doi:10.1002/pip.3228

Green, M. A., Hishikawa, Y., Dunlop, E. D., Levi, D. H., Hohl-Ebinger, J., & Ho-Baillie, A. W. Y. (2018). Solar cell efficiency tables (version 52). *Progress in Photovoltaics: Research and Applications*, *26*(7), 427–436. doi:10.1002/pip.3040

Heo, J. H., Im, S. H., Noh, J. H., Mandal, T. N., Lim, C.-S., Chang, J. A., Lee, Y. H., Kim, H., Sarkar, A., Nazeeruddin, M. K., Grätzel, M., & Seok, S. I. (2013). Efficient inorganicorganic hybrid heterojunction solar cells containing perovskite compound and polymeric hole conductors. *Nature Photonics*, *7*(6), 486–491. doi:10.1038/nphoton.2013.80

Hong, X. L., Loy, D. J. J., Dananjaya, P. A., Tan, F., Ng, C. M., & Lew, W. S. (2018). Oxide-based RRAM materials for neuromorphic computing. *Journal of Materials Science*, *53*(12), 8720–8746. doi:10.1007/s10853-018-2134-6

Huang, J., Yuan, Y., Shao, Y., & Yan, Y. (2017). Understanding the physical properties of hybrid perovskites for photovoltaic applications. *Nature Reviews. Materials*, *2*(7), 17042. doi:10.1038/natrevmats.2017.42

Im, J.-H., Lee, C.-R., Lee, J.-W., Park, S.-W., & Park, N.-G. (2011). 6.5% efficient Perovskite quantum-dot-sensitized solar cell. *Nanoscale*, *3*(10), 4088–4093. doi:10.1039/c1nr10867k PMID:21897986

Ippili, S., Jella, V., Kim, J., Hong, S., & Yoon, S. G. (2018). Enhanced piezoelectric output performance via control of dielectrics in Fe_{2+}-incorporated $MAPbI_3$ perovskite thin films: Flexible piezoelectric generators. *Nano Energy*, *49*, 247–256. doi:10.1016/j.nanoen.2018.04.031

Ippili, S., Jella, V., Thomas, A. M., Yoon, C., Jung, J. S., & Yoon, S. G. (2021). ZnAl–LDH-induced electroactive β-phase and controlled dielectrics of PVDF for a high-performance triboelectric nanogenerator for humidity and pressure sensing applications. *Journal of Materials Chemistry. A, Materials for Energy and Sustainability*, *9*(29), 15993–16005. doi:10.1039/D1TA02966E

Jella, V., Ippili, S., Eom, J.-H., Pammi, S. V. N., Jung, J.-S., Tran, V.-D., Nguyen, V. H., Kirakosyan, A., Yun, S., Kim, D., Sihn, M. R., Choi, J., Kim, Y.-J., Kim, H.-J., & Yoon, S.-G. (2019). A comprehensive review of flexible piezoelectric generators based on organic-inorganic metal halide perovskites. *Nano Energy*, *57*, 74–93. doi:10.1016/j.nanoen.2018.12.038

Jella, V., Ippili, S., & Yoon, S. G. (2020). Halide (Cl/Br)-incorporated organic–inorganic metal trihalide perovskite films: Study and investigation of dielectric properties and mechanical energy harvesting performance. *ACS Applied Electronic Materials*, *2*(8), 2579–2590. doi:10.1021/acsaelm.0c00473

Jena, A. K., Kulkarni, A., & Miyasaka, T. (2019). Halide perovskite photovoltaics: Background, status, and future prospects. *Chemical Reviews*, *119*(5), 3036–3103. doi:10.1021/acs.chemrev.8b00539 PMID:30821144

Jeon, N. J., Noh, J. H., Kim, Y. C., Yang, W. S., Ryu, S. C., & Seok, S. I. (2014). Solvent engineering for high-performance inorganic–organic hybrid perovskite solar cells. *Nature Materials*, *13*(9), 897–903. doi:10.1038/nmat4014 PMID:24997740

Jeong, D. S., Thomas, R., Katiyar, R. S., Scott, J. F., Kohlstedt, H., Petraru, A., & Hwang, C. S. (2012). Emerging memories: Resistive switching mechanisms and current status. *Reports on Progress in Physics*, *75*(7), 076502. doi:10.1088/0034-4885/75/7/076502 PMID:22790779

Juarez-Perez, E. J., Ono, L. K., & Qi, Y. (2019). Thermal degradation of formamidinium based lead halide perovskites into *sym*-triazine and hydrogen cyanide observed by coupled thermogravimetry-mass spectrometry analysis. *Journal of Materials Chemistry. A, Materials for Energy and Sustainability*, *7*(28), 16912–16919. doi:10.1039/C9TA06058H

Kagan, C., Mitzi, D., & Dimitrakopoulos, C. (1999). Organic-inorganic hybrid materials as semiconducting channels in thin-film field-effect transistors. *Science*, *286*(5441), 945–947. doi:10.1126/science.286.5441.945 PMID:10542146

Kazim, S., Nazeeruddin, M. K., Grätzel, M., & Ahmad, S. (2014). Perovskite as light harvester: A game changer in photovoltaics. *Angewandte Chemie International Edition*, *53*(11), 2812–2824. doi:10.1002/anie.201308719 PMID:24519832

Kieslich, G., Sun, S., & Cheetham, A. K. (2014). Solid-state principles applied to organic–inorganic perovskites: New tricks for an old dog. *Chemical Science (Cambridge)*, *5*(12), 4712–4715. doi:10.1039/C4SC02211D

Kim, D. S., Kim, Y. H., Lee, C. E., & Kim, Y. T. (2006). Colossal electroresistance mechanism in a Au/Pr0.7Ca0.3MnO$_3$/Pt sandwich structure: Evidence for a Mott transition. *Physical Review B: Condensed Matter and Materials Physics*, *74*(17), 174430. doi:10.1103/PhysRevB.74.174430

Kim, H. D., Ohkita, H., Benten, H., & Ito, S. (2016). Photovoltaic performance of perovskite solar cells with different grain sizes. *Advanced Materials*, *28*(5), 917–922. doi:10.1002/adma.201504144 PMID:26639125

Kim, H.-S., Lee, C.-R., Im, J.-H., Lee, K.-B., Moehl, T., Marchioro, A., Moon, S.-J., Humphry-Baker, R., Yum, J.-H., Moser, J. E., Grätzel, M., & Park, N.-G. (2012). Lead iodide perovskite sensitized all-solid-state submicron thin film mesoscopic solar cell with efficiency exceeding 9%. *Scientific Reports*, *2*(1), 591. doi:10.1038/srep00591 PMID:22912919

Kim, H. S., Lee, J. W., Yantara, N., Boix, P. P., Kulkarni, S. A., Mhaisalkar, S., Gratzel, M., & Park, N. G. (2013). High efficiency solid-state sensitized solar cell-based on submicrometer rutile TiO_2 nanorod and $CH_3NH_3PbI_3$ perovskite sensitizer. *Nano Letters*, *13*(6), 2412–2417. doi:10.1021/nl400286w PMID:23672481

Kim, Y. J., Dang, T. V., Choi, H. J., Park, B. J., Eom, J. H., Song, H. A., Seol, D., Kim, Y., Shin, S. H., Nah, H., & Yoon, S.-G. (2016). Piezoelectric properties of $CH_3NH_3PbI_3$ perovskite thin films and their applications in piezoelectric generators. *Journal of Materials Chemistry. A, Materials for Energy and Sustainability*, *4*(3), 756–763. doi:10.1039/C5TA09662F

Kitazawa, N., Watanabe, Y., & Nakamura, Y. (2002). Optical properties of $CH_3NH_3PbX_3$ (X = halogen) and their mixed-halide crystals. *Journal of Materials Science*, *37*(17), 3585–3587. doi:10.1023/A:1016584519829

Knutson, J. L., Martin, J. D., & Mitzi, D. B. (2005). Tuning the band gap in hybrid tin iodide perovskite semiconductors using structural templating. *Inorganic Chemistry*, *44*(13), 4699–4705. doi:10.1021/ic050244q PMID:15962978

Kojima, A., Teshima, K., Shirai, Y., & Miyasaka, T. (2009). Organometal Halide Perovskites as Visible-Light Sensitizers for Photovoltaic Cells. *Journal of the American Chemical Society*, *131*(17), 6050–6051. doi:10.1021/ja809598r PMID:19366264

Koutselas, I. B., Ducasse, L., & Papavassiliou, G. C. (1996). Electronic properties of three- and low- dimensional semiconducting materials with Pb halide and Sn halide units. *Journal of Physics Condensed Matter*, *8*(9), 1217–1227. doi:10.1088/0953-8984/8/9/012

Lee, M. M., Teuscher, J., Miyasaka, T., Murakami, T. N., & Snaith, H. J. (2012). Efficient Hybrid Solar Cells Based on Meso-Superstructured Organometal Halide Perovskites. *Science*, *338*(6107), 643–647. doi:10.1126/science.1228604 PMID:23042296

Li, M., Sun, R., Chang, J., Dong, J., Tian, Q., Wang, H., Li, Z., Yang, P., Shi, H., Yang, C., Wu, Z., Li, R., Yang, Y., Wang, A., Zhang, S., Wang, F., Huang, W., & Qin, T. (2023). Orientated crystallization of FA-based perovskite via hydrogen-bonded polymer network for efficient and stable solar cells. *Nature Communications*, *14*(1), 573. doi:10.1038/s41467-023-36224-6 PMID:36732540

Li, Z., Yang, M., Park, J. S., Wei, S. H., Berry, J., & Zhu, K. (2016). Stabilizing perovskite structures by tuning tolerance factor: Formation of formamidinium and cesium lead iodide solid-state alloys. *Chemistry of Materials*, *28*(1), 284–292. doi:10.1021/acs.chemmater.5b04107

Lin, K., Xing, J., Quan, L. N., de Arquer, F. P. G., Gong, X., Lu, J., Xie, L., Zhao, W., Zhang, D., Yan, C., Li, W., Liu, X., Lu, Y., Kirman, J., Sargent, E. H., Xiong, Q., & Wei, Z. (2018). Perovskite Light-Emitting Diodes with External Quantum Efficiency Exceeding 20 Per Cent. *Nature*, *56*(7726), 245–248. doi:10.1038/s41586-018-0575-3 PMID:30305741

Lin, K., Xing, J., Quan, L. N., García de Arquer, F. P., Gong, X., Lu, J., Xie, L., Zhao, W., Zhang, D., & Yan, C. (2018). Perovskite light-emitting diodes with external quantum efficiency exceeding 20 per cent. *Nature*, *562*(7726), 245–248. doi:10.1038/s41586-018-0575-3 PMID:30305741

Liu, M., Johnston, M. B., & Snaith, H. J. (2013). Efficient planar heterojunction perovskite solar cells by vapour deposition. *Nature*, *501*(7467), 395–398. doi:10.1038/nature12509 PMID:24025775

Maculan, G., Sheikh, A. D., Abdelhady, A. L., Saidaminov, M. I., Haque, M. A., Murali, B., Alarousu, E., Mohammed, O. F., Wu, T., & Bakr, O. M. (2015). $CH_3NH_3PbCl_3$ single crystals: Inverse temperature crystallization and visible-blind UV-Photodetector. *The Journal of Physical Chemistry Letters*, *6*(19), 3781–3786. doi:10.1021/acs.jpclett.5b01666 PMID:26722870

Mashiyama, H., & Kurihara, Y. (1998). Disordered cubic Perovskite structure of CH_3 NH_3PbX_3 (X = Cl, Br, I). *Journal of the Korean Physical Society*, *32*, 156–158.

Meijer, G. I., Staub, U., Janousch, M., Johnson, S. L., Delley, B., & Neisius, T. (2005). Valence states of Cr and the insulator-to-metal transition in Cr-doped $SrTiO_3$. *Physical Review B: Condensed Matter and Materials Physics*, *72*(15), 155102. doi:10.1103/PhysRevB.72.155102

Meng, L., You, J., & Yang, Y. (2018). Addressing the stability issue of perovskite solar cells for commercial applications. *Nature Communications*, *9*(1), 5265. doi:10.1038/s41467-018-07255-1 PMID:30532038

Mitzi, D. B. (1999). *Synthesis, Structure, and roperties of Organic-Inorganic Perovskites and Related Materials*. John Wiley & Sons.

Mitzi, D. B. (2001). Templating and structural engineering in organic–inorganic perovskites. *Dalton Transactions (Cambridge, England)*, *1*, 1–12.

Mitzi, D. B., Chondroudis, K., & Kagan, C. R. (2001). Organic-inorganic electronics. *IBM Journal of Research and Development*, *45*(1), 29–45. doi:10.1147/rd.451.0029

Mitzi, D. B., Feild, C. A., Harrison, W. T. A., & Guloy, A. M. (1994). Conducting tin halides with a layered organic-based perovskite structure. *Nature*, *369*(6480), 467–469. doi:10.1038/369467a0

Mitzi, D. B., Prikas, M. T., & Chondroudis, K. (1999). Thin film deposition of organic-inorganic hybrid materials using a single source thermal ablation technique. *Chemistry of Materials*, *11*(3), 542–544. doi:10.1021/cm9811139

National Renewable Energy Laboratory. (n.d.). *Best Research-Cell Efficiencies*. Available online: https://www.nrel.gov/pv/assets/pdfs/best-research-cell efficiencies.20200925.pdf

Nature News Features. (2013). 365 days: Nature's 10. *Nature*, *504*(7480), 357–365. doi:10.1038/504357a PMID:24352276

Nenon, D. P., Christians, J. A., Wheeler, L. M., Blackburn, J. L., Sanehira, E. M., Dou, B., Olsen, M. L., Zhu, K., Berry, J. J., & Luther, J. M. (2016). Structural and chemical evolution of methylammonium lead halide perovskites during thermal processing from solution. *Energy & Environmental Science*, *9*(6), 2072–2082. doi:10.1039/C6EE01047D

Orrea-Baena, J.-P., Saliba, M., Buonassisi, T., Grätzel, M., Abate, A., Tress, W., & Hagfeldt, A. (2017). Promises and challenges of perovskite solar cells. *Science*, *358*(6364), 739–744. doi:10.1126/science.aam6323 PMID:29123060

Park, N. G. (2015). Perovskite solar cells: An emerging photovoltaic technology. *Materials Today*, *18*(2), 65–72. doi:10.1016/j.mattod.2014.07.007

Qin, P., Tanaka, S., Ito, S., Tetreault, N., Manabe, K., Nishino, H., Nazeeruddin, M. K., & Grätzel, M. (2014). Inorganic hole conductor-based lead halide perovskite solar cells with 12.4% conversion efficiency. *Nature Communications*, *5*(1), 3834. doi:10.1038/ncomms4834 PMID:24815001

Rakita, Y., Bar-Elli, O., Meirzadeh, E., Kaslasi, H., Peleg, Y., Hodes, G., Lubomirsky, I., Oron, D., Ehre, D., & Cahen, D. (2017). Tetragonal $CH_3NH_3PbI_3$ is ferroelectric. *Proceedings of the National Academy of Sciences of the United States of America*, *114*(28), E5504–E5512. doi:10.1073/pnas.1702429114 PMID:28588141

Röhm, H., Leonhard, T., Hoffmann, M. J., & Colsmann, A. (2017). Ferroelectric domains in methylammonium lead iodide perovskite thin-films. *Energy & Environmental Science*, *10*(4), 950–955. doi:10.1039/C7EE00420F

Rong, Y., Hu, Y., Mei, A., Tan, H., Saidaminov, M. I., Seok, S. I., McGehee, M. D., Sargent, E. H., & Han, H. (2018). Challenges for commercializing perovskite solar cells. *Science*, *361*(6408), 8235. doi:10.1126/science.aat8235 PMID:30237326

Saidaminov, M. I., Abdelhady, A. L., Murali, B., Alarousu, E., Burlakov, V. M., Peng, W., Dursun, I., Wang, L., He, Y., Maculan, G., Goriely, A., Wu, T., Mohammed, O. F., & Bakr, O. M. (2015). Highquality bulk hybrid perovskite single crystals within minutes by inverse temperature crystallization. *Nature Communications*, *6*(1), 7586. doi:10.1038/ncomms8586 PMID:26145157

Saski, M., Prochowicz, D., Marynowski, W., & Lewiński, J. (2019). Mechanosynthesis, optical, and morphological properties of MA, FA, Cs-SnX_3 (X = I, Br) and phase-pure mixed-halide $MASnI_xBr_{3-x}$ perovskites. *European Journal of Inorganic Chemistry*, *22*(22), 2680–2684. doi:10.1002/ejic.201801506

. (2013). Science News. *Science*, *342*, 1438–1439.

Shi, D., Adinolfi, V., Comin, R., Yuan, M., Alarousu, E., Buin, A., Chen, Y., Hoogland, S., Rothenberger, A., Katsiev, K., Losovyj, Y., Zhang, X., Dowben, P. A., Mohammed, O. F., Sargent, E. H., & Bakr, O. M. (2015). Low trap-state density and long carrier diffusion in organolead trihalide perovskite single crystals. *Science*, *347*(6221), 519–522. doi:10.1126/science.aaa2725 PMID:25635092

Snaith, H., & Gratzel, M. (2006). Enhanced charge mobility in a molecular hole transporter via addition of redox inactive ionic dopant: Implication to dye-sensitized solar cells. *Applied Physics Letters*, *89*(26), 262114. doi:10.1063/1.2424552

Snaith, H. J. (2010). Estimating the maximum attainable efficiency in dye-sensitized solar cells. *Advanced Functional Materials*, *20*(1), 13–19. doi:10.1002/adfm.200901476

Song, Z., McElvany, C. L., Phillips, A. B., Celik, I., Krantz, P. W., Watthage, S. C., Liyanage, G. K., Apul, D., & Heben, M. J. (2017). A technoeconomic analysis of perovskite solar module manufacturing with low-cost materials and techniques. *Energy & Environmental Science*, *10*(6), 1297–1305. doi:10.1039/C7EE00757D

Steirer, K. X., Schulz, P., Teeter, G., Stevanovic, V., Yang, M., Zhu, K., & Berry, J. J. (2016). Defect tolerance in methylammonium lead triiodide perovskite. *ACS Energy Letters*, *1*(2), 360–366. doi:10.1021/acsenergylett.6b00196

Stoumpos, C. C., & Kanatzidis, M. G. (2015). The Renaissance of Halide Perovskites and Their Evolution as Emerging Semiconductors. *Accounts of Chemical Research*, *48*(10), 2791–2802. doi:10.1021/acs.accounts.5b00229 PMID:26350149

Stoumpos, C. C., Malliakas, C. D., & Kanatzidis, M. G. (2013). Semiconducting tin and lead iodide perovskites with organic cations: Phase transitions, high mobilities, and near-infrared photoluminescent properties. *Inorganic Chemistry*, *52*(15), 9019–9038. doi:10.1021/ic401215x PMID:23834108

Stranks, S. D., Eperon, G. E., Grancini, G., Menelaou, C., Alcocer, M. J. P., Leijtens, T., Herz, L. M., Petrozza, A., & Snaith, H. J. (2013). Electron-hole diffusion lengths exceeding 1 micrometer in an organometal trihalide perovskite absorber. *Science*, *342*(6156), 341–344. doi:10.1126/science.1243982 PMID:24136964

Stranks, S. D., & Snaith, H. J. (2015). Metal-halide perovskites for photovoltaic and light-emitting devices. *Nature Nanotechnology*, *10*(5), 391–402. doi:10.1038/nnano.2015.90 PMID:25947963

Subbiah, A. S., Halder, A., Ghosh, S., Mahuli, N., Hodes, G., & Sarkar, S. K. (2014). Inorganic hole conducting layers for perovskite-based solar cells. *The Journal of Physical Chemistry Letters*, *5*(10), 1748–1753. doi:10.1021/jz500645n PMID:26270378

Takahashi, Y., Obara, R., Lin, Z.-Z., Takahashi, Y., Naito, T., Inabe, T., Ishibashi, S., & Terakura, K. (2011). Charge transport in tin-iodide perovskite: Origin of high conductivity. *Dalton Transactions (Cambridge, England)*, *40*(20), 5563–5568. doi:10.1039/c0dt01601b PMID:21494720

Teshima, K., Shirai, Y., & Miyasaka, T. (2009). Organometal halide perovskites as visible-light sensitizers for photovoltaic cells. *Journal of the American Chemical Society*, *131*(17), 6050–6051. doi:10.1021/ja809598r PMID:19366264

Walsh, A., Scanlon, D. O., Chen, S., Gong, X. G., & Wei, S.-H. (2015). Self-regulation mechanism for charged point defects in hybrid halide perovskites. *Angewandte Chemie International Edition*, *54*(6), 1791–1794. doi:10.1002/anie.201409740 PMID:25504875

Wang, H., & Kim, D. H. (2017). Perovskite-based photodetectors: Materials and devices. *Chemical Society Reviews*, *46*(17), 5204–5236. doi:10.1039/C6CS00896H PMID:28795697

Wang, H., Wang, Y., Xuan, Z., Chen, T., Zhang, J., Hao, X., Wu, L., Constantinou, I., & Zhao, D. (2021). Progress in perovskite solar cells towards commercialization—A review. *Materials (Basel)*, *14*(21), 6569. doi:10.3390/ma14216569 PMID:34772092

Wang, K., Wang, S., Xiao, S., & Song, Q. (2018). Recent advances in perovskite micro- and Nanolasers. *Advanced Optical Materials*, *6*(18), 1800278. doi:10.1002/adom.201800278

Wang, R., Mujahid, M., Duan, Y., Wang, Z.-K., Xue, J., & Yang, Y. (2019). A review of perovskites solar cell stability. *Advanced Functional Materials*, *29*(47), 1808843. doi:10.1002/adfm.201808843

Wang, Y., Mahmoudi, T., & Hahn, Y.-B. (2020). Highly stable and Efficient Perovskite Solar Cells Based on FAMA-Perovskite-Cu:NiO Composites with 20.7% Efficiency and 80.5% Fill Factor. *Advanced Energy Materials*, *10*(27), 2000967. doi:10.1002/aenm.202000967

Weber, D. (1978). $CH_3NH_3SnBrxI_{3-x}$ (x = 0-3), a Sn(II)-system with the cubic perovskite structure. *Zeitschrift für Naturforschung Part B*, *33*, 1443–1445. doi:10.1515/znb-1978-1214

Wei, Z., & Xing, J. (2019). The rise of perovskite light-emitting diodes. *The Journal of Physical Chemistry Letters*, *10*(11), 3035–3042. doi:10.1021/acs.jpclett.9b00277 PMID:31117692

Wu, C., Zhang, Q., Liu, Y., Luo, W., Guo, X., Huang, Z., Ting, H., Sun Zhong, X., Wei, S., & … . (2018). The dawn of lead-free perovskite solar cell: Highly stable double perovskite $Cs_2AgBiBr_6$ film. *Advancement of Science*, *5*(3), 1700759. doi:10.1002/advs.201700759 PMID:29593974

Xiao, Z., Dong, Q., Bi, C., Shao, Y., Yuan, Y., & Huang, J. (2014). Solvent annealing of perovskite-induced crystal growth for photovoltaic-device efficiency enhancement. *Advanced Materials*, *26*(37), 6503–6509. doi:10.1002/adma.201401685 PMID:25158905

Xiao, Z., Yuan, Y., Wang, Q., Shao, Y., Bai, Y., Deng, Y., Dong, Q., Hu, M., Bi, C., & Huang, J. (2016). Thin-film semiconductor perspective of organometal trihalide perovskite materials for high-efficiency solar cells. *Materials Science and Engineering R Reports*, *101*, 1–38. doi:10.1016/j.mser.2015.12.002

Xing, G., Mathews, N., Sun, S., Lim, S. S., Lam, Y. M., Grätzel, M., Mhaisalkar, S., & Sum, T. C. (2013). Long-range balanced electron- and hole-transport lengths in organic-inorganic $CH_3NH_3PbI_3$. *Science*, *342*(6156), 344–347. doi:10.1126/science.1243167 PMID:24136965

Xu, X., Wang, Z., Yu, J., Li, L., & Yan, X. (2019). Phase engineering for highly efficient quasi-two-dimensional all-inorganic perovskite light-emitting diodes via adjusting the ratio of Cs cation. *Nanoscale Research Letters*, *14*(255), 1–8. doi:10.1186/s11671-019-3076-x PMID:31352526

Yadegarifard, A., Lee, H., Seok, H. J., Kim, I., Ju, B. K., Kim, H. K., & Lee, D. K. (2023). FA/Cs-based mixed Pb–Sn perovskite solar cells: A review of recent advances in stability and efficiency. *Nano Energy*, *112*, 108481. doi:10.1016/j.nanoen.2023.108481

Yamada, Y., Nakamura, T., Endo, M., Wakamiya, A., & Kanemitsu, Y. (2014). Photocarrier recombination dynamics in perovskite $CH_3NH_3PbI_3$ for solar cell applications. *Journal of the American Chemical Society*, *136*(33), 11610–11613. doi:10.1021/ja506624n PMID:25075458

Yang, F., Jang, D., Dong, L., Qiu, S., Distler, A., Li, N., Brabec, C. J., & Egelhaaf, H. J. (2021). Upscaling Solution-Processed Perovskite Photovoltaics. *Advanced Energy Materials*, *11*(42), 2101973. doi:10.1002/aenm.202101973

Yang, T. Y., Gregori, G., Pellet, N., Grätzel, M., & Maier, J. (2015). The significance of ion conduction in a hybrid organic-inorganic lead-iodide-based perovskite photosensitizer. *Angewandte Chemie International Edition*, *54*(27), 7905–7910. doi:10.1002/anie.201500014 PMID:25980541

Yang, W. S., Noh, J. H., Jeon, N. J., Kim, Y. C., Ryu, S., Seo, J., & Seok, S. I. (2015). High-performance photovoltaic perovskite layers fabricated through intramolecular exchange. *Science*, *348*(6240), 1234–1237. doi:10.1126/science.aaa9272 PMID:25999372

Yang, Z., Surrente, A., Galkowski, K., Miyata, A., Portugall, O., Sutton, R. J., Haghighirad, A. A., Snaith, H. J., Maude, D. K., Plochocka, P., & Nicholas, R. J. (2017). Impact of the halide cage on the electronic properties of fully inorganic cesium lead halide perovskites. *ACS Energy Letters*, *2*(7), 1621–1627. doi:10.1021/acsenergylett.7b00416

Yin, W.-J., Shi, T., & Yan, Y. (2014). Unusual defect physics in $CH_3NH_3PbI_3$ perovskite solar cell absorber. *Applied Physics Letters*, *104*(6), 063903. doi:10.1063/1.4864778

Yin, X., Chen, P., Que, M., Xing, Y., Que, W., Niu, C., & Shao, J. (2016). Highly efficient flexible perovskite solar cells using solution-derived NiOx hole contacts. *ACS Nano*, *10*(3), 3630–3636. doi:10.1021/acsnano.5b08135 PMID:26958704

Yongping, F. (2015). Solution growth of single crystal Methylammonium Lead halide Perovskite nanostructures for optoelectronic and photovoltaic applications. *Journal of the American Chemical Society*, *137*(17), 5810–5818. doi:10.1021/jacs.5b02651 PMID:25871732

Yoo, J. J., Wieghold, S., Sponseller, M. C., Chua, M. R., Bertram, S. N., Hartono, N. T. P., Tresback, J. S., Hansen, E. C., Correa-Baena, J.-P., Bulović, V., Buonassisi, T., Shin, S. S., & Bawendi, M. G. (2019). An interface stabilized perovskite solar cell with high stabilized efficiency and low voltage loss. *Energy & Environmental Science*, *12*(7), 2192–2199. doi:10.1039/C9EE00751B

Yu, H., Ryu, J., Lee, J. W., Roh, J., Lee, K., Yun, J., Lee, J., Kim, Y. K., Hwang, D., Kang, J., Kim, S. K., & Jang, J. (2017). Large grain-based hole-blocking layer free planar-type perovskite solar cell with best efficiency of 18.20%. *ACS Applied Materials & Interfaces*, *9*(9), 8113–8120. doi:10.1021/acsami.6b15710 PMID:28211274

Zhao, Y., Tan, H., Yuan, H., Yang, Z., Fan, J. Z., Kim, J. H., Voznyy, O., Gong, X., Quan, L. N., Tan, C. S., Hofkens, J., Yu, D., Zhao, Q., & Sargent, E. H. (2018). Perovskite seeding growth of formamidinium-lead-iodide-based perovskites for efficient and stable solar cells. *Nature Communications*, *9*(1), 1607. doi:10.1038/s41467-018-04029-7 PMID:29686304

Zhao, Y., & Zhu, K. (2016). Organic–inorganic hybrid lead halide perovskites for optoelectronic and electronic applications. *Chemical Society Reviews*, *45*(3), 655–689. doi:10.1039/C4CS00458B PMID:26645733

Zhou, H., Chen, Q., Li, G., Luo, S., Song, T. B., Duan, H. S., Hong, Z., Jingbi You, J., Liu, Y., & Yang, Y. (2014). Interface engineering of highly efficient perovskite solar cells. *Science*, *345*(6196), 542–546. doi:10.1126/science.1254050 PMID:25082698

Chapter 7
Piezoelectric Nanogenerators

E. Devaraj
CMR University, India

ABSTRACT

Piezoelectric nanoparticles offer a chance for advancement toward self-powered systems through energy harvesting. The use of a system or device that performs a function without the requirement for an external power source, such as a battery or any other type of source, is made possible by the newly emerging technology known as self-powered systems. For instance, this technology can capture energy from nearby sources and transform it into electric energy utilizing the piezoelectric effect. The creation of some self-powered systems based on nano-devices would be facilitated by the development of a nanogenerator (NG) to transform environmental energy into electric energy. Zinc oxide (ZnO) has been employed in a wide range of applications, including optoelectronics, chemical sensors, piezoelectric transducers, and varistors, due to its many functions. Consequently, ZnO-based piezoelectric composites would be a different option for flexible and small implantable sensors that may immediately detect and monitor mechanical stimulation or pressure changes occurring in the patient's organs.

1. INTRODUCTION TO PIEZOELECTRIC EFFECT

The discovery that compressing and stretching quartz can produce macroscopic polarization and, subsequently, electric charges was made in 1880 by Pierre and Jacques Curie (Curie & Curie, 1880). The term "piezoelectricity," which means "electricity resulting from the press," was used to describe this phenomenon (Taylor, 1985). The Greek word "piezo" means "to press or squeeze." The first official use of piezoelectric material wasn't made until 1921 when quartz crystal was used in a transducer to find submarines via echolocation. Since the development of artificial piezoelectric materials, the emergence of piezoelectricity has opened up new avenues for research in the field of crystal physics and has quickly generated an industry market. The study and use of piezoelectric materials in modern engineering were hastened in particular by the discovery of ferroelectricity (Valasek, 1921). In 1969, it was discovered that polyvinylidene fluoride polymer had a high piezoelectric effect and met the criteria for mechanical flexibility in piezoelectric materials (Murayama et al., 1976). Piezoelectric materials, such as single

DOI: 10.4018/979-8-3693-1306-0.ch007

Figure 1. Direct piezoelectric effect: (a) at zero stress, (b) at applied compression, and (c) at applied tension. Converse piezoelectric effect: (d) at zero voltage, (e) at applied voltage, and (f) at applied reversed voltage.

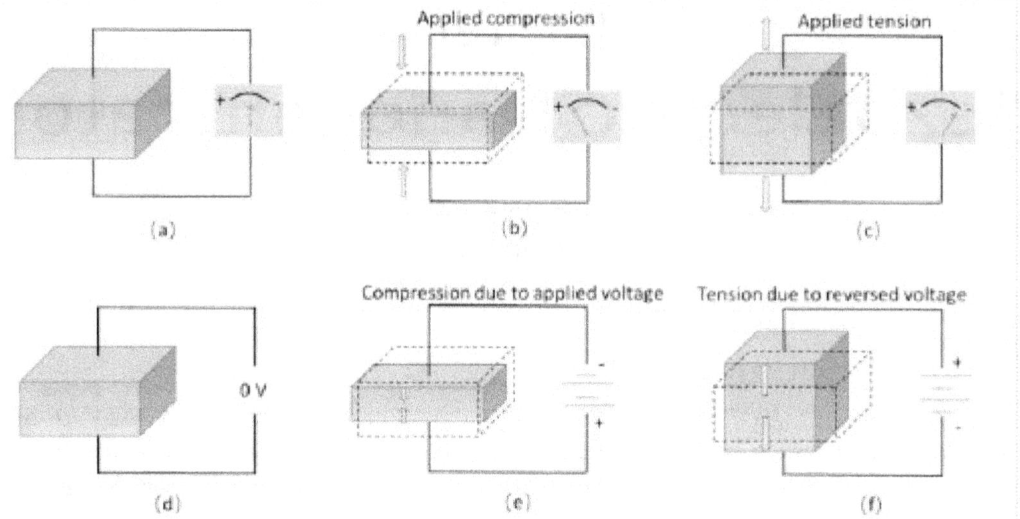

crystal, ceramic, and polymer, have several uses in the fields of medical, telecommunications, information, and the military up to this point.

The generation of an electric charge in response to a mechanical stimulus is known as the direct piezoelectric effect. As demonstrated in Figures 1 (a) to (c), if compressive stress is applied, the result is a positive electric voltage. If tension is applied, the result is a negative electric voltage. As roughly depicted in Figure 1 (d) to (f), the inversed piezoelectric effect is the generation of mechanical deformation in the material in response to electric stimulation. When a voltage is applied to a piezoelectric material, cocompressive deformation results, but a reversed voltage results in tension.

Indeed, the straightforward molecular model shown in Figure 2 can be used to illustrate the fundamentals of piezoelectricity. Centrosymmetric and non-centrosymmetric materials can both be categorized as crystals. A centrosymmetric crystal cannot produce electric polarization in response to external stimulation, as shown in Figure 2(a). Positive and negative charge centers coincide in the initial condition, absent any external force, creating an electrical neutral in the basic unit cell.

Due to the presence of the center of symmetry, the polarization dipole pi always cancels one another after the application of an external force, resulting in a null total dipole moment. As a result, there is no internal net polarization (P), proving that centrosymmetric crystals are nonpolar and non-piezoelectric. The unperturbed non-centrosymmetric unit cell, on the other hand, starts out in an electrically neutral condition, just like a centrosymmetric crystal (Figure 2(b)). Because a unit cell lacks a center of symmetry, when external stress is applied, the internal structure is deformed, causing the separation of the positive and negative centers to produce a dipole moment. Inside the substance, even nearby half-dipoles carrying the opposing charge indications cancel one another out. However, this substance polarized and surface net charges that produce an electrical field elaborate the genesis of piezoelectricity (Arnau & Soares, 2008).

Figure 2. (a) A centrosymmetric molecular and (b) a non-centrosymmetric molecular model subjected to an external force (Arnau & Soares, 2008)

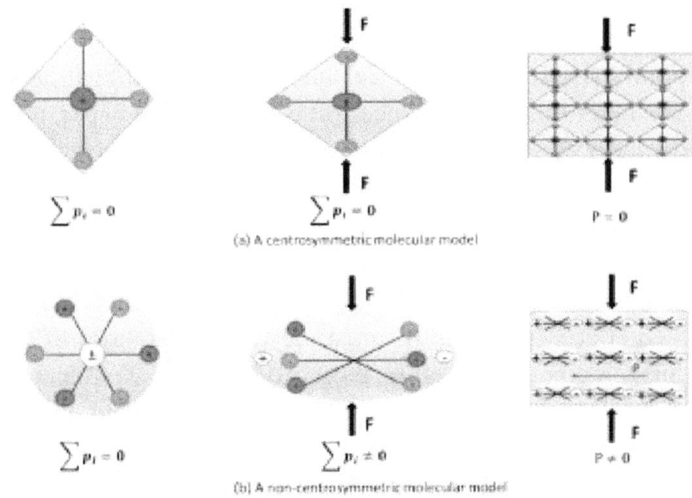

1.1 Nanotechnology

The concept of nanotechnology was first articulated in a lecture by physicist Richard Feynman entitled "There is Plenty of Room at the bottom" given at the California Institute of Technology in December 1959. In that speech, Feynman described a technique that helps a scientist modify and adjust certain atoms and molecules. The word "nanotechnology" was brought up by Professor Norio Taniguchi more than ten years following the speech. The advent of the scanning tunneling microscope in 1981 made it possible to "see" individual atoms, which marked the beginning of contemporary nanotechnology. The invention of the scanning electron microscope made it possible to observe and work with particular atoms and molecules.

All branches of research, including physics, biology, chemistry, materials science, medicine, and engineering, involve nanotechnology, which is the study and utilization of incredibly small objects . In many applications today, researchers are taking advantage of the advantages that nanostructures have over their bulk counterparts.

1.2 Semiconductor Nanostructures

Semiconductor nanostructures (NSs) are distinctive as the structural components of numerous different nanoscale devices. ZnO, GaN, CdS, and potentially ZnS are examples of semiconductor NS materials that may play a big role in collaborating with objects in relation to energy challenges. It is crucial to understand the semiconducting, piezoelectric, and pyroelectric properties of various semiconducting NSs. Such NSs essentially have optoelectronics, sensors, and actuators as their primary study fields. In general, some NSs have a high surface area to volume ratio, are low toxicity, environmentally friendly, have great chemical stability, and, in the case of some metal oxide NSs, may even be biocompatible. Due to ZnO's distinct physical and chemical characteristics, research on metal oxide NSs has been successful for the past 20 years. The "top-down" and "bottom-up" approaches are the two approaches that can be used to create an NS. Although the top-down method, which uses lithographic techniques, is best for some sspecific applications, such as photonic crystals, it is quite expensive and time-consuming.

The bottom-up strategy is further separated into physical and chemical components. The physical paths are typically hot (a few hundred degrees Celsius). Though chemical pathways are typically low-temperature (100^{0}C). The latter is appealing since soft substrates like paper, plastic, textiles, etc. can be used. The vapor-liquid-solid (VLS) technique and chemical vapor deposition (CVD) are examples of common physical techniques. While electrochemical deposition (ED) and hydrothermal techniques are frequently utilized in chemical processes. The aqueous chemical growth (ACG) techniques used to create metal oxide NSs have recently gained a lot of attention from scientists. This is a result of the method's ease of use, low temperature (100°C), affordability, and environmental friendliness.

Additionally, it is theoretically possible to adjust the various qualities of any material by changing the preparation procedures, recipe, temperature, etc. . Additionally, ZnO has distinctive and appealing morphologies in zero-dimensional configurations like nanoparticles and quantum dots, as well as one-dimensional configurations like nanowires (NWs) and nanorods (NRs), as well as two-dimensional configurations like, for example, nanosheets and nanowalls, as well as three-dimensional configurations like nanotubes, nano leaves, nanoflowers, and nano tetrapods, among others.For future applications, they are advantageously beneficial for fabricating and developing nanodevices for sensing, personal electronics, optoelectronics, medicinal devices, and battery-free system.

1.3 Nanogenerators (NGs)

A nanogenerator (NG) is an element that uses nanostructured materials to transform energy into usable electrical energy. Piezoelectric, triboelectric, and pyroelectric NG are the three distinct techniques of an NG.

A piezoelectric new generation (NG) is a device that uses active materials that produce charges when they are physically pressured, turning mechanical energy into electricity. The term "NG," however, is specifically used to refer to a piezoelectric-based device that uses kinetic energy and was first demonstrated in 2006. It can refer to any type of device that is based on an NS and converts any type of environmental energy, such as airflow, wind, ambient noise, vibration, and human body movements. It is crucial to remember that NG research is still in its infancy; the demonstration was hailed as a milestone due to the technology's promise as an energy harvester. The advancement of NGs is predicted to pave the way for easy integration with other energy harvesters that convert various forms of energy, such as solar energy converters.

With less worry about the energy source, such energy harvesting systems can be employed to power portable electronic gadgets . However, the link between piezoelectric and semiconducting qualities is connected to the idea of the NG. When an external force applies stress, ZnO NWs growing parallel to the c-axis are compressed uniaxially. The top and bottom sides of the ZnO NWs will each have a piezoelectric potential with a distinct polarity. Then, the transient current travels through the external circuit before flowing from the top to the bottom of the NW . The result is recognized as an electric pulse. The piezoelectric potential in the NWs will cease to exist once the external force is removed. The result is the harvesting of an AC current.

Triboelectric NGs make use of the electrostatic charges that develop when two distinct materials come into physical contact with one another. Triboelectric displays charges and discharges when force is applied as a result of the contact and separation of the two materials. As a result, electrons began to flow between the material's two electrodes. Because of the friction and movement of various clouds on top of one another on cloudy days, the triboelectric phenomenon is the cause of the lightning. Z. L. Wang et al. released the first report on contemporary triboelectric NG in 2012.

Using the pyroelectric effect, a pyroelectric NG transforms thermal energy into electrical energy. A pyroelectric NG typically relies heavily on the Seebeck effect, which uses temperature fluctuations at the device's end to drive the diffusion of charge carriers .

1.4 Self-Powered Nano-Systems

Self-powering with long-term operation, realistic characterizations, and a wide range of wireless components, such as those used for space monitoring, are crucial. In addition to numerous diverse biomedical uses, it is also utilized for control and diagnostics in a variety of industries, including the military, business, industrial, and space research, for example. Despite having a relatively lengthy operational life, small devices (such as nano and micro devices) are difficult to function when the energy level is low. Finding solutions for the ongoing functional shortfall will therefore be aided by the discovery of sustainable energy sources, such as the collecting of ambient energy.

The commercial and academic sectors are paying increasingly close attention to the nanodevices created to demonstrate wireless nanosystems. Additionally, these nanodevices are crucial for sensing, medical research, and other fields. It may be absolutely necessary for biomedical equipment that operates wirelessly and without the need for storage. This describes self-powered nanosystems precisely. Small components in personal electronics, such as mobile phones, may be powered by self-powered systems. Therefore, for electronic gadgets to operate sustainably and without maintenance, novel power sources are urgently needed. This technique is extremely inventive and can be applied to various sensor systems, home security applications, nano/microelectromechanical (NMEMS) systems, etc.

2. PIEZOELECTRIC MATERIAL

It is important to understand the relationship between dielectric and piezoelectric before discussing the many types of piezoelectric materials. materials that are ferroelectric and pyroelectric, as seen in Figure 3.

An electrical insulator that can become polarized by an external electric field is referred to as a dielectric substance. One such class of anisotropic dielectrics is the piezoelectric material, which has the capacity to produce an electric charge in response to the application of stress or strain or, conversely, mechanical deformation in response to an applied electric field. Pyroelectric materials are a subclass of piezoelectrics, showing that polarization can alter in response to temperature changes (Webster, 1999). The positive and negative charge centers are moved away from one another as a result of temperature variations, which alter the crystal's polarization. A specific type of pyroelectric material is ferroelectric material. An opposing electric field can turn around the spontaneous electric polarization that ferroelectrics exhibit (Lines & Glass, 1979).

The polarization direction is included inside several domains that make up the grains in ferroelectric crystals. In order to align all dipoles in the same direction and create a macroscopic net polarization, an electric field poling procedure is required. Piezoelectric materials can be broadly divided into two

Figure 3. Classification of piezoelectricity (Khanbareh, 2016)

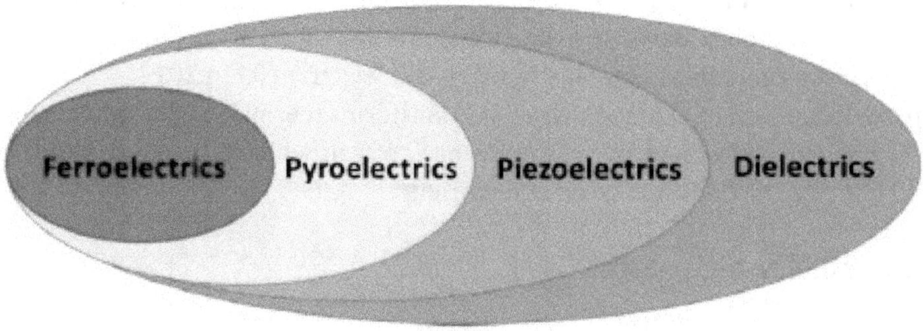

groups: natural materials and synthetic materials manufactured by humans, such as single crystals, ceramics, and polymers.

- **Nature material:** Natural crystals with non-centrosymmetric crystal lattices can act as piezoelectric materials. Materials used to make crystals include quartz, topaz, sucrose, Rochelle salt, and some organic materials like silk and wood. Rochelle salt has been used for many different types of transducers; however, it has drawbacks such as being susceptible to fluids and having a high-temperature limit (Uchino, 2012). The low piezoelectric constant is the primary disadvantage of natural piezoelectric materials.
- **Single crystal:** Artificial piezoelectric materials have been created in order to overcome the drawbacks of natural piezoelectric materials and improve piezoelectric capabilities. Single crystals without a center of symmetry for piezoelectricity experience the piezoelectric effect. According to their point groups, the 32 crystal classes can be broken down into seven groups: cubic, hexagonal, trigonal, tetragonal, orthorhombic, monoclinic, and triclinic. Of the 32 classes, only 20 have piezoelectric characteristics. Ten of them are polar, showing spontaneous polarization without mechanical excitation, whereas the other ten are non-polar, showing polarization only in response to mechanical stress. The common single crystal, quartz, is extensively used in transducers, high-frequency filters, and frequency-controlling oscillators . It has great qualities such as a mechanical strength of 2 to 3 GPa, thermal stability, and high voltage sensitivity, but its dielectric constant and piezoelectric capabilities are poor (Wilson, 2005).
- **Ceramics:** At room temperature, a ceramic is an inorganic solid substance. In fact, most ferroelectrics with a polycrystalline structure made up of ragged, tiny, irregular grains are ferroelectrics, including piezoelectric ceramics. The process of making piezoelectric ceramics involves heating thin powders containing a mixture of metal oxides to create a homogeneous powder. To create structural elements, the powders are then mixed with an organic binder, and finally sintered to create a thick crystal structure. Typically, binary or ternary combinations with a perovskite crystal structure make up piezoelectric ceramics. The perovskite crystal structure has the chemical formula ABO3, where A is a large metal ion and B is a smaller metal ion. After 1945, barium titanate (BaTiO3) was first produced as a piezoelectric material (Megaw, 1945).

Due to their higher piezoelectric constant than quartz, piezoelectric ceramics like lead titanate ($PbTiO_3$) and lead zirconate titanate (PZT) have considerable production potential. The majority of ceramics, however, are expensive, fragile, have a large electric loss, and have low stability, which limits their use in high-frequency systems (Jiansirisomboon et al., 2006). Additionally, lead oxide, a component of PZT, has been shown in numerous studies to be hazardous and harmful to the environment (Panda, n.d.).

- **Polymer:** A polymer is a big molecule, sometimes known as a macromolecule, made up of many compounds together (Polymer, n.d.). These ferroelectric polymers, which primarily belong to that family, display the piezoelectric effect by drawing and repelling long chains that are entangled in an electric field. Kawai made the initial discovery of polyvinylidene fluoride's (PVDF) piezoelectricity in 1969 (Kawai, 1969). Due to its elastic nature and chemical biocompatibility, PVDF is frequently used in sensors, acoustic ultrasound measurement, and biomedical fields (Lee et al., 2008). Due to their excellent flexibility and environmentally benign characteristics, piezoelectric polymers are therefore becoming more and more popular. However, these materials are constrained by their low dielectric and relatively low piezoelectric charge constant (Ohigashi, 2008; Wang et al., 1988).

3. ZnO-BASED COMPOSITES

Composites are made up of at least two parts that, when put together, result in a substance with considerably altered physical or chemical properties. A unique technique for creating a new class of composite with improved functionality and a wide range of applications in various fields is one of them: embedding fillers into the polymer (Della Schiava et al., 2018; Thetpraphi et al., 2019). Over the past 20 years, more well-known researchers have worked to investigate and market those multicomponent polymer systems (Furusawa et al., 1994; Duguet et al., 2000; Hayashi et al., 2000). The advantages of the polymer matrix, including its superior processability, flexibility, and durability, are combined with the special electric, magnetic, optical, thermal, and piezoelectric capabilities of inorganic particles at the nano- or microscale level (Le et al., 2014; Le et al., 2015). The choice of polymer and fillers is determined by the intended use. The chain geometry of a polymer determines whether it is polar or non-polar. Under an alternating field, nonpolar polymers' dipole moments cancel one another, resulting in low polarisation. Thus, nonpolar polymers like polytetrafluoroethylene (PTFE), polyethylene (PE), and polystyrene (PS) typically have low dielectric constants (less than 3). of contrast, the dipole moments of polar polymers, such as poly (methyl methacrylate) (PMMA) and polyvinyl chloride (PVC), reinforce one another and produce a high dielectric constant. In addition to taking the dielectric constant into account, the selection of the polymer matrix should carefully examine stiffness, viscosity, transparency, and curing conditions. In addition to polymers, fillers can also be magnetic (Tatarenko et al., 2007), nonconductive (Roy et al., 2012), or conductive (Gong et al., 2011).

Conductive fillers are chosen to raise the effective dielectric constant of composites at low concentrations whereas nonconductive fillers serve as an insulator. Additionally, the shapes, sizes, and concentration of the fillers can be used to tune the properties of composite materials, which the experiment should confirm.

High piezoelectric response, high thermal stability, and low mechanical loss piezoelectric ceramics are suppressed by their brittleness and complicated procedure by sintering at very high temperature

Figure 4. Schematic representation of piezoelectric composites

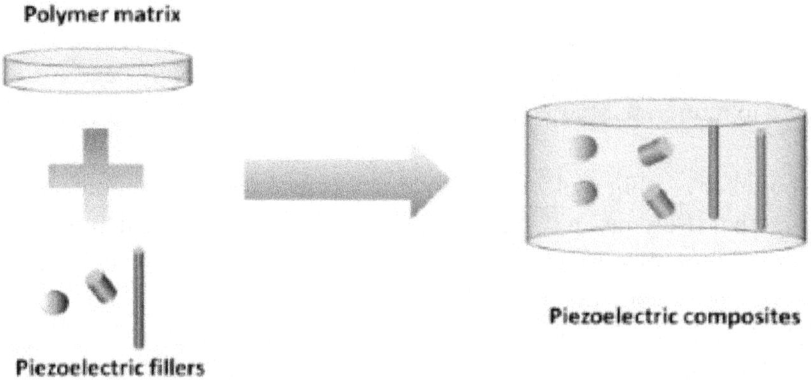

(D'Ambrogio et al., 2020; Grinberg et al., 2019). Despite being more flexible and eco-friendly, the piezoelectric polymer is constrained by a low dielectric constant and poor electromechanical coupling (Ganet et al., 2015; Liu et al., 2019). Therefore, piezoelectric fillers are produced and added to polymers to create two phase systems, as shown in Figure 4, in order to address these disadvantages. They have customised qualities and are ductile, flexible, and simple to shape.

Zinc oxide (ZnO) comes in a variety of sizes and geometries, including nano- and microparticles, microrods, and nanowires. By utilising the potent direct piezoelectric effect, ZnO, which has a non-centrosymmetric wurtzite structure, is frequently employed in piezoelectric applications. A broad direct band gap of 3.39 eV, a high excitation binding energy of roughly 60 meV at ambient temperature, and strong electron mobility make ZnO one of the most important II-VI semiconductor materials (Look, 2001; Ozgur et al., 2005; Xu & Wang, 2011). ZnO has significant applications in optoelectronic, electromechanical, electrochemical, and photovoltaic devices, including light-emitting diodes, solar cells, ultraviolet lasers, and piezoelectric nanogenerators (Arnold et al., 2003; Huang et al., 2001; Govorov et al., 2007; Park & Yi, 2004; Law et al., 2005). This is due to the connection of semiconducting and piezoelectric characteristics. In comparison to PZT, BaTiO3, and PVDF films, ZnO does not require a high-voltage poling technique when it comes to the piezoelectric behaviour. In addition, ZnO is a biocompatible, non-toxic material, which broadens its possibilities in the healthcare industry, particularly for in vivo biosensing and bio-detection tools (Li et al., 2008; Zhou et al., 2006). Polydimethylsiloxane (PDMS) is the polymer of choice for the lab-fabricated composites because to its stability, excellent transparency, low cost, and—most importantly—thermosetting capability. When using the dielectrophoresis process, polymers must have a thermosetting action.

3.1 Zinc Oxide (ZnO) Crystal

Schematically, the Wurtzite, Zinc Blende, and Rocksalt crystal formations make up the ZnO crystal. The thermodynamically stable phase in an ambient environment is the non-centrosymmetric crystal hexagonal Wurtziite (Kulkarni et al., 2011). Figure 5 (a) illustrates the hexagonal unity cell of the ZnO wurtzite structure, which has an axial lattice parameter of 0.52 mm perpendicular to the basal plane and a lattice parameter of 0.33 mm in the basal plane (Morkoç & Özgür, 2009). In the wurtzite structure, a

Figure 5. (a) Wurtzite structure model of ZnO where shaded gray and black spheres represent zinc and oxygen atoms, respectively. (b) Tetrahedral coordination between Zn and O in Wurtzite structure without external force, (c) the distortion of tetrahedral unit under an external compressive force (Özgür et al., 2005).

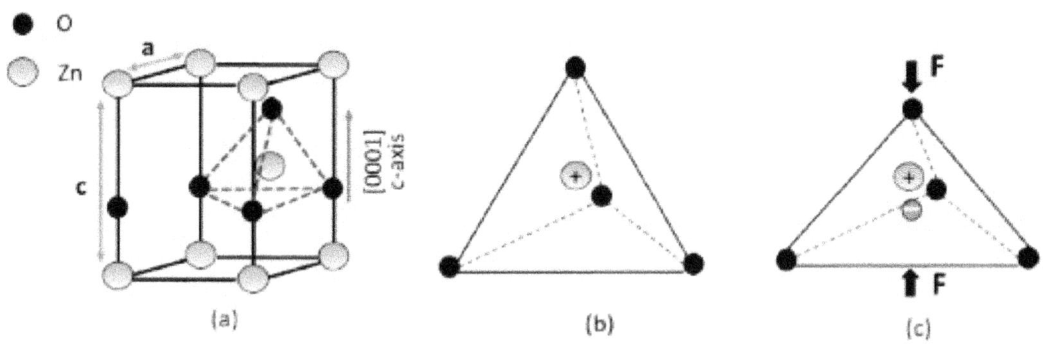

basal plane is formed by four tetrahedrally coordinated Zn cations or oxygen anions that stack alternately along the c-side direction.

The most frequent instance is ZnO growing in the direction (Curie & Curie, 1880) along the c-axis. A polar surface forms on the basal plane as a result of the surface plane being completely terminated by either a negatively charged oxygen polar plane (0001) or a positively charged Zn polar plane (0001). The distribution of surface charges effects the developing nanostructure with the dominance of polar surfaces by lowering the crystal's electrostatic energy (Wang, 2009). The fundamental unit is a tetrahedral unit made up of four oxygen anions and one Zn cation, despite the growth orientation, as shown in Figure 5 (b). An electrically neutral unity cell and zero net charge in the crystal are produced when the negative charge centres of the four oxygen anions locate at an equal distance from the positive Zn cation. The distortion of the tetrahedral unite caused by the displacement of the positive charge centre from the negative charge centre, as shown in Figure 5 (c), results in the creation of an electric dipole when a physical compressive force is applied to the unite cell. As a result, a macroscopic dipole forms and the crystal as a whole exhibits internal electric polarisation.In that situation, the charge carriers, which demonstrate remarkable piezoelectric ability, transform mechanical energy into electrical energy. In fact, the electronegativity of zinc and oxygen atoms differ, leading to a shift in the electron cloud, which causes the repeating unit cell to show a nonzero charge in steady state. Additionally, the positively charged Zn poplar and negatively charged O polar surfaces along the c-axis are classified as Type III by Tasker's classification system (Tasker, 1979). In that situation, a typical dipole moment is produced along the c-axis by a spontaneous polarisation field.

3.2 Piezoelectricity in ZnO Nanostructure

ZnO nanostructures can now be produced in a variety of shapes, including nanoparticles, nanowires, belts, rings, bows, and more thanks to the advancement of nanotechnology synthesis in recent years. Materials with at least one dimension less than 100 nm and special features permitting novel applications are included in the definition of nanotechnology (Consonni & Lord, 2021). Due to surface and quantum confinement effects, mechanical, electromechanical, electrical, and optical characterizations at the nanoscale differ significantly from those of bulk materials (What It Is and How It Works,). Additionally, nanostructures offer the potential for miniaturisation, particularly in implantable biological nanodevices.

One-dimensional (1D) nanostructures with a high aspect ratio, such as nanowire, nanotubes, and nanofibers, are particularly well-received by the scientific communities due to their better properties. According to reports, 1D nanowires (NWs) exhibit electrical and mechanical features that are diameter dependent (Fu et al., 2014; Wang, 2004). Young's modulus in (Curie & Curie, 1880) oriented ZnO NWs with diameter less than 120 nm (Figure 6 a) has been shown by Chen et al. to grow considerably with decreasing diameter. This is due to the significant role of surface elasticity, which tends to the value of bulk ZnO (140 GPa) with large diameter (>120 nm) (He et al., 2009) and occurs at large surface-to-volume ratio structures. When the diameter of the ZnO NWs was reduced from 475 to 225 nm, it was discovered that the fracture strain rose from 5% to 15% (Chen et al., 2006).

A theoretical computing approach has been used to determine the variation tendency in the piezoelectric coefficient's size dependency. Regarding the first principle-based density functional theory calculations, Agrawal et al. have shown that the volume reduction caused by local polarisation and surface relaxation was responsible for the enormous piezoelectric size effect for both ZnO and gallium nitrides (GaN) NWs (Desai & Haque, 2007). With respect to an increase in diameter and coverage to bulk values with diameter higher than 1.5 nm and 2.5 nm, respectively (Figure 6 b), the piezoelectric coefficients of ZnO and GaN NWs drop.For ferroelectric PZT nanowires, size reduction increases both the piezoelectric coefficient and the dielectric constant, which are attributed to the improvement in the dielectric constant and the decrease in lateral interaction, respectively (Agrawal & Espinosa, 2011). When the diameter is greater than 2.4 nm, the piezoelectric coefficient for BaTiO3 NWs rises and approaches the bulk value (Figure 6 c) (Hong & Fang, 2008). Additionally, the piezoelectric capabilities are significantly influenced by the NW's aspect ratio in terms of radius and length. The output voltage with a constant diameter of 50 nm (Figure 6 d) increases as the aspect ratio grows and starts to fall above an aspect ratio of 80 as a result of excessive deflection, according to the FEM simulation of a deflected ZnO NW (Zhang et al., 2010). As a result, the size-dependences of piezoelectricity in various 1D nanomaterials differ.

Therefore, it was thought that the ZnO NWs should have higher piezoelectric constants than the bulk ZnO. Experimental research on the piezoelectric size impact of 1D nanostructures is expected to be difficult at such a nanoscale, nevertheless. These challenges include complex nanoscale manipulation and the need for high sensitivity when measuring the generated voltage and current at the same time. The first experimental measurement of piezoelectric production from bent ZnO NWs using atomic force microscopy (AFM) technology was published by Wang et al. (Gao et al., 2007; Riaz et al., 2011). On the other hand, the piezoelectric force microscope (PFM), which measures the sample's displacement by applying an electric field on the order of a picometer (Zhao et al., 2004), has evolved into a typical technique to investigate the piezoelectric effect at the nanoscale.

3.3 Connectivity Pattern of Composites

In a multiphase composite system, Newnham et al. (Newnham et al., 1978) showed that the connection patterns of the constituent phases can be used to control electromechanical coupling and piezoelectric properties. There are ten possible connectives in a two-phase system if each phase can be self-connected into zero, one, two, and three dimensions: 0-0, 1-0, 2-0, 3-0, 1-1, 2-1, 3-1, 2-2, 3-2, and 3-3. The first and second numbers relate to the connection pattern of the filler and the polymer, respectively.Cubic is

Figure 6. (a) Diameter dependence of effective Young's modulus in bended ZNO nanowires) (Agrawal & Espinosa, 2011); (b) piezoelectric coefficient of ZnO and GaN nanowires as a function of the diameter (Reddy et al., 2019); (c) the size dependent piezoelectric coefficient of $BaTiO_3$ (Wang & Song, 2006); (d) output voltage as a function of aspect ratio of ZnO NWs controlled at a constant diameter of 50 nm (Zhang et al., 2010)

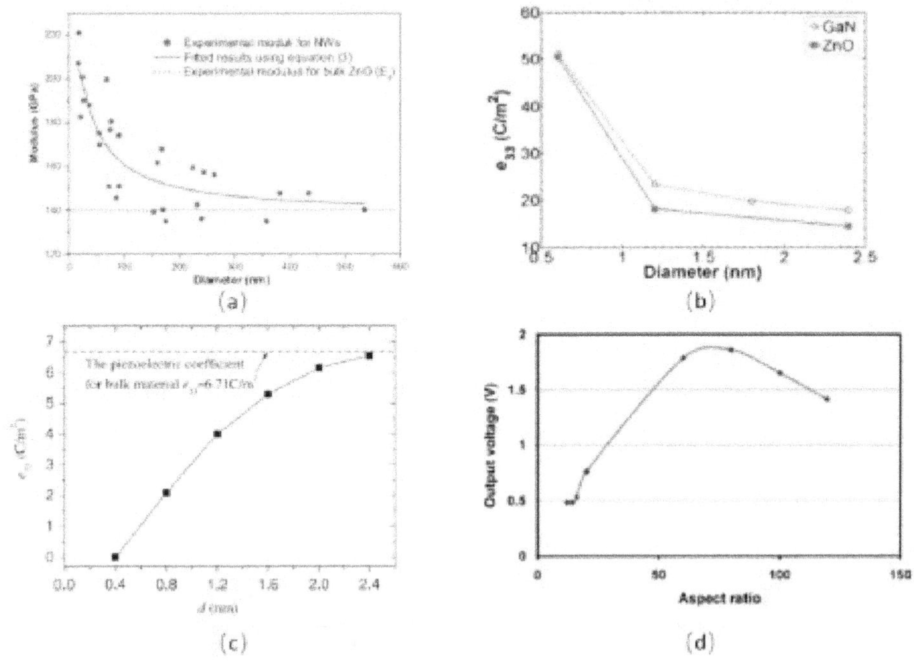

used as the fundamental building component in Figure 7, and the colour white denotes the filler phase. In a fundamental 0-0 connectivity pattern, two phases are connected in zero dimensions, which is not conceivable in particle composites. The disconnected particles are randomly spread in a completely connected matrix according to the 0-3 connectivity pattern, which is thought to be the simplest type. This pattern features unshaded phase self-connections in the zero-direction and shaded phase self-connections in three-dimensional directions. A composite with 1-3 connectivity phases has a one-dimensionally linked filler and a three-dimensionally connected matrix, which means that fibres or rods with continuous phases are aligned in the polymer matrix.

4. TRIBOELECTRIC NANOGENERATORS

A novel and renewable energy device called the triboelectric nanogenerator (TENG), which is based on the triboelectric phenomenon, was introduced in 2012 for capturing ambient mechanical energy. Figure 8 shows the first triboelectric nanogenerator in action.

The first triboelectric nanogenerator to be reported is multilayered. The periodic contact and separation between PET and Kapton, caused by outside mechanical motion, alternately drives the induced electrons between electrodes due to the interaction between triboelectric effect and electrostatic induction. Also

Figure 7. Representation of connectivity patterns for two-phase composites, where arrows indicate the connection directions (Newnham et al., 1978)

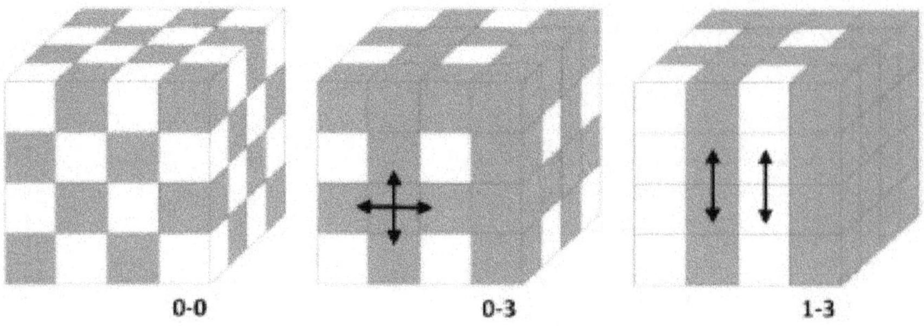

Figure 8. A schematic illustration of the device structure and working principle of the first triboelectric generator. (a) The structure of the triboelectric nanogenerator in bending and releasing process and experimental setup for electrical measurement. (b) An illustration of the working principle of the first triboelectric nanogenerator (Fan et al., 2012).

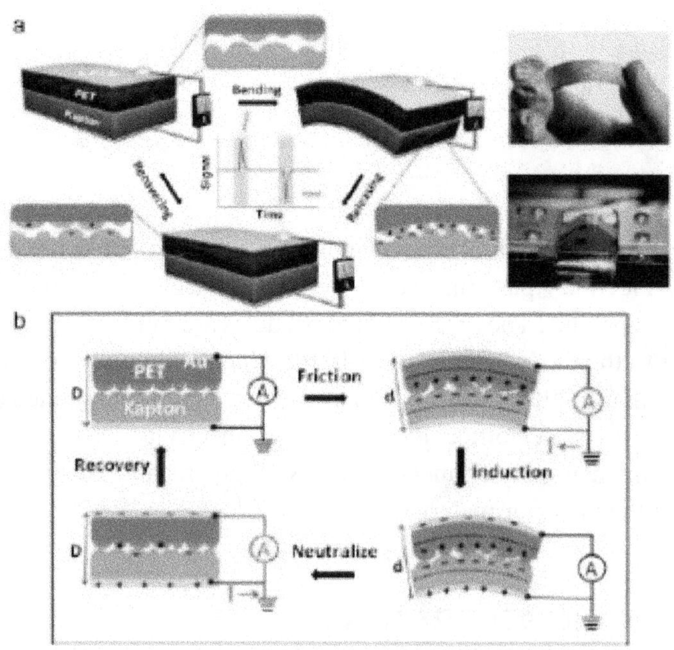

supplied were a peak voltage of 3.3 V, a peak current of 0.6 A, and a peak power density of 10.4 mW/ cm^3. The mechanical energy is successfully transferred to electricity throughout this procedure.

Figure 9. The working mechanism of four modes of triboelectric nanogenerator

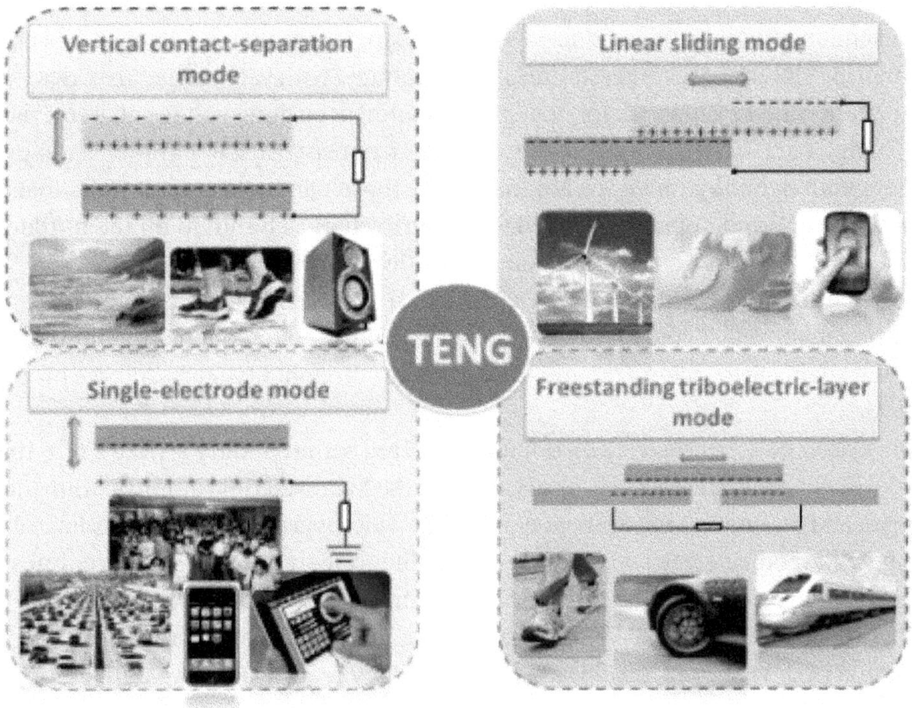

Since the introduction of the first triboelectric nanogenerator, it has gained increasing scientific interest and been demonstrated to be a reliable, affordable, and straightforward method for harvesting mechanical energy. Over the last three years, triboelectric nanogenerator-based self-powered devices and systems have made a tremendous amount of development (Wang et al., 2015).

As schematically depicted in Figure 9, the triboelectric nanogenerator currently has four fundamental operating modes, which are the vertical contact-separation mode, in-plane linear sliding mode, single electrode mode, and freestanding triboelectric-layer mode (Yang et al., 2013; Zhu et al., 2013).

5. POTENTIAL APPLICATIONS

Piezoelectric nanogenerators are machines that, at the nanoscale, transform mechanical energy into electrical energy. Because they may derive energy from a variety of mechanical vibrations or movements, these devices offer a wide range of possible uses. The following are some possible uses for piezoelectric nanogenerators:

5.1 Application-Based Electrical Properties

Applications based on the electrical properties of materials are widespread and diverse, as these properties are fundamental to modern technology. Here are some key applications based on electrical properties:

5.1.1 ZnO Varistor

In terms of the semiconducting and conducting mechanism, traditional ZnO varistor is created by high temperature sintering sintered of polycrystalline ZnO combined with various metal oxides. A "diode-like" electrical component called a ZnO varistor, it displays nonlinear current-voltage characteristics above the switch field (Zhang et al., 2020). The occurrence of barriers between ZnO grains is the primary cause. Potential barriers with Schottky junctions are formed by those nanometer-thick walls that comprise lattice mismatches, defects, and dopants (He, 2019). Electric charges build up on the borders when a low voltage is applied, and the junctions experience very little reverse leakage current.

When the switching field is exceeded by the enforced voltage, the assembled charges avalanche to produce a significant current. Typically, ZnO varistor is employed to conduct the high current to guard against overvoltage damage to electric circuits or electronic components, such as surge protection devices and surge arresters.

A composite-based polymer matrix filled with powdered semi-conductor (like ZnO) makes up a microvaristor. A device like this is frequently used to reduce the localised field amplification that could cause partial discharges and corona in high voltage equipment. Microvaristor compounds have a considerable ability to change and modify electrical characteristics due to their advantageous qualities. To generate a stereotyped pattern in traditional varistors, a high sintering temperature (about 1000 °C) is necessary (Ishibe et al., 2014; Saadeldin et al., 2019). On the other hand, ZnO-based microvaristor composites allow for the construction of devices on flexible substrates and only require a curing temperature of about 100 °C. In high voltage applications such stator coils, polymer insulators, electric-field grading of bushings, and cable accessories, ZnO microvaristors are therefore frequently used (Eda, 1989; Donzel et al., 2011; Weida et al., 2011; Donnelly & Varlow, 2003). When the filler concentration is over the percolation threshold, ZnO-epoxy, ZnO-Linear low density PE, and ZnO-Low density PE exhibit electric nonlinear field-dependent conductivity (Chiu, 2014; Chiu et al., 2009; Lin et al., 2008).

5.1.2 Pulsed Field Ablation Technique

The technique of catheter ablation is used to treat cardiac arrhythmias that are brought on by early or inappropriate myocardial activation due to electrical triggers or faulty conduction patterns in the myocardium (Yang & Zu, 2016). Radiofrequency microwave, high intensity focused ultrasound, lasers, and cryogenics are a few of the technologies that have been developed over the past three decades for cardiac ablation. These technologies, however, frequently fall short in closing lesion gaps, which might cause the arrhythmia to reoccur and make the treatment plan more challenging. Endocardial pulsed field ablation (PFA) stands out among them because of its non-thermal ablative action, which is very advantageous to myocardial tissue.

It is known as "electroporation" when a procedure like this produces an immediate electrical field that causes microscopic doors (nano-pores) in surrounding cells to open. All other modern thermal energy sources have the potential to cause collateral damage, which PFA has the capacity to fully prevent. Cardiomyocytes (heart tissues) have a particularly lower irreversible threshold of electroporation than other types of cells, such as the liver (0.6 V/m), kidney (0.7 V/m), vascular smooth muscle cells (1750 V/m), and nervous system (3800 V/m), according to the review by F. D. Ramirez et al. (Kramar et al., n.d.). As a result, PFA turns into a selective agent that only affects cardiomyocytes while sparing nearby non-heart tissue like nerve cells or the oesophagus.

The PFA catheter also differs from conventional ablation energy sources in that it simply needs to be close to the tissue being ablated rather than making direct contact with it. As a result, the patients are not left with scarring or char formation. This essentially means that PFA is significantly safer than other modern methods.

For this treatment, the business Farapulse developed a unique catheter, which has been CE-marked since October 2020 (Ramirez et al., 2020). Compared to hours for radiofrequency, the invented device could generate a suitable electric field in seconds or minutes. Different catheters can be configured to deliver therapy using the Farapulse generator, which generates the waveform. PFA is now extremely quick and precise because to this. It's interesting how many optimisations may be made to the first functional Farapulse gadget in order to forward the advancement of such a novel method. Additionally, the creation of multifunctional materials is a key innovation axis for PFA devices. A number of factors would be taken into account when designing a smart catheter-based PFE, including 1) the bending angles the structure is capable of, 2) the potential to change the shape from a 2D to a 3D structure, 3) the degree of rotation required and the design's ability to be miniaturised, and 4) the safety of the patients in the event of an electrical failure.

This section's goal is to provide information on the possibilities of ZnO composite for the creation of a smart catheter for PFA. Figure 10, where several electrodes are deposited above the composite surface, shows the basic design of the proposed device. A setup like this enables the delivery of a succession of extremely brief electrical pulses to destroy heart tissue. The following factors were the primary limitations of the manufactured materials: 1) mechanical flexibility and biocompatibility; 2) patient electric protection; 3) regulating electric field level; and 4) contact force sensing.

Regarding the first issue, it has been shown that PDMS and ZnO have biocompatible qualities (Chen et al., 2015; Jing & Guo, 2019; Reddy et al., 2019) and mechanical rigidity that is in line with PFA specifications (Ateş et al., 2000; Blanc et al., 2017; Jakmuangpak et al., 2020). Additionally, by adjusting the particle concentration, the mechanical characteristics of the created composites can be somewhat modified. The second point focuses on patient safety and states that in the event that the composite composite has dielectric breakdown, electric discharge within the heart tissue must be avoided. In fact, ZnO powder with voltage-dependent electrical conductivity can be successfully used as a microvaristor. This effect is frequently employed to reduce the localised field amplification that could result in partial

Figure 10. Principle scheme of catheter integrated with random or aligned ZnO/PDMS composite

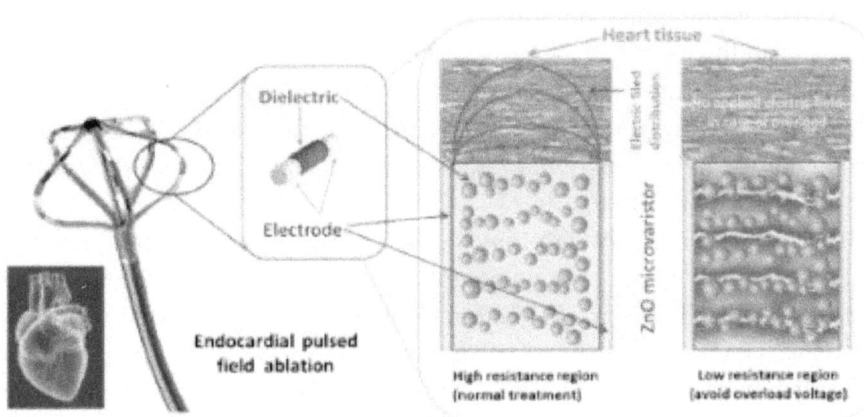

discharges and corona during the PFA procedure in high voltage equipment (Kramar et al., n.d.). Figure 10 illustrates how ZnO particle-based composites act as dielectrics with a high resistive region close to the heart tissue, where an electric field is imposed to carry out a typical ablation procedure, under low electric fields. ZnO composites function as a variable conductor to draw a big current through the heart tissue once the imposed electric field is too strong for the tissue, leaving little electric field in nearby tissue.

The third problem tries to demonstrate how easily several key parameters can be changed to adjust the electric field that is applied to the treated tissue. It should be proven that various composite parameters, including particle concentration, percolation threshold, electric and dielectric properties, affect the performance of microvaristor. External factors including temperature, frequency, and voltage excitation amplitude all play a significant role. Thus, it is highly advantageous to optimise the material process by dielectrophoresis, which could result in a comparable performance by employing lower particle concentration and voltage stimulation. Last but not least, the goal is to create a smart PFA with several functions, such as sensing capabilities in addition to electroporation effect. Performing the contact force (CF) between the ablation electrode and the atrial wall, which is a crucial factor in lesion size and durability, is one of the essential aspects here (Shirasaki et al., 2001).

5.2 Application-Based Piezoelectric Properties

In terms of the piezoelectric property, composites based on ZnO have been investigated in a variety of applications. Actually, there are two uses for the electrical signals that were transformed from the mechanical input. One method is to gather mechanical energy and store the resulting electric energy as power sources. Another one functions as sensors that measure mechanical deformation's frequency and magnitude based on corresponding in-time electric signals.

5.2.1 Energy Harvesting

In actuality, all are wasting ambient energy every day, including thermal, mechanical, and solar energy. Green energy from the environment must be harvested and scavenged in order to lessen the burden and consumption of energy from fossil fuels and coal. In the environment, mechanical energy is one of them and is widespread. By way of the direct piezoelectric effect, piezoelectric materials, for example, can capture vibration from machines, air flow, body motion, etc. Particularly, ZnO nanostructure-based systems are unquestionably becoming attractive alternatives because of their high harvesting effectiveness and ability to adapt to tiny scale vibration (Aryana et al., 2016; Falconi et al., 2009). ZnO NWs-based devices have the ability to collect mechanical vibration over a broad frequency range, in contrast to cantilever-based microelectromechanical systems, which can only operate at a few high resonance frequencies determined by the length of the cantilever. Li and his coworkers have described crystalline ZnO NWs with integrated micro-energy harvesters that produce 22 mW/cm^3 of volume-normalized power at 1.4 g at 40 Hz (Li et al., 2016).

Additionally, its capacity for energy gathering aids in the revolutionary creation of power sources. In order to satisfy the low energy needs of nanosystems or nanodevices, the piezoelectric nanogenerator was created (Lee et al., 2012; Tao et al., 2017). First, Wang et al. hypothesised that by deflecting ZnO NWs with a conductive AFM tip (Liu et al., 2019), nanoscale mechanical energy may be transformed into electric energy . In such instance, a single NW/nanotube-based device can be driven with an efficiency of up to 30% by a nanogenerator composed of ZnO NW arrays of size 10 m 10 m. Additionally, Yang

Figure 11. (a) Generating electricity by deforming a piezoelectric NW with a conductive AFM tip (Liu et al., 2019); (b) a flexible power generator based on cyclic stretching-releasing of a piezoelectric fine wire create an oscillating output voltage up to ~ 50 mV (Yang, Qin, Dai et al, 2009); (c) a three-layer integration of the vertical nanowire array integrated nanogenerator enhance the output voltage up to 0.243 V (Xu et al., 2010); (d) nanogenerators with a structure of a cantilever with five layers was integrated onto a tire's inner surface harvest energy from the motion of automobile (Hu et al., 2011)

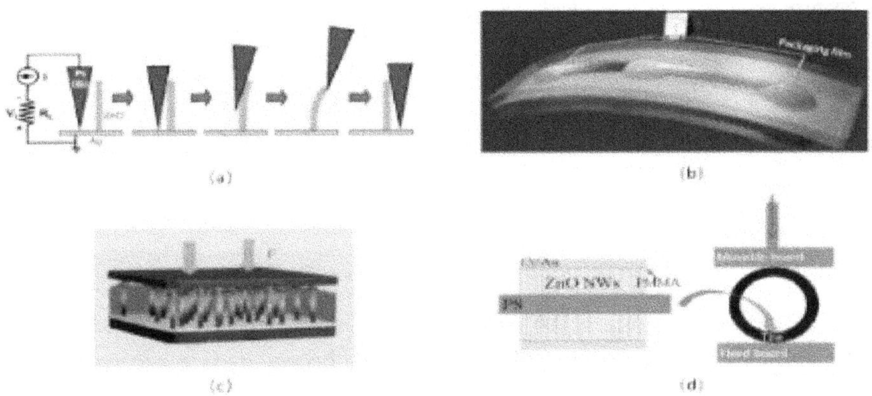

et al. (Yang, Qin, Dai et al, 2009) have worked on the creation of flexible laterally integrated nanogenerators based on cyclic stretching-releasing movement with both ends fixed. The vertically integrated nanogenerators (VING) (Figure 11 c), which have one end linked to the substrate and the other free to move, have been reported to have excellent energy harvesting properties (Xu et al., 2010). Hu et al. have shown how a nanogenerator could be used to capture energy from the rolling of car tyres (Figure 11 d) (Hu et al., 2011). These new generations of power sources can provide sustainable and long-term assistance without polluting the environment once they are used up, in contrast to conventional power sources.

5.2.2 Pressure Sensor in Biomedical Field

Our intended use for the created sensing device is to assess the relevance of arterial narrowing, notably in coronary stenoses, using the fractional flow reserve (FFR) technique. A stenosis in the arteries limits the amount of blood that can flow through the heart's blood vessels in a cardiac patient. It probably causes breathing and/or heart discomfort, or even a heart attack. Additionally, accurate artery narrowing diagnosis can aid in determining the severity of the condition and the best course of treatment. For instance, tight obstructions could be handled medically rather than surgically in order to save money. The pressure-derived FFR methodology, when compared to visual angiography, is a reliable invasive method for assessing coronary artery disease (Pijls & De Bruyne,).

The FFR index is calculated by dividing the maximum myocardial blood flow while stenosis is present by the theoretical maximum flow while the stenosis is absent (Pijls et al., 2009). If a pressure sensor guidewire is introduced into a heart channel to concurrently measure the pressure on both sides of the arterial obstruction, as shown in Figure 12(a), the results will be displayed.

The three pressure sensor technologies now used in FFR marketing are piezo-resistive sensor, optical fibre technology, and microcatheter (Sharma et al.,). The fiber-optic sensor must be used with opti-

Figure 12. (a) Schematic illustration of FFR sensor implemented inside the vessel. Pa is the proximal coronary pressure while Pd is the distal coronary pressure; (b) fabricated sensor implemented on a catheter was inserted into a cardiac simulation arm (D'Ambrogio et al., 2021).

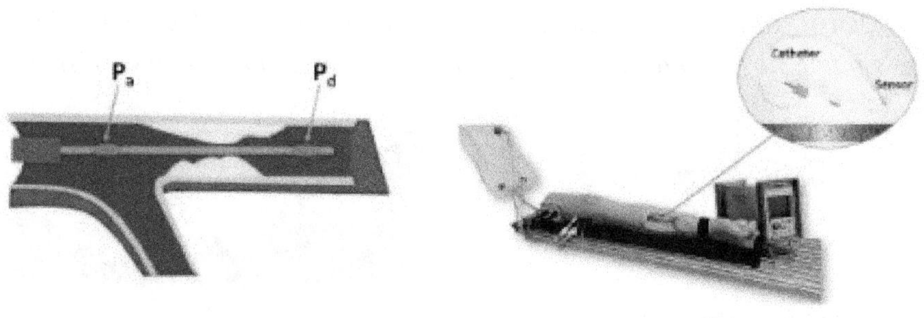

(a) (b)

cal fibre and microcatheter technologies. An optical sensor has advantages including small size, quick response, and sensitivity to fibre bending. The optic fibre, however, is quite brittle and constrained by the minimum bend radius in the case of convoluted vessels. As opposed to a pressure wire, it has been shown that FFR assessed with a microcatheter overestimates the severity of stenosis (Pouillot et al., 2018). The piezo-resistive sensor has some shortcomings, including a high power need, long-term instability in dynamic field settings, and non-linearity. However, it offers good temperature sensing and mechanical stability.Barium titanate nanoparticles/PDMS composite film incorporated into a catheter has been effectively simulated as an FFR sensor in a cardiac simulation arm, according to D'Ambrogio et al.'s paper (D'Ambrogio et al., 2021). This piezoelectric film exhibits a comparable and acceptable sensitivity around 0.75 mV kPa^{-1} under an arterial pulse-shape stress, donated as 12 kPa pressure with a 66-bpm heart rate, demonstrating high potential employed in the FFR tool with additional merits such as high flexibility, and ease of process and integration (D'Ambrogio et al., 2021). As a result, we think that ZnO-based piezoelectric composites would be a more practical alternative as a flexible and sensitive pressure sensor, making it simpler to integrate one on a catheter for the measurement of FFR pressure.

Additionally, the therapy of vascular illness has frequently utilised endovascular technology (Reuben et al., 2007; Rheaume et al., 2010). Remote vascular access, in particular when treating arterial vascular injuries, could speed up healing and cause less tissue damage and contamination, which would increase patient survival when compared to traditional open surgery. Pressure sensors that are attached to the catheter can be used for a thorough and accurate monitoring of the blood vessels by enabling the detection of various pressure locations. Internal catheter space must be preserved for interventional equipment due to the working environment. In that situation, it would be necessary to integrate a thin, flexible pressure sensor onto the curved surfaces of catheters.Consequently, the ability of ZnO vertically aligned NWs composites to achieve ultrathin thickness, even to several micrometres, would make them suitable and well-liked, while the flexibility could be tuned by the material characteristics of the encapsulating polymer matrix or the volume ratio between the NWs and the polymer. The developed ZnO sensor is also environmentally friendly and biologically compatible, demonstrating its tremendous potential for implantable sensing under in vivo physiological settings.

5.2.3 Self-Powered Systems Adopted in the Biomedical Field

Once ZnO NW-based piezoelectric nanogenerators are used in the biological environment, from natural to in vitro and in vivo working as self-powered systems, it becomes much more appealing on a larger scale. These devices have potential benefits for use as medical implants or portable electronics, including no need for an external power source, no need for replacement or recharging, ability to be miniaturised, biocompatibility, and non-toxicity (Wang, 2011). Particularly, there are many mechanical energy sources in living things, including muscle stretching, body movement, and blood flow and heartbeat in vivo environments. The movement of a human index finger and a live running hamster have been used to detect and harvest biomechanical energy using a single wire generator with both ends fixed (Yang, Qin, Li et al, 2009). To track different swimming techniques, a portable self-powered biosensor based on ZnO NWs has been developed (Mao et al., 2021). Additionally, sensors based on NWs are utilised to monitor the health situation by detecting human activities linked to skin deformation. In order to characterise the movements of the eyeball beneath the thin film of aligned ZnO nanowires utilised by Lee et al., it was possible to record sleep behaviour, fatigue, and brain activity (Lee et al., 2014). What's more, ZnO NWs can be implanted into living things without harm in order to capture low-frequency mechanical energy. In order to harvest the mechanical energy produced by a live rat's breathing and heartbeat, Li et al. have already described the viability of implanting a single ZnO wire generator in the animal (Li et al., 2010). Self-powered systems can also use human interior vibrations to power smart nanodevices that distribute drugs or monitor those vibrations to speed up diagnostics.

In addition, it is generally known that ZnO can be investigated as a multifunctional material due to its diverse features, including those involving the optical, piezoelectric, electric, semiconductor, and more. The concept of piezotronics and piezo-photonics in wurtzite structured material was initially introduced by Wang et al. (Wang, 2012). Numerous coupling effects may be produced as a result of piezoelectricity, piezoelectricity, and photoexcitation overlap. When designing electronic devices, such as piezoelectric field effect transistors and piezoelectric gated diodes based on a single ZnO nanowire, one interpretation of piezotronics, which is based on the piezoelectricity-semiconductor coupling, is to use the piezoelectric potential as a gate to tune the charge transport process.A piezoelectricity-semiconductor-photoexcitation coupling is known as piezo-photonics, and it allows for the control of electro-optical processes via a strain-induced piezopotential. The piezoelectric effect has been discovered to increase the light emission efficiency of ZnO micro-based LEDs (He et al., 2007; Wang et al., 2006; Yang et al., 2011).

6. CONCLUSION

The ambient world is filled with mechanical motion on a huge diversity of scales and in many different forms. It can be obtained from human body as well. Harvesting mechanical energy is a crucial step in generating affordable, sustainable, and clean electric energy. Over the past few decades, more research has been put into developing renewable energy technology that can turn the motions of the environment into power.

The investigation of improving implantable biomedical sensor applications' potential piezoelectric performance of ZnO-based composites. The optimum tuning of material properties of individual phases as well as structural design were used to develop the optimisation guidelines in order to achieve flexible, tiny, and effective sensors adopted to low-frequency biological media.

ZnO has attracted a great deal of interest for a wide range of electrical, optoelectronic, chemical, and biological sensing applications because it is a biocompatible semiconductor made of plentiful elements. Due to its high crystalline quality, intriguing physical features (dielectric and semiconductor), cost-effective application in flexible electronics, and potential for recycling, this material has a number of positive traits. As a result, research into the incorporation of ZnO into piezoelectric devices such as nanogenerators, pressure/strain sensors, and varistors has been ongoing over the past ten years. Furthermore, technological issues like stack architecture, polymer matrix, alignment of ZnO fillers, etc. have been significantly taken into account in order to optimize the electric potential generated under mechanical stress. To establish the potential and dependability of the created materials, complete characterizations of ZnO composites in terms of dielectric, electrical, mechanical, and piezoelectric properties are required.

REFERENCES

Agrawal, R., & Espinosa, H. D. (2011). Giant piezoelectric size effects in zinc oxide and gallium nitride nanowires. A first principles investigation. *Nano Letters, 11*(2), 786–790. doi:10.1021/nl104004d PMID:21241064

Arnau, A., & Soares, D. (2008). *Fundamentals of piezoelectricity*. Piezoelectric Transducers Appl. doi:10.1007/978-3-540-77508-9_1

Arnold, M. S., Avouris, P., Pan, Z. W., & Wang, Z. L. (2003). Field-effect transistors based on single semiconducting oxide nanobelts. *The Journal of Physical Chemistry B, 107*(3), 659–663. doi:10.1021/jp0271054

Aryana, A., Singh, S. M., Mugnai, G., de Asmundis, C., Kowalski, M., Pujara, D. K., Cohen, A. I., Singh, S. K., Fuenzalida, C. E., Prager, N., Bowers, M. R., O'Neill, P. G., Brugada, P., d'Avila, A., & Chierchia, G. B. (2016). Pulmonary vein reconnection following catheter ablation of atrial fibrillation using the second-generation cryoballoon versus open-irrigated radiofrequency: Results of a multicenter analysis. *Journal of Interventional Cardiac Electrophysiology, 47*(3), 341–348. doi:10.1007/s10840-016-0172-z PMID:27475949

Ateş, Y., Yücesoy, C. A., Ünlü, M. A., Saygin, B., & Akkaş, N. (2000). The Mechanical Properties of Intact and Traumatized Epidural Catheters. *Anesthesia and Analgesia, 90*(2), 393–399. doi:10.1213/00000539-200002000-00029 PMID:10648328

Blanc, L., Delchambre, A., & Lambert, P. (2017). Flexible medical devices: Review of controllable stiffness solutions. *Actuators, 6*(3), 23. Advance online publication. doi:10.3390/act6030023

Chen, Shi, Zhang, Zhu, & Yan. (2006). *Size Dependence of Young's Modulus in ZnO Nanowires*. . doi:10.1103/PhysRevLett.96.075505

Chen, Y., Tse, W. H., Chen, L., & Zhang, J. (2015). Ag nanoparticles-decorated ZnO nanorod array on a mechanical flexible substrate with enhanced optical and antimicrobial properties. *Nanoscale Research Letters, 10*(1), 106. doi:10.1186/s11671-014-0712-3 PMID:25852401

Chiu. (2014). *A Review on Conduction Mechanisms in Dielectric Films*. Academic Press.

Chiu, F. C., Lee, C. Y., & Pan, T. M. (2009). Current conduction mechanisms in Pr2 O3 /oxynitride laminated gate dielectrics. *Journal of Applied Physics*, *105*(7), 074103. doi:10.1063/1.3103282

Consonni, V., & Lord, A. M. (2021). Polarity in ZnO nanowires: A critical issue for piezotronic and piezoelectric devices. *Nano Energy*, *83*, 105789. doi:10.1016/j.nanoen.2021.105789

Curie, J., & Curie, P. (1880). Développement par compression de l'électricité polaire dans les cristaux hémièdres à faces inclinées, Bull. *La Société Minéralogique Fr.*, *3*(4), 90–93. doi:10.3406/bulmi.1880.1564

D'Ambrogio, G., Zahhaf, O., Bordet, M., Le, M. Q., Della Schiava, N., Liang, R., Cottinet, P.-J., & Capsal, J.-F. (2021). Structuring BaTiO3 /PDMS Nanocomposite via Dielectrophoresis for Fractional Flow Reserve Measurement. *Advanced Engineering Materials*, *2100341*(10), 2100341. Advance online publication. doi:10.1002/adem.202100341

D'Ambrogio, G., Zahhaf, O., Hebrard, Y., Le, M. Q., Cottinet, P., & Capsal, J. (2020). Micro-Structuration of Piezoelectric Composites Using Dielectrophoresis: Toward Application in Condition Monitoring of Bearings. *Advanced Engineering Materials*, *2000773*. Advance online publication. doi:10.1002/adem.202000773

Della Schiava, Thetpraphi, Le, Lermusiaux, Millon, Capsal, & Cottinet. (2018). Enhanced Figures of Merit for a High-Performing Actuator in Electrostrictive Materials. *Polym.*, *10*, 263. . doi:10.3390/polym10030263

Desai, A. V., & Haque, M. A. (2007). Mechanical properties of ZnO nanowires. *Sensors and Actuators. A, Physical*, *134*(1), 169–176. doi:10.1016/j.sna.2006.04.046

Donnelly, K. P., & Varlow, B. R. (2003). Non-linear dc and ac conductivity in electrically insulating composites. *IEEE Transactions on Dielectrics and Electrical Insulation*, *10*(4), 610–614. doi:10.1109/TDEI.2003.1219645

Donzel, L., Greuter, F., & Christen, T. (2011). Nonlinear resistive electric field grading part 2: Materials and applications. *IEEE Electrical Insulation Magazine*, *27*(2), 18–29. doi:10.1109/MEI.2011.5739419

Duguet, E., Abboud, M., Morvan, F., Maheu, P., & Fontanille, M. (2000). PMMA encapsulation of alumina particles through aqueous suspension polymerisation processes. *Macromolecular Symposia*, *151*(1), 365–370. doi:10.1002/1521-3900(200002)151:1<365::AID-MASY365>3.0.CO;2-T

Eda, K. (1989). Zinc Oxide Varistors. *IEEE Electrical Insulation Magazine*, *5*(6), 28–30. doi:10.1109/57.44606

Falconi, C., Mantini, G., D'Amico, A., & Wang, Z. L. (2009). Studying piezoelectric nanowires and nanowalls for energy harvesting. *Sensors and Actuators. B, Chemical*, *139*(2), 511–519. doi:10.1016/j.snb.2009.02.071

Fan, F.-R., Tian, Z.-Q., & Wang, L. (2012). Z. L. *Nano Energy*, *1*, 328–334. doi:10.1016/j.nanoen.2012.01.004

Fu, M., Pan, D., Yang, Y., Shi, T., Zhang, Z., Zhao, J., Xu, H. Q., & Chen, Q. (2014). Electrical characteristics of field-effect transistors based on indium arsenide nanowire thinner than 10 nm. *Applied Physics Letters*, *105*(14), 143101. doi:10.1063/1.4897496

Furusawa, K., Nagashima, K., & Anzai, C. (1994). Synthetic process to control the total size and component distribution of multilayer magnetic composite particles. *Colloid & Polymer Science*, *272*(9), 1104–1110. doi:10.1007/BF00652379

Ganet, F., Le, M. Q., Capsal, J. F., Gérard, J. F., Pruvost, S., Duchet-Rumeau, J., Livi, S., Lermusiaux, P., Millon, A., & Cottinet, P. J. (2015). Haptic feedback using an all-organic electroactive polymer composite. *Sensors and Actuators. B, Chemical*, *220*, 1120–1130. doi:10.1016/j.snb.2015.06.071

Gao, P. X., Song, J., Liu, J., & Wang, Z. L. (2007). Nanowire piezoelectric nanogenerators on plastic substrates as flexible power sources for nanodevices. *Advanced Materials*, *19*(1), 67–72. doi:10.1002/adma.200601162

Gong, H., Zhang, Y., Quan, J., & Che, S. (2011). Preparation and properties of cement based piezoelectric composites modified by CNTs. *Current Applied Physics*, *11*(3), 653–656. doi:10.1016/j.cap.2010.10.021

Govorov, A. O., Richardson, H. H., Soci, C., Zhang, A., Xiang, B., Dayeh, S. A., Aplin, D. P. R., Park, J., Bao, X. Y., Lo, Y. H., & Wang, D. (2007). ZnO nanowire UV photodetectors with high internal gain. *Nano Letters*, *7*(4), 30–38. doi:10.1021/nl070111x PMID:17358092

Grinberg, D., Siddique, S., Le, M. Q., Liang, R., Capsal, J. F., & Cottinet, P. J. (2019). 4D Printing based piezoelectric composite for medical applications, J. Polym. Sci. Part B Polym. *Journal of Polymer Science. Part B, Polymer Physics*, *57*(2), 109–115. doi:10.1002/polb.24763

Hayashi, S., Fujiki, K., & Tsubokawa, N. (2000). Grafting of hyperbranched polymers onto ultrafine silica: Postgraft polymerization of vinyl monomers initiated by pendant initiating groups of polymer chains grafted onto the surface. *Reactive & Functional Polymers*, *46*(2), 193–201. doi:10.1016/S1381-5148(00)00053-5

He, J. (2019). *Introduction of Varistor Ceramics, Met.* Oxide Varistors. doi:10.1002/9783527684038.ch1

He, J. H., Hsin, C. L., Liu, J., Chen, L. J., & Wang, Z. L. (2007). Piezoelectric Gated Diode of a Single ZnO Nanowire. *Advanced Materials*, *19*(6), 781–784. doi:10.1002/adma.200601908

He, M.-R., Shi, Y., Zhou, W., Chen, J. W., Yan, Y. J., & Zhu, J. (2009). Diameter dependence of modulus in zinc oxide nanowires and the effect of loading mode: In situ experiments and universal core-shell approach. *Applied Physics Letters*, *95*(9), 091912. doi:10.1063/1.3205102

Hong, J., & Fang, D. (2008). Systematic study of the ferroelectric properties of Pb(Zr0.5Ti0.5)O3 nanowires. *Journal of Applied Physics*, *104*(6), 064118. doi:10.1063/1.2982090

Hu, Y., Xu, C., Zhang, Y., Lin, L., Snyder, R. L., & Wang, Z. L. (2011). A nanogenerator for energy harvesting from a rotating tire and its application as a self-powered pressure/speed sensor. *Advanced Materials*, *23*(35), 4068–4071. doi:10.1002/adma.201102067 PMID:21812038

Huang, Mao, Feick, Yan, Wu, Kind, Weber, Russo, & Yang. (2001). Room-temperature ultraviolet nanowire nanolasers. *Science, 292*, 1897–1899. . doi:10.1126/science.1060367

Ishibe, Mori, & Kozako. (2014). *A New Concept Varistor With Epoxy / Microvaristor Composite*. Academic Press.

Jakmuangpak, S., Prada, T., Mongkolthanaruk, W., Harnchana, V., & Pinitsoontorn, S. (2020). Engineering Bacterial Cellulose Films by Nanocomposite Approach and Surface Modification for Biocompatible Triboelectric Nanogenerator. *ACS Applied Electronic Materials*, *2*(8), 2498–2506. doi:10.1021/acsaelm.0c00421

Jiansirisomboon, S., Songsiri, K., Watcharapasorn, A., & Tunkasiri, T. (2006). Mechanical properties and crack growth behavior in poled ferroelectric PMN-PZT ceramics. *Current Applied Physics*, *6*(3), 299–302. doi:10.1016/j.cap.2005.11.004

Jing, X., & Guo, Z. (2019). Fabrication of biocompatible super stable lubricant-immobilized slippery surfaces by grafting a polydimethylsiloxane brush: Excellent boiling water resistance, hot liquid repellency and long-term slippery stability. *Nanoscale*, *11*(18), 8870–8881. doi:10.1039/C9NR01556F PMID:31012900

Kawai, H. (1969). The Piezoelectricity of Poly (vinylidene Fluoride). *Japanese Journal of Applied Physics*, *8*(7), 975–976. doi:10.1143/JJAP.8.975

KhanbarehH. (2016). Expanding the Functionality of Piezo-Particulate Composites. doi:10.4233/UUID:AAB2497D-6EEC-4956-8C15-8F5FDEDC94F3

Kramar, P., Miklavčič, D., & Mir, L. M. (n.d.). Electroporation-based Technologies and Treatments. *Proceedings of the Electroporation-based Technologies and Treatments*. www.ebtt.org

Kulkarni, S. B., Patil, U. M., Salunkhe, R. R., Joshi, S. S., & Lokhande, C. D. (2011). Temperature impact on morphological evolution of ZnO and its consequent effect on physico-chemical properties. *Journal of Alloys and Compounds*, *509*(8), 3486–3492. doi:10.1016/j.jallcom.2010.12.036

Law, M., Greene, L. E., Johnson, J. C., Saykally, R., & Yang, P. (2005). Nanowire dye-sensitized solar cells. *Nature Materials*, *4*(6), 455–459. doi:10.1038/nmat1387 PMID:15895100

Le, M. Q., Belhora, F., Cornogolub, A., Cottinet, P. J., Lebrun, L., & Hajjaji, A. (2014). Enhanced magnetoelectric effect for flexible current sensor applications. *Journal of Applied Physics*, *115*(19), 194103. doi:10.1063/1.4876910

Le, M. Q., Capsal, J. F., Galineau, J., Ganet, F., Yin, X., Yang, M. D., Chateaux, J. F., Renaud, L., Malhaire, C., Cottinet, P. J., & Liang, R. (2015). All-organic electrostrictive polymer composites with low driving electrical voltages for micro-fluidic pump applications. *Scientific Reports*, *5*(1), 1–13. doi:10.1038/srep11814 PMID:26139015

Lee, H., Cooper, R., Wang, K., & Liang, H. (2008). Nano-Scale Characterization of a Piezoelectric Polymer (Polyvinylidene Difluoride, PVDF). *Sensors (Basel)*, *8*(11), 7359–7368. doi:10.3390/s8117359 PMID:27873933

Lee, K. Y., Kumar, B., Seo, J. S., Kim, K. H., Sohn, J. I., Cha, S. N., Choi, D., Wang, Z. L., & Kim, S. W. (2012). P-type polymer-hybridized high-performance piezoelectric nanogenerators. *Nano Letters*, *12*(4), 1959–1964. doi:10.1021/nl204440g PMID:22409420

Lee, S., Hinchet, R., Lee, Y., Yang, Y., Lin, Z.-H., Ardila, G., Montès, L., Mouis, M., & Wang, Z. L. (2014). Ultrathin Nanogenerators as Self-Powered/Active Skin Sensors for Tracking Eye Ball Motion. *Advanced Functional Materials*, *24*(8), 1163–1168. doi:10.1002/adfm.201301971

Li, Y., Çelik-Butler, Z., & Butler, D. P. (2016). An integrated piezoelectric zinc oxide nanowire micro-energy harvester. *Nano Energy*, *26*, 456–465. doi:10.1016/j.nanoen.2016.05.008

Li, Z., Yang, R., Yu, M., Bai, F., Li, C., & Wang, Z. L. (2008). Cellular level biocompatibility and bio-safety of ZnO nanowires. *The Journal of Physical Chemistry. C, Nanomaterials and Interfaces*, *112*(51), 20114–20117. doi:10.1021/jp808878p

Li, Z., Zhu, G., Yang, R., Wang, A. C., & Wang, Z. L. (2010). Muscle-driven in vivo nanogenerator. *Advanced Materials*, *22*(23), 2534–2537. doi:10.1002/adma.200904355 PMID:20446305

Lin, C. C., Lee, W. S., Sun, C. C., & Whu, W. H. (2008). A varistor-polymer composite with nonlinear electrical-thermal switching properties. *Ceramics International*, *34*(1), 131–136. doi:10.1016/j.ceramint.2006.09.018

Lines, M., & Glass, A. (1979). *Principles and applications of ferroelectrics and related materials*. Clarendon Press.

Liu, Q., Le, M. Q., Richard, C., Liang, R., Cottinet, P. J., & Capsal, J. F. (2019). Enhanced pseudo-piezoelectric dynamic force sensors based on inkjet-printed electrostrictive terpolymer. *Organic Electronics*, *67*, 259–271. doi:10.1016/j.orgel.2019.01.028

Look, D. C. (2001). Recent advances in ZnO materials and devices. *Materials Science and Engineering B*, *80*(1-3), 383–387. doi:10.1016/S0921-5107(00)00604-8

Mao, Y., Zhu, Y., Zhao, T., Jia, C., Bian, M., Li, X., Liu, Y., & Liu, B. (2021). A portable and flexible self-powered multifunctional sensor for real-time monitoring in swimming. *Biosensors (Basel)*, *11*(5), 1–11. doi:10.3390/bios11050147 PMID:34066654

Megaw. (1945). Crystal structure of barium titanate. *Nature*, *155*(3938), 484–485. doi:10.1038/155484b0

Morkoç & Özgür. (2009). *General Properties of ZnO*. . doi:10.1002/9783527623945.ch1

Murayama, N., Nakamura, K., Obara, H., & Segawa, M. (1976). The strong piezoelectricity in polyvinylidene fluroide (PVDF). *Ultrasonics*, *14*(1), 15–24. doi:10.1016/0041-624X(76)90067-6

Newnham, R. E., Skinner, D. P., & Cross, L. E. (1978). Connectivity and piezoelectric-pyroelectric composites. *Materials Research Bulletin*, *13*(5), 525–536. doi:10.1016/0025-5408(78)90161-7

Ohigashi, H. (2008). Electromechanical properties of polarized polyvinylidene fluoride films as studied by the piezoelectric resonance method. *Journal of Applied Physics*, *47*(3), 949–955. doi:10.1063/1.322685

Özgür, Ü., Alivov, Y. I., Liu, C., Teke, A., Reshchikov, M. A., Doğan, S., Avrutin, V., Cho, S.-J., & Morkoç, H. (2005). A comprehensive review of ZnO materials and devices. *Journal of Applied Physics*, *98*(4), 041301. doi:10.1063/1.1992666

Panda. (n.d.). *Review: Environmental friendly lead-free piezoelectric materials*. . doi:10.1007/s10853-009-3643-0

Park, W. I., & Yi, G.-C. (2004). Electroluminescence in n-ZnO Nanorod Arrays Vertically Grown on p-GaN. *Advanced Materials*, *16*(1), 87–90. doi:10.1002/adma.200305729

Pijls & De Bruyne. (n.d.). *Coronary pressure measurement and fractional flow reserve*. Academic Press.

Pijls, N. H. J., de Bruyne, B., Peels, K., van der Voort, P. H., Bonnier, H. J. R. M., Bartunek, J., & Koolen, J. J. (2009). *Measurement of Fractional Flow Reserve to Assess the Functional Severity of Coronary-Artery Stenoses*. doi:10.1056/NEJM199606273342604

Polymer. (n.d.). https://www.britannica.com/science/polymer

Pouillot, C., Fournier, S., Glasenapp, J., Rambaud, G., Bougrini, K., Vi Fane, R., Geyer, C., & Adjedj, J. (2018). Pressure wire versus microcatheter for FFR measurement: A head-to-head comparison. *Archives of Cardiovascular Diseases. Supplements*, *10*(1), 25–26. doi:10.1016/j.acvdsp.2017.11.048 PMID:28804057

Ramirez, F. D., Reddy, V. Y., Viswanathan, R., Hocini, M., & Jaïs, P. (2020). Emerging Technologies for Pulmonary Vein Isolation. *Circulation Research*, *127*(1), 170–183. doi:10.1161/CIRCRESAHA.120.316402 PMID:32716722

Reddy, V. Y., Neuzil, P., Koruth, J. S., Petru, J., Funosako, M., Cochet, H., Sediva, L., Chovanec, M., Dukkipati, S. R., & Jais, P. (2019). Pulsed Field Ablation for Pulmonary Vein Isolation in Atrial Fibrillation. *Journal of the American College of Cardiology*, *74*(3), 315–326. doi:10.1016/j.jacc.2019.04.021 PMID:31085321

Reuben, Whitten, Sarfati, & Kraiss. (2007). Increasing use of endovascular therapy in acute arterial injuries: Analysis of the National Trauma Data Bank. *J. Vasc. Surg.*, *46*, 1222-1226. doi:10.1016/j.jvs.2007.08.023

Rheaume, P., Chen, J., & Casey, P. (2010). Open vs endovascular repair of blunt traumatic thoracic aortic injuries. *Journal of Vascular Surgery*, *51*(3), 763–769. doi:10.1016/j.jvs.2009.12.014 PMID:20206817

Riaz, M., Song, J., Nur, O., Wang, Z. L., & Willander, M. (2011). Study of the piezoelectric power generation of ZnO nanowire arrays grown by different methods. *Advanced Functional Materials*, *21*(4), 628–633. doi:10.1002/adfm.201001203

Roy, A. K., Ahmad, Z., Prasad, A., & Prasad, K. (2012). Concentration dependent dielectric properties of Barium Titanate / Polyvenylidene Fluoride (PVDF) and. *Advanced Materials Research*, *1*(4), 285–297. doi:10.12989/amr.2012.1.4.285

Saadeldin, M. M., Desouky, O. A., Ibrahim, M., Khalil, G. E., & Helali, M. Y. (2019). Investigation of structural and electrical properties of ZnO varistor samples doped with different additives. *NRIAG J. Astron. Geophys.*, *7*(2), 201–207. doi:10.1016/j.nrjag.2018.06.002

Sharma, Aroom, Naik, Gill, & Zhang. (n.d.). *Flexible Thin-Film PVDF-TrFE Based Pressure Sensor for Smart Catheter Applications*. doi:10.1007/s10439-012-0708-z

Shirasaki, Yoshizawa, & Tateishi. (2001). Changes in the mechanical properties of silicone catheter in degrading process. *Nihon Reoroji Gakkaishi.*, *29*, 9–13. doi:10.1678/rheology.29.9

Tao, R., Parmar, M., Ardila, G., Oliveira, P., Marques, D., Montès, L., & Mouis, M. (2017). Performance of ZnO based piezo-generators under controlled compression. *Semiconductor Science and Technology*, *32*(6), 64003. doi:10.1088/1361-6641/aa691f

Tasker, P. W. (1979). The stability of ionic crystal surfaces. *Journal of Physics. C. Solid State Physics, 12*(22), 4977–4984. doi:10.1088/0022-3719/12/22/036

Tatarenko, Gheevarughese, Srinivasan, Antonenkov, & Bichurin. (2007). Microwave magnetoelectric effects in ferrite—piezoelectric composites and dual electric and magnetic field tunable filters. *J. Electroceramics, 241*(24), 5–9. . doi:10.1007/s10832-007-9382-1

Taylor, G. W. (1985). Piezoelectricity. Academic Press.

Thetpraphi, K., Le, M. Q., Houachtia, A., Cottinet, P. J., Petit, L., Audigier, D., Kuhn, J., Moretto, G., & Capsal, J. F. (2019). Surface Correction Control Based on Plasticized Multilayer P(VDF-TrFE-CFE) Actuator—Live Mirror. *Advanced Optical Materials, 7*(13), 1900210. doi:10.1002/adom.201900210

Uchino, K. (2012). *Piezoelectric ceramics for transducers, Ultrason.* Transducers Mater. Des. Sensors, Actuators Med. Appl., doi:10.1533/9780857096302.1.70

Valasek, J. (1921). Piezo-Electric and Allied Phenomena in Rochelle Salt. *Physical Review, 17*(4), 475–481. doi:10.1103/PhysRev.17.475

Wang, Herbert, & Glass. (1988). The Applications of ferroelectric polymers. *Blackie, 160.*

Wang, X., Zhou, J., Song, J., Liu, J., Xu, N., & Wang, Z. L. (2006). Piezoelectric Field Effect Transistor and Nanoforce Sensor Based on a Single ZnO Nanowire. *Nano Letters, 6*(12), 2768–2772. doi:10.1021/nl061802g

Wang, Z. (2011). *Nanogenerators for self-powered devices and systems.* http://smartech.gatech.edu/handle/1853/39262

Wang. (2004). *Zinc oxide nanostructures: growth, properties and applications.* . doi:10.1088/0953-8984/16/25/R01

Wang. (2009). ZnO nanowire and nanobelt platform for nanotechnology. *Mater. Sci. Eng. R Reports, 64*, 33–71. doi:10.1016/j.mser.2009.02.001

Wang, Z. L. (2012). Progress in piezotronics and piezo-phototronics. *Advanced Materials, 24*(34), 4632–4646. doi:10.1002/adma.201104365 PMID:22331639

Wang, Z. L., Chen, J., & Lin, L. (2015). Progress in triboelectric nanogenerators as a new energy technology and self-powered sensors. *Energy & Environmental Science, 8*(8), 2250–2282. doi:10.1039/C5EE01532D

Wang & Song. (2006). Piezoelectric nanogenerators based on zinc oxide nanowire arrays. *Science, 312*, 242–246. . doi:10.1126/science.1124005

Webster. (1999). *The measurement, instrumentation, and sensors handbook.* Academic Press.

Weida, D., Richter, C., & Clemens, M. (2011). Design of ZnO microvaristor material stress-cone for cable accessories. *IEEE Transactions on Dielectrics and Electrical Insulation, 18*(4), 1262–1267. doi:10.1109/TDEI.2011.5976125

What It Is and How It Works. (n.d.). National Nanotechnology Initiative. https://www.nano.gov/nanotech-101/what

Wilson, J. S. (2005). *Sensor Technology Handbook*. Elsevier Inc. doi:10.1016/B978-0-7506-7729-5.X5040-X

Xu, S., Qin, Y., Xu, C., Wei, Y., Yang, R., & Wang, Z. L. (2010). Self-powered nanowire devices. *Nature Nanotechnology*, *5*(5), 366–373. doi:10.1038/nnano.2010.46 PMID:20348913

Xu, S., & Wang, Z. L. (2011). One-dimensional ZnO nanostructures: Solution growth and functional properties. *Nano Research*, *4*(11), 1013–1098. doi:10.1007/s12274-011-0160-7

Yang, Q., Wang, W., Xu, S., & Wang, Z. L. (2011). Enhancing Light Emission of ZnO Microwire-Based Diodes by Piezo-Phototronic Effect. *Nano Letters*, *11*(9), 4012–4017. doi:10.1021/nl202619d PMID:21823605

Yang, R., Qin, Y., Dai, L., & Wang, Z. L. (2009). Power generation with laterally packaged piezoelectric fine wires. *Nature Nanotechnology*, *4*(1), 34–39. doi:10.1038/nnano.2008.314 PMID:19119280

Yang, R., Qin, Y., Li, C., Zhu, G., & Wang, Z. L. (2009). Converting biomechanical energy into electricity by a muscle-movement- driven nanogenerator. *Nano Letters*, *9*(3), 1201–1205. doi:10.1021/nl803904b PMID:19203203

Yang, Y., Zhang, H., Chen, J., Jing, Q., Zhou, Y. S., Wen, X., & Wang, Z. L. (2013). Single-Electrode-Based Sliding Triboelectric Nanogenerator for Self-Powered Displacement Vector Sensor System. *ACS Nano*, *7*(8), 7342–7351. doi:10.1021/nn403021m PMID:23883397

Yang, Z., & Zu, J. (2016). Comparison of PZN-PT, PMN-PT single crystals and PZT ceramic for vibration energy harvesting. *Energy Conversion and Management*, *122*, 321–329. doi:10.1016/j.enconman.2016.05.085

Zhang, X., Le, M. Q., Zahhaf, O., Capsal, J. F., Cottinet, P. J., & Petit, L. (2020). Enhancing dielectric and piezoelectric properties of micro-ZnO/PDMS composite-based dielectrophoresis. *Materials & Design*, *192*, 108783. doi:10.1016/j.matdes.2020.108783

Zhang, Y., Hong, J., Liu, B., & Fang, D. (2010). Strain effect on ferroelectric behaviors of BaTiO3 nanowires: A molecular dynamics study. *Nanotechnology*, *21*(1), 015701. Advance online publication. doi:10.1088/0957-4484/21/1/015701 PMID:19946153

Zhao, Wang, & Mao. (2004). *Piezoelectric Characterization of Individual Zinc Oxide Nanobelt Probed by Piezoresponse Force Microscope*. doi:10.1021/nl035198a

Zhou, J., Xu, N., & Wang, Z. L. (2006). Dissolving behavior and stability of ZnO wires in biofluids: A study on biodegradability and biocompatibility of ZnO nanostructures. *Advanced Materials*, *18*(18), 2432–2435. doi:10.1002/adma.200600200

Zhu, G., Chen, J., Liu, Y., Bai, P., Zhou, Y. S., Jing, Q., Pan, C., & Wang, Z. L. (2013). Linear-Grating Triboelectric Generator Based on Sliding Electrification. *Nano Letters*, *13*(5), 2282–2289. doi:10.1021/nl4008985 PMID:23577639

Chapter 8
Status and Prospects of GdIG Garnet Ferrites for Energy Storage Devices: A Review

Anjori Sharma
Lovely Professional University, India

Dipesh
Lovely Professional University, India

Hamnesh Mahajan
Lovely Professional University, India

A. K. Srivastavaa
Lovely Professional University, India

ABSTRACT

Energy storage devices are essential parts of contemporary energy networks because they allow for the effective use and integration of renewable energy sources and provide a steady supply of electricity. Due to their distinct features and ability to improve the efficiency of various energy storage systems, gadolinium iron garnet GdIG ($Gd_3Fe_5O_{12}$) garnet ferrites have emerged as interesting options for energy storage applications among a variety of other materials. This review study provides a thorough examination of GdIG garnet ferrites for energy storage devices, including information on their structural characteristics, and uses in a range of energy storage systems. Additionally, it analyses the difficulties and constraints these materials confront and investigates new fields of study, including improvements in customising material properties, and nanoscale applications.

DOI: 10.4018/979-8-3693-1306-0.ch008

1. INTRODUCTION

1.1 Background and Motivation

Energy storage systems need to be effective and dependable due to the world's rising energy demand as well as environmental sustainability and climate change concerns (Evliya, 2007, pp. 87-99; Hadjipaschalis et al., 2009). By storing excess energy during times of low demand and releasing it when demand exceeds supply, energy storage devices are essential for addressing the intermittent nature of renewable energy sources like solar and wind. The energy grid must be balanced in order to provide a steady and uninterrupted supply of power.

1.2 Significance of Energy Storage Devices in 21st Century

The focus of modern energy systems is shifting to be more decentralised and renewable. Renewable energy sources, however, are erratic by nature, making grid stability and energy management more difficult. The gap between energy production and consumption can be closed with the help of energy storage devices. They provide a number of advantages, such as (Prakash et al., 2022; Li & Wang, 2019; Al-Ghussain et al., 2021):

- Energy storage devices aid in regulating the frequency, voltage, and power quality of the grid, which helps ensure a steady and dependable supply of energy.
- Peak demand management: Energy storage systems make it possible to store excess energy during times of low demand and release it during times of high demand, hence avoiding the need for pricey peak power generating.
- Energy storage makes it easier to integrate intermittent renewable energy sources effectively, resulting in a cleaner and more sustainable energy mix.
- Energy storage offers backup power in the event of grid outages or blackouts, boosting energy security and resilience.
- Load Shifting: Energy storage systems help flatten load profiles by storing excess energy during off-peak hours and providing it during peak demand, maximising energy utilisation.

1.3 Overview to Gadolinium Iron Garnet Ferrite GdIG

Based on their structure, ferrites can be classified into three types (Tolani et al., 2019; Goldman, 2006; Tahanian et al; 2020): cubic structure ferrites, including spinel and garnets ferrite, and hexagonal structure ferrites, known as hexaferrite. Despite sharing Fe_2O_3 as the basic formula, these ferrites differ not only in structure but also in their properties and applications. This review will specifically focus on garnet ferrites. Figure 1 illustrates the Scopus data displaying the number of publications on garnet ferrites in the last 10 years. The basic rare-earth-based gadolinium iron garnet exhibits excellent resistivity, dielectric properties, high energy storage capability, large Faraday rotation, high initial permeability, thermal and chemical stability, high remanence, and low propagation losses (Ganne et al., 2007; Sekijima et al., 2001; Xu & Yang, 2007; Xu et al., 2008; Sharma et al., 2022; Li, 2020 pp. 15408-15416).

These materials exhibit excellent qualities that make them very appealing for use in energy storage applications. These ferrites, which are members of the rare-earth iron garnet family, have special qualities like:

- High Curie Temperature: GdIG garnet ferrites have a Curie temperature that is higher than room temperature, which ensures their performance and stability under a variety of operating circumstances.
- Low Magnetic Losses: GdIG garnet ferrites offer great efficiency and reduced energy dissipation in energy storage devices as a result of their low magnetic loss properties.
- GdIG garnet ferrites have magneto-strictive capabilities, which allow them to transform magnetic energy into mechanical vibrations that can be used in energy harvesting applications.
- Excellent chemical stability of these ferrites guarantees their long-term robustness and dependability in energy storage devices.
- Modifiable Properties: The properties of GdIG garnet ferrites can be modified to satisfy certain specifications for various energy storage applications by modifying the chemical structure and crystal structure.

Figure 1. Previous 10-year Scopus data of no. of publication on garnet ferrites and its application

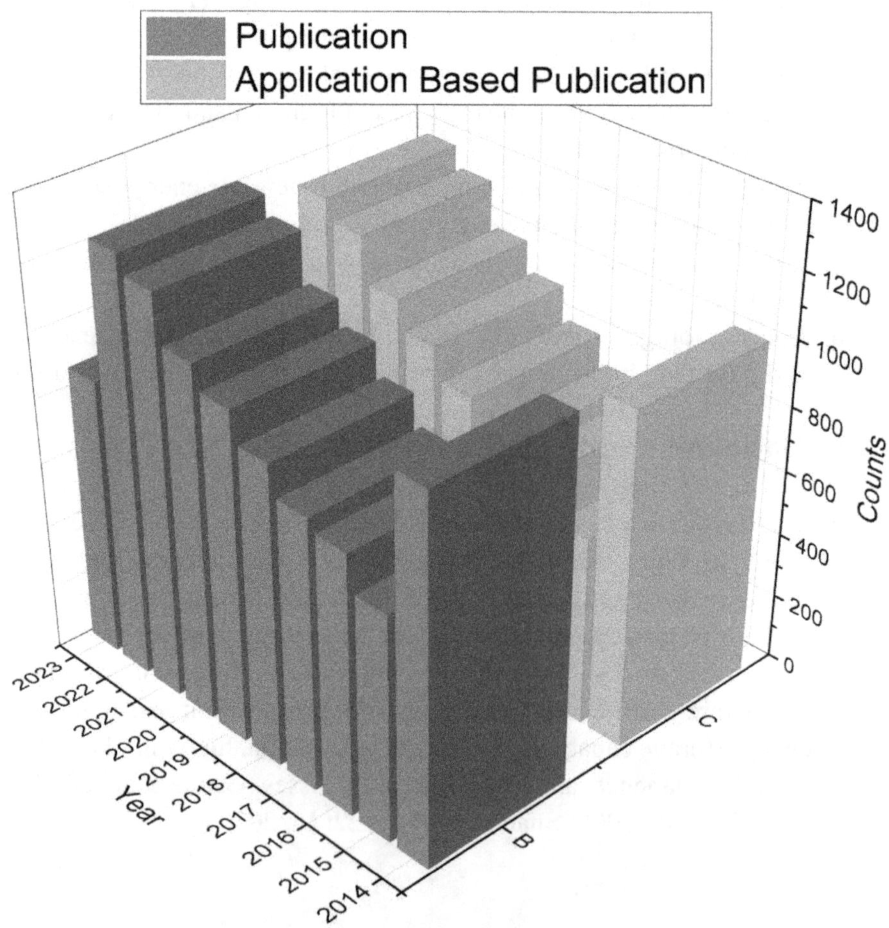

1.4 Aim of the Review

This review paper's main goal is to provide a thorough analysis of GdIG garnet ferrites for use in energy storage technologies. This essay seeks to:

- Summarise, contrast, and discuss the effects of the various synthesis techniques used to create GdIG garnet ferrites.
- Examine the structural and magnetic characteristics of GdIG garnet ferrites, focusing on how rare-earth dopants and synthesis variables affect them.
- Assess GdIG garnet ferrites' applicability in various energy-storage devices, including lithium-ion batteries, supercapacitors, magneto-strictive energy harvesters, and magnetic inductors
- Talk about the difficulties and restrictions involved in using GdIG garnet ferrites in actual energy storage devices.
- Examine recent developments in research and potential future applications to improve the functionality and use of GdIG garnet ferrites in the constantly changing field of energy storage technologies.
- Examine current research breakthroughs and prospective new uses to enhance the performance and applicability of GdIG garnet ferrites in the dynamic field of energy storage technologies.

2. CRYSTALLOGRAPHIC ORGANISATION AND LATTICE PARAMETERS

The family of rare-earth iron garnets includes GdIG garnet ferrites, which have a cubic crystal structure. Garnet ferrites possess a cubic crystal structure characterized by a stoichiometric formula of $C_3A_2D_3O12$, where C, A, and D represent the distribution of cations in the dodecahedral, octahedral, and tetrahedral sites within the crystal lattice. The cations are situated in the interspace of oxygen atoms (O^{2-}). In total, a single unit cell of garnet ferrite contains 160 atoms, comprising 8 formula units. By substituting various rare-earth elements or altering the synthesis conditions, the lattice characteristics of GdIG garnet ferrites can be changed, resulting in tuneable qualities for certain energy storage applications.

2.1 Magnetic Properties and Hysteresis (Ji et al., 2022)

Due to its distinctive crystal structure, rare-earth and iron ions, and unusual magnetic properties, GdIG garnet ferrites display remarkable magnetic behaviour and hysteresis loops. Since these substances have both spontaneous magnetization and a net magnetic moment, they are ferrimagnetic. The aligned magnetic moments of the Fe^{3+} ions at the B sites and the Gd^{3+} ions at the A sites interact to produce the magnetic behaviour. Hysteresis loops, which describe the relationship between the applied magnetic field and the resulting magnetization, can be used to quantify and visualise the magnetic behaviour.

In a hysteresis loop, the magnetic moments of the material align with the field as an external magnetic field is applied and gradually increased, leading to an increase in magnetization. The magnetization does not, however, revert to zero when the external field is lowered to zero, indicating the presence of remanent magnetization. Remanent magnetization is the difference between the magnetism at saturation and remanence, and this property is essential for magnetic storage applications, such as in magnetic inductors.

2.2 Rare-Earth Dopants' Effect on Material Properties (Sharma et al., 2022)

The structural and magnetic characteristics of GdIG garnet ferrites are significantly influenced by the addition of various rare-earth dopants, such as Gd, Y, Dy, and La. The material's crystal structure, lattice properties, and magnetic behaviour are all impacted by the dopant choice. For instance, the Curie temperature, magnetic anisotropy, and magneto-strictive characteristics of GdIG garnet ferrites can be changed by replacing Gd with other rare-earth elements. The thermal stability, magnetic loss, and electrical properties are likewise impacted by rare-earth dopants, making them essential variables to customise the material for certain energy storage applications.

2.3 Analysis of Mössbauer Spectroscopy

A potent method for examining the local surroundings and magnetic characteristics of atoms in a substance is Mössbauer spectroscopy. Mössbauer spectroscopy is used to look at the hyperfine interactions between the Fe^{3+} ions at the B sites and the Gd^{3+} ions at the A sites in the setting of GdIG garnet ferrites. Important new information about the coordination, oxidation, and magnetic ordering of the iron and rare-earth ions is provided by this technique. By understanding the distribution of Fe^{3+} ions in various magnetic sub-lattices and the effect of rare-earth doping on the hyperfine interactions, researchers may design and develop GdIG garnet ferrites for energy storage devices.

2.4 Analysis via X-Ray Diffraction (Sharma et al., 2022)

For the analysis of a material's crystal structure and phase purity, X-ray diffraction (XRD), in particular for GdIG garnet ferrites, is a common technique. By analysing the X-ray diffraction patterns, scientists may identify the sample's crystallographic phases and quantify the lattice parameters. The cubic crystal structure of the GdIG garnet ferrites is confirmed by XRD analysis, which also establishes how much phase creation took place throughout the manufacturing process. It is an essential tool for ensuring the synthesised material's purity and crystallinity, which directly affects its magnetic and structural properties.

2.5 Techniques for Microstructural Characterization (Sharma et al., 2022; Sharma et al., 2023)

The microstructure and morphology of GdIG garnet ferrites are investigated at the nanoscale using microstructural characterisation methods including scanning electron microscopy (SEM) and transmission electron microscopy (TEM). High-resolution surface pictures from SEM can be used to determine the distribution of rare-earth dopants as well as the size and porosity of the surface's grains. Using TEM, scientists may examine crystal flaws and structure at the atomic level. These methods are crucial for comprehending the interactions between GdIG garnet ferrites' microstructure and properties, especially in energy storage devices, where the material's microstructure has a big impact on its performance and dependability.

In a nutshell it is essential to have a thorough grasp of the structural and magnetic characteristics of GdIG garnet ferrites in order to maximise their use in energy storage systems. Fundamental characteristics of the material are determined by its crystal structure and lattice parameters, and their appropriateness for magnetic storage applications is indicated by their magnetic behaviour and hysteresis loops. Mössbauer spectroscopy and X-ray diffraction studies help investigate how the choice of rare-earth dopants affects the material's properties. Additionally, the nanoscale characteristics that affect the overall performance of GdIG garnet ferrites in energy storage devices can be understood by microstructural characterisation approaches.

3. APPLICATIONS OF GdIG GARNET FERRITES FOR ENERGY STORAGE

3.1 Garnet Ferrite as a Solid-State Electrolyte for Lithium-Ion Batteries (Huang et al., 2013)

3.1.1 Improved Stability and Safety

Due to their great energy density, lithium-ion batteries are widely used in a variety of applications. However, the presence of liquid electrolytes raises safety issues, particularly in situations with high temperatures. Ferrites made up of GdIG garnet can demonstrated potential as solid-state electrolytes for lithium-ion batteries (Ren et al., 2016; Thangadurai et al., 2014). They are a safer alternative to liquid electrolytes due to their exceptional chemical stability and non-flammability. For use in next-generation, high-performance lithium-ion batteries, GdIG-based solid-state electrolytes provide improved safety by lowering the possibility of electrolyte leakage, flammability, and thermal runaway.

3.1.2 Increase in Conductivity

Getting high ionic conductivity to match liquid electrolyte performance is one of the main obstacles to using solid-state electrolytes in lithium-ion batteries. The ionic conductivity of GdIG garnet ferrites can be improved further by tuning the crystal structure and doping concentrations. The goal of research has been to improve grain boundary conductivity using different processing methods, including grain boundary engineering and spark plasma sintering. GdIG-based solid-state electrolytes can enable quicker charge-discharge rates and enhance the general performance of lithium-ion batteries by enhancing the ionic conductivity.

3.2 Garnet Ferrites in Supercapacitors (Shiri et al., 2015)

3.2.1 Design and Production of Electrodes

The energy storage technology that fills the space between conventional capacitors and batteries is known as a supercapacitor, also referred to as an ultracapacitor. The potential of GdIG garnet ferrites as supercapacitor electrode materials has been researched. To incorporate GdIG garnet ferrites into the electrode structure, researchers have investigated several electrode designs and fabrication methods. These initiatives aim to maximise the energy storage capacity and charge-discharge rates of supercapacitors by optimising the electrode's active surface area, porosity, and electrical conductivity.

3.2.2 Increased Capacitance and cycle Stability

GdIG garnet ferrites' special qualities, such as their tuneable magnetization and low magnetic losses, have the potential to improve the capacitance and cycle stability of supercapacitors. Due to their steady electrochemical behaviour, GdIG garnet ferrites can be used as electrode materials to increase charge storage capacity and lengthen cycle life. GdIG garnet ferrites can also reduce the effects of polarisation and improve the overall energy efficiency of supercapacitors.

3.3 Magneto-Strictive Energy Harvesters Based on GdIG (Thirumalasetty et al., 2023)

3.3.1 Design Considerations and the Working Principle

Magneto-strictive energy harvesters use the magneto-strictive characteristics of specific materials to transform mechanical vibrations or strain into electrical energy. These uses benefit from the distinctive magneto-strictive properties of GdIG garnet ferrites. The GdIG garnet ferrite is subjected to mechanical vibrations in the working principle, which alters its magnetic characteristics and induces electrical voltage across a coil. In order to maximise the efficiency of energy conversion, design factors include optimising the geometry, size, and coupling between the mechanical vibrations and the GdIG garnet ferrite.

3.3.2 GdIG Garnet Ferrites' Magneto-Strictive Qualities

GdIG garnet ferrites have magneto-strictive qualities, which implies that when they are exposed to an external magnetic field, they alter in size. In magneto-strictive energy harvesters, the conversion of mechanical vibrations into electrical energy depends on this magneto-strictive effect. GdIG garnet ferrites are excellent candidates for energy harvesting applications because to their special combination of high Curie temperature and magneto-strictive properties, which enables efficient energy harvesting across a wide range of temperatures and environmental conditions.

3.4 GdIG Garnet Ferrites for Magnetic Inductors (Wang et al., 2016; Mallmann et al., 2013; Eckel et al., 2009)

Magnetic inductors are crucial parts of power electronics and electrical circuits. They reduce core losses and increase efficiency. Magnetic inductors' efficiency is constrained by core losses, which are mostly caused by hysteresis and eddy current losses. In comparison to conventional magnetic materials like silicon steel and ferrites, GdIG garnet ferrites have lower magnetic losses and a higher Curie temperature. GdIG garnet ferrites are advantageous in high-frequency and high-power applications because they can reduce core losses and boost energy efficiency when used as the magnetic inductors' core material.

3.4.1 Design and Performance Optimisation

For magnetic inductors, design optimisation entails adjusting the material's microstructure and composition to obtain a desired magnetic property. The magnetic characteristics of GdIG garnet ferrites can be tailored to meet the needs of magnetic inductors for various applications by adjusting the rare-earth dopants and synthesis conditions. Performance of the inductor is also greatly influenced by the core geometry and winding configuration. Magnetic inductors with improved performance characteristics, smaller size, and lighter weight can be produced with the proper design and selection of GdIG garnet ferrites.

4. LIMITATIONS AND CHALLENGES

4.1 Bulk Production and Economically

Energy storage systems that use GdIG garnet ferrites must be able to be produced using processes that are both affordable and scalable. High-quality GdIG garnet ferrites are currently difficult to produce in large quantities, which has an impact on their commercial viability. These materials' widespread use is constrained by the sophisticated and pricey methods frequently used in the synthetic techniques for creating them. To get over this restriction, scientists need to investigate more effective and affordable synthesis methods that preserve the material's desirable features without sacrificing quality.

4.2 Interface Compatibility in Energy Storage Devices

GdIG garnet ferrites need to have strong interface compatibility with other components used in the device's construction in order to be employed effectively in energy storage devices. Interfacial resistances or chemical processes that impair device performance can be introduced at interfaces between GdIG garnet ferrites and other parts, such as electrodes or separators. In order to reduce interfacial losses and increase overall device efficiency, good interfacial qualities require careful material selection and manufacturing condition optimisation.

GdIG garnet ferrites may have brittleness and fragility, which might provide difficulties during the manufacture and integration of devices. Particularly in applications involving mechanical stress or vibrations, like magneto-strictive energy harvesters, mechanical strength and resilience are critical factors. To increase the durability and dependability of GdIG garnet ferrites in energy storage devices, researchers must look into ways to boost their mechanical properties, such as through composite architectures or doping schemes.

4.3 Issues With Thermal and Electrical Conductivity

In order to ensure effective heat dissipation and minimal energy loss, thermal and electrical conductivity are crucial factors in energy storage systems. When compared to other typical materials, GdIG garnet ferrites may have comparatively poorer thermal and electrical conductivities, which could influence their ability to dissipate heat and the effectiveness of their charge transport. By adding conductive additives or creating composite structures, it may be possible to increase thermal and electrical conductivity while maintaining the necessary magnetic properties of GdIG garnet ferrites.

4.4 Scalability and Commercial Viability

The ability of GdIG garnet ferrites to be scaled up for use in energy storage systems is directly related to their commercial viability. Although promising at the laboratory scale, it is difficult to scale up manufacturing of high-quality GdIG garnet ferrites to satisfy industrial demand. To guarantee consistent and repeatable performance of GdIG-based energy storage devices, issues with batch-to-batch consistency, uniformity, and quality control must be addressed.

4.5 Environmental and Safety Considerations

The environmental and safety implications of GdIG garnet ferrites must be carefully considered, as with any material used in energy storage systems. Despite the fact that GdIG garnet ferrites are typically regarded as secure and reliable substances, the probable inclusion of rare-earth elements in the raw materials raises questions about their potential effects on the environment and their ethical sourcing. For sustainable use and safe disposal, thorough research on their long-term stability and environmental consequences is required. GdIG garnet ferrites must have their environmental and safety implications thoroughly assessed, just like any other material utilised in energy storage systems.

The performance, dependability, and scalability of GdIG garnet ferrites in energy storage applications should be improved through the development of novel synthesis techniques, material property optimisation, and investigation of novel composite structures. In addition, overcoming these obstacles and realising the full potential of GdIG garnet ferrites in the energy storage environment require interdisciplinary cooperation between material scientists, engineers, and industry stakeholders.

5. FUTURE PERSPECTIVES AND EMERGING RESEARCH

5.1 Synthesis Techniques Advancements

One important topic of new research is the creation of inventive and enhanced synthesis methods for GdIG garnet ferrites. Innovative techniques that provide better control over material characteristics, crystal structure, and phase purity are constantly being investigated by researchers. The manufacturing of GdIG garnet ferrites could be streamlined and made more commercially viable by improvements in sol-gel, solid-state reaction, and hydrothermal procedures as well as the application of cutting-edge processing methods including microwave-assisted synthesis. The development of scalable and affordable techniques for the mass manufacture of high-quality GdIG garnet ferrites in this field would likely make it easier to incorporate them into energy storage devices.

5.2 Tailoring Material Properties for Specific Applications

This interesting area of research involves modifying the characteristics of GdIG garnet ferrites to satisfy the demands of various energy storage applications. You can use controlled rare-earth doping, crystal structure manipulation, grain size and morphology optimisation, and controlled rare-earth doping to get the appropriate magnetic, electrical, and mechanical properties. Researchers can maximise the performance of GdIG garnet ferrites in different energy storage technologies by tailoring them for particular uses like high-power magnetic inductors, solid-state lithium-ion batteries, or energy harvesters.

5.3 Advanced Characterization Methods

Understanding the underlying characteristics and behaviour of GdIG garnet ferrites at the nanoscale and beyond requires the use of advanced characterisation techniques. The microstructure, flaws, and magnetic interactions of the material can be studied using methods including nuclear magnetic resonance (NMR) spectroscopy, atomic force microscopy (AFM), and transmission electron microscopy (TEM). In order to improve the performance and functionality of GdIG garnet ferrites in energy storage devices, it may be helpful to apply these cutting-edge characterisation approaches in order to gain a deeper understanding of the links between their structure and properties.

5.4 GdIG Garnet Based Thin Films

A promising new field for energy storage is the study of thin films and nanoscale GdIG garnet ferrites. Nanoscale materials are superior in a range of applications due to their special properties and enhanced surface-to-volume ratios. Researchers are investigating technologies including pulsed laser deposition, molecular beam epitaxy, and sol-gel techniques in order to produce GdIG garnet ferrite thin films with particular properties. Nanoscale GdIG garnet ferrites may be useful in advanced energy storage technologies like high-performance capacitors, nanoscale inductors, and tiny energy harvesters.

5.5 Integration With Other Energy Storage Materials

It is being researched to see if integrating more materials with beneficial properties would enhance energy storage systems. GdIG garnet ferrites can be combined with other energy storage elements such lithium-ion cathode elements, supercapacitor electrodes, or piezoelectric elements in energy harvesting devices. Device performance can be enhanced and versatile energy storage systems can be produced by combining GdIG garnet ferrites with different materials. Current research focuses on developing hybrid or composite structures that make use of the unique properties of GdIG garnet ferrites and other materials.

5.6 Application in Future Energy Storage Technologies

GdIG garnet ferrites are expected to play an important role in next-generation devices as energy storage technologies continue to advance. It is possible to create energy storage solutions that are safer, more effective, and ecologically friendly by making improvements to their synthesis, customization, and integration. By enhancing energy storage capacity, stability, and dependability, GdIG garnet ferrites can aid in the development of smart grids, portable electronics, electric cars, and renewable energy systems.

6. CONCLUSION

A promising class of materials for energy storage applications has developed in the form of GdIG garnet ferrites. These materials display a distinctive blend of magnetic, electrical, and mechanical properties.

This review study has investigated their potential in a wide range of energy storage technologies, including solid-state lithium-ion batteries, supercapacitors, magneto-strictive energy harvesters, and magnetic inductors. GdIG garnet ferrites have limitations and drawbacks despite their remarkable qualities, such as high manufacturing costs, mechanical brittleness, and interface compatibility. However, with continuous study and advancements in synthesis techniques, alterations to material properties, and integration with other materials, these challenges can be solved to fully achieve the potential of GdIG garnet ferrites in energy storage devices. Emerging research fields have a lot of potential to improve the performance and adaptability of these materials, including nanoscale GdIG garnet ferrites, enhanced characterisation methods, and their use in next-generation energy storage devices. We can anticipate the widespread use of these materials, which will contribute to a more sustainable and effective energy future, as researchers continue to investigate and optimise GdIG garnet ferrites for certain energy storage applications. GdIG garnet ferrites have exciting possibilities for energy storage in the future and have the power to completely alter how we store and use energy in the years to come.

REFERENCES

Al-Ghussain, L., Ahmad, A. D., Abubaker, A. M., & Mohamed, M. A. (2021). An integrated photovoltaic/wind/biomass and hybrid energy storage systems towards 100% renewable energy microgrids in university campuses. *Sustainable Energy Technologies and Assessments*, *46*, 101273. doi:10.1016/j.seta.2021.101273

Eckel, S., Sushkov, A. O., & Lamoreaux, S. K. (2009). Magnetic susceptibility and magnetization fluctuation measurements of mixed gadolinium-yttrium iron garnets. *Physical Review B: Condensed Matter and Materials Physics*, *79*(1), 014422. doi:10.1103/PhysRevB.79.014422

Evliya, H. (2007). Energy Storage for Sustainable Future—A Solution to Global Warming. In *Thermal Energy Storage for Sustainable Energy Consumption: Fundamentals, Case Studies and Design* (pp. 87–99). Springer Netherlands. doi:10.1007/978-1-4020-5290-3_5

Ganne, J. P., Lebourgeois, R., Paté, M., Dubreuil, D., Pinier, L., & Pascard, H. (2007). The electromagnetic properties of Cu-substituted garnets with low sintering temperature. *Journal of the European Ceramic Society*, *27*(8-9), 2771–2777. doi:10.1016/j.jeurceramsoc.2006.11.054

Goldman, A. (2006). *Modern ferrite technology*. Springer Science & Business Media.

Hadjipaschalis, I., Poullikkas, A., & Efthimiou, V. (2009). Overview of current and future energy storage technologies for electric power applications. *Renewable & Sustainable Energy Reviews*, *13*(6-7), 1513–1522. doi:10.1016/j.rser.2008.09.028

Huang, B., Ren, R., Zhang, Z., & Zheng, S. (2013). The improvement of dispersibility of YIG precursor prepared via chemical coprecipitation. *Journal of Alloys and Compounds*, *558*, 56–61. doi:10.1016/j.jallcom.2013.01.037

Ji, X., Zhou, K., Zhao, Y., Sun, M., Dong, S., Zhang, H., Cao, H., Zheng, H., Wu, Q., & Zhang, Y. (2022). Crystal Structure, Magnetic, Dielectric and Ferromagnetic Resonance Properties of Pr-Zn-Zr Co-Doped Yttrium Iron Garnet. *Journal of Electronic Materials*, *51*(3), 1–9. doi:10.1007/s11664-021-09382-w

Li, H. (2020). Investigation of the structural, magnetic, impedance properties in samarium-doped yttrium iron garnet. *Ceramics International*, *46*(10), 15408–15416. doi:10.1016/j.ceramint.2020.03.086

Li, X., & Wang, S. (2019). Energy management and operational control methods for grid battery energy storage systems. *CSEE Journal of Power and Energy Systems*, *7*(5), 1026–1040.

Mallmann, E. J. J., Sombra, A. S. B., Goes, J. C., & Fechine, P. B. A. (2013). Yttrium iron garnet: Properties and applications review. *Diffusion and Defect Data, Solid State Data. Part B, Solid State Phenomena*, *202*, 65–96. doi:10.4028/www.scientific.net/SSP.202.65

Prakash, K., Ali, M., Siddique, M. N. I., Chand, A. A., Kumar, N. M., Dong, D., & Pota, H. R. (2022). A review of battery energy storage systems for ancillary services in distribution grids: Current status, challenges and future directions. *Frontiers in Energy Research*, *10*, 971704. doi:10.3389/fenrg.2022.971704

Ren, Y., Liu, T., Shen, Y., Lin, Y., & Nan, C. W. (2016). Chemical compatibility between garnet-like solid state electrolyte Li6. 75La3Zr1. 75Ta0. 25O12 and major commercial lithium battery cathode materials. *Journal of Materiomics*, *2*(3), 256–264. doi:10.1016/j.jmat.2016.04.003

Sekijima, T., Itoh, H., Fujii, T., Wakino, K., & Okada, M. (2001). Influence of growth atmosphere on solubility limit of Ce3+ ions in Ce-substituted fibrous yttrium iron garnet single crystals. *Journal of Crystal Growth*, *229*(1-4), 409–414. doi:10.1016/S0022-0248(01)01191-5

Sharma, A., Godara, S. K., Maji, P. K., & Srivastava, A. K. (2022). Influence of Temperature on Structural and Magnetic Properties of Gd3AlxFe5− xO12 (x= 2). *Crystal Research and Technology*, *57*(2), 2100109. doi:10.1002/crat.202100109

Sharma, A., Godara, S. K., & Srivastava, A. K. (2022). Effect of Y3+, Bi3+, La3+ substitution on structural, optical and magnetic properties of gadolinium iron garnets. *Materials Today: Proceedings*, *50*, 1867–1877. doi:10.1016/j.matpr.2021.09.234

Sharma, A., Godara, S. K., & Srivastava, A. K. (2022). Influence of composition variation on structural, magnetic and dielectric properties of Gd3Fe5O12 (x)/MgFe2O4 (1− x) composite. *Indian Journal of Physics*, *96*(14), 4173–4184. doi:10.1007/s12648-022-02365-5

Sharma, A., Mahajan, H., Mohammed, I., Sinha, S., Godara, S. K., & Srivastava, A. K. (2023). Effect of Sintering Temperature on the Structural, Dielectric, and Magnetic Properties of Garnet-Spinel Ferrite Composites for Use in L-Band Devices. *Materials Performance and Characterization*, *12*(1), 64–78. doi:10.1520/MPC20220084

Sharma, A., Mohammed, I., Godara, S. K., & Srivastava, A. K. (2022). Low dielectric losses and enhanced magnetic property of ErIG (x)/YIG (1− x)(x= 0.5) composite for antenna applications. *Bulletin of Materials Science*, *45*(4), 248. doi:10.1007/s12034-022-02810-5

Shiri, H. M., Ehsani, A., & Shayeh, J. S. (2015). Synthesis and highly efficient supercapacitor behavior of a novel poly pyrrole/ceramic oxide nanocomposite film. *RSC Advances*, *5*(110), 91062–91068. doi:10.1039/C5RA19863A

Tahanian, H., Aliahmadi, M., & Faiz, J. (2020). Ferrite permanent magnets in electrical machines: Opportunities and challenges of a non-rare-earth alternative. *IEEE Transactions on Magnetics*, *56*(3), 1–20. doi:10.1109/TMAG.2019.2957468

Thangadurai, V., Narayanan, S., & Pinzaru, D. (2014). Garnet-type solid-state fast Li ion conductors for Li batteries: Critical review. *Chemical Society Reviews*, *43*(13), 4714–4727. doi:10.1039/c4cs00020j PMID:24681593

Thirumalasetty, A. B., Krishnan, T., Kota Venkata, S. K., Adiraj, S., Xavier, D., Venkatachalam, S., Kesava Reddy, R. G., Dudekula, Z., & Wuppulluri, M. (2023). Strategic Design and Analysis of Energy Storage and ME Applications of Low Temperature Hybrid Microwave Sintered Electroceramic Composites. *ACS Applied Electronic Materials*, *5*(6), 3120–3129. doi:10.1021/acsaelm.3c00190

Tolani, S. C., Golhar, A. R., & Rewatkar, K. G. (2019, May). A review of morphological, structural behaviour and technological applications of ferrites. In AIP Conference Proceedings (Vol. 2104, No. 1). AIP Publishing. doi:10.1063/1.5100459

Wang, G., Liu, H., Wu, H., Li, X., Qiu, H., Yang, Y., Qu, B., Ren, T.-L., Han, X., Zhang, R., & Wang, H. (2016). Epitaxial yttrium iron garnet film for fabrication of high frequency on-chip inductors. *Applied Physics Letters*, *109*(16), 162405. doi:10.1063/1.4964642

Xu, H., & Yang, H. (2007). Magnetic properties of Y3Fe5O12 nanoparticles doped Bi and Ce ions. *Materials and Manufacturing Processes*, *23*(1), 1–4. doi:10.1080/10426910701524212

Xu, H., Yang, H., Xu, W., & Feng, S. (2008). Magnetic properties of Ce, Gd-substituted yttrium iron garnet ferrite powders fabricated using a sol–gel method. *Journal of Materials Processing Technology*, *197*(1-3), 296–300. doi:10.1016/j.jmatprotec.2007.06.061

Chapter 9
Supercapacitors:
Revolutionizing Energy Storage

Hamnesh Mahajan
Lovely Professional University, India

Anjori Sharma
Lovely Professional University, India

Ajeet Kumar Srivastava
Lovely Professional University, India

ABSTRACT

This chapter explores supercapacitors as advanced energy storage solutions in the context of sustainable and efficient energy. It introduces the need for such technologies, details supercapacitor construction and working principles, covers various types and applications, and discusses crucial materials and quantitative modeling. The chapter also offers insights into the future of supercapacitor research. In summary, it provides a comprehensive overview of supercapacitors for researchers, engineers, and enthusiasts, encapsulating their potential in the energy storage landscape.

1. INTRODUCTION

Energy is incredibly important to our daily lives. From 7,323 TWh in 1980 to 23,845 TWh in 2019, global electricity consumption has expanded dramatically in recent decades. According to the World Energy and Climate Statistics - Year Book 2021, worldwide energy usage fell by 1.1% in 2020, the first drop since 2009. The global economy is largely reliant on energy generation, storage, and transportation. The investment in global energy, as per the International Energy Agency, is delineated in Table 1 spanning the years 2015 to 2023. The International Energy Agency asserts that the utilization of non-renewable fossil fuels including gas, oil, and coal has become vital for the generation of energy (Castro-Gutiérrez et al., 2020). Furthermore, it is widely accepted that the increase in greenhouse gas concentrations in the atmosphere during the 1950s is largely due to the combustion of fossil fuels, which is driving global

DOI: 10.4018/979-8-3693-1306-0.ch009

Table 1. Investment in global energy, as per the International Energy Agency (Gould, 2023)

Year	Billion USD	
	Clean Energy	Fossil Fuels
2015	1074	1319
2016	1132	1105
2017	1129	1114
2018	1137	1109
2019	1225	1066
2020	1259	839
2021	1408	914
2022	1617	1002
2023	1740	1050

climate change (Lazarus & van Asselt, 2018). Population growth, the depletion of fossil fuel reserves, and the harmful effects of greenhouse gases on the ecosystem and climate all pose a severe danger in the future decades. As a result, switching from fossil fuels to renewable energy sources and ecologically friendly technology is critical for long-term growth (Månberger, 2021). Renewable sources of energy including wind and solar energy, among others, cannot be employed for commercial or residential use since supply interruptions might result in huge losses (Saikia et al., 2020). As a result, the comprehensive approach toward clean energy emerges in modern civilization, as does the necessity to investigate new storage and conversion technologies as the global economy grows (Guan et al., 2019). Energy-storage systems are becoming increasingly important in this respect, and have been hailed as the most promising among numerous renewable energy storage technologies due to their high efficiency, adaptability, and flexibility.

2. ENERGY STORAGE DEVICES

The currently used energy storage and conversion devices as mentioned in Figure 1 are fuel cells, batteries, capacitors, and supercapacitors (SCs) (Krishan & Suhag, 2019)(Sun et al., 2019). Unlike fuel cell and batteries which converts chemical energy into electrical energy, conventional capacitor deals with the physical separation of electric charge across the dielectric medium such as ceramic, polymer film, or oxide layer (Tao et al., 2007). The limited surface area of conventional capacitors makes it challenging to achieve maximum energy storage. Consequently, researchers are placing significant emphasis on discovering devices with superior energy storage capacity and high power delivery (Nikzad et al., 2009). In this context, Supercapacitors (SCs) emerge as highly promising options for meeting these energy storage demands.

Due to their higher energy density (E), fuel cells and batteries are commonly utilized in high-energy applications. Meanwhile, they're also plagued by deficiencies that involve low power density (P), an inadequate cycle life, high cost, complex manufacturing processes, and safety concerns. On the other hand, conventional capacitors exhibit high power density and long cycle life but have lower level of

Figure 1. Energy storage and conversion devices

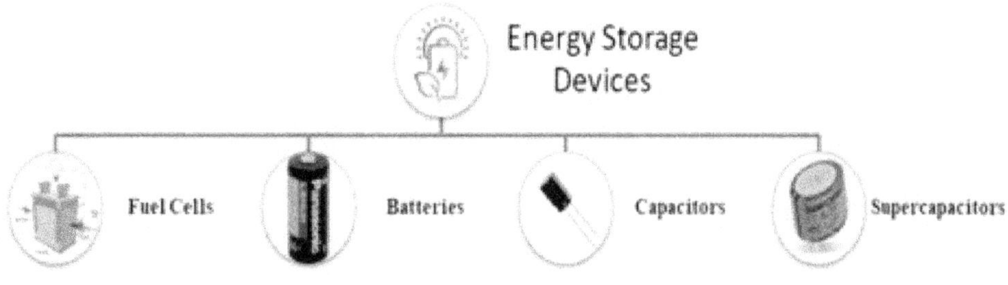

energy density when it's juxtaposed with fuel cells and batteries. As a result, electrochemical supercapacitors (ESCs) are considered a fascinating energy storage technology capable of bridging the gap amid batteries and conventional capacitors, as illustrated in Figure 2 Ragone plot (Nagao et al., 2016).

The electrochemical supercapacitors (ESCs) have superior energy and power density, excellent cyclic life, rapid charge-discharge rate, superb reversibility, longer shelf life, fast dynamic response, minimum equivalent series resistance (ESR), economic, featherweight, pliable, highly efficient, operational safety, and eco-friendly character (Winter & Brodd, 2004)(Yu & Chen, 2020)(Perera et al., 2015). Owing to the above-mentioned characteristics of the ESCs it seems to have its application in mobile phones, laptops, notebook computers, smart watches, memory backup, digital cameras, electric as well as hybrid electric vehicles, urban rail transit, energy power systems, toys, headphones, calculators, etc. (Hall et al., 2010). Because ESCs have magnificent power density, they may quickly release stored energy, making them ideal for short-term applications that require high electricity bursts (Pell et al., 1999).

3. SUPERCAPACITOR

In comparison with conventional capacitors, supercapacitors (SCs) are specifically constructed capacitors with large capacitance and energy density values. Relatively small, the SCs can provide capacitance in the tens, hundreds, or even thousands of farad range. They not only can hold a lot of charges, but they can also withstand thousands of charge-discharge cycles without breaking down. That's why SCs are usually called ultracapacitors (R. R. Zhang et al., 2015). Due to its high capacitance value, it can store much more electric energy between its plates. The SCs are more like an electrochemical device similar to the electrolytic capacitor but in this, no chemical reactions are involved for the storage of electric energy. As an electrostatic field ($\vec{E}$), the SCs store a huge amount of charge.

3.1. Construction of the Supercapacitor

The construction of the SCs is somewhat like electrolytic capacitors and it consists of the porous electrodes (cathode as well as the anode), electrolyte, and separator as depicted in Figure 3. These porous electrodes have the ability to attract a large quantity of charge. The electrolytes employed in SCs can be solid, liquid, or gel, and they play an important part in the SCs redox processes (Du et al., 2020). Supercapacitors (SCs) electrodes are separated by a permeable separator, typically a paper membrane. This separator allows the flow of charge carriers while ensuring there is no direct short circuit between the

Figure 2. Ragone plot of specific power versus specific energy for supercapacitors (Ayornu et al., 2022)

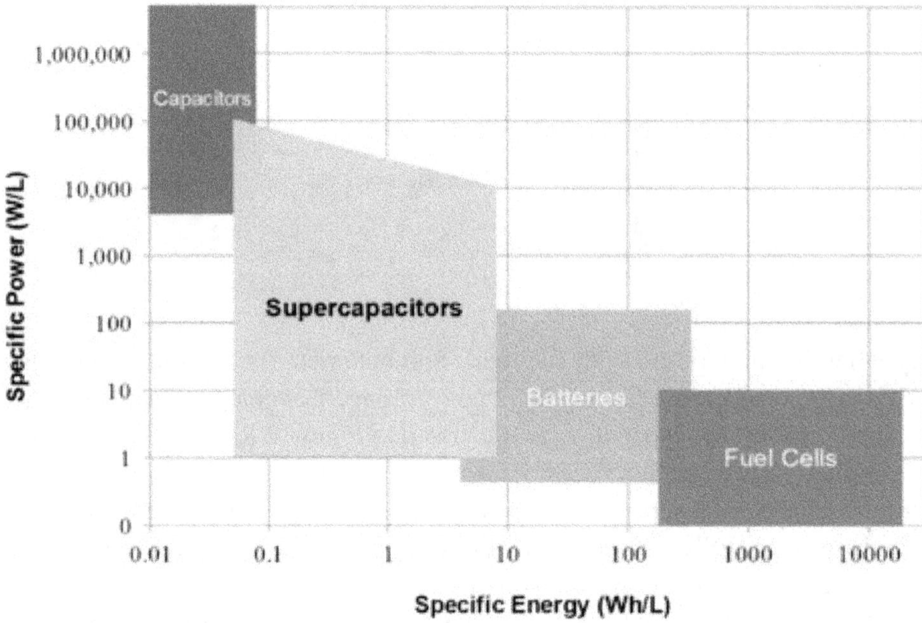

electrodes. SCs store charge in a completely different way than conventional capacitors. The polarization of the dielectric material is responsible for charge storage in conventional capacitors. The electrolyte of SCs, on the other hand, does not act as a dielectric. It is solely responsible for supplying charge carriers to the electrodes. Supercapacitors (SCs) inevitably store the charge on account of a proliferation of opposing charge carriers on the surfaces of the electrodes. The electrolyte plays a crucial role in influencing the internal or equivalent series resistance (ESR) of the SCs. As per Eq. (1), the higher the power density (P) of the SCs, the lower the resistance (R_{ES}) given by the electrolyte (Yassine & Fabris, 2017).

$$P = \frac{V^2}{4R_{ES}} \tag{1}$$

Here P, V, and R_{ES} signify the power density, applied voltage, and equivalent series resistance respectively (Hu & Fisher, 2018).

3.2. Principle and Working of the Supercapacitor

In a supercapacitor (SC), the energy storage principle can either be pure charge storage on the interface of electrode-electrolyte electrochemically or it can be a transfer of the charge to the redox molecule layers on the surface of the electrode (Forouzandeh et al., 2020).

The supercapacitor (SC) works similar to the conventional capacitor, both electrostatic and electrolytic, with only minor differences in technical features. It comprises two exceptionally permeable or porous electrodes, a paper membrane that isolates positive and negative plates called a separator. Utiliz-

Figure 3. Construction of the supercapacitor (Qi & Koenig, 2017)

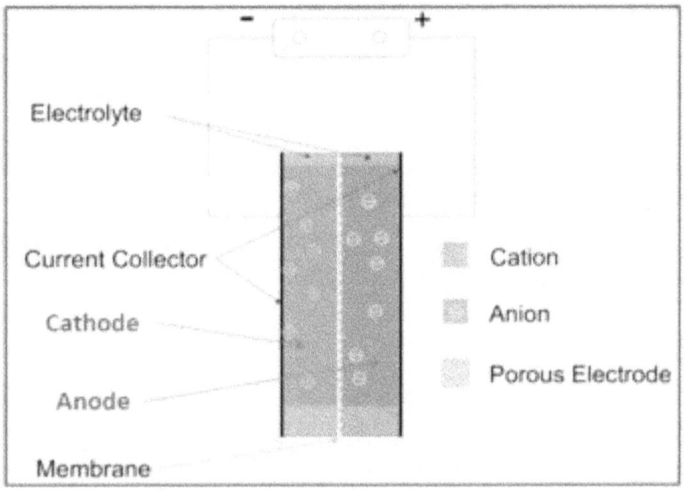

ing an ionic liquid affectionately referred to as the electrolyte, these electrodes are electrically linked. Positive ions are drawn from an electrolyte when voltage is hooked up to the negative electrode, and vice versa. These ions accumulate on the electrode surface, thereby reducing the distance (l) between the electrodes (Pal et al., 2019). As a result, the capacitance (C) of the SC is extremely high. Because the C of the SC is inversely related to the l between the electrodes $\left(C = \frac{A}{l}\right)$, where A and l are the cross-sectional areas and separation distance between the two electrodes respectively.

Figure 4. Working of the supercapacitor (Ferrer et al., 2017)

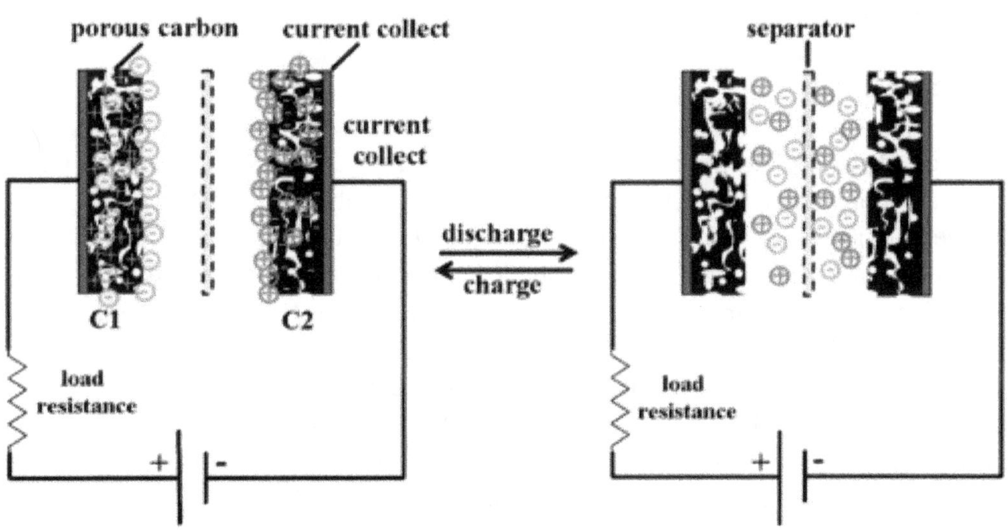

An electrostatic field ($\vec{E}$) is created between the current collectors due to the buildup of opposing charges. The SC is charged until the $\vec{E}$ between the current collectors equals and opposes the applied voltage. Till the supply voltage lowers or reverses polarity, the charge carriers are held by the current collectors. When the supplied voltage is reduced, the current collectors return a corresponding number of charge carriers to the electrolyte. During such an operation, an equivalent current travels in the opposite direction through the SC. The SC goes through a similar charging and discharging cycle when the polarity changes, as seen in Figure 4 (Bharti et al., 2021). The SC has a ten-times higher specific capacitance (Cs) and energy density (E) than an electrolytic capacitor and a hundred-times higher Cs and E than an electrostatic capacitor (Abruña et al., 2008).

3.3. Types of the Supercapacitors

Electrical double-layer capacitors, pseudocapacitors, and hybrid capacitors are currently the three primary categories into which supercapacitors (SCs) may be classified (Vangari et al., 2013). This categorization is based on the charge storage strategy and active material utilized in each type.

3.3.1. Electric Double-Layer Capacitor

An electrode, an electrolyte, and a separator make up an electric double-layer capacitor (EDLC). Carbon-based substances, which involve graphene, carbon aerogel, carbon nanotubes, activated carbon, carbon fiber cloth, and among others, are deployed as active materials or components for the electrode in EDLC. A non-faradaic or electrostatic process, in which there is absolutely no conveyance of charge amid the electrolyte and the electrode's surface has been employed for storing charge in the EDLC (Kiamahalleh et al., 2012)(Jayalakshmi & Balasubramanian, 2008). The moment an electrode has been subjected to a voltage, charge builds up on its surface. Ions from the electrolyte solution eventually permeate through the separator and transmit the opposing charge into the electrode's pores. A double-layered layer of charge forms at each electrode as a result of the meticulously planned electrode design, which inhibits ion recombination. In contrast to a conventional capacitor, the EDLC achieves a greater energy density (E) through the formation of this double layer, expanding the surface area, and lessening the distance (d) that exists amid the two electrodes (Choi & Yoon, 2015).

3.3.2. Pseudocapacitor

The construction of the pseudocapacitor (PC) is similar to EDLC but uses conducting polymer [poly (3,4-ethylenedioxythiophene) (PEDOT), polypyrrole (PPy), polyaniline (PANI) and Polythiophene (PTh)] and metal oxides (Fe_2O_3, Co_3O_4, ZnO_2, MnO_2, RuO_2, NiO, V_2O_5, SnO_2, $CoFe_2O_4$, etc.) as an active material or component for the SC electrode. Charge storage in PC is mostly accomplished by a faradic mechanism, where a charge is transferred between the electrolyte and electrode. The pseudocapacitor (PC) achieves higher energy densities (E) and specific capacitance (C_s) compared to an EDLC by utilizing under potential deposition, intercalation, and reduction-oxidation processes to store charge. These mechanisms allow the PC to store charge more efficiently, resulting in enhanced performance. The reduction-oxidation reaction occurs in the PC, just as it does in batteries, resulting in inferior cycle stability and inadequate power density (P) (S. M. Chen et al., 2014)(T. Chen & Dai, 2013).

3.3.3. Hybrid Capacitor

To boost the characteristics of performance, the hybrid capacitors endeavor to merge the beneficial features of EDLC and PC while balancing out their relative drawbacks. While using both Faradaic and non-Faradaic approaches for storing charge, hybrid capacitors deliver power and energy densities that surpass EDLCs without making the trade-offs in stability over cycles and expenses that have impeded the development of pseudocapacitors. Owing to the nature of the electrode arrangement, the three primary types of hybrid capacitors - composite, asymmetric, and battery-type - have garnered an overwhelming amount of research emphasis (Sayed et al., 2020).

a. Composite

In an attempt to fabricate a single electrode that incorporates both physical and chemical storage of charge procedures, the composite electrodes amalgamate carbon-based substances with conducting polymer or metal oxide. The carbon-based substances enable for a capacitive double-layer of charge in conjunction with developing an underpinning with a substantial amount of surface area thereby improving interactions amid the electrolyte and the coated pseudocapacitive substances. The composite electrode's capacitance may be further increased by the pseudocapacitive materials using Faradaic processes (Frackowiak et al., 2001). Successful composite electrodes made of conducting polymer polypyrrole and carbon nanotubes have been developed. This electrode may obtain better capacitances than an electrode made only of a polypyrrole polymer or a carbon nanotube, according to several investigations (Frackowiak et al., 2006). The reason for this is owing to the accessible nature of the intertwined mat structure that enables a three-dimensional charge distribution and a homogeneous coating of polypyrrole. Furthermore, it has been researched that the integrity of the structure that exists in the intertwined mat lessens the wear and tear caused by the simultaneous addition and removal of ions from the coated polypyrrole. As a result, these composites, as opposed to conducting polymers, were consequently proven to accomplish a cyclic stability that is on par with EDLCs (Jurewicz et al., 2001).

b. Asymmetric

To integrate non-Faradaic and Faradaic procedures, asymmetric hybrids make use of an EDLC electrode and a PC electrode. The association that exists between an activated carbon-based negative electrode and a conducting polymer based positive electrode has prompted an enormous amount of focus (Arbizzani et al., 2001). In light of the dearth of an effective negative-charge in the conducting polymer, the conducting polymer pseudocapacitor hasn't proven to be as beneficial as they would have potentially been. An electrode that's constructed from charge-negative activated carbon has been employed to navigate through the problem at hand. The capacitances and resistances of conducting polymer electrodes are often larger than those of activated carbon electrodes, conducting polymer electrodes frequently have bigger capacitances and resistances than activated carbon electrodes, albeit they additionally possess inferior cyclic stability and voltage (Laforgue et al., 2003). With the objective of accomplishing more effective power and energy densities than corresponding EDLC, asymmetrical hybrid capacitors are used that interconnect these two electrodes to ameliorate the severity of this compromise. Additionally, they exhibit superior cycle stability compared to similar pseudocapacitors (Liangliang & Chunyang, 2010).

c. Battery-Type

Comparable to asymmetrical hybrids in which they link two distinct electrodes, battery-type hybrids distinguished apart owing to the inclusion of a supercapacitor electrode. Such a distinctive combination encompasses the desire for superior energy supercapacitors and more efficient batteries by amalgamating the power, cycle life, and recharge times of supercapacitors along with the energy features associated with batteries. The overwhelming majority of studies comprised the use of lead-dioxide, and nickel-hydroxide to serve as one electrode while activated carbon as a second electrode (H. Li et al., 2005) (Wang & Zheng, 2004). The scientific evidence at hand indicates that battery-type hybrids might have the potential to eliminate disconnect amid batteries and supercapacitors, yet there doesn't exist as much evidence from experiments accessible regarding them as for different sorts of supercapacitors. Despite the encouraging outcomes, it is generally agreed that further study will be required to fully understand possibilities of battery-type hybrids (Du Pasquier et al., 2003)(Pell & Conway, 2004).

4. MATERIALS FOR THE SUPERCAPACITOR ELECTRODE

The electrode materials have a major impact on the supercapacitors (SCs) electrochemical performance. Owing to the types of the SCs as discussed earlier, three distinct types of electrode materials are being used for energy storage applications in the SCs: carbon-based, metal oxides, and conducting polymer. Electrical conductivity, pore structure and shape, pore size distribution, specific surface area, and surface functionality are all important material properties that impact the electrochemical performance of the SCs (Yang et al., 2017).

4.1. Carbon-Based Materials

The active carbon-based material viz. graphene, carbon fiber cloth, activated carbon, carbon aerogel, and carbon nanotubes etc. have been frequently employed as electrode materials in supercapacitor (SC) fabrication. The reasons for this include (a) minimal cost, (b) large surface-to-volume ratio, (c) well-established electrode fabrication technology, and (d) easy availability. A double electrochemical layering emerges at the interface amid the electrodes and the electrolytes as the storage procedure for the material made from carbon. With regards to this, the entire surface area that the ions from electrolytes may access in carbon-based materials is largely what determines capacitance (Bo et al., 2012). A considerable capacity for charge accumulation at the electrode-electrolyte interface is seen in carbon materials when their substantial specific surface area is present. When improving specific capacitance (Cs) for carbon materials, surface functionalization must also be taken into consideration (Singh et al., 2015), along with pore size and a substantial amount of surface area. Extremely stable and conductive SC electrodes made of carbon generally have a high power density (P) and excellent cycle stability. Moreover, the energy density (E) of carbon-based electrodes for SCs is often low due to the limits of the energy storage technique.

4.2. Metal Oxides

The potential for energy storage was investigated for a number of metal oxides, including Fe_2O_3, Co_3O_4, ZnO_2, MnO_2, RuO_2, NiO, V_2O_5, SnO_2, etc. Owing to their magnificent activity and superior inherent electrochemical characteristics, these metal oxides have outstanding electrochemical performance in the supercapacitors (SCs). However, they still face issues like scarcity of natural resources, poor conductivity, and exorbitant costs. RuO_2 was found to be the most remarkable active material for pseudocapacitor (PC) electrodes among the aforementioned metal oxides because it has exceptional specific capacitance (C_s), a very high electrical conductivity, and reverse charge-discharge characteristics. RuO_2 can be used as an active material for electrodes, however there are some challenges with this because it is both very poisonous and costly (Mohd Abdah et al., 2020). Additionally, there is a requirement to increase the electrical conductivity and Cs of single transition metal oxides. Cobaltite (MCo_2O_4), aluminate (MAl_2O_4), spinel ferrite (MFe_2O_4), and other mixed metal oxides can be used to solve the aforementioned problems. Due to their outstanding electrochemical performance brought on by the synergistic interaction of several metal cations, these mixed metal oxides have been thought of as potential rivals for improved SCs (Arsalani et al., 2018). Spinel ferrite (MFe_2O_4) is one of the best-known mixed metal oxides because it has a variety of unexpected features, including electrochemical stability, multiple oxidation states, electrical, magnetic, optical, and catalytic (Liu et al., 2015). These spinel ferrites are moreover widely accessible, inexpensive, extraordinarily productive, ecologically safe, and simple to synthesize on an industrial scale.

4.3. Conducting Polymers

Numerous polymers with conducting properties have been extensively investigated as supercapacitor (SC) materials for electrodes because of how simple and inexpensive they are to synthesize (Paper et al., 2009). Conducting polymers are superior to carbon-based electrode materials in terms of conductivity, specific capacitance (Cs), and equivalent series resistance (ESR). Different layouts or combinations of the electrodes were used for conducting polymers. The arrangement of the n-p type yields substantial energy densities (E). However, lack of n-doped electrodes in the conducting polymer has hindered pseudocapacitors (PC) from reaching their maximum potential (Iro et al., 2016). Charge was stored and released in the conducting polymers via the reversible redox process. Ions get transported onto the polymer's backbone when oxidation, frequently alluded to by the term doping, takes place. While ions get restored within the electrolyte when reduction, frequently alluded to by the term de-doping, takes place (Sharma & Bhatti, 2010). In conducting polymer the mechanical stress caused by redox reactions restricts the stability of the material during numerous charge-discharge cycles (Liangliang & Chunyang, 2010). Owing to the minimal cost, high conductivity, excellent energy storage capacity, and ease of synthesis, polyaniline (PANI) is perhaps the most encouraging SC electrode material out of the various existing conducting polymers [viz. polypyrrole (PPy), poly (3,4-ethylenedioxythiophene) (PEDOT), polythiophene (PTh), and polyaniline (PANI)] (Cheng et al., 2013). Swelling and shrinking of PANI are prone to a substantial loss in performance owing to recurrent charge-discharge cycles. When coupled with metal oxides or carbon-based materials, PANI has been shown to be more stable, which also raises the value of Cs (J. Li et al., 2011).

5. QUANTITATIVE MODELING OF SUPERCAPACITORS

During the preceding section 3.3, we dealt with the numerous varieties of supercapacitors, which are contingent on an extensive assortment of materials followed by a diverse stretch in terms of performance attributes. The community of scientists has leveraged quantitative modeling to foresee the functionality attributes of supercapacitors in an effort to shorten the time and expense for manufacturing and physical investigation. This has provided insight into how to create supercapacitors that perform more closely to their theoretical potential. Equivalent circuit models are of great importance. In studies on the quantitative modeling of supercapacitors, equivalent circuit models have been used to simulate the behaviour of porous electrodes and to investigate the empirical correlations between capacitance, ESR, pore size, and surface area. The theoretical limitations of supercapacitors with various architectures and compositions have also been established using these models (L. Zhang et al., 2018).

5.1. Equivalent Circuit Models

To identify the sole purpose of portraying advanced electrochemical processes, equivalent circuit models deploy mathematical procedures or computerized simulations that mimic fundamental electrical circuitry components like the capacitors and the resistors. For a long time, porous electrode performance characteristics have been predicted using straightforward equivalent circuits (Helseth, 2019). Strategies to mimic the behavior that occurs within the double-layer across the interface that exists amid the solution of electrolytes and the electrode pores are overwhelmingly achieved via these equivalent circuits. Numerous Faradaic anomalies encountered in pseudocapacitors were just lately captivated by equivalent circuits (Song et al., 2012).

Figure 5 showcases the hierarchical arrangement of the equivalent circuits that have been built to mimic porous electrodes. The equivalent circuit that utilizes a porous electrode (1.5 e) can be accomplished in this arrangement through the introduction of a basic capacitor (1.5 a) once a time. As the ions from the solution of the electrolyte propagate in the direction of the electrode's surface at the final stage of the equivalent circuit (1.5 e), subsequently referred to as the line of transmission, the dispersed resistances portray the ESR built-in to each pore. The dispersed capacitances are a representation of each pore's

Figure 5. Hierarchy of equivalent circuits for porous electrodes: (a) capacitor, (b) capacitor with series resistance, (c) simple double-layer, (d) simple pseudocapacitor, (e) transmission line model

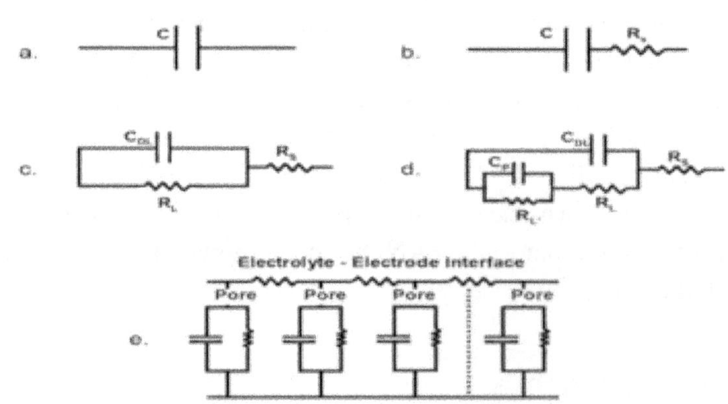

non-Faradaic double-layer capacitance. The Faradaic pore equivalent circuit (1.5 d) may be added to this equivalent circuit in order to simulate a porous pseudocapacitor electrode, which is another crucial point to make (Fletcher et al., 2014)(Pourkheirollah et al., 2023).

5.2. Empirical Relationships

The empirical correlations between pore size, surface area, and capacitance have also been the subject of much investigation. Although surface area and capacitance have a proportionate connection in principle, early practical experimentation revealed that the two quantities were unrelated. To explain this difference between theory and experiment, two rival mathematical models have been created (Ellenbogen, 2006). According to the first model, the capacitance per surface area of the microspores and the capacitance per surface area of the external medium must be determined independently because of the particular electrosorption behaviour present in micropores. The second, now-commonly-accepted model postulates that electrolyte ions cannot penetrate into holes smaller than a certain size threshold, and as a result, the surface area of such pores cannot contribute to capacitance. There have been attempts to ascertain the ideal pore size and size distribution required to maximize ion accessibility when taking the second model into consideration. As a consequence, it has also been shown that there is an inverse link between pore size and ESR (Bharti et al., 2021).

5.3. Theoretical Limits

The theoretical upper and lower bounds of supercapacitor power and energy densities have also been estimated via quantitative modeling. The study has also produced fresh insights on ways to optimize the design of supercapacitors by identifying the limiting parameters that prohibit supercapacitors from reaching their theoretical limit (Lewandowski & Galinski, 2007). However, theoretical models imply that the energy densities of supercapacitors are frequently constrained by the breakdown voltage and ion concentration of the electrolyte (Zheng, 2005). This is true even though there has always been interest in the development of new electrode materials to boost energy densities. Additionally, new research suggests that the electrolyte may also be able to constrain the power densities of supercapacitors. The study emphasizes that in order to achieve energy and power densities that are closer to the theoretical limits of supercapacitors, electrolyte optimization is just as crucial as electrode optimization (Ellenbogen, 2006).

6. PROSPECTUS ON THE FUTURE OF SUPERCAPACITOR RESEARCH AND DEVELOPMENT (R&D)

Research and development (R&D) on supercapacitors has recently concentrated on initiatives to raise the capacitance of electrode materials and to create better quantitative models. The development of supercapacitors, however, may be moving towards new domains, according to recent research trends. Research and development initiatives in particular pertaining to hybrid capacitors, equivalent series resistance, electrolyte optimization, and self-discharge are projected to grow and enable significant performance improvements in supercapacitors (Navarro et al., 2021).

6.1. Hybrid Capacitors

It has been verified that hybrid capacitors display a mix of performance traits that were previously impossible. They create a single supercapacitor by combining the finest qualities of EDLCs and pseudocapacitors. Although hybrid capacitors have received less attention than EDLCs or pseudocapacitors, the available research indicates that they may be able to outperform these devices. As a result, R&D efforts to fabricate better hybrid capacitors and to create more precise quantitative models of hybrid capacitors have continued to grow. Hybrid capacitors are quickly overtaking EDLCs as the most promising type of supercapacitors due to their incredible flexibility in modifying the design and performance and the growing interest in creating high cycle life, high-energy supercapacitors (Muzaffar et al., 2019).

6.2. Equivalent Series Resistance (ESR)

Supercapacitors can't reach power densities that are as near to the theoretical limits as they could due to the ESR. As a result, figuring out how to reduce the ESR of supercapacitors is turning into a key field of R&D. There are already a number of techniques for lowering the ESR, such as chemically attaching the electrode to the current collector, polishing the current collector's surface, and utilizing colloidal thin film suspensions. Additionally, studies have been done to determine the intrinsic ESR of different electrolytes and to define the link between pore size and ESR in electrode materials. As these R&D projects advance, supercapacitors should be able to reach power densities that are closer to their theoretical limitations (Redondo et al., 2020).

6.3. Electrolyte Optimization

The crucial first step towards bettering supercapacitors has repeatedly been emphasized in the scientific literature as electrolyte optimization. While an electrolyte's resistance can restrict power density, a supercapacitors operating voltage and ion concentration can limit energy density. Even though the performance of supercapacitors is influenced by the electrolyte, research and development (R&D) efforts to improve electrolytes have not yet been as rigorous or successful as those made to improve electrodes. However, the authors contend that in order to refine electrolytes and increase the synergy between electrolyte and electrode, additional R&D efforts must be supported given the significance of electrolyte optimization and the literature's attention on this topic (Wong et al., 2020).

6.4. Ameliorating Self-Discharge

In order for supercapacitors to live up to their promise, it is also necessary to ameliorate their propensity for self-discharge. Supercapacitors are under thermodynamic pressure to discharge because they are in a greater state of potential energy when charged than when discharged. Self-discharge, which usually transpires once a capacitor inwardly discharges through a circuit that is open, is an unfavorable occurrence that occasionally results from this pressure. Self-discharge is a fundamental property of all electrochemical energy storage devices, such as batteries and capacitors. Supercapacitors, however, experience it more frequently. Therefore, self-discharge usually has a worse outcome for them. The reverse process in systems based on chemical processes tends to be hampered via thermodynamic or kinetic obstacles in the dearth of an external link with the electrodes; however, the supercapacitor does not comprise an intrinsic obstacle to operate in reverse. Supercapacitors also frequently have very significant potential differences between their electrodes despite their close proximity. The difference in the potential in EDLC could render it substantially more difficult to sustain in contrast to what is in the battery as a result of these several reasons. Although there are many various ways that self-discharge occurs, they frequently are the consequence of uncontrollable Faradaic events, including the reduction and oxidation of contaminants in the material of the electrode. The rate of self-discharge in supercapacitors can therefore be decreased in part by increasing material purity (Kowal et al., 2011).

7. CONCLUSION

Supercapacitors don't appear likely to displace batteries as the primary power storage option, according to the study. This is mainly due to the fact that currently proposed supercapacitor technologies cannot store as much energy as batteries.

Supercapacitors, however, may be configured to do tasks that electrochemical batteries have not been perfectly suitable to perform owing to the way they adapt. Supercapacitors are also well suited for specific roles and applications that enhance the advantages of batteries due to some inherent qualities. Particularly for applications that need a prolonged shelf life, excellent cyclic stability, rapid recharging, and outstanding power, supercapacitors offer a lot of potential.

Supercapacitors ought to subsequently be revealed to be among the superior energy sources to feed an assortment of specific application prerequisites. Particularly, there has been a lot of interest in creating supercapacitors for backup and emergency power sources, pulse power applications, and hybrid electric vehicle power systems.

Supercapacitors have benefits in these specialized fields, but up until now, manufacture and use have been restricted. This lack of commercial penetration might seems brought on by an assortment of aspects such as high cost, problematic packing, and self-discharge. At least some of these problems may be resolved, according to recent study.

Supercapacitors may eventually become a practical, widely accessible power source for a growing variety of applications as R&D activities' outcomes continue to evolve. In addition to acting as a springboard for creating future applications, it is anticipated that this study will further encourage the R&D necessary to achieve this result.

REFERENCES

Abruña, H. D., Kiya, Y., & Henderson, J. C. (2008). Batteries and electrochemical capacitors. *Physics Today*, *61*(12), 43–47. doi:10.1063/1.3047681

Arbizzani, C., Mastragostino, M., & Soavi, F. (2001). New trends in electrochemical supercapacitors. *Journal of Power Sources*, *100*(1-2), 164–170. doi:10.1016/S0378-7753(01)00892-8

Arsalani, N., Tabrizi, A. G., & Ghadimi, L. S. (2018). Novel PANI/MnFe2O4 nanocomposite for low-cost supercapacitors with high rate capability. *Journal of Materials Science Materials in Electronics*, *29*(7), 6077–6085. doi:10.1007/s10854-018-8582-6

Ayornu, H., Anwar, S., & Thakur, Y. (2022). Fabrication and Characterization of Semi-Crystalline and Amorphous Dielectric Polymer Films for Energy Storage. *Materials Sciences and Applications*, *1*(11), 559–568. doi:10.4236/msa.2022.1311034

Bharti, K., Kumar, A., Ahmed, G., Gupta, M., Bocchetta, P., Adalati, R., Chandra, R., & Kumar, Y. (2021). Theories and models of supercapacitors with recent advancements: Impact and interpretations. *Nano Express*, *2*(2), 022004. doi:10.1088/2632-959X/abf8c2

Bo, Z., Wen, Z., Kim, H., Lu, G., Yu, K., & Chen, J. (2012). One-step fabrication and capacitive behavior of electrochemical double layer capacitor electrodes using vertically-oriented graphene directly grown on metal. *Carbon*, *50*(12), 4379–4387. doi:10.1016/j.carbon.2012.05.014

Castro-Gutiérrez, J., Celzard, A., & Fierro, V. (2020). Energy Storage in Supercapacitors: Focus on Tannin-Derived Carbon Electrodes. *Frontiers in Materials*, *7*(July), 1–25. doi:10.3389/fmats.2020.00217

Chen, S. M., Ramachandran, R., Mani, V., & Saraswathi, R. (2014). Recent advancements in electrode materials for the high-performance electrochemical supercapacitors: A review. *International Journal of Electrochemical Science*, *9*(8), 4072–4085. doi:10.1016/S1452-3981(23)08076-8

Chen, T., & Dai, L. (2013). Carbon nanomaterials for high-performance supercapacitors. *Materials Today*, *16*(7-8), 272–280. doi:10.1016/j.mattod.2013.07.002

Cheng, Q., Tang, J., Shinya, N., & Qin, L. C. (2013). Polyaniline modified graphene and carbon nanotube composite electrode for asymmetric supercapacitors of high energy density. *Journal of Power Sources*, *241*, 423–428. doi:10.1016/j.jpowsour.2013.04.105

Choi, H., & Yoon, H. (2015). Nanostructured electrode materials for electrochemical capacitor applications. *Nanomaterials (Basel, Switzerland)*, *5*(2), 906–936. doi:10.3390/nano5020906 PMID:28347044

Du, H., Wu, Z., Xu, Y., Liu, S., & Yang, H. (2020). Poly(3,4-ethylenedioxythiophene) based solid-state polymer supercapacitor with ionic liquid gel polymer electrolyte. *Polymers*, *12*(2), 297. Advance online publication. doi:10.3390/polym12020297 PMID:32024287

Du Pasquier, A., Plitz, I., Menocal, S., & Amatucci, G. (2003). A comparative study of Li-ion battery, supercapacitor and nonaqueous asymmetric hybrid devices for automotive applications. *Journal of Power Sources*, *115*(1), 171–178. doi:10.1016/S0378-7753(02)00718-8

Ellenbogen, J. C. (2006). *Supercapacitors : A Brief Overview*. The MITRE Corporation. https://www.mitre.org/tech/nanotechSponsor

Ferrer, P. R., Mace, A., Thomas, S. N., & Jeon, J. W. (2017). Nanostructured porous graphene and its composites for energy storage applications. *Nano Convergence*, *4*(1), 29. Advance online publication. doi:10.1186/s40580-017-0123-0 PMID:29152447

Fletcher, S., Black, V. J., & Kirkpatrick, I. (2014). A universal equivalent circuit for carbon-based supercapacitors. *Journal of Solid State Electrochemistry*, *18*(5), 1377–1387. doi:10.1007/s10008-013-2328-4

Forouzandeh, P., Kumaravel, V., & Pillai, S. C. (2020). Electrode materials for supercapacitors: A review of recent advances. *Catalysts*, *10*(9), 1–73. doi:10.3390/catal10090969

Frackowiak, E., Jurewicz, K., Delpeux, S., & Béguin, F. (2001). Nanotubular materials for supercapacitors. *Journal of Power Sources*, *97-98*, 822–825. doi:10.1016/S0378-7753(01)00736-4

Frackowiak, E., Khomenko, V., Jurewicz, K., Lota, K., & Béguin, F. (2006). Supercapacitors based on conducting polymers/nanotubes composites. *Journal of Power Sources*, *153*(2), 413–418. doi:10.1016/j.jpowsour.2005.05.030

Gould, T. (2023). *World Energy Investment 2023*. https://www.iea.org/reports/world-energy-investment-2023

Guan, L., Pan, L., Peng, T., Qian, T., Huang, Y., Li, X., Gao, C., Li, Z., Hu, H., & Wu, M. (2019). Green and scalable synthesis of porous carbon nanosheet-assembled hierarchical architectures for robust capacitive energy harvesting. *Carbon*, *152*, 537–544. doi:10.1016/j.carbon.2019.06.059

Hall, P. J., Mirzaeian, M., Fletcher, S. I., Sillars, F. B., Rennie, A. J. R., Shitta-Bey, G. O., Wilson, G., Cruden, A., & Carter, R. (2010). Energy storage in electrochemical capacitors: Designing functional materials to improve performance. *Energy & Environmental Science*, *3*(9), 1238–1251. doi:10.1039/c0ee00004c

Helseth, L. E. (2019). Modelling supercapacitors using a dynamic equivalent circuit with a distribution of relaxation times. *Journal of Energy Storage*, *25*(August), 100912. doi:10.1016/j.est.2019.100912

Hu, Y., & Fisher, T. S. (2018). Suggested standards for reporting power and energy density in supercapacitor research. *Bulletin of Materials Science*, *41*(5), 10–13. doi:10.1007/s12034-018-1641-z

Iro, Z. S., Subramani, C., & Dash, S. S. (2016). A brief review on electrode materials for supercapacitor. *International Journal of Electrochemical Science*, *11*(12), 10628–10643. doi:10.20964/2016.12.50

Jayalakshmi, M., & Balasubramanian, K. (2008). Simple capacitors to supercapacitors - An overview. *International Journal of Electrochemical Science*, *3*(11), 1196–1217. doi:10.1016/S1452-3981(23)15517-9

Jurewicz, K., Delpeux, S., Bertagna, V., Béguin, F., & Frackowiak, E. (2001). Supercapacitors from nanotubes/polypyrrole composites. *Chemical Physics Letters*, *347*(1-3), 36–40. doi:10.1016/S0009-2614(01)01037-5

Kiamahalleh, M. V., Zein, S. H. S., Najafpour, G., Sata, S. A., & Buniran, S. (2012). Multiwalled carbon nanotubes based nanocomposites for supercapacitors: A review of electrode materials. *Nano*, *7*(2), 1–27. doi:10.1142/S1793292012300022

Kowal, J., Avaroglu, E., Chamekh, F., Šenfelds, A., Thien, T., Wijaya, D., & Sauer, D. U. (2011). Detailed analysis of the self-discharge of supercapacitors. *Journal of Power Sources*, *196*(1), 573–579. doi:10.1016/j.jpowsour.2009.12.028

Krishan, O., & Suhag, S. (2019). An updated review of energy storage systems: Classification and applications in distributed generation power systems incorporating renewable energy resources. *International Journal of Energy Research*, *43*(12), 6171–6210. doi:10.1002/er.4285

Laforgue, A., Simon, P., Fauvarque, J. F., Mastragostino, M., Soavi, F., Sarrau, J. F., Lailler, P., Conte, M., Rossi, E., & Saguatti, S. (2003). Activated Carbon/Conducting Polymer Hybrid Supercapacitors. *Journal of the Electrochemical Society*, *150*(5), A645. doi:10.1149/1.1566411

Lazarus, M., & van Asselt, H. (2018). Fossil fuel supply and climate policy: Exploring the road less taken. *Climatic Change*, *150*(1-2), 1–13. doi:10.1007/s10584-018-2266-3

Lewandowski, A., & Galinski, M. (2007). Practical and theoretical limits for electrochemical double-layer capacitors. *Journal of Power Sources, 173*(2), 822–828. doi:10.1016/j.jpowsour.2007.05.062

Li, H., Cheng, L., & Xia, Y. (2005). A hybrid electrochemical supercapacitor based on a 5 V Li-ion battery cathode and active carbon. *Electrochemical and Solid-State Letters*, *8*(9), A433. Advance online publication. doi:10.1149/1.1960007

Li, J., Xie, H., Li, Y., Liu, J., & Li, Z. (2011). Electrochemical properties of graphene nanosheets/polyaniline nanofibers composites as electrode for supercapacitors. *Journal of Power Sources*, *196*(24), 10775–10781. doi:10.1016/j.jpowsour.2011.08.105

Liangliang, T., & Chunyang, J. (2010). Conducting polymers as electrode materials for supercapacitors. *Huaxue Jinzhan*, *22*(8), 1610–1618.

Liu, S., Hu, L., Xu, X., Al-Ghamdi, A. A., & Fang, X. (2015). Nickel Cobaltite Nanostructures for Photoelectric and Catalytic Applications. *Small, 11*(34), 4267–4283. doi:10.1002/smll.201500315 PMID:26121217

Månberger, A. (2021). Reduced Use of Fossil Fuels can Reduce Supply of Critical Resources. *Biophysical Economics and Sustainability*, *6*(2), 1–15. doi:10.1007/s41247-021-00088-5

Mohd Abdah, M. A. A., Azman, N. H. N., Kulandaivalu, S., & Sulaiman, Y. (2020). Review of the use of transition-metal-oxide and conducting polymer-based fibres for high-performance supercapacitors. *Materials & Design*, *186*, 108199. doi:10.1016/j.matdes.2019.108199

Muzaffar, A., Ahamed, M. B., Deshmukh, K., & Thirumalai, J. (2019). A review on recent advances in hybrid supercapacitors: Design, fabrication and applications. *Renewable and Sustainable Energy Reviews, 101*(October), 123–145. doi:10.1016/j.rser.2018.10.026

Nagao, M., Suzuki, K., Imade, Y., Tateishi, M., Watanabe, R., Yokoi, T., Hirayama, M., Tatsumi, T., & Kanno, R. (2016). All-solid-state lithium–sulfur batteries with three-dimensional mesoporous electrode structures. *Journal of Power Sources*, *330*, 120–126. doi:10.1016/j.jpowsour.2016.09.009

Navarro, G., Torres, J., Blanco, M., Nájera, J., Santos-Herran, M., & Lafoz, M. (2021). Present and future of supercapacitor technology applied to powertrains, renewable generation and grid connection applications. *Energies*, *14*(11), 3060. Advance online publication. doi:10.3390/en14113060

Nikzad, L., Alibeigi, S., Vaezi, M. R., Yazdani, B., & Rahimipour, M. R. (2009). Synthesis of a graphite-polyaniline nanocomposite and evaluation of its electrochemical properties. *Chemical Engineering & Technology*, *32*(6), 861–866. doi:10.1002/ceat.200800470

Pal, B., Yang, S., Ramesh, S., Thangadurai, V., & Jose, R. (2019). Electrolyte selection for supercapacitive devices: A critical review. *Nanoscale Advances*, *1*(10), 3807–3835. doi:10.1039/C9NA00374F PMID:36132093

Paper, C., Anodic, S., Wang, D., Li, F., Zhao, J., Ren, W., Chen, Z., Tan, J., Wu, Z., Gentle, I., Lu, G. Q., & Cheng, H. (2009). Fabrication of Graphene / Polyaniline. *ACS Nano*, *3*(7), 1745–1752. PMID:19489559

Pell, W. G., & Conway, B. E. (2004). Peculiarities and requirements of asymmetric capacitor devices based on combination of capacitor and battery-type electrodes. *Journal of Power Sources, 136*(2), 334–345. doi:10.1016/j.jpowsour.2004.03.021

Pell, W. G., Conway, B. E., Adams, W. A., & De Oliveira, J. (1999). Electrochemical efficiency in multiple discharge/recharge cycling of supercapacitors in hybrid EV applications. *Journal of Power Sources, 80*(1), 134–141. doi:10.1016/S0378-7753(98)00257-2

Perera, S. D., Ding, X., Bhargava, A., Hovden, R., Nelson, A., Kourkoutis, L. F., & Robinson, R. D. (2015). Enhanced Supercapacitor Performance for Equal Co-Mn Stoichiometry in Colloidal Co 3-x Mn x O 4 Nanoparticles, in Additive-Free Electrodes. *Chemistry of Materials, 27*(23), 7861–7873. doi:10.1021/acs.chemmater.5b02106

Pourkheirollah, H., Keskinen, J., Mäntysalo, M., & Lupo, D. (2023). Simplified exponential equivalent circuit models for prediction of printed supercapacitor's discharge behavior - Simulations and experiments. *Journal of Power Sources, 567*(February), 232932. Advance online publication. doi:10.1016/j.jpowsour.2023.232932

Qi, Z., & Koenig, G. M. Jr. (2017). Review Article: Flow battery systems with solid electroactive materials. *Journal of Vacuum Science & Technology B. Journal of Vacuum Science and Technology. B, Nanotechnology & Microelectronics : Materials, Processing, Measurement, & Phenomena : JVST B, 35*(4), 040801. doi:10.1116/1.4983210

Redondo, E., Fevre, L. W. L., Fields, R., Todd, R., Forsyth, A. J., & Dryfe, R. A. W. (2020). Enhancing supercapacitor energy density by mass-balancing of graphene composite electrodes. *Electrochimica Acta, 360*. doi:10.1016/j.electacta.2020.136957

Saikia, B. K., Benoy, S. M., Bora, M., Tamuly, J., Pandey, M., & Bhattacharya, D. (2020). A brief review on supercapacitor energy storage devices and utilization of natural carbon resources as their electrode materials. *Fuel, 282*(July), 118796. doi:10.1016/j.fuel.2020.118796

Sayed, D. M., Taha, M. M., Ghanem, L. G., El-Deab, M. S., & Allam, N. K. (2020). Hybrid supercapacitors: A simple electrochemical approach to determine optimum potential window and charge balance. *Journal of Power Sources, 480*(September), 229152. doi:10.1016/j.jpowsour.2020.229152

Sharma, P., & Bhatti, T. S. (2010). A review on electrochemical double-layer capacitors. *Energy Conversion and Management, 51*(12), 2901–2912. doi:10.1016/j.enconman.2010.06.031

Singh, A. P., Tiwari, N. K., Karandikar, P. B., & Dubey, A. (2015). Effect of electrode shape on the parameters of supercapacitor. *2015 International Conference on Industrial Instrumentation and Control, ICIC 2015, Icic*, 669–673. 10.1109/IIC.2015.7150826

Song, M. K., Cheng, S., Chen, H., Qin, W., Nam, K. W., Xu, S., Yang, X. Q., Bongiorno, A., Lee, J., Bai, J., Tyson, T. A., Cho, J., & Liu, M. (2012). Erratum: Anomalous pseudocapacitive behavior of a nanostructured, mixed-valent manganese oxide film for electrical energy storage. *Nano Letters, 12*(8), 4416. https://doi.org/ doi:10.1021/nl300984y

Sun, H., Zhu, J., Baumann, D., Peng, L., Xu, Y., Shakir, I., Huang, Y., & Duan, X. (2019). Hierarchical 3D electrodes for electrochemical energy storage. *Nature Reviews. Materials*, *4*(1), 45–60. doi:10.1038/s41578-018-0069-9

Tao, F., Zhao, Y. Q., Zhang, G. Q., & Li, H. L. (2007). Electrochemical characterization on cobalt sulfide for electrochemical supercapacitors. *Electrochemistry Communications*, *9*(6), 1282–1287. doi:10.1016/j.elecom.2006.11.022

Vangari, M., Pryor, T., & Jiang, L. (2013). Supercapacitors: Review of Materials and Fabrication Methods. *Journal of Energy Engineering*, *139*(2), 72–79. doi:10.1061/(ASCE)EY.1943-7897.0000102

Wang, X., & Zheng, J. P. (2004). The Optimal Energy Density of Electrochemical Capacitors Using Two Different Electrodes. *Journal of the Electrochemical Society*, *151*(10), A1683. doi:10.1149/1.1787841

Winter, M., & Brodd, R. J. (2004). What are batteries, fuel cells, and supercapacitors? *Chemical Reviews*, *104*(10), 4245–4269. doi:10.1021/cr020730k PMID:15669155

Wong, S. I., Lin, H., Sunarso, J., Wong, B. T., & Jia, B. (2020). Optimization of ionic-liquid based electrolyte concentration for high-energy density graphene supercapacitors. *Applied Materials Today*, *18*, 100522. doi:10.1016/j.apmt.2019.100522

Yang, H., Kannappan, S., Pandian, A. S., Jang, J. H., Lee, Y. S., & Lu, W. (2017). Graphene supercapacitor with both high power and energy density. *Nanotechnology*, *28*(44), 445401. Advance online publication. doi:10.1088/1361-6528/aa8948 PMID:28854156

Yassine, M., & Fabris, D. (2017). Performance of commercially available supercapacitors. *Energies*, *10*(9), 1340. Advance online publication. doi:10.3390/en10091340

Yu, L., & Chen, G. Z. (2020). Supercapatteries as High-Performance Electrochemical Energy Storage Devices. *Electrochemical Energy Reviews*, *3*(2), 271–285. doi:10.1007/s41918-020-00063-6

Zhang, L., Hu, X., Wang, Z., Sun, F., & Dorrell, D. G. (2018). A review of supercapacitor modeling, estimation, and applications: A control/management perspective. *Renewable and Sustainable Energy Reviews, 81*(February), 1868–1878. doi:10.1016/j.rser.2017.05.283

Zhang, R. R., Xu, Y. M., Harrison, D., Fyson, J., Qiu, F. L., & Southee, D. (2015). Flexible strip supercapacitors for future energy storage. *International Journal of Automation and Computing*, *12*(1), 43–49. doi:10.1007/s11633-014-0866-6

Zheng, J. P. (2005). Theoretical Energy Density for Electrochemical Capacitors with Intercalation Electrodes. *Journal of the Electrochemical Society*, *152*(9), A1864. doi:10.1149/1.1997152

Chapter 10
Nanowires for Bio-Sensing Applications

Nipun Jain
https://orcid.org/0009-0008-8270-9449
Indian Institute of Science, India

Ranjit Barua
https://orcid.org/0000-0003-2236-3876
Indian Institute of Engineering Science and Technology, India

Yusuf Olatunji Waidi
Indian Institute of Science, India

Sudipto Datta
https://orcid.org/0000-0003-2161-6878
Indian Institute of Science, India

ABSTRACT

Nanomaterials like nanowires, carbon nanotubes, quantum dots, and nanoparticles have breakthrough applications in various engineering domains including the various sensor applications in biomedical engineering. Because of its huge ratio of surface area to volume compared to the macro materials, it is widely used. In this chapter, the authors describe briefly the fabrication, the working principle, as well as the various applications like biomarker detection, RNA and virus detection, and drug discovery of these nanomaterials. Lastly, they briefly highlight its advancement which includes long-term stability, high ionic strength solvent, sensitivity, reusability, and self-powering. Nanowire is going to bring huge innovations in biomedical sensors in the upcoming days.

DOI: 10.4018/979-8-3693-1306-0.ch010

1. INTRODUCTION

Biomedical sensors with exceptional sensitivity hold the potential to revolutionize disease detection, increasing the likelihood of life-saving intervention during the early stages of illnesses. Take, for instance, breast cancer: when identified in its localized state, the five-year survival rate surpasses 90%, but plunges to around 20% once it progresses to an advanced stage (Etzioni et al., 2003). There remains a strong demand for highly sensitive biosensors dedicated to early cancer detection (Wulfkuhle et al., 2003). While extensive research has been conducted to enhance biomedical sensor sensitivity, recent strides in nanotechnology promise groundbreaking solutions in the field of biomedical sensing (Kim et al., 2016). Nanotechnology encompasses a wide range of scientific and engineering disciplines that explore materials and structures on a scale of approximately 100 nm or less (Patolsky et al., 2006). At this nanoscale, material properties are profoundly influenced by surface-to-volume ratios and quantum size effects, yielding properties distinctly different from those at the macroscopic level. The vast surface area-to-volume ratio at the nanoscale renders nanostructures extremely responsive to surface changes. The electron behavior within these nanostructures is constrained by quantization effects, giving rise to quantum size effects where the device's energy levels are size-dependent. Moreover, the reduced scale enhances the excited energy of semiconductors at their lowest state and the strength of volume-normalized oscillators (Wang et al., 1991), resulting in nanostructures with high energy conversion efficiency and relatively low thermal noise (Koo et al., 2015). This is particularly advantageous in the context of biomedical sensors, as biomolecules fall within the same size range as synthetic nanostructures, spanning from approximately 10 to 100 nm.

Biomedical sensors primarily serve to identify and characterize chemical and biological substances, with applications ranging from disease diagnosis to drug development. Employing nanomaterials such as nanowires, carbon nanotubes, and nanoparticles, these crucial tasks can be readily accomplished (Kurkina et al., 2012). For example, semiconductor crystalline nanoparticles have been employed to detect labeled disease markers and other biological entities (Ballou et al., 2004), while colloidal gold has found use in optical imaging and magnetic resonance (Perez et al., 2004). Semiconductor nanowires have made it feasible to electrically detect a variety of substances without the need for labels (Datta et al., 2023). These nanowires are crafted from semiconductor materials (Waidi et al., 2023), and their surfaces can be easily tailored to enhance sensitivity to chemical and biological species (Datta et al., 2023). On the other hand, carbon nanotubes, another potential option for biomedical sensors, are produced by blending semiconductor materials and metals, necessitating additional purification steps (Cui et al., 2003). Furthermore, the binding protocols for various analytes to nanotubes have not been thoroughly established (Datta et al., 2023) (Barua et al., 2018). Consequently, nanostructures constructed with nanowires emerge as the most suitable choice for biomedical sensors, offering high sensitivity, uniformity, reproducibility, and scalability through a relatively straightforward fabrication process (Kaushik et al., 2014). This study delves into the fundamental principles underpinning nanowire-based biomedical sensors and explores an array of applications in the realms of biology and medicine, expanding upon our previous work (Barua et al., 2020). The content in this review avoids duplicating the material found in our prior reports. Additionally, we delve into advances related to long-term stability, sensitivity in high-ionic-strength solvents, reusability, and self-powering capabilities, all of which hold significant promise in overcoming the present limitations associated with nanowire-based biomedical sensors.

2. NANOWIRE FIELD-EFFECT SENSORS

The working principles of nanowire sensors, specifically the nanowire field-effect transistor (FET), offer insight into their operation. Nanowire FETs are an evolution of standard planar FETs, which traditionally consist of a gate, source, drain, and body. In the body, the source and drain are micro- or nanoscale structures made of a metallic material. A gate with a critically thin isolation layer, typically SiO2, is positioned between the source and drain. This isolation layer alters the electric potential on the gate, thereby regulating the conductivity between the source and drain (Sze et al., 2021). In addition to the standard potential changes in the gate caused by external voltage application, chemically or biologically charged species can modify the potential (and consequently the conductivity) in the gate by binding to these charged species. The concept of electrical detection through the accumulation of charged species has been proposed for several decades (Bergveld., 1972). However, the requirement for substantial samples for detection (i.e., low sensitivity) limits the impact of planar gate FET sensors. Compared to planar FET sensors, the nanowire FET replaces the doped channel with a nanowire and employs receptors as a 'gate'. Although the FET structure changes, the core concept remains the same: external alterations caused by charged species can influence the nanowire's conductivity. The semiconducting nanowire can be composed of silicon or other materials (Das et al., 2023). Silicon and silica nanowires are widely used due to their compatibility with standard CMOS (complementary metal–oxide–semiconductor) technology, the natural growth of silicon oxidation (the isolation layer) on the silicon surface, and the ease with which the surfaces of silicon and silica can be modified. The native surfaces of silicon and silica do not respond to changes in biomolecules. Even though biomolecules contain charges, their accumulation near the isolation layer does not affect the nanowires' conductivity (Perez et al., 2004). To render the surface responsive, it must be functionalized and bound with receptors capable of capturing specific charged species like DNA, RNA, viruses, and so on. The functionalization process establishes a chemical connection between the top surface of the isolation layer (as well as the nanowires) and biomolecules. The receptors effectively control the 'gate' of the nanowire-based FET. Through alterations in the number of collected species, the receptors generate an electric field within the nanowire through the thin isolation oxidation layer, thereby determining the nanowire's conductivity. Consequently, if we consider the species as the input signal, the function of the receptors closely resembles that of the gate, as they convert the input signal into changes in the device's conductivity. The most common method for receptor attachment is covalent binding, given that the oxide layer can easily form on the silicon surface. After exposure to air or plasma, the silicon nanowire forms silicon oxide to passivate the silicon surface. Covalent binding aims to modify the silicon surface with a silane derivative monolayer. For instance, 3-aminopropoyl-triethoxysilane (APTES) is a receptor used for DNA, peptide nucleic acid (PNA), and antibodies (Perez et al., 2004). When negatively charged species in an aqueous solution bind to the receptor, they induce positive charges in the Si nanowire FET surface through the receptor. This generation of positive charges can be seen as either the disappearance of electrons (negatively charged carriers) or the emergence of holes (positively charged carriers). Consequently, the conductivity (or current) between the source and drain is altered. In the case of an n-type Si nanowire, the conductivity decreases because electrons serve as charge carriers. Conversely, for a p-type Si nanowire, the current increases since holes are the charge carriers. The Si nanowire FET can be electrically connected to a semiconductor parameter analyzer or other measurement and logging equipment to enable real-time conductivity monitoring (Datta et al., 2023). Unlike FETs used in electronic circuits, nanowire FETs typically remain in the ON state, allowing DC currents to continuously flow due to their sensing nature. Multiple or an array of nanowires are typically used in a single sensing device, as multiple nanowires are more sensitive to species accumulation and experience reduced noise due to averaging (Wang et al., 1991).

3. FABRICATION PROCESS

For the creation of nanowire sensors, both top-down and bottom-up techniques are used. The bottom-up techniques are applied to generate top-notch nanowires frequently on Si wafers (Kim et al., 2014). Even though most nanowires are cylindrical, other variants (such as round, square, and triangular) can be created using current bottom-up procedures (Li et al., 2003). Si nanowires are first fabricated utilizing the chemical vapour deposition (CVD) technique (Patolsky et al., 2006). Vapor-Liquid-Solid process in the CVD procedure enables the catalytic growth of Si nanowires. The silicon nanowires saturated with ethanol are then applied to a silicon substrate via deposition. The generated nanowires are subsequently spin-coated with a photoresist, and electrodes are designed using a lift-off system. Bottom-up production is completed by surface alteration and passivation with receptor affinity (Patolsky et al., 2006). It is simple to create a separation on the surface by subjecting it to an air atmosphere. This method also offers the disadvantage of random orientation, resulting in a non-uniform distribution. In addition, realignment processes during manufacturing, namely, Langmuir-Blodgett, microfluidics (Huang et al., 2001), contact printing (Datta et al., 2023), and electric field (Freer et al., 2010), are necessary to improve the orientation. Nevertheless, the large-scale manufacturing of positioned sensors is challenging since the common CMOS manufacturing method is incompatible with these alignment techniques.

Contrary to these techniques, a top-down approach is established from the microfabrication of silicon-on-insulator or single-crystalline wafers capable of producing aligned nanowires. This technique remains consistent with the conventional CMOS technology. The manufacturing process begins by doping boron or phosphorous on the silicon layer of the wafer (Figure 1). The source and drain connections are then defined by substantial density doping on the patterned region, followed by reactive ion etching (RIE), to create electrodes having a size of a few micrometers. Then, nanometer-sized Si nanowires are created using electron-beam (E-beam) lithography, and thermal evaporation creates the metal contact leads. Similar to the bottom-up approach, passivation, and surface modification mark the end of the fabrication process. (Vu et al., 2009). Although well-oriented nanowires produced by top-down strategies are extremely compatible with CMOS technologies, their diameter is greater than that of bottom-up methods. The devices produced using this have nearly a similar design pattern despite its variations in the orientation and materials used. The fluid exchange system constitutes an important element in nanowire-based sensors. The fluid exchange system sends fluids or analytes to the nanowire sensor's surface or very near it. Polydimethylsiloxane (PDMS) channels are frequently utilized to feed analytes to nanowire sensors. Due to its great biocompatibility, PDMS is also advantageous in biosensing applications. (Barua et al., 2023). This microfluidic technology does have some restrictions, though, as the particles inside the channel find it challenging to contact the surface because the flow at the microscale level is mainly laminar. Additionally, because PDMS can absorb biomolecules, there is a chance that the sensor's sensitivity would decrease (Barua et al., 2022). An acrylic chamber has been created as a fluid exchange mechanism to get around these drawbacks of PDMS microfluidic channels (Datta et al., 2018).

Figure 1. The conceptual overview of field-effect transistors (FETs), with illustrations and explanations: (A) schematics illustrating the differences between standard and nanowire field-effect transistors; (B) an overview of the functionalization process of the nanowire, based on 3-aminopropoyltriethoxysilane (APTES); (C) illustrates the operating principles of nanowire sensors; (D) shows nanowire sensors fabricated using the bottom-up approach; (E) depicts nanowire sensors fabricated using the top-down approach (Wang et al., 2016)

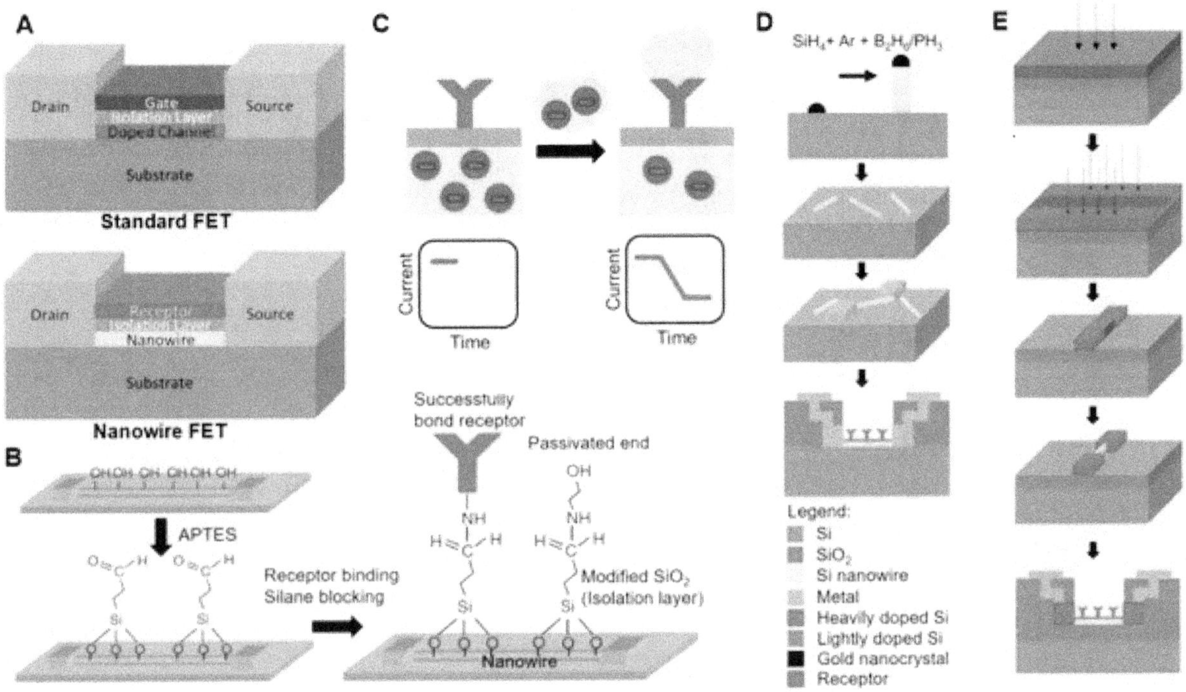

4. DETECTION OF VARIOUS BIOLOGICAL AGENTS

4.1 Detection of DNA and RNA

Numerous methods, such as electrochemical, optical fiber sensors, and surface plasmon resonance bio-sensors strategies, have experienced a lot of development in DNA sensors owing to the large spectrum of possible uses for DNA and RNA detection methods, which includes recognizing of clinical diagnosis, the identification of gene sequences, nano-engineering, and in the discipline of food (Russell et al., 2014) (Tran et al., 2023). Successful pathology diagnosis is particularly critical for scientists working in the healthcare sciences (Shariati., 2018). Using indium tin oxide nanowires, Shariati et al. created a transistor-based DNA sensor for the hepatitis B virus (Shariati., 2018). Due to the strong conductance and functionally modified surface of ITO nanowires, probe immobilization, and target hybridization have increased. An HRTEM measurement revealed that the crystalline diameter of ITO nanowires was not more than 50 nm. The single-stranded viral DNA was immobilized as a probe on the modified nanowires in which the targets have a range of 1 fM to 10 M. The hybridization procedure for a particular single strand takes 90 minutes with a switching ratio 1.1. For biosensor sensitivity, DNA oligonucleotide sequences that are

Figure 2. The generation of a nanowire pattern on a silicon-on-insulator (SOI) wafer, along with the use of a field effect transistor (FET) based on silicon nanowires to detect GABA (Gamma-Aminobutyric Acid), a neurotransmitter (Lee et al., 2019)

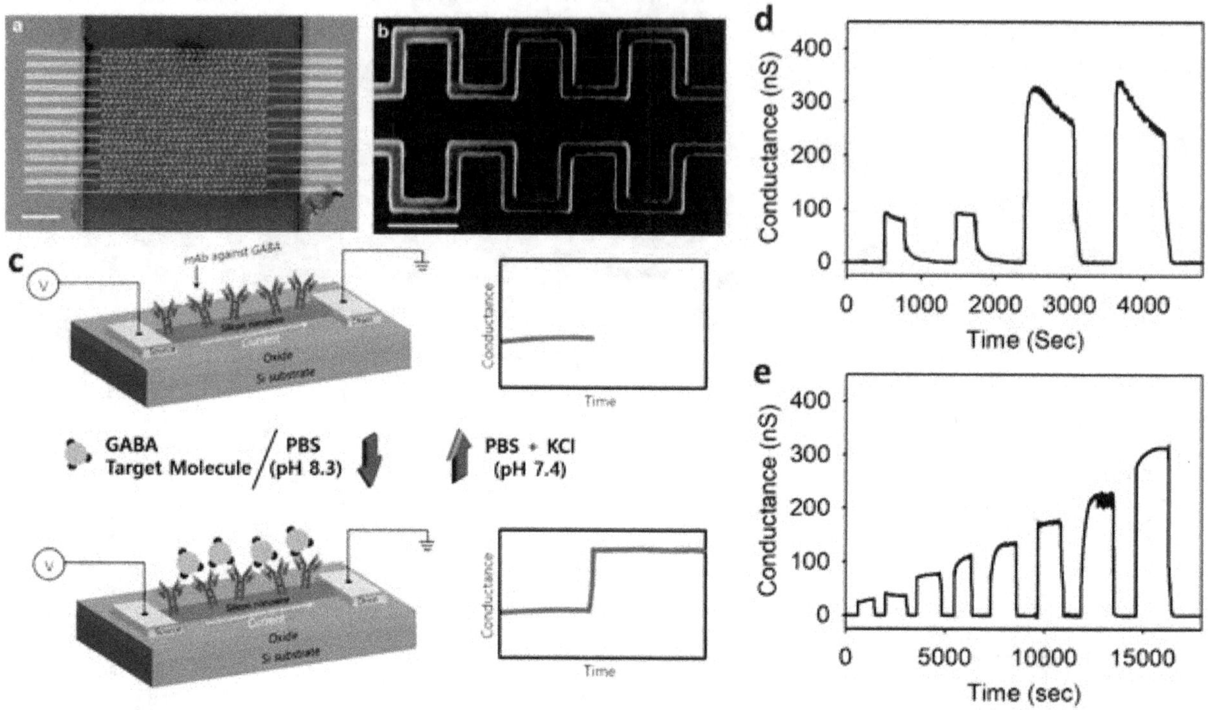

mismatched, complementary, and non-complementary can be easily distinguished. Wang et al. created ultrasensitive sensors of PEG-polypyrrole to give better antifouling characteristics (Wang et al., 2019). PPy nanowire arrays were created by electrochemically polymerizing pyrrole at a constant voltage on a carbon electrode. The surfaces were adorned with 4-armed PEG by electrochemically oxidizing the amine groups. These novel PEG-modified nanowires combine their highly effective antifouling properties with remarkable conductivity. Because they are essential for the growth of cancer and several other diseases, microRNAs (miRNAs) are possible markers as diagnostic and prognostic markers. It can effectively construct an ultrasensitive biosensor by immobilizing DNA probes on these nanowires. DNA/RNA hybridization was measured by tracking differences in the methylene blue redox signal.

4.2 Viruses

Since most viruses inflict serious illnesses in humans and can be utilized as biological weapons, effectively detecting viruses serves as one of the greatest challenges for nanowire sensors in maintaining the safety and health of our communities (Stadler et al., 2003). Many harmful viral diseases, such as influenza, dengue (Zhang et al., 2010), and HIV (Inci et al., 2013), have been detected recently using these nanowire-based sensors. Antibodies binding precisely to the virus of interest are functionalized onto the sensor surface, which alters the substrate conductivity. For example, a silicon NW sensor was created to identify up to 29 flu viruses per milliliter (mL) in exhaled breath condensate samples (EBC).

(Shen et al., 2012). Si nanowires have been developed by Ibarlucea et al. that can identify the Ebola VP40 (Ibarlucea et al., 2018). After 30 minutes of incubation, the minimum detection was discovered to be approximately 6.25 nM, surpassing the ELISA method six times. In the coming time, nanowire biosensors will be used more frequently at the point-of-care diagnosis of widespread diseases because of their benefits, such as speed, accuracy, and motility.

5. DETECTION OF PROTEIN

Biosensors for protein detection are employed to recognize specific proteins in biological samples. When a specific analysis is recognized, a cell, protein (including enzymes), or any other biocomponent can convert the biological signal into a marker signal. An electrochemical biosensor that recognizes target proteins using specific peptide sequences with high affinity and selectivity is known as a peptide-based biosensor. Another type of biosensor used for protein detection is the aptamer biosensor. Aptamers, which are biological recognition elements, can replace antibodies in several functions. CRISPR/Cas-based technologies are another type of biosensor used to find proteins. In order to detect prostate-specific antigen (PSA) with extreme precision and focus, an innovative aptasensing substrate driven by SERS was developed by Awatef. The transducing SERS substrate is composed of vertically aligned silver nanoparticle-coated silicon nanowires (SiNWs). The aptamer is immobilized using thiol chemistry, and the silver nanoparticles improve the Raman signal. The formation of amide bond vibration modes in response to PSA depends on the aptamer's presence on the surface of Si NWs. With a detection limit of 0.1 gL1, the equipment detects in a wide concentration spectrum, up to 20 gL1. It encompasses the samples from healthy as well as diagnosed patients (Ouhibi et al., 2021). Kim et al., 2019 investigated fiber optic-based, 3D nanomaterial-based sensors for localized surface plasmon resonance. The sensors in question were developed for highly accurate sensing using ZnO and Au nanoparticles. The development of a sensor is notable for the following aspects: An increased sensing area, the effect of NWs on light entrapment, and simple optics derived from the optical fiber. The 3D nanomaterials were produced hydrothermally using AuNPs on the NWs and ZnO nanowires on the fiber cross sections. The suggested sensor responded to changes in refractive index by producing a constant value. The 3D FO-LSPR demonstrated a large elevation in the localized SPR sensitivity for large-scale refractive index variations compared to 2D FO-LSPR sensors, which comprise a single layer of nanoparticles linked to an optical fiber. Additionally, employing these multiple biosensors decreased prostate-specific antigen detection limits to 0.51 pg/mL. This biomarker becomes useful for increased detection of prostate cancer. N, S-graphene quantum dots @Au-polyaniline (N, S-GQDs@Au-PANI NWs) were created in order to quantify carcinoembryonic antigen (CEA). The N, S-GQDs, and Au-PANI were made via straightforward interfacial polymerization and hydrothermal pyrolysis. A multifunctional sensor is created to evaluate an elevation in electrochemical functioning by linking Au-thiol to Au-PANI nanowires. The biomedical applications allow for the impedimetric calculation of CEA based on assigned immunoassay. The substances that expedite electron movement include N, S-GQDs@Au-PANI NWs. The non-tagged immunosensor detects a very low range and demonstrates impressive solidity and precision response against numerous carcinogens (Ganganboina et al., 2019).

6. RECENT ADVANCES IN NANOWIRE BIOSENSORS

6.1 Integration With Paper-Based Devices (PADs)

Researchers are using nanowires to make PADs that can detect biomarkers in a portable way. PADs have several advantages over silicon-based chips, such as being cheaper and more biodegradable (Song et al., 2017). Li et al. were the first to use ZnO nanowires in small, portable devices to measure glucose levels (Li et al., 2015). They used a conductive ink to print the entire PAD, a device that measures glucose levels. They then used this PAD to grow ZnO nanowires from deposited ZnO nanoparticles. The ZnO nanowires were deposited quickly, efficiently, and scalable (hydrothermal). This method also allows for precise ability to adjust the size of the ZnO nanowires, which is important for downstream purposes. The researchers showed how to use the sensor to identify glucose concentrations. The developed sensor could detect glucose levels as low as 94.7 μM and up to 15 mM. This sensor is not as sensitive as conventional sensors, which can detect very low glucose levels (0.5 μM). However, this sensor is still promising for remote biosensing applications. A few challenges must be addressed before this sensor can be used in real-world applications. First, many proteins need enzymes to be detected. Enzymes are unstable at room temperature and can lose their activity over time, which is a problem that needs to be solved for PADs with nanowires to be successful. Second, the rough surface of PADs makes it difficult to detect low concentrations. Researchers have recently developed a new type of paper made from nanofibrillated cellulose (NFC) (Missoum et al., 2013) that could improve the sensitivity of zinc oxide photoacoustic devices (ZnO PADs). NFC paper has a rough surface on the nanometer scale, which could help scatter light more effectively, leading to stronger photoacoustic signals. Finally, to make these devices suitable for remote sensing, they need to be wireless and not need any power. This can be done using a technology that allows the device to work without batteries or other external power sources.

6.2 Signal Processing and Data Analysis

To diagnose a disease, doctors usually must detect and analyze various markers. Many multiplexing biosensors are being developed, but most are still in the early stages of development, and testing them on spiked samples is the standard method. For multivariable and multiplexing biosensors to be used in clinical practice, the signal processing techniques used to analyze their output data must accurately identify and extract useful information about the diagnosed disease or condition. Scientists started using signal processing to improve the data from nanowire sensor arrays in gas sensors (Potyrailo., 2016). For instance, Cho et al. used platinum, indium, copper, and nickel nanowire sensor arrays to identify vapors from explosive materials such as octane, acetone, nitrotoluene, and nitrobenzene (Cho et al., 2011). The researchers evaluated the performance of the decision tree on a separate test set of data and achieved an accuracy of over 90%. This means that the decision tree could correctly predict the concentration of explosive precursors in over 90% of the samples. The error rate was less than 1%, meaning the decision tree made few mistakes. Shehada et al. 2016 applied the hypothesis of applying signal administration to improve data quality from nanowire sensor arrays to diagnose diseases. They developed an array of silicon nanowire biosensors to detect the unique mix of chemicals in the breath of people with cancer. The researchers first fabricated and functionalized silicon nanowires to create an array of biosensors. They then characterized the array using a modeled breath containing a combination associated with eleven disease markers: lung cancer, chronic obstructive pulmonary disease, gastric cancer, and asthma.

Once the array was successfully characterized, the researchers built an artificial neural network (ANN) and trained it on a dataset of sensing characteristics and established disease conditions. The researchers used the instructed ANN and an array of nanowire biosensors to probe 374 patients' breath samples. The researchers found that the array with ANN could differentiate between almost all pairs of diseases with more than 80% accuracy. Conversely, the small sample size hindered the study. This means that the array could not learn the patterns of all diseases as well as it could have. As a result, the accuracy rate for exact diseases with controlled samples was low, at around 60%. In a later study, the researchers used nanobiosensors to create a large clinical data collection on the exhaled molecules of 1404 participants with 17 different diseases (Nakhleh et al., 2017). This finding found 13 new chemicals in the breath of people with certain diseases. This suggests that nano biosensors could diagnose diseases by detecting these chemicals in the breath. Additionally, new machine-learning techniques (Silver et al., 2017) could be used to improve the accuracy of nano biosensors. Self-powering nanowire biosensors can be powered by triboelectric nanogenerators (TENGs) or biofuels. TENGs convert mechanical energy into electrical energy. Scientists have demonstrated that a voltage of 1.26 volts can be generated by applying a mechanical force to a well-aligned array of zinc oxide (ZnO) nanowires connected in series (Wang et al., 2006). This powers a nanowire-based pH sensor (Cao et al., 2018) and a wireless data transmission technique (Hu et al., 2011). Nevertheless, TENGs can only generate power when there is significant physical motion. This limits their use in vivo to muscle, lung, and cardiac tissues (103, 104). Biofuels are a preferred power source for nanowire biosensors used inside the body. Biofuels are chemical fuels that can be produced from biological sources, such as glucose from blood or lactate from sweat. Biofuel cells generate electricity from the chemical energy stored in biofuels. Researchers have explored two glucose-mediated approaches for powering nanowire biosensors in vivo: glucose/O2 (Yang et al., 2014) and glucose/air (Zhao et al., 2015). Both strategies use enzymes to convert glucose into electricity. However, the glucose/O2 strategy is more efficient because it uses oxygen and glucose to generate electricity. Hansen et al. have shown that creating a self-powered nanowire biosensor device inside the body can harness the biofuels and piezoelectric effect (Hansen et al., 2010). This system converts the mechanical energy of breathing and the heartbeat into electrical energy using poly(vinylidene fluoride) nanofibers. The system also uses a tunable enzymatic biofuel cell to generate electricity from the chemicals in body fluids. Combining the two methods allows the system to generate more power and operate for longer times.

7. CONCLUSION AND PERSPECTIVE

Nanowires have been used to develop biosensors for various applications, including detecting pathogens, DNA, bacteria, protease, RNA, molecular molecules, and neurotransmitters. Nanowires have significantly improved biosensing technology over the past decade due to advances in the design, miniaturization, and development of nanostructured tools and new systems for creating nanowires. Using nanowires in biosensors has improved biosensing technology's sensitivity, dynamic range, and durability. Additionally, using multiple nanowires with different properties in biosensors has greatly improved the transduction process. How well nanowire biosensors perform will depend on how they compare to current gold standards, like ELISA and PCR, in terms of ease of use, sensitivity, specificity, and reliability. Although nanowire biosensors have exceptionally superior sensitivity, their analytical signal strength is not yet high enough to be corrupted by significant background signals, notably in vivo settings. The low sensitivity of nanowire biosensors and the challenges of fabricating them may be solved by advances in receptor binding techniques and the availability of commercially available materials at a lower cost. However, some challenges still need to be addressed before nanowire biosensors can be widely used in biomedical applications. These challenges include (i) The difficulty of producing nanostructures. (ii) The toxicity of nanowires varies depending on the material type and physical characteristics. (iii) The sustainability of nanostructures in sensor applications has not been well-studied. Nanowire biosensors are still in their early stages of development, and most of them require a large sample to identify a target. This can lead to inaccurate results, such as false-positive or false-negative diagnoses. Few nanowire biosensors have been commercialized, but researchers believe that companies will soon translate current academic research into commercially viable prototypes. Nanowires are expected to play a key role in developing wearable biosensors, which could be used to monitor people's health in real time.

CONFLICTS OF INTEREST

The authors declare that there are no conflicts of interest regarding the publication of this chapter.

FUNDING

No funding has been provided for this work.

ETHICAL APPROVAL

Not required.

ACKNOWLEDGMENT

The authors would like to thank IIEST Shibpur, and IISc Bangalore.

REFERENCES

Ballou, B., Lagerholm, B. C., Ernst, L. A., Bruchez, M. P., & Waggoner, A. S. (2004). Noninvasive imaging of quantum dots in mice. *Bioconjugate Chemistry*, *15*(1), 79–86. doi:10.1021/bc034153y PMID:14733586

Barua, R., Das, S., Roy Chowdhury, A., & Datta, P. (2022). Experimental and simulation investigation of surgical needle insertion into soft tissue mimic biomaterial for minimally invasive surgery (MIS). *Proceedings of the Institution of Mechanical Engineers, Part H: Journal of Engineering in Medicine*. 10.1177/09544119221143860

Barua, R., Datta, S., Chowdhury, A. R., & Datta, P. (2018). Advances in MEMS and Micro-Scale technologies for application in controlled Drug-Dosing systems: MEMS-Based drug delivery systems. In *Design and Development of Affordable Healthcare Technologies* (pp. 165–179). IGI Global. doi:10.4018/978-1-5225-4969-7.ch007

Barua, R., Datta, S., & Das, J. (2020). Application of Nanotechnology in Global Issues. In M. Merviö (Ed.), *Global Issues and Innovative Solutions in Healthcare, Culture, and the Environment* (pp. 292–300). IGI Global. doi:10.4018/978-1-7998-3576-9.ch015

Barua, R., Sarkar, A., & Datta, S. (2023). Modern Lab-on-Chip Biosensors Application on Infectious COVID-19 Detection. In R. Singh, R. Phanden, B. Sikarwar, & J. Davim (Eds.), *Advances in MEMS and Microfluidic Systems* (pp. 258–270). IGI Global. doi:10.4018/978-1-6684-6952-1.ch013

Bergveld, P. (1972). Development, operation, and application of the ion-sensitive field-effect transistor as a tool for electrophysiology. *IEEE Transactions on Biomedical Engineering*, *19*(5), 342–351. doi:10.1109/TBME.1972.324137 PMID:5038390

Cao, R., Wang, J., Zhao, S., Yang, W., Yuan, Z., Yin, Y., Du, X., Li, N.-W., Zhang, X., Li, X., Wang, Z. L., & Li, C. (2018). Self-powered nanofiber-based screen-print triboelectric sensors for respiratory monitoring. *Nano Research*, *11*(7), 3771–3779. doi:10.1007/s12274-017-1951-2

Cho, J., Li, X., Gu, Z., & Kurup, P. U. (2011). Recognition of explosive precursors using nanowire sensor array and decision tree learning. *IEEE Sensors Journal*, *12*(7), 2384–2391. doi:10.1109/JSEN.2011.2182042

Cui, Y., Zhong, Z., Wang, D., Wang, W. U., & Lieber, C. M. (2003). High performance silicon nanowire field effect transistors. *Nano Letters*, *3*(2), 149–152. doi:10.1021/nl0258751

Das, S., Datta, S., Barman, A., & Barua, R. (2023). Smart Biodegradable and Bio-Based Polymeric Biomaterials for Biomedical Applications. In A. Kumar, P. Kumar, A. Srivastava, & V. Goyat (Eds.), *Modeling, Characterization, and Processing of Smart Materials* (pp. 56–82). IGI Global. doi:10.4018/978-1-6684-9224-6.ch003

Datta, S., & Barua, R. (2023). Fluorescent Nanomaterials and Its Application in Biomedical Engineering. In A. Rakha, A. Munawar, V. Khanna, & S. Bansal (Eds.), *Modeling and Simulation of Functional Nanomaterials for Forensic Investigation* (pp. 164–186). IGI Global. doi:10.4018/978-1-6684-8325-1.ch009

Datta, S., Barua, R., Sarkar, R., Barui, A., Chowdhury, A. R., & Datta, P. (2018, September). Design and development of alginate: Poly-l-lysine scaffolds by 3D bio printing and studying their mechanical, structural and cell viability properties. *IOP Conference Series. Materials Science and Engineering*, *402*, 012113. doi:10.1088/1757-899X/402/1/012113

Etzioni, R., Urban, N., Ramsey, S., McIntosh, M., Schwartz, S., Reid, B., Radich, J., Anderson, G., & Hartwell, L. (2003). The case for early detection. *Nature Reviews. Cancer*, *3*(4), 243–252. doi:10.1038/nrc1041 PMID:12671663

Freer, E. M., Grachev, O., Duan, X., Martin, S., & Stumbo, D. P. (2010). High-yield self-limiting single-nanowire assembly with dielectrophoresis. *Nature Nanotechnology*, *5*(7), 525–530. doi:10.1038/nnano.2010.106 PMID:20526324

Ganganboina, A. B., & Doong, R. A. (2019). Graphene quantum dots decorated gold-polyaniline nanowire for impedimetric detection of carcinoembryonic antigen. *Scientific Reports*, *9*(1), 7214. doi:10.1038/s41598-019-43740-3 PMID:31076624

Hansen, B. J., Liu, Y., Yang, R., & Wang, Z. L. (2010). Hybrid nanogenerator for concurrently harvesting biomechanical and biochemical energy. *ACS Nano*, *4*(7), 3647–3652. doi:10.1021/nn100845b PMID:20507155

Hu, Y., Zhang, Y., Xu, C., Lin, L., Snyder, R. L., & Wang, Z. L. (2011). Self-powered system with wireless data transmission. *Nano Letters*, *11*(6), 2572–2577. doi:10.1021/nl201505c PMID:21604749

Huang, Y., Duan, X., Wei, Q., & Lieber, C. M. (2001). Directed assembly of one-dimensional nanostructures into functional networks. *Science*, *291*(5504), 630–633. doi:10.1126/science.291.5504.630 PMID:11158671

Ibarlucea, B., Fawzul Akbar, T., Kim, K., Rim, T., Baek, C. K., Ascoli, A., Tetzlaff, R., Baraban, L., & Cuniberti, G. (2018). Ultrasensitive detection of Ebola matrix protein in a memristor mode. *Nano Research*, *11*(2), 1057–1068. doi:10.1007/s12274-017-1720-2

Inci, F., Tokel, O., Wang, S., Gurkan, U. A., Tasoglu, S., Kuritzkes, D. R., & Demirci, U. (2013). Nanoplasmonic quantitative detection of intact viruses from unprocessed whole blood. *ACS Nano*, *7*(6), 4733–4745. doi:10.1021/nn3036232 PMID:23688050

Kaushik, A., Kumar, R., Huey, E., Bhansali, S., Nair, N., & Nanir, M. (2014). Silica Nanowires: Growth, Integration, and Sensing Applications. *Mikrochimica Acta*, *181*(15-16), 1759–1780. doi:10.1007/s00604-014-1255-0 PMID:25382871

Kim, D. M., & Jeong, Y. H. (Eds.). (2014). *Nanowire Field Effect Transistors: Principles and Applications* (Vol. 43). Springer. doi:10.1007/978-1-4614-8124-9

Kim, H. M., Park, J. H., & Lee, S. K. (2019). Fiber optic sensor based on ZnO nanowires decorated by Au nanoparticles for improved plasmonic biosensor. *Scientific Reports*, *9*(1), 15605. doi:10.1038/s41598-019-52056-1 PMID:31666617

Kim, K., Park, C., Kwon, D., Kim, D., Meyyappan, M., Jeon, S., & Lee, J. S. (2016). Silicon nanowire biosensors for detection of cardiac troponin I (cTnI) with high sensitivity. *Biosensors & Bioelectronics*, *77*, 695–701. doi:10.1016/j.bios.2015.10.008 PMID:26496224

Koo, K. I., & Seo, J. M. (2015). A review of nanotechnology for highly sensitive photodetectors for vision sensors of insect-like robots. *Sensors and Materials*, *27*(6), 465–474.

Kurkina, T., & Balasubramanian, K. (2012). Towards in vitro molecular diagnostics using nanostructures. *Cellular and molecular life sciences. Cellular and Molecular Life Sciences*, *69*(3), 373–388. doi:10.1007/s00018-011-0855-7 PMID:22009454

Lee, J. H., Chae, E. J., Park, S. J., & Choi, J. W. (2019). Label-free detection of γ-aminobutyric acid based on silicon nanowire biosensor. *Nano Convergence*, *6*(1), 1–6. doi:10.1186/s40580-019-0184-3 PMID:31041617

Li, C. P., Lee, C. S., Ma, X. L., Wang, N., Zhang, R. Q., & Lee, S. T. (2003). Growth direction and cross-sectional study of silicon nanowires. *Advanced Materials*, *15*(7-8), 607–609. doi:10.1002/adma.200304409

Li, X., Zhao, C., & Liu, X. (2015). A paper-based microfluidic biosensor integrating zinc oxide nanowires for electrochemical glucose detection. *Microsystems & Nanoengineering*, *1*(1), 1–7. doi:10.1038/micronano.2015.14

Missoum, K., Belgacem, M. N., & Bras, J. (2013). Nanofibrillated Cellulose Surface Modification: A Review. *Materials (Basel)*, *6*(5), 1745–1766. doi:10.3390/ma6051745 PMID:28809240

Nakhleh, M. K., Amal, H., Jeries, R., Broza, Y. Y., Aboud, M., Gharra, A., Ivgi, H., Khatib, S., Badarneh, S., Har-Shai, L., Glass-Marmor, L., Lejbkowicz, I., Miller, A., Badarny, S., Winer, R., Finberg, J., Cohen-Kaminsky, S., Perros, F., Montani, D., ... Haick, H. (2017). Diagnosis and Classification of 17 Diseases from 1404 Subjects via Pattern Analysis of Exhaled Molecules. *ACS Nano*, *11*(1), 112–125. doi:10.1021/acsnano.6b04930 PMID:28000444

Ouhibi, A., Raouafi, A., Lorrain, N., Guendouz, M., Raouafi, N., & Moadhen, A. (2021). Functionalized SERS substrate based on silicon nanowires for rapid detection of prostate specific antigen. *Sensors and Actuators. B, Chemical*, *330*, 129352. doi:10.1016/j.snb.2020.129352

Patolsky, F., Zheng, G., & Lieber, C. M. (2006). Nanowire sensors for medicine and the life sciences. *Nanomedicine (London)*, *1*(1), 51–65. doi:10.2217/17435889.1.1.51 PMID:17716209

Perez, J. M., Josephson, L., & Weissleder, R. (2004). Use of magnetic nanoparticles as nanosensors to probe for molecular interactions. *ChemBioChem*, *5*(3), 261–264. doi:10.1002/cbic.200300730 PMID:14997516

Potyrailo, R. A. (2016). Multivariable Sensors for Ubiquitous Monitoring of Gases in the Era of Internet of Things and Industrial Internet. *Chemical Reviews*, *116*(19), 11877–11923. doi:10.1021/acs.chemrev.6b00187 PMID:27602947

Russell, C., Welch, K., Jarvius, J., Cai, Y., Brucas, R., Nikolajeff, F., Svedlindh, P., & Nilsson, M. (2014). Gold nanowire based electrical DNA detection using rolling circle amplification. *ACS Nano*, *8*(2), 1147–1153. doi:10.1021/nn4058825 PMID:24433087

Shariati, M. (2018). The field effect transistor DNA biosensor based on ITO nanowires in label-free hepatitis B virus detecting compatible with CMOS technology. *Biosensors & Bioelectronics*, *105*, 58–64. doi:10.1016/j.bios.2018.01.022 PMID:29355779

Shehada, N., Cancilla, J. C., Torrecilla, J. S., Pariente, E. S., Brönstrup, G., Christiansen, S., Johnson, D. W., Leja, M., Davies, M. P., Liran, O., Peled, N., & Haick, H. (2016). Silicon Nanowire Sensors Enable Diagnosis of Patients via Exhaled Breath. *ACS Nano*, *10*(7), 7047–7057. doi:10.1021/acsnano.6b03127 PMID:27383408

Shen, F., Wang, J., Xu, Z., Wu, Y., Chen, Q., Li, X., Jie, X., Li, L., Yao, M., Guo, X., & Zhu, T. (2012). Rapid flu diagnosis using silicon nanowire sensor. *Nano Letters*, *12*(7), 3722–3730. doi:10.1021/nl301516z PMID:22731392

Silver, D., Schrittwieser, J., Simonyan, K., Antonoglou, I., Huang, A., Guez, A., Hubert, T., Baker, L., Lai, M., Bolton, A., Chen, Y., Lillicrap, T., Hui, F., Sifre, L., van den Driessche, G., Graepel, T., & Hassabis, D. (2017). Mastering the game of Go without human knowledge. *Nature*, *550*(7676), 354–359. doi:10.1038/nature24270 PMID:29052630

Song, P., Wang, Y. H., & Liu, X. (2017). Flexible physical sensors made from paper substrates integrated with zinc oxide nanostructures. *Flexible and Printed Electronics*, *2*(3), 034001. doi:10.1088/2058-8585/aa765d

Stadler, K., Masignani, V., Eickmann, M., Becker, S., Abrignani, S., Klenk, H. D., & Rappuoli, R. (2003). SARS—Beginning to understand a new virus. *Nature Reviews. Microbiology*, *1*(3), 209–218. doi:10.1038/nrmicro775 PMID:15035025

Sze, S. M., Li, Y., & Ng, K. K. (2021). *Physics of semiconductor devices*. John wiley & sons.

Tran, V. A., Vo, G. N., Vo, T., Doan, V. D., Vo, V., & Le, V. T. (2023). Recent Applications and Prospects of Nanowire-Based Biosensors. *Processes (Basel, Switzerland)*, *11*(6), 1739. doi:10.3390/pr11061739

Vu, X. T., Eschermann, J. F., Stockmann, R., GhoshMoulick, R., Offenhäusser, A., & Ingebrandt, S. (2009). Top-down processed silicon nanowire transistor arrays for biosensing. *Physica Status Solidi (A), 206*(3), 426-434.

Waidi, Y. O., Barua, R., & Datta, S. (2023). Metals, Polymers, Ceramics, Composites Biomaterials Used in Additive Manufacturing for Biomedical Applications. In A. Kumar, P. Kumar, A. Srivastava, & V. Goyat (Eds.), *Modeling, Characterization, and Processing of Smart Materials* (pp. 165–184). IGI Global. doi:10.4018/978-1-6684-9224-6.ch008

Wang, J., & Hui, N. (2019). Electrochemical functionalization of polypyrrole nanowires for the development of ultrasensitive biosensors for detecting microRNA. *Sensors and Actuators. B, Chemical*, *281*, 478–485. doi:10.1016/j.snb.2018.10.131

Wang, Y., & Herron, N. (1991). Nanometer-sized semiconductor clusters: Materials synthesis, quantum size effects, and photophysical properties. *Journal of Physical Chemistry*, *95*(2), 525–532. doi:10.1021/j100155a009

Wang, Z., Lee, S., Koo, K., & Kim, K. (2016). Nanowire-Based Sensors for Biological and Medical Applications. *IEEE Transactions on Nanobioscience*, *15*(3), 186–199. doi:10.1109/TNB.2016.2528258 PMID:26978831

Wang, Z. L., & Song, J. (2006). Piezoelectric nanogenerators based on zinc oxide nanowire arrays. *Science*, *312*(5771), 242–246. doi:10.1126/science.1124005 PMID:16614215

Wulfkuhle, J. D., Liotta, L. A., & Petricoin, E. F. (2003). Proteomic applications for the early detection of cancer. *Nature Reviews. Cancer*, *3*(4), 267–275. doi:10.1038/nrc1043 PMID:12671665

Yang, L., Zhang, Y., Chu, M., Deng, W., Tan, Y., Ma, M., Su, X., Xie, Q., & Yao, S. (2014). Facile fabrication of network film electrodes with ultrathin Au nanowires for nonenzymatic glucose sensing and glucose/O2 fuel cell. *Biosensors & Bioelectronics*, *52*, 105–110. doi:10.1016/j.bios.2013.08.038 PMID:24035853

Zhang, G. J., Zhang, L., Huang, M. J., Luo, Z. H. H., Tay, G. K. I., Lim, E. J. A., Kang, T. G., & Chen, Y. (2010). Silicon nanowire biosensor for highly sensitive and rapid detection of Dengue virus. *Sensors and Actuators. B, Chemical*, *146*(1), 138–144. doi:10.1016/j.snb.2010.02.021

Zhao, M., Gao, Y., Sun, J., & Gao, F. (2015). Mediatorless glucose biosensor and direct electron transfer type glucose/air biofuel cell enabled with carbon nanodots. *Analytical Chemistry*, *87*(5), 2615–2622. doi:10.1021/acs.analchem.5b00012 PMID:25666266

Chapter 11
Lead-Free Alkali Niobate Piezoelectric Materials:
Synthesis and Characterizations

Haobam Samananda Singh
https://orcid.org/0000-0002-9931-096X
National Institute of Technology, Manipur, India

Irom Monika Aniz
National Institute of Technology, Manipur, India

Chiranjit Chaliha
National Institute of Technology, Manipur, India

Mamata Maisnam
National Institute of Technology, Manipur, India

ABSTRACT

Piezoceramics with attractive structural and electrical properties are important for the development of energy harvesters and energy conservation devices. The lead-based piezoelectric ceramics need to be replaced by lead-free piezoceramics because of the toxicity of the lead. Barium titanate (BT), sodium bismuth titanate (BNT), potassium bismuth titanate (KNT), and sodium potassium niobate (KNN) are some lead-free piezo materials. Among the lead-free piezo materials, KNN has a high Curie temperature and attractive electric and piezoelectric properties. The attractive properties of KNN-based ceramics are due to the polymorphic phase boundary observed in KNN ceramics which is similar to the MPB as observed in PZTs. Therefore, KNN has the potential to replace the PZTs. However, it is difficult to get high-density ceramics due to limitations in sintering at high temperatures that can volatilize the alkali elements components. The structural and electrical properties of the material depend on chemical compositions and processing techniques. The addition of oxides as sintering aids, different sintering techniques, and chemical modification improves the density and electrical properties.

DOI: 10.4018/979-8-3693-1306-0.ch011

1. INTRODUCTION

The piezoelectric effect is the phenomenon where an electric charge is generated in a material when mechanical stress is applied to it. It was first developed by Pierre Curie and Jacques Curie in 1880 (Cady, 1963; Curie & Curie, 1880). The materials exhibiting piezoelectric effect are called piezoelectric materials. Quartz, tourmaline, Rochelle salt, etc. are natural crystals while PZT, BT, KNN etc. are synthetic polycrystalline materials exhibiting piezoelectric effect. During First World War I, Paul Langevin developed a sonar device to detect submarines and became the first practical application of the piezoelectric effect (Duck, 2022). The development of this sonar device leads to the development of transducer, sensor, actuator industries, etc. The natural piezoelectric crystals are hygroscopic while the synthetics are non-hygroscopic and have better piezoelectric properties. Therefore, a mass study of synthetic piezoelectric materials especially PZTs is done and dominates the market of piezoelectric materials. Recently, the toxicity of lead which is the main component of PZT become aware and therefore must be replaced by lead-free piezoelectric materials.

2. APPLICATION OF PIEZOELECTRIC MATERIALS

The piezoelectric materials are widely used in many applications because of its direct piezoelectric effect. Some of the applications of piezoelectric materials are discussed below.

2.1. Automobile Industries

PZTs are used in automobile industries to produce power for low-power-consuming electrical gadgets, tyre pressure sensors, knock sensors, fuel injectors, etc (Kulkarni et al., 2018). The mechanical force that comes into play when the tyre is in contact with the ground is converted into electrical energy by using benders made of PZTs attached to the tyre. The power produced using benders of PZTs is small and used in running low-power sensors in the automobile. The power produced may be increased by stacking the layers of PZT benders for more useful applications. Proper monitoring of the tyre pressure of the vehicle can be done using a piezoelectric disc over the surface of the rim of the wheel. The internal pressure developed due to the reduction in the tyre pressure is taped by the piezoelectric disc and an electrical signal is produced due to the piezoelectric effect. When the tyre pressure reduces, an alarm will raise in the dashboard and a critical alarm will raise if the tyre gets punctured. The knock sensor at the head of the engine block of vehicles is used to prevent the detonation and spark knock. Piezoelectric materials are also used as piezoelectric fuel injectors. These injectors are faster than the standard injector solenoid.

2.2. Aerospace Industries

PZTs can be used as power harvesting devices for running the important subsystems of aircraft viz engine fire system, cockpit system and avionics systems (Elahi et al., 2020). The extraction of energy from the turbine for running these subsystems reduces its important output for driving the compressor through the shaft. Therefore, PZTs can be used to convert the energy lost in the surrounding environment into a useful form of energy for running the subsystems. Moreover, piezoelectric sensors can detect damages in various parts of aircraft viz engine blades, brackets and fittings, bulkheads and frames, etc. and can sense environmental changes.

2.3. Electrical Equipment

Piezoelectric actuators are used in the printer heads of dot-matrix printers and inkjet printers. Cell phones, earbuds, sound-producing toys, etc use piezoelectric speakers because of low power consumption for producing large sound volumes. Piezoelectric materials are also used as humidifiers where a piezoelectric transducer is used to transmit ultrasonic sound into water to atomize the fine water from the surface of the water.

2.4. Medical Applications

Ultrasonic equipment for medical purposes uses a piezoelectric transducer. They are used to monitor the pregnancy, break kidney stones, destroy malignant tissue, cardiac batteries etc.

2.5. Daily Life Applications

The commonly used ignitors for gas stoves, blowtorches etc. are made of piezoelectric materials. The sparks due to the voltage produced in the piezoelectric ceramics when sudden mechanical stress is applied have been used to ignite fuels. Piezoelectric sensors and actuators are used to integrate useful devices like gyroscopes, accelerometers etc into chip-sized packages (Sekhar et al., 2021).

3. LEAD ZIRCONIUM TITANATE (PZT)

Lead zirconium titanate (PZT) is a solid solution of ferroelectric $PbTiO_3$ and antiferroelectric $PbZrO_3$ which exhibit a ferroelectric nature at room temperature. It exhibits the piezoelectric effect and belongs to the perovskite family with the chemical formula $Pb[Zr_xTi_{1-x}]O_3$ (Bhide et al., 1968; Haertling & Buchanan, 1991; Hao et al., 2014; Jaffe, 1958). The similarity in the ionic size of the Pb^{2+} (1.32Å) ion and O^{2+} (1.4Å) leads to the face-centred cubic unit cell where the larger divalent Pb^{2+} ion (1.32Å) occupies the corner A- site of the unit cell while the smaller tetravalent Zr^{4+}(0.79Å) or Ti^{4+}(0.68Å) occupies the body centre which is B-site of the unit cell (Panda & Sahoo, 2015). The tetravalent ion is surrounded by six oxygen atoms at the face centre of the unit cell forming BO_6 octahedral as shown in fig 1 (Bochenek et al., 2011).

PZT shows various phases at different Zr: Ti ratios and different temperatures. A typical phase diagram for PZTs is shown in Figure 2. At room temperature, PZT exhibits two ferroelectric phases- Rhombohedral with space group *R3c* and *R3m* and tetragonal phase with space group *P4mm* depending on the Zr/Ti ratio (Noheda et al., 2000; Ragini et al., 2002). Ti-rich PZT exhibits a tetragonal phase while a rhombohedral phase for the Zr-rich PZT where the *R3c* rhombohedral phase is found at lower temperatures and the *R3m* rhombohedral phase at higher temperatures (Corker et al., 1998). The phase diagram as shown in Figure 2 indicates that the Curie temperature also varies with the Zr: Ti ratio indicating an increasing nature with the increase of Ti content. The microscopic compositional fluctuations lead to the formation of tetragonal or rhombohedral-rich small regions. Therefore, MPB between the two ferroelec-

Figure 1. Unit cell structure of PZT (M. Mir et al.)

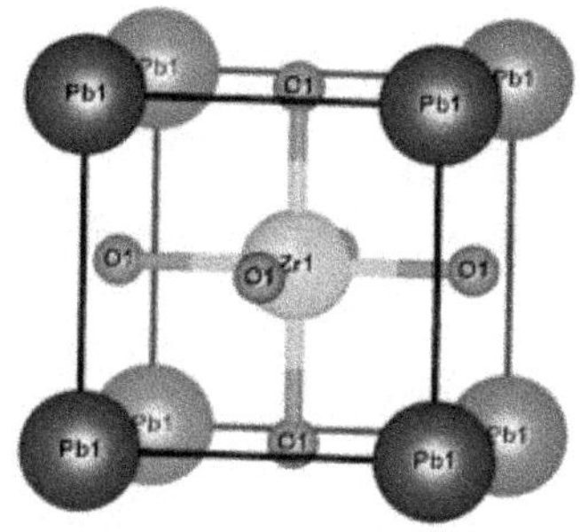

Figure 2. Phase diagram of PZT (A. Bouzid et al.)

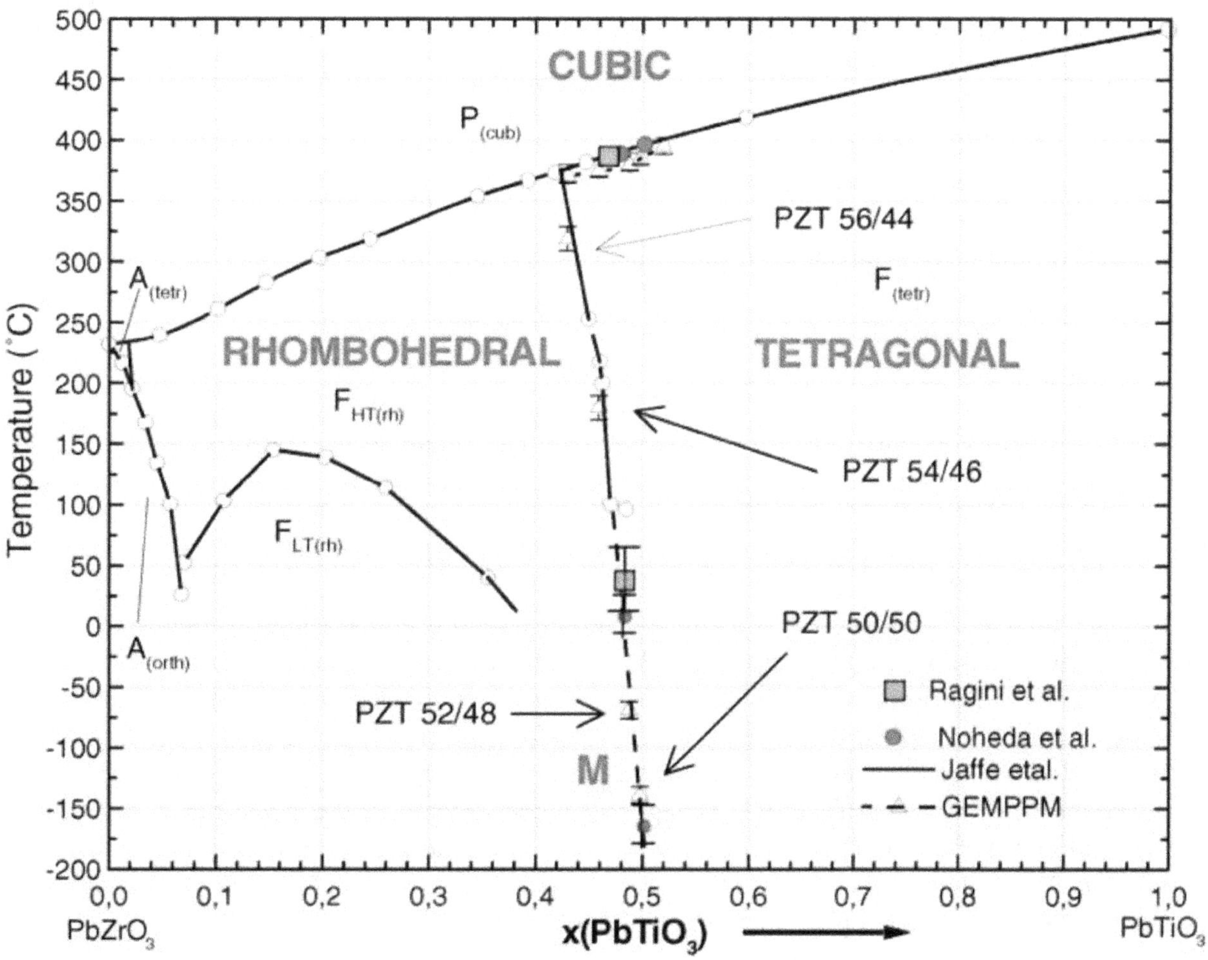

Figure 3. Phase transition of PZT at different temperatures

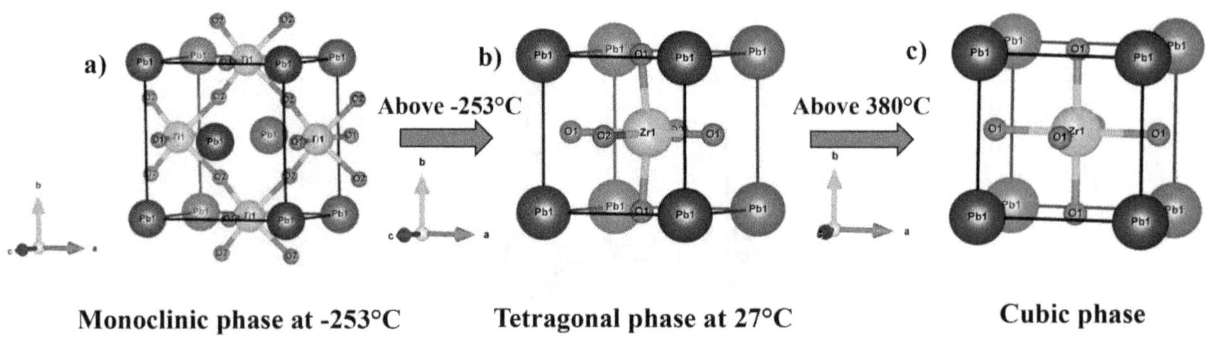

tric phases - rhombohedral and tetragonal is observed in the composition, Zr/Ti ~ 50/50 (Bouzid et al., 2005; Joseph et al., 2000; Vasiliu et al., 1983). When the Zr: Ti ratio changes from 50:50, distortion of rhombohedral and tetragonal appears and the same pattern of distortion due to the variation of the Zr: Ti ratio is observed with the variation of sintering temperatures (Vasiliu et al., 1983). Moreover, *B. Noheda et al.* reported that PZT with the chemical formula $PbZr_{0.52}Ti_{0.48}O_3$ exhibits a ferroelectric monoclinic phase with subgroup *Cm* at temperature -253°C. With the increase in temperature, the Pb atom displaces along one of the (110) axes and becomes ferroelectric tetragonal of space group *P4mm* at 27°C and further changed to the paraelectric cubic phase at a temperature above 380°C (Noheda et al., 2000). The phase transition of PZT at different temperatures as reported by *B. Noheda et al.* is shown in Figure 3.

Figure 3 (c) shows the symmetric structure where there is no distortion in the BO_6 octahedral. When the temperature reduces below 380°C the Zr-O bond lengths change along the c axis and the structure of BO_6 octahedral becomes distorted as shown in Figure 3 (b). This structural distortion may be responsible for the spontaneous electrical polarisation that makes the PZT ferroelectric in nature at room temperature. However, the structure of the PZT perovskite can be modified by substituting either the Pb-site, Zr-site or Ti-site ions of the pure PZT perovskite. The difference in the ionic size is responsible for the structural distortion and the degree of distortion is given by the tolerance factor which is given by $t = \frac{(R_A + R_O)}{\sqrt{2}(R_B + R_O)}$ where R_A, R_B and R_O are the radii of the respective ions (Panda & Sahoo, 2015).

PZT has attractive ferroelectric, electrical and piezoelectric properties. There are 14 different directions of polarisations; 6 from the tetragonal and 8 from the rhombohedral phase of the PZTs in MPB (Newnham, 1997). These directions of polarisations are oriented through the polling process and enhance the dielectric and piezoelectric properties of PZTs. Therefore, the morphotropic phase boundary (MPB) is responsible for its excellent electrical and piezoelectric properties (Panda & Sahoo, 2015). The electrical and piezoelectric properties of PZTs change when the structure of the perovskite changes due to site substitutions. Substitution by iso-valent dopants may lower the Curie temperature and increase the permittivity. Donor dopants facilitate the domain wall motion and make the PZT soft by creating cation vacancies whereas acceptor dopants make the PZT hard due to the creation of oxygen vacancies (Panda & Sahoo, 2015). The effects of some dopants on the dielectric, piezoelectric and ferroelectric properties of PZTs are shown in Table 1.

Table 1. Electrical, piezoelectric and ferroelectric properties of PZTs

Sl. No	Composition	Dielectric Constant, ε	Dielectric Loss	Piezoelectric Constant, d_{33} (pC/N)	Remnant Polarisation, P_r (μC/cm²)	*Ref.*
1	$Pb_{0.07}La_{0.03}Zr_{0.53}Ti_{0.47}O_3$	1100	0.019	325	18.03	(Sahoo & Panda, 2013)
2	$Pb_{0.08}La_{0.02}Zr_{0.53}Ti_{0.47}O_3$	700	0.018	250	12.50	(Sahoo & Panda, 2013)
3	0.5()-0.5$PbZrO_3$	1000	0.05	260	31.20	(Liu et al., 2020)
4	PLSZTN	2384	0.043	498	32.66	(Ramam & Lopez, 2007)
5	PSLZT	1800	0.015	640	19.50	(Kalem et al., 2011)
6	$Pb(Zr_{0.52}Ti_{0.48})_{0.976}Nb_{0.024}O_3$	1178	0.02	-	-	(Mirzaei et al., 2016)
7	PNZT	-	-	535	17.00	(van der Veer et al., 2021)
8	PZT + 0.4 wt%CuOLi_2CO_3	1881	0.013	474	37.00	(Zhao & Bian, 2023)
9	0.2PZN-PZT	1232	0.030	347	37.90	(Du et al., 2023)
10	PNB-PZT +2 mol% Li_2CO_3	7000	0.024	936	27.00	(Pu et al., 2022)

4. NEED FOR REPLACEMENT OF LEAD-BASED PIEZOELECTRIC MATERIALS

PZTs have excellent piezoelectric properties and therefore they have been extensively used in many applications. However, the lead which is the main constituent of PZT is toxic. After arsenic (As), lead is the most toxic element whose natural level should be below 50 $mgKg^{-1}$. Large lead concentration delays puberty and reduces hearing capacity in children. It also causes the kidney, central nervous system, fertility problems and hampers foetal growth in the early stage of pregnancy. In plants, there might be impaired chlorophyll production, cell divisions, seed germinations etc. Lead causes morphological, physiological and biochemical toxic effects on living organisms (Kumar & Kumar, 2020). Therefore, European nations and some other developed countries started banning the use of lead-containing materials. Although there are no equivalent substitutes for PZTs in various applications (Ringgaard & Wurlitzer, 2005), replacement of lead-based piezoelectric materials with non-toxic piezoelectric material is required. Therefore, a scientific study is the need of the present for the search of substitutes.

5. LEAD-FREE PIEZOELECTRIC MATERIALS

Any piezoelectric materials where lead is not a constituent element are termed lead-free piezoelectric materials. They may be either perovskite having the chemical formula ABO_3 or non-perovskite. Barium titanate (BT), bismuth sodium titanate (BNT), bismuth sodium titanate (BKT), potassium sodium niobate (KNN) etc. are perovskite lead-free piezoelectric materials and bismuth layer structured ferroelectrics (BLSFs), tungsten bronze structured ferroelectrics etc are non-perovskite piezoelectric materials. The large anisotropic piezoelectric properties of the perovskite group as compared to the non-perovskite

Figure 4. Structure of barium titanate (N.B. Mahmood et al.)

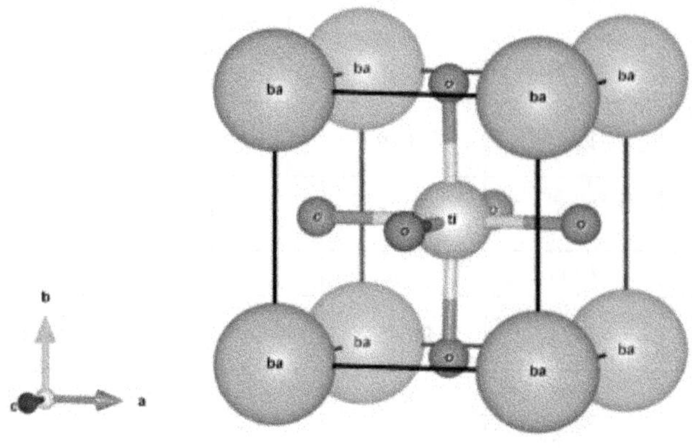

group make perovskite piezoelectric materials a hopeful candidate for replacement of the PZTs (Panda, 2009). Non-perovskite materials are found suitable for ceramics filter and resonator applications.

5.1. Barium Titanate (BT)

Barium titanate with the chemical formula $BaTiO_3$ is the first known ferroelectric perovskite material with spontaneous polarisation ~ 30 μC/cm². It is also the first polycrystalline piezoelectric material with a piezoelectric strain constant, d_{33} ~ 190pC/N (Takenaka, 2013). The unit cell structure of barium titanate is shown in Figure 4.

Figure 4 shows that the A-site Ba atoms having 12 coordination numbers are located at the corner and the B-site Ti atom occupies the body centre of the unit cell. The B-site Ti atom has 6 coordination numbers and is surrounded by six oxygen atoms located at the face centre of the unit cell (Rached et al., 2021). Barium Titanate undergoes a phase transformation at different temperatures. At temperatures below -90°C, it is found to have a rhombohedral phase and changes to an orthorhombic phase above-90°C. The orthorhombic phase changes to tetragonal when the temperature is above 0°C and further changes to the paraelectric cubic phase above a temperature of 120°C (Al-Shakarchi & Mahmood, 2011; Nayak et al., 2014; Pullar et al., 2009). The tetragonal phase with space group *P4mm* predominates when prepared using the sodium hydroxide-ethanol chemical method while the cubic phase with space group *Pm-3m* predominates when prepared using the oxalate and sodium hydroxide chemical methods. The cubic phase of BT (calcined at 600°C for 6h) observed in the oxalate chemical method changes to the tetragonal phase when calcined at 800°C for 6h. The structural phase transition of barium titanate at different temperatures is shown below:

It is seen that the structure of BT is symmetric above 120°C and there is a relative displacement of centres of positive and negative charges within the sub-lattice at temperatures below 120°C due to the distorted structures. Therefore, barium titanate is paraelectric in nature at temperatures above 120°C and becomes ferroelectric in nature below 120°C. The ferroelectric properties of BTs are related to the domain structure and domain motion formed during the phase transition. Uniform and stable ferroelectric properties are obtained for uniform and small-grained microstructure (Kumar et al., 2009). The domain structure of barium titanate is shown in Fig 6.

Figure 5. Phase transition of barium titanate at different temperatures (Khastgir D. et al.)

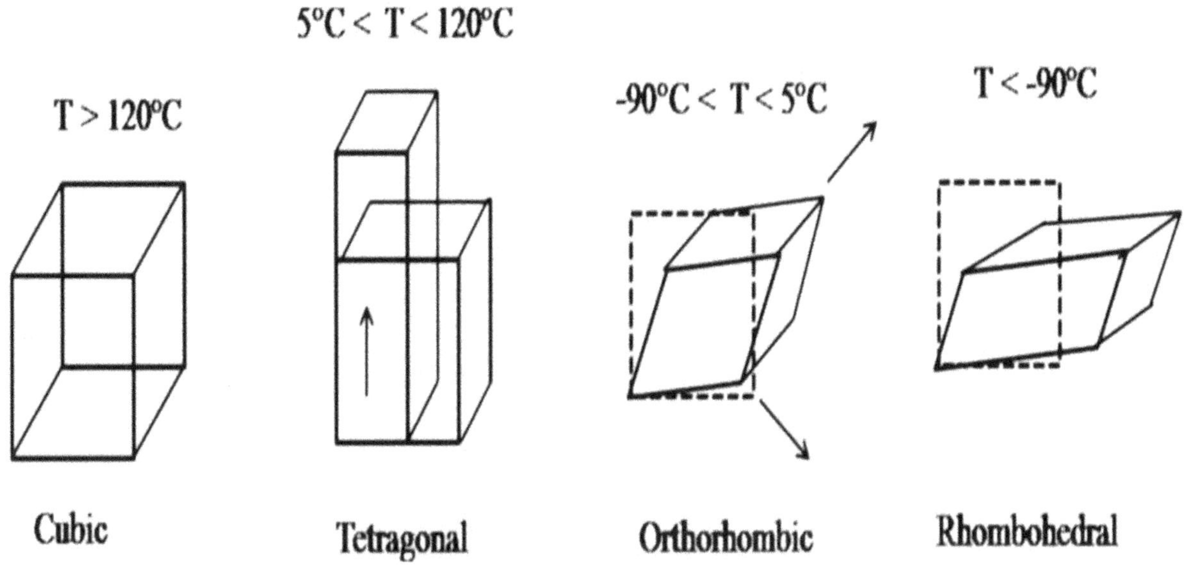

It is seen that BT has two domain walls viz 90°-wall separating domains which are polarised perpendicularly and 180°-wall separating domains which are polarised antiparallel to each other. The microstructure of the BT changes when the Ba atom or Ti atom or both are substituted and correspondingly change the domain structures. Therefore, the ferroelectric properties of barium titanate depend on the

Figure 6. Domain structure of single crystal barium titanate (N.T. Tsou et al.)

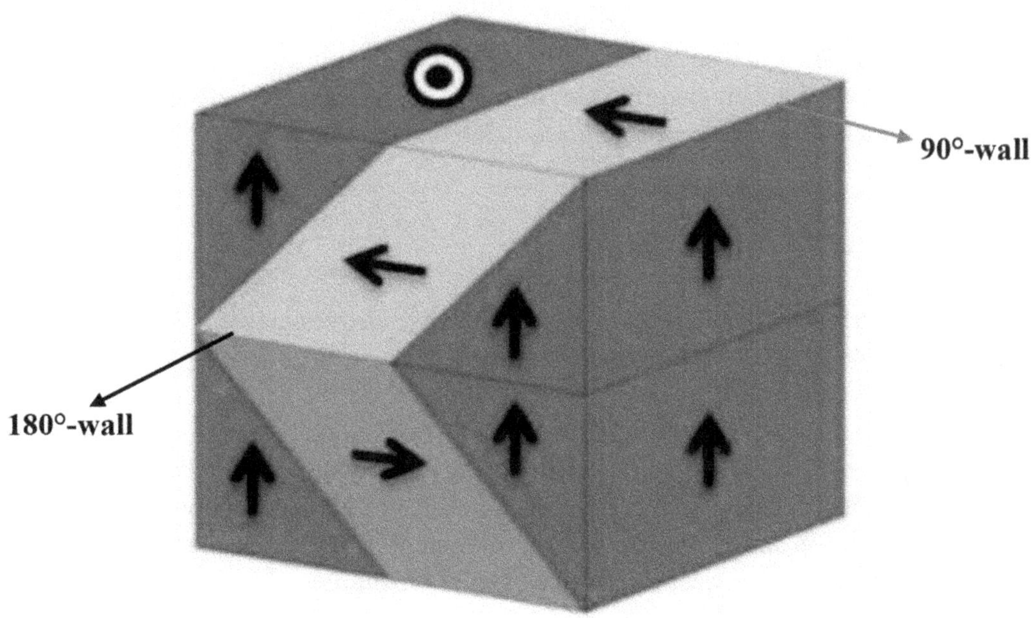

dopants and the processing techniques. In studying the ferroelectric properties of zirconium-substituted barium titanate *P. Kumar et al* conclude that the ferroelectric properties of barium titanate correlate with the material's grain size and porosity (Kumar et al., 2009). The ratio of remanent polarisation and spontaneous polarisation is found to be decreasing till 2 mol% Zirconium substitution in the Ti-site followed by the increase with the 4 mol% and 6 mol% substitutions. The lattice incompatibility due to the different ionic radii of the dopants as compared to that of the Ba or Ti fluctuates the ionic states of the dopants. Therefore, there is the creation of cationic or anionic vacancies depending on the nature of the dopants (iso-valent, donor, acceptor or amphoteric). These vacancies affect the B-site octahedral, which results in changes in the ferroelectric, piezoelectric and dielectric properties of BT (Behera et al., 2022; Rached et al., 2021). Moreover, the electrical properties of BTs are correlated with the degree of space charge polarization in them. The degree of polarization is also affected by external conditions like temperature and frequency of the applied alternating field which in turn affect the electrical properties of the BTs (Rached et al., 2021).

BTs have attractive electrical and piezoelectric properties viz high dielectric constant, low dielectric loss, and high piezoelectric strain constant which enable them to be used for capacitors, multilayer capacitors, and energy storage devices. However, the temperature range for the actual piezoelectric application is narrow due to low Curie temperature (Zhao et al., 2018). Therefore, many efforts have been made to increase the Curie temperature of BTs. Yang et al. observed a decrease in the Curie temperature to about 60°C when Ca substitutes the B-site Ti atom by 3 mol%. However, the substitution of the A-site Ba atom by 3 mol% of Ca increases the Curie temperature to about 130°C (Yang et al., 2018). *Su et al.* report that the structural, microstructural and electrical properties of BTs are sensitive to calcination and sintering temperatures. There is an increase in the particle size of strontium-substituted barium titanate with the increase in calcination time while incomplete calcination results in abnormal grain growth and uniform grain size distribution is observed for those samples calcined for a lower time. Moreover, the dielectric constant increases with the increase of sintering temperature and sintered density increase whereas the dielectric loss decreases with the increase in sintering temperature (Su et al., 2002).

Therefore, the processing method is very important for the attractive structural, microstructural, electrical, ferroelectric and piezoelectric properties of BT. S. *Yun et al.* confirm the importance of the processing method. The microstructural evolution of BT ceramics is observed when prepared by two different methods viz reaction-sintering and mixed oxide method. The electrical and ferroelectric properties are higher for BT prepared by the mixed oxide method as compared to the reaction sintering method (Yun et al., 2008).

5.2. Bismuth Sodium Titanate (BNT)

Bismuth Sodium Titanate, commonly referred to as BNT, is a lead-free perovskite material characterized by the chemical formula $Bi_{0.5}Na_{0.5}TiO_3$. It exhibits significant ferroelectric and piezoelectric characteristics at room temperature, with a spontaneous polarization of about 40 μC/cm2 *(Smolenskii et al., 1960)* (Jones & Thomas, 2002). The phase configurations of BNT are complex and undergo temperature-dependent phase transitions (Dorcet et al., 2008; Jones & Thomas, 2002; Trolliard & Dorcet, 2008; Zhou et al., 2021). At temperatures below 200°C, BNT displays a ferroelectric rhombohedral phase characterized by the *R3c* space group and an antiphase $a^-a^-a^-$ tilt system. However, as the temperature surpasses 200°C, the material experiences a decrease in remanent polarization and becomes an intermediate antiferroelectric orthorhombic phase, featuring the *Pnma* space group and $a^-b^+c^-$ - octahedral tilting. When

Figure 7. Phase transition of bismuth sodium titanate (BNT) at different temperatures

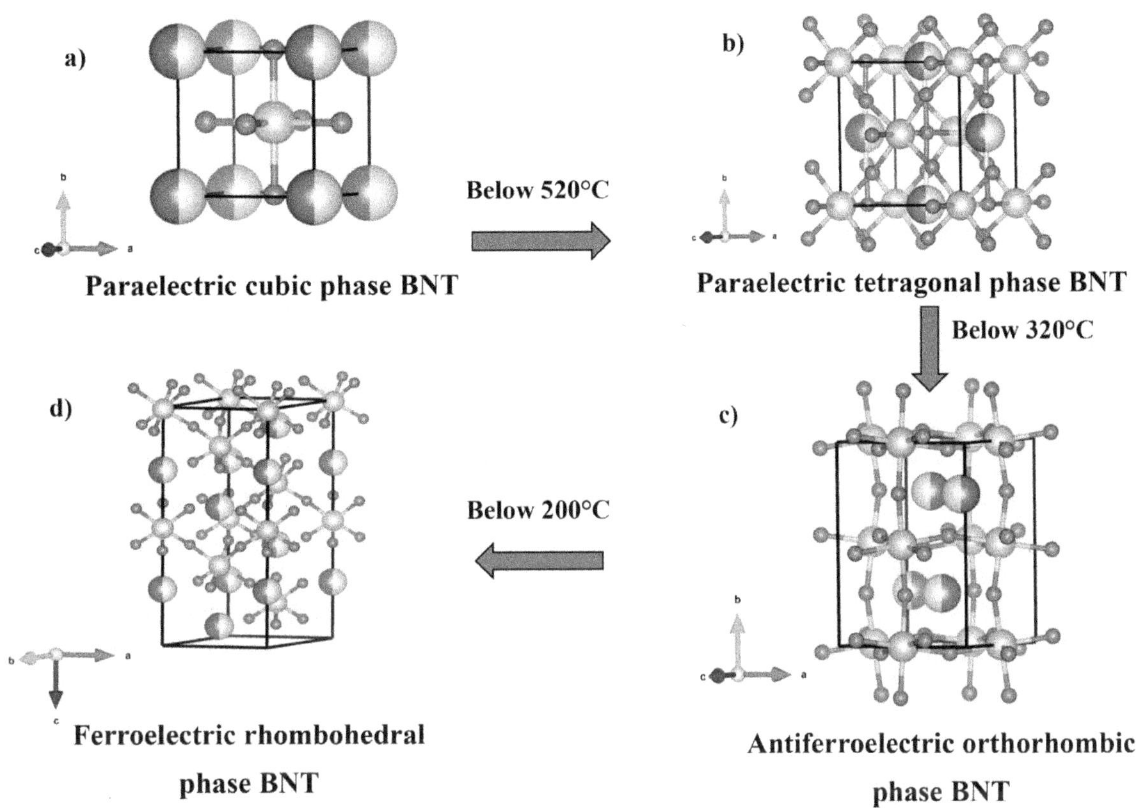

the temperature rises over 320°C, the antiferroelectric phase becomes a paraelectric tetragonal phase, characterized by the P4bm space group and an in-phase $a°a°c+$ tilt system. Subsequently, the material transits into cubic phase with $Pm\overline{3}m$ space group and $a°a°a°$ zero tilt system when the temperature is above 520°C (Zhou et al., 2021). The point at which the decrease in spontaneous polarization initiates, which occurs at 200°C, is referred to as the depolarization temperature whereas the point where the ferroelectric phase changes to the paraelectric phase is referred to as curie temperature. The unsymmetrical unit cell structures due to cation displacement and tilting of oxygen octahedral at lower temperatures as shown in Fig 6. are responsible for the phase transitions of BNT at different temperatures. The phase transitions of BNT at different temperatures are shown in the Fig.6.

However, *Aksel et al.* found BNT to exist in the ferroelectric monoclinic phase, which is distinguished by the Cc space group, at room temperature (Aksel et al., 2013; Ogino et al., 2014). *B.N. Rao et al.* report the simultaneous existence of rhombohedral and monoclinic ferroelectric phases at room temperature (Rao, Fitch, & Ranjan, 2013). However, it is important to remember that the rhombohedral phase is essentially an averaged monoclinic structure, which results from strain and structural inhomogeneities that can be corrected and reversed using heating and poling procedures. (Beanland & Thomas, 2014; Rao,

Datta, Chandrashekaran et al, 2013; Rao, Fitch, & Ranjan, 2013). Therefore, the structural properties of BNT at room temperature depend on mechanical, electrical and thermal treatments.

The electrical properties of BNTs are attractive because of their significant spontaneous polarization at room temperature. Pure BNTs are known to possess a piezoelectric coefficient (d_{33}) of approximately 57 pC/N, a dielectric constant of around 240, and a minimal dielectric loss of approximately 0.011 (Zhou & Liu, 2008). However, the substantial coercive field, which stands at approximately 73 kV/cm, presents a challenge for effective poling due to the elevated conductivity. Therefore, complex perovskite solid solutions that include BNT are investigated. *C. Zhou et al.* studied the solid solutions (1-x)($Bi_{1/2}Na_{1/2}$)TiO_3- xBi($Mg_{2/3}Nb_{1/3}$)O_3 and showed improvement in the piezoelectric and dielectric properties of BNT. This may be due to the compensation of the Bi^{3+} ion of BNT that volatilized during high-temperature sintering by the Bi^{3+} ion of Bi($Mg_{2/3}Nb_{1/3}$)O_3 and the creation of oxygen vacancies due to the substitution of Ti^{4+} ion by $(Mg_{2/3}Nb_{1/3})^{3+}$ ion (Zhou & Liu, 2008). The electrical properties of the BNT depend on various factors, including temperature, the frequency of the applied electric field, the type of dopants involved, and the presence of space charge polarization. Furthermore, the electrical properties are influenced by the methodology employed during processing, as well as other factors such as sintering and calcination temperatures and sintering techniques. The respective site disorders created by the dopants due to different ionic sizes affect the polarisation in the material and the nature of polarisation is different at different frequency ranges (Chen et al., 2020; Pattipaka et al., 2018; Zhou & Liu, 2008). Therefore, modification of electrical properties can be achieved through A or B-site substitutions, as well as by creating solid solutions with other ferroelectric materials. BNT-BKT-BT ceramics prepared using the solid-state reaction method exhibit superior characteristics, including higher piezoelectric constant (d_{33}), dielectric constant, mechanical quality factor, and electromechanical coupling factor, compared

Table 2. Electrical, piezoelectric and ferroelectric properties of BNTs

Sl. No	Composition	Dielectric Constant, ε	Dielectric Loss	Piezoelectric Constant, d_{33} (pC/N)	Remnant Polarisation, P_r (μC/cm²)	*Ref.*
1	BNT	240	0.011	57	-	(Zhou & Liu, 2008)
2	BNT - KNNG	1074	0.059	108	12.50	(Pattipaka et al., 2018)
3	()0.06BT	3500	0.06	-	22.50	(Chen et al., 2020)
4	BNT - 0.3 wt% Bi_2O_3	406	-	93.4	-	(Nagata et al., 2012)
5	BNT - HP	422	-	97.5	-	(Nagata et al., 2012)
6	BNKT - ST	1500	0.05	-	30.00	(Malik et al., 2014)
7	BNKLT	1600	0.06	250	22.50	(Han et al., 2020)
8	BNKLT - 0.015Ta_2O_5	1500	0.06	25	10.00	(Han et al., 2020)
9	87.5BNT + 10BKT +2.5BT	1506	0.05	109	34.40	(Fernandez-Benavides et al., 2018)
10	BNT - KNN	1030	0.05	-	20.00	(Chandrasekhar & Kumar, 2016)

to ceramics prepared via the hydrothermal method (Liu et al., 2011). Moreover, microwave-sintered 0.92(BNT-BKT-BT)-0.08NN ceramics exhibit electrical properties that are superior to those of conventionally sintered ceramics and highlight the utility of microwave sintering as a more practical method for producing ceramics appropriate for a variety of uses (Chandrasekhar et al., 2023; Rahman et al., 2015).

The effects of dopants, and complex solid solutions of BNTs on the dielectric, piezoelectric and ferroelectric properties of BNT at room temperature are summarised in Table 2.

5.3. Potassium Sodium Niobate (KNN)

Sodium potassium niobate, often abbreviated as KNN, is a piezoelectric material that is made when ferroelectric $KNbO_3$ and antiferroelectric $NaNbO_3$ are combined. Its chemical composition can be expressed as $[K_xNa_{1-x}]NbO3$, where the values of x vary to produce various solid solution compositions. As a result of the strict laws that the European countries have imposed on the use of dangerous materials like lead in electrical appliances, alkali niobates which have the general formula ANbO3, where A is an alkali metal have become a competitive alternative piezoelectric ceramic (Ringgaard & Wurlitzer, 2005). Before 2000, there was no intensive study on the alkali niobates. However, the European LEAF project from 2001 to 2004 focused on KNN. KNN undergoes distinct phase transitions at various temperature ranges, as reported by *B. Orayech et al.* According to their results, KNN exhibits a rhombohedral phase with a space group of *R3c* below -138°C and an orthorhombic phase with a space group of *Amm2* above -138°C. (Zhang et al., 2009; Orayech et al., 2015). At about 200°C, the orthorhombic phase transforms to the tetragonal phase, which is followed by a gradual transition into the cubic phase with space group $Pm\ \overline{3}m$ at about 420°C (Orayech et al., 2015). In the composition of $K_{0.5}Na_{0.5}O_3$, a coexistence of monoclinic and orthorhombic phases has been detected at room temperature (Kumar et al., 2013; Leontsev & Eitel, 2010), which is similar to the situation in PZT. Additionally, at temperatures of about -90°C, *N. Zhang et al.* have also reported the simultaneous presence of orthorhombic and monoclinic phases in $K_{0.30}Na_{0.70}NbO_3$ (Zhang et al., 2009).

In Fig 8, the phase diagram of the KNN solid solution is depicted, highlighting distinct phase transitions occurring at different temperatures and the morphotropic phase boundary (MPB) between the ferroelectric phases. Beyond MPB, the transition temperature for the phase shift from the ferroelectric orthorhombic phase to the ferroelectric tetragonal phase is lowered to room temperature, a phenomenon referred to as the polymorphic phase transition (PPT). Consequently, KNN ceramics also exhibit another phase boundary known as the polymorphic phase boundary (PPB) (Dwivedi et al., 2018; Fu et al., 2017; Skidmore & Milne, 2007; Yan et al., 2019).

The KNN ceramic has good ferroelectric properties, such as a high Curie temperature of more than 400°C, a remnant polarization of about 33 μC/cm^2, and a significant piezoelectric constant, d_{33}, of about 80 pC/N (Panda, 2009). KNN can be doped with different elements including Li^+, Zn^{2+}, and Ta^{5+} to enhance its electric, piezoelectric, and ferroelectric properties. This enhancement occurs by reducing the energy barrier between different ferroelectric phases at the morphotropic phase boundary (MPB) or polymorphic phase boundary (PPB). Consequently, both intrinsic lattice distortion and extrinsic domain wall motion are enhanced in regions with MPB or PPT upon the introduction of suitable dopants. The coexistence of multiple phases in KNN ceramics is responsible for the enriched piezoelectric characteristics that result (Gao et al., 2021).

Figure 8. Phase diagram of KNN ceramics (Jaffe et.al)

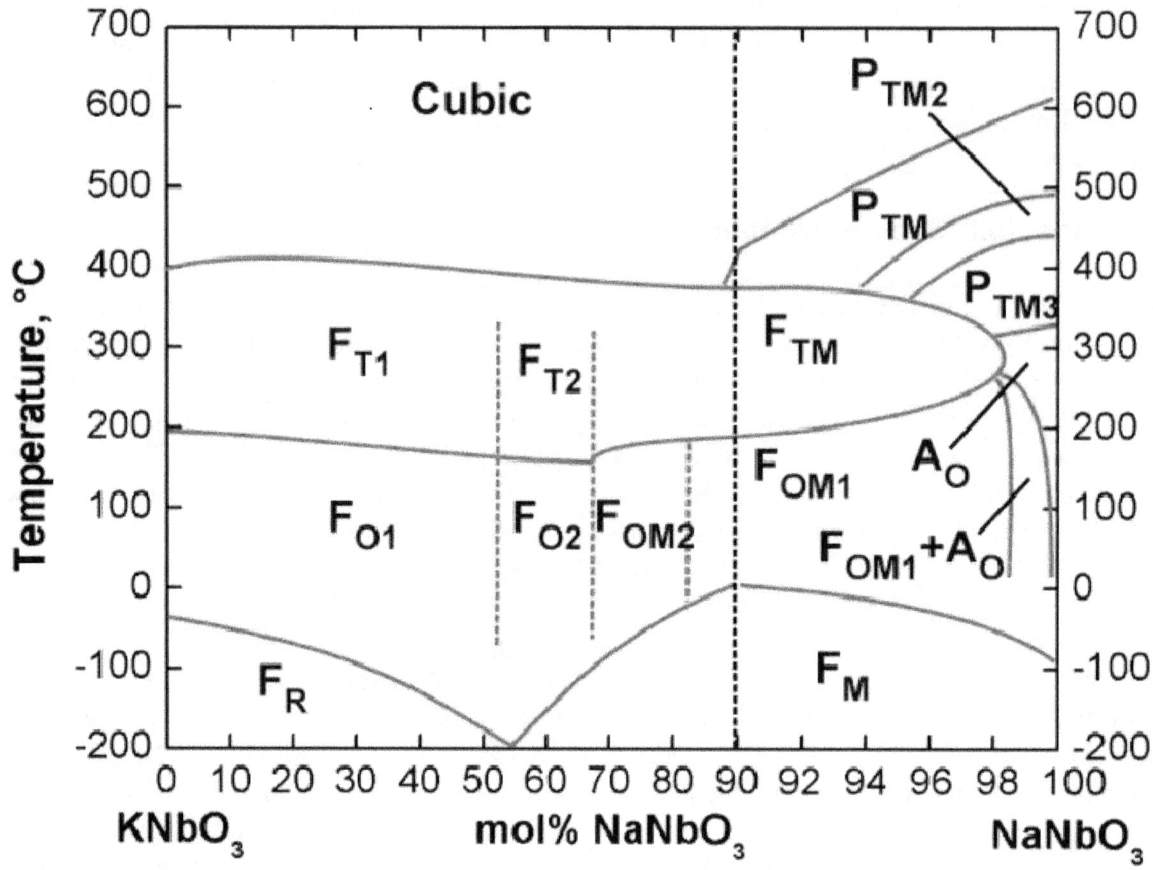

An analysis of the effects of various dopants on the electrical, piezoelectric, and ferroelectric properties of KNN ceramics is shown in Table 3.

KNN provides a wide temperature range that is appropriate for piezoelectric applications due to its high curie temperature. However, its fabrication poses challenges due to the volatility of alkali elements at elevated sintering temperatures and the complexities involved in preserving the material's stoichiometry. Chemical modifications, adding oxides as sintering aids and different sintering techniques are possible ways to solve the problems of KNN ceramics.

6. SYNTHESIS PROCESS AND CHARACTERISATIONS

The synthesis process of advanced ceramics is completely different from that of traditional ceramics. The first step to synthesis advanced ceramics is to synthesis fine powders followed by sintering to obtain a highly dense ceramic. There are many ways to prepare fine powders viz mechanical method, chemical method etc. and different sintering method viz conventional sintering, microwave sintering, hot pressing etc. Among the various synthesis processes solid state chemical reaction method and sol-gel method

Table 3. Electrical, piezoelectric, and ferroelectric properties of KNN ceramics

Sl. No	Composition	Dielectric Constant, ε	Dielectric Loss	Piezoelectric Constant, d_{33} (pC/N)	Remnant Polarisation, P_r (μC/cm²)	*Ref.*
1	KNN	316	0.08	95	26	(Alkoy & Berksoy, 2010)
2	KNN + 1 mol% CuO	335	0.01	82	20	(Alkoy & Berksoy, 2010)
3	$K_{0.465}Na_{0.465}Li_{0.07}NbO_3$	560	0.04	155	28	(Alkoy & Berksoy, 2010)
4	$K_{0.465}Na_{0.465}Li_{0.07}NbO_3$ +1mol% CuO	554	0.04	84	08	(Alkoy & Berksoy, 2010)
5	KNN + 0.5 mol% ZnO	359	0.13	95	10	(Ramajo et al., 2014)
6	KNN + 1 mol % ZnO	408	0.10	90	05	(Ramajo et al., 2014)
7	$KNLNST_{0.06}$	1577	0.03	270	60	(Shao et al., 2012)
8	0.96KNN + 0.04 mol% MnO_2	3000	-	-	28	(Tian & Du, 2014)
9	KNN - KCN	430	0.04	81	-	(Murutoğlu et al., n.d.)
10	KNN- KCN - Sb	550	0.02	104	-	(Murutoğlu et al., n.d.)

are commonly used due to simplicity in process, low cost of installation and the solid-state chemical reaction is more efficient for the large-scale productions.

6.1. Powder Synthesis

6.1.1. Solid-State Chemical Reaction Method

It is a common method for synthesising polycrystalline ceramics. This method is used to produce powders from solid reactants. During heating, chemical reactions between the solid reactants occur and there is transport of atoms, ions or molecules through the phase boundaries by several mechanisms. The rate of reaction depends on several factors viz nature of the reactants and products, shape and size of the particles, uniformity of mixing, reaction atmosphere, temperature and time. The reaction rate decreases with the increase in particle size due to the increase in the diffusion distance while the rate changes with temperature according to the Arrhenius equation: $K = Ae^{\frac{-Q}{RT}}$, where K is the reaction rate, Q is the activation energy, R is the universal gas constant, A is the frequency factor and T is the temperature. There might be an undesirable phase due to an incomplete reaction when the solid reactants are poorly mixed. Therefore, grinding and uniformity in mixing are important to produce powders with better characteristics. However, it should be taken care not to contaminate the powders due to impurities during milling (Reitz, 1997).

6.1.2. Sol-Gel Process

The sol-gel process is a wet chemical reaction technique used to generate metal oxides from metal alkoxides or metal salt precursors. In this procedure, a sol, a liquid suspension of colloidal particles, is made, and after that, a gel is formed. In this method, the precursors are mixed with a solvent and metal salts are dissociated into metal cations and anions. The solvated metal cations are deprotonated by the hydrolysis process to form a hydroxo ligand when one proton is deprotonated or an oxo ligand when two protons are deprotonated. The hydrolysed cations undergo a polymerisation reaction to form polynuclear complexes consisting of two or more metals. This process of polymerisation is called condensation. A solid phase will precipitate beyond the solubility limit and the precipitate observed is dried and calcined to get powders (Pierre, 2020; Reitz, 1997).

6.2. Sintering

The powders, whether obtained through the solid-state chemical reaction method or the sol-gel process, are compacted into pellets of the desired dimensions and subsequently sintered to produce highly dense ceramics. The ferroelectric, electric and piezoelectric effects can be affected by the sintering method. Conventional sintering, multistep sintering, microwave sintering, hot press sintering, and spark plasma sintering are a few popular sintering techniques. Although there is evidence that hot press sintering and spark plasma sintering greatly enhance ceramic properties, these processes are complex and require specialized equipment. Alternatively, ceramic properties can be improved by using the easier-to-use and more accessible multistep sintering process (Zhang, Zheng, & Wu, 2020). Notably, two-step and three-step sintering methods have been reported for the preparation of ceramics with highly attractive characteristics (Wang et al., 2009; Zhang, Zheng, & Wu, 2020; Zhang, Zhai, & Xue, 2020).

6.3. Characterisations

Certain characterization techniques are used in order to thoroughly evaluate the electrical, ferroelectric, piezoelectric, structural, and microstructural properties of the prepared ceramics. Techniques like X-ray diffraction (XRD) and Fourier transform infrared spectroscopy (FTIR) are used for structural analysis. Transmission electron microscopy (TEM) and scanning electron microscopy (SEM) are used for microstructural assessment. Dielectric measurements, including the investigation of room temperature frequency-dependent parameters like dielectric constant, dielectric loss, and AC conductivity, are conducted using a Keysight-Impedance Analyzer. In order to gain insight into the performance of the ceramics, additional properties related to ferroelectricity and piezoelectricity, such as polarization-electric (PE) loops and piezoelectric coefficients, are measured.

6.4. Case Study of Preparation and Characterisation of KNN Ceramics

Alkali niobate ceramics with the chemical formula $K_{0.47}Na_{0.47}Li_{0.06}NbO_3$ (KNLN) are prepared using a solid-state chemical reaction method and two-step sintering technique. The effects of five different oxides are studied. The preparation process, XRD patterns and room temperature variation of various electrical properties of the ceramics with frequency in the range 20Hz to 5MHz are shown in figures 8, 9, and 10 respectively and observations are tabulated in Table 4 (Singh & Maisnam, 2023). The observed structural

Figure 9. Synthesis process of KNN ceramics

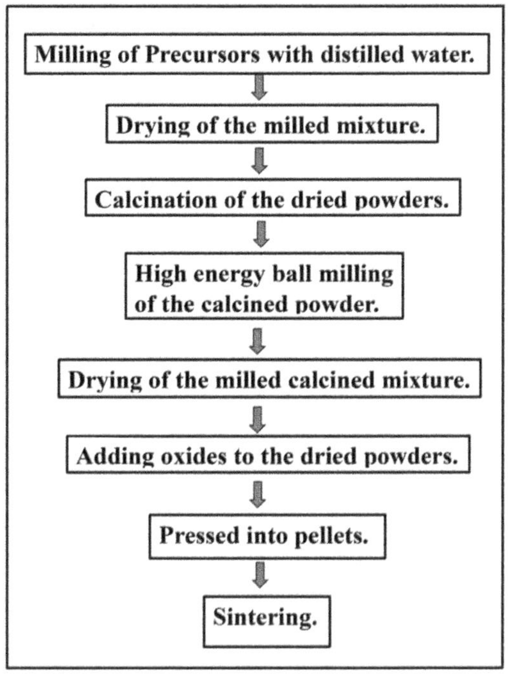

Figure 10. Synthesis process of KNN ceramics

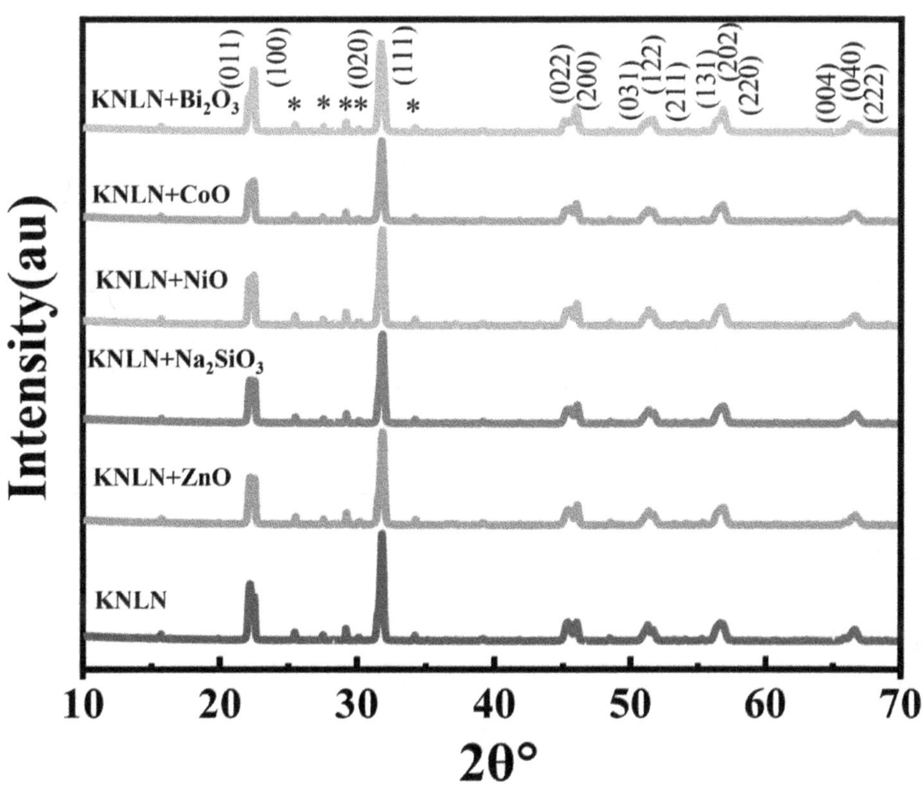

Figure 11. a) Nyquist plot of KNN ceramics, b) the variation of ac conductivity with frequency at room temperature, c) the variation of dielectric constant with frequency at room temperature, and d) the variation of dielectric loss with frequency at room temperature of KNN ceramics

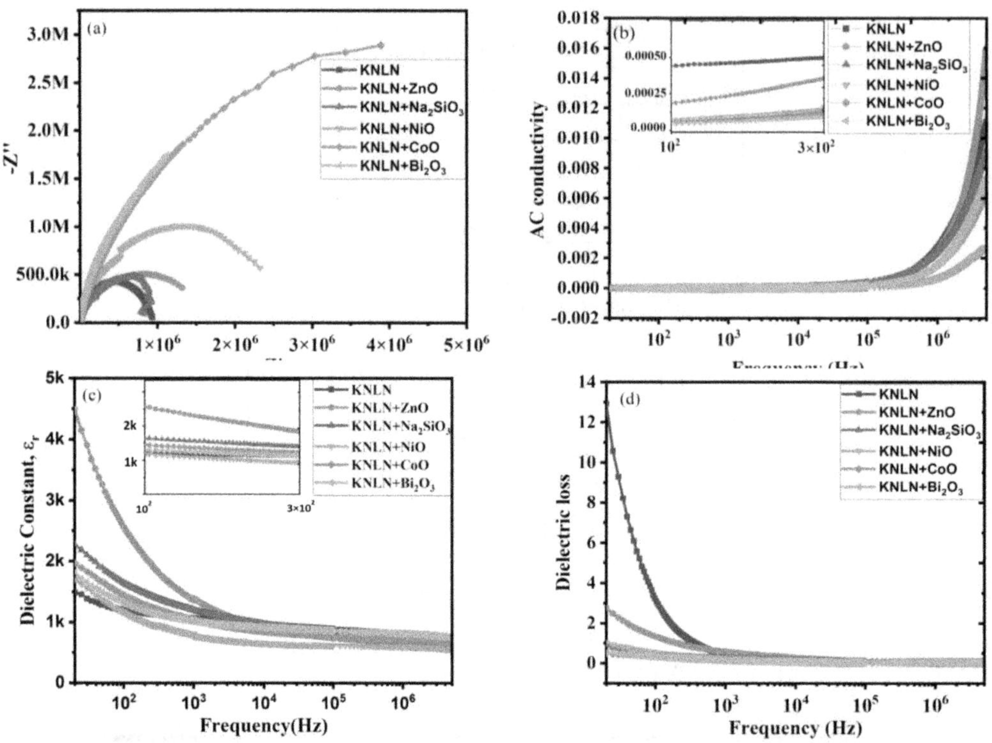

Table 4. Structural and electrical properties of KNLN ceramics

Properties	KNLN	KNLN + ZnO	KNLN+Na_2SiO_3	KNLN + NiO	KNLN + CoO	KNLN+Bi_2O_3
Phase	Orthorhombic+ Tetragonal	Orthorhombic+ Tetragonal	Orthorhombic+ Tetragonal	Orthorhombic+ Tetragonal	Orthorhombic+ Tetragonal	Orthorhombic +Tetragonal
Symmetry	*Amm2+P4mm*	*Amm2+P4mm*	*Amm2+ P4mm*	*Amm2+P4mm*	*Amm2+P4mm*	*Amm2+P4mm*
a (Å)	3.94954	3.94317	3.94223	3.94830	3.95017	3.95101
b (Å)	5.63107	5.61965	5.62708	5.60740	5.62033	5.64234
c (Å)	5.68178	5.67454	5.67426	5.67748	5.67716	5.63924
X-ray density (g/cc)	4.48	4.50	4.49	4.50	4.49	4.50
Crystallite size(nm)	33.47	43.72	44.00	47.30	50.58	43.58
Lattice strain	0.110	0.085	0.090	0.078	0.069	0.067
Dislocation density($\times 10^{15}nm^{-2}$)	1.7853	1.0463	1.0330	0.8939	0.7817	0.5265
Dielectric constant at 1kHz	1070	1390	1190	780	1030	1010
Dielectric loss at 1kHz	0.5	0.5	0.2	0.2	0.2	0.1

and electrical properties of the ceramics confirm that adding sintering aids increases the density of the ceramics, thereby improving their dielectric properties.

7. CONCLUSION

The lead-free piezoelectric ceramics have advantages in some applications but limitations in others. For instance, BT has a significant dielectric constant and advantageous ferroelectric properties. However, it is not appropriate for high-temperature applications due to its low Curie temperature of 120°C. Bismuth Sodium Titanate (BNT) is a material with a dielectric constant of notable value and a higher Curie temperature of 320°C. The substantial conductivity that results from its strong remnant polarization makes the polling difficult. As a result, BNT can only be used in certain piezoelectric applications. On the other hand, Potassium Sodium Niobate (KNN) has excellent ferroelectric, piezoelectric, and electric properties in addition to having a high Curie temperature. The PPT and MPB found in KNN-based ceramics are similar to those found in PZT-based ceramics. Therefore, by removing their issues through chemical modifications, adopting a variety of sintering techniques, and incorporating of oxides as sintering aids, KNN ceramics present a promising candidate with the potential to replace PZT.

ACKNOWLEDGMENT

The author (Haobam Samananda Singh) acknowledged DST (Department of Science and Technology), New Mehrauli Road, Technology Bhavan, New Delhi-110 016, Govt. of India for giving financial support in the form of INSPIRE Fellowship (IF200544).

REFERENCES

Ahtee, M., & Hewat, A. W. (1978). Structural phase transitions in sodium–potassium niobate solid solutions by neutron powder diffraction. *Acta Crystallographica. Section A, Crystal Physics, Diffraction, Theoretical and General Crystallography*, *34*(2), 309–317. doi:10.1107/S056773947800056X

Aksel, E., Forrester, J. S., Nino, J. C., Page, K., Shoemaker, D. P., & Jones, J. L. (2013). Local atomic structure deviation from average structure of Na 0.5 Bi 0.5 TiO 3: Combined x-ray and neutron total scattering study. *Physical Review B: Condensed Matter and Materials Physics*, *87*(10), 104113. doi:10.1103/PhysRevB.87.104113

Al-Shakarchi, E. K., & Mahmood, N. B. (2011). Three techniques used to produce BaTiO 3 fine powder. *Journal of Modern Physics*.

Alkoy, E. M., & Berksoy, A. (2010, August). Effect of Cu and Li addition on the electrical properties of KNN ceramics. In *2010 IEEE International Symposium on the Applications of Ferroelectrics (ISAF)* (pp. 1-4). IEEE.

Beanland, R., & Thomas, P. A. (2014). Symmetry and defects in rhombohedral single-crystalline Na 0.5 Bi 0.5 TiO 3. *Physical Review B: Condensed Matter and Materials Physics*, *89*(17), 174102. doi:10.1103/PhysRevB.89.174102

Behera, C., Patel, P., Pradhan, N., & Choudhary, R. N. P. (2022). Studies of structural, dielectric and electrical characteristics of nickel-modified barium titanate for device applications. *Journal of Materials Science Materials in Electronics*, *33*(3), 1657–1669. doi:10.1007/s10854-022-07709-9

Bhide, V. G., Hegde, M. S., & Deshmukh, K. G. (1968). Ferroelectric properties of lead titanate. *Journal of the American Ceramic Society*, *51*(10), 565–568. doi:10.1111/j.1151-2916.1968.tb13323.x

Bochenek, D., Skulski, R., Wawrzała, P., & Brzezińska, D. (2011). Dielectric and ferroelectric properties and electric conductivity of sol–gel derived PBZT ceramics. *Journal of Alloys and Compounds*, *509*(17), 5356–5363. doi:10.1016/j.jallcom.2011.02.046

Bouzid, A., Bourim, E. M., Gabbay, M., & Fantozzi, G. (2005). PZT phase diagram determination by measurement of elastic moduli. *Journal of the European Ceramic Society*, *25*(13), 3213–3221. doi:10.1016/j.jeurceramsoc.2004.07.018

Cady, W. G. (1963). Piezoelectricity and ultrasonics. *Sound: Its Uses and Control, 2*(1), 46-52.

Chandrasekhar, M., & Kumar, P. (2016). Processing and characterizations of BNT-KNN ceramics for actuator applications. *Processing and Application of Ceramics*, *10*(2), 73–77. doi:10.2298/PAC1602073C

Chandrasekhar, M., Negi, R. R., Nanda, D., Mallam, M., & Kumar, P. (2023). Microwave versus conventional sintering of NaNbO3 modified BNT–BT–BKT ceramics. *Journal of Materials Science Materials in Electronics*, *34*(6), 475. doi:10.1007/s10854-023-09918-2

Chen, B., Niu, M., Peng, Z., Qiao, X., Liang, P., Yang, Z., & Chao, X. (2020). Dielectric properties of lead-free BNT-based ferroelectric ceramics near the morphotropic phase boundary. *Materials Chemistry and Physics*, *256*, 123639. doi:10.1016/j.matchemphys.2020.123639

Corker, D. L., Glazer, A. M., Whatmore, R. W., Stallard, A., & Fauth, F. (1998). A neutron diffraction investigation into the rhombohedral phases of the perovskite series. *Journal of Physics Condensed Matter*, *10*(28), 6251–6269. doi:10.1088/0953-8984/10/28/007

Curie, J., & Curie, P. (1880). Bulletin de la Société Minéralogique de France. *Bulletin de la Societe Mineralogique de France*, *3*, 90–93. doi:10.3406/bulmi.1880.1564

Dorcet, V., Trolliard, G., & Boullay, P. (2008). Reinvestigation of phase transitions in Na0. 5Bi0. 5TiO3 by TEM. Part I: First order rhombohedral to orthorhombic phase transition. *Chemistry of Materials*, *20*(15), 5061–5073. doi:10.1021/cm8004634

Du, J., Yang, C., Li, Y., Yan, J., Qiu, L., Wang, L., & Zhu, K. (2023). Microstructural, dielectric, piezoelectric and ferroelectric properties of x PZN–PZT ternary ceramics. *Journal of Materials Science Materials in Electronics*, *34*(9), 780. doi:10.1007/s10854-023-10152-z

Duck, F. (2022). Paul Langevin, U-boats, and ultrasonics. *Physics Today*, *75*(11), 42–48. doi:10.1063/PT.3.5122

Dwivedi, S., Pareek, T., & Kumar, S. (2018). Structure, dielectric, and piezoelectric properties of K 0.5 Na 0.5 NbO 3-based lead-free ceramics. *RSC Advances*, *8*(43), 24286–24296. doi:10.1039/C8RA04038A PMID:35539216

Elahi, H., Munir, K., Eugeni, M., Abrar, M., Khan, A., Arshad, A., & Gaudenzi, P. (2020). A review on applications of piezoelectric materials in aerospace industry. *Integrated Ferroelectrics*, *211*(1), 25–44. doi:10.1080/10584587.2020.1803672

Fernandez-Benavides, D. A., Gutierrez-Perez, A. I., Benitez-Castro, A. M., Ayala-Ayala, M. T., Moreno-Murguia, B., & Muñoz-Saldaña, J. (2018). Comparative study of ferroelectric and piezoelectric properties of BNT-BKT-BT ceramics near the phase transition zone. *Materials (Basel)*, *11*(3), 361. doi:10.3390/ma11030361 PMID:29494486

Fu, Z., Yang, J., Lu, P., Zhang, L., Yao, H., Xu, F., & Li, Y. (2017). Influence of secondary phase on polymorphic phase transition in Li-doped KNN lead-free ceramics. *Ceramics International*, *43*(15), 12893–12897. doi:10.1016/j.ceramint.2017.06.185

Gao, X., Cheng, Z., Chen, Z., Liu, Y., Meng, X., Zhang, X., Wang, J., Guo, Q., Li, B., Sun, H., Gu, Q., Hao, H., Shen, Q., Wu, J., Liao, X., Ringer, S. P., Liu, H., Zhang, L., Chen, W., ... Zhang, S. (2021). The mechanism for the enhanced piezoelectricity in multi-elements doped (K, Na) NbO3 ceramics. *Nature Communications*, *12*(1), 881. doi:10.1038/s41467-021-21202-7 PMID:33564001

Haertling, G. I., & Buchanan, R. C. (1991). *Ceramic materials for electronics*. Buchanan.

Han, J., Yin, J., & Wu, J. (2020). BNT-based ferroelectric ceramics: Electrical properties modification by Ta2O5 oxide addition. *Journal of the American Ceramic Society*, *103*(1), 412–422. doi:10.1111/jace.16752

Hao, X., Zhai, J., Kong, L. B., & Xu, Z. (2014). A comprehensive review on the progress of lead zirconate-based antiferroelectric materials. *Progress in Materials Science*, *63*, 1–57. doi:10.1016/j.pmatsci.2014.01.002

Jaffe, H. (1958). Piezoelectric ceramics. *Journal of the American Ceramic Society*, *41*(11), 494–498. doi:10.1111/j.1151-2916.1958.tb12903.x

Jones, G. O., & Thomas, P. A. (2002). Investigation of the structure and phase transitions in the novel A-site substituted distorted perovskite compound Na0. 5Bi0. 5TiO3. *Acta Crystallographica. Section B, Structural Science*, *58*(2), 168–178. doi:10.1107/S0108768101020845 PMID:11910154

Joseph, J., Vimala, T. M., Sivasubramanian, V., & Murthy, V. R. K. (2000). Structural investigations on Pb (ZrxT1− x) O3 solid solutions using the X-ray Rietveld method. *Journal of Materials Science*, *35*(6), 1571–1575. doi:10.1023/A:1004778223721

Kalem, V., Cam, I., & Timuçin, M. (2011). Dielectric and piezoelectric properties of PZT ceramics doped with strontium and lanthanum. *Ceramics International*, *37*(4), 1265–1275. doi:10.1016/j.ceramint.2010.12.003

Kulkarni, H., Zohaib, K., Khusru, A., & Aiyappa, K. S. (2018). Application of piezoelectric technology in automotive systems. *Materials Today: Proceedings*, *5*(10), 21299–21304. doi:10.1016/j.matpr.2018.06.532

Kumar, A., & Kumar, A. (2020). Lead toxicity: Health hazards, influence on food chain, and sustainable remediation approaches. *International Journal of Environmental Research and Public Health, 17*(7), 2179. doi:10.3390/ijerph17072179 PMID:32218253

Kumar, P., Pattanaik, M., & Sonia. (2013). Synthesis and characterizations of KNN ferroelectric ceramics near 50/50 MPB. *Ceramics International, 39*(1), 65–69. doi:10.1016/j.ceramint.2012.05.093

Kumar, P., Singh, S., Juneja, J. K., Prakash, C., & Raina, K. K. (2009). Ferroelectric properties of substituted barium titanate ceramics. *Physica B, Condensed Matter, 404*(12-13), 1752–1756. doi:10.1016/j.physb.2009.02.016

Leontsev, S. O., & Eitel, R. E. (2010). Progress in engineering high strain lead-free piezoelectric ceramics. *Science and Technology of Advanced Materials, 11*(4), 044302. doi:10.1088/1468-6996/11/4/044302 PMID:27877343

Liu, J., Fei, L. J., Chen, C. F., Han, X. L., & Luo, Y. (2011, December). Effect of preparation method on properties of BNT-BKT-BT piezoelectric ceramics. In *2011 Symposium on Piezoelectricity, Acoustic Waves and Device Applications (SPAWDA)* (pp. 279-281). IEEE. 10.1109/SPAWDA.2011.6167243

Liu, Z., Yuan, Y., Luo, Z., Wan, H., Gao, P., Wu, H., Zhuang, J., Zhang, J., Zhang, N., Liu, H., Ren, W., & Ye, Z. G. (2020). Effects of antiferroelectric substitution on the structure and ferroelectric properties of a complex perovskite solid solution. *Journal of Materials Chemistry. C, Materials for Optical and Electronic Devices, 8*(17), 5795–5806. doi:10.1039/C9TC05084A

Malik, R. A., Kang, J. K., Hussain, A., Ahn, C. W., Han, H. S., & Lee, J. S. (2014). High strain in lead-free Nb-doped Bi1/2 (Na0. 84K0. 16) 1/2TiO3–SrTiO3 incipient piezoelectric ceramics. *Applied Physics Express, 7*(6), 061502. doi:10.7567/APEX.7.061502

Mirzaei, A., Bonyani, M., & Torkian, S. (2016). Effect of Nb doping on sintering and dielectric properties of PZT ceramics. *Processing and Application of Ceramics, 10*(3), 175–182. doi:10.2298/PAC1603175M

Murutoğlu, M., Akça, E., & Yilmaz, H. (n.d.). *Effects of Antimony Doping and Sintering Atmosphere on the Properties of K4CuNb8O23 Modified K0. 5Na0. 5NbO3 (KNN-KCN) Lead-free Piezoceramics.* Academic Press.

Nagata, H., Shinya, T., Hiruma, Y., Takenaka, T., Sakaguchi, I., & Haneda, H. (2012). Piezoelectric properties of bismuth sodium titanate ceramics. *Developments in Dielectric Materials and Electronic Devices, 167*, 213-221.

Nayak, S., Sahoo, B., Chaki, T. K., & Khastgir, D. (2014). Facile preparation of uniform barium titanate (BaTiO 3) multipods with high permittivity: Impedance and temperature dependent dielectric behavior. *RSC Advances, 4*(3), 1212–1224. doi:10.1039/C3RA44815K

Newnham, R. E. (1997). Molecular mechanisms in smart materials. *MRS Bulletin, 22*(5), 20–34. doi:10.1557/S0883769400033170

Noheda, B., Gonzalo, J. A., Cross, L. E., Guo, R., Park, S. E., Cox, D. E., & Shirane, G. (2000). Tetragonal-to-monoclinic phase transition in a ferroelectric perovskite: The structure of PbZr 0.52 Ti 0.48 O 3. *Physical Review B: Condensed Matter, 61*(13), 8687–8695. doi:10.1103/PhysRevB.61.8687

Ogino, M., Noguchi, Y., Kitanaka, Y., Miyayama, M., Moriyoshi, C., & Kuroiwa, Y. (2014). Polarization rotation and monoclinic distortion in ferroelectric (Bi0. 5Na0. 5) TiO3–BaTiO3 single crystals under electric fields. *Crystals*, *4*(3), 273–295. doi:10.3390/cryst4030273

Orayech, B., Faik, A., López, G. A., Fabelo, O., & Igartua, J. M. (2015). Mode-crystallography analysis of the crystal structures and the low-and high-temperature phase transitions in Na0. 5K0. 5NbO3. *Journal of Applied Crystallography*, *48*(2), 318–333. doi:10.1107/S1600576715000941

Panda, P. K. (2009). Environmental friendly lead-free piezoelectric materials. *Journal of Materials Science*, *44*(19), 5049–5062. doi:10.1007/s10853-009-3643-0

Panda, P. K., & Sahoo, B. (2015). PZT to lead free piezo ceramics: A review. *Ferroelectrics*, *474*(1), 128–143. doi:10.1080/00150193.2015.997146

Pattipaka, S., James, A. R., & Dobbidi, P. (2018). Enhanced dielectric and piezoelectric properties of BNT-KNNG piezoelectric ceramics. *Journal of Alloys and Compounds*, *765*, 1195–1208. doi:10.1016/j.jallcom.2018.06.138

Pierre, A. C. (2020). *Introduction to sol-gel processing*. Springer Nature. doi:10.1007/978-3-030-38144-8

Pu, T., Chen, H., Xing, J., Luo, Y., Fan, S., Liu, H., Chen, Q., & Zhu, J. (2022). High piezoelectricity of low-temperature sintered Li2CO3-added PNN–PZT relaxor ferroelectrics. *Journal of Materials Science Materials in Electronics*, *33*(8), 4819–4830. doi:10.1007/s10854-021-07671-y

Pullar, R. C., Zhang, Y., Chen, L., Yang, S., Evans, J. R., Salak, A. N., Kiselev, D. A., Kholkin, A. L., Ferreira, V. M., & Alford, N. M. (2009). Dielectric measurements on a novel Ba 1− x Ca x TiO 3 (BCT) bulk ceramic combinatorial library. *Journal of Electroceramics*, *22*(1-3), 245–251. doi:10.1007/s10832-008-9430-5

Rached, A., Wederni, M. A., Khirouni, K., Alaya, S., Martín-Palma, R. J., & Dhahri, J. (2021). Structural, optical and electrical properties of barium titanate. *Materials Chemistry and Physics*, *267*, 124600. doi:10.1016/j.matchemphys.2021.124600

Ragini, R., Ranjan, R., Mishra, S. K., & Pandey, D. (2002). Room temperature structure of Pb (Zr x Ti 1− x O 3) around the morphotropic phase boundary region: A Rietveld study. *Journal of Applied Physics*, *92*(6), 3266–3274. doi:10.1063/1.1483921

Rahman, J. U., Hussain, A., Maqbool, A., Malik, R. A., Kim, M. S., & Kim, M. H. (2015). Effect of sintering temperature on the electromechanical properties of 0.945 Bi 0.5 Na 0.5 TiO 3-0.055 BaZrO 3 ceramics. *Journal of the Korean Physical Society*, *66*(7), 1072–1076. doi:10.3938/jkps.66.1072

Ramajo, L. A., Taub, J., & Castro, M. S. (2014). Effect of ZnO addition on the structure, microstructure and dielectric and piezoelectric properties of K0. 5Na0. 5NbO3 ceramics. *Materials Research*, *17*(3), 728–733. doi:10.1590/S1516-14392014005000048

Ramam, K., & Lopez, M. (2007). Effect of strontium doping on dielectric, ferroelectric and piezoelectric properties of PLZTN ceramics. *Materials Science and Engineering B*, *145*(1-3), 41–47. doi:10.1016/j.mseb.2007.09.085

Rao, B. N., Datta, R., Chandrashekaran, S. S., Mishra, D. K., Sathe, V., Senyshyn, A., & Ranjan, R. (2013). Local structural disorder and its influence on the average global structure and polar properties in Na 0.5 Bi 0.5 TiO 3. *Physical Review B: Condensed Matter and Materials Physics*, *88*(22), 224103. doi:10.1103/PhysRevB.88.224103

Rao, B. N., Fitch, A. N., & Ranjan, R. (2013). Ferroelectric-ferroelectric phase coexistence in Na 1/2 Bi 1/2 TiO 3. *Physical Review B: Condensed Matter and Materials Physics*, *87*(6), 060102. doi:10.1103/PhysRevB.87.060102

Reitz, W. (1997). Ceramic processing and sintering by MN Rahaman. *Materials and Manufacturing Processes*, *12*(3), 555–556. doi:10.1080/10426919708935164

Ringgaard, E., & Wurlitzer, T. (2005). Lead-free piezoceramics based on alkali niobates. *Journal of the European Ceramic Society*, *25*(12), 2701–2706. doi:10.1016/j.jeurceramsoc.2005.03.126

Sahoo, B., & Panda, P. K. (2013). Effect of lanthanum, neodymium on piezoelectric, dielectric and ferroelectric properties of PZT. *Journal of Advanced Ceramics*, *2*(1), 37–41. doi:10.1007/s40145-013-0039-z

Sekhar, B. C., Dhanalakshmi, B., Rao, B. S., Ramesh, S., Prasad, K. V., Rao, P. S., & Rao, B. P. (2021). Piezoelectricity and its applications. *Multifunct. Ferroelectr. Mater, 71*.

Shao, B., Qiu, J., Zhu, K., Pang, X., & Meng, Q. (2012). Tantalum influence on electrical properties of lead-free (K 0.4425 Na 0.52 Li 0.0375)(Nb 0.93− x Ta x Sb 0.07) O 3 piezoelectric ceramics. *Journal of Materials Science Materials in Electronics*, *23*(4), 846–850. doi:10.1007/s10854-011-0504-9

Singh, H. S., & Maisnam, M. (2023, December 10). Study of structural and electrical properties of oxides added K0. 47Na0. 47Li0. 06NbO3 ceramics. *Ferroelectrics*, *617*(1), 83–92. doi:10.1080/00150193.2023.2271138

Skidmore, T. A., & Milne, S. J. (2007). Phase development during mixed-oxide processing of a [Na0. 5K0. 5NbO3] 1− x–[LiTaO3] x powder. *Journal of Materials Research*, *22*(8), 2265–2272. doi:10.1557/jmr.2007.0281

Su, B., Holmes, J. E., Cheng, B. L., & Button, T. W. (2002). Processing effects on the microstructure and dielectric properties of barium strontium titanate (BST) ceramics. *Journal of Electroceramics*, *9*(2), 111–116. doi:10.1023/A:1022850205284

Takenaka. (2013). Lead-Free Piezoelectric Ceramics. Handbook of Advanced Ceramics.

Tian, A., & Du, H. (2014). Enhanced ferroelectric property of manganese doped lead-free (K0. 5Na0. 5) NbO3 Ceramics. *Ferroelectrics*, *463*(1), 72–77. doi:10.1080/00150193.2014.891921

Trolliard, G., & Dorcet, V. (2008). Reinvestigation of phase transitions in Na0. 5Bi0. 5TiO3 by TEM. Part II: Second order orthorhombic to tetragonal phase transition. *Chemistry of Materials*, *20*(15), 5074–5082. doi:10.1021/cm800464d

van der Veer, E., Noheda, B., & Acuautla, M. (2021). Piezoelectric properties of PZT by an ethylene glycol-based chemical solution synthesis. *Journal of Sol-Gel Science and Technology*, *100*(3), 517–525. doi:10.1007/s10971-021-05651-6

Vasiliu, F., Lucuta, P. G., & Constantinescu, F. (1983). Structure-sintering temperature relationship in PZT solid solutions at the morphotropic phase boundary. *Physica Status Solidi (A), 80*(2), 637-642.

Wang, D., Zhu, K., Ji, H., & Qiu, J. (2009). Two-step sintering of the pure K0. 5Na0. 5NbO3 lead-free piezoceramics and its piezoelectric properties. *Ferroelectrics, 392*(1), 120–126. doi:10.1080/00150190903412622

Yan, K., Wang, F., Chen, X., Wu, D., & Zhu, K. (2019). Effects of A-site ionic size on the phase transition behavior of lead-free niobate ceramics. *Ceramics International, 45*(16), 20323–20330. doi:10.1016/j.ceramint.2019.07.005

Yang, Y., Hao, H., Zhang, L., Chen, C., Luo, Z., Liu, Z., Yao, Z., Cao, M., & Liu, H. (2018). Structure, electrical and dielectric properties of Ca substituted BaTiO3 ceramics. *Ceramics International, 44*(10), 11109–11115. doi:10.1016/j.ceramint.2018.03.111

Yun, S., Wang, X., & Xu, D. (2008). Influence of processing parameters on the structure and properties of barium strontium titanate ceramics. *Materials Research Bulletin, 43*(8-9), 1989–1995. doi:10.1016/j.materresbull.2007.10.003

Zhang, N., Glazer, A. M., Baker, D., & Thomas, P. A. (2009). Structures of K0. 05Na0. 95NbO3 (50–300 K) and K0. 30Na0. 70NbO3 (100–200 K). *Acta Crystallographica. Section B, Structural Science, 65*(3), 291–299. doi:10.1107/S0108768109011057 PMID:19461138

Zhang, N., Zheng, T., & Wu, J. (2020). Lead-free (K, Na) NbO3-based materials: Preparation techniques and piezoelectricity. *ACS Omega, 5*(7), 3099–3107. doi:10.1021/acsomega.9b03658 PMID:32118126

Zhang, Y., Zhai, J., & Xue, S. (2020). Effect of three step sintering on piezoelectric properties of KNN-based lead free ceramics. *Chemical Physics Letters, 758*, 137906. doi:10.1016/j.cplett.2020.137906

Zhao, C., Wu, B., Thong, H. C., & Wu, J. (2018). Improved temperature stability and high piezoelectricity in lead-free barium titanate-based ceramics. *Journal of the European Ceramic Society, 38*(16), 5411–5419. doi:10.1016/j.jeurceramsoc.2018.08.004

Zhao, S. Y., & Bian, J. J. (2023). Effects of Li2CO3–CuO addition on the sintering behavior, dielectric and piezoelectric properties of PZT ceramics. *Journal of Materials Science Materials in Electronics, 34*(15), 1202. doi:10.1007/s10854-023-10658-6

Zhou, C., & Liu, X. (2008). Dielectric and piezoelectric properties of bismuth-containing complex perovskite solid solution of Bi 1/2 Na 1/2 TiO 3– Bi (Mg 2/3 Nb 1/3) O 3. *Journal of Materials Science, 43*(3), 1016–1019. doi:10.1007/s10853-007-2246-x

Zhou, X., Xue, G., Luo, H., Bowen, C. R., & Zhang, D. (2021). Phase structure and properties of sodium bismuth titanate lead-free piezoelectric ceramics. *Progress in Materials Science, 122*, 100836. doi:10.1016/j.pmatsci.2021.100836

Chapter 12
Improving Water Quality Through Mathematical Modeling Adsorption With Carbonaceous Materials

Luis Alfonso Cavazos-Cuello
Universidad Autónoma de Nuevo León, Mexico

Jacob Josafat Salazar-Rábago
Universidad Autónoma de Nuevo León, Mexico

Nancy E. Dávila-Guzmán
https://orcid.org/0000-0002-9362-4458
Universidad Autónoma de Nuevo León, Mexico

Eduardo Soto-Regalado
Universidad Autónoma de Nuevo León, Mexico

Jesús Botello-González
https://orcid.org/0000-0003-4710-5785
Universidad Autónoma de Nuevo León, Mexico

ABSTRACT

Carbonaceous materials (CMTs) have applications across diverse fields, from medical to food industry; their most pivotal use lies in wastewater treatment and purification through tertiary processes, particularly through adsorption, which is the process where liquids, gasses, or dissolved molecules will adhere to the surface of the solid material through either weak or strong intermolecular forces. Adsorption processes are versatile since they can be performed in batch or continuous regime and have been used to purify water that has been polluted with a wide variety of contaminants, such as dyes, pesticides, phenolic compounds, among many others. The aim of this chapter is to explore the different schools of mathematical modeling in adsorption systems both in batch and continuous regime as well as the benefits it provides in the large-scale removal of pollutants in aqueous phase using both traditional and novel carbonaceous materials.

DOI: 10.4018/979-8-3693-1306-0.ch012

INTRODUCTION

Carbonaceous materials are a class of solids primarily comprised of carbon molecules arranged in different configurations, ranging from amorphous to crystalline structures. CMTs have captivated the attention of many researchers and industries globally due to their diverse applications and the versatility of their properties, which can be tailored using several processing techniques to meet specific demands depending on their intended use.

CMTs can be broadly separated into two main categories: rigid and flexible materials. Rigid materials obstruct molecule transfer through their solid matrix, only allowing species transport across their surface. One example of a widely used rigid CMT is activated carbon, which consists of amorphous graphene layers and generally has an elevated surface area as well as amphoteric characteristics allowing it to interact with molecules of varying sizes and structures. Other examples of rigid materials are carbon blacks, biochars, among many others.

Contrastingly, flexible materials do allow species transfer within their solid matrix, the main example of this type of material are lignocellulosic residues, which, as their name implies are materials which consist of varying proportions of lignin, cellulose, and hemicellulose. Flexible CMTs are becoming an increasingly popular alternative to other materials mainly due to their reduced cost of application and high availability.

LITERATURE REVIEW

Carbon Nanotubes

A carbon nanotube (CNT) is a structure composed of carbon atoms aligned in a hexagonal pattern. These structures resemble tubes or cylinders, with diameters measuring in the nanometer scale around a few nanometers and lengths that can range from a few nanometers to several millimeters. CNTs have attracted attention due to their characteristics and the wide range of applications across different fields. They possess a surface area enabling the attachment of components for biosensing purposes (Ferrier & Honeychurch, 2021). CNTs are widely regarded as one of the types of nanomaterials based on carbon and have undergone extensive research for their potential uses in various applications, including water treatment (Dong et al., 2021). There are two categories of carbon nanotubes, which are classified based on the number of layers of carbon they contain. Single walled carbon nanotubes (SWCNTs) consist of a layer of graphene and have a diameter that ranges from 0.4 to 2 nm (Jeon et al., 2019). They usually occur in bundles that are hexagonally packed. On the other hand, multiwalled carbon nanotubes (MWCNTs) comprise two or more cylinders each made up of sheets. The diameter of these tubes varies from 1 to 3 nm. (Jha et al., 2020)

In recent years the synthesis of carbon nanotubes has witnessed some developments. One notable trend involves the utilization of machine learning algorithms to optimize synthesis parameters and predict the properties of CNTs (Hwang et al., 2019). Another advancement lies in the creation of wafer scale CNT synthesis techniques, such as strategies based on solution processing and dip coating methods (Jeon et al., 2019). Moreover, there have been strides in synthesizing SWCNTs with specific chirality by designing catalysts and manipulating growth conditions. Additionally, the production of MWCNTs has been enhanced through methods like nebulized pyrolysis and arc discharge using hydrogen gas (Jha et al., 2020). These emerging trends in CNT synthesis aim to improve the quality, yield and control over CNT properties opening possibilities for their application, in fields including electronics, energy storage and biomedical devices.

Carbon Aerogels

Carbon aerogels are materials that are created using a chemical process called sol-gel chemistry. These materials have some properties such as high surface area, electrical conductivity, and chemical inertness (Gan et al., 2019). They are produced by polymerizing organic precursors, such as resorcinol and formaldehyde, in aqueous solution, followed by supercritical drying and pyrolysis in an inert atmosphere (Lee & Park, 2020). The resulting carbon aerogels have a three-dimensional hierarchical morphology with ultrafine cell sizes. They can be made into shapes like monoliths or thin films and their structure can be customized well (Zubizarreta et al., 2008). Carbon aerogels find applications in fields including energy storage, desalination, catalysis, and customized well (Zubizarreta et al., 2008). One of their uses is in the removal of pollutants from water (Ptaszkowska-Koniarz et al., 2018a, 2018b), although its high cost has limited their application.

ACTIVATION PROCESSES OF AGRO-INDUSTRIAL WASTES TO PRODUCE CARBONOUS MATERIALS

Carbonous materials can be produced from agro-wastes, which are widely available because these are by-products of agricultural industrial wastes that includes agricultural residues and industrial residues, in Table 1 the main classification of these residues is shown (Sadh et al., 2018). Because their lignin, cellulose and hemicellulose contents, these residues can be used to prepare activated carbon or biochar, in other words carbonous materials.

In Table 2 the composition of some agro-industrial wastes is denoted (Sadh et al., 2018). The relative chemical composition of every agro-industrial waste affects the textural properties because cellulose, hemicellulose, and lignin have different thermal performance under non-oxidating conditions. Boundzanga el al. studied the contribution of hemicellulose, cellulose, and lignin over the porous characteristics of biochar produced from tomato plants, poplar wood, olive pomace, and miscanthus, the chemical activation process was carried out with H_3PO_4 under argon atmosphere at 500 °C for 40 min. According to their findings, the most important mass loss was 249 °C, 354 °C, 375 °C for hemicellulose, cellulose,

Table 1. Type of agricultural residues waste

Agricultural Residues		Industrial Residues
Field Residues	**Process Residues**	
Stems	Husks	Potato peel
Stalks	Seeds	Orange peel
Leaves	Roots	Cassava peel
Seed pods	Bagasse	Soybean oil cake

Note. Agricultural residues

Table 2. Composition of agro-industrial wastes

Agro-Industrial Wastes	Chemical Composition (% w/w)		
	Cellulose	Hemicellulose	Lignin
Sugarcane bagasse	30.2	56.7	13.4
Rice straw	39.2	23.5	36.1
Corn stalks	61.2	19.3	6.9
Sawdust	45.1	28.1	24.2
Sugar beet waste	26.3	18.5	2.5
Barley straw	33.8	21.9	13.8
Cotton stalks	58.2	14.4	21.5
Oat straw	39.5	27.1	17.5
Soya stalks	34.5	24.8	19.8
Sunflower stalks	42.1	29.7	13.4
Wheat straw	32.9	24.0	8.9

Note. Some Agro-industrial waste composition

and lignin. The yield of every biochar was not significantly affected by the type of biomass and was ranged from 68.9 to 74.3%; however, the textural properties were different, and was range from 0.29 to 0.47 cm^3/g and 396 to 1009 m^2/g for total pore volume and surface area, respectively (Boundzanga et al., 2022).

Carbonous materials such as biochar or activated carbon is typically produced by pyrolysis, gasification, and torrefaction, to thermochemical decompose the biomass (agro-wastes); pyrolysis is carried out in a temperature range of 250 to 900 °C, to prevent its complete oxidation these processes are conducted in furnaces under inert atmosphere, the products of these process comprise biochar (solid), bio-oil (liquid), and gaseous products. Gasification is conducted at elevated temperatures (650–900 °C) under atmosphere of oxygen, air, or steam; the main products are syngas and low amounts of hydrocarbons. Torrefaction is an incomplete pyrolysis and is an innovative process to prepare carbonous materials or biochar, this process is carried out in atmosphere free of oxygen at low temperatures (300 °C) (Dhyani & Bhaskar, 2019).

Chemical Activation

Although the produced carbonous materials have textural properties to be considered as an adsorbent material, the above-mentioned properties can be improved by chemical or physical activation. Chemical activation is carried out in a single stage at temperatures ranged from 450 to 900 °C and comprises carbonization and activation; prior these processes the biomass is impregnated with chemical substances, such as HCl, $ZnCl_2$, $MgCl_3$, HNO_3, H_2SO_4, KOH, among other, these substances are useful to dehydrant the biomass and to prevent the tar production; thus, both the porosity and surface are enhanced (H. Liu et al., 2019). In recent years several research groups have studied the chemical activation of agro-wastes using of above-mentioned chemical substances; in Table 3 some agro-wastes used as precursor to produce biochar or activated carbons, in addition their textural properties, the studied adsorbate, and the maximum adsorption capacities are shown.

Table 3. Agro-wastes used as precursor to produce biochar or activated carbons by chemical activation

Precursor	Adsorbate	T (°C)	Time (h)	Activating Agent	Surface Area (m^2/g)	Total Pore Volume (cm^3/g)	Qmax (mg/g)	Ref.
Ascophyllum nodosum	Cu(II)	300, 500, and 700	120	None	19.815	0.021	222.27	(Katiyar et al., 2021)
Stem of khat	Malachite Green	600	60	H_2SO_4	1,735	---	5.62	(Abate et al., 2020)
Sesame straw	Pb Cd Cr Cu Zn	700	360	None	289.2	0.0411	102 86 65 55 34	(Park et al., 2016)
Barley Husks	Phenol	300, 500, 700	20, 100, 180	$ZnCl_2$	811.44	0.263	132.9	(Loredo-Cancino et al., 2016) (Loredo-Cancino et al., 2013)
Corn pericarp	Methylene organge	350, 500, 650	20, 60, 100	KOH	23.31	---	141.13	(Loya-González et al., 2019)
Soybean shells	Iodine*	450, 550, 650	30, 105, 180	$ZnCl_2$	---	---	1061	(Sepúlveda-Cervantes et al., 2018)
Pine sawdust	Carbamazepine bisphenol A	200, 350, 500, 650	120	H_3PO_4	1,148	1.1	215 225	(G. Chu et al., 2018)

Note. *This substance is used as an indicator of surface area and was measured according to the ASTM: Standard Method for Determination of Iodine Number of Activated Carbon (ASTM, 2006).

It is important to note that these carbonous materials are useful to remove by adsorption process several pollutants, such as heavy metals, dyes, and organic molecules, among others, it is not possible to compare the adsorption capacities because in every study the adsorbent was different and their textural properties, initial concentrations, adsorbent dosages, experimental conditions (pH solution, stirring speed, temperature, etc.), as well their surface chemistry are different; nevertheless, for heavy metals the maximum adsorption capacity was achieved with the activated carbon produced from Ascophyllum nodosum, on the other hand the maximum adsorption capacity for organic molecules was reported for the activated carbon prepared from pine sawdust.

Physical Activation

Physical activation is used to enhance the textural properties of biochar or activated carbons prepared from agro-wastes; however, this process is carried out mainly in two stages, first, the biomass is pyrolyzed or carbonized under inert gas flow, for instance, N_2 or another inert gas, afterwards the activation process is conducted at high temperatures (700–900 °C) using CO_2 or steam as atmosphere into the furnace. This activation process produces biochar or activated carbons with more porous because during this process excessive volatile matter is released, then the yield of biochar is lower in comparison to the yield of chemical activation process; moreover, when an agro-waste is activated by physical mode the textural properties (surface area and pore volume distribution) are better compared with the carbonous materials produced by chemical activation; nevertheless, physical activation requires more power consumption due to this process is carried out at higher temperatures in the two required stages (Eduardo et al., 2022). When steam at high temperature is used to activate physically a precursor to produce biochar or activated carbons (carbonous materials), this gas flow is forced to pass through the solid matrix, therefore its structural porosity is improved due to this flow gas contributes to remove the products of incomplete pyrolysis of the carbonization or pyrolysis stage (Sizmur et al., 2017).

Menya et al. determined the optimal conditions to produce char from rice husks, which was used as precursor to prepare biochar by pyrolysis. The production of char was carried out into a furnace, the operational conditions were established by an experimental design, as follows temperature (400 - 600 °C), heating rate (10 - 25 °C/min), and heating time 60-120 °C), a maximum surface area of 123.9 m^2/g was achieved (Menya et al., 2020). Corncob was used as precursor to produce biochar by pyrolysis under air with low O_2 content at 200 °C during 2 h, thereafter the material was activated at 350, 450, 550, and 650 °C, the best textural properties were found to be 25.67 m^2/g, 0.0085 cm^3/g, and 3.38 nm, for surface area, micropore volume, and average pore width, respectively. This biochar was used to adsorb Atrazine, the maximum adsorption capacity was 2.65 mg/g (Hao et al., 2013). The pyrolysis of biomass is a thermal process to prepared biochar or activated carbons, which could be including three stages. The first one is carried out within a temperature range of 50 to 150 °C, during this stage about 10% of lost mass is recorded as humidity releasing; the second is denoted as pyrolytic decomposition, and can be registered at a range of temperature of 200 to 500 °C, within this range the pyrolytic decomposition of cellulose and hemicellulose is observed into the reactor; thus volatile matter is released, the third stage (carbonization) is due to the condensation reaction and further decomposition of char and inorganic matter (X. Liu et al., 2014). Li. J. et al studied the physical activation of rice husk, first the biomass was dried, and then the biomass was carbonized at 550 °C for 30 min; thereafter the char was activated by steam at 750 °C with a mass ratio of 3:1 (char/steam). The best surface area and total pore volume were 351.23 m^2/g and 0.315 cm^3/g. This adsorbent material was used to adsorb hexane, and the saturated adsorption capacities were reported as 29.33 and 5.02 mg/g for activated and non-activated biochar, respectively (Li et al., 2019).

CHEMICAL TREATMENT OF CARBONOUS MATERIALS TO ENHANCE ITS ADSORPTION BEHAVIOR

Although by traditional activation processes, such as chemical or physical activation it is possible to produce carbonous materials with particularly good textural properties and good chemical surface composition; recently several authors have reported the enhancing of the above-mentioned characteristics to improve the behavior of the adsorbent materials. These modifications can be developed chemical substances. These treatments can be carried out by acid modification, alkaline modification, salt impregnation, and oxidation by hydrogen peroxide. The modification with acids is useful to enhance the textural properties of biochar, it is possible because this treatment removes impurities from surface and more microporous are developed, thus the surface area increases, the typical used acids are HCl, H_2SO_4, H_3PO_4, and HNO_3. On the other hand, the modification by alkalis (KOH or NH_4OH) increases BET

(Brunauer Emmett Teller) surface area and porosity, moreover, the content of oxygenated functional groups increases it is important because this carbonous materials will be useful to adsorb cationic species (Tomczyk et al., 2023).

Simulation and Mathematical Modeling with Carbonaceous Materials

One of the most widely employed methods for achieving sustainability and eco-friendliness in the removal of contaminants at low concentrations is through adsorption with carbonaceous materials, but the adsorption process is not instantaneous; thus, the diluted molecules must transport from the bulk of the solution to the active sites of the CMTs through three mass transfer stages: Film layer transport, transport through the solid pores or permeation (depending if the CMT is rigid or flexible) and finally deposition on the active sites of the material.

The stages previously mentioned are analyzed using mathematical modeling and process simulation, which are an essential tool in the holistic design of adsorption separation processes, since they allow the manipulation of the operational conditions to maximize the pollutant removal efficiency, with the added benefit of enhancing the environmental and economical sustainability if the process. There are two main types of modeling in the adsorption field, the first one is the so called 'semi-empirical' modeling, which consists of algebraic models that are obtained through considering specific stages of the adsorption process as unimportant, such is the case of the Crank model, demonstrated by Equation (1), this model only considers intra-particle diffusion as a meaningful step, considering film transport and active site adsorption insignificant.

$$\frac{q}{q_{\infty}} = 1 - \frac{6}{\pi^2}\sum_{n=1}^{\infty}\frac{1}{n^2}\exp\left(-\frac{D_s n^2 \pi^2 t}{R^2}\right) \tag{1}$$

The Crank model has been widely used in the description of adsorption systems involving a variety of carbonaceous materials, however, this model is also based on the assumption that the solute concentration on the material surface remains constant, which is rarely the case in real adsorption systems, limiting the theoretical accuracy of the description provided by the model.

The second type of modeling in adsorption systems is the diffusional modeling, this consists of employing models that consider every stage of the adsorption process, these models consist of mass balances in a differential element of the CMT surface as well as the liquid (or gas) bulk resulting in a system of Partial Differential Equations (PDEs). The correct derivation of said PDEs depend on two factors: The CMT's geometry and its classification as rigid or flexible.

The geometry of the material defines the coordinate system used during the formulation of the mass balance in the solid. Meaning that if the CMT used in an adsorption process is a granular activated carbon, spherical coordinates can be assumed, in the case of pelletized activated carbon it would be fair to assume cylindrical coordinates and so on. This indicates that determining the coordinate system to be used in the diffusional models requires a comprehensive characterization of the CMT using complementary techniques such as Scanning Electron Microscopy as well as Helium pycnometry in the case of porous materials.

On the other hand, the category of the material influences the phenomena occurring after the film layer transport stage of the process. Liquid molecules diffusing through the internal structure of a rigid carbon material, such as an active carbon or carbon xerogel, will have to navigate through the pores of the material through two parallel diffusion mechanisms, effective pore volume diffusion and surface diffusivity.

On the contrary, for a long time it was thought that no type of diffusion occurred in the internal structure of flexible materials like lignocellulosic CMTs, however, recent research reports a combination of diffusion through the hollow structure of lignocellulosic solids as well as permeation through the cell walls of the flexible material. This also significantly impacts the mass balance on the solid material.

The most used diffusional model in the description of adsorption systems is the Pore Volume Surface Diffusion Model (PVSDM), it was originally derived for spherical rigid CMTs such as active carbons and it supposes that the adsorption in the active sites occurs instantaneously. The model is presented by the Equations (2) and (4) as well as their corresponding initial and boundary conditions (Equations (3, 5-7)).

$$V\frac{dC_A}{dt} = -mSk_L\left(C_A - C_A|_{r=R}\right) \tag{2}$$

$$t = 0; C_A = C_{A0} \tag{3}$$

$$\varepsilon_P\frac{\partial C_{Ar}}{\partial t} + \rho_p\frac{\partial q}{\partial t} = \frac{1}{r^2}\frac{\partial}{\partial r}\left[r^2\left(\mathcal{D}_{ep}\frac{\partial C_{Ar}}{\partial t} + \mathcal{D}_S\rho_p\frac{\partial q}{\partial r}\right)\right] \tag{4}$$

$$t = 0; C_{Ar} = 0; 0 \le r \le R \tag{5}$$

$$\left(\frac{\partial C_{Ar}}{\partial r}\right)_{r=0} = 0 \tag{6}$$

$$\varepsilon_P\left(\frac{\partial C_{Ar}}{\partial t}\right)_{r=R} + \rho_p\left(\frac{\partial q}{\partial r}\right)_{r=R} = k_L\left(C_A - C_A|_{r=R}\right) \tag{7}$$

Equation 2 along with its initial condition (Eq (3)) represent the mass balance on the bulk of the liquid or gas phase, meanwhile, Equation (4) and its associated conditions (Eqs (5-7)) correspond to the mass balance performed on a spherical differential element of an Active Carbon sphere, where the two intraparticle diffusion mechanisms, pore volume and surface diffusion, can be found. The model can be simplified to either the Surface Diffusion Model (SDM) or the Pore Volume Diffusion Model (PVDM), when considering the contribution of either mechanism as negligible ($\mathcal{D}_{ep} = 0 \; or \; \mathcal{D}_S = 0$).

For the case of flexible materials, the only diffusional model used so far in the description of adsorption systems is the Diffusion-Permeation model, it was originally developed for flexible rectangular sawdust flakes, and it makes similar assumptions as the PVSDM but considering the CMT's geometry. The model is described by Equations (8, 10) and their respective initial and boundary conditions given by the Equations (9, 11-13).

$$V\frac{dC_A}{dt} = -mSk_L(C_A - C_{A_z}|_{z=L}) \tag{8}$$

$$t = 0; C_A = C_A \tag{9}$$

$$\frac{\partial q}{\partial t} = \frac{\partial}{\partial z}\left(P_M \frac{\partial q}{\partial z}\right) \tag{10}$$

$$t = 0; C_{Az} = 0; 0 \leq z \leq L \tag{11}$$

$$-P_M \rho_p \left(\frac{\partial q}{\partial z}\right)_{z=0} = 0 \tag{12}$$

$$P_M \rho_p \left(\frac{\partial q}{\partial z}\right)_{z=0} = k_L\left(C_A - C_A\big|_{z=L}\right) \tag{13}$$

where P_M is the permeability coefficient in cm^2/s. As previously mentioned, both diffusional described models assume that the adsorption on the active site stage is instantaneous, meaning that this stage can be modeled using the adsorption isotherms of the liquid molecules on the CMT, indicating that a deep understanding of the thermodynamic behavior of the equilibrium between both phases is also required to employ the diffusional models.

While using semi-empirical models to describe adsorption systems provides an accurate representation of experimental adsorption data, diffusional modeling not only allows a deeper description of the adsorption system, but also allows the calculation and optimization of design parameters used in the diffusional modeling of continuous adsorption systems meant for large-scale purification of wastewater, making the diffusional modeling of adsorption systems an invaluable asset for the environment.

LITERATURE REVIEW

The Sorption Process

The adsorption or biosorption process (simply referred to as 'sorption') is a solid-liquid or solid-gas separation process consisting of the retention of molecules over the surface of a solid material, in this case, a carbonaceous material (Calvert, 1990). The sorption process is widely used in the purification of water systems due to its ease of implementation and low operational costs. The sorption process can also be implemented in both batch and transient regimes, making it a versatile option in the purification of polluted water.

Additionally, since the sorption process consists of the interaction between a solid and a fluid phase, the solid to be used can have a wide variety of structures. Moreover, in the case of carbonaceous materials, their surface properties can be easily tuned during their production stage to maximize the removal of contaminants (Turk Sekulic et al., 2019; T. Wang et al., 2020; You et al., 2023) (Sanchez-Silva et al., 2022).

Stages of Sorption Studies Using Carbonaceous Materials

The sorption process can be analyzed from different perspectives. For instance, since the process involves the equilibrium between a solid and fluid phase, it can be analyzed from a thermodynamic perspective. Conversely, since the equilibrium is not attained instantly, a kinetic perspective can be studied. Both perspectives are studied in a batch regime, therefore the last perspective involves studying the process in a continuous regime.

However, to fully understand the interaction between a carbonaceous material and the solute to be removed, all the perspectives previously mentioned must be studied, this is summarized in Figure 1, where sorption studies are usually structured like a pyramid, mainly since, to advance to the next stage of the study, data from the previous stages is required.

Figure 1. Stages of a sorption study

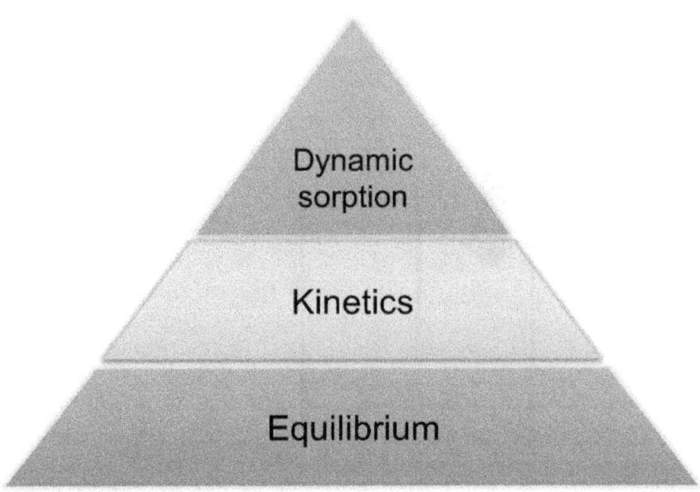

Mathematical Modeling in the Sorption Process

The mathematical modeling of the sorption process is categorized according to the study stage: The equilibrium stage involves mostly algebraic models, whilst the kinetics and dynamic sorption stages are both divided in algebraic and diffusional models.

Sorption Equilibrium Models

The mathematical interpretation of the sorption equilibrium process begins with a material balance performed on a batch sorption tank such as the one presented in Figure 2, the result of this is presented in the equation (14).

$$q_e = \frac{V}{m}\left(C_0 - C_e\right) \tag{14}$$

where:

q_e: Equilibrium sorption capacity, [mg/g].
V : Solution volume, [L].
m : Mass of the carbonaceous material, [g].
C_0: Inlet concentration of the solute, [mg/L].
C_e : Outlet concentration of the solute, [mg/L].

In a laboratory-scale sorption equilibrium study, 4 or more batch sorption containers are used, using equation 1 for each container. Doing so will yield a C_e and q_e data pair for each container. This set of data must be fitted to sorption isotherm models. Sorption isotherm models describe the relation between the solute's equilibrium concentration and the carbonaceous material's sorption capacity at equilibrium. A wide variety of sorption isotherm models have been reported in literature (Al-Ghouti & Da'ana, 2020;

Figure 2. Batch sorption tank

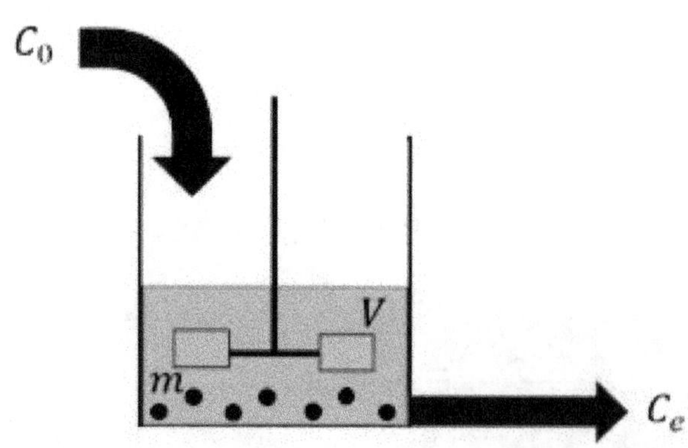

Foo & Hameed, 2010), however, the most reported sorption isotherms are the Freundlich and Langmuir isotherms:

$$q_e = kC_e^{1/n} \tag{15}$$

$$q_e = \frac{q_m K C_e}{1 + K C_e} \tag{16}$$

where k and n are fitting parameters, while q_m is the maximum sorption capacity from the Langmuir isotherm in mg/g and K is the Langmuir isotherm equilibrium constant in L/mg. The experimental sorption data is fitted to the two (or more) models employed and an error measurement is calculated to determine which model represents the experimental data better.

The Freundlich isotherm has been commonly regarded to have no physical meaning, mainly due to it being an exponential function, indicating that the surface of the carbonaceous material will continue to accumulate solute molecules regardless of the equilibrium concentration, which is not the case, since the surface of the solid will eventually reach saturation. Due to this, the Freundlich isotherm has commonly been associated with multi-layer sorption, where solute molecules will stack on top of previously adsorbed molecules.

Additionally, graphic representation of the sorption isotherm depends on the value of the n parameter, Figure 3 presents the possible shapes of the Freundlich isotherm in correspondence to the n value, where values of n smaller than one represent non-favorable sorption, this could be due to experimental conditions hindering the sorption process, such as the solution pH. Values of n bigger than one are considered as favorable sorption.

The Langmuir isotherm is founded on the equilibrium condition that the sorption and desorption rates remain constant, as well as assuming homogeneous sorption energy as well as non-lateral interaction among adsorbed molecules, the characteristic shape of a Langmuir isotherm is that of a plateau such as the one presented in Figure 4.

Figure 3. Possible shapes of the Freundlich isotherm

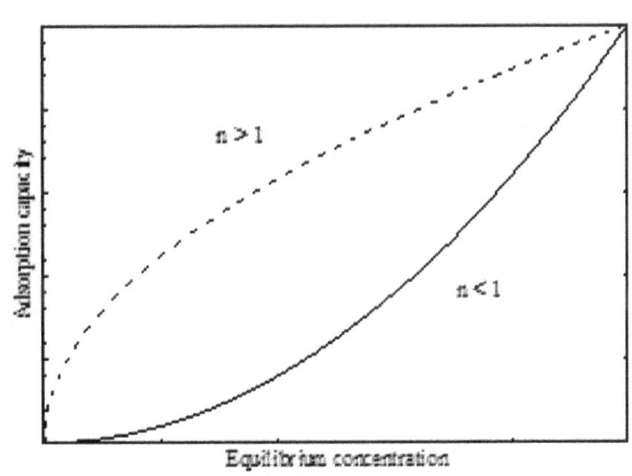

Figure 4. Typical shape of a Langmuir isotherm

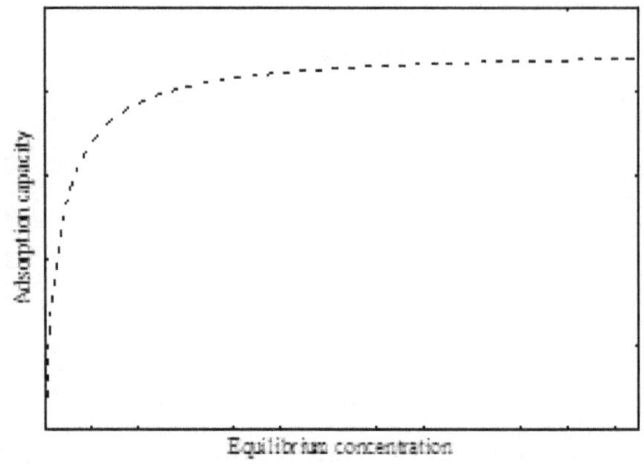

Sorption Kinetics Models

Similarly, to the equilibrium sorption stage, the interpretation of sorption kinetic data begins with a mass balance on a batch sorption container such as the one presented in Figure 2, however, the solute concentration is determined at specific time intervals rather than just the beginning and equilibrium states of the operation. The resulting mass balance is presented by equation (17).

$$q_t = \frac{V}{m}\left(C_0 - C_t\right) \tag{17}$$

where q_t and C_t are the sorption capacity and solute concentration at a given time t. As previously mentioned, sorption kinetics models can be categorized in two: algebraic (also known as reaction kinetic) and phenomenological models,

Algebraic models only consider one of the stages of the sorption process described in advance, namely the active site sorption stage, while considering the rest of the stages as non-significant. Many algebraic kinetic models have been used in literature for the description of sorption kinetics data, such as the Elovich model (Qiu et al., 2009; J. Wang & Guo, 2020), however, the most reported models in literature are the Pseudo First Order (PFO) model and the Pseudo Second Order (PSO) models:

$$q_t = q_e\left(1 - e^{-k_1 t}\right) \tag{18}$$

$$q_t = \frac{q_e^2 k_2 t}{1 + q_e k_2 t} \tag{19}$$

where:

q_t : Sorption capacity at a given time, [mg/g].
q_e : Sorption capacity at equilibrium, [mg/g]
k_1 : First order kinetic constant, [min^{-1}]
k_2 : Second order kinetic constant, [mg^{-1}min^{-1}]
t : Time, min.

The PFO and PSO models have been used in the description of a wide variety of sorption systems, however, they still present inconsistencies in the analysis derived from their analysis, for example, the PSO model was originally derived in the description of the sorption process of bivalent metal ions on peat (Ho & McKay, 1998), however it still has been used to successfully describe sorption kinetics data associated with the physical sorption of organic pollutants. Due to these theoretical inconsistencies, the description of the kinetic sorption system usually consists of comparing the fit of both models to experimental data as well as comparing kinetic constant values.

In contrast, diffusional modeling is based upon mass transport in the bulk of the fluid phase and the solid particle of the carbonaceous material and consists of performing mass balances on both the fluid and solid phase. Using diffusional models presents the advantage of allowing the deep knowledge of the mechanisms governing the sorption process in contrast to algebraic models that don't have their theoretical basis stablished enough. Additionally, diffusional models allow the obtention of parameters which are of vital importance in the design of dynamic sorption columns, which will be discussed further on.

The most used diffusional model is the Pore Volume Surface Diffusion Model (PVSDM), originally derived by Geankoplis and his Ph.D. Student, Roberto Leyva-Ramos (Leyva-ramos, 2007; Ocampo-Pérez et al., 2013), this model considers the following conditions:

- The carbonaceous material geometry is specific. This indicates that the coordinate system of the model will be adjusted according to the solid geometry.
- The carbonaceous material is porous.
- The process occurs through Fick diffusion.
- The active site sorption stage occurs instantaneously.

The result of performing mass balances on the solute bulk and a differential element of the carbonaceous material is described by equations (20, 22) alongside their initial and boundary conditions.

$$V\frac{dC_A}{dt} = -mSk_L\left(C_A - (C_{Ar})_{r=R}\right) \tag{20}$$

$$t = 0; C_A = C_{A0} \tag{21}$$

$$\varepsilon_p\frac{\partial C_{Ar}}{\partial t} + \rho_p\frac{\partial q}{\partial t} = \frac{1}{r^2}\frac{\partial}{\partial r}\left[r^2\left(\mathcal{D}_{ep}\frac{\partial C_{Ar}}{\partial r} + \mathcal{D}_S\rho_p\frac{\partial q}{\partial r}\right)\right] \tag{22}$$

$$t = 0; C_{Ar} = 0; 0 \leq r \leq R \tag{23}$$

$$\left(\frac{\partial C_{Ar}}{\partial r}\right)_{r=0} = 0 \tag{24}$$

$$\varepsilon_p \left(\frac{\partial C_{Ar}}{\partial t}\right)_{r=R} + \rho_p \left(\frac{\partial q}{\partial t}\right)_{r=R} = k_L \left(C_A - \left(C_{Ar}\right)_{r=R}\right) \tag{25}$$

$$q = f\left(C_{Ar}\right) \tag{26}$$

where:

ε_p = Void fraction of the carbonaceous material, dimensionless

ρ_p = Particle density, [g/cm^3]

D_{ep} = Pore volume diffusivity, [cm^2/s]

D_S = Surface diffusivity, [cm^2/s]

Equation (20) describes the bulk mass balance, where the film layer transport stage is represented by the film layer coefficient k_L. By contrast, equation (9) describes the mass balance on the differential element of the carbonaceous material, where two intraparticle diffusion mechanisms are introduced, pore volume diffusion and surface diffusion:

- Pore volume diffusion indicates that the solute will diffuse through the pores of the carbonaceous materials without ever contacting the pore walls and the pore volume diffusivity can be calculated as a function of the solute's diffusion coefficient as well as geometric factors of the carbonaceous material:

$$\mathcal{D}_{ep} = \frac{\mathcal{D}_{AB}\varepsilon_p}{\tau} \tag{27}$$

where $\mathcal{D}_{ep}$ is the solute's pore volume diffusion coefficient in cm^2/s, $\mathcal{D}_{AB}$ is the solute's molecular diffusivity in cm^2/s, ε_p and τ are the void fraction and tortuosity of the carbonaceous material, which can be determined through pycnometric characterization techniques.

- In the surface diffusion mechanism, the solute molecules will adhere to the surface of the pores inside the carbonaceous material and then diffuse along them.

Finally, the boundary condition described by the equation (25) indicates that all the solute molecules lost by the bulk during the film layer diffusion stage will transfer to the pores of the carbonaceous material through intraparticle diffusion.

One versatility of the PVSDM model is that the contribution of each mass transfer mechanism can be evaluated by considering that either one has null contribution ($\mathcal{D}_{ep}$ or $\mathcal{D}_S = 0$). However, to truly quantify the contributions of both mechanisms during a sorption kinetics experiment using carbonaceous materials, the following equations must be used:

$$N_{AS,total} = N_{AS} + N_{AP} = -\mathcal{D}_{ep}\frac{\partial C_{Ar}}{\partial r} - \mathcal{D}_S \rho_p \frac{\partial q}{\partial r} \tag{28}$$

$$\%CN_{As} = \frac{N_{As}}{N_{A,\ Total}} = \frac{\mathcal{D}_S \rho_p \frac{\partial q}{\partial r}}{\mathcal{D}_S \rho_p \frac{\partial q}{\partial r} + \mathcal{D}_{ep}\frac{\partial C_{Ar}}{\partial r}} \tag{29}$$

where N_{AS} and N_{AP} represent the flow densities of solute molecules through surface and pore volume diffusion respectively and $\%\ CN_{As}$ represents the percentage contribution of surface diffusivity towards the total intraparticle diffusion.

The literature for diffusional sorption studies of organic molecules on carbonaceous materials such as activated carbons has demonstrated a predominance of surface diffusivity over pore volume diffusion. Cavazos-Cuello et al reported the sorption of chlorpheniramine maleate on the surface of a wood based activated carbon, reporting surface diffusion contributions ranging from 96 to 56% when the equilibrium uptake of the active carbon increased, this is presented in Figure 5.

Ocampo-Pérez reported the sorption of ibuprofen, acetaminophen, and metronidazole, three widely used pharmaceutical compounds, on the surface of three activated carbons, revealing surface diffusivity contributions of 95% for all three compounds (Díaz-Blancas et al., 2018; Fröhlich et al., 2018; Ocampo-Perez et al., 2017).

However, surface diffusion is not always the predominant mechanism. Recently, Leyva-Ramos reported the sorption of heavy metal cations over the surface of activated carbon fibers. In this study, surface diffusivity could not be contemplated due to the heavy metals having average molecular sizes bigger than the average pore diameter of the activated carbons, resulting in what is known as hindered diffusion (Valdez-garcía & Leyva-ramos, 2023).

Figure 5. Contributions to surface diffusivity in the sorption of chlorpheniramine as reported (Cavazos-Cuello et al., 2023)

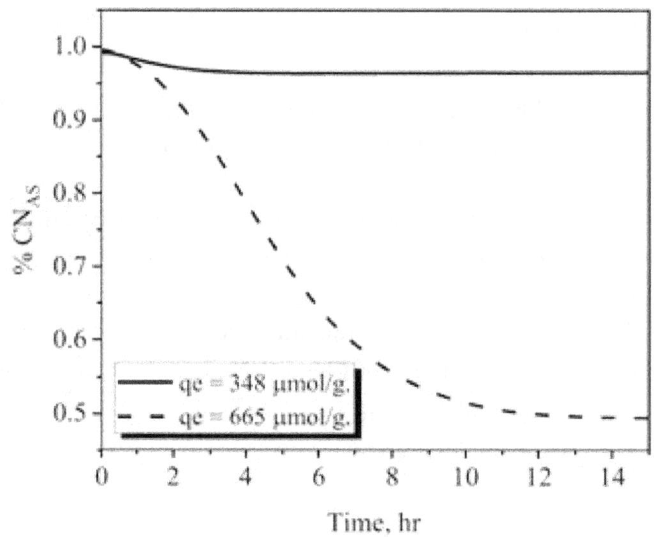

It is important to note that the PVSDM is used in the case of carbonaceous materials that are rigid in their structure, meaning that they don't allow species transfer along their solid matrix. However, in the case of materials that do allow transfer species through their matrix, which is the case for lignocellulosic biomasses, the PVSDM model cannot be applied, this is due to biomasses not containing an internal porous structure.

Historically speaking, mass transfer phenomena has not been reported in the description of sorption kinetic processes using lignocellulosic biomasses, however, recent research has revealed that there is a combination of diffusion in the canalized structure of the components of lignocellulosic biomasses as well as permeation through the polymeric matrix of the material. This has led to the derivation of a diffusional model for its application in sorption kinetics systems which use flexible biomasses, the permeation model. Derived originally by Leyva-Ramos and his Ph.D. Student Salazar-Rábago for the biosorption of dyes on wood sawdust, this model contemplates the non-porous nature of lignocellulosic biomass as well as the biosorbent geometry and makes similar considerations as the PVSDM (Leyva-Ramos et al., 2021).

$$V\frac{dC_A}{dt} = -mSk_L\left(C_A - (C_{Az})_{z=L}\right) \tag{30}$$

$$t = 0; C_A = C_{A0} \tag{31}$$

$$\frac{\partial q}{\partial t} = \frac{\partial}{\partial z}\left(P_M \frac{\partial q}{\partial z}\right) \tag{32}$$

$$t = 0; C_{Az} = 0; 0 \leq z \leq L \tag{33}$$

$$-P_M \rho_P \left(\frac{\partial q}{\partial z}\right)_{z=0} = 0 \tag{34}$$

$$P_M \rho_P \left(\frac{\partial q}{\partial z}\right)_{z=L} = k_L\left(C_A - \left(C_{Ar}\right)_{z=L}\right) \tag{35}$$

$$q = f\left(C_{Az}\right) \tag{36}$$

where P_M is the permeation coefficient in cm^2/s.

It is worth noting that both equations (13) and (23) describe the instantaneous sorption of the solute in the surface of the carbonaceous material, meaning that this stage is and can be modeled using the sorption isotherm parameters obtained in the previous stage of the study, emphasizing the pyramid structure of sorption studies. Moreover, the determination of the coordinate system to use in the resolution of both the PVSDM and the permeation model requires a thorough textural characterization of the carbonaceous material, making this a preliminary stage on the sorption pyramid.

Currently the only study where the permeation of a solute through a biosorbent is analyzed is the one published by Leyva-Ramos and his Ph.D. Student, due to the derivation of this model being very recent. In this study, permeation coefficients with values ranging between 0.76 and 1.89 × 10^{-9} cm^2/s, which is smaller than the molecular diffusivity of methylene blue (5.05 × 10^{-6} cm^2/s).

When solving diffusional models, one must verify the physical consistency of the parameters obtained, one way of verifying this is through the order of magnitude of the intraparticle diffusion and permeation coefficients ($\approx 10^{-8} - 10^{-10}$ cm^2/s) with relation to the molecular diffusivity of the solute ($\approx 10^{-6}$ cm^2/s). If the diffusion or permeation coefficients are greater that the solute's molecular diffusivity it would indicate that the solute diffuses faster in the internal structure of the carbonaceous materials than on the bulk of the fluid phase, which is physically possible.

Dynamic Sorption and Its Mathematical Models

Dynamic sorption presents a very different perspective from the previous study stages, both in operational and mathematical terms, Figure 6 represents a lab scale dynamic sorption system, consisting of a fixed bed and a peristaltic pump. In this system, the solute solution is fed through the bottom of the column at

Figure 6. Experimental continuous batch sorption system

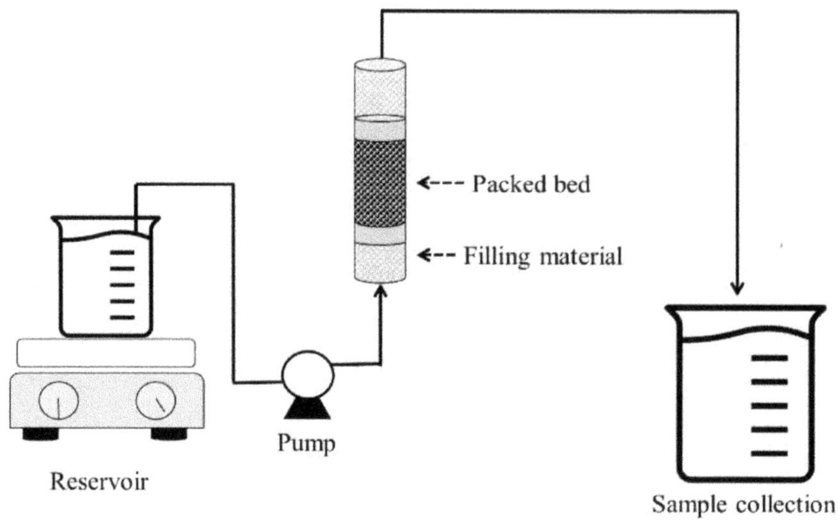

a given initial concentration to maximize contact time between the solute and the carbonaceous material and the outlet of the column is then free of solute.

The main component of the dynamic sorption process is the breakthrough curve. A breakthrough curve is a graphic representation of the outlet solute relative concentration as a function of time, this is shown in Figure 7. As the column outlet will remain free of solute often the objective of a dynamic sorption study is to maximize the time in which remotion of the solute is also maximized, as well as predicting the time it will take for the concentration to let solute molecules to pass through the fixed bed, this is known as the breakthrough time, and it is determined by operational or purity requirements of the bed's outlet.

The outlet solute concentration will then gradually increase until it reaches the inlet concentration, meaning that sorption is not occurring anymore, indicating the saturation of the carbonaceous material, this point is thereby called the saturation point. In dynamic sorption studies, the sorption capacity is calculated as follows (Worch, 2012):

$$q = \frac{QC_{A0}}{m} \int_{0}^{t} \left(1 - \frac{C_A}{C_{A0}} \right) \tag{37}$$

where Q is the volumetric flow rate, usually in cm^3/min. The dynamic sorption capacity can be calculated at both the breakthrough and saturation points simply by changing the upper integration limit to either the breakthrough time t_b or the saturation time t_s.

In a similar form to batch sorption studies, the mathematical modeling of the dynamic sorption process consists of algebraic and phenomenological models. Algebraic models are derived from mass balances performed in the volume of the packed bed and carbonaceous material similarly to the diffusional modeling previously discussed, however, these models are simplified through specific considerations to obtain algebraic models.

Figure 7. A typical breakthrough curve

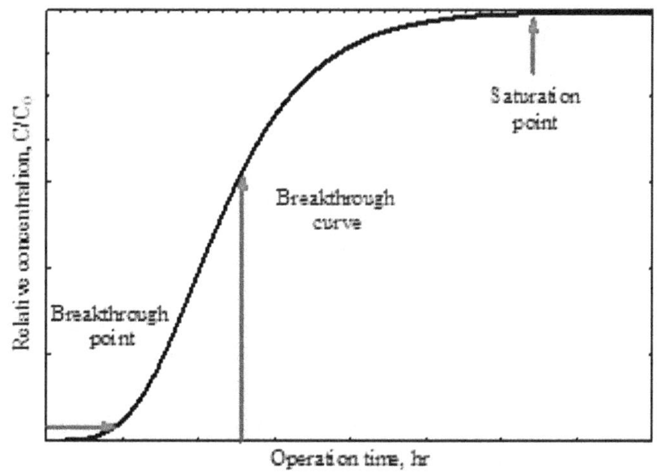

The most reported breakthrough models in literature are the Thomas, Yoon-Nelson (YN) model and, however, there are many more reported (K. H. Chu, 2020; Omitola et al., 2022; Xu et al., 2013). The Thomas model is derived from the assumption that the sorption equilibrium between the solute and the carbonaceous material is described by a monolayer using the Langmuir isotherm and that the sorption kinetics are described by the PSO model, ignoring the film layer and intraparticle transport stages of the sorption process and its mathematical form is as follows (Ashton et al., 2004):

$$\ln\left(\frac{C_0}{C_t}-1\right)=\frac{k_{Th}q_0w}{Q}-k_{Th}C_0t \tag{38}$$

where:

k_{Th} = Thomas model constant [mL/min×mg]

q_0 = Sorption capacity, [mg/g]

w = Mass of the carbonaceous material, [g]

Although the Thomas model has been widely used in the description of dynamic water purification systems using carbonaceous materials (Abdolali et al., 2017; Ahmad & Hameed, 2010; Lakshmipathy & Sarada, 2015), it also presents inconsistencies in its theoretical derivation, mainly due to the equilibrium and kinetic sorption models used for the derivation, since the Langmuir isotherm is commonly associated with physical sorption whereas the PSO model has been erroneously associated with chemical sorption (Ho & McKay, 1998).

Figure 8. Plug flow and axially dispersed flow

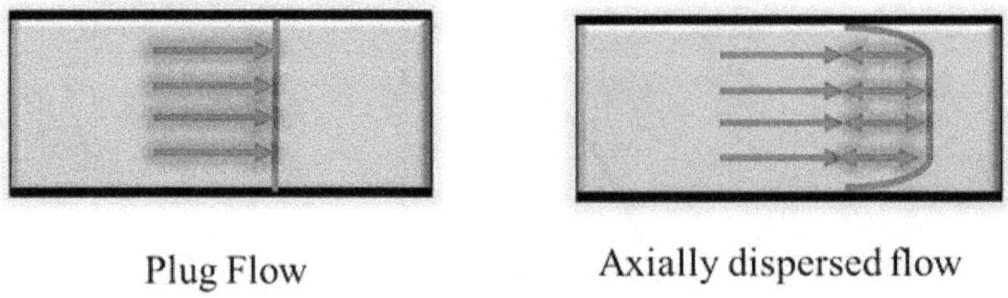

On the other hand, the YN model is a probabilistic model originally derived for the description of gas sorption in medical applications and has also been widely use in the description of many sorption systems (Abdolali et al., 2017; K. H. Chu, 2020; Gupta & Garg, 2019), this model presents the advantage of not requiring any information regarding the characterization of the carbonaceous material as well as not requiring a specific set of equilibrium and kinetic sorption models as a basis. It additionally allows the knowledge of the exact time where 50% of the fixed bed has been saturated, the so-called stochiometric time.

$$\ln\left(\frac{C_t}{C_0 - C_t}\right) = k_{YN}t - \tau k_{YN} \tag{39}$$

where:

k_{YN} = Yoon-Nelson's model velocity constant, [cm/s]

τ = Stochiometric time, [min]

t = Time, [min]

Diffusional dynamic sorption is commonly modeled through the axial dispersion model, which is similarly obtained through performing mass balances in the bulk of the solution flowing through the packed bed and it is described by equation (40) alongside its initial and boundary conditions. Axial dispersion is a material transfer phenomenon which occurs in tubular systems which involves the dispersion of the solute molecules alongside the axial flow direction because of laminar flow regimes, this is graphically presented in Figure 8, where a comparison between plug flow and axially dispersed flow is shown.

$$\frac{\partial C_A}{\partial t} = -\nu_z \frac{\partial C_A}{\partial z} + \mathcal{D}_z \frac{\partial^2 C_A}{\partial z^2} - k_L \frac{S}{\varepsilon_b}\left(C_A - C_{Ar}\Big|_{r=R}\right) \tag{40}$$

$$t = 0; z \geq 0; C_{Ar} = 0; C_A = 0. \tag{41}$$

$$t > 0; z = 0; C_A = C_{A0} \tag{42}$$

$$t > 0; z = L_b; \frac{\partial C_A}{\partial z} = 0 \tag{43}$$

where v_z is the surface velocity of the solute, L_b is the packed bed length in cm, ε_b is the packed bed void volume and $\mathcal{D}_z$ is the axial dispersion coefficient in cm^2/s. The model previously described must be solved alongside the intraparticle balance described by the equation (22) alongside its initial and boundary conditions, transforming it into the Axial Dispersion Model – Pore Volume Surface Diffusion Model (ADM-PVSDM).

The use of the ADM-PVSDM in the description of dynamic sorption systems allows the quantification of the axial dispersion present in the sorption column through the fitting of the dynamic sorption data to the model. Axial dispersion is a function of the volumetric flow rate of the solute solution. Figure 9 presents the effect of increasing the volumetric flow rate (therefore, the axial dispersion) on a typical breakthrough curve, where a significant decrease in breakthrough time is observed.

Increasing the volumetric flow rate of the solute through the packed bed decreases the contact time between the carbonaceous material and the solute molecules, allowing said molecules to literally break-through the fixed bed.

Similarly, the intraparticle diffusion mechanism also impacts the breakthrough time, as previously mentioned, surface diffusion is commonly found in the description of sorption kinetics systems involving carbonaceous materials and organic pollutants, so it will be used to exemplify its effect on the breakthrough time, which is shown in Figure 10. Decreasing the surface diffusion coefficient also decreases the breakthrough time of dynamic sorption systems. Surface diffusivity is itself a function of the equilibrium sorption capacity, which is in term a function of the inlet concentration on the fixed bed.

Figure 9. Effect of the volumetric flow rate on the breakthrough time

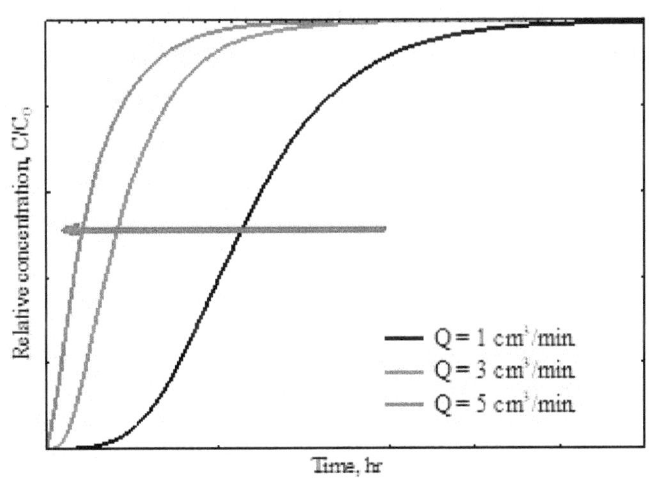

Figure 10. Effect of the surface diffusivity on the breakthrough time

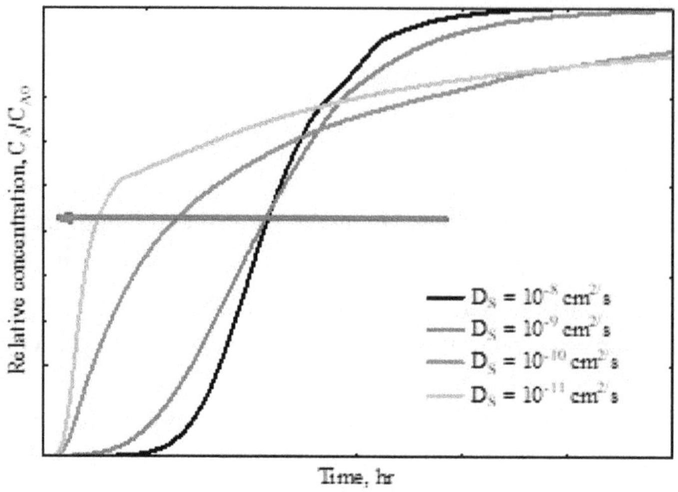

The use of the axial dispersion model is not limited to fitting experimental sorption data to obtain the axial dispersion, one of the most interesting applications of this model resides in the prediction of the breakthrough behavior of sorption systems, this is an application highlighted in a recent study performed by Aguirre- Contreras et al (Aguirre-Contreras et al., 2023), who studied the remotion of 8-Hydroxy-quinoleine (8HQ) using Norit commercial activated carbon. In this study, parameters such as the film layer coefficient and the axial dispersion coefficient were calculated through correlations to predict the breakthrough behavior of 8HQ (Worch, 2012; Xu et al., 2013), obtaining accurate prediction like those of the algebraic breakthrough models.

CONCLUSION

Overall, the use of phenomenological modeling in the description of water purification systems using carbonaceous materials presents a clear advantage over traditional algebraic modeling, due to them allowing the determination of the transport mechanisms involved in the process as well as allowing the design and prediction of the behavior presented in dynamic sorption systems, whereas algebraic models only allow fitting of experimental sorption data to determine an sorption kinetic rate or the stochiometric time of a given dynamic sorption process.

REFERENCES

Abate, G. Y., Alene, A. N., Habte, A. T., & Getahun, D. M. (2020). Adsorptive removal of malachite green dye from aqueous solution onto activated carbon of Catha edulis stem as a low cost bio-adsorbent. *Environmental Systems Research*, *9*(1), 29. Advance online publication. doi:10.1186/s40068-020-00191-4

Abdolali, A., Ngo, H. H., Guo, W., Zhou, J. L., Zhang, J., Liang, S., Chang, S. W., Nguyen, D. D., & Liu, Y. (2017). Application of a breakthrough biosorbent for removing heavy metals from synthetic and real wastewaters in a lab-scale continuous fixed-bed column. *Bioresource Technology*, *229*, 78–87. doi:10.1016/j.biortech.2017.01.016 PMID:28110128

Aguirre-Contreras, S., Leyva-Ramos, R., Ocampo-Pérez, R., Aguilar-Madera, C. G., Flores-Cano, J. V., & Medellín-Castillo, N. A. (2023). Mathematical modeling of breakthrough curves for 8-hydroxyquinoline removal from fundamental equilibrium and adsorption rate studies. *Journal of Water Process Engineering*, *54*(April), 103967. doi:10.1016/j.jwpe.2023.103967

Ahmad, A. A., & Hameed, B. H. (2010). Fixed-bed adsorption of reactive azo dye onto granular activated carbon prepared from waste. *Journal of Hazardous Materials*, *175*(1–3), 298–303. doi:10.1016/j.jhazmat.2009.10.003 PMID:19883979

Al-Ghouti, M. A., & Da'ana, D. A. (2020). Guidelines for the use and interpretation of adsorption isotherm models: A review. *Journal of Hazardous Materials*, *393*(February), 122383. doi:10.1016/j.jhazmat.2020.122383 PMID:32369889

Ashton, D., Hilton, M., & Thomas, K. V. (2004). Investigating the environmental transport of human pharmaceuticals to streams in the United Kingdom. *The Science of the Total Environment*, *333*(1–3), 167–184. doi:10.1016/j.scitotenv.2004.04.062 PMID:15364527

ASTM. (2006). Standard Test Method for Determination of Iodine Number of Activated Carbon 1. *ASTM International, 94*(Reapproved), 1–5. http://compass.astm.org.acces.bibl.ulaval.ca/download/D4607.6656.pdf

Boundzanga, H. M., Cagnon, B., Roulet, M., de Persis, S., Vautrin-Ul, C., & Bonnamy, S. (2022). Contributions of hemicellulose, cellulose, and lignin to the mass and the porous characteristics of activated carbons produced from biomass residues by phosphoric acid activation. *Biomass Conversion and Biorefinery*, *12*(8), 3081–3096. doi:10.1007/s13399-020-00816-9

Calvert, J. G. (1990). Glossary of atmospheric chemistry terms. *Pure and Applied Chemistry*, *62*(11), 2167–2219. doi:10.1351/pac199062112167

Cavazos-Cuello, L. A., Ocampo-Pérez, R., Tututi-Ávila, S., Dávila-Guzmán, N. E., Soto-Regalado, E., & Salazar-Rábago, J. J. (2023). Comparative phenomenological modeling of the overall adsorption rate of drugs onto activated carbon. The case of the chlorpheniramine. *Chemical Engineering Research & Design*, *199*, 152–161. doi:10.1016/j.cherd.2023.09.043

Chu, G., Zhao, J., Huang, Y., Zhou, D., Liu, Y., Wu, M., Peng, H., Zhao, Q., Pan, B., & Steinberg, C. E. W. (2018). Phosphoric acid pretreatment enhances the specific surface areas of biochars by generation of micropores. *Environmental Pollution*, *240*, 1–9. doi:10.1016/j.envpol.2018.04.003 PMID:29729563

Chu, K. H. (2020). Breakthrough curve analysis by simplistic models of fixed bed adsorption: In defense of the century-old Bohart-Adams model. *Chemical Engineering Journal, 380*(June), 122513. doi:10.1016/j.cej.2019.122513

Dhyani, V., & Bhaskar, T. (2019). Pyrolysis of biomass. *Biomass, Biofuels, Biochemicals: Biofuels: Alternative Feedstocks and Conversion Processes for the Production of Liquid and Gaseous Biofuels,* 217–244. doi:10.1016/B978-0-12-816856-1.00009-9

Díaz-Blancas, V., Ocampo-Pérez, R., Leyva-Ramos, R., Alonso-Dávila, P. A., & Moral-Rodríguez, A. I. (2018). 3D modeling of the overall adsorption rate of metronidazole on granular activated carbon at low and high concentrations in aqueous solution. *Chemical Engineering Journal, 349*(May), 82–91. doi:10.1016/j.cej.2018.05.076

Dong, Y. D., Zhang, H., Zhong, G. J., Yao, G., & Lai, B. (2021). Cellulose/carbon Composites and their Applications in Water Treatment – a Review. *Chemical Engineering Journal, 405*, 126980. doi:10.1016/j.cej.2020.126980

Eduardo, S. R., Dávila-Guzmán, N. E., & Margarita, L. C. (2022). Agro-wastes as precursors of biochar, a cleaner adsorbent to remove pollutants from aqueous solutions. In *Biomass-Derived Materials for Environmental Applications* (pp. 349–376). Elsevier. doi:10.1016/B978-0-323-91914-2.00001-5

Ferrier, D. C., & Honeychurch, K. C. (2021). Carbon nanotube (CNT)-based biosensors. *Biosensors (Basel), 11*(12), 1–33. doi:10.3390/bios11120486 PMID:34940243

Foo, K. Y., & Hameed, B. H. (2010). Insights into the modeling of adsorption isotherm systems. *Chemical Engineering Journal, 156*(1), 2–10. doi:10.1016/j.cej.2009.09.013

Fröhlich, A. C., Ocampo-Pérez, R., Diaz-Blancas, V., Salau, N. P. G., & Dotto, G. L. (2018). Three-dimensional mass transfer modeling of ibuprofen adsorption on activated carbon prepared by sonication. *Chemical Engineering Journal, 341*(January), 65–74. doi:10.1016/j.cej.2018.02.020

Gan, G., Li, X., Fan, S., Wang, L., Qin, M., Yin, Z., & Chen, G. (2019). Carbon Aerogels for Environmental Clean-Up. *European Journal of Inorganic Chemistry, 2019*(27), 3126–3141. doi:10.1002/ejic.201801512

Gupta, A., & Garg, A. (2019). Adsorption and oxidation of ciprofloxacin in a fixed bed column using activated sludge derived activated carbon. *Journal of Environmental Management, 250*(August), 109474. doi:10.1016/j.jenvman.2019.109474 PMID:31505384

Hao, F., Zhao, X., Ouyang, W., Lin, C., Chen, S., Shan, Y., & Lai, X. (2013). Molecular structure of corncob-derived Biochars and the mechanism of Atrazine sorption. *Agronomy Journal, 105*(3), 773–782. doi:10.2134/agronj2012.0311

Ho, Y. S., & McKay, G. (1998). Sorption of dye from aqueous solution by peat. *Chemical Engineering Journal, 70*(2), 115–124. doi:10.1016/S0923-0467(98)00076-1

Hwang, S. I., Franconi, N. G., Rothfuss, M. A., Bocan, K. N., Bian, L., White, D. L., Burkert, S. C., Euler, R. W., Sopher, B. J., Vinay, M. L., Sejdic, E., & Star, A. (2019). Tetrahydrocannabinol Detection Using Semiconductor-Enriched Single-Walled Carbon Nanotube Chemiresistors. *ACS Sensors, 4*(8), 2084–2093. doi:10.1021/acssensors.9b00762 PMID:31321969

Jeon, I., Xiang, R., Shawky, A., Matsuo, Y., & Maruyama, S. (2019). Single-Walled Carbon Nanotubes in Emerging Solar Cells: Synthesis and Electrode Applications. *Advanced Energy Materials*, *9*(23), 1–27. doi:10.1002/aenm.201801312

Jha, R., Singh, A., Sharma, P. K., & Fuloria, N. K. (2020). Smart carbon nanotubes for drug delivery system: A comprehensive study. *Journal of Drug Delivery Science and Technology*, *58*(May), 101811. doi:10.1016/j.jddst.2020.101811

Katiyar, R., Patel, A. K., Nguyen, T.-B., Singhania, R. R., Chen, C.-W., & Dong, C.-D. (2021). Adsorption of Copper (II) in aqueous solution using biochars derived from Ascophyllum nodosum seaweed. *Bioresource Technology*, *328*(January), 124829. doi:10.1016/j.biortech.2021.124829 PMID:33618185

Lakshmipathy, R., & Sarada, N. C. (2015). A fixed bed column study for the removal of Pb2+ ions by watermelon rind. *Environmental Science. Water Research & Technology*, *1*(2), 244–250. doi:10.1039/C4EW00027G

Lee, J. H., & Park, S. J. (2020). Recent advances in preparations and applications of carbon aerogels: A review. *Carbon*, *163*, 1–18. doi:10.1016/j.carbon.2020.02.073

Leyva-ramos, R. (2007). Importancia y aplicaciones de la adsorción en fase líquida. Sólidos Porosos, Preparación. *Caracterización y Aplicaciones.*, *1*, 160–170.

Leyva-Ramos, R., Salazar-Rábago, J. J., & Ocampo-Pérez, R. (2021). A novel intraparticle mass transfer model for the biosorption rate of methylene blue on white pine (Pinus durangensis) sawdust. Diffusion-permeation. *Chemical Engineering Research & Design*, *172*, 43–52. doi:10.1016/j.cherd.2021.05.029

Li, J., Li, Q., Qian, C., Wang, X., Lan, Y., Wang, B., & Yin, W. (2019). Volatile organic compounds analysis and characterization on activated biochar prepared from rice husk. *International Journal of Environmental Science and Technology*, *16*(12), 7653–7662. doi:10.1007/s13762-019-02219-4

Liu, H., Wang, L., & Lei, M. (2019). Positive impact of biochar amendment on thermal balance during swine manure composting at relatively low ambient temperature. *Bioresource Technology*, *273*, 25–33. doi:10.1016/j.biortech.2018.10.033 PMID:30399607

Liu, X., Zhang, Y., Li, Z., Feng, R., & Zhang, Y. (2014). Characterization of corncob-derived biochar and pyrolysis kinetics in comparison with corn stalk and sawdust. *Bioresource Technology*, *170*, 76–82. doi:10.1016/j.biortech.2014.07.077 PMID:25125195

Loredo-Cancino, M., Soto-Regalado, E., Cerino-Córdova, F. J., García-Reyes, R. B., García-León, A. M., & Garza-González, M. T. (2013). Determining optimal conditions to produce activated carbon from barley husks using single or dual optimization. *Journal of Environmental Management*, *125*, 117–125. doi:10.1016/j.jenvman.2013.03.028 PMID:23651918

Loredo-Cancino, M., Soto-Regalado, E., García-Reyes, R. B., Cerino-Córdova, F. de J., Garza-González, M. T., Alcalá-Rodríguez, M. M., & Dávila-Guzmán, N. E. (2016). Adsorption and desorption of phenol onto barley husk-activated carbon in an airlift reactor. *Desalination and Water Treatment*, *57*(2), 845–860. doi:10.1080/19443994.2014.970579

Loya-González, D., Loredo-Cancino, M., Soto-Regalado, E., Rivas-García, P., Cerino-Córdova, F. de J., García-Reyes, R. B., Bustos-Martínez, D., & Estrada-Baltazar, A. (2019). Optimal activated carbon production from corn pericarp: A life cycle assessment approach. *Journal of Cleaner Production, 219*, 316–325. doi:10.1016/j.jclepro.2019.02.068

Menya, E., Olupot, P. W., Storz, H., Lubwama, M., Kiros, Y., & John, M. J. (2020). Optimization of pyrolysis conditions for char production from rice husks and its characterization as a precursor for production of activated carbon. *Biomass Conversion and Biorefinery*, *10*(1), 57–72. doi:10.1007/s13399-019-00399-0

Ocampo-Perez, R., Aguilar-Madera, C. G., & Díaz-Blancas, V. (2017). 3D modeling of overall adsorption rate of acetaminophen on activated carbon pellets. *Chemical Engineering Journal*, *321*, 510–520. doi:10.1016/j.cej.2017.03.137

Ocampo-Pérez, R., Leyva-Ramos, R., Sanchez-Polo, M., & Rivera-Utrilla, J. (2013). Role of pore volume and surface diffusion in the adsorption of aromatic compounds on activated carbon. *Adsorption*, *19*(5), 945–957. doi:10.1007/s10450-013-9502-y

Omitola, O. B., Abonyi, M. N., Akpomie, K. G., & Dawodu, F. A. (2022). Adams-Bohart, Yoon-Nelson, and Thomas modeling of the fix-bed continuous column adsorption of amoxicillin onto silver nanoparticle-maize leaf composite. *Applied Water Science*, *12*(5), 1–9. doi:10.1007/s13201-022-01624-4

Park, J. H., Ok, Y. S., Kim, S. H., Cho, J. S., Heo, J. S., Delaune, R. D., & Seo, D. C. (2016). Competitive adsorption of heavy metals onto sesame straw biochar in aqueous solutions. *Chemosphere*, *142*, 77–83. doi:10.1016/j.chemosphere.2015.05.093 PMID:26082184

Ptaszkowska-Koniarz, M., Goscianska, J., & Pietrzak, R. (2018a). Removal of rhodamine B from water by modified carbon xerogels. *Colloids and Surfaces. A, Physicochemical and Engineering Aspects*, *543*(January), 109–117. doi:10.1016/j.colsurfa.2018.01.057

Ptaszkowska-Koniarz, M., Goscianska, J., & Pietrzak, R. (2018b). Synthesis of carbon xerogels modified with amine groups and copper for efficient adsorption of caffeine. *Chemical Engineering Journal*, *345*(March), 13–21. doi:10.1016/j.cej.2018.03.132

Qiu, H., Lv, L., Pan, B. C., Zhang, Q. J., Zhang, W. M., & Zhang, Q. X. (2009). Critical review in adsorption kinetic models. *Journal of Zhejiang University. Science A*, *10*(5), 716–724. doi:10.1631/jzus.A0820524

Sadh, P. K., Duhan, S., & Duhan, J. S. (2018). Agro-industrial wastes and their utilization using solid state fermentation: a review. In Bioresources and Bioprocessing (Vol. 5, Issue 1). Springer Science and Business Media Deutschland GmbH. doi:10.1186/s40643-017-0187-z

Sanchez-Silva, J. M., Collins-Martínez, V. H., Padilla-Ortega, E., Aguilar-Aguilar, A., Labrada-Delgado, G. J., Gonzalez-Ortega, O., Palestino-Escobedo, G., & Ocampo-Pérez, R. (2022). Characterization and transformation of nanche stone (Byrsonima crassifolia) in an activated hydrochar with high adsorption capacity towards metformin in aqueous solution. *Chemical Engineering Research & Design*, *183*, 580–594. doi:10.1016/j.cherd.2022.05.054

Sepúlveda-Cervantes, C. V., Soto-Regalado, E., Rivas-García, P., Loredo-Cancino, M., Cerino-Córdova, F. J., & García Reyes, R. B. (2018). Technical-environmental optimisation of the activated carbon production of an agroindustrial waste by means response surface and life cycle assessment. *Waste Management & Research, 36*(2), 121–130. doi:10.1177/0734242X17741680 PMID:29189111

Sizmur, T., Fresno, T., Akgül, G., Frost, H., & Moreno-Jiménez, E. (2017). Biochar modification to enhance sorption of inorganics from water. In *Bioresource Technology* (Vol. 246, pp. 34–47). Elsevier Ltd. doi:10.1016/j.biortech.2017.07.082

Tian, L., Li, Z., Wang, P., Zhai, X., Wang, X., & Li, T. (2021). Carbon quantum dots for advanced electrocatalysis. *Journal of Energy Chemistry, 55*, 279–294. doi:10.1016/j.jechem.2020.06.057

Tomczyk, A., Kondracki, B., & Szewczuk-Karpisz, K. (2023). Chemical modification of biochars as a method to improve its surface properties and efficiency in removing xenobiotics from aqueous media. In *Chemosphere* (Vol. 312). Elsevier Ltd. doi:10.1016/j.chemosphere.2022.137238

Turk Sekulic, M., Boskovic, N., Slavkovic, A., Garunovic, J., Kolakovic, S., & Pap, S. (2019). Surface functionalised adsorbent for emerging pharmaceutical removal: Adsorption performance and mechanisms. *Process Safety and Environmental Protection, 125*, 50–63. doi:10.1016/j.psep.2019.03.007

Valdez-garcía, G. D., & Leyva-ramos, R. (2023). Hindered diffusion of heavy metal cation in the adsorption rate on activated carbon fiber. *Chemical Engineering Research & Design, 196*, 556–567. Advance online publication. doi:10.1016/j.cherd.2023.07.005

Wang, J., & Guo, X. (2020). Adsorption kinetic models: Physical meanings, applications, and solving methods. In *Journal of Hazardous Materials* (Vol. 390). Elsevier B.V. doi:10.1016/j.jhazmat.2020.122156

Wang, T., Lü, L., & Shi, L. (2020). Surface functionalization and shape tuning of carbon fiber monofilament via direct microplasma scanning for ultramicroelectrode application. *Applied Surface Science, 531*(July), 147414. Advance online publication. doi:10.1016/j.apsusc.2020.147414

Worch, E. (2012). Adsorption Technology in Water Treatment. doi:10.1515/9783110240238

Xu, Z., Cai, J. G., & Pan, B. C. (2013). Mathematically modeling fixed-bed adsorption in aqueous systems. In Journal of Zhejiang University: Science A (Vol. 14, Issue 3, pp. 155–176). doi:10.1631/jzus.A1300029

You, Z., Zhao, L., Zhao, K., Liao, H., Wen, S., Xiao, Y., Cheng, B., & Lei, S. (2023). Highly tunable three-dimensional porous carbon produced from tea seed meal crop by-products for high performance supercapacitors. *Applied Surface Science, 607*(September), 155080. doi:10.1016/j.apsusc.2022.155080

Zubizarreta, L., Arenillas, A., Domínguez, A., Menéndez, J. A., & Pis, J. J. (2008). Development of microporous carbon xerogels by controlling synthesis conditions. *Journal of Non-Crystalline Solids, 354*(10–11), 817–825. doi:10.1016/j.jnoncrysol.2007.08.015

Chapter 13
Exploring X-Ray Techniques for Comprehensive Material Characterization and Analysis

Sangeeta Prasher
https://orcid.org/0000-0002-0699-8372
Kanya Maha Vidyalaya Jalandhar, India

Mukesh Kumar
Lovely Professional University, India

ABSTRACT

The use of X-rays for material analysis has revolutionized our understanding of various materials, providing invaluable insights. This chapter delves into X-ray applications, including XRD, XRF, XPS, AES, and SIMS. It begins with X-ray production and then details the principles, operations, and applications of these techniques, offering deep insights into material properties. XRD is vital for deciphering solid crystalline structures and atomic arrangements, revealing lattice spacing, symmetry, and phases. XRF excels in elemental analysis, offering both qualitative and quantitative results by generating distinctive fluorescent X-rays when X-rays interact with a sample. XPS focus on surface analysis, determining elemental composition and chemical states. SIMS enables high-sensitivity elemental and isotopic analysis of surfaces. These techniques find diverse applications. This chapter enhances readers' understanding of these techniques, helping them choose the right approach for material characterization, fostering scientific progress and technological innovation.

INTRODUCTION

X-rays were a breakthrough discovery in 1895 by the German physicist Wilhelm Conrad Röentgen, who accidently discovered them and given the name "X-rays" because their true nature remained a mystery during their initial revelation. X-rays displayed remarkable properties, far surpassing those of visible light such as an ability to penetrate through human body, dense wood, and even thick metal objects

DOI: 10.4018/979-8-3693-1306-0.ch013

(Meadowcroft, 1896), which were conventionally considered "opaque". The fascinating aspect of X-rays was that knowing their underlying principles was not a prerequisite for utilizing these rays effectively. Almost immediately following their discovery, physicians eagerly embraced X-rays as a valuable tool for medical diagnosis without dissections and engineers recognized their potential in examining the internal structures of materials. The process involved positioning an X-ray source on one side of the object of interest and placing a photographic film on the opposite side a shadow or radiograph could be created. This method proved immensely valuable in locating fractures in bones and identifying cracks within metal castings. However, it wasn't until the year 1907 that the true nature of X-rays was finally elucidated (Edsall, 1907). During 1912, scientists discovered the phenomenon of X-ray diffraction by crystals. This revelation introduced a revolutionary method to unveil internal structures of matter with astonishing precision, reaching down to sizes as minuscule as 10^{-10} meters. It is this profound phenomenon, and its applications in solving materials-related challenges, that forms the core focus of this chapter. X-ray-based material characterization techniques play a vital role in various fields, including materials science for developing customized materials, geology and archaeology for uncovering Earth's history, the semiconductor industry for quality control, and pharmaceuticals for enhancing drug formulations through the analysis of crystalline structures(Kalam et al., 1978).

The chapter is structured to provide a comprehensive understanding of the utilization of X-rays in material characterization offering insights into the generation of X-rays through various sources and methods. Subsequently, the chapter delves into specific X-ray techniques, such as X-ray diffraction (XRD), X-ray fluorescence spectroscopy (XRF), X-ray photoelectron spectroscopy (XPS), Auger electron spectroscopy (AES) and Secondary Ion Mass Spectrometry (SIMS). These techniques are reviewed in detail, elucidating their principles, instrumentation, and applications. In essence, this chapter serves as a comprehensive guide to empower readers with the knowledge needed to select the most suitable approach for characterizing materials of interest, ultimately fostering scientific advancement and technological innovation.

X-RAYS GENERATION

X-rays are produced when charged particles possessing sufficient kinetic energy experience rapid deceleration, but not all electrons lose speed at the same rate. Some cease abruptly during a single impact, releasing all their energy at once, while others undergo multiple deflections by the target's atoms, progressively losing fractions of their total kinetic energy until it is entirely dissipated (Seibert, 1997). Hence, at the point of impact, X-rays comprising a mixture of different wavelengths are generated and disperse in all directions. This phenomenon takes place within a specialized device known as an X-ray tube (Figure 1), housing a source of electrons, a filament, usually made of tungsten, which emits electrons when heated and an anode made of copper or tungsten, which enables the controlled emission of X-rays for a myriad of scientific and medical applications (Redler et al., 2018). Maintaining a high voltage, typically in the range of tens of thousands of volts, between these electrodes, leads to emission of a stream of electrons due to thermionic emission. The emitted electrons are focused into a narrow beam by the focusing cup and accelerated toward the anode and strike the target area on the anode, usually a small, specific spot known as the focal spot, imparting sufficient energy to some of the inner-shell electrons of the anode atoms, causing ejections of electrons and creating vacancies in the core orbits. To regain stability, electrons from outer shells of the atom make transitions into the lower energy levels, filling the

Figure 1. X-rays tube casing a source of electrons, a filament, and an anode

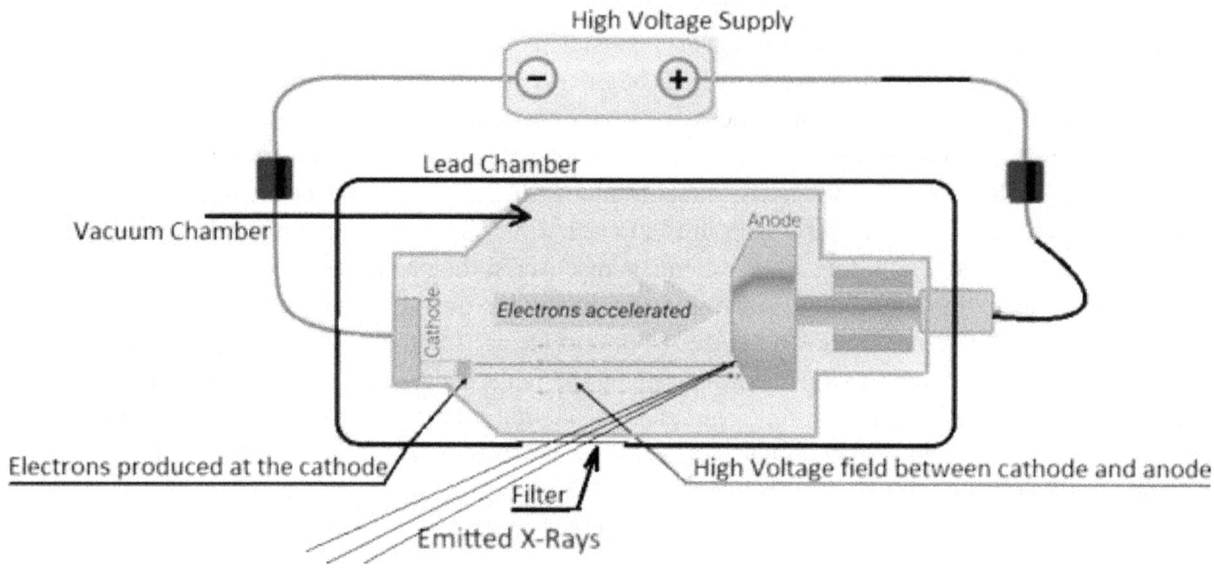

vacancies created by the ejection of inner-shell electrons releasing excess energy in the form of X-ray photons emitted in all directions. To focus the X-rays into a specific direction, a lead chamber with an aperture is used to guide the X-ray beam to the desired direction. These X-ray photons are characteristic of the anode material and have specific energy levels corresponding to the energy differences between electron shells. These factors are integral to both continuous and pulsed X-ray production methods, each of which exhibits unique characteristics and applications. Continuous X-ray production involves the continuous acceleration of electrons from the cathode towards the anode through a high-voltage potential difference, leading to a continuous stream of X-rays through the process of Bremsstrahlung radiation (Seibert & Boone, 2005). The kinetic energy lost by the electrons is converted into X-rays of various energies, creating a continuous spectrum of X-rays. This method is commonly used in medical X-ray machines and other applications where a continuous and stable X-ray beam is required for imaging bones, soft tissues, and organs. They are also used in industrial testing, airport security scanners, and material science applications. On the other hand, pulsed X-ray production involves generating X-rays in short bursts or pulses to generate high energy precisely timed pulsed electron beams. This method is used in applications that require precise control over X-ray exposure, such as computed tomography (CT) scanning and certain research experiments.

ROLE OF X-RAYS IN MATERIAL CHARACTERIZATION

After Following the elucidation of the fundamental principles and methodologies involved in the generation of X-rays, this discourse delves into the profound significance of X-rays within the realm of material characterization. However, prior to their deployment in the arena of material characterizations and other associated domains, the next section is structured to unveil the mechanisms governing the interactions between X-rays and matter.

Interaction of X-Rays With Matter

X-rays are high-energy electromagnetic waves that may get scattered or absorbed by the material leading to photoelectric effect, Compton or coherent scattering. The Photoelectric Effect, involves X-rays knocking electrons out of the material they strike. Whereas Compton Scattering involves the collision of X-rays with electrons in the material. During this interaction, X-rays lose some of their energy and change direction. Another important phenomenon is Coherent Scattering, where X-rays interact with the atoms of the material without significant energy loss. These interaction mechanisms are strategically harnessed for investigating and unveiling the material composition, atomic structure, crystallography, composition and lattice structure of the material (Janssens, 2004). Subsequent sections delve into a comprehensive exploration of various material characterization techniques, such as XRD, XRF, XPS, AES and SIMS, utilizing interactions of x-rays with materials, elucidating their fundamental principles and requisite instrumentation working and applications in materials structure or composition elucidation.

X-RAY DIFFRACTION (XRD)

X-ray diffraction (XRD) is a sophisticated analytical technique that allows the scientific community to probe the atomic and molecular structures of crystalline materials elucidating its crystallographic structure, lattice parameters, grain sizes at the atomic and molecular levels (Ali et al., 2022) and even the presence of defects or impurities (Snyder, 1999). Rooted in the principles of constructive interference of monochromatic x-rays scattered from the material atoms, XRD has proven essential across material characterization. X-rays strike a crystalline sample and scatter in various directions due to the periodic arrangement of atoms within the crystal lattice. The scattered x-rays interfere constructively or destructively depending on the angle of scattering and spacing between the atoms in the crystal and produce a distinctive intensity distribution known as the diffractogram. The angles and intensities of the intensity peaks provide crucial information about the material's crystal structure and spacing between lattice planes. The condition for constructive interference is given by Bragg's law, formulated by British physicists William Henry Bragg and his son William Lawrence Bragg in 1913 (Bragg & Bragg, 1913). It states that when X-rays encounter a crystalline sample at a particular angle θ, the waves reflected from different atomic planes reinforce each other, resulting in a constructive interference if:

$$n\lambda = 2d\sin\theta$$

In this equation, 'n' represents the order of diffraction and has an integer value consistent with sin θ not exceeding unity. 'λ' stands for the wavelength of the incident monochromatic X-rays, and 'd' symbolizes the interplanar separation within the crystal lattice. For a given combination of 'λ' and 'd,' multiple angles of incidence ('θ') can lead to diffraction, corresponding to different values of 'n'. For angles of incidence that do not adhere to Bragg's Law, there is no significant scattering because the scattered X-rays interfere destructively and cancel each other out in directions that deviate from the law. The spatial arrangements of atoms within the confines of a unit cell exert a profound influence on the amplitudes of diffracted beams, while maintaining the invariance of their directional attributes.

Figure 2. Labelled diagram of X-rays diffraction instrumentation

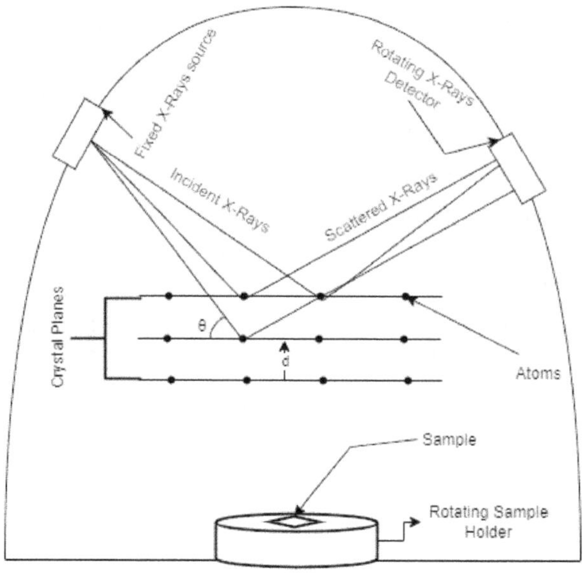

XRD Instrumentation

To carry out XRD experiments, specialized instruments called X-ray diffractometers are used. These instruments consist of several key components, including an X-ray source, a sample holder, a goniometer, and a detector (Figure 2). The X-rays emitted by the x-ray tube utilized in the setup are directed towards the sample. The sample holder securely positions the material, ensuring a consistent and precise orientation. The goniometer is a mechanical device that allows controlled rotation of the sample along multiple axes, making it possible to collect diffraction data from various crystallographic orientations. Finally, the detector records the intensity of diffracted X-rays at specific angles. Modern XRD detectors have significantly advanced the field, offering high sensitivity, speed, and precision. Some common types of detectors used in XRD include scintillation detectors, and area detectors (Chauhan, 2014). Scintillation detectors rely on scintillating materials that emit flashes of light when struck by X-rays. This light is then converted into electrical signals that could be directly read by a computer. The conversion primarily occurs through the absorption of photons by the detector material, leading to ionization and the generation of electron-ion pairs. The number of ionizations (N) for each absorbed photon is directly related to the photon's energy (E) and is influenced by the average energy (e) required to create an electron-ion pair in the detector material. Because scintillation detectors generally measure the scattered intensity at only one angle at a time, hence a collimation is required between the sample and the detector, often similar to that found between the source and the detector.

Two-dimensional area detectors gained prominence in X-ray diffraction experiments. These detectors encompass a range of technologies, all sharing a fundamental function resembling that of electronic film analogous to traditional photographic film. The area detectors are capable of capturing and recording the intensity distribution across an entire surface. However, instead of producing a tangible image, the resulting data is directly transmitted to the data acquisition computer in the form of an array of intensity values. Area detectors amalgamate many of the advantages associated with scintillation detectors and traditional film-based methods. They exhibit a high level of sensitivity, facilitating rapid data acquisition, while simultaneously allowing for the measurement of diffraction patterns at numerous angles. Consequently, the utilization of area detectors significantly escalates the volume of generated data.

X-ray diffraction operates in various modes to extract distinct information about the material being studied. The most common modes include Laue method, rotating crystal method and powder XRD. A brief overview of the three methods is discussed.

The Laue Method

The Laue method, named after Max von Laue, was the first diffraction technique employed in the history of crystallography. In the method, a broad spectrum of white radiation generated by an X-ray tube is directed at a stationary single crystal specimen fixed at Bragg angle, which corresponds to specific crystallographic planes within the crystal lattice (Amoros, 2012). Consequently, only specific wavelengths that satisfy Bragg's law for the given crystal parameters, such as interplanar spacing (d) and the incident angle (θ), undergo diffraction. Each resulting diffracted beam possesses a distinct wavelength. The Laue method can be categorized into two variations, each contingent on the relative positioning of the X-ray source, crystal specimen, and photographic film. In the transmission Laue method, the film is positioned behind the crystal, capturing the diffracted beams in the forward direction. The nomenclature "transmission" is derived from the partial transmission of diffracted beams through the crystal. On the contrary, the back-reflection Laue method involves positioning the film between the crystal and the x-ray source. The incident beam passes through an aperture in the film, and the resulting diffracted beams are recorded in a backward direction. However, for the production of noticeable diffraction pattern, the method relies on the order and periodicity of atoms within a large sized single crystal to ensure the incident X-ray beam encounters a consistent crystal structure.

Rotating-Crystal Method

The rotating-crystal method represents another essential approach in crystallography, wherein a single crystal specimen is oriented with one of its axes or a critical crystallographic direction perpendicular to a monochromatic X-ray beam. A cylindrical film surrounds the crystal, and the crystal is systematically rotated about the selected direction, with the film's axis aligning with the crystal's axis of rotation (Bunaciu et al., 2015). During this rotation, at specific instances, a particular set of lattice planes aligns precisely to meet the Bragg condition for diffraction of the incident monochromatic beam and results in diffracted beams. These diffracted beams are situated on imaginary cones, with their axes coinciding with the axis of rotation. The rotating-crystal method limited by its inability to cover all possible Bragg angles (ranging from 0° to 90°) for every set of crystallographic planes and only nearly planes perpendicular to the rotation axis will contribute to diffraction. Despite these limitations, the primary utility of the rotating-crystal method lies in the determination of unknown crystal structures.

Powder Method

The powder method represents a versatile approach to crystallography, particularly suitable for materials reduced to fine powders or composed of loose or consolidated microscopic grains. The specimen, in an appropriate holder, is exposed to a monochromatic x-ray beam. Within the powder, individual particles act as tiny crystallites or aggregates of smaller crystals, each randomly oriented concerning the incident beam (Swanson, 1955).

Remarkably, due to this random orientation, some crystallites will, by chance, align perfectly for specific reflections, allowing them to diffract the incident beam. Consequently, the powder method ensures that every set of lattice planes is capable of producing diffraction. One can envision the mass of powder as akin to a single crystal undergoing rotation, not about a single axis, but around all feasible axes. In summary, the powder method offers a robust means of examining crystalline materials, especially those existing in a powdered or granular state, by exploiting the random orientation of individual particles to achieve comprehensive diffraction information. This method has proved invaluable in the characterization and analysis of crystalline structures, contributing significantly to the advancement of crystallography.

Applications in Material Characterization

X-ray diffraction (XRD) has evolved into an indispensable tool, finding its application across diverse scientific and industrial domains, notably in materials science and chemistry. This technique's enduring relevance stems from ongoing advancements in instrumentation and analytical methodologies, perpetually enhancing our comprehension of the atomic realm and diverse materials. Key to XRD's efficacy are Bragg's Law and the Scherrer Formula (Scherrer, 1918), serving as foundational principles. Bragg's Law facilitates the determination of interplanar spacing (d-spacing) between crystallographic planes, while the Scherrer Formula, expressed as:

$$\beta\left(2\theta\right) = \frac{k\lambda}{L\cos\theta}$$

correlates diffraction peak broadening (β) to the size of the crystallites responsible for diffraction. In instances where organic or polymeric materials exhibit helical or spiral crystal structures, XRD can be instrumental in recognizing these phases, employing the relation:

$$Spiral\ orientations\ in\ a\ crystal = \frac{\beta}{2}\cos\theta$$

Quantitative phase analysis within XRD entails the assessment of relative phase proportions within a sample, a task often addressed by Rietveld refinement method introduced by Rietveld in 1969 (Rietveld, 1969). This technique extends its utility beyond lattice parameter determination, encompassing the occupancy of various atomic sites. Rietveld refinement excels in tracking structural variations within materials, including phase transitions and defects, under divergent conditions such as doping, irradiation, and temperature shifts.

Subsequent to the introduction of Rietveld's original algorithms for XRD profile generation, a diverse array of academic and commercial software has emerged and proliferated. These software platforms enable precise lattice parameter determinations, encompassing unit cell dimensions and angles, by iteratively optimizing these parameters to minimize disparities between observed and calculated diffraction

Figure 3. Mechanism of production of x-rays fluorescence

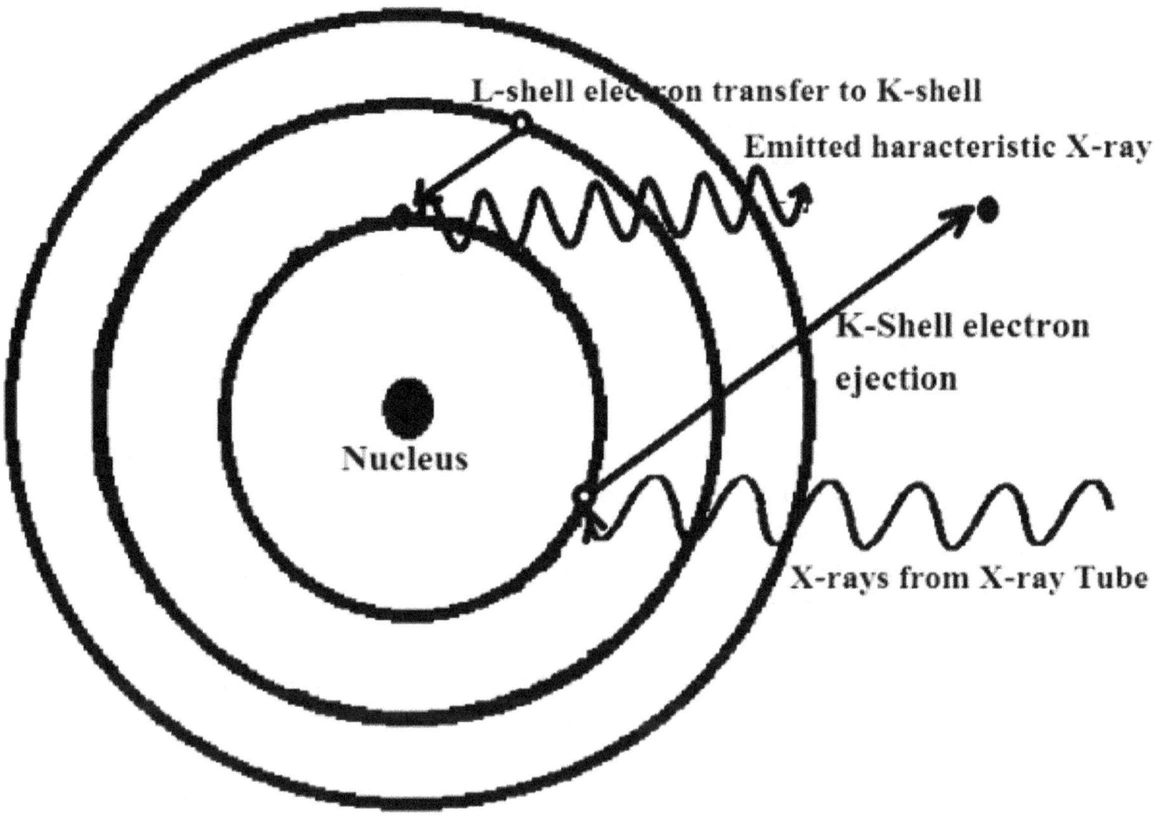

patterns. While XRD is a powerful technique, it is primarily limited to crystalline materials and may struggle with amorphous or disordered structures.

X-RAY FLUORESCENCE SPECTROSCOPY (XRF)

X-ray fluorescence (XRF) spectrometry is a non-destructive powerful and widely used analytic technique for elemental analysis in numerous scientific and industrial applications. It operates on the fundamental principle of exciting the atoms in a sample using a primary x-ray source. These x-rays knock out the inner-shell electrons from their stable orbits, creating temporary vacancies in these inner electron shells. The electrons from the upper shells move into vacancies created in the innermost shells. During this transition, excess energy is released in the form of characteristic fluorescent X-rays with discrete energy levels unique to each element (Nasrazadani & Hassani, 2016) as shown in figure 3. The intensity and energy of the emitted secondary x-rays can be detected with the help of a detector positioned at a specific angle to the sample to record in a spectrum. The energy spectrum of the detected X-rays is then meticulously analyzed to identify the elements present in the sample. The technique is used to analyze the surface layers of only about 100 µm thickness and is highly sensitive to detect elements present in the sample at concentrations as low as microgram-per-gram with high degree of precision and accuracy (Simon, 2012; Streli et al., 1999). It can analyze a wide range of materials, including solids, liquids, and powders, and can simultaneously analyze multiple elements across the periodic table having atomic number greater than 11. The intensity of the spectral line is proportional to the concentration of the element in the sample (Mantler, 1986). XRF can be performed using either energy-dispersive X-ray fluorescence (EDXRF) or wavelength-dispersive X-ray fluorescence (WDXRF) spectrometry based on the technology used for the separation (dispersion), identification and intensity of a sample's X-ray fluorescence spectrum (Sitko & Zawisz, 2012). In EDXRF, the sample is excited with a broad range of X-ray energies and the emitted X-rays have different energies corresponding to the elements from which they originate. The energy-dispersive detector is capable of separating the emitted X-rays based on the energies associated with them. The spectrum is generated by measuring the intensity of detected X-rays associated with various energies. Each peak in the spectrum corresponds to a specific element present in the sample. The technique offers faster analysis times and can analyze wide range of samples.

In WDXRF, a crystal monochromator disperses the emitted X-rays, separating them by their wavelengths, and the detector is positioned to detect only a specific wavelength of X-ray emission at a time. As the crystal is rotated, different wavelengths of X-rays are detected, and a spectrum is generated that represents the elemental composition of the sample. It offers better detection limits and can be used to analyze light elements and elements with low atomic numbers, making it ideal for high-precision applications. The crystal monochromator in WDXRF provides narrow spectral peaks, resulting in higher resolution and better differentiation between closely spaced X-ray lines. In EDXRF, the energy resolution is generally lower due to the broader spectral peaks obtained from the solid-state detector. The detector is typically a scintillation counter or a proportional counter in WDXRF. The main disadvantage of WDXRF is its relatively slow analysis time compared to EDXRF. WDXRF instruments typically require more complex and specialized components, such as a crystal monochromator and high-voltage power supplies for the X-ray tube, compared to EDXRF instruments, which use solid-state detectors. This can make WDXRF instruments more expensive and require more maintenance compared to EDXRF instruments.

X Ray Fluorescence Spectrometry Instrumentation

XRF instrumentation comprises several essential components, such as X-ray source, a sample holder, a detector and a data analysis system (Figure 4).

X-Ray Source

The XRF instruments can employ either an X-ray tube or a radioisotope as a source of X-rays to knockout the inner shell electrons of the sample, depending on the specific application and portability requirements. X-ray tubes are commonly used in laboratory-based XRF instruments. They generate intense and controllable X-ray beams, making them suitable for precise elemental analysis. In some field applications and handheld XRF devices, a radioisotope source such as cadmium-109 or americium-241 is employed. These sources emit X-rays that can interact with the sample, allowing for elemental analysis in remote or non-laboratory settings.

Figure 4. Block diagram of X-rays fluorescence spectroscopy

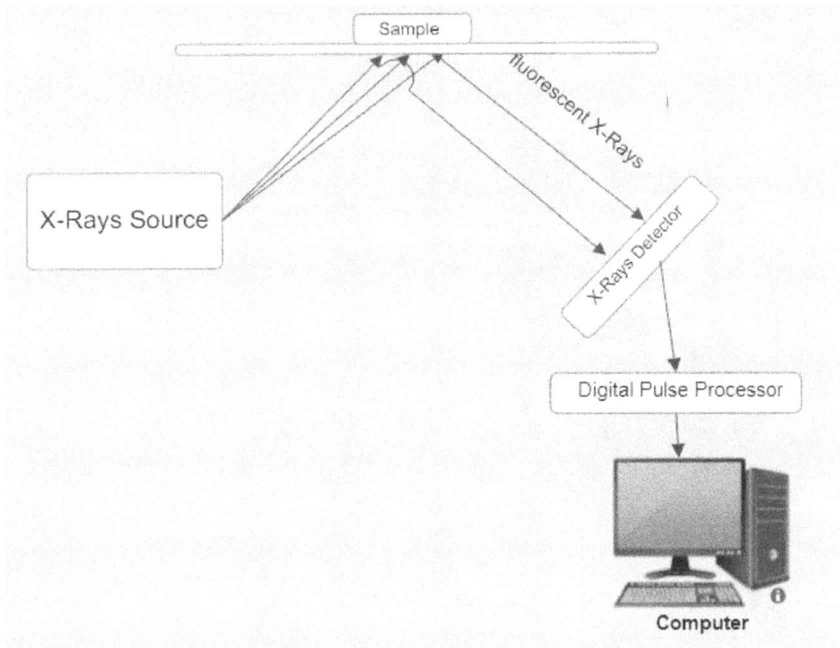

Sample Holder and Sample Chamber

The sample holder is designed to ensure uniform exposure of the sample surface carefully positioned in the path of the incident X-rays. Some advanced XRF instruments include automated sample changers for high-throughput analysis. The sample chamber is where the sample holder is placed. It usually has a window made of a material transparent to X-rays, such as beryllium or polypropylene, to allow the X-rays generated by the X-ray tube to pass through and interact with the sample.

Detectors

The XRF detectors works on principle of interaction between incident X-rays and the detector material. The characteristic X-rays emitted by the sample interact with the detector material, ionizing its atoms and creating electron-hole pairs. These electron-hole pairs then migrate toward the electrodes within the detector material. This migration generates an electrical signal that is directly proportional to the energy of the incident X-rays. The voltage signal produced during this process is subsequently converted into a digital signal, which is further transformed into a spectrum (Streli et al., 1999). Common types of detectors used in XRF instruments include solid-state detectors, such as silicon drift detectors (SDD) or germanium detectors, and gas-filled detectors, such as proportional counters or scintillation detectors. Silicon drift detectors and germanium detectors are solid-state detectors fabricated from high-purity silicon or germanium. An extensive contact area on one side faces the incoming X-rays, while a central anode contact on the opposite side is surrounded by concentric drift electrodes (Figure 5).

When a bias voltage is applied to the SDD detector chip, and it is exposed to X-rays, each incident X-ray is converted into an electron cloud with a charge proportional to its characteristic energy. These

Figure 5. Interface of silicon drift detectors

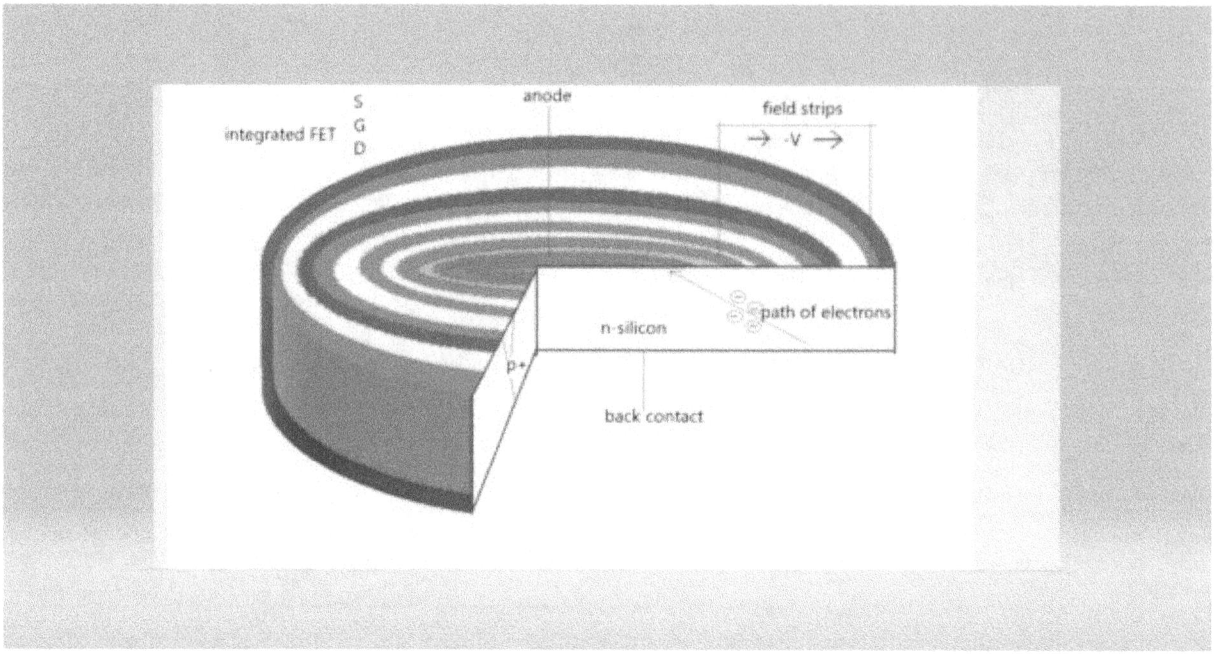

electrons are elevated into the conduction band of the silicon semiconductor, leaving behind holes that assume the role of free positive charges. Subsequently, the electrons 'drift' along a field gradient, maintained by the drift electrodes, and are eventually collected at the anode. This collection process transforms the emitted X-rays into electrical signals for processing and analysis (Strüder et al., 1998). These detectors are capable of detecting X-rays in the range of 5-30 keV and 2-100 keV, respectively and have good energy resolution of 0.1-0.2 keV. Silicon detectors are also compact and require minimal cooling. However, germanium detectors are more expensive than silicon detectors and require cooling with liquid nitrogen to achieve their best performance. One of the main advantages of silicon and germanium detectors is their high sensitivity to X-rays, i.e., they can distinguish between different X-ray energies with great accuracy. Gas-filled detectors such as proportional counters and scintillation detectors are alternative options in XRF instrumentation. Although less commonly used than solid-state detectors, they possess unique characteristics that make them suitable for specific applications.

Filters and Collimators

Filters and collimators are used to control the energy and direction of the X-rays emitted by the sample and detected by the detectors. Filters are used to selectively filter out certain X-ray energies, allowing only the desired X-ray energies to reach the detectors. Collimators are used to limit the direction of X-ray emission and ensure that only X-rays emitted from the sample at a specific angle are detected, reducing background noise and improving measurement accuracy.

Pulse Processor

A pulse processor is used to process the electrical signals generated by the X-ray detector. It typically includes an amplifier, a pulse height analyzer (PHA), and a multi-channel analyzer (MCA) to convert the incoming signals into digital data that can be analyzed.

Crystal Monochromator

The crystal monochromator is a key component in WDXRF instruments, and it is used to disperse the emitted X-rays based on their wavelength. The crystal monochromator selects specific X-ray wavelengths corresponding to the elements of interest and directs them to the X-ray detector. Common types of crystals used in WDXRF instruments include germanium (Ge), lithium fluoride (LiF), and quartz (SiO2), among others.

Qualitative and Quantitative Analysis

XRF spectra often contain background signals, such as Compton scattering and the Bremsstrahlung effect, which can interfere with peak identification and quantification. Data analysis software applies background corrections to isolate the true characteristic X-ray peaks. X-ray Fluorescence (XRF) spectroscopy can be used for both qualitative and quantitative analysis of materials, making it a versatile analytical technique. Qualitative XRF analysis is particularly valuable in scenarios where the composition of a sample is unknown, such as in material identification, archaeology, and forensic investigations. This is achieved by comparing the energy levels and intensities of the emitted fluorescent X-rays to known reference spectra for each element. The characteristic X-ray peaks serve as unique "fingerprints" that allow for accurate element identification. In addition to qualitative analysis, XRF is widely utilized for quantitative analysis, where the concentration of elements within a sample is precisely determined. This quantitative capability is achieved through a calibration process involving standard reference materials with well-known elemental compositions (Sitko, 2009). The standardized calibration curves establish the relationship between X-ray intensity and elemental concentration. To analyze an unknown sample, the XRF spectrum is compared to the calibration curves for the determination of the sample's elemental composition and concentration. The applications of XRF are vast and impactful, ranging from archaeology and geology to environmental science, material science, and beyond. XRF has proven its worth in understanding the composition of ancient artifacts, exploring mineral deposits, assessing environmental pollution, and ensuring the quality of materials in numerous industries.

X-RAY PHOTOELECTRON SPECTROSCOPY (XPS)

X-ray Photoelectron Spectroscopy (XPS), also known as Electron Spectroscopy for Chemical Analysis (ESCA), stands as a formidable analytical technique, yielding profound insights into the elemental composition, chemical states, and structural attributes of a material's surface (Baer & Engelhard, 2010). Its foundation rests on the photoelectric effect, a phenomenon initially illuminated by Hertz in 1887. Subsequently, K. Siegbahn and his research team at Uppsala University, Sweden, in the mid-1960s, extended this principle to surface analysis, which eventually earned Siegbahn the Nobel Prize in Physics

Figure 6. Photoemission of electrons from inner core of atom due to highly energetic X-rays

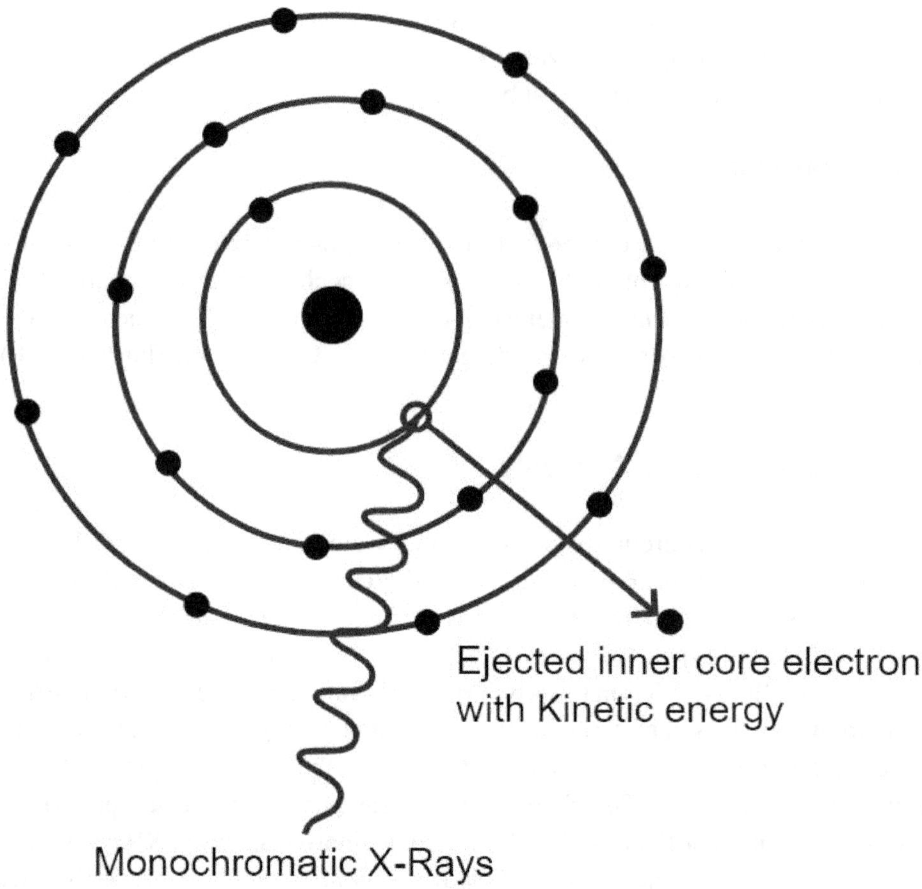

in 1981. Here a sample is exposed to a beam of monochromatic X-rays to excite the core level electrons of the elements. This makes it possible to detect not only several elements but also their oxidation states, as each electron has a specific binding energy. However, XPS probes to a depth of 10 nm only and the sample is analyzed at ultra-high vacuum of around 10^{-9} millibar as at a pressure of 10^{-6} mbar (Seah, 1984), background gas constituents (O2, H2O, etc.) can oxidize or reduce the sample molecules.

Principle

The key concept in XPS is the characteristic binding energy of the inner core electrons of elements. When an X-ray photon with sufficient energy strikes inner core electrons of an atom, the electrons get ejected with a kinetic energy equal to the difference between the energy of the incident X-ray photon and the binding energy of the electron in the atom (Figure 6). The ejected photoelectron's kinetic energy is measured using an energy analyzer and the XPS spectrum is observed between the intensity of ejected electrons and their kinetic energies (Benninghoven, 1973). The intensity of the XPS peaks is directly related to the concentration of the corresponding element, thus making it a valuable technique for identification and quantification of elemental composition of the sample's surface (except hydrogen and helium). The binding energy, E_b, of the core-level electron is calculated from the kinetic energy, E_k, of the ejected electrons using the simple relationship:

$$E_k = hv - E_b$$

XPS is particularly surface-sensitive because it mainly probes the top few nanometers of a material as within the bulk of the sample, the mean free path of the ejected electrons is very small due to interaction with the surrounding atoms (Salmeron & Schlogl, 2008). But electrons expelled from atoms in the sample's top layer have a high chance of leaving without further interaction. Different chemical environments, such as chemical bonds or oxidation states, result in slightly different binding energies. By analyzing the fine structure of the XPS peaks, one can gain insights into the chemical composition and state of the material's surface. The mechanism of photoionization is utilized to eject electrons from surfaces as shown in figure.

X-Ray Photoelectron Spectroscopy Instrumentation

The XPS instrument encompassing several critical components are listed as the following.

Ultra-High Vacuum (UHV) Sample Chamber

X-ray Photoelectron Spectroscopy is a powerful analytical tool conducted in an Ultra-High Vacuum System. This exceptionally low pressure is achieved using specialized vacuum pumps designed for different purposes. Initially, a Roughing Pump evacuates the chamber to reduce the pressure to a lower but still substantial level. The Turbo Pump takes over to further reduce the pressure, creating the necessary UHV conditions. It operates at high speeds, effectively removing residual gases. Ion Pumps, which are part of the UHV setup, continuously remove any remaining trace gases to maintain the extremely low pressure required for accurate XPS measurements. The UHV system allows a longer path length for photoelectrons, enhancing their detectability and accuracy. Second, it keeps the sample surfaces clean by preventing contamination, which can interfere with the analysis by producing spurious X-ray signals.

X-Ray Source of XPS

The X-ray source in XPS is typically an X-ray emission line of a particular element, such as magnesium (Mg Kα), aluminum (Al Kα), or chromium (Cr Kα) generated by bombarding these material with high-energy (10-15 keV) electrons resulting in ejection of inner-shell electrons from the atoms of the target (Dolcet & Gross, 2021). Subsequently, outer-shell electrons fill these vacancies, releasing energy in the form of X-rays with distinct energies corresponding to the specific element involved. These monochromatic X-rays are then directed onto the sample being analyzed in the XPS instrument. When they interact with the sample, they can excite core-level electrons, initiating the photoelectron emission process, which forms the basis of XPS analysis.

Electrons Lens

The electron lens collects the photoelectrons depending on their initial binding energies and the energy of the incident X-rays and direct them towards the energy analyzer. It also plays a role in focusing the incoming photoelectrons into a well-defined beam for maintaining the precision of the energy measurements, ensuring that the electrons follow a specific trajectory. Once the focused photoelectrons enter the energy analyzer, they pass through a magnetic field or an electrostatic field. This field causes the electrons to disperse based on their kinetic energies.

Energy Analyzer

The signal strength of the photoelectrons emitted relies on the energy and hence influence the response of energy analyzer. The phenomenon can be utilized to precisely detect the energy of photoelectrons before being detected by the detector (Cortenraad et al., 2000). To do this correctly, one needs to accurately understand the influence of electrons' energy on the performance of the analyzer. Energy analyzer is, thus, a critical component of the XPS instrument and is used to measure the energy and intensity of the emitted electrons. There are several types of energy analyzers including the hemispherical analyzer (HA) or cylindrical mirror analyzer (CMA). HA is based on the principle of electrostatic lenses and use a combination of electrostatic fields to focus and deflect emitted electrons onto a detector, allowing for energy analysis. Hemispherical analyzers are known for their high energy resolution and are capable of providing detailed information about the electronic structure of materials. Whereas CMAs consist of a cylindrical mirror that focuses the emitted electrons onto a detector, while deflecting electrons with different kinetic energies to different angles. These analyzers use magnetic or electric field to separate and measure the kinetic energy of the emitted electrons.

Detector

The detector identifies and translates the kinetic energy of electrons into electrical signals. A multichannel display detector which is equivalent to a photographic plate allows simultaneous detection of the kinetic energy of the ejected electrons. It converts the kinetic energy of the electrons into electrical signals, which are then processed to obtain the XPS spectrum. Prevalent detector types encompass Electron Multipliers (EM), Delay-Line Detectors, and Faraday Cups, each boasting distinct advantages in terms of sensitivity and energy resolution.

Electron Multipliers (EM): Electron multipliers, such as photomultiplier tube used in scintillation counter, are based on the principle of secondary electron emission, where incoming electrons striking a surface release multiple secondary electrons, leading to electron multiplication. EMs are highly sensitive and can provide high gain, making them suitable for detecting low-intensity XPS signals. They are typically used in combination with hemispherical analyzers in XPS instruments.

Delay-Line Detectors: Delay-line detectors, also known as time-of-flight (TOF) detectors, measure the time taken by the emitted electrons to travel to the detector from the sample surface. They are based on the principle of detecting the time-of-flight of electrons, which is proportional to their kinetic energy. Delay-line detectors can provide high energy resolution and are commonly used in combination with cylindrical mirror analyzers in XPS instruments. The basic principle of a delay-line detector involves the collection of emitted electrons and accelerating them towards a positive sensitive detector, such as microchannel plate (MCP) by the application of high voltage. They generate electrical signals that are proportional to the position and time or arrival of the electrons, which is used to calculate their kinetic energy based on the known distance between the sample surface and the detector. These detectors provide high energy resolution, which allows for precise determination of binding energies and detailed analysis of chemical states.

Faraday Cups: Faraday cups are metal cups that collect and measure the charge generated by the incoming electrons. They are simple and robust detectors that can be used for detecting high-energy electrons in XPS instruments. Faraday cups detectors consist of a conductive material such as gold or stainless steel, which collects the emitted electrons. These electrons transfer their kinetic energy to the metal surface, leading to the generation of a current or charge with a magnitude proportional to the number of electrons collected by the Faraday cup. It is used in combination with non-monochromatic X-ray sources, where the XPS spectra are acquired using the energy information of the emitted electrons. In these cases, the energy resolution of the detector is not a critical factor, as the X-ray source provides a broad range of X-ray energies. These detectors are used to study bulk materials and thick films.

Application of X-Ray Photoelectron Spectroscopy

XPS is a versatile and powerful analytical technique widely employed in various scientific and industrial fields due to its ability to provide detailed information about the composition and chemical state of a material's surface. It is used extensively to analyze the elemental composition of the surface of materials, including metals, polymers, ceramics, and semiconductors. It provides insights into the adsorption and desorption of molecules on surfaces, crucial for understanding surface reactions and gas-surface interactions. XPS is not used in material characterization but is also a useful tool in studying the chemical composition of cell membranes and biomolecular interactions at cell surfaces (Ojeda et al., 2008; van der Mei et al., 2000).

The electron energy levels of an atom can be divided into two types, core levels, which are bound to the nucleus, and valence levels, which are only weakly bound. The core-level electrons of an atom have energies that are nearly independent of the chemical species in which the atoms are bound since they are not involved in the bonding process. Thus, the identification of Core level binding energies provides unique signatures of the elements. XPS is also capable of quantifying the chemical shift, the difference in binding energy of same atom in two chemically distinct states in the same molecule or in two separate compounds, characteristic of the nature and the number of atoms surrounding the ionized atoms of the sample. It can be used to study the nature of chemical bonding of atoms in a molecule.

AUGER ELECTRON SPECTROSCOPY (AES)

Auger Electron Spectroscopy is a surface analysis technique used to determine the elemental composition and chemical state of the top few atomic layers of a material (Chang, 1971) and was discovered by Liese Meitner, an Austrian physicist who reported that the primary β-rays can eject outer electrons from the daughter atom, leading to the excitation of Kα radiation. This, in turn, can induce the ejection of L-, M-, or N-electrons from the same atom. Then Pierre Auger, a French physicist in the 1925 observed

Figure 7. Auger electron ejection process

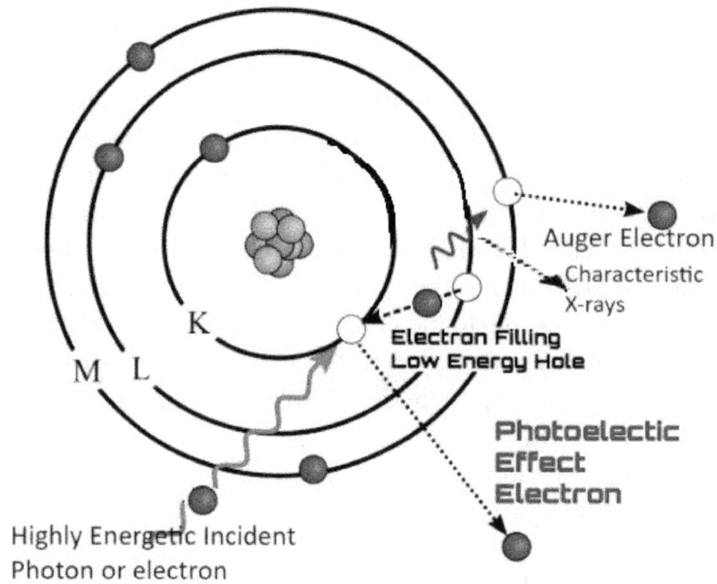

that when an inner-shell electron is ejected from an atom, it can be followed by the emission of another electron (Figure 7), which became known as the Auger electron (Auger, 1975) and in the late 1930s, Dutch physicist J. H. DeBoer proposed a quantum mechanical theory to explain the effect. Nearly two decades later, in 1953, JJ Lander suggested the idea of using stimulated Auger signals for surface analysis. This concept marked a groundbreaking development in the field of surface analysis by introducing the notion of enhancing the sensitivity and selectivity of AES measurements through external excitation or stimulation. It wasn't until 1967 that Larry Harrison demonstrated a practical method for enhancing Auger signals. The first commercial Auger electron spectrometer became available in 1969 (Hardouin Duparc, 2009). It relies on the emission of Auger electrons, which are low-energy electrons ejected from an atom when an electron from a higher energy level fills the core hole created by the removal of an initial electron.

Principle of AES

In AES, a high-energy electron beam with an energy typically ranging from 1 to 10 keV strikes an atom on the material's surface, it imparts its energy to one of the inner core electrons, causing it to be ejected from the atom. This ejection creates a core hole in the atom, leaving it in an ionized and excited state. To fill this core hole, an electron from an outer shell make transition to the inner shell and releases energy. This energy is passed to another outer shell electron, causing it to leave the atom in the form of an Auger electron (Bishop & Rivière, 1970). The emitted Auger electrons are collected and their energies are analyzed using a spectrometer. The energy of the Auger electron is specific to the elements involved and can be measured accurately. By analyzing the energies of Auger electrons and referencing them to known energy levels for each element, AES can identify the elements present on the sample's surface

(Vrakking & Meyer, 1971). The spectrum can also provide information about the chemical states of the elements and hence, is useful to determine the chemical bonding and oxidation states of the elements in the analyzed material.

Auger Electron Spectroscopy Instrumentation

AES instruments are equipped with sophisticated components designed to operate under a high-vacuum environment to facilitate precise and accurate measurements. The essential components of an AES spectrometer are

Ultra-High Vacuum (UHV) Environment

The Ultra high vacuum is required to reduce the chances of low energy electrons being scattered by gas molecules on their way to reach the detector, and to keep the sample surface free from contamination from the gas molecules. Low energy Auger electrons are easily scattered by the gas molecules; scattering reduces the signal intensity and increases background noise in the spectra (Frank Ogletree et al., 2009). The Ultra high vacuum chamber is commonly prepared using stainless steel and joints of chamber parts are made from crushed copper gaskets. UHV can be achieved using diffusion pumps, sputter ion pumps or turbo molecular pumps (Carlson, 2013). Magnetic shielding is also required for the chamber to change the trajectory of signal electrons; since electrons are affected by the magnetic field.

Electron Gun

The electron gun in Auger Electron Spectroscopy (AES) stands as a fundamental apparatus responsible for the generation of a high-energy electron beam that interacts with the sample. The nature of the electron gun used for AES depends upon 3 factors including the speed of analysis, the desired special resolution and beam induced changes to the sample surface. A selection of distinct electron sources is harnessed depending on the desired energy range for the electron beam and the level of brightness needed or better resolution and performance. Conventional sources encompass the tungsten thermionic emitter and the lanthanum hexaboride thermionic emitter, both esteemed for their robustness and constancy. Nevertheless, contemporary AES instruments may also incorporate cold and hot field emitters, offering superior brilliance and narrower energy distributions. The current of electrons produced by the electron gun is a vital factor in AES analysis. It affects spatial resolution, analysis speed, and sensitivity. Lower current provides better spatial resolution for detailed surface analysis, while higher current speeds up the analysis and boosts sensitivity.

Electron Energy Analyzer

At the heart of AES instrumentation is the energy analyzer, which serves to selectively capture and quantify the kinetic energies of Auger electrons emitted from the sample. The energy-selecting element in the electron energy analyzer is responsible for filtering the emitted electrons based on their energy. It also disperses the secondary emitted electrons according to their energies. This element typically consists of an electric or magnetic field that can selectively deflect electrons with different energies, allowing only a narrow range of energies to pass through to the detector. This allows for the energy distribution of the Auger electrons to be measured with high precision. The most commonly used analyzer is the CHA also called hemispherical sector analyzer. CHA is made up of two metal hemispheres, one inside the other and arranged in a symmetrical fashion. The inner one, called the hemispherical analyzer, has a small aperture at its center through which the electrons pass. The outer one, called the electrostatic lens, has a larger aperture that lets electrons enter the analyzer. When an electron beam hits the CHA, it gets a speed boost from an electric field between the hemispheres. But as the electrons go through the hemispherical analyzer, they start to slow down because of a retarding electric field that works against them. By adjusting the retarding potential on the hemispherical analyzer, electrons with different energies can be selected.

Electron Detector

In the Auger Electron Spectroscopy (AES) process, once the Auger electrons have exited the analyzer, they proceed to the detector, where their signals are amplified and counted using a device known as an electron multiplier. AES instruments employ various types of detectors, including scintillation detectors, semiconductor detectors, and electron multipliers. These detectors are essential for capturing the energy of the Auger electrons and converting it into an electrical signal that can be further processed and analyzed.

One of the detector types used in AES is the single channel detector. This electrostatic device incorporates a continuous dynode surface, which is essentially a thin, conductive layer lining the inside of a tubular channel. This detector requires only two electrical connections to establish the conditions for electron multiplication. The output of the single channel detector consists of a series of pulses, which are then directed into a pulse amplifier or discriminator before being sent to a computer for analysis. Multichannel detectors are also utilized for auger electron detection. In this type of detector, a multiple detection system is added at the output of the analyzer. The system maybe in the form of a few multiple, parallel, equivalent detector chains or position sensitive detectors spread across the whole of the analyzer output slit plane.

Data Acquisition and Analysis Software

Modern AES instruments are equipped with computer-controlled data acquisition systems. Specialized software collects, processes, and analyzes the data, allowing for the identification of elements and chemical states on the sample surface.

SECONDARY ION MASS SPECTROMETRY (SIMS)

Secondary Ion Mass Spectrometry (SIMS) stands as a critical analytical technique that has significantly evolved over the decades, enabling researchers and scientists to gain valuable insights into the composition, structure, and distribution of elements and isotopes on solid surfaces. It offers unique capabilities that have revolutionized various scientific domains, from materials science to environmental research. SIMS finds its roots in the broader field of mass spectrometry, which originated in the early 20th century with the pioneering work of scientists like J.J. Thomson, Francis W. Aston, and Arthur J. Dempster. These early developments laid the foundation for SIMS. The first dedicated SIMS instruments were constructed in the 1960s, initially for the analysis of moon rocks during the Apollo missions. This marked the beginning of SIMS as a standalone technique. Throughout the 1970s, significant strides were made in SIMS technology, leading to the introduction of the first commercial SIMS spectrometers by companies like Cameca and VG Instruments.

Principle of SIMS

SIMS operates on the fundamental principle of bombarding a sample's surface with high-energy primary ions (ranging from keV to MeV) generated from an ion gun. These primary ions penetrate the top few nanometers of the sample, causing atoms or molecules to be sputtered from the surface. These sputtered species include secondary ions carrying chemical and isotopic information. An electric field extracts and accelerates these secondary ions towards a mass spectrometer, where they are separated based on their mass-to-charge ratio (m/z) (Heide, 2014). The resulting mass spectrum provides information about the relative abundance and spatial distribution of elements and isotopes on the sample's surface.

Instrumentation

A typical SIMS instrument comprises several key components.

Primary Ion Source

The primary ion gun generates high-energy primary ions. Various types of ion guns can be employed, including electron impact ion sources, plasma ion sources, and liquid metal ion guns (LMIGs), depending on the required current, beam dimensions, and sample characteristics.

Primary Ion Column

This component focuses and directs the primary ion beam onto the sample surface, causing sputtering and the emission of secondary ions. The primary ion column often includes elements such as electrostatic lenses, ion optics, and deflectors, which are designed to precisely focus and steer the primary ion beam towards the target area of the sample. In some instruments, there is an option to separate primary ion species or pulse the beam. The advanced SIMS instruments primary ion column may include a Wien filter or other mass-selection devices to separate these primary ions based on their mass-to-charge ratios (m/z) to selectively choose primary ion species for specific analysis.

High Vacuum Sample Chamber

A high vacuum (below 10^{-4} Pa) is essential to prevent collisions of secondary ions with background gases, ensuring accurate measurements and limiting surface contamination. The primary objective of achieving and maintaining a high vacuum is to prevent collisions between the emitted secondary ions and background gas molecules. When secondary ions are sputtered from the sample surface, they need to travel from the sample to the mass analyzer without colliding with other particles, as these collisions can alter their trajectories and energies. Such disruptions can lead to erroneous measurements and affect the reliability of the data collected.

Mass Analyzer

This device separates the secondary ions according to their mass-to-charge ratio (m/z). Various types of mass analyzers are used, including quadrupole, time-of-flight (TOF), magnetic sector, ion trap, or others, each with unique principles of operation. Each type has unique principles and applications, ranging from routine analysis to high-precision isotopic measurements, structural analysis, and tandem mass spectrometry. Quadrupole mass analyzers utilize a combination of radiofrequency (RF) and direct current (DC) electric fields to select ions based on their m/z values. These analyzers allow ions with specific m/z ratios to pass through while filtering out others. TOF mass analyzers measure the time it takes for ions to travel from the ion source to the detector. Since all ions possess the same kinetic energy, their time of flight is directly related to their m/z ratio. Lighter ions reach the detector faster than heavier ones. Magnetic sector analyzers employ a combination of magnetic fields to bend the path of ions based on their m/z ratios. The degree of bending is proportional to the mass of the ion, allowing for precise separation. Ion trap analyzers use electromagnetic fields to trap ions within a confined space. Ions are then selectively ejected from the trap based on their m/z values. The trapping and ejection cycles are controlled to scan across the mass range of interest. The choice depends on specific analytical requirements such as mass resolution, sensitivity, and sample complexity.

Detector

The detector records the separated secondary ions, and various types of detectors, such as Faraday cups, electron multipliers, or microchannel plate detectors, can be used based on the specific requirements of the analysis.

Data Analysis and Interpretation

Detected ion signals are processed and analyzed to generate a mass spectrum, which represents the relative abundance of ions as a function of their m/z values. Interpretation of the mass spectrum involves comparing the observed m/z values and ion abundances with reference databases, spectra, or theoretical calculations to identify the molecules or elements present in the sample. Quantitative analysis can also be performed by comparing ion abundances with known standards or using internal or external calibration methods.

Modes of Operation

SIMS spectroscopy can be conducted in different modes, tailored to specific analytical needs.

Static SIMS

In this mode, a low-intensity primary ion beam bombards the sample surface for a short duration. This results in the sputtering of only a thin layer of the sample, the emitted secondary ions are collected and analyzed to determine the sample's composition and structure. Static SIMS is primarily used for surface characterization and elemental analysis. Unlike dynamic SIMS, it does not focus on depth profiling or time-dependent changes but offers high sensitivity and spatial resolution for surface studies.

Dynamic SIMS (DSIMS)

DSIMS first emerged in the early 1980s and quickly found significant applications in the semiconductor industry. DSIMS is especially prominent in the study of thin films, where it can precisely determine layer composition and thickness, aiding in the development of advanced materials and coatings (Guerquin-Kern et al., 2005). In polymer research, DSIMS helps elucidate polymer structures, revealing information about molecular weight, branching, and composition. Additionally, in the field of catalyst surface chemistry, DSIMS is instrumental in understanding catalytic processes and identifying surface species, contributing to advancements in catalysis.

Imaging SIMS

This mode involves mapping and imaging the elemental or isotopic distribution across a sample surface with high spatial resolution, down to the nanometer scale. It provides valuable insights into the distribution of elements and isotopes in complex structures. This mode facilitates a deep understanding of how elements and isotopes are distributed within intricate structures.

CONCLUSION

In conclusion, the utilization of X-rays in material characterization has indeed sparked a revolution in the analysis and understanding of various materials, offering invaluable insights into their properties. This chapter has highlighted the historical significance of X-rays, from their accidental discovery by Wilhelm Conrad Röntgen in 1895 to their role in revolutionizing medical diagnosis and material analysis. The generation of X-rays within specialized devices like X-ray tubes was explained, and the distinction between continuous and pulsed X-ray production methods was made clear.

This chapter has also provided a comprehensive overview of the applications of X-rays in techniques XRD, XRF, XPS, AES, and SIMS. These techniques have been explored in depth, shedding light on their working principles, operations, and applications. XRD emerged as a powerful tool for deciphering the crystalline structure of solids and inspecting atomic arrangements within crystallites. XRF stands out as a versatile technique for elemental analysis. XPS and SIMS enable high-sensitivity elemental and isotopic analysis of material surfaces. These techniques find applications in diverse fields, from mineralogy and pharmaceuticals to the semiconductor industry and environmental science.

REFERENCES

Ali, A., Chiang, Y. W., & Santos, R. M. (2022). X-ray Diffraction Techniques for Mineral Characterization: A Review for Engineers of the Fundamentals, Applications, and Research Directions. *Minerals (Basel)*, *12*(2), 205. doi:10.3390/min12020205

Amoros, J. L. (2012). *The Laue Method*. Elsevier Science.

Auger, P. (1975). The Auger effect. *Surface Science*, *48*(1), 1–8. doi:10.1016/0039-6028(75)90306-4

Baer, D. R., & Engelhard, M. H. (2010). XPS analysis of nanostructured materials and biological surfaces. *Journal of Electron Spectroscopy and Related Phenomena, 178–179*, 415–432. doi:10.1016/j.elspec.2009.09.003

Benninghoven, A. (1973). New developments in the surface analysis of solids. *Applied Physics (Berlin), 1*(1), 3–16. doi:10.1007/BF00886799

Bishop, H. E., & Rivière, J. C. (1970). Characteristic ionization losses observed in auger emission spectroscopy. *Applied Physics Letters, 16*(1), 21–23. doi:10.1063/1.1653016

Bragg, W. H., & Bragg, W. L. (1913). The reflection of X-rays by crystals. *Proceedings of the Royal Society of London. Series A, Containing Papers of a Mathematical and Physical Character, 88*(605), 428–438. doi:10.1098/rspa.1913.0040

Bunaciu, A. A., Udriştioiu, E., & Aboul-Enein, H. Y. (2015). X-Ray Diffraction: Instrumentation and Applications. *Critical Reviews in Analytical Chemistry, 45*(4), 289–299. doi:10.1080/10408347.2014.949616 PMID:25831472

Carlson, T. (2013). *Photoelectron_and_Auger_Spectroscopy*. Springer US.

Chang, C. C. (1971). Auger electron spectroscopy. *Surface Science, 25*(1), 53–79. doi:10.1016/0039-6028(71)90210-X

Chauhan, A. (2014). Powder XRD Technique and its Applications in Science and Technology. *Journal of Analytical & Bioanalytical Techniques, 5*(6). Advance online publication. doi:10.4172/2155-9872.1000212

Cortenraad, R., Denier van der Gon, A. W., & Brongersma, H. H. (2000). Influence of analyser transmission and detection efficiency on the energy dependence of low-energy ion scattering signals. *Surface and Interface Analysis, 29*(8), 524–534. doi:10.1002/1096-9918(200008)29:8<524::AID-SIA897>3.0.CO;2-J

Dolcet, P., & Gross, S. (2021). Electronic and Chemical Properties: X-Ray Absorption and Photoemission. In O. Diwald & T. Berger (Eds.), *Metal Oxide Nanoparticles* (pp. 383–434). Wiley., doi:10.1002/9781119436782.ch11

Edsall, D. L. P. R. (1907). The nature of the general toxic reaction following exposure to the x-rays. *The American Journal of the Medical Sciences (1827-1924), 133*(3), 426.

Frank Ogletree, D., Bluhm, H., Hebenstreit, E. D., & Salmeron, M. (2009). Photoelectron spectroscopy under ambient pressure and temperature conditions. *Nuclear Instruments & Methods in Physics Research. Section A, Accelerators, Spectrometers, Detectors and Associated Equipment, 601*(1–2), 151–160. doi:10.1016/j.nima.2008.12.155

Guerquin-Kern, J.-L., Wu, T.-D., Quintana, C., & Croisy, A. (2005). Progress in analytical imaging of the cell by dynamic secondary ion mass spectrometry (SIMS microscopy). Biochimica et Biophysica Acta (BBA) -. *Biochimica et Biophysica Acta. G, General Subjects, 1724*(3), 228–238. doi:10.1016/j.bbagen.2005.05.013 PMID:15982822

Hardouin Duparc, O. (2009). Pierre Auger – Lise Meitner: Comparative contributions to the Auger effect. *International Journal of Materials Research, 100*(9), 1162–1166. doi:10.3139/146.110163

Janssens, K. (2004). *X-ray based methods of analysis*. doi:10.1016/S0166-526X(04)80008-4

Kalam, M. A., Hussam, A., Khaliquzzaman, M., Khan, A. H., Islam, M. M., Zaman, M. B., & Husain, M. (1978). Elemental analysis in solutions by radioisotope excited X-ray fluorescence. *Journal of Radioanalytical Chemistry*, *46*(2), 285–297. doi:10.1007/BF02519895

Mantler, M. (1986). X-ray fluorescence analysis of multiple-layer films. *Analytica Chimica Acta*, *188*, 25–35. doi:10.1016/S0003-2670(00)86026-2

Meadowcroft, W. H. (1896). The abc of x-rays.

Nasrazadani, S., & Hassani, S. (2016). Modern analytical techniques in failure analysis of aerospace, chemical, and oil and gas industries. Handbook of Materials Failure Analysis with Case Studies from the Oil and Gas Industry, 39–54. doi:10.1016/B978-0-08-100117-2.00010-8

Ojeda, J. J., Romero-González, M. E., Bachmann, R. T., Edyvean, R. G. J., & Banwart, S. A. (2008). Characterization of the Cell Surface and Cell Wall Chemistry of Drinking Water Bacteria by Combining XPS, FTIR Spectroscopy, Modeling, and Potentiometric Titrations. *Langmuir*, *24*(8), 4032–4040. doi:10.1021/la702284b PMID:18302422

Redler, G., Jones, K. C., Templeton, A., Bernard, D., Turian, J., & Chu, J. C. H. (2018). Compton scatter imaging: A promising modality for image guidance in lung stereotactic body radiation therapy. *Medical Physics*, *45*(3), 1233–1240. doi:10.1002/mp.12755 PMID:29360151

Rietveld, H. M. (1969). A profile refinement method for nuclear and magnetic structures. *Journal of Applied Crystallography*, *2*(2), 65–71. doi:10.1107/S0021889869006558

Salmeron, M., & Schlogl, R. (2008). Ambient pressure photoelectron spectroscopy: A new tool for surface science and nanotechnology. *Surface Science Reports*, *63*(4), 169–199. doi:10.1016/j.surfrep.2008.01.001

Scherrer, P. (1918). Bestimmung der Grosse und inneren Struktur von Kolloidteilchen mittels Rontgenstrahlen. *Nach Ges Wiss Gottingen*, *2*, 8–100.

Seah, M. P. (1984). A review of the analysis of surfaces and thin films by AES and XPS. *Vacuum*, *34*(3–4), 463–478. doi:10.1016/0042-207X(84)90084-8

Seibert, J. A. (1997). The AAPM/RSNA physics tutorial for residents. X-ray generators. *Radiographics*, *17*(6), 1533–1557. doi:10.1148/radiographics.17.6.9397462 PMID:9397462

Seibert, J. A., & Boone, J. M. (2005). X-ray imaging physics for nuclear medicine technologists. Part 2: X-ray interactions and image formation. *Journal of Nuclear Medicine Technology*, *33*(1), 3–18. PMID:15731015

Simon, A. H. (2012). *Sputter Processing. Handbook of Thin Film Deposition: Techniques* (3rd ed.). Processes, and Technologies. doi:10.1016/B978-1-4377-7873-1.00004-8

Sitko, R. (2009). Quantitative X-ray fluorescence analysis of samples of less than 'infinite thickness': Difficulties and possibilities. *Spectrochimica Acta. Part B, Atomic Spectroscopy*, *64*(11–12), 1161–1172. doi:10.1016/j.sab.2009.09.005

Sitko, R., & Zawisz, B. (2012). Quantification in X-Ray Fluorescence Spectrometry. In S. K. Sharma (Ed.), *X-Ray Spectroscopy*. InTech. doi:10.5772/29367

Snyder, R. L. (1999). X-Ray Diffraction. In X-ray Characterization of Materials (pp. 1–103). Wiley. doi:10.1002/9783527613748.ch1

Streli, C., Wobrauschek, P., & Kregsamer, P. (1999). X-ray Fluorescence Spectroscopy, Applications. Encyclopedia of Spectroscopy and Spectrometry, 2478–2487. doi:10.1006/rwsp.2000.0337

Strüder, L., Lechner, P., & Leutenegger, P. (1998). Silicon drift detector – the key to new experiments. *Naturwissenschaften*, *85*(11), 539–543. doi:10.1007/s001140050545

Swanson, H. E. (1955). *Standard X-ray Diffraction Powder Patterns* (Vol. 5–10). U.S. Department of Commerce, National Bureau of Standards.

van der Heide, P. (2014). *Secondary Ion Mass Spectrometry An Introduction to Principles and Practices* (1st ed.). Wiley. doi:10.1002/9781118916780

van der Mei, H. C., de Vries, J., & Busscher, H. J. (2000). X-ray photoelectron spectroscopy for the study of microbial cell surfaces. *Surface Science Reports*, *39*(1), 1–24. doi:10.1016/S0167-5729(00)00003-0

Vrakking, J. J., & Meyer, F. (1971). Auger electron spectroscopy made quantitative by ellipsometric calibration. *Applied Physics Letters*, *18*(6), 226–228. doi:10.1063/1.1653640

Chapter 14
Analysis on Various Aspects of Internet of Nano-Things (IoNT), Its Integration in Machine Learning, and Its Diverse Applications

Preeti Singh Bahadur
Amity University, India

ABSTRACT

The internet of things (IoT) concept has long been regarded as a crucial driver of the Fourth Industrial Revolution, having the potential to change the way we live. However, the incorporation of nanotechnology promises to increase its impact significantly more. The proliferation of intelligent settings implies application interconnection and the use of the internet. As a result, the concept of the internet of things (IoT) has emerged. The internet of nano things (IoNT) is made possible by the development of the IoT idea, a new communication networks paradigm based on nanotechnology and IoT, that is, a paradigm capable of interconnecting nano-scale devices via existing networks. The concept of the internet of nano things (IoNT) arose from the connecting of these nano devices with the internet. The internet of nano-things (IoNT) is a network of nanoconnected devices, objects, or organisms with unique IDs that wirelessly transport data to the cloud over a computer or cellular network. Data delivery, caching, and energy usage are three of the most important topics in IoNT today.

1. INTRODUCTION

Everybody's daily life has changed dramatically as a result of the internet of things (Perwej et al., 2019). The goal of the highly connected global network known as the Internet is to connect digital and analogue devices. Many IoT applications have recently been created and put into use (Parwej et al., 2018). By connecting various devices and objects from many sectors, the Internet of things (IoT) broadens the

DOI: 10.4018/979-8-3693-1306-0.ch014

scope of the internet. Big data (Perwej et al., 2017) analytics and the Internet of Things (IoT), a new dawn technology that depicts how data, people, and connected physical objects act depending on communicated information, have been used by a variety of domains for a variety of goals. Big data and the Internet of Things (IoT) are enormous, complicated concepts. They are separate even though they are connected. The Internet of Things (IoT) is made up of millions of information-gathering and -sharing devices, whereas big data (Akhtar et al., 2017) covers a far wider area. Although big data (Perwej et al., 2019) and IoT are two independent concepts, their success ultimately depends on one another. The amount of data available for analysis by all types of organisations will significantly increase as a result of the IoT. However, before the benefits of knowledge are fully realised, there are significant restrictions that must be solved sooner rather than later. Big data and the IoT are undoubtedly closely related because, by definition, billions of internet-connected 'things' will produce enormous amounts of data (Parwej et al., 2018). The most popular technology subjects in recent years have been big data, the Internet of Things (IoT), and the Internet of Nano Things (IoNT) (Akyildiz et al., 2015). Nanotechnology is the primary component of the Internet of Nano Things (IoNT). Because of the unique characteristics of materials at the nanoscale, nanotechnology (science on the size of individual atoms and molecules) has been dubbed the second Industrial Revolution (Tabassum et al., 2020). The IoNT uses a large number of nanosensors to deliver more accurate and thorough information about a specific object in order to improve our comprehension of object behaviour. By embedding nano-sensors in the devices, IoNT expands the scope of IoT and enables connections and communication between it and the internet via a nanotechnology network. The IoNT contains nanotechnology, which enables seamless data transmission and communication over a specific range of activities and is deployed in desired devices within the nanotechnology radius. The quantity and variety of linked objects, as well as the technologies for data collection, processing, and sharing, have all significantly altered this vision and paradigm. The IoNT infrastructure (Akyildiz et al., 2010) enables numerous combinations of nano-sensor network technologies, nano cameras, nanophones, nanothings objects, and more.

The powerful IoT is being significantly improved by the rapidly growing IoNT. By pooling resources and radically expanding the scope and dependability of the ensuing services, these "nano things" in the IoNT infrastructure will realise, investigate, and learn to utilise one another's data. The idea of the IoNT is presented as a sort of IoT in which conventional networks are coupled to nano-devices with dimensions that may range from 1 to 100 nm (Tabassum et al., 2020). This results in new networking paradigms. By combining nanotechnology with a number of other technologies including the Internet of Things (IoT), Sensor Networks, Cloud Computing, Big Data Analytics, etc., Internet of Nano Things (IoNT) communications can be formed (Akyildiz et al., 2011). The healthcare industry as well as homeland security and environmental protection will be significantly impacted by this new networking paradigm. The rest of the paper is organized as follows: Section 2 provides reviewed literature, section 3 provides machine learning, section 4 provides internet of nanothing and its intergration on nanotechnology, section 5 provides about the internet of nanothings, section 6 provides functioning of internet of nanothings, section 7 provides the main reason of internet of nanothings, Section 8 provides internet of nanothings and its market trends, section 9 provides application of internet of nanothings and section 10 provides future scope and conclusion.

2. REVIEWED LITERATURE

A brand-new technology that is developing more quickly than ever and has the potential to address many of the most important problems facing the world is the Internet of Nano Things. It functions similarly to how we connect devices in an IoT environment, but a key distinction is that it can connect nano components, something that IoT cannot do. As a result, it brings about a state-of-the-art revolution in nanoscale device electromagnetic communication. This section covers works related to the Internet of Nano Things. Using channel modelling, information encoding, and networking protocols for nano-devices based on IoNT, Ian F. Akyildiz et al. concentrate on electromagnetic communication in nanoscale devices. Regarding communication in nanoscale devices and nanonetworks, the information and communication technology (ICT) society should offer some fresh alternatives (Akyildiz et al., 2010). A. Nayyar et al. talk about The definition of a nan machine is a basic functional unit that incorporates nano-components to carry out crucial functions like detecting or igniting. Nano-machines can also be used as a foundation for the development of nano-memory, nano-processors, nano-clocks, and nano-bots. In addition to researching various application areas and improving IoNT devices based on nanotechnology, new security and protection mechanisms should be investigated with regard to the data being gathered by nano sensors (Nayyar et al., 2017). The (Kulakowski et al. 2017) concentrated on signal routing using nano-networks and Forster resonance energy transfer (FRET), which was discovered to be a technology with a very high signal propagation speed. Based on the biological characteristics of particular molecules, they proposed five new routing mechanisms and empirically verified one of them. Due to the rise in internet users, more bandwidth is required, according to K. Agarwal et al. The Terahertz band is a frequency range that can be used to address this problem. Large amounts of channel capacity are made available for very short ranges (>1m) thanks to this technology. The various Internet of Nano-Things (IoNT) network models and the architectural specifications for implementation were described by (Ali et al. 2016). They emphasised the primary uses for IoNT as well as the substantial difficulties encountered while putting this technology into use in the medical field. In addition, they examined two paradigms, layer-based and non-layer-based models, as well as the comprised layers in the layered model, to explain the communication and networking elements of IoNT. provided a variety of fundamental concepts, architecture, communication classifications, communication issues, applications, benefits, security, and future research directions in order to provide in-depth knowledge about Internet of Nano Things (IoNT) and Industrial Internet of Things (IIoT) technologies. By reviewing theoretical and experimental ideas, (Kuscu et al. 2016) developed a thorough architectural view of Nano communication that was centred on its core principles and design requirements. Under the umbrella of the "Internet of Molecular Things," they provided an overview of the networking options provided by the inherent properties of fluorophores. IoT operational models are categorised by the Internet Architecture Board (IAB) depending on how they are networked and used. These models include IoT Device-to-Device, IoT Device-to-Cloud, IoT Device-to-Gateway, and IoT Back-End-Sharing Data. The primary potential and difficulties associated with implementing IoT technology were examined and analysed. The Internet of Nano Things' (IoNT) development is primarily focused on enhancing processing power, delivering more storage space at a reduced cost, and expanding the function of communication terminals. The IoNT infrastructure developed by Yusuf Perwej et al. can be implemented in a variety of eco-systems, including radio frequency identification (RFID), nano antenna, Wi-Fi, Li-Fi, RedTacton (Perwej et al., 2016), and Wi-Fi. The current state of IoT, technology supporting IoT, its applications and difficulties, recent advancements and development through an in-depth review have been provided in this paper by (Whitmore et al..2014). Depending on the manufacturing process for the Nano-machines and the intended application, different communication paradigms can be employed in Nano-networks. The most recent methods for creating nanoscale objects and significant difficulties in identifying IoMNTs are covered by (Simonjan et. al., 2018) in their discussion. Depending on the manufacturing process for the Nano-machines and the intended application, different communication paradigms can be employed in Nano-networks. The most recent methods for creating nanoscale objects and significant difficulties in identifying IoMNTs are covered in their discussion. A novel QoS aware cross-layer communication module, addressing methods, neighbour finding and routing mechanisms, and novel security solutions for the IoMNT were also suggested by the researchers. Looked explored the difficulties and possibilities of integrating body-worn nano-devices with body area networks and other external gateways. The application needs were supported by a novel network architecture that was developed, and in particular, simulation-based performance evaluation and security challenges were also discovered. IoNT network models and architectural requirements for use in healthcare applications such medicine delivery and disease detection.

3. MACHINE LEARNING

A subset of artificial intelligence known as machine learning allows machines to mimic the way humans learn (Mahdavinejad et. al., 2018). It aids a machine's usage of information and algorithms as well as its comprehension of behavioural and technological patterns, eventually enhancing machine accuracy.

After his famous IBM checkers game, which was shown on television, Arthur Samuel, an American pioneer in the field of artificial intelligence research, came up with the word. Millions of people witnessed for the first time how a machine defeated renowned checkers player Robert Nealey in a video game. This little accomplishment was a significant turning point for machine learning and artificial intelligence.

Machine learning has subsequently merged perfectly with data science and artificial intelligence. Machines are educated to predict and find important insights with the aid of statistical techniques and algorithms. These insights consequently improve business and application decision-making, hopefully having an impact on crucial growth indicators. Applications for machine learning are being used in all major sectors.

3.1. Incorporation of Machine Learning Into IoT

The first smart refrigerator with internet access and video calling was released by LG in 2000. This was the first time the Internet of Things was ever commercialised, and by 2008, there were more IoT devices than there were people on the planet. Since then, IoT usage has been inescapable. But it was noted that the IoT needs to become more compassionate in order to advance. On this need, machine learning delivers.

IoT's use of machine learning is revolutionary since it allows for some degree of behaviour prediction. Through the use of advanced algorithms, it aids in the extraction of accurate information from vast amounts of raw data acquired from IoT devices. IoT application development devices must be able to comprehend unstructured, multimodal data and respond intelligently to both users and environmental changes. This is where machine learning comes into play. This enhances intelligent decision-making systems that are effective and more humane.

3.2. How can we Benefit From Machine Learning in IoT?

- Applying learning algorithms to IoT devices can be done in a variety of ways. In addition to making them understandable, this approach makes them more ergonomic and sustainable by requiring fewer workers and less time to complete a task. Some of the ways that machine learning inference in IoT can help us are listed below.
- In utilities, energy can be conserved by individually forecasting usage and then dynamically allocating. Given the volume of data generated by this, machine learning might be applied to smart metres to further map data and patterns of energy consumption and distribute resources accordingly, saving manpower, energy, and money.
- Manufacturing suffers significant losses as a result of subpar processes and goods. Any operational defect or abnormality may be quickly identified and addressed with the aid of machine learning inferred IoT devices, minimising the waste of both human and financial resources.

Smart and targeted marketing: To enhance the performance of B2B sales teams, many marketing firms are integrating Machine Learning with their IoT platforms. To do this, all sales-related calls, web conferences, and emails are recorded, transcribed, and their contents are examined. The financial behemoth JPMorgan employs an AI copywriter who creates content that is better geared towards the needs of its clients.

Apart from these, practically any industry can profit from the fusion of ML and IoT technology. This combination has produced automation that is far more effective and time-saving. (Bzai et al., 2022), (Fischer et. al., 2022).

4. INTEGRATION OF NANOTECHNOLOGY INTO INTERNET OF THINGS

Long seen as a major driver of the Fourth Industrial Revolution with the potential to revolutionise how we live our lives, the internet of things (IoT) paradigm. Yet the incorporation of nanotechnology promises to significantly increase its impact. The Internet of Things (IoT) is a network of interconnected physical things that use sensors, antennas, CPUs, software, and other technologies to exchange pertinent data online. The range of these devices, from pharmaceuticals to guided missiles, is enormous and is expected to expand; estimates place the peak number of IoT-connected devices at 75 billion in 2025, with tens or perhaps hundreds of zettabytes of created data.

The development of enabling technologies (such as cloud computing and big data analytics) and numerous communication modalities, referred to as IoT protocols, has made such significant projections possible. These protocols allow information to be sent between sensors and the following piece of gear in a networked environment. They include low-power wide-area (LPWA) and 5G for long distances, as well as Bluetooth, Wi-Fi, ZigBee, and near field communication (NFC) for short distances. One of the most intriguing advancements nowadays, in my opinion, is the incorporation of nanotechnology. The internet of Nano-Things (IoNT) is a brand-new IoT derivative that claims to fully extend the IoT concept using nanodevices.

4.1. Nano-Devices

IoT devices that incorporate nanomaterials can take advantage of their exceptional qualities to improve functionality, energy efficiency, and accuracy while shrinking the size of the devices. Nanosensors have proven to be the most useful nanodevices within IoT endpoints, although nanoantennas, nanoprocessors, and nanobatteries are further examples of IoT nanodevices currently in use or being developed.

4.2. Nano-Sensors

To offer pertinent information for further analysis, IoT sensors must monitor particular phenomena in sensing environments. This is accomplished using a variety of nanomaterials in nanosensors, which can also monitor physical, chemical, and biological processes.

For instance, (Tang et al., 2019) created a flexible ammonia (NH3) monitoring sensor based on nanowires. Because of the nanowires' extraordinarily high surface area to volume ratio, the sensor—which was designed to be utilised in a watch-type device—displayed a lower detection limit and a quicker response time than conventional NH3 sensors.

The impressively low power consumption (as low as 3W) and scalable soft lithography production approach provide additional evidence for the potential for nanomaterials to improve IoT sensors.

Non-invasive biosensors for continuous blood glucose monitoring as well as for chemical, microbial, and other analyte monitoring in drinking water have all benefited from similar nano-based advantages.

4.3. Nano-Antennas

IoT antennas handle information reception, decoding, and transmission through a variety of wave types to enable wireless communication between IoT devices. Nanoantennas, which are frequently made of graphene, perform this role largely by radiating in the terahertz frequency range. They can even be combined with nanosensors that can both sense and signal using carbon nanotubes, making them considerably smaller than conventional antennas. The method of manufacture may provide a further intriguing nano-based benefit. Any object may easily turn into a smart Internet of Things device thanks to a titanium carbide nanoantenna that Drexel University researchers have created. This nanoantenna can be sprayed directly onto any stiff or flexible object in a single step without adding any weight or electronics.

4.4. Nano-Processors

Data from IoT endpoints must be processed by an IoT processor by making the appropriate computations. They are predominantly silicon-based and comprise millions of transistors that work as binary switches inside of groups of gates that represent logic operations. The first programmable carbon nanotube processor was created by a team of MIT researchers just a few years ago in 2019. Nanoprocessors are still largely in the laboratory. Realising the higher efficiencies and speeds of carbon nanotubes in such a function may still be many years away given that it only has 14000 transistors.

4.5. Nano-Batteries

Unsurprisingly, a wireless system of smart devices has high power requirements compatible with rechargeable, long-lasting batteries with high energy densities. Currently, lithium-ion models are the most widely used.

Nanomaterials are used in nanobatteries' electrodes or membranes to lower self-discharge rates, boost energy density, and speed up charging. Carbon nanotubes, lithium titanate oxide nanoparticles, and germanium nanowires among a few examples of nanomaterials that have been effectively used in this way, according to (Soliman et al., 2021).

5. ABOUT INTERNET OF NANO THINGS (IoNT)

Nanotechnology, the Internet of Things (IoT), and Industry 4.0 all come together at the Internet of Nano Things (IoNT). The IoNT essentially represents a nanoscale version of the IoT, which is its basic idea. However, the IoNT is a manifestation of small-scale IoT systems that is the appropriate answer for remote environmental monitoring and medical applications. These regions also merge inside sensors that may be employed in regular IoT systems (Perwej et al., 2019). In order to offer more accurate and thorough information about a specific object that allows greater comprehension of object behaviour, the IoNT uses a considerable number of nanosensors.

In a study titled "The Internet of Nano- Things" from 2010, Ian Akyildiz and Josep Jornet presented the initial IoNT concept. The Internet of Nano Things (IoNT) is an emerging networking paradigm that's further referred to as the integration of nanoscale devices with current communication networks and finally the Internet. IoNT is the latest evolution that combines nanotechnology and IoT. Synthetic biology and nanotechnology techniques that enable the building of biological integrated computing devices are the foundation of the Internet of Nano Things. Due to its useful applications in computing systems, medicine, agriculture, and the environment, nanosensors have emerged as one of the key areas of research in the nanoscience community (Akyildiz et al., 2008). The wireless communication systems currently in use might be coupled with these nanosensors, creating a new domain known as the "internet of nano-things" (IoNTs). Typically, nanosensors fall under one of the following three categories: chemical (atomic and molecular energies), biological (antibody and antigen contact, DNA interaction, or enzymatic interaction), or physical (mechanical, acoustic, thermal and radiation, optical, and magnetic). Figure 1 depicts methods for creating and integrating nanosensors.

The Internet of Nano-Things (IoNT) offers a wide range of nano-applications for the real-time data, including rehabilitation, diagnostics, and preventive care. IoNT creates important potential and problems for research on wearable sensor-based big data processing. Traditional algorithms are unable to handle such massive amounts of heterogeneous data, therefore appropriate mechanisms for data analysis are required to keep up with the demands of managing, processing, and responding while maintaining the accuracy of the data. The Terahertz Electromagnetic Nano-Communication (Akyildiz et al., 2010), which is regarded as the transmission and reception of electromagnetic radiation from components, and the molecular Communication, which is regarded as the transmission and reception of information encoded in molecules, are the two main communication channels used by IoNT. By combining nanotechnology with other well-known technologies including big data, wireless sensor networks, cloud computing, and grid computing, IoNT infrastructure may be put into practise.

Figure 1. The outlook for the fabrication and integration of nanosensors

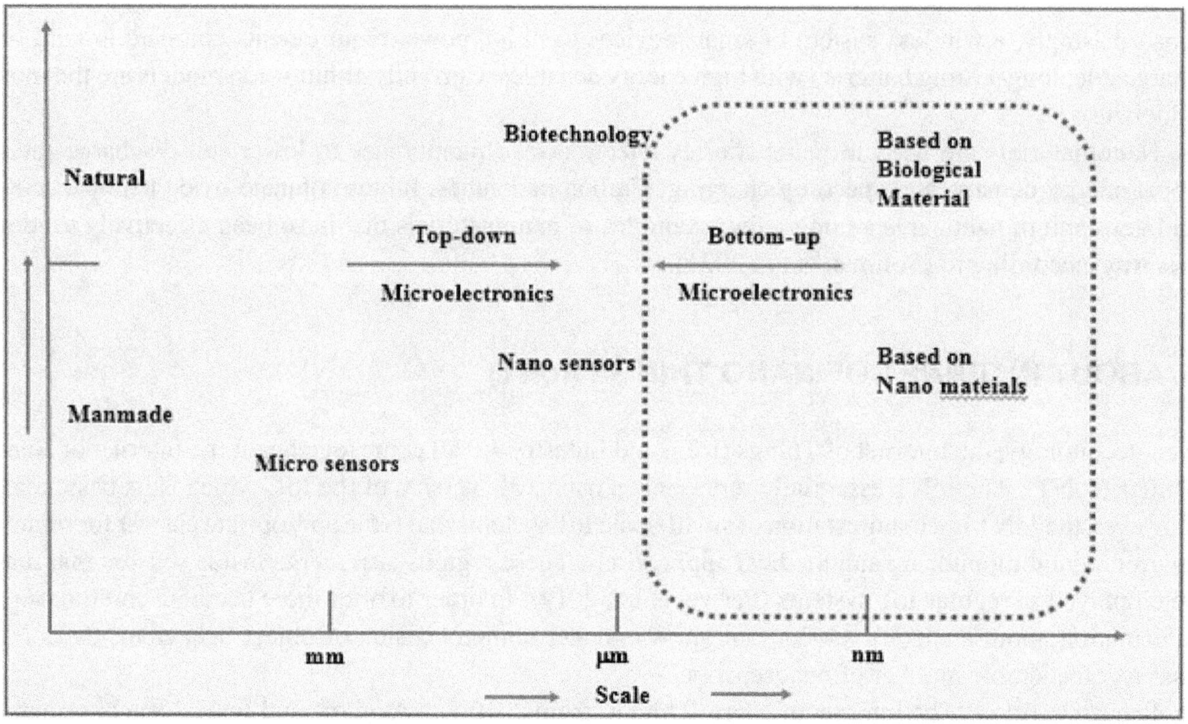

New network designs must be developed in order to connect nano devices to current communication networks like the Internet. IoNT development and adoption are dependent on processing power, affordable large-scale storage, smart antennas, and RFID tag technology. The Internet of Nanotechnology Things (IoNT) has given rise to new domains including the Internet of Bio-Nano Things (IoBNT) and the Internet of Multimedia-Nano Things (IoMNT), which can bring innovative advances to the healthcare and multimedia fields. Numerous nanosensors that function as wireless nano antennae that can sense and signal have also been created using non-biological materials like carbon nanotubes. Nanosensors can gather data from millions of different sites since they are so small. The information can then be combined by external devices to produce extraordinarily accurate maps that depict even the smallest variations in light, vibration, electrical currents, magnetic fields, chemical concentrations, and other environmental factors. The IoNT might offer considerably more accurate, reasonably priced, and current images of our cities, homes, factories, and even our bodies. Today, devices like wearables, security cameras, and traffic lights are being connected to the Internet (Parwej et al., 2019). Next, many nanosensors will capture enormous volumes of real-time data and transmit it to the cloud. One of the newest research areas has received a lot of attention is IoT using nanomachines.

6. FUNCTIONING OF INTERNET OF NANO THINGS (IoNT)

The IoT and nanotechnology can be used to create a physical network made of nanomaterials that makes data sharing easier by allowing various components to communicate with one another at the nanoscale.

Figure 2. The functioning of internet of nano things (IoNT)

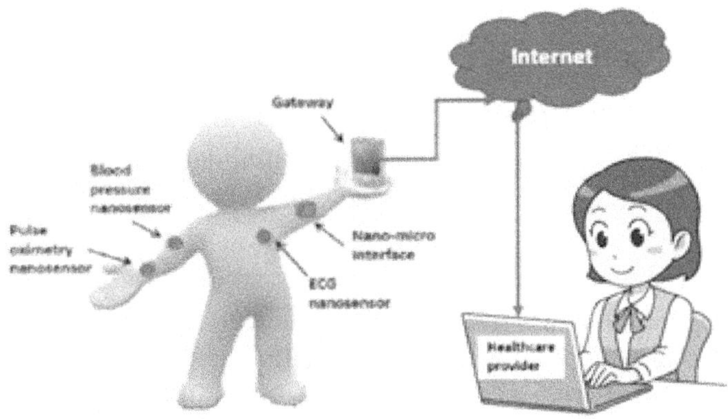

The Internet of Nano Things (IoNT) refers to this (Agarwal et al., 2017). It has not yet reached the same level of development as other IoT systems, but the communication and medical industries are showing interest in it. One such instance is in applications that require remote sensing in the field or while measuring several places on a human body. In order to build better sensor networks and more automated processes, we must first look at the traditional IoT systems that are currently being implemented. The Internet of Things (IoT) is a collection of sensor networks, data collectors, and transmitters that communicate data across the cloud from various entry points to a centralised place. This makes it possible for the Internet of Things (IoT) to function autonomously without requiring for human interaction—unless the system notifies an operator of a problem that it discovers through analysis. The Internet of Things (IoNT) is essentially a miniaturised version of these systems, which use tiny sensors and data network hubs to transfer data far distances. IoNT systems currently do not have the same level of sophistication as their IoT counterparts, but their capacity to collect data from such tiny sensor points makes them ideal for applications that cannot use other (bulkier) sensor networks. The IoNT network has a number of different parts that communicate with one another in order to send data over great distances.

There are various components in every system, and the IoNT is no exception. These components frequently communicate with one another through electromagnetic nano-communication (the transmission and reception of electromagnetic waves) as well as through molecular communication (the encoding of information in molecules). Nano nodes, nano-routers, nano-micro interface devices, and gateways are the four key components of the Internet of Things that aid in the transport of information. Figure 2 depicts the four essential parts of an IoNT system. Nano nodes, nano-routers, nano-micro interface devices, and gateways are the names given to these. The nano node is the smallest component (Miller et al., 2012). These are essentially simple nano machines and are commonly referred to as sensors in traditional IoT networks. The actions they can carry out and the range at which they can send data are constrained by their small size and limited internal capacity. Numerous nano nodes can link to a number of nano-routers, similar to how sensors relay localised data to a local hub before distributing it over great distances. Since nano-routers are much bigger than nano nodes and have more processing capacity, they can gather all the data from the nearby nano nodes, combine it, and send it across a distance to the nano-micro interface device. The simplest and smallest part of the IoNT (Miraz et al., 2015) configuration, a nano node is regarded as a fundamental nano machine. These tiny nano devices are utilised for data transmission and simple computations. They have a relatively little internal memory, and their small size (and energy) restricts how far they can send data. However, they can be positioned in a certain area and communicate information to a bigger nano-router, which then sends it over wider distances. As a result, the system's true sensor component may frequently be the nano nodes. The nano-router, a nano machine with a significantly bigger computing power, receives the data from the nano nodes. They serve as an aggregator for all the nearby nano nodes that collect the initial data because they have a significantly higher processing capacity. The information can then be sent to the nano-micro interface device and they can control the exchange of commands between the nano nodes. These interface devices collect all the data from the nano-routers and send it (Perwej et al., 2017) to the micro scale (and vice versa) utilising a blend of conventional network protocols and nano-communication techniques. The gateway then controls the entire system and makes the data accessible from anywhere using the internet. As a result, the IoNT (Balasubramaniam et al., 2013) does demonstrate some parallels with how IoT systems function, but due to the small size of the components, some hubs must be placed closer together.

7. THE MAIN REASON OF INTRNET OF NANO THINGS (IoNT)

Nanotechnology, nanomachines, and the Internet of Nano Things (IoNT) development will have a significant impact on technological development in nearly every industry in the near future (Ali et al., 2015). A new networking paradigm known as the Internet of Nano Things is defined by the connectivity of nanoscale devices with current communication networks and ultimately the Internet.

Figure 3. The nano communication in internet of nano things

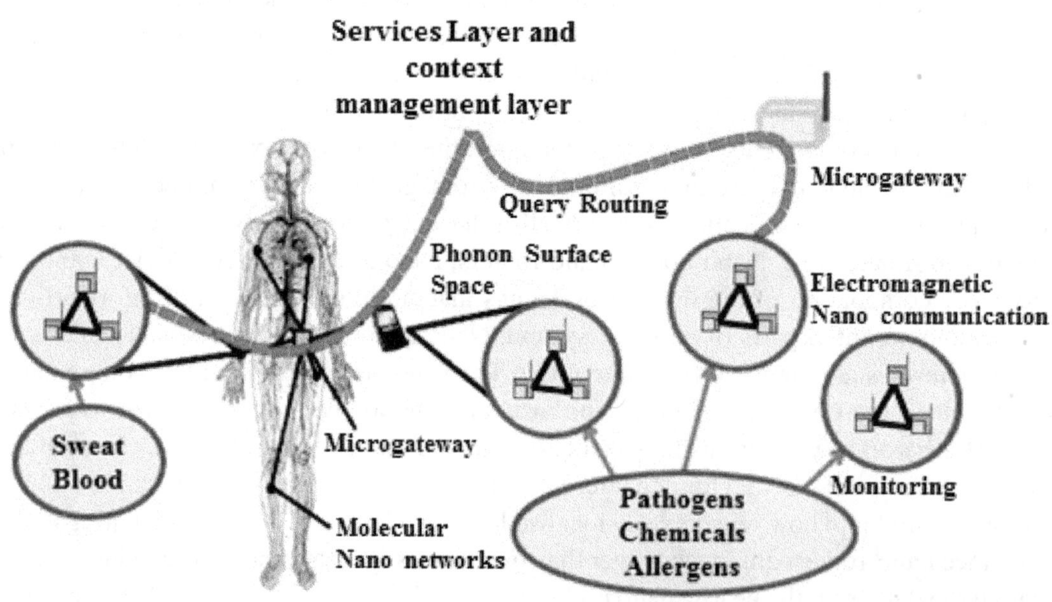

The IoNT is an extension of the Internet of Everything, but it allows for the use of nano-sensors and nano-networks in diverse items. In other words, the IoNT's motivation stems from its ability to connect various types of nanoscale devices in a communication network, enabling the collection of data in inaccessible locations (Akyildiz et al., 2015). Figure 3 depicts the interconnection that is created between various devices, as well as between nanosensors and nano-networks, with the objective of delivering crucial information in difficult-to-access locations. For instance, environmental nano-sensors might gather data on infections and allergens in a specific area, while on-body nano-sensors could provide electrocardiographic and other crucial signals. Nano-networks are not just a scaled-down version of conventional communication networks. They represent an entirely new paradigm in communication, where the majority of communication processes are motivated by biological systems found in nature.

8. INTERNET OF NANO THINGS (IoNT) MARKET TRENDS

The relatively recent series of nano-scale devices and the systems that are constructed utilising them can be used to describe the worldwide internet of nano things market. In order to better direct, gather, and analyse data, his global internet of nano things market indicates a business devoted to offering interconnectivity amongst nano-scale systems. Additionally, it makes it easier to share the data with different end users. Retail, media and entertainment, energy and utilities, transportation and logistics, manufacturing, and healthcare are just a few of the sectors that now make use of the worldwide internet of nano things market. The IoNT consists of a nanoscale network of various physical items that communicate with one another via nanotechnology. According to recent research, the IoNT market is anticipated to increase from roughly USD 5 billion in 2016 to USD 10 billion by 2020, as shown in figure 4. This growth is predicted to occur at a projected rate of more than 24.12% for the present forecast period of 2016 to 2025 (https://www.mordorintelligence.com/industry-reports/internet-of-nano-things-market). Nanotechnology with a number of other technologies, including IoT, sensors networks, cloud computing, and big data analytics, among others, can be combined to create IoNT infrastructure. The IoNT infrastructure is dependent on the operational region and necessary bandwidth for a particular application. Processing power, massive amounts of inexpensive storage, and sophisticated RFID tag technology are all necessary for the development and acceptance of IoNT.

The rising rate of demand for ubiquitous connection is a key factor in the global market for the internet of nano things. The internet of things concept and, by extension, the IoNT market are driven by the growing need for increased interconnection across linked and computing devices. Companies have been able to significantly increase the speed of data transfer between systems and reduce the complexity

Figure 4. The market growth rate internet of nano things (IoNT) by region (2020 - 2025)

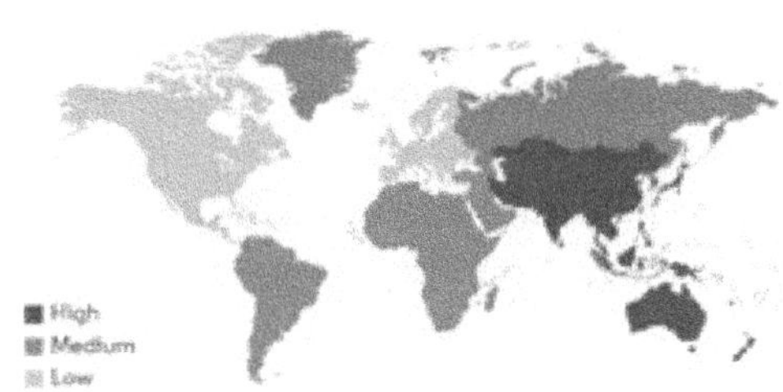

of various communications systems because to the worldwide market for the internet of nano things. The demand for and knowledge of IoNT has also grown as a result of the government's assistance for the development of IoNT technology for the healthcare industry (Ali et al., 2015). However, because to privacy and security concerns, the IoNT market's expansion confronts several difficulties. Concerns about the security of the data have grown as a result of the communication of sensitive data between devices over the internet. The significant capital outlay necessary for the advancement of nanotechnology is another factor impeding the growth of the IoNT market. The participants in the worldwide network of nano things are anticipated to concentrate on developed economies in North America and Europe, which possess a wider range of IoT and IoNT usage. In contrast to developing economies, which have only recently shown a greater level of interest in the industry, there are more players in these regions of the global IoNT market. Applications of IoNT across a variety of industries, including healthcare, transportation, logistics, manufacturing, media, entertainment, and retail, have revealed the market's enormous development potential. The IoNT's scope and uses are expanding as a result of technical developments, according to research. Major corporations are continually researching nanotechnology and creating nano systems with a greater range of uses. Among the leading companies in the IoNT market are SAP SE and Siemens AG in Germany, Cisco Systems Inc., Qualcomm Incorporated, Juniper Networks, and IBM Corporation in the United States, Schneider Electric, and Alcatel-Lucent S.A. in France, and Intel Corporation.

9. INTERNET OF NANO THINGS (IoNT) APPLICATIONS

The Internet of Nano Things (IoNT) will undoubtedly have a significant impact on the nanotechnology industry. In contrast to IoT, the opportunities provided by the new methods for collecting finely detailed information by IoNT allow for the expansion of already-existing applications as well as forays into new ones (Najah et al., 2016). Numerous industries, including oil and gas, defence, agriculture, smart cities, healthcare, and many more, are using IoT. The different IoNT applications that can enhance the advantages of nanotechnology in several fields are covered in this section (Kethineni., 2017).

9.1. Fertilizers and Pesticides Delivery in Agriculture

The agriculture industry could see significant changes as a result of nanotechnology. Among these difficulties are the rising dangers of plant-related diseases and threats to agricultural production. Tools developed using nanotechnology will help the agriculture industry tremendously by enabling quick disease detection, enhancing plant nutrient uptake, and advancing molecular disease therapy.

9.2. Drug Delivery Systems

The Internet of Bio-Nano Things (IoBNT) uses nanotechnology to create regulator implants that could be used to treat metabolic illnesses like diabetes. Nanosensors and intelligent glucose reservoirs work together to assist the glucose level mechanisms. Utilising a drug delivery system to administer particular medications or neuro transmitters to the nervous system can eradicate the effects of neurodegenerative illnesses (Freitas 2006).

9.3. Environmental Monitoring

Nanosensors are used in public places like bus stations, airports, hotels, etc. for environmental monitoring, along with real-time traffic and temperature monitoring. Nanotechnology could be used to monitor temperature, humidity, and air quality in intelligent industries and farms. IoNT is utilised in precision agricultural applications, which promotes effective crop development, environmental monitoring, and animal monitoring. The farm animals are being watched over and tracked using it.

9.4. Smart Cities

When a smart city is implemented, household appliances, monitoring sensors, observation cameras, actuators, buses, autos, and other devices can interact and communicate with each other. The geographical setting and daily routines of people influence the smart city's roadmap. No of their financial position, everybody can use the communication technology. The Internet of Things has already been used to construct smart cities. We can utilise nanosensors to track and identify the location of litter found in high-absorption areas of the air, and then to activate nanosensors to clean up that specific area. Additionally, we may use the countless nanosensors' assistance to gather a vast amount of data in real time to enhance living conditions and provide new services and applications (Jarmakiewicz et al., 2016).

9.5. Animals and Biodiversity Control

In natural settings, nano networks can manage a variety of animal species. Nano networks might create pheromones or messages that cause particular animals to behave in a certain way. As a result, it would be feasible to manage which animal species are present in a given area.

9.6. High Speed Data Transfer and Cellular

IoNT integrated electromagnetic wave nature qualities for Pico and Fem to cells-based networking over the Terahertz range (0.1THz -10THz), for communication among sensors, in the telecommunications domain. Due to the larger capacity and available transmission rate of THz-based frequencies, it is possible to convey information quickly using these frequencies in cutting-edge cell systems and military-related applications (Lee et al., 2002).

9.7. Biomedical Applications

Molecules, proteins, DNA, organelles, and the main parts of living cells all naturally exist at the nanoscale. As a result, the biomedical industry has a very large variety of applications for nano networks. For instance, nanomaterial-based biological nanosensors (Yonzon et al., 2005) can be used externally (like a tattoo) or even internally (like a pill or intramuscular injection) to track blood sugar, sodium, and cholesterol levels, to find the presence of infectious agents (Tallury et al., 2010), or to recognise particular cancer types. To gather data and send it to a healthcare professional, a wireless interface between these nano devices and a micro device, like a mobile phone or medical equipment, may be used. Better health monitoring and treatment systems that combine chemical and biological nanosensors, ultra-high-resolution nanocameras, and ultrasonic nanophones are being developed for the early detection and treatment of diseases like cancer. Parents who want to monitor their child's use of addictive drugs are another example. They can do this by placing specific nanosensors inside their child's body. These sensors will offer insightful, unbiased information about the occasions and settings of drug use. Since visible range sensors placed outside of bodies could be removed, nanosensors are preferable in this circumstance.

9.8. Functionalized Materials and Fabrics

Nano networks can be used in the production of new, high-tech fabrics and materials to enhance specific capabilities. Nano functionalized materials have been used to make items like fabrics that are antibacterial and stain-repellent (Ravindra et al., 2010). Advanced smart fabrics' improved airflow is achieved by controlling the reaction using nano actuators and nanosensors in communication.

9.9. Oil and Gas

We can increase the oil's rate of repossession by employing the nanosensor. We can find the oil linked to the rocks by using nanosensors, which can pass through the gaps in the rock. Although a cross well imaging and seismic instrument has a greater impact on this region, the firmness they provide is quite modest. Nanosensors collaborate and connect with one another through molecular communication in the Internet of Nano Things. Additionally, the neighbouring gateway can be used to transmit the data collected in real time. Consequently, without the necessity for an accurate magnetic source and receiver, the oil position can be easily plotted.

9.10. Industrial and Consumer Goods Applications

Nanotechnology is being used to create a variety of new industrial and consumer products, from stretchable and flexible electronics to textiles that are anti-microbial and self-cleaning thanks to newly developed functionalized nanoparticles (Rogers et al., 2009). Additionally, practically everything in our everyday lives—from kitchen appliances to every component of our workplace—will be connected as nano devices with communication capabilities are integrated into every single object.

9.11. Multimedia

It focuses on creating high-resolution multimedia content by building devices like photo-detectors and acoustic nano-transducers (El-din et al., 2017). Focus areas for the nano-multimedia systems include forensic research, industrial process control, biological attacks, and health. It is difficult to increase the clarity and accuracy of visual and acoustic information, but nano-cameras and nano-phones can solve this problem by providing larger computing and storage capacity, higher quality image and audio sensing capabilities, and higher energy efficiency.

9.12. Military and Defense Applications

Two examples of the military applications made possible by nano networks include sophisticated nuclear, biological, and chemical (NBC) defences and complex damage detection systems for civil structures, soldier armour, and military vehicles. A network of nanosensors, for instance, can be used to detect dangerous chemicals and biological weapons with unprecedented accuracy and promptness in a variety of settings, such as airport lobbies or conference rooms (where they might be hidden in the wall paint or deployed from an unmanned vehicle that is invisible to the human eye) (https://www.shiksha.com/online-courses/articles/applications-of-internet-of-things-iot-in-defence-and-military/).

9.13. Food Packaging

Regarding clever packaging Over the past several years, there has been a sharp increase in the amount of research being done in the subject of nanotechnology. Several businesses are now concentrating on developing new types of nanoscale materials. The food industry is one that is starting to take notice of this, which is not surprising given that the public's preference for natural food products has historically prevented the adoption of cutting-edge food technologies. Undoubtedly, wrapping is the most recent frontier of food nano science investigation and expansion (Chaudhary et al., 2020). The sealing of food packaging can be considerably improved after the introduction of nanomaterial compounds like $SiO2$, $TiO2$, and $KMnO4$.

9.14. Body Sensor Network (BSN)

This comprises of in body nano sensors playing a crucial role in collecting and monitoring patient's biological activities. The nano sensors being used in BSN to provide real time data on a wearable device. They can be placed in the body and a gateway to communicate with Internet (Lueken et al., 2020).

9.15. Immune System Support

IoBNT can be utilized to support the immune system to identify and control foreign and pathogen elements in the human body. Several nano devices such as sensors and actuators collaborate with each other in macro, micro, and nano systems to protect organism against diseases. Implementing nano devices can advance the medical field by utilizing theses nano devices to predict, detect, and eliminate certain procedures based on localization of malicious agents and cells, such as cancer cells. This will minimize the risk of developing such disease and provide treatments less aggressive and invasive compared to the existing ones (Chen et al., 2005).

10. CONCLUSION AND FUTURE SCOPE

Extensive research is already being done to address the remaining issues the IoT and IoNT are encountering. Future research is expected to be diverse and abundant, ranging from switched-beam nanoantennas made of graphene to multi-layer blockchain security models. Energy harvesting, which involves obtaining energy from outside sources and transforming it into usable electrical energy, may be the answer to the problems associated with sensor and nanosensor power supplies. Once more, nanotechnology is anticipated to play a significant role in the future of this solution. In addition to other examples, (Phillips 2021) demonstrated how nanowires in piezoelectric nanogenerators and quantum dots in thermoelectric generators could completely replace IoT and IoNT endpoint batteries, allowing for a rapid uptake of the technology while being both environmentally and commercially viable.

REFERENCES

Agarwal, Agarwal, & Agarwal. (2017). Evolution of Internet of Nano Things (IoNT). *International Journal of Engineering Technology Science and Research, 4*(7).

Agarwal, K., Agarwal, K., & Agarwal, S. (2017). Evolution of Internet of Nano Things (IoNT). *Int. J. Eng. Technol. Sci. Res., 4*(7), 274–277.

Akhtar, Parwej, & Perwej. (2017). A Perusal of Big Data Classification and Hadoop Technology. *International Transaction of Electrical and Computer Engineers System (ITECES), 4*(1), 26-38. DOI:. doi:10.12691/iteces-4-1-4

Akyildiz, I., & Jornet, J. (2010). The Internet of nano-things. *IEEE Wireless Communications, 17*(6), 58–63. doi:10.1109/MWC.2010.5675779

Akyildiz, I., Pierobon, M., Balasubramaniam, S., & Koucheryavy, Y. (2015). The internet of Bio-Nano things. *IEEE Communications Magazine, 53*(3), 32–40. doi:10.1109/MCOM.2015.7060516

Akyildiz, I. F., Brunetti, F., & Blázquez, C. (2008). Nanonetworks: A new communication paradigm. *Computer Networks, 52*(12), 2260–2279. doi:10.1016/j.comnet.2008.04.001

Akyildiz, I. F., & Jornet, J. M. (2010). Electromagnetic wireless nanosensor networks. *Nano Communication Networks, 1*(1), 3–19. doi:10.1016/j.nancom.2010.04.001

Akyildiz, I. F., Jornet, J. M., & Pierobon, M. (2011). Nanonetworks. *Communications of the ACM, 54*(11), 84–89. doi:10.1145/2018396.2018417

Al-Shargabi, B., & Sabri, O. (2017). Internet of Things: An exploration study of opportunities and challenges. *2017 International Conference on Engineering & MIS (ICEMIS).* https://doi.org/10.1109/ICEMIS.2017.8273047

Ali, N. A., & Abu-Elkheir, M. (2015). Internet of nano-things healthcare applications: Requirements, opportunities, and challenges. *2015 IEEE 11th International Conference on Wireless and Mobile Computing, Networking and Communications (WiMob).* 10.1109/WiMOB.2015.7347934

Ali, N. A., Aleyadeh, W., & AbuElkhair, M. (2016). Internet of Nano-Things network models and medical applications. *2016 International Wireless Communications and Mobile Computing Conference (IWCMC).* 10.1109/IWCMC.2016.7577059

AnandNayyar. (2017). Internet of Nano Things(IoNT): Next Evolutionary Step in Nanotechnology. *Nanoscience and Nanotechnology, 7*(1), 4–8. doi:10.5923/j.nn.20170701.02

Balasubramaniam, S., & Kangasharju, J. (2013). Realizing the Internet of Nano Things: Challenges, Solutions, and Applications. *Computer, 46*(2), 62–68. doi:10.1109/MC.2012.389

Bzai, J., Alam, F., Dhafer, A., Bojović, M., Altowaijri, S. M., Niazi, I. K., & Mehmood, R. (2022). Machine Learning-Enabled Internet of Things (IoT): Data, Applications, and Industry Perspective. *Electronics (Basel), 11*(17), 2676. doi:10.3390/electronics11172676

Chaudhary, P., Fatima, F., & Kumar, A. (2020). Relevance of Nanomaterials in Food Packaging and its Advanced Future Prospects. *Journal of Inorganic and Organometallic Polymers and Materials*, *30*(12), 5180–5192. doi:10.1007/s10904-020-01674-8 PMID:32837459

Ching-Jen Chen, Haik, Y., & Chatterjee, J. (2005). Development of nanotechnology for biomedical applications. Conference. *Emerging Information Technology*, *2005*, 9–12. Advance online publication. doi:10.1109/EITC.2005.1544329

Dressler, F., & Fischer, S. (2015). Connecting in-body nano communication with body area networks: Challenges and opportunities of the Internet of Nano Things. *Nano Communication Networks*, *6*(2), 29–38. doi:10.1016/j.nancom.2015.01.006

Ezz El-Din, H., & Manjaiah, D. H. (2017). Internet of Nano Things and Industrial Internet of Things. *Internet of Things: Novel Advances and Envisioned Applications*, 109–123. . doi:10.1007/978-3-319-53472-5_5

Fischer, H., Seidenstricker, S., Berger, T., & Holopainen, T. (2022). *Artificial intelligence in B2B sales: Impact on the sales process.* AHFE International. doi:10.54941/ahfe1001456

Freitas, R. A. Jr. (2006). Pharmacytes: An Ideal Vehicle for Targeted Drug Delivery. *Journal of Nanoscience and Nanotechnology*, *6*(9), 2769–2775. doi:10.1166/jnn.2006.413 PMID:17048481

Ghder Soliman, W., Swathi, Ch., Yasasvi, T., Keerthi Priya, B., & Akhila Reddy, D. (2021). Review on poly(ethylene oxide)-based electrolyte and anode nanomaterials for the internet of things node-level lithium-ion batteries. *Materials Today: Proceedings*, *42*, 429–435. doi:10.1016/j.matpr.2020.09.806

https://www.mordorintelligence.com/industry-reports/internet
-of-nano-things-market

https://www.shiksha.com/online-courses/articles/applications
-of-internet-of-things-iot-in-defence-and-military/

Jarmakiewicz, J., Parobczak, K., & Maslanka, K. (2016). On the Internet of Nano Things in healthcare network. *2016 International Conference on Military Communications and Information Systems (ICMCIS).* 10.1109/ICMCIS.2016.7496572

Jornet, J. M., & Akyildiz, I. F. (2012). The Internet of Multimedia Nano-Things. *Nano Communication Networks*, *3*(4), 242–251. doi:10.1016/j.nancom.2012.10.001

Kethineni, P. (2017). Applications of internet of nano things: A survey. *2017 2nd International Conference for Convergence in Technology (I2CT).* 10.1109/I2CT.2017.8226154

Kulakowski, P., Solarczyk, K., & Wojcik, K. (2017). Routing in FRET-Based Nanonetworks. *IEEE Communications Magazine*, *55*(9), 218–224. doi:10.1109/MCOM.2017.1600760

Kuscu, M., & Akan, O. B. (2016). The Internet of Molecular Things Based on FRET. *IEEE Internet of Things Journal*, *3*(1), 4–17. doi:10.1109/JIOT.2015.2439045

Lee, S.-W., Mao, C., Flynn, C. E., & Belcher, A. M. (2002). Ordering of Quantum Dots Using Genetically Engineered Viruses. *Science*, *296*(5569), 892–895. doi:10.1126/science.1068054 PMID:11988570

Lueken, M., Mueller, L., Decker, M. G., Bollheimer, C., Leonhardt, S., & Ngo, C. (2020). Evaluation and Application of a Customizable Wireless Platform: A Body Sensor Network for Unobtrusive Gait Analysis in Everyday Life. *Sensors (Basel), 20*(24), 7325. doi:10.3390/s20247325 PMID:33419278

Mahdavinejad, M. S., Rezvan, M., Barekatain, M., Adibi, P., Barnaghi, P., & Sheth, A. P. (2018). Machine learning for internet of things data analysis: A survey. *Digital Communications and Networks, 4*(3), 161–175. doi:10.1016/j.dcan.2017.10.002

Miller & Kearnes. (2012). *Nanotechnology, Ubiquitous Computing and the Internet of Things: Challenges to Rights to Privacy and Data Protection*. Draft Report to the Council of Europe.

Miraz, M. H., Ali, M., Excell, P. S., & Picking, R. (2015). *A review on Internet of Things (IoT), Internet of Everything (IoE) and Internet of Nano Things (IoNT). 2015 Internet Technologies and Applications*. ITA. doi:10.1109/ITechA.2015.7317398

Parwej, Akhtar, & Perwej. (2018). A Close-Up View About Spark in Big Data Jurisdiction. *International Journal of Engineering Research and Application, 8*(1), 26-41. DOI: doi:10.9790/9622-0801022641

Parwej, D. (2019). An empirical analysis of web of things (WOT). *International Journal of Advanced Research in Computer Science, 10*(3), 32–40. doi:10.26483/ijarcs.v10i3.6434

Perwej, Omer, Sheta, Harb, & Adrees. (2019). The Future of Internet of Things (IoT) and Its Empowering Technology. *International Journal of Engineering Science and Computing, 9*(3), 20192–20203.

Perwej, Y. (2016). A Literature Review of the Human Body as a Communication Medium using RedTacton. *Communications on Applied Electronics, 4*(9), 7–17. doi:10.5120/cae2016652161

Perwej, Y. (2017). The Next Generation of Wireless Communication Using Li-Fi (Light Fidelity) Technology. *Journal of Computer Networks, 4*(1), 20–29. doi:10.12691/jcn-4-1-3

Perwej, Y. (2017, March). An Experiential Study of the Big Data. *Science and Education Publishing, 4*(1), 14–25. doi:10.12691/iteces-4-1-3

Perwej, Y. (2019). The Hadoop Security in Big Data: A Technological Viewpoint and Analysis. *International Journal of Scientific Research in Computer Science and Engineering, 7*(3), 1–14. doi:10.26438/ijsrcse/v7i3.1014

Perwej, Y., Ahmed, M., Kerim, B., & Ali, H. (2019). An Extended Review on Internet of Things (IoT) and its Promising Applications. *Communications on Applied Electronics, 7*(26), 8–22. doi:10.5120/cae2019652812

Perwej, Y., Haq, K., Parwej, F., & M, M. (2019). The Internet of Things (IoT) and its Application Domains. *International Journal of Computer Applications, 182*(49), 36–49. doi:10.5120/ijca2019918763

Perwej, Y., Kerim, B., Adrees, M. S., & Sheta, O. E. (2017). An Empirical Exploration of the Yarn in Big Data. *International Journal of Applied Information Systems, 12*(9), 19–29. doi:10.5120/ijais2017451730

Perwej, Y., Parwej, F., Mohamed Hassan, M. M., & Akhtar, N. (2019). The Internet-of-Things (IoT) Security : A Technological Perspective and Review. *International Journal of Scientific Research in Computer Science, Engineering and Information Technology*, 462–482. . doi:10.32628/CSEIT195193

Phillips, J. D. (2021). Energy Harvesting in Nanosystems: Powering the Next Generation of the Internet of Things. *Frontiers in Nanotechnology*, *3*, 633931. Advance online publication. doi:10.3389/fnano.2021.633931

Ravindra, S., Murali Mohan, Y., Narayana Reddy, N., & Mohana Raju, K. (2010). Fabrication of antibacterial cotton fibres loaded with silver nanoparticles via "Green Approach.". *Colloids and Surfaces. A, Physicochemical and Engineering Aspects*, *367*(1–3), 31–40. doi:10.1016/j.colsurfa.2010.06.013

Rogers, J. A. (2009). Materials and mechanics for stretchable electronics - from electronic eye cameras to conformal brain monitors. *TRANSDUCERS 2009 - 2009 International Solid-State Sensors, Actuators and Microsystems Conference*. 10.1109/SENSOR.2009.5285765

Simonjan, J., Jornet, J. M., Akyildiz, I. F., & Rinner, B. (2018). Nano-cameras. *Proceedings of the 5th ACM International Conference on Nanoscale Computing and Communication*. 10.1145/3233188.3233209

Stelzner, M., Dressler, F., & Fischer, S. (2016). Function Centric Networking. *Proceedings of the 3rd ACM International Conference on Nanoscale Computing and Communication*. 10.1145/2967446.2967479

Tabassum, N. (2020). An empirical exploration of the nanotechnology. *International Journal of Advanced Research*, *8*(7), 885–915. doi:10.21474/IJAR01/11352

Tallury, P., Malhotra, A., Byrne, L. M., & Santra, S. (2010). Nanobioimaging and sensing of infectious diseases. *Advanced Drug Delivery Reviews*, *62*(4–5), 424–437. doi:10.1016/j.addr.2009.11.014 PMID:19931579

Tang, N., Zhou, C., Xu, L., Jiang, Y., Qu, H., & Duan, X. (2019). A Fully Integrated Wireless Flexible Ammonia Sensor Fabricated by Soft Nano-Lithography. *ACS Sensors*, *4*(3), 726–732. doi:10.1021/acssensors.8b01690 PMID:30793588

Whitmore, A., Agarwal, A., & Da Xu, L. (2014). The Internet of Things—A survey of topics and trends. *Information Systems Frontiers*, *17*(2), 261–274. doi:10.1007/s10796-014-9489-2

Yonzon, C., Stuart, D., Zhang, X., McFarland, A., Haynes, C., & Vanduyne, R. (2005). Towards advanced chemical and biological nanosensors—An overview. *Talanta*, *67*(3), 438–448. doi:10.1016/j.talanta.2005.06.039 PMID:18970187

Chapter 15
Modeling Tools for Evaluating Materials in CO_2 Capture:
Performance and Sustainability

Ilse María Hernández-Romero
https://orcid.org/0000-0003-3787-1603
Tecnológico de Monterrey, Mexico

Michael Pérez-Rodríguez
Tecnológico de Monterrey, Mexico

Karina Hernández-Romero
Universidad Michoacana de San Nicolás de Hidalgo, Mexico

Lucy T. González
Tecnológico de Monterrey, Mexico & Tecnologico de Monterrey, Centro del Agua, Monterrey, Mexico

ABSTRACT

This chapter examines the importance of materials in capturing carbon dioxide (CO_2) and the criteria for selecting these materials using modeling tools. The discussion starts by analyzing the adsorption process of CO_2 capture as a phenomenon and selecting suitable adsorbent materials under different considerations. After that, an exploration of important modeling tools that can be used to simulate the CO_2 process and to evaluate the performance of these materials is conducted. In addition, a brief overview of the relationship between theoretical mathematical models and experimental methods to validate and improve both is addressed. Additionally, it discusses the integration of mathematical programming and optimization approaches in the selection and performance of materials in the CO_2 capture process and related technologies. Finally, it examines the economic and sustainability considerations when selecting materials, highlighting the need for technology to be economically feasible and environmentally sustainable.

DOI: 10.4018/979-8-3693-1306-0.ch015

1. INTRODUCTION

The global fight against climate change requires capturing and storing carbon dioxide (CO_2) to reduce greenhouse gas emissions (Gür, 2022). To achieve this goal, technology for capturing CO_2 has become an essential part of efforts to lower CO_2 levels in the atmosphere and combat global warming. One crucial aspect of the technology is selecting and evaluating absorbent materials that can effectively trap CO_2 from industrial and energy sources (Lu et al., 2015). Capturing CO_2 has thus become a crucial strategy for practical solutions to address climate change, entailing the extraction and retention of this greenhouse gas from such sources to reduce atmospheric emissions. In this way, the effectiveness and sustainability of absorbent materials are critical factors for the successful implementation of CO_2 capture (Wang et al., 2014). In order to address this issue, modeling tools have been developed to assess the performance and sustainability of these materials, offering a comprehensive and multidisciplinary approach to the fight against climate change (Papadopoulos & Seferlis, 2017).

In this sense, one of the primary concerns in CO_2 capture is the efficiency, the sustainability, and the cost of absorbent materials (Yu et al., 2012). Therefore, mathematical and modeling tools allow for simulating how these materials interact with CO_2 for capture and how to maximize their absorption capacity. It includes approaches such as quantum mechanics, molecular dynamics, among others which provide a detailed understanding of the capture process and their molecular interactions. On the other hand, by simulating and optimizing these processes, derived from fundamental modeling, it becomes possible to consider their effects in different technologies (Hernández-Romero et al., 2022). Thus, it leads to the proposal of optimal operating conditions that not only maximize CO_2 capture but also minimize energy consumption, reduce environmental impact, and optimize costs (Chatziasteriou et al., 2022). In this way, mathematical modeling plays a crucial role in evaluating and predicting fundamental aspects such as stability, performance, and regeneration of these materials in adsorbent materials (Murge & Roy, 2019). On the other hand, it is essential to ensure the longevity and reusability of these materials to maximize the efficiency of CO_2 capture processes. The longevity of absorbent materials is fundamental because their life cycle must be as long as possible to justify their investment and environmental impact and it can be considered in the modeling of this process. The reuse of these materials not only reduces the costs associated with the production and disposal of discarded materials and reduces the need to produce new absorbent materials. It, in turn, reduces the carbon footprint and waste generated by the industry while promoting sustainability.

In addition, mathematical modeling of adsorbent materials for CO_2 capture has become an interdisciplinary discipline that spans chemistry, physics, engineering, and materials science. This tool has become crucial for designing and optimizing materials with specific properties that maximize CO_2 capture efficiency (Berger & Bhown, 2013). The mathematical models can be used to predict how materials interact with CO_2, understand the underlying mechanisms, and optimize operating conditions. The mathematical modeling techniques employed in this field vary in complexity and scope. At the most fundamental level, calculations based on electron density function theory (DFT) allow the simulation of molecular interactions between adsorbent materials and CO_2 at the atomic level (Zhang et al., 2021). These approaches provide detailed information on the electronic structure of materials and chemical properties, which is essential for understanding CO_2 adsorption. On the other hand, molecular dynamics (MD) simulations allow to study how adsorbent materials behave under realistic and dynamic conditions (Imteyaz et al., 2023). These simulations provide information on the diffusion of CO_2 molecules in the structure of the materials and how this affects the trapping capacity.

For example, one of the most important models consider in the CO_2 capture process are the thermodynamic models, that are necessary to predict the CO_2 adsorption equilibrium at different temperature and pressure conditions (Hernández-Romero et al., 2022). It is essential to evaluate an adsorbent material's capacity in various practical scenarios. Moreover, predicting and adjusting these variables is crucial to ensure the optimal performance of adsorbent materials over time, extending their lifetime and CO_2 capture efficiency. By considering all aspects together, developing sustainable and efficient CO2 capture processes is possible. The integration of mathematical modeling in evaluating absorbent materials contributes to creating solutions that maximize capture capacity, reduce operating costs, minimize environmental impact, and ensure the long-term reuse of these materials (Dziejarski et al., 2023).

On the other hand, it is crucial to ensure the sustainability of adsorbent materials to meet specific economic and sustainable standards. One important aspect is to consider modeling tools for a comprehensive assessment. Life cycle models provide a view of environmental effects, evaluating the environmental impact of absorbent materials from manufacturing to final disposal (González-Olmos et al., 2022). Using these models, it can identify the most sustainable materials and make informed decisions regarding their use. Minimizing the carbon footprint associated with producing and using these materials is a step towards adopting an eco-friendly approach and contributes to mitigating climate change.

Furthermore, economic modeling tools are essential to cost-benefit evaluations, and they are also used to evaluate the efficiency of CO_2 capture strategies. These models assess the production costs, capture efficiency, and benefits related to emission reductions (Zhao et al., 2018). With the help of accurate economic models, it is possible to choose sustainable and economically feasible absorbent materials. By balancing costs and benefits, it becomes possible to achieve a CO_2 capture strategy that is both effective in reducing emissions and economically justifiable. Moreover, modeling significantly evaluates the materials considering multiple objectives, such as efficiency, cost, energy sustainability, performance, and other vital aspects (Zamarripa et al., 2018). Integrating these multiple factors into the modeling process can provide a more complete and balanced view of the selection and optimization of sorbent materials for CO_2 capture (Romero-García et al., 2022). This approach aims to increase capture capacity and reduce operating costs, energy consumption, and environmental impact, resulting in sustainable solutions to combat climate change.

This chapter provides an overview of the current modeling tools used for assessing materials in the context of CO_2 capture. The discussion is focused on three key factors: the significance of the material adsorbent, their efficiency and capture capacity, the interaction between experimental and modeling approaches to develop a framework for selecting the appropriate material, and lastly, the environmental and economic evaluation to ensure the sustainability of the process.

2. CO_2 CAPTURE AND MATERIAL SELECTION

2.1 CO_2 Capture Process by Adsorption

The CO_2 capture process through adsorption involves an intricate interaction between carbon dioxide molecules and the surface of a solid material known as the adsorbent (see **Figure 1**). In this way, adsorption is a process in which molecules or particles from a gas or liquid adhere to the surface of a solid or liquid material (Chaffee et al., 2007). This adherence occurs due to attractive forces between the adsorbate (the substance being adsorbed) and the surface of the adsorbent (the material on which

Figure 1. Absorption and adsorption dynamics: A comparative diagram

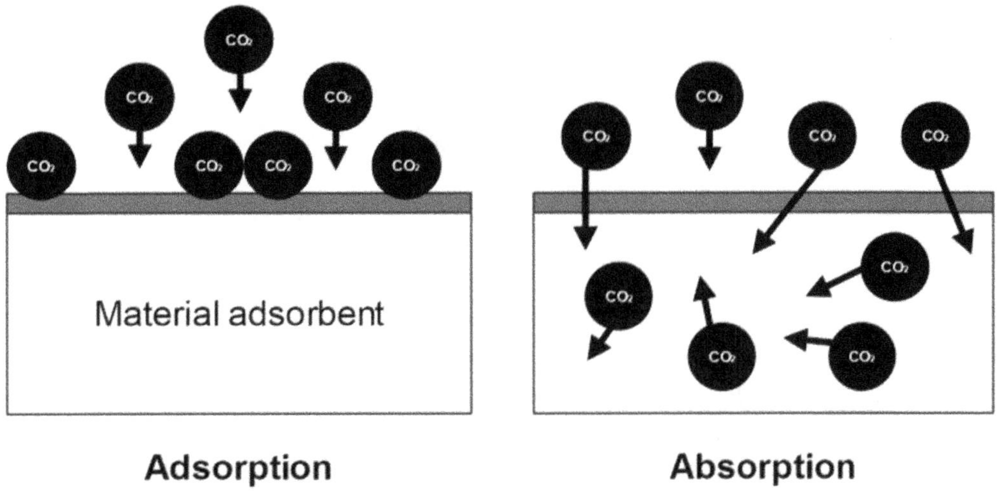

adsorption takes place). Unlike absorption, where another substance takes up a substance throughout its volume, adsorption involves only the surface layer. This phenomenon, referred to as adsorption, is influenced by several factors determining its efficiency and effectiveness. Adsorption isotherms play a crucial role in establishing the quantitative relationship between the amount of adsorbed CO_2 and the equilibrium pressure in the gas phase, allowing for the characterization of adsorption capacity under different conditions (Chaffee et al., 2017). The surface chemistry of the adsorbent plays a decisive role as functional groups interact with CO_2 through mechanisms such as physisorption or chemisorption, depending on the nature of the adsorbent. The material's pore structure is also essential as it influences the accessibility of CO_2 molecules to adsorption sites (Kayan, 2019).

Additionally, temperature and pressure impact the kinetics and adsorption capacity, with adsorption often favored at lower temperatures and higher pressures (Raganati & Ammendola, 2020). The process is dynamic, involving continuous adsorption of CO_2 in contact with the gas stream and subsequent desorption under different conditions. Competitive adsorption, where other components in the gas stream interact with the adsorbent, and the adsorbent regeneration after capture are crucial aspects for the overall efficiency of the CO_2 capture system. In summary, detailed optimization of these factors is essential to designing CO_2 capture systems that are efficient and sustainable in the long term.

2.2 Importance of Adsorbent Material

Thorough consideration of materials is necessary for effective CO_2 capture from emission sources, as shown in Figure 2.

Different materials have different properties that can be used for CO_2 capture. These materials can be categorized into sorbents, membranes, and adsorbents. However, each material type has its benefits and challenges. It makes it difficult to select the best material for optimal CO_2 capture efficiency. For instance,

- Sorbents

Figure 2. Overview of high-performance adsorbent materials for CO_2 capture

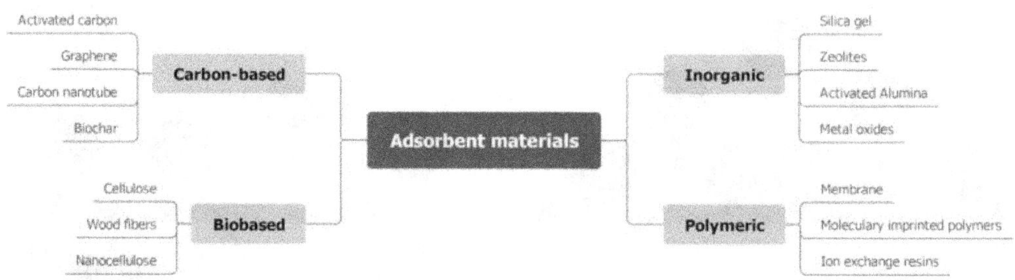

Sorbents play a crucial role in capturing CO_2 through absorption processes. These materials have a natural affinity for CO_2 molecules, making capturing and removing them easy. The most common types of sorbents include amine-based compounds, metal-organic frameworks (MOFs), and activated carbon. Amine-based sorbents react chemically with CO_2 to form stable compounds, while MOFs have adjustable structures that allow for tailored adsorption capacities (Trickett et al., 2017). The selection of sorbent materials depends on their capacity, selectivity, regenerability, and stability under operational conditions.

- Membranes

Carbon dioxide capture can be achieved using membrane-based technologies, which utilize semi-permeable materials for selective gas permeation (Han & Ho, 2018). The three most used types of membranes are polymeric, mixed-matrix, and zeolite. Polymeric membranes use the selective permeability of polymers to CO_2, while mixed-matrix membranes incorporate nanoparticles to enhance performance. Zeolite membranes have ordered microporous structures that enable molecular sieving. The selection of materials for membranes is based on their permeability, selectivity, durability, and scalability.

- Adsorbents

Adsorbents such as activated carbon and zeolites can capture CO_2 by physically adsorbing it onto their surfaces. Activated carbon is known for its high surface area and porosity, which allows it to undergo reversible adsorption-desorption cycles. Conversely, zeolites are crystalline aluminosilicates that exhibit tunable structures and have specific adsorption sites (Boer et al., 2023). The choice of adsorbent materials depends on adsorption capacity, kinetics, thermal stability, and regenerability (Dziejarski et al., 2023).

2.3 Selecting Optimal Materials: Criteria and Considerations

Selecting materials for CO_2 capture is a complex process that requires careful consideration of different critical criteria (Majchrzak & Nowak, 2017), as Table 1 shows. These criteria ensure that the chosen materials are efficient, sustainable, and economically viable, for instance:

1. **Efficiency:** The ability of a material to capture CO_2 depends on two fundamental factors: capacity and selectivity. Capacity refers to how much CO2 a material can absorb or adsorb, which is crucial for ensuring the efficiency of the capture process. Selectivity, on the other hand, describes

Table 1. Criteria for evaluating materials in CO_2 capture process (Espinal et al., 2013)

Criteria	Sorbents	Membranes	Adsorbents
Capacity	High, dependent on chemical reactions	Moderate to High, varies with material	High, influenced by porosity
Selectivity	High for target gas	Based on permeability	Based on adsorption sites
Stability and Durability	Sensitive to chemical degradation	Depends on material resilience	Resilient to physical changes
Regenerability	Feasible through regeneration	Varies; regeneration is possible	Feasible through desorption
Cost-effectiveness	Cost of amines, MOFs.	Cost of polymerization, fillers	Cost of synthesis and production
Material Adaptability	Subject to chemical reactions	Adaptable to temperature and pressure	Influenced by adsorption conditions

the material's ability to differentiate CO_2 from other gases in the emission stream, which helps to maintain the purity of the captured gas.

2. **Stability and Durability:** In CO_2 capture processes, materials need to be stable and durable, even under harsh and dynamic conditions. It is crucial that materials can resist chemical degradation, physical changes, and thermal fluctuations to ensure their longevity and sustained operational performance over multiple capture and release cycles (Oschatz & Antonietti, 2018).
3. **Regenerability:** The ability of a material to go through capture and release cycles without considerable damage is known as regenerability. If a material can quickly regenerate and efficiently release the CO_2 it has absorbed, it can aid in the economic viability of CO_2 capture technologies. Maintaining performance over multiple cycles is a critical aspect of regenerability.
4. **Cost-effectiveness:** When selecting materials for CO_2 capture technologies, the cost of material synthesis, deployment, and maintenance is a crucial factor to consider. It is crucial to balance performance and cost-effectiveness to ensure the widespread adoption of these technologies. This criterion evaluates the expenses associated with material synthesis, integration into capture systems, and any required regeneration processes.
5. **Material Adaptability and Conditions:** Compatibility with the specific CO_2 capture conditions is crucial for optimal performance and reliability of materials used. This includes temperature, pressure, and chemical environment considerations.
6. **Sustainable Impact:** In the context of sustainability, the ecological impact of materials is becoming increasingly important. When choosing materials for CO_2 capture, it is crucial to consider their environmental impact throughout their life cycle, from synthesis to disposal. To achieve broader sustainability goals, it is essential to minimize environmental consequences such as energy consumption and waste generation.

Besides carefully choosing materials, using advanced modeling tools can improve CO_2 capture technologies. These computational tools can better understand material behaviors, predict their performance in different situations, and help optimize capture systems. Combining material selection with robust modeling can advance CO_2 capture technologies, improve their performance, and significantly contribute to global efforts to reduce emissions. This approach prioritizes sustainability in carbon capture by combining material selection and modeling tools.

Figure 3. Conceptual framework for mathematical modeling in CO_2 capture

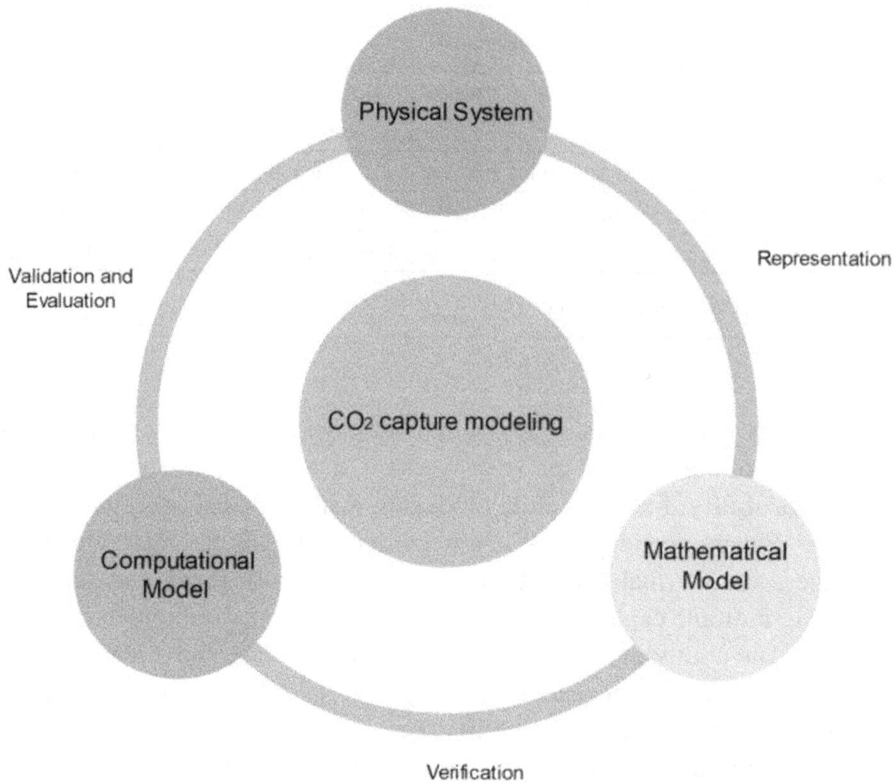

3. MODELING PERSPECTIVES: A COMPREHENSIVE OVERVIEW

Before exploring the modeling tools and optimization strategies for CO_2 capture, it is important to have a fundamental understanding of mathematical modeling, as illustrated in the Figure 3. This section aims to guide the transition from the physical system to the realm of computational precision, emphasizing the crucial role of mathematical models in solving the complexities of CO_2 capture processes (Miller et al., 2017). Since the CO_2 capture process involves a dynamic and multiscale physical system, comprehending the intricacies of the process, from molecular interactions in materials to the intricate dynamics of the capture process, is essential (Thierry & Biegler, 2019). However, it is challenging to capture the essence of the involved physical phenomena systematically. Mathematical modeling can function as a bridge that transcends the tangible complexities of the physical system. Mathematical models represent the underlying physical laws and behaviors governing CO_2 capture, such as mass and energy balances, physical constraints, cost analysis, and environmental considerations in expressions (Li et al., 2018).

Furthermore, these models capture the dynamic relationships, thermodynamic properties, and intricate interactions that define the CO_2 capture process. By using mathematical abstractions, the complexities of the physical system become manageable, allowing for precise analysis, prediction, and optimization of CO_2 capture strategies. Among the mathematical modeling techniques, computational models are powerful tools for analysis and prediction. Thus, by algorithms and numerical techniques, these models transform abstract representations of the physical system into computationally executable frameworks.

In this sense, evaluating material performance at crucial points is important in pursuing refining carbon dioxide (CO_2) capture technologies. This section presents a systematic exploration of various modeling tools and techniques. It begins with molecular simulations, followed by process modeling and empirical analyses such as adsorption isotherms, and concludes with optimization methodologies. This sequential approach forms the foundation for informed material selection and system design in CO_2 capture by delving into the microscopic intricacies, comprehending macroscopic system dynamics, and strategically optimizing.

In this way, different modeling tools contribute to the nuanced evaluation of materials in CO_2 capture processes. These tools offer insights into material behaviors under diverse conditions. Some critical modeling approaches include:

a. Molecular Simulations

Molecular simulations use computational methods to model the behavior of individual molecules, providing insights into the interactions between materials and CO_2 at the molecular level (Liu & Wilcox, 2013). Techniques such as molecular dynamics and Monte Carlo simulations play a crucial role in predicting thermodynamic properties, structural changes, and diffusion kinetics. Molecular simulations enable a deep understanding of how materials interact with CO_2 on a molecular scale, thus aiding in the development and optimization of materials for improved capture performance.

b. Process Modeling

In order to improve the CO_2 capture system, a process modeling approach is used to simulate the entire process, from molecular interactions to macroscopic dynamics. Process models analyze factors like temperature, pressure, and flow rates to gain comprehensive insights into the system's performance under real-world conditions. Based on this knowledge, efficient, scalable, and sustainable capture processes can be developed.

c. Adsorption Isotherms

Adsorption isotherms are crucial for determining the quantity of CO_2 that a material can adsorb at constant temperature based on empirical principles (Serafin & Dziejarski, 2023). This empirical approach is essential for understanding the adsorption capacity and kinetics of a material. Adsorption isotherms serve as a fundamental tool in material selection, as they help in designing materials that have optimal capture characteristics by delineating the adsorption behavior under varying conditions.

d. Mathematical Programming and Optimization

On the other hand, Mathematical Programming has emerged as a crucial tool for developing robust models in CO_2 capture (Thierry & Biegler, 2019). It is an optimization technique that uses mathematical models to identify the optimal combination of system parameters and materials while considering specific constraints, for instance, as Figure 4 shows. The technique has proven to be particularly useful in this field as it can navigate complex solution spaces and identify configurations that maximize capture efficiency under predefined conditions. Advanced modeling tools utilize Mathematical Programming, integrating

Figure 4. Optimal material selection for CO_2 capture: A multi-objective objective approach

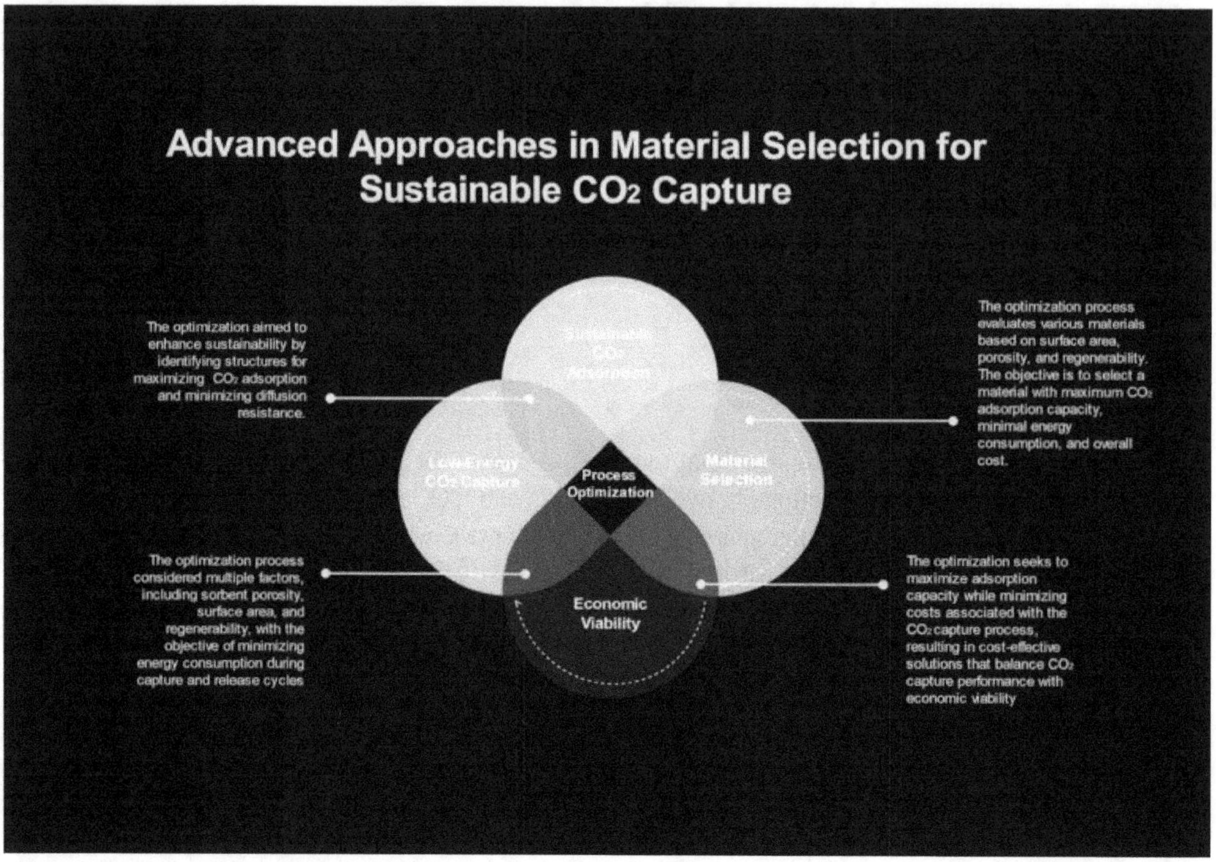

optimization techniques to enhance CO_2 capture technology efficiency. It has multiple applications in material selection for CO_2 capture and can help identify the most suitable materials based on specific criteria. Therefore, it is essential to acknowledge the indispensable role of Mathematical Programming in refining material selection and design strategies and its potential to contribute significantly to the development of efficient CO_2 capture technologies.

In conclusion, integrating modeling tools and optimization techniques is crucial for a comprehensive evaluation and enhancement of carbon dioxide (CO_2) capture materials. These methodologies aid in understanding material behavior in capture contexts and provide practical and effective solutions for designing efficient and sustainable CO_2 capture systems. Modeling tools provide a detailed analysis that helps to solve the complexities of the capture processes and identify critical points and opportunities to improve system efficiency. By applying optimization techniques, we can explore solution spaces thoroughly and identify optimal materials and system parameter combinations. This approach contributes to continuous improvement and lays the groundwork for designing more efficient and sustainable CO_2 capture systems.

4. MODELING AND EXPERIMENTAL IN MATERIAL EVALUATION

Integrating both modeled and experimental approaches is essential to effectively capture carbon dioxide. This strategic approach requires careful consideration of models, optimization, and experimental designs. The priority is to validate theoretical models with empirical data and continuously improve efficiency in identifying optimal materials and conditions for CO_2 capture (Park et al., 2019).

The design of experiments (DoE) becomes a fundamental tool for structuring experimental research efficiently. It defines how data will be collected, enabling the systematic identification of factors affecting the CO_2 capture process (Akinola et al., 2022). For example, the acquisition of adsorption isotherms greatly benefits from well-planned experimental designs, allowing the systematic exploration of different experimental conditions to understand how adsorption quantities vary under various combinations of temperature, pressure, and material type. Furthermore, model optimization relies on experimental data to refine predictions and adjust parameters. Information gathered in experiments, such as adsorption isotherms, becomes valuable benchmarks for calibrating mathematical models. Optimization becomes an iterative process where models are adjusted to accurately reflect real-world conditions, continuously enhancing predictive capabilities, and understanding material behavior.

On the other hand, the optimization process extends beyond parameter adjustments; it also addresses strategies to improve the efficiency of CO_2 capture processes. For instance, by varying experimental conditions in the design of experiments, local and global optima in a material's adsorption capacity can be identified. These results, in turn, guide the selection of ideal operational conditions for CO_2 capture, contributing to the system's overall efficiency. Thus, the interaction between model optimization and the design of experiments establishes a cycle of iterative validation and continuous improvement. Information collected in experiments not only validates existing models but also informs the need for adjustments and enhancements. The constant feedback between theory and practice drives the continuous evolution of models and CO_2 capture strategies.

This strategic approach has significant applications in material selection for CO_2 capture. By combining experimental data on adsorption isotherms with theoretical models, materials with optimal properties for efficient CO_2 capture can be identified (Moubarak et al., 2023). Optimization refines operational conditions, maximizing adsorption capacity and minimizing associated costs. By optimizing models through integration with experimental data, the development of more sustainable and efficient CO_2 capture technologies is encouraged. The ability to identify promising materials and optimal operational conditions translates into more effective and economical systems, contributing to the effective and sustainable reduction of CO_2 emissions. However, this approach has challenges. Obtaining precise experimental data and dealing with the complexity of some mathematical models can pose challenges. Nevertheless, these challenges offer opportunities for additional research and methodological improvements that will further strengthen the integration of modeled and experimental approaches.

An example is presented to highlight the importance of interaction between mathematical modeling and optimization in the context of the adsorption process. Now, to address the adsorption process as an optimization problem. We will consider maximizing the adsorbed amount of CO_2 subject to certain constraints. The objective of the optimization problem is to maximize the adsorbed amount of CO_2 (Q). It is the quantity of CO_2 the adsorbent material captures, and maximizing it aligns with the goal of efficient CO_2 capture.

$$Max(Q) \tag{1}$$

The modeling includes some constraints such as Langmuir isotherm, mass, and energy balances, among others (see Table 2). In this way, The Langmuir isotherm equation is a fundamental equation in adsorption, describing the relationship between the adsorbed amount (Q) and the concentration of CO_2 in the gas stream (C). This constraint describes the adsorption of a gas onto a solid surface. Moreover, K is the Langmuir constant. It is often used to model the adsorption of CO_2 onto a porous material.

$$Q = \frac{Q_{max}KC}{1+KC} \tag{2}$$

On the other way, the temperature-dependent Langmuir isotherm equation can be expressed as:

$$Q(T) = \frac{Q_{max}b_0 e^{\left(-\frac{E_a}{RT}\right)}}{1+b_0 e^{\left(-\frac{E_a}{RT}\right)}} \tag{3}$$

where b_0 is the Langmuir constant at a reference temperature, E_a is the activation energy, R is the gas constant, and T is the temperature.

On the other hand, the mass balance constraint relates the rate of change of adsorbed amount $\left(\frac{dQ}{dt}\right)$ to the flow rate of the gas stream (*F*), initial concentration of CO_2 (C_0), and the surface area of the adsorbent (*A*). This constraint ensures that the adsorbed amount is influenced by the characteristics of the gas stream and the adsorbent surface.

$$\frac{dQ}{dt} = AF(C_0 - C) \tag{4}$$

Finally, the energy balance constraint connects the rate of temperature change $\left(\frac{dT}{dt}\right)$ to the heat of adsorption (ΔH), density of the adsorbent (ρ), and the heat capacity of the adsorbent (Cp). This constraint reflects the fact that the adsorption process can involve heat exchange and temperature changes.

$$\frac{dT}{dt} = \frac{\Delta H}{\rho Cp}\frac{dQ}{dt} \tag{5}$$

The decision variables are the parameters that can be adjusted in the process. In this case, they are the initial concentration of CO_2 (C_0) and the flow rate of the gas stream (*F*). These are the variables that the optimization algorithm will adjust to maximize the objective function.

Table 2. CO_2 capture optimization formulation summary

CO_2 Capture Optimization Problem	
Objective Function	Maximize Q (Adsorbed amount of CO2)
Decision Variables	C_0 - Initial concentration of CO_2 F - Flow rate of the gas stream
Constraints	Langmuir Isotherm: $Q = \frac{Q_{max} \cdot C}{1+K \cdot C}$ Mass Balance: $\frac{dQ}{dt} = F \cdot C_0 - A \cdot \frac{dQ}{dt}$ Energy Balance: $\frac{dT}{dt} = \frac{\Delta H \cdot \rho}{C_p} \cdot \frac{dQ}{dt}$

These equations provide a basis for modeling and understanding the adsorption process for CO_2 capture. Depending on the specific conditions and assumptions of the system, other equations and models may also be employed. Thus, the optimization problem seeks to find the optimal values of the decision variables (initial concentration and flow rate) that maximize the adsorbed amount of CO_2 while satisfying the constraints imposed by the Langmuir isotherm, mass balance, and energy balance. Solving this optimization problem can guide the design and operation of the CO_2 capture system for enhanced performance.

Through the optimization model, it is crucial to identify the most effective conditions for maximizing CO_2 adsorption capacity in the capture process. The primary objective is to systematically pinpoint the optimal combination of material properties (e.g., pore size, surface area) and process conditions (e.g., temperature, pressure). This ensures the highest possible efficiency in capturing CO_2 while considering real-world constraints and resource utilization. The optimization model aims to strike a balance between maximizing adsorption capacity and addressing practical limitations, contributing to the development of cost-effective, sustainable, and environmentally friendly CO_2 capture processes.

4.1 Synergy Between the Mathematical Model and the Experiment

The development of CO_2 capture technologies requires a collaborative approach between experimental data and mathematical modeling, for example, the obtention of the Langmuir isotherm. The process begins with collecting data through experiments that measure CO_2 adsorption at different concentrations and temperatures (see Figure 5). Langmuir isotherm modeling is then utilized to create a theoretical framework for understanding adsorption, where the parameters are determined through model fitting to align the theoretical model with experimental data.

After that, calibration and validation of Langmuir parameters optimize model alignment with experimental findings. It is a crucial step in ensuring the accuracy of Langmuir model predictions against additional experimental datasets. The optimization process involves identifying optimal conditions for maximizing CO_2 adsorption capacity and adjusting model parameters for peak efficiency in CO_2 capture. The integration of the Langmuir model and experimental data provides a more nuanced understanding of the adsorption process, leading to insightful observations. This collaboration helps identify limitations and refine the Langmuir model to make it more accurate.

Figure 5. CO_2 adsorption capacity profiles from experimental data (simulated example)

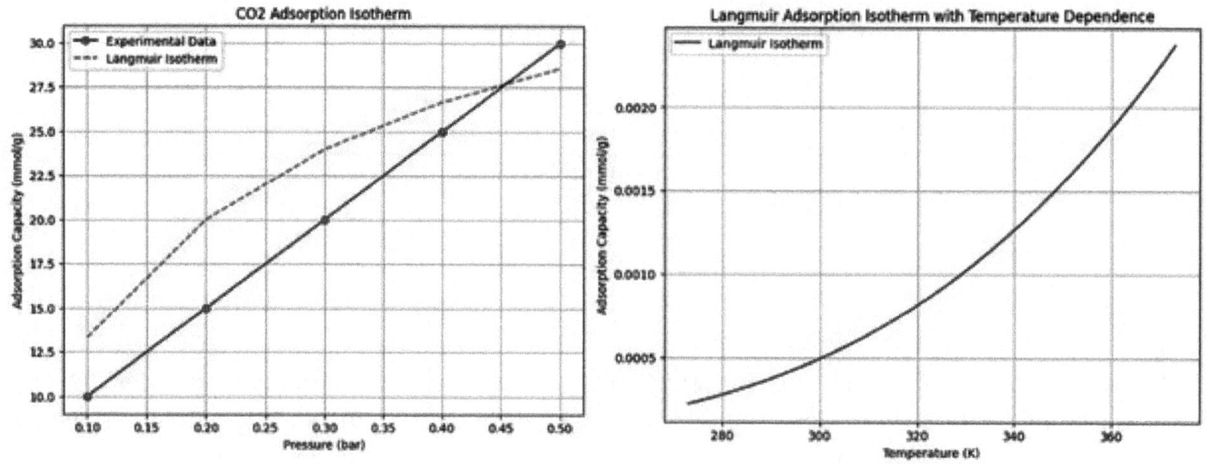

5. MODEL VALIDATION AND BENCHMARKING

In CO_2 capture technologies, it is crucial to validate theoretical models with empirical data. Although theoretical models can provide a framework for understanding complex processes, their practical applications depend on how well they align with real-world observations. Experimental validation serves as a crucial checkpoint by allowing researchers to assess the accuracy of theoretical predictions against tangible, measured outcomes. Therefore, the importance of validating modeling efforts with experimental data cannot be overstated. This iterative calibration and validation process improves the parameters of the model to accurately represent the details of CO_2 capture under different conditions.

The relationship between modeling and experimental data plays a key role in validating theoretical foundations, identifying potential discrepancies, and identifying areas for improvement. This dynamic interaction contributes to the continuous improvement of models, resulting in a more accurate representation of material behaviors and system dynamics. When modeling is robustly validated with experimental data, it instills confidence in the predictive capabilities of theoretical frameworks. It helps guide informed decision-making and drives advancements in CO_2 capture technologies built on a solid foundation of real-world efficacy.

5.1 Calibration for Precision

Model validation is crucial in refining theoretical models for CO_2 capture processes. The process involves an iterative loop of model calibration that acknowledges the dynamic nature of the process and allows for fine-tuning model parameters based on empirical feedback to achieve precision. The aim is to capture the nuances of molecular interactions, system dynamics, and the overarching process. One of the critical elements in model validation is benchmarking against experimental data. This process thoroughly compares the theoretical model with tangible outcomes from practical experiments. Benchmarking serves as a stringent test, verifying accuracy and highlighting areas for improvement. It provides a quantitative basis for evaluating the model's performance against real-world conditions.

5.2 Modeling and Experimental Insight

The relationship between theoretical modeling and experimental insights becomes evident during model validation. This collaborative approach ensures that the theoretical foundation faithfully represents the material behaviors, system dynamics, and intricate processes inherent in CO_2 capture. The continuous feedback loop between modeling and experiments enriches the model, making it a more accurate and reliable tool for understanding and predicting CO_2 capture phenomena. Establishing performance benchmarks is critical beyond validation. Performance benchmarks provide a comparative yardstick for evaluating different CO_2 capture models and technologies. Setting standards and benchmarks drives continuous improvement, pushing the boundaries of efficiency and sustainability in CO_2 capture solutions.

5.3 Addressing Model Limitations

Finally, recognizing and addressing the limitations of the developed model is essential for guiding its application and informing future research directions. Transparently addressing limitations contributes to the evolution of more robust and accurate models, fostering a culture of continuous improvement. The validation and benchmarking process creates a feedback loop fundamental to continuous improvement. Insights gained from experimental data inform and shape the modeling process, fostering an environment of ongoing refinement and evolution. This iterative approach ensures that CO_2 capture models stay agile, adaptive, and aligned with the latest scientific and technological advancements.

6. ECONOMIC AND SUSTAINABILITY CONSIDERATIONS IN MATERIALS SELECTION

For CO_2 capture technologies to be successful and widely adopted, it is vital to consider economic viability and sustainability goals while selecting materials. While high-performance materials are crucial for effective CO_2 capture, economic factors are also essential to ensure scalability and economic feasibility (Peh et al., 2023). The focus on economic viability aligns with the intent to make these technologies accessible on a large scale. Additionally, it is essential to carefully evaluate materials with sustainability in mind to address broader environmental concerns (Parekh et al., 2023). The efficiency and long-term environmental impact of CO_2 capture materials depend on their ecological footprint. To promote a sustainable and economically viable future, adopting a dual optimization approach that combines economic evaluation with sustainability objectives is crucial. This approach will help us develop materials that excel technically and contribute to a greener future while aligning technological progress with societal and environmental goals.

7. CONCLUSION AND FUTURE DIRECTIONS

This chapter provides a valuable understanding of materials selection used in carbon dioxide (CO_2) capture processes from the model development perspective. It starts by discussing the significance of CO_2 capture and material selection in achieving an effective process. The modeling overview section analyzes various tools and techniques utilized for evaluating the performance of materials in CO_2 capture scenarios. The integration of modeling and experimental approaches is an essential element of the chapter, emphasizing the relationship between mathematical modeling and experimental data. It provides valuable insights into optimization formulation and the relationship between mathematical models and experimental results. The chapter also highlights the importance of robust validation methodologies to enhance the reliability and accuracy of the models through model validation and comparison.

Additionally, the chapter highlights the need to align technological advancements with economic viability and environmental sustainability when considering economic and sustainability considerations in material selection. It is crucial to consider economic feasibility and sustainability when developing CO_2 capture technologies for global scalability and accessibility. The relevance of modeling tools as mathematical programming and optimization in selecting materials for CO_2 capture becomes significant when considering multiple objectives. These tools enable a thorough analysis of various materials and their properties and facilitate strategic decision-making in addressing divergent and often conflicting objectives.

In this way, mathematical modeling provides a precise and structured representation of the relationships among different variables and properties of materials, allowing project performance of such materials under various conditions. This predictive capability becomes an essential component for optimization, as it enables the evaluation of how different combinations of properties affect desired objectives, such as CO_2 capture efficiency, sustainability, and economic viability. On the other hand, mathematical programming and optimization go beyond addressing the inherent complexity in material selection, especially when considering multiple objectives. These tools allow the identification of optimal or Pareto-optimal solutions that balance different objectives, such as maximizing capture efficiency, minimizing costs, and satisfying sustainable considerations. Finally, considering multiple objectives drives the need for an integrated approach in developing mathematical models and optimization strategies. By emphasizing the ability of these tools to manage the complexity of decision-making efficiently and effectively in material selection, their essential role is strengthened in the quest for balanced and sustainable solutions to the challenges associated with CO_2 capture.

REFERENCES

Akinola, T. E., Prado, P. L. B., & Wang, M. (2022). Experimental studies, molecular simulation and process modelling\simulation of adsorption-based post-combustion carbon capture for power plants: A state-of-the-art review. *Applied Energy*, *317*, 119156. doi:10.1016/j.apenergy.2022.119156

Berger, A. H., & Bhown, A. S. (2013). Optimizing solid sorbents for CO2 capture. *Energy Procedia*, *37*, 25–32. doi:10.1016/j.egypro.2013.05.081

Boer, D. G., Langerak, J., & Pescarmona, P. P. (2023). Zeolites as Selective Adsorbents for CO2 Separation. *ACS Applied Energy Materials*, *6*(5), 2634–2656. doi:10.1021/acsaem.2c03605

Chaffee, A. L., Knowles, G. P., Liang, Z., Zhang, J., Xiao, P., & Webley, P. A. (2007). CO2 capture by adsorption: Materials and process development. *International Journal of Greenhouse Gas Control*, *1*(1), 11–18. doi:10.1016/S1750-5836(07)00031-X

Chaffee, N., Ebelegi, A. N., & Wankasi, D. (2017). Modelling and interpretation of adsorption isotherms. *Journal of Chemistry*, •••, 2017.

Chatziasteriou, C. C., Kikkinides, E. S., & Georgiadis, M. C. (2022). Recent advances on the modeling and optimization of CO2 capture processes. *Computers & Chemical Engineering*, *165*, 107938. doi:10.1016/j.compchemeng.2022.107938

Dziejarski, B., Serafin, J., Andersson, K., & Krzyżyńska, R. (2023). CO2 capture materials: a review of current trends and future challenges. *Materials Today Sustainability*, 100483.

Espinal, L., Poster, D. L., Wong-Ng, W., Allen, A. J., & Green, M. L. (2013). Measurement, standards, and data needs for CO2 capture materials: A critical review. *Environmental Science & Technology*, *47*(21), 11960–11975. doi:10.1021/es402622q PMID:24060087

Gonzalez-Olmos, R., Gutierrez-Ortega, A., Sempere, J., & Nomen, R. (2022). Zeolite versus carbon adsorbents in carbon capture: A comparison from an operational and life cycle perspective. *Journal of CO2 Utilization, 55*, 101791.

Gür, T. M. (2022). Carbon dioxide emissions, capture, storage and utilization: Review of materials, processes and technologies. *Progress in Energy and Combustion Science*, *89*, 100965. doi:10.1016/j.pecs.2021.100965

Han, Y., & Ho, W. W. (2018). Recent advances in polymeric membranes for CO2 capture. *Chinese Journal of Chemical Engineering*, *26*(11), 2238–2254. doi:10.1016/j.cjche.2018.07.010

Hernández-Romero, I. M., Flores-Tlacuahuac, A., Nápoles-Rivera, F., Esquivel-Patiño, G. G., & García-Pardo, M. L. (2022). Multi objective optimization of the amines-CO2 capture absorption-desorption process by a non-equilibrium rate model. *Chemical Engineering Research & Design*, *187*, 93–104. doi:10.1016/j.cherd.2022.08.050

Imteyaz, S., Suresh, C. M., Kausar, T., & Ingole, P. P. (2023). Carbon dioxide capture and its electrochemical reduction study in deep eutectic solvent (DES) via experimental and molecular simulation approaches. *Journal of CO2 Utilization, 68*, 102349.

Kayan, A. (2019). Inorganic-organic hybrid materials and their adsorbent properties. *Advanced Composites and Hybrid Materials*, *2*(1), 34–45. doi:10.1007/s42114-018-0073-y

Li, S., Deng, S., Zhao, L., Zhao, R., Lin, M., Du, Y., & Lian, Y. (2018). Mathematical modeling and numerical investigation of carbon capture by adsorption: Literature review and case study. *Applied Energy*, *221*, 437–449. doi:10.1016/j.apenergy.2018.03.093

Liu, Y., & Wilcox, J. (2013). Molecular simulation studies of CO2 adsorption by carbon model compounds for carbon capture and sequestration applications. *Environmental Science & Technology*, *47*(1), 95–101. doi:10.1021/es3012029 PMID:22747244

Lu, X., Jin, D., Wei, S., Wang, Z., An, C., & Guo, W. (2015). Strategies to enhance CO 2 capture and separation based on engineering absorbent materials. *Journal of Materials Chemistry. A, Materials for Energy and Sustainability*, *3*(23), 12118–12132. doi:10.1039/C4TA06829G

Majchrzak, A., & Nowak, W. (2017). Separation characteristics as a selection criteria of CO2 adsorbents. *Journal of CO2 Utilization, 17*, 69-79.

Miller, D. C., Agarwal, D., Bhattacharyya, D., Boverhof, J., Chen, Y., Eslick, J., ... & Zitney, S. E. (2017). Innovative computational tools and models for the design, optimization and control of carbon capture processes. *Process Systems and Materials for CO2 Capture: Modelling, Design, Control and Integration*, 311-342.

Moubarak, E., Moosavi, S. M., Charalambous, C., Garcia, S., & Smit, B. (2023). A Robust Framework for Generating Adsorption Isotherms to Screen Materials for Carbon Capture. *Industrial & Engineering Chemistry Research*, *62*(26), 10252–10265. doi:10.1021/acs.iecr.3c01358 PMID:37425135

Murge, P., Dinda, S., & Roy, S. (2019). Zeolite-based sorbent for CO2 capture: Preparation and performance evaluation. *Langmuir*, *35*(46), 14751–14760. doi:10.1021/acs.langmuir.9b02259 PMID:31661624

Oschatz, M., & Antonietti, M. (2018). A search for selectivity to enable CO 2 capture with porous adsorbents. *Energy & Environmental Science*, *11*(1), 57–70. doi:10.1039/C7EE02110K

Papadopoulos, A. I., & Seferlis, P. (Eds.). (2017). *Process systems and materials for CO2 capture: modelling, design, control and Integration*. John Wiley & Sons. doi:10.1002/9781119106418

Parekh, A., Chaturvedi, G., & Dutta, A. (2023). Sustainability analyses of CO2 sequestration and CO2 utilization as competing options for mitigating CO2 emissions. *Sustainable Energy Technologies and Assessments*, *55*, 102942. doi:10.1016/j.seta.2022.102942

Park, J., Rubiera Landa, H. O., Kawajiri, Y., Realff, M. J., Lively, R. P., & Sholl, D. S. (2019). How well do approximate models of adsorption-based CO2 capture processes predict results of detailed process models? *Industrial & Engineering Chemistry Research*, *59*(15), 7097–7108. doi:10.1021/acs.iecr.9b05363

Peh, S. B., Farooq, S., & Zhao, D. (2023). Techno-economic analysis of MOF-based adsorption cycles for postcombustion CO2 capture from wet flue gas. *Chemical Engineering Science*, *268*, 118390. doi:10.1016/j.ces.2022.118390

Raganati, F., Chirone, R., & Ammendola, P. (2020). CO2 capture by temperature swing adsorption: Working capacity as affected by temperature and CO2 partial pressure. *Industrial & Engineering Chemistry Research*, *59*(8), 3593–3605. doi:10.1021/acs.iecr.9b04901

Romero-García, A. G., Ramírez-Corona, N., Sánchez-Ramírez, E., Alcocer-García, H., De Blasio, C., & Segovia-Hernández, J. G. (2022). Sustainability assessment in the CO2 capture process: Multi-objective optimization. *Chemical Engineering and Processing*, *182*, 109207. doi:10.1016/j.cep.2022.109207

Serafin, J., & Dziejarski, B. (2023). Application of isotherms models and error functions in activated carbon CO2 sorption processes. *Microporous and Mesoporous Materials*, *354*, 112513. doi:10.1016/j.micromeso.2023.112513

Thierry, D., & Biegler, L. T. (2019). Dynamic real-time optimization for a CO2 capture process. *AIChE Journal. American Institute of Chemical Engineers*, *65*(7), e16511. doi:10.1002/aic.16511

Trickett, C. A., Helal, A., Al-Maythalony, B. A., Yamani, Z. H., Cordova, K. E., & Yaghi, O. M. (2017). The chemistry of metal–organic frameworks for CO2 capture, regeneration and conversion. *Nature Reviews. Materials*, *2*(8), 1–16. doi:10.1038/natrevmats.2017.45

Wang, J., Huang, L., Yang, R., Zhang, Z., Wu, J., Gao, Y., Wang, Q., O'Hare, D., & Zhong, Z. (2014). Recent advances in solid sorbents for CO 2 capture and new development trends. *Energy & Environmental Science*, *7*(11), 3478–3518. doi:10.1039/C4EE01647E

Yu, C. H., Huang, C. H., & Tan, C. S. (2012). A review of CO2 capture by absorption and adsorption. *Aerosol and Air Quality Research*, *12*(5), 745–769. doi:10.4209/aaqr.2012.05.0132

Zamarripa, M. A., Eslick, J. C., Matuszewski, M. S., & Miller, D. C. (2018). Multi-objective optimization of membrane-based CO2 capture. *Computer-Aided Chemical Engineering*, *44*, 1117–1122. doi:10.1016/B978-0-444-64241-7.50181-6

Zhang, X., Wang, J., Song, Z., & Zhou, T. (2021). Data-driven ionic liquid design for CO2 capture: Molecular structure optimization and DFT verification. *Industrial & Engineering Chemistry Research*, *60*(27), 9992–10000. doi:10.1021/acs.iecr.1c01384

Zhao, R., Deng, S., Wang, S., Zhao, L., Zhang, Y., Liu, B., Li, H., & Yu, Z. (2018). Thermodynamic research of adsorbent materials on energy efficiency of vacuum-pressure swing adsorption cycle for CO2 capture. *Applied Thermal Engineering*, *128*, 818–829. doi:10.1016/j.applthermaleng.2017.09.074

Chapter 16
AI's Function in Sustainable Development's Renewable Energy Planning

Sabyasachi Pramanik
https://orcid.org/0000-0002-9431-8751
Haldia Institute of Technology, India

ABSTRACT

The integration of artificial intelligence (AI) in the renewable energy sector has emerged as a disruptive force that has the potential to fundamentally alter the ways in which clean energy is produced, distributed, and used. Artificial intelligence (AI) technologies, such as machine learning, predictive analytics, optimization algorithms, and smart control systems, have enabled unprecedented improvements in renewable energy systems. These advancements include optimizing energy storage systems to increase grid dependability and stability as well as optimizing solar and wind turbine efficiency via the use of adaptive control and real-time data analysis. Artificial intelligence-driven predictive maintenance has decreased operational costs and downtime, extending the lifespan of renewable infrastructure. AI-powered energy management solutions encourage an energy-saving culture by giving consumers the ability to make informed decisions about how much energy they use.

1. INTRODUCTION

The relationship between energy production and consumption is crucial for assessing the level of economic growth in a particular country. Energy supply and use are closely related to urgent worldwide issues, such as environmental problems and global warming. These difficulties include a broad spectrum of problems, including greenhouse gas emissions, acid rain, deforestation, air pollution, ozone depletion, water and land usage, biodiversity loss, and radioactive material discharge (Salim et al., 2018). Humanity must face the challenges that emerge in the areas of environmental, economic, and social consequences if it is to attain a more optimistic and sustainable energy future. With certain restrictions, there has been a worldwide movement toward the use of clean energy resources to mitigate the nega-

DOI: 10.4018/979-8-3693-1306-0.ch016

tive effects of conventional energy sources like coal, oil, and natural gas, which are known to impede sustainable development. The only practical and affordable substitute for conventional energy sources is renewable energy, which includes hydroelectric power, geothermal energy, tidal energy, solar energy, and wind energy (Ediger, 2019). It is possible to address social, economic, and environmental issues by using renewable energy sources. These substitutes are acknowledged for their capacity to promote community development, especially in developing and rural areas, cut power costs, create jobs, improve public health, and ease the adoption of environmentally friendly technology. Additionally, they aid in the decrease or eradication of dangerous pollutants such as carbon dioxide, sulfur dioxide, and carbon monoxide (Kumar, 2020).

In addition to offering a long-lasting and sustainable energy supply, the use of renewable energy sources has the potential to increase the industry's diversification and reduce emissions on a local and worldwide scale. Particularly in developing nations and rural regions, it may provide economically appealing solutions for meeting specific electrical service demands. It may also create opportunities for local manufacturing. Numerous facets of renewable energy, including as design, development, assessment, operation, distribution, and regulation, heavily rely on artificial intelligence (AI). This study's major goal is to clarify the artificial intelligence techniques used in the renewable energy industry (Asif & Muneer, 2007). Artificial intelligence (AI) is becoming more and more integrated into many areas of renewable energy systems (REs) as a result of the expanding availability of computer power, sophisticated tools, and enhanced data gathering techniques. The design, control, and maintenance approaches now used in the energy business have shown a propensity to provide outcomes that are somewhat erroneous. Furthermore, the use of artificial intelligence (AI) in carrying out these duties has led to increased degrees of precision and accuracy, establishing it as a front-runner in this field of technology. Because it may improve the efficiency and quality of automated systems, artificial intelligence (AI) has been a popular area of research in recent decades (Bryson, 2019). This system uses sophisticated training methods and gives them a complete set of instructions so they may learn, reason, and make judgments in a way that is similar to human cognition.

Renewable energy is expected to overtake coal as the world's leading source of electricity production by the early 2025, making it the best source in this area. The estimate indicates that their part of the power mix will increase by 10 percentage points over the course of the predicted period, to reach a projected share of 38% by 2027. Only the renewable energy sector is expected to expand in the electrical generating industry; the proportions of coal, natural gas, nuclear, and oil generation are expected to decline. It is anticipated that the amount of electricity generated worldwide by solar photovoltaic (PV) and wind sources would more than double over the next five years. By 2027, it is projected that this trend would result in wind and solar PV together producing close to 20% of the world's total electricity production. Throughout the course of the projection period, the aforementioned variable technologies are expected to account for 80% of the increase in global renewable energy. As such, further paths for improving power system flexibility must be investigated. Dispatched renewable energy sources including hydropower, bioenergy, geothermal, and concentrated solar power have a vital role in incorporating solar photovoltaic (PV) and wind technologies into global electricity grids, but their growth is still limited (IEA, 2022).

Artificial intelligence (AI) has many interesting applications in the field of renewable energy and resources. These include the ability to learn on its own, simple implementation, effective computation after learning, generalization skills, and major contributions to decision-making processes. Using AI approaches more successfully is primarily justified by the capacity to integrate these capabilities into decision-making, control, and modeling processes in response to changing requirements.

Table 1. Share of cumulative power capacity by source of generation, 2010-2027(IEA (2022))

	2010	2011	2012	2013	2014	2015	2016	2017	2018	2019	2020	2021	2022	2023	2024	2025	2026	2027
Solor Power	0.8	1.3	1.8	2.3	2.9	3.5	4.5	5.7	6.9	8.1	9.4	10.9	12.8	14.7	16.5	18.4	20.2	22.2
Wind Power	3.5	4.1	4.7	5.1	5.7	6.5	7	7.4	7.8	8.3	9.4	10.1	10.8	11.4	12.1	12.9	13.7	14.4
Hydrogen Power	19.8	19.5	19.3	19.4	19.2	18.9	18.6	18.2	17.9	17.5	16.9	16.6	16.2	15.7	15.2	14.9	14.5	14.1
Bioenergy	1.4	1.4	1.5	1.6	1.6	1.6	1.7	1.7	1.8	1.8	1.9	1.9	2	2	2	2	2	2
Coal	31.2	31.3	31.2	31	30.8	30.6	30.3	29.8	28.9	28.4	27.5	26.7	25.6	24.7	23.8	22.8	21.9	20.9
Natural Gas	26.7	26.4	26.3	26	25.8	25.2	24.6	24.4	24.3	24	23.3	22.6	22.1	21.4	20.8	20.2	19.7	19.1

Figure 1. Cumulative power generation share vs. year of analysis/prediction (IEA, 2022)

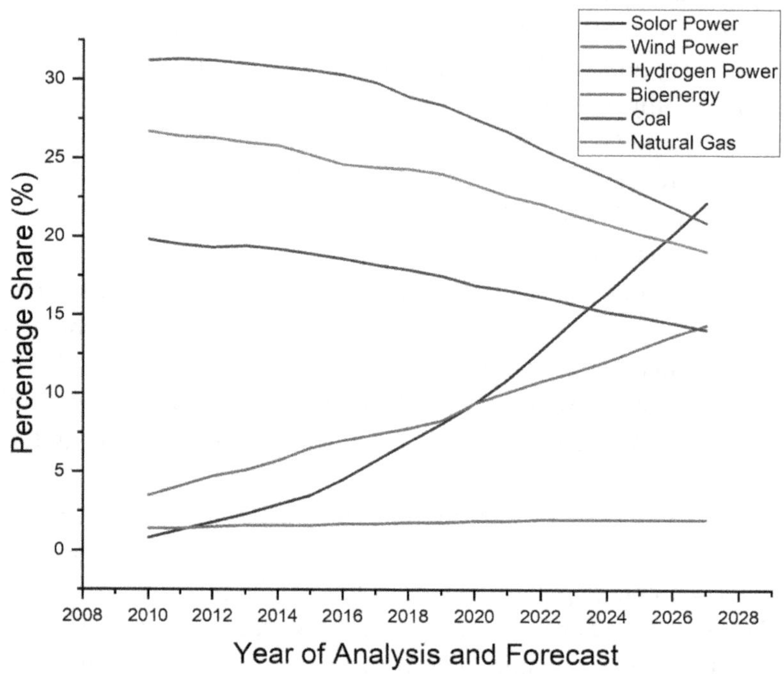

2. AI'S PLACE IN THE USE OF RENEWABLE ENERGY

2.1 Sunlight Power

Using solar energy, which is produced by the sun's endless and plentiful rays, is essential to current sustainable development. Solar energy has the potential to improve energy security, lessen carbon emissions, and mitigate climate change since it promises to provide clean, renewable electricity. There are two main ways to use solar power: directly using photovoltaic (PV) cells to convert sunlight into electrical power, or indirectly utilizing solar energy concentration to create steam. After that, a turbine is spun by this steam to produce energy. By helping electrons advance to a higher energy level via their contact with light photons, a phenomenon known as the photovoltaic effect is used to directly transform solar

Figure 2. Net Solar PV electricity capacity additions by country or region, 2022-2024(IEA, 2023)

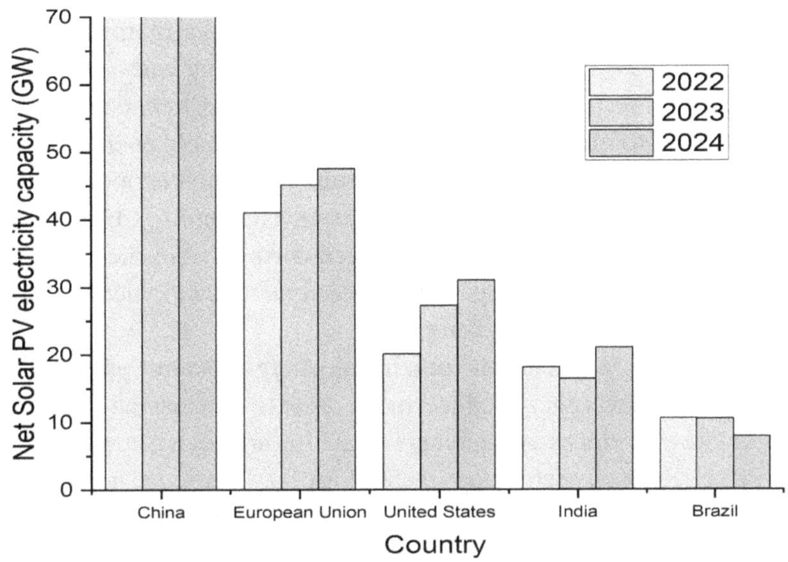

radiation into electrical energy. Originally used for spaceship propulsion, photovoltaics have several practical uses. These include electric mobility, roadside emergency phones, remote sensing, water usage pumps, and grid-independent homes.

In the years 2023 and 2024, it is projected that China will contribute more to the increase of the world's renewable energy capacity, reaffirming its leadership position as the world's largest renewable energy user. China accounted for more than half of the growth in the world's capacity to generate renewable energy in 2022. By 2024, forecasts indicate that the nation will have deployed an astonishing 55% of the world's annual renewable capacity. By 2024, it is anticipated that more than half of all solar photovoltaic projects would be implemented in China (IEA, 2023).

The efficiency and effect of solar energy management have been substantially enhanced by the use of Artificial Intelligence (AI). Artificial intelligence (AI)-powered solutions make it possible to measure solar irradiance in real time, optimize solar panel orientation for optimal energy collection, forecast trends of energy production, and simplify smart grid integration. In line with the objectives of a cleaner and greener future, these developments not only improve the viability and dependability of solar energy but also open the door to a more sustainable energy landscape. Bose offers a thorough explanation of the metaheuristic techniques used in photovoltaic (PV) systems to monitor maximum power point (Bose, 2017). (Mellit et al., 2009), a paper that focuses on the use of PV particularly, discusses the usage of AI methods in PV systems. In the realm of solar energy, artificial intelligence techniques called artificial neural networks are used to improve prediction and thermal load optimization at the design stage. A number of meteorological parameters, including temperature, solar radiation, relative humidity, atmospheric pressure, wind speed, precipitation, temperature range, and day length, have shown to be amenable to prediction and optimization using artificial neural network techniques.

2.2 Biofuel

The use of bioenergy, which is produced from organic resources like biomass and biofuels, offers a complex approach to achieving sustainable development and lowering carbon emissions. It includes a broad range of sources, such as forestry waste, energy crops specifically intended for agricultural use, and agricultural wastes. In addition to offering a dependable and sustainable energy source, bioenergy has the ability to help with organic waste management and promote circular economy principles. There are two primary types of bioenergy: contemporary and conventional technology. Using liquid biofuels made from bagasse plants (bio-refiners, biogas) that are produced during the anaerobic digestion of wastes is a component of modern bioenergy technology. Conventional methods include burning biomass—such as wood, charcoal, and animal dung—to produce energy.

Artificial Intelligence (AI) is essential for maximizing bioenergy resource production, conversion, and consumption in the context of today's energy concerns. In order to forecast the sugar yields of pretreated biomass throughout the hydrolysis process, researchers created an artificial neural network (ANN) application tool (Sarker et al., 2021). Raman spectroscopy technology may be used to augment the understanding of the structural alterations that transpire in biomass throughout the pyrolysis and combustion processes (Kumbhar et al., 2021). Artificial intelligence (AI)-powered solutions improve biogas generation from organic waste via process optimization, enable precision farming methods, allow real-time monitoring of biofuel production processes, and help with predictive modeling for efficient biomass output. The combination of artificial intelligence (AI) with bioenergy highlights the need for diverse and sustainable energy portfolios, which in turn makes ecologically responsible energy solutions possible. Fuzzy logic was used to build a gasification model. According to Nassif et al. (2020), there was a noteworthy 75% rise in the conversion rate and an amazing 85.51% gasification rate under ideal circumstances. In the bioenergy industry, artificial intelligence (AI) is frequently used to anticipate several factors, including the characteristics of biomass feedstock, the thermochemical and biochemical conversion processes, and the end-user application. Artificial intelligence (AI) has a big impact on biomass quality prediction, which affects both the end product's overall quality and the viability of biomass conversion operations from an operational standpoint. The calculation of greater heating values via data analysis from final analysis also makes use of artificial intelligence (AI). Because they are more economical and efficient than final analysis data, proximate analysis data are often used to estimate greater heating values.

2.3 Wind Power

Utilizing the kinetic energy of air masses in motion, wind energy is a notable symbol of the shift towards sustainable and clean energy sources. Tall onshore wind turbines and offshore wind farms are examples of wind power facilities that have the potential to transform the energy generating industry while reducing the negative impacts of climate change. Because wind energy can be naturally renewed, it is regarded as a renewable energy source. It has attributes like cost, accessibility, and purity. Every day, people all around the globe use wind turbines to capture the kinetic energy in the air and transform it into electrical energy. Given that wind energy provides a clean and sustainable option for supplying energy to our society, its use is becoming more important in the context of the world's energy production. Wind power has been used for a long time as a major energy source. Its kinetic energy is captured and transformed into electrical energy by using windmills and wind turbines (Burton et al., 2011; Robert &

Figure 3. Net onshore wind electricity capacity additions by country or region, 2022-2024 (IEA, 2023)

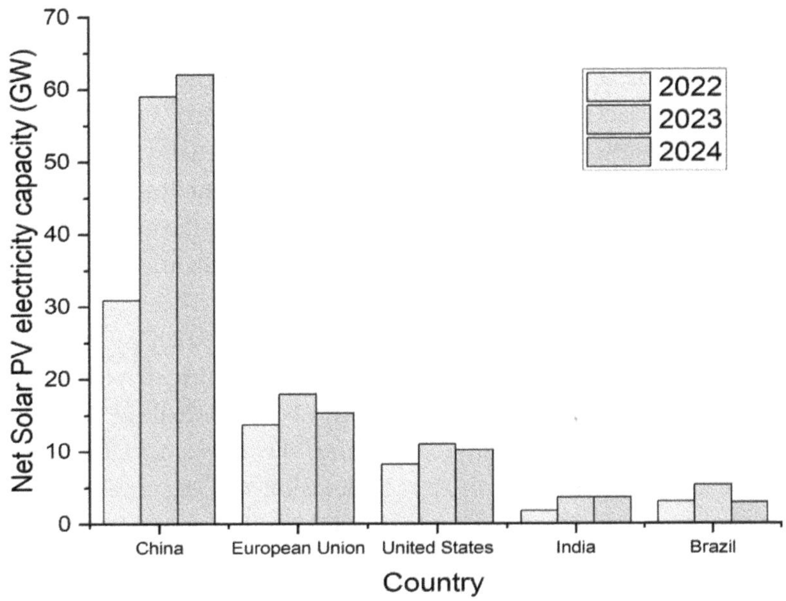

Brown, 2006). By 2024, more than 70% of new offshore wind projects and more than 60% of onshore wind projects globally are expected to be implemented by China (IEA, 2023).

The efficiency, dependability, and flexibility of wind power systems are all improved by the use of Artificial Intelligence (AI) into wind energy management. Predictive models powered by AI allow for accurate wind speed forecasts, maximizing power production and turbine performance. AI algorithms also make it easier to monitor and manage wind turbine operations in real time, identifying irregularities and reducing downtime with predictive maintenance. (Damousis & Dokopoulos, 2001) Developed fuzzy approaches using several Genetic algorithms to estimate wind power and speed. The use of wireless modems made it possible to get wind energy data from a distance. the application of methods for artificial intelligence. The study's use of renewable energy was its main emphasis, and the fuzzy technique was utilized to evaluate the data. According to the findings, accuracy rates employing the fuzzy approach were 29.7% and 39.8% higher within the designated time period than those achieved using the permanent technique. In a different research, (Carolin Mabel & Fernandez, 2008) used data gathered from seven different wind farms to estimate wind power over a three-year period using a neural network architecture with feed-forward backpropagation. The test set's and training set's root mean square errors, which were determined to be 0.0065 and 0.0070, respectively, show how accurate the BPNN's predictions were. AI strengthens the basis for a sustainable energy environment by promoting innovation that is in line with a cleaner, more sustainable future as it continues to improve wind energy operations. In the realm of wind energy, artificial intelligence (AI) is used to anticipate wind power, calculate the operating lifetime of wind turbines, and improve energy efficiency by reducing energy losses.

2.4 Thermoelectric Power

Earth's subsurface layers are where geothermal energy is generated and stored (Archer, 2020). Temperature differences between the Earth's surface and core give birth to the geothermal gradient. The progressive breakdown of radioactive elements found inside the Earth's crust produces lava. Because of the movement of tectonic plates, lava disintegrates, resulting in the development of large geothermal reservoirs that are the main source of geothermal energy. Geothermal systems provide a consistent and dependable energy source with no effect on the environment by using the natural heat that is stored under the surface of the Earth. For use in commercial, industrial, and residential settings, these systems may provide energy, heating, and cooling solutions.

The effectiveness and profitability of geothermal energy management systems have been markedly improved by the use of Artificial Intelligence (AI). Predictive modeling driven by artificial intelligence (AI) improves reservoir management by helping to identify the best drilling locations and characterize reservoirs. The exploration of geothermal resources requires the assessment and management of several uncertainties, which poses challenges when it comes to decision-making on operations and investments. Remote sensing (RS), machine learning (ML), and artificial intelligence (AI) together show promise for solving the difficult problems of geothermal research. To provide a preliminary evaluation of geothermal potential, researchers use machine learning, artificial intelligence, and remote sensing methods. This evaluation is based on well-established markers for geothermal areas, including deformation, faults, surface temperatures, and mineral markers. The research demonstrated how the concept was applied to Desert Peak and Brady, two geothermal locations. Even though these sites are close together, they have different features. Brady has surface manifestations that are obvious, and Desert Peak is a blind site. Several satellite images and geographic data are used to analyze temperature, deformation, faults, and mineral markers (Nassef et al., 2020). Recent advances in machine learning techniques might improve systems built with expert judgment. When expert opinions are not available, the XGBoost method shows more agreement with the results than linear logistic regression. This is explained by the fact that XGBoost is non-linear, which enables it to capture the non-linearity brought about by the expert recommendations. Notably, the assessment only used linear techniques, but the expert judgments made the whole strategy non-linear (Mordensky et al., 2023).

Furthermore, real-time monitoring and management of geothermal power plants are made easier by AI-driven algorithms, which maximize resource use and reduce downtime. The combination of AI with geothermal energy is a monument to innovation and sustainability, propelling a more sustainable and resilient energy future as geothermal energy continues to gain popularity in the world energy mix.

2.5 Marine Energy

The energy that comes from the sea is also categorized as hydroelectric energy, albeit it is also often called marine energy. There is a phenomena that occurs at large depths in the water where temperature differences have enough energy to produce thermal energy. This phenomenon is known as water thermal energy (OTE), which is the use of temperature differences in the water to produce electricity. Wave energy is captured by means of the use of floating devices. Things that move in an elliptical pattern when affected by waves. The use of ocean energy, which is produced by the powerful forces of tides, waves, and currents, is a viable option for a dependable and large-scale renewable energy source. Considering that the seas make up over 70% of the Earth's surface, ocean energy systems have a great deal of potential for contributing significantly to the production of sustainable energy. These systems use a variety of methods to capture the kinetic energy found in marine motions, such as tidal turbines and wave energy converters. Even though renewable energy still has a lot of untapped potential, the ocean energy sector plays a big part in helping coastal communities and countries greatly increase their access to power (Esteban & Leary, 2012). The development of computer approaches for wave production, together with updating and forecasting systems, are examples of research opportunities in the energy resources sector. Furthermore, special methods for enhancing energy converter control are also highlighted. The pursuit of sustainable and environmentally friendly renewable energy sources is reflected in the current research endeavors. It is important to note, too, that studies with an engineering emphasis sometimes overlook the possible long-term effects of climate change. The use and development of computational engineering techniques remain relevant, especially when dealing with continuum mechanics issues. Still, scholars are emphasizing machine learning technologies more and more (Tavakoli et al., 2023).

A new era of increased efficiency and optimization has begun with the use of artificial intelligence (AI) to ocean energy management. By adjusting their location and operation to achieve maximum power extraction, artificial intelligence (AI)-based algorithms significantly improve the efficiency of maritime energy devices by forecasting oceanic conditions. An innovative method for forecasting the daily energy of ocean waves in Queensland State, Australia, was presented. This method combines a Cascaded Forward Neural Network (CFNN) and Boruta-Extreme Gradient Boosting (BXGB) feature selection with a multi-stage Multivariate Variational Mode Decomposition (MVMD). The goal of this integrated system, known as MVMD-BXGB-CFNN, is to improve wave energy predictions' precision and dependability. MVMD-BXGB-MGGP, MVMD-BXGB-LSSVM, and MVMD-BXGB-GBDT are the names of the three other robust intelligence-based alternatives to the modeling results: Multigene Genetic Programming (MGGP), Least Square Support Machine (LSSVM), and Gradient Boosted Decision Tree (GBDT) models that were hybridized with MVMD and BXGB. Furthermore included for comparison were the standalone models of GBDT, MGGP, LSSVM, and CFNN (Jamei et al., 2022).

Artificial intelligence (AI) is also essential for enhancing data processing and enabling real-time equipment monitoring, which helps with predictive maintenance and lengthens the operating life of maritime energy projects. As artificial intelligence (AI) continues to grow, ocean energy is poised to become an increasingly important part of the renewable energy landscape. This will contribute significantly to the development of robust and sustainable energy systems in the future.

2.6 Vital Hydrogen

Hydrogen energy offers a versatile and emission-free method of reaching sustainable energy solutions. It is produced by the ecologically beneficial conversion of hydrogen gas into fuel or power. As an energy carrier, hydrogen has the ability to store and transfer large amounts of renewable energy efficiently. This feature makes it possible to reduce sporadic problems and makes decarbonization of many industries, such as transportation and manufacturing, easier. The fundamental component of all fuel products is hydrogen. Hydrogen is produced during the electrolysis process, which also includes the biological activity of bacteria and the breakdown of other composites created from biomass, such as organic wastes. Further use of this hydrogen as a renewable energy source for burning to produce power is possible (Kothari et al., 2008; McDowall, 2012).

Significant improvements are produced when hydrogen and artificial intelligence (AI) are combined, enhancing the hydrogen energy lifecycle at every level. Algorithms based on artificial intelligence (AI) are used to increase the productivity of hydrogen production processes, such as reforming and electrolysis, in order to maximize energy efficiency and minimize total costs. In addition, artificial intelligence (AI) is essential to the distribution and storage of hydrogen since it makes it easier to detect leaks, monitor in real time, and maximize use efficiency. (Dufo-López et al., 2007) provide a novel way for controlling stand-alone hybrid renewable power systems with hydrogen storage via the use of an optimized genetic algorithm. The integration of artificial intelligence (AI) with hydrogen energy is a groundbreaking endeavor towards the advancement of a sustainable and eco-friendly energy framework, hence hastening the transition to a future where hydrogen is the primary energy source.

3. HOW AI CONTRIBUTES TO SUSTAINABLE DEVELOPMENT

The management of renewable energy has been greatly impacted by the development of artificial intelligence (AI), which has also acted as a driver for sustainable growth. Artificial intelligence (AI) offers a wide range of technologies that may improve the efficiency, reliability, and integration of renewable energy systems in the face of climate change and the urgent need to transition to more sustainable energy sources. The amalgamation of artificial intelligence (AI) with renewable energy sources presents a plausible resolution to tackle ecological issues and has the capability to revolutionize the techniques used for energy generation, distribution, and consumption.

Solar, wind, hydro, geothermal, and ocean energy are examples of renewable energy sources that are vital to reducing carbon emissions and maintaining energy security. However, these energy sources' erratic and intermittent nature makes it difficult to keep up a consistent and dependable energy supply. In order to bridge the gap between the inherent unpredictability of renewable energy sources and the need for a dependable energy system, artificial intelligence (AI) is essential. The ability of artificial intelligence (AI) to analyze vast amounts of data effectively and in real time makes it easier to anticipate energy production patterns precisely, which improves the efficient use of existing resources. Artificial intelligence (AI) systems, for instance, can forecast wind or solar irradiance, which makes it possible to arrange solar panels or wind turbines optimally to maximum energy collection.

Photovoltaic systems are now intelligent entities capable of adjusting their performance to different environmental circumstances as a consequence of the integration of artificial intelligence (AI) into solar energy management. Intelligent inverters with AI capabilities can adapt dynamically to changes in voltage and satisfy grid needs, resulting in increased energy efficiency. In order to maximize the performance of solar arrays and increase total energy output, machine learning algorithms are used to analyze historical data (Hannan et al., 2021). Additionally, the smooth integration of solar energy into intelligent power grids is made possible by artificial intelligence (AI), which creates a two-way communication channel between energy providers and consumers. Demand response, load balancing, and energy storage optimization are made easier by the integration of many technologies, which increases the grid's overall stability (Som, 2021).

Similar to this, using AI-powered technology has benefited wind energy as it greatly raises the general efficacy and efficiency of wind farms and turbines. By using artificial intelligence (AI) to drive predictive maintenance, mechanical abnormalities may be identified in a timely manner, allowing for the proactive execution of repair methods and reducing the likelihood of operational downtime. Moreover, by considering the unique wind patterns and topographical features of a particular site in order to optimize energy output, artificial intelligence (AI) plays a critical role in supporting the optimal placing of wind turbines. Artificial intelligence (AI) integration may significantly expand the portfolio of renewable energy sources by improving the economic viability, efficiency, and dependability of wind energy plants (Kumar et al., 2022; Som, 2021).

The use of geothermal energy, which captures the natural heat of the Earth, is encouraged even more by advances in artificial intelligence (AI). Predicting reservoir characteristics, allocating resources optimally, and determining the best places to drill are all made possible by artificial intelligence (AI). In order to maximize operational efficiency and enable early issue detection, geothermal power facilities must employ real-time monitoring and control systems. This eventually results in lower maintenance costs and better system performance as a whole. In addition to optimizing energy extraction, the integration of artificial intelligence (AI) with geothermal energy management also helps to increase the operational lifespan of geothermal systems (He et al., 2022; Som, 2021).

The integration of artificial intelligence projects has led to a greater acknowledgment of hydrogen as a highly appreciated energy carrier. The efficiency of producing hydrogen from water, a crucial process for storing excess renewable energy, is increased by the use of AI-optimized electrolysis methods (Sai Ramesh et al., 2023). Algorithms for artificial intelligence (AI) are used to regulate the distribution and storage of hydrogen in order to improve efficiency and safety. Artificial intelligence (AI)-produced hydrogen is a renewable energy source that can power a variety of industries, including transportation and industrial. This reduces reliance on fossil fuels and accelerates the transition to a sustainable and commercially feasible energy system (Jha et al., 2017).

While the benefits of artificial intelligence (AI) in managing renewable energy sources are undeniable, challenges still need to be overcome. To guarantee exact modeling and accurate forecasts, the effective integration of artificial intelligence (AI) with renewable energy requires the availability of extensive and dependable data (Goralski & Tan, 2020). Furthermore, a certain set of abilities and expertise is required for the use of artificial intelligence (AI) technology. Data security and privacy issues must be taken into critical consideration in order to maintain the integrity of energy systems. Furthermore, while thinking about the broader deployment of AI-driven solutions, it is critical to examine their cost-effectiveness and scalability.

4. AI'S LIMITED USE IN SUSTAINABLE DEVELOPMENT

The integration of Artificial Intelligence (AI) with sustainable development has ushered in a new era of opportunity, capable of addressing complex problems across several industries. However, it is imperative to acknowledge that artificial intelligence (AI) is not a panacea; rather, it comes with a unique set of limitations and moral dilemmas that need careful management. In order to encourage a more equitable and morally responsible use of technology, it is critical to recognize these limitations when examining how artificial intelligence (AI) affects sustainable development.

4.1 Fairness and Bias in Data

Data bias is one of the main challenges facing the artificial intelligence community. Machine learning algorithms learn on historical data, and if historical data contains biased information, artificial intelligence models may reinforce and amplify inherent prejudices. In the context of sustainable development, biased algorithms have the potential to result in unfair resource allocation, unequal access to opportuni-

Figure 4. Limitation of the role of AI in sustainable development

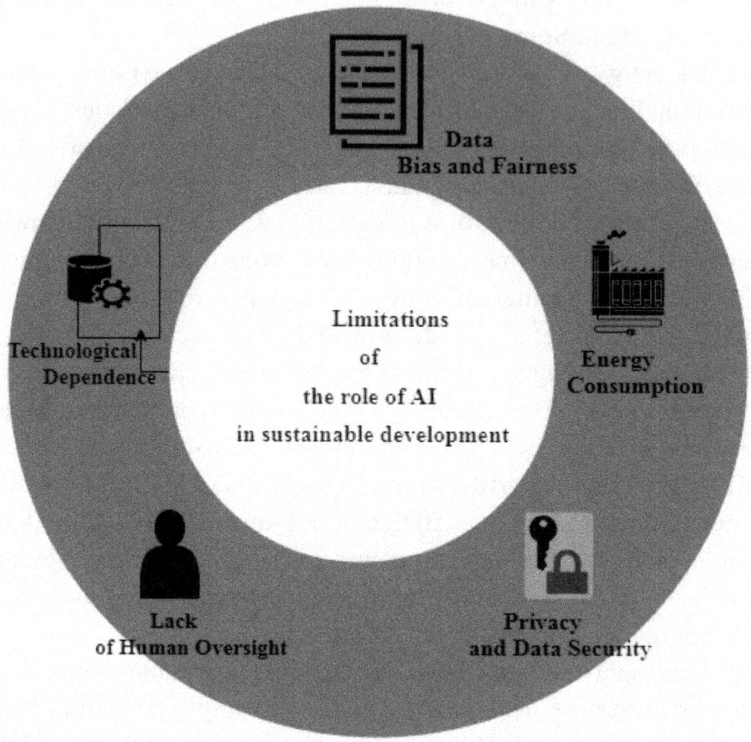

ties, and the maintenance of existing inequities. For example, artificial intelligence (AI) systems used in employment processes have the capacity to inadvertently maintain prejudice against certain groups of people based on historical hiring records. The prevention of social inequality being exacerbated by AI-driven sustainable development initiatives requires the reduction of data bias and the advancement of algorithmic fairness.

4.2 Usage of Energy

While artificial intelligence (AI) has great potential to improve energy efficiency and facilitate sustainable energy management, creating and implementing complex AI models may be energy-intensive. When artificial intelligence (AI) technologies that are supposed to encourage sustainability inadvertently lead to higher energy use, that is when irony might appear. The deployment of deep learning and other large-scale artificial intelligence (AI) applications requires the use of massive computer resources. As a consequence, there is now a greater need for energy resources. Maintaining a balanced relationship between the benefits of artificial intelligence (AI) and the energy it uses is necessary to avoid inadvertently weakening the core sustainability principles that AI aims to support.

4.3 Security of Data and Privacy

Systems using artificial intelligence (AI) work best when they can access and process large amounts of personal data. Concerns about data security and privacy protection are raised by these conditions. In order to support social programs, health treatments, and environmental monitoring, AI-driven systems must be able to efficiently handle the gathering and processing of personal data within the context of sustainable development. But doing so requires striking a delicate balance between achieving desirable results and respecting people's basic right to privacy. In order to apply AI responsibly, it is essential to integrate extensive data protection protocols and get individuals' informed permission.

4.4 Absence of Human Supervision

Deep learning and neural network-based artificial intelligence (AI) systems have the property of being opaque, meaning that their findings are difficult for humans to understand or explain. When it comes to sustainable growth, making decisions only via AI without human oversight may have unforeseen consequences. Attaining a state of balance between automation and human interaction is crucial for maintaining ethical decision-making, accountability, and transparency in the context of AI applications.

4.5 Dependence on Technology

As artificial intelligence (AI) becomes more and more integrated into sustainable development strategies, worries about an over-reliance on technology are raised. The only reliance on artificial intelligence (AI) solutions might lead to a diminished value placed on human abilities and traditional knowledge. A variety of academic fields are involved in sustainable development, which calls for the participation of many stakeholders. Within this framework, artificial intelligence (AI) need to be seen as a way to complement human intellect and cooperative endeavors, as opposed to replacing them.

5. THE DIFFICULTIES WITH AI IN RENEWABLE ENERGY

This section discusses a few of the hot-button research issues and challenges surrounding the practical implementation of artificial intelligence algorithms in renewable energy applications.

Artificial intelligence (AI) technology integration into renewable energy systems may need significant upfront costs for hardware, software, and training. It's also crucial to remember that setting up and implementing renewable energy infrastructures may need a large financial outlay. The monetary consequences linked to these levies might pose a substantial challenge for smaller businesses and developing countries with limited resources. The integration of artificial intelligence (AI) with renewable energy systems requires large amounts of data to be collected, analyzed, and stored. When sensitive data is involved, such as customer energy use patterns or grid vulnerabilities, concerns about data security and privacy may arise. The complex dynamics and uncertainties present in renewable energy systems, such as extreme weather events and abrupt changes in grid conditions, may be too difficult for the current AI algorithms to handle. The creation of more resilient algorithms is essential to solving the issues raised by the continuous developments in AI technology. The integration of artificial intelligence (AI) with renewable energy systems may lead to job displacement in some areas, since task automation eliminates the need for manual labor. Moreover, there may be a disparity in skill sets in the labor market given that the industry requires people with extensive expertise in both artificial intelligence (AI) and renewable energy. Implementing targeted educational and training programs that help workers adapt to the changing job market is crucial for successfully addressing these challenges.

Artificial intelligence (AI) systems are seen by many as essential forces behind social and economic advancement, leading governments all over the globe to prioritize their development. Deploying artificial intelligence approaches for large-scale renewable energy production requires a thorough grasp of both AI expertise and industry-specific information. This calls for a systematic approach to talent development, with an emphasis on long-term training programs that acquaint people with the essential components and operating framework of certain businesses (Hassan et al., 2023).Artificial intelligence (AI) systems are seen by many as essential forces behind social and economic advancement, leading governments all over the globe to prioritize their development. Deploying artificial intelligence approaches for large-scale renewable energy production requires a thorough grasp of both AI expertise and industry-specific information. This necessitates a systematic approach to talent development, with an emphasis on long-term training programs that acquaint people with the essential components and operating framework of certain sectors. The use of artificial intelligence (AI) techniques to large-scale renewable energy generation is still in its infancy, and there are still few financially feasible and scientifically supported policies in place (Liu et al., 2022) (Matschoss et al., 2019).

6. CONCLUSION

Artificial Intelligence (AI) integration in the renewable energy industry has become a disruptive force that might completely change how we produce, distribute, and use clean energy. Unprecedented breakthroughs in renewable energy systems have been made possible by artificial intelligence (AI) technologies, which include machine learning, predictive analytics, optimization algorithms, and smart control systems. These developments include improving energy storage systems for improved grid stability and dependability as well as increasing the efficiency of solar panels and wind turbines using real-time data analysis and adaptive control techniques. Predictive maintenance powered by AI has reduced downtime and operating expenses, increasing the longevity of renewable infrastructure. AI-driven energy management systems also enable users to make knowledgeable choices about how much energy they use, which promotes an energy-saving culture. The integration of artificial intelligence (AI) with renewable energy has enabled the development of virtual power plants capable of intelligently managing dispersed energy supplies, resulting in a grid that is more flexible and robust. Nonetheless, issues including algorithmic biases, data privacy concerns, and the difficulty of integrating AI into current energy infrastructures must be resolved. AI's contribution to renewable energy is expected to increase as it develops, accelerating the shift to a sustainable energy future by maximizing the potential of renewable resources, cutting waste, and eventually lowering our carbon footprint.

REFERENCES

Archer, R. (2020). Geothermal Energy. In Future Energy (pp. 431–445). Elsevier. doi:10.1016/B978-0-08-102886-5.00020-7

Asif, M., & Muneer, T.ASIF. (2007). Energy supply, its demand and security issues for developed and emerging economies. *Renewable & Sustainable Energy Reviews*, *11*(7), 1388–1413. doi:10.1016/j.rser.2005.12.004

Bose, B. K. (2017). Artificial Intelligence Techniques in Smart Grid and Renewable Energy Systems—Some Example Applications. *Proceedings of the IEEE*, *105*(11), 2262–2273. doi:10.1109/JPROC.2017.2756596

Bryson, J. J. (2019). *Towards a New Enlightenment? A Transcendent Decade - The Past Decade and the Future of Globalization*. OpendMind BBVA.

Burton, T., Jenkins, N., Sharpe, D., & Bossanyi, E. (2011). *Wind Energy Handbook*. Wiley., doi:10.1002/9781119992714

Carolin Mabel, M., & Fernandez, E. (2008). Analysis of wind power generation and prediction using ANN: A case study. *Renewable Energy*, *33*(5), 986–992. doi:10.1016/j.renene.2007.06.013

Damousis, I. G., & Dokopoulos, P. (2001). A fuzzy expert system for the forecasting of wind speed and power generation in wind farms. *IEEE Power Industry Computer Applications Conference*, 63–69. 10.1109/PICA.2001.932320

Dufo-López, R., Bernal-Agustín, J. L., & Contreras, J. (2007). Optimization of control strategies for stand-alone renewable energy systems with hydrogen storage. *Renewable Energy*, *32*(7), 1102–1126. doi:10.1016/j.renene.2006.04.013

Ediger, V. Ş. (2019). An integrated review and analysis of multi-energy transition from fossil fuels to renewables. *Energy Procedia*, *156*, 2–6. doi:10.1016/j.egypro.2018.11.073

Esteban, M., & Leary, D. (2012). Current developments and future prospects of offshore wind and ocean energy. *Applied Energy*, *90*(1), 128–136. doi:10.1016/j.apenergy.2011.06.011

Goralski, M. A., & Tan, T. K. (2020). Artificial intelligence and sustainable development. *International Journal of Management Education*, *18*(1), 100330. doi:10.1016/j.ijme.2019.100330

Hannan, M. A., Al-Shetwi, A. Q., Ker, P. J., Begum, R. A., Mansor, M., Rahman, S. A., Dong, Z. Y., Tiong, S. K., Mahlia, T. M. I., & Muttaqi, K. M. (2021). Impact of renewable energy utilization and artificial intelligence in achieving sustainable development goals. *Energy Reports*, *7*, 5359–5373. doi:10.1016/j.egyr.2021.08.172

Hassan, Q., Sameen, A. Z., Salman, H. M., Al-Jiboory, A. K., & Jaszczur, M. (2023). *The role of renewable energy and articial intelligence towards environmental sustainability and net zero*. Academic Press.

He, Z., Guo, W., & Zhang, P. (2022). Performance prediction, optimal design and operational control of thermal energy storage using artificial intelligence methods. *Renewable & Sustainable Energy Reviews*, *156*, 111977. doi:10.1016/j.rser.2021.111977

IEA. (2022). *Renewables 2022 - December*. https://www.iea.org/reports/renewables-2022

IEA. (2023). *Renewable Energy Market Update - June*. https://www.iea.org/reports/renewable-energy-market-update-june-2023

Jamei, M., Ali, M., Karbasi, M., Xiang, Y., Ahmadianfar, I., & Yaseen, Z. M. (2022). Designing a Multi-Stage Expert System for daily ocean wave energy forecasting: A multivariate data decomposition-based approach. *Applied Energy*, *326*, 119925. doi:10.1016/j.apenergy.2022.119925

Jha, S., Bilalovic, J., Jha, A., Patel, N., & Zhang, H. (2017). Renewable energy: Present research and future scope of Artificial Intelligence. *Renewable & Sustainable Energy Reviews*, *77*, 297–317. doi:10.1016/j.rser.2017.04.018

Kothari, R., Buddhi, D., & Sawhney, R. L. (2008). Comparison of environmental and economic aspects of various hydrogen production methods. *Renewable & Sustainable Energy Reviews*, *12*(2), 553–563. doi:10.1016/j.rser.2006.07.012

Kumar, M. (2020). Social, Economic, and Environmental Impacts of Renewable Energy Resources. In Wind Solar Hybrid Renewable Energy System. IntechOpen. doi:10.5772/intechopen.89494

Kumar, N., Kumar, D., Layek, A., & Yadav, S. (2022). Renewable energy and sustainable development. In *Artificial Intelligence for Renewable Energy Systems* (pp. 305–328). Elsevier. doi:10.1016/B978-0-323-90396-7.00011-0

Kumbhar, D., Palliyarayil, A., Reghu, D., Shrungar, D., Umapathy, S., & Sil, S. (2021). Rapid discrimination of porous bio-carbon derived from nitrogen rich biomass using Raman spectroscopy and artificial intelligence methods. *Carbon*, *178*, 792–802. doi:10.1016/j.carbon.2021.03.064

Liu, Z., Sun, Y., Xing, C., Liu, J., He, Y., Zhou, Y., & Zhang, G. (2022). Artificial intelligence powered large-scale renewable integrations in multi-energy systems for carbon neutrality transition: Challenges and future perspectives. *Energy and AI*, *10*, 100195. doi:10.1016/j.egyai.2022.100195

Matschoss, P., Bayer, B., Thomas, H., & Marian, A. (2019). The German incentive regulation and its practical impact on the grid integration of renewable energy systems. *Renewable Energy*, *134*, 727–738. doi:10.1016/j.renene.2018.10.103

McDowall, W. (2012). Technology roadmaps for transition management: The case of hydrogen energy. *Technological Forecasting and Social Change*, *79*(3), 530–542. doi:10.1016/j.techfore.2011.10.002

Mellit, A., Kalogirou, S. A., Hontoria, L., & Shaari, S. (2009). Artificial intelligence techniques for sizing photovoltaic systems: A review. *Renewable & Sustainable Energy Reviews*, *13*(2), 406–419. doi:10.1016/j.rser.2008.01.006

Mordensky, S. P., Lipor, J. J., DeAngelo, J., Burns, E. R., & Lindsey, C. R. (2023). When less is more: How increasing the complexity of machine learning strategies for geothermal energy assessments may not lead toward better estimates. *Geothermics*, *110*, 102662. doi:10.1016/j.geothermics.2023.102662

Nassef, A. M., Sayed, E. T., Rezk, H., Inayat, A., Yousef, B. A. A., Abdelkareem, M. A., & Olabi, A. G. (2020). Developing a fuzzy-model with particle swarm optimization-based for improving the conversion and gasification rate of palm kernel shell. *Renewable Energy*, *166*, 125–135. doi:10.1016/j.renene.2020.11.037

Robert, B., & Brown, E. B. (2006). Book Review: Wind Energy Explained: Theory, Design and Application. In Wind Engineering (Vol. 30, Issue 2). doi:10.1260/030952406778055054

Sai Ramesh, A., Vigneshwar, S., Vickram, S., Manikandan, S., Subbaiya, R., Karmegam, N., & Kim, W. (2023). Artificial intelligence driven hydrogen and battery technologies – A review. *Fuel*, *337*, 126862. doi:10.1016/j.fuel.2022.126862

Salim, H. K., Padfield, R., Hansen, S. B., Mohamad, S. E., Yuzir, A., Syayuti, K., Tham, M. H., & Papargyropoulou, E. (2018). Global trends in environmental management system and ISO14001 research. *Journal of Cleaner Production*, *170*, 645–653. doi:10.1016/j.jclepro.2017.09.017

Sarker, T. R., Pattnaik, F., Nanda, S., Dalai, A. K., Meda, V., & Naik, S. (2021). Hydrothermal pretreatment technologies for lignocellulosic biomass: A review of steam explosion and subcritical water hydrolysis. *Chemosphere*, *284*, 131372. doi:10.1016/j.chemosphere.2021.131372 PMID:34323806

Som, T. (2021). Sustainability in Energy Economy and Environment: Role of AI Based Techniques. doi:10.1007/978-3-030-72929-5_31

Tavakoli, S., Khojasteh, D., Haghani, M., & Hirdaris, S. (2023). A review on the progress and research directions of ocean engineering. *Ocean Engineering*, *272*, 113617. doi:10.1016/j.oceaneng.2023.113617

Chapter 17
Recent Trends in Photonic Crystals

Kannadhasan Suriyan
https://orcid.org/0000-0001-6443-9993
Study World College of Engineering, India

R. Nagarajan
https://orcid.org/0000-0002-4990-5869
Gnanamani College of Technology, India

ABSTRACT

This chapter explains the fascinating realm of photonic crystals. With a period on the order of an optical wavelength, these are artificial multidimensional periodic structures. They are a lot like solid state crystals in many ways. The band of photons is the most significant one since it provides a potent hypothesis for comprehending how light behaves in a challenging photonic crystal structure. We can produce the photonic bandgap and localise light thanks to it. They have considerable promise for cutting-edge uses in fields including acoustics, bio-photonics, quantum engineering, optoelectronics, waves, and optics. 2-D photonic crystals may provide integrated optics the much-needed miniaturisation. A modest selection of measurements on several kinds of waveguides on this size have already been provided. Straight waveguides alone are not sufficient for good routing or for creating more complicated circuits.

1. INTRODUCTION

All across the natural world, from the shifting colors of an opal held up to the light to the patterns on a butterfly's wings, periodic structures' optical characteristics may be seen. Although photonic crystals have been used by nature for millions of years, humans have only just begun to recognise their potential. Thin film stacks, which are one-dimensional periodic structures, had been researched for many years, but 1987 theory of three-dimensional photonic crystals was the first to do so. According to periodic dielectric structures in three dimensions may have an electromagnetic bandgap, or a range of frequencies below which light cannot pass through the structure in either direction. Additionally, he said that it is

DOI: 10.4018/979-8-3693-1306-0.ch017

possible to avoid undesired spontaneous emission in semiconductors by arranging the material such that the emission frequencies fall inside a photonic bandgap; as no propagating states exist at that frequency, emission is in fact prohibited. John demonstrated that a lot of the characteristics of PhCs continue to exist even when the periodic lattice is disordered. Strong light localization is still possible in such structures if the index contrast is great enough, analogous to the electrical bandgaps of amorphous semiconductors. Interpretation of the cavity modes that may be inserted into a periodic structure by producing a defect or phase slip was perhaps more pertinent to much of the PhC research that followed and to the subject of this paper A.Chelnokov,1997-A Blanco 2000

This method had earlier been used to establish resonant cavities in distributed feedback lasers, proved that these modes could be localised in three dimensions and explained the phenomenon in terms of defect states in the photonic bandgap. The idea of regulating light using periodic structures has quickly grown into a field of study on a global scale as a result of this finding and the earliest recommendations for restricting spontaneous emission. In bibliometric analysis, science mapping is crucial for capturing the discipline's state of growth and current circumstances. For the scientific mapping technique in this research, VOSviewer software, created by the Centre for Science and Technology Studies at Leiden University in the Netherlands, was utilised. To see the data, VOSviewer shows the size of the nodes. Symbolizing the input data is a node. The word size, which is greater than the node and represents the frequency of the data, demonstrates how uniformly distributed and broadly linked the data is to study Yablonovitch 1987- Dubey 2021. To make the search process easier, the cluster contains nodes with similar colour values. A line between two or more keywords indicates that they were used together in the article. The likelihood of simultaneous occurrence increases with line thickness. To facilitate searching, a cluster is made up of nodes that share the same colour. Two or more keywords are connected by a line to indicate that they have featured in publications together. The performance of conventional optical fibres is excellent in both telecom and non-telecom applications, but they have a number of structural limitations. PCFs have an extremely versatile design. Lattice pitch, air hole form and diameter, glass's refractive index, and lattice type are among the variables that may be changed. Since there is no wavelength and single mode fibres are single mode in all optical ranges, one may acquire them indefinitely thanks to design freedom. In addition, PCF has two guiding mechanisms: the photonic bandgap mechanism and an index guiding mechanism that is comparable to the one found in traditional optical fibres. It is feasible to create the fiber's desired dispersion qualities by modifying the structure. It is possible to design and construct PCFs with zero, low, or anomalous dispersion at visible wavelengths. A very broad range may also be used to attenuate the dispersion Arafa 2017- A.Cucinotta 2002.

Outstanding nonlinear fibres are produced when anomalous dispersion and tiny mode regions are combined. On the other hand, it is possible to create massive, solid or air core single mode fibres. A rod was used in lieu of the central air capillary in a lattice structure.A photonic bandgap may be obtained if the centre defect is created by introducing a central air capillary with a diameter that is different from other capillaries (typically larger) (PBG). The electron conduction process in materials with an energy-band structure is therefore analogous to light guiding in solid state physics. The demonstration of light guiding in an air fault occurred in 1997. (hollow core PGB guidance). A hexagonal lattice had some of its core capillaries removed, creating a sizable hole that was filled with air White RC 2001 – J.C.Knight 1998.

2. PHOTONICS CRYSTAL

An artificial 2D photonic crystal structure with a lattice constant close to the wavelength of light may be created using periodically dispersed air cores. Photonic bandgaps in the 2D crystal formations hinder the transmission of light in a certain frequency range. A distinct area with optical characteristics different from the photonic crystal is generated if the periodicity of the structure is disrupted by a defect (absence of air cores or big air core). Modes with frequencies within the photonic bandgap may be supported in the defect zone, which prevents them from penetrating the photonic crystal. The modes go through the fibre being strongly attracted to the flaws. It is not necessary for the defect zone to have a higher refractive index than the surrounding area since photonic bandgap is what conducts light in the core.

The initial generation of such devices used a straightforward method, which included removing holes from the required locations in the PhC lattice. Unfortunately, such devices show extremely limited bandwidth, which may be a design-inherent property, as well as poor transmission, which improves with fabrication technique and input coupler design. Numerous methods for increasing this bandwidth have been suggested by taking into account the field patterns in such devices using both the finite-difference time-domain (FDTD) and eigenmode expansion (EME) modelling approaches. To put it somewhat more simply, Y-splitters can be made better by adding holes to minimise the size of the junction cavity, and bends can be made better by using a mirror plane at the bend's point. For a number of these manually created devices, results from simulation and experimentation have both been reported. The most recent of them has extremely little bandwidth, about 30 nm, which amply indicates the necessity for a more complex numerical technique.Device transmission in PhC circuits is heavily influenced by the substrate shape, and it is particularly difficult to calculate the excess loss per component in Y-splitter devices. The greatest overall performance of a Y-splitter for the heterostructure-type substrate investigated in this study, at the time of writing, is a total transmission of 80% with regard to a similar W1 waveguide. Please take notice that we report a comparable number in this study for the transmission of a full microcircuit made up of a Y-junction, bends, and injectors. This is considerably more notable when we compare against a 5-m-wide ridge waveguide as opposed to the standard comparison versus a W1.

Access to PhC devices on the submicron scale must also be taken into account. Research is ongoing to determine the optimal way to link the access waveguide to the PhC. The key challenge is the mismatch at the waveguide-PhC interface and the subsequent loss from reflections and scattering. The adiabatic taper used in the conventional methods needs a length of at least 100 microns in order to function properly. In sharp contrast to the small size of PhC devices, this length. Therefore, there is strong motive to shorten PhC device access times.The advantages of numerical optimization approaches to further enhance the transmission and bandwidth of such injectors, bends, and Y-splitters have recently generated a lot of attention in the PhC community. Here, we provide the experimental findings from a fully optimisedPhC system that includes each of these components. The material heterostructure employed in our tests is fully taken into consideration by the optimization procedure. In our tests, we made use of a single-mode AlGaAsheterostructure (250 nm Al Ga As/488 nm Al Ga As / 55 nm Al Ga As / 2000 nm Al Ga As), which has previously shown strong experimental performance. Using a maximum length of 10 m as the primary design constraint, an injector from a 5- m access guide to the PhC W1 waveguide is optimised using a global approach in conjunction with 2-D FDTD and 2-D EME.

Similar to how a 60 bend is improved, a Y-splitter is also improved by modifying the size and location of a tuning hole in the Y-splitter. The design process's specifics, which describe a Y-splitter on a membrane with a refractive index of 2.5, have already been published. The design procedure was then carried out again for our wafer and an operating point of 1.5 m. The transmission of the improved design is 97%.

In order to guarantee that 3-D phenomena such out-of-plane diffraction do not completely dominate the response, 3-D FDTD was utilised to confirm the appropriateness of the designs. In each instance, the device's peak transmission for the 3-D FDTD is 76%. The bandwidth at 3 dB is around 120 nm. Due to memory constraints, this simulation does not include an output taper, therefore a little greater transmission in the simulation than in the experiment is to be anticipated—there will be no output losses caused by interfaces. The output taper is not anticipated to interfere with the taper's 300 nm bandwidth, as shown by Felici. The outcome of a 3-D FDTD simulation for a Y-splitter with an improved injector but a "basic" PhC is also provided for comparison to highlight the value of such optimization. The first PCF development work began in 1997 with a proposal for single mode PCF using sensors, interferometers, and mode filtering. In 1999, waveguide-structured PCF with a photonic bandgap air core was presented. The year 2000 saw the introduction of highly birefringent PCF, which comprises of various air hole diameters with an asymmetric core design or two orthogonal axes. The same year, Super continuum generation in PCF was also produced because of the wavelength's zero dispersion and strong nonlinearity. These were mostly used in pulse compression and laser sound spectroscopy. Bragg fibre fabrication was launched in 2000 for widespread use in optical sensors and fibre lasers. Along with ytterbium doping, double cladding was also launched in that year. Ultra-flattened dispersion was first presented in the year 2002. Bragg fibre with an air core and silica was invented in 2003. A brand-new chalcogenide PCF was launched in 2004 and demonstrated a very high refractive index coefficient. Later, in 2005, the kagome lattice PCF, which has a greater contribution to gas sensing, was produced. Physical sensors are useful because of their flexibility and interior structure, which may be filled with analytes to interact with light. Physical sensors are primarily used to measure certain desired parameters, such as pressure, torsion, refractive index, electric field, magnetic field, temperature, displacement, stress, and strain for a variety of applications in health monitoring, bio-medical sensing, civil structure monitoring, and remote sensing, in order to identify the problem in its initial state and prevent accidents or abnormal conditions. In the medical profession, sensing is increasingly important.

DNA hybridization, blood sensing, illness detection, pathogen detection, antibody and antigen detection, and glucose detection. To broaden the sensitivity range, several architectures are added. Because they are stable structures, PCF structures are employed for sensing. When chemicals are present as a gas, vapour, or liquid, certain sensors are employed to find them. They are used in the chemical industry to feel different chemicals and find chemical leaks and concentrations. Blood constituents and hazardous substances in the body are detected in the medical profession. To detect dangerous additions, food components are also felt. Following the initial work of (among others) Purcell (spontaneous emission in microcavities) in the late 1940s and Bykov (spontaneous emission in periodic structures) in the early 1970s, the modification of spontaneous emission by the surrounding dielectric structure has been studied throughout the second half of the 20th century. Yablonovitch used the phrase "photonic bandgap" in his influential 1987 study on "Inhibited spontaneous emission in solid state physics and electronics." The fundamentals of Bragg reflection that underlie a photonic bandgap were well understood in the discipline, but Yablonovitch was the first to link them to an electronic bandstructure and its associated vocabulary, providing fresh perspectives on how periodic structures in optics function.

The main characteristic of these structures that attracted attention was the suppression of light propagation, which sparked early investigation into new light emitters and zero-threshold lasers based on the suppression of undesirable spontaneous emission. As a result of the significant confinement of light caused by the suppression of propagation, Joannopoulos was motivated to promote photonic crystals for use in integrated optics in the middle of the 1990s. A new class of devices and circuits that are evocative of the large-scale integration attained by the microelectronics industry are made possible by this strong confinement's ability to facilitate high-density integration in photonics.

While the USA hosted the majority of the early research on photonic crystals (PhCs), the situation now paints a much more even regional picture. Since 1999, the USA, Japan, and Europe have alternated hosting the International Symposium on Photonic and Electromagnetic Crystal Structures (PECS). 2005 will see the holding of PECS-6 in Creta. The establishment of a European photonic crystal research community and the growth of the subject as a whole have unquestionably been significantly influenced by the IST-programme under the 5th Framework Programme (FP5) in Europe. With a greater participation of corporate partners and an expansion to new research subjects in nanophotonics and molecular photonics, this research is currently developing under the 6th Framework Programme (FP6). The reader is directed to additional information about the IST-programma. Holography, distributed feedback lasers, periodic waveguides, and x-ray diffraction are examples of devices that use standard electromagnetic gratings, which are made up of mild perturbations around a mean refractive index. These weak gratings are theoretically treated using a perturbation-based methodology that results in coupled mode equations for a limited number of grating plane waves. Weak gratings are not present in all traditional periodic structures. Large one-dimensional PBGs may be found in multilayer dielectric stacks that are used to cover optical components with high reflection and antireflection coatings.

Here, one may use the one-dimensional systems' simplicity to one's advantage in order to explain the PBG's physical genesis.

Layers with changing dielectric constants make up a one-dimensional photonic crystal. With a period an equal to the order of the light wavelength, this system repeats itself in the z-direction. At the interface of two media, a plane wave travelling in the z-direction along the line of periodicity will be dispersed. Within the structures, this results in waves that propagate both forward and backward. Standing waves will be created when these waves collide.the light's dispersion It is helpful to return to actual space and think about the electric field just above and below the gap, where k = /a, to have a physical understanding of this bandgap development. Standing waves with a wavelength of 2 an are the modes in this situation. The nodes of these modes may be found in either the layers with a high or low dielectric constant, giving rise to only two conceivable configurations for these modes within the structure.

3. CONCLUSION

The mode focused in the high index material will have a shorter effective distance, and as a result a lower frequency, than the mode concentrated in the low index material, even if both modes have the same wavelength. The energy gap in the photonic dispersion relation is provided by this difference in frequency, which is similar to the energy gap between the valence and conduction bands in a semiconductor material. The lower frequency range is often referred to as the dielectric band and the higher frequency band as the air band. The photonic system may be expressed in Bloch form, which consists of a plane wave modulated by a function derived from the periodicity of the lattice, since the two systems are in fact nearly identical. This method is very helpful when thinking about how light is coupled into two- and three-dimensional photonic crystals.

REFERENCES

Arafa, S., Bouchemat, M., Bouchemat, T., Benmerkhi, A., & Hocini, A. (2017). Infiltrated photonic crystal cavity as a highly sensitive platform for glucose concentration detection. *Optics Communications*, *384*, 93–100. doi:10.1016/j.optcom.2016.10.019

Birks, T. A., Knight, J. C., & Russell, P. St. J. (1997). Endlessly single-mode photonic crystal fiber. *Optics Letters*, *22*(13), 961. doi:10.1364/OL.22.000961 PMID:18185719

Chelnokov, A., Rowson, S., Lourtioz, J. M., Duvillaret, L., & Coutaz, J. L. (1997). Terahertzcharacterisation of mechanically machined 3D photonic crystal. *Electronics Letters*, *33*(23), 1981–1983. doi:10.1049/el:19971337

Cucinotta, A., Selleri, S., Vincetti, L., & Zoboli, M. (2002). Holey fiber analysis through the finite-element method. *IEEE Photonics Technology Letters*, *14*(11), 1530–1532. doi:10.1109/LPT.2002.803375

Dubey, S. K., Kumar, A., Kumar, A., Pathak, A., & Srivastava, S. K. (2021). A study of highly sensitive d-shaped optical fiber surface plasmon resonance based refractive index sensor using grating structures of Ag-TiO2 and Ag-SnO2. *Optik (Stuttgart)*, *252*, 168527. doi:10.1016/j.ijleo.2021.168527

Ferrando, A., Silvestre, E., Miret, J. J., Andres, P., & Andres, M. V. (1999). Full-vector analysis of a realistic photonic crystal fiber. *Optics Letters*, *24*(5), 276. doi:10.1364/OL.24.000276 PMID:18071478

Fleming, J. G., & Lin, S. Y. (1999). Three-dimensional photonic crystal with a stop band from 1.35 to 1.95 μm. *Optics Letters*, *24*(1), 49–51. doi:10.1364/OL.24.000049 PMID:18071404

Ge, R., Xie, J., Yan, B., Liu, E., Tan, W., & Liu, J. (2018). Refractive index sensor with high sensitivity based on circular photonic crystal. *Journal of the Optical Society of America. A, Optics, Image Science, and Vision*, *35*(6), 992–997. doi:10.1364/JOSAA.35.000992 PMID:29877344

Ho, M. M. (1998). Sigalas, W. Zubrzycki, S. R. Kurtz, and J. Bur, "A three-dimensional photonic crystal operating at infrared wavelengths,". *Nature*, *394*(6690), 251–253. doi:10.1038/28343

Jensen, M. O., & Brett, M. J. (2005). Square spiral 3D photonic bandgap crystals attelecommunications frequencies. *Optics Express*, *13*(9), 3348–3354. doi:10.1364/OPEX.13.003348 PMID:19495237

John, S. (1987). Strong localization of photons in certain disordered dielectric superlattices. *Physical Review Letters*, *58*(23), 2486–2489. doi:10.1103/PhysRevLett.58.2486 PMID:10034761

Knight, J. C., Birks, T. A., Russel, P. St. J., & de Sandro, J. P. (1998). Properties of photonic crystal fiber and the effective index model. *Journal of the Optical Society of America. A, Optics, Image Science, and Vision*, *15*(3), 748. doi:10.1364/JOSAA.15.000748

Mahdavi, A., Sarau, G., Xavier, J., Paraïso, T. K., Christiansen, S., & Vollmer, F. (2016). Maximizing photoluminescence extraction in silicon photonic crystal slabs. *Scientific Reports*, *6*(1), 1–6. doi:10.1038/srep25135 PMID:27113674

Mogilevtsev, D., Birks, T. A., & Russell, P. St. J. (2001). Localized function method for modeling defect modes in 2-D photonic crystals. *Journal of Lightwave Technology*, *17*(11), 2078–2081. doi:10.1109/50.802997

Noda, S., Tomoda, K., Yamamoto, N., & Chutinan, A. (2000). Full three-dimensionalphotonic bandgap crystals at near-infrared wavelengths. *Science*, *289*(5479), 604–606. doi:10.1126/science.289.5479.604 PMID:10915619

Scarmozzino, R., Gopinath, A., Pregla, R., & Helfert, S. (2000). Numerical techniques for modeling guided-wave photonic devices. *IEEE Journal of Selected Topics in Quantum Electronics*, *6*(1), 150–162. doi:10.1109/2944.826883

van Driel, H. M. (2000). Large-scale synthesis of a silicon photonic crystal with a completethree-dimensional bandgap near 1.5 micrometres. *Nature*, *405*(6785), 437–440. doi:10.1038/35013024 PMID:10839534

White, P., McPhedran, R. C., Botten, L. C., Smith, G. H., & de Sterke, C. M. (2001). Calculations of air-guided modes in photonic crystal fibers using the multipole method. *Optics Express*, *9*(13), 721. doi:10.1364/OE.9.000721 PMID:19424312

Yablonovitch, E. (1987). Inhibited spontaneous emission in solid-state physics and electronics. *Physical Review Letters*, *58*(20), 2059.

Zhi, W., Guobin, R., Shuqin, L., & Shuisheng, J. (2003). Supercell lattice method for photonic crystal fibers. *Optics Express*, *11*(9), 980. doi:10.1364/OE.11.000980 PMID:19465960

Compilation of References

Abate, G. Y., Alene, A. N., Habte, A. T., & Getahun, D. M. (2020). Adsorptive removal of malachite green dye from aqueous solution onto activated carbon of Catha edulis stem as a low cost bio-adsorbent. *Environmental Systems Research*, *9*(1), 29. Advance online publication. doi:10.1186/s40068-020-00191-4

Abdallah, M., El-Rayes, K., & Liu, L. (2016). Economic and GHG emission analysis of implementing sustainable measures in existing public buildings. *Journal of Performance of Constructed Facilities*, *30*(6), 4016055. doi:10.1061/(ASCE)CF.1943-5509.0000911

Abdolali, A., Ngo, H. H., Guo, W., Zhou, J. L., Zhang, J., Liang, S., Chang, S. W., Nguyen, D. D., & Liu, Y. (2017). Application of a breakthrough biosorbent for removing heavy metals from synthetic and real wastewaters in a lab-scale continuous fixed-bed column. *Bioresource Technology*, *229*, 78–87. doi:10.1016/j.biortech.2017.01.016 PMID:28110128

Abedin, A. H. (2011, August). A Critical Review of Thermochemical Energy Storage Systems. *The Open Renewable Energy Journal*, *4*(1), 42–46. doi:10.2174/1876387101004010042

Abraham, K. M. (2020). How comparable are sodium-ion batteries to lithium-ion counterparts? *ACS Energy Letters*, *5*(11), 3544–3547. doi:10.1021/acsenergylett.0c02181

Abruña, H. D., Kiya, Y., & Henderson, J. C. (2008). Batteries and electrochemical capacitors. *Physics Today*, *61*(12), 43–47. doi:10.1063/1.3047681

Abu Nayem, S. M., Ahmad, A., Shaheen Shah, S., Saeed Alzahrani, A., Saleh Ahammad, A. J., & Aziz, M. A. (2022). High Performance and Long-cycle Life Rechargeable Aluminum Ion Battery: Recent Progress, Perspectives and Challenges. *Chemical Record (New York, N.Y.)*, *22*(12), e202200181. doi:10.1002/tcr.202200181 PMID:36094785

Abulencia, A. B., Villoria, M. B. D., Libre, R. G. D. Jr, Quiatchon, P. R. J., Dollente, I. J. R., Guades, E. J., Promentilla, M. A. B., Garciano, L. E. O., & Ongpeng, J. M. C. (2021). Geopolymers as sustainable material for strengthening and restoring unreinforced masonry structures: A review. *Buildings*, *11*(11), 532. doi:10.3390/buildings11110532

Achenbach, G. D., James, W. J., & Gerson, R. (1967). Preparation of single-phase polycrystalline BiFeO3. *Journal of the American Ceramic Society*, *50*(8), 437–437. doi:10.1111/j.1151-2916.1967.tb15153.x

Adam, G. O., Sharker, S. M., & Ryu, G. (2022). Emerging biomedical applications of carbon dot and polymer composite materials. *Applied Sciences (Basel, Switzerland)*, *12*(20), 10565. doi:10.3390/app122010565

Agarwal, Agarwal, & Agarwal. (2017). Evolution of Internet of Nano Things (IoNT). *International Journal of Engineering Technology Science and Research, 4*(7).

Agarwal, K., Agarwal, K., & Agarwal, S. (2017). Evolution of Internet of Nano Things (IoNT). *Int. J. Eng. Technol. Sci. Res.*, *4*(7), 274–277.

Agrawal, R., & Espinosa, H. D. (2011). Giant piezoelectric size effects in zinc oxide and gallium nitride nanowires. A first principles investigation. *Nano Letters*, *11*(2), 786–790. doi:10.1021/nl104004d PMID:21241064

Aguirre-Contreras, S., Leyva-Ramos, R., Ocampo-Pérez, R., Aguilar-Madera, C. G., Flores-Cano, J. V., & Medellín-Castillo, N. A. (2023). Mathematical modeling of breakthrough curves for 8-hydroxyquinoline removal from fundamental equilibrium and adsorption rate studies. *Journal of Water Process Engineering*, *54*(April), 103967. doi:10.1016/j.jwpe.2023.103967

Agyapong-Fordjour, F. (2021). Substitutional Vanadium SulfideNanodispersed in MoS2 Film for Pt-Scalable Catalyst. *Advanced Science, 8*(16).

Ahmad, A. A., & Hameed, B. H. (2010). Fixed-bed adsorption of reactive azo dye onto granular activated carbon prepared from waste. *Journal of Hazardous Materials*, *175*(1–3), 298–303. doi:10.1016/j.jhazmat.2009.10.003 PMID:19883979

Ahmed, A. (2014). Energy smart buildings: Potential for conservation and efficiency of energy. *Pakistan Development Review*, *53*(4II), 371–380. doi:10.30541/v53i4IIpp.371-381

Ahmed, S. F., Rafa, N., Mehnaz, T., Ahmed, B., Islam, N., Mofijur, M., Hoang, A. T., & Shafiullah, G. M. (2022, September). Integration of phase change materials in improving the performance of heating, cooling, and clean energy storage systems: An overview. *Journal of Cleaner Production*, *364*, 132639. Advance online publication. doi:10.1016/j.jclepro.2022.132639

Ahtee, M., & Hewat, A. W. (1978). Structural phase transitions in sodium–potassium niobate solid solutions by neutron powder diffraction. *Acta Crystallographica. Section A, Crystal Physics, Diffraction, Theoretical and General Crystallography*, *34*(2), 309–317. doi:10.1107/S056773947800056X

Akhtar, Parwej, & Perwej. (2017). A Perusal of Big Data Classification and Hadoop Technology. *International Transaction of Electrical and Computer Engineers System (ITECES), 4*(1), 26-38. DOI:. doi:10.12691/iteces-4-1-4

Akinola, T. E., Prado, P. L. B., & Wang, M. (2022). Experimental studies, molecular simulation and process modelling\ simulation of adsorption-based post-combustion carbon capture for power plants: A state-of-the-art review. *Applied Energy*, *317*, 119156. doi:10.1016/j.apenergy.2022.119156

Akkerman, Q. A., & Manna, L. (2020). What defines a halide perovskite? *ACS Energy Letters*, *5*(2), 604–610. doi:10.1021/acsenergylett.0c00039 PMID:33344766

Akkerman, Q. A., Rainò, G., Kovalenko, M. V., & Manna, L. (2018). Genesis, Challenges and Opportunities for Colloidal Lead Halide Perovskite Nanocrystals. *Nature Materials*, *17*(5), 394–405. doi:10.1038/s41563-018-0018-4 PMID:29459748

Aksel, E., Forrester, J. S., Nino, J. C., Page, K., Shoemaker, D. P., & Jones, J. L. (2013). Local atomic structure deviation from average structure of Na 0.5 Bi 0.5 TiO 3: Combined x-ray and neutron total scattering study. *Physical Review B: Condensed Matter and Materials Physics*, *87*(10), 104113. doi:10.1103/PhysRevB.87.104113

Akyildiz, I. F., Brunetti, F., & Blázquez, C. (2008). Nanonetworks: A new communication paradigm. *Computer Networks*, *52*(12), 2260–2279. doi:10.1016/j.comnet.2008.04.001

Akyildiz, I. F., & Jornet, J. M. (2010). Electromagnetic wireless nanosensor networks. *Nano Communication Networks*, *1*(1), 3–19. doi:10.1016/j.nancom.2010.04.001

Akyildiz, I. F., Jornet, J. M., & Pierobon, M. (2011). Nanonetworks. *Communications of the ACM*, *54*(11), 84–89. doi:10.1145/2018396.2018417

Akyildiz, I., & Jornet, J. (2010). The Internet of nano-things. *IEEE Wireless Communications*, *17*(6), 58–63. doi:10.1109/MWC.2010.5675779

Akyildiz, I., Pierobon, M., Balasubramaniam, S., & Koucheryavy, Y. (2015). The internet of Bio-Nano things. *IEEE Communications Magazine*, *53*(3), 32–40. doi:10.1109/MCOM.2015.7060516

Alam, S. N., Sharma, N., & Kumar, L. (2017). Synthesis of graphene oxide (GO) by modified hummers method and its thermal reduction to obtain reduced graphene oxide (rGO). *Graphene (Irvine, Calif.)*, *6*(1), 1–18. doi:10.4236/graphene.2017.61001

Al-Ghouti, M. A., & Da'ana, D. A. (2020). Guidelines for the use and interpretation of adsorption isotherm models: A review. *Journal of Hazardous Materials*, *393*(February), 122383. doi:10.1016/j.jhazmat.2020.122383 PMID:32369889

Al-Ghussain, L., Ahmad, A. D., Abubaker, A. M., & Mohamed, M. A. (2021). An integrated photovoltaic/wind/biomass and hybrid energy storage systems towards 100% renewable energy microgrids in university campuses. *Sustainable Energy Technologies and Assessments*, *46*, 101273. doi:10.1016/j.seta.2021.101273

Alhassan, I., Park, J., & Tang, X. (2023). Characterizing a Local Sewage Sludge Ash for Developing Alkali-Activated Geopolymer Mortar. *Transportation Research Record: Journal of the Transportation Research Board*, *2677*(9). doi:10.1177/03611981231159112

Ali, N. A., & Abu-Elkheir, M. (2015). Internet of nano-things healthcare applications: Requirements, opportunities, and challenges. *2015 IEEE 11th International Conference on Wireless and Mobile Computing, Networking and Communications (WiMob).* 10.1109/WiMOB.2015.7347934

Ali, A., Chiang, Y. W., & Santos, R. M. (2022). X-ray Diffraction Techniques for Mineral Characterization: A Review for Engineers of the Fundamentals, Applications, and Research Directions. *Minerals (Basel)*, *12*(2), 205. doi:10.3390/min12020205

Ali, B. A., & Allam, N. K. (2019). A first-principles roadmap and limits to design efficient supercapacitor electrode materials. *Physical Chemistry Chemical Physics*, *21*(32), 17494–17511. doi:10.1039/C9CP02614B PMID:31372620

Ali, N. A., Aleyadeh, W., & AbuElkhair, M. (2016). Internet of Nano-Things network models and medical applications. *2016 International Wireless Communications and Mobile Computing Conference (IWCMC).* 10.1109/IWCMC.2016.7577059

Alkoy, E. M., & Berksoy, A. (2010, August). Effect of Cu and Li addition on the electrical properties of KNN ceramics. In *2010 IEEE International Symposium on the Applications of Ferroelectrics (ISAF)* (pp. 1-4). IEEE.

Allen, M. J., Tung, V. C., & Kaner, R. B. (2010). Honeycomb carbon: A review of graphene. *Chemical Reviews*, *110*(1), 132–145. doi:10.1021/cr900070d PMID:19610631

Al-Shakarchi, E. K., & Mahmood, N. B. (2011). Three techniques used to produce BaTiO 3 fine powder. *Journal of Modern Physics.*

Al-Shargabi, B., & Sabri, O. (2017). Internet of Things: An exploration study of opportunities and challenges. *2017 International Conference on Engineering & MIS (ICEMIS).* https://doi.org/10.1109/ICEMIS.2017.8273047

Alva, G., Lin, Y., & Fang, G. (2018). An overview of thermal energy storage systems. *Energy*, *144*, 341–378. doi:10.1016/j.energy.2017.12.037

Alves, C., Pelisser, F., Labrincha, J., & Novais, R. (2023). Effect of Hydrogen Peroxide on the Thermal and Mechanical Properties of Lightweight Geopolymer Mortar Panels. *Minerals (Basel)*, *13*(4), 542. doi:10.3390/min13040542

Alzahrani, A. S., Otaki, M., Wang, D., Gao, Y., Arthur, T. S., Liu, S., & Wang, D. (2021). Confining sulfur in porous carbon by vapor deposition to achieve high-performance cathode for all-solid-state lithium–sulfur batteries. *ACS Energy Letters*, *6*(2), 413–418. doi:10.1021/acsenergylett.0c01956

Amoros, J. L. (2012). *The Laue Method*. Elsevier Science.

Amran, M., Debbarma, S., & Ozbakkaloglu, T. (2021). Fly ash-based eco-friendly geopolymer concrete: A critical review of the long-term durability properties. *Construction & Building Materials*, *270*, 121857. doi:10.1016/j.conbuildmat.2020.121857

AnandNayyar. (2017). Internet of Nano Things(IoNT): Next Evolutionary Step in Nanotechnology. *Nanoscience and Nanotechnology*, *7*(1), 4–8. doi:10.5923/j.nn.20170701.02

An, G.-H., Ahn, H.-J., & Hong, W.-K. (2015). Electrochemical properties for high surface area and improved electrical conductivity of platinum-embedded porous carbon nanofibers. *Journal of Power Sources*, *274*, 536–541. doi:10.1016/j.jpowsour.2014.10.086

Aniket & Akhade. (n.d.). *Review of Scheffler reflector*. Academic Press.

Ansari. (2022). Critical aspects of various techniques for synthesizing metal oxides and fabricating their composite-based supercapacitor electrodes: a review. *Nanomaterials, 12*(11).

ANSI/IEEE Std 176-1987. (1987). IEEE standard on piezoelectricity.

Arafa, S., Bouchemat, M., Bouchemat, T., Benmerkhi, A., & Hocini, A. (2017). Infiltrated photonic crystal cavity as a highly sensitive platform for glucose concentration detection. *Optics Communications*, *384*, 93–100. doi:10.1016/j.optcom.2016.10.019

Aravindan, Lee, & Madhavi. (2017). Best Practices for Mitigating Irreversible Capacity Loss of Negative Electrodes in Li-Ion Batteries. *Advanced Energy Materials, 7*(17).

Arbizzani, C., Mastragostino, M., & Soavi, F. (2001). New trends in electrochemical supercapacitors. *Journal of Power Sources*, *100*(1-2), 164–170. doi:10.1016/S0378-7753(01)00892-8

Archer, R. (2020). Geothermal Energy. In Future Energy (pp. 431–445). Elsevier. doi:10.1016/B978-0-08-102886-5.00020-7

Arivazhagan, R., Geetha, N. B., Sivasamy, P., Kumaran, P., Gnanamithra, M. K., Sankar, S., Loganathan, G. B., & Arivarasan, A. (2020). Review on performance assessment of phase change materials in buildings for thermal management through passive approach. *Materials Today: Proceedings*, *22*, 419–431. doi:10.1016/j.matpr.2019.07.616

Arnau, A., & Soares, D. (2008). *Fundamentals of piezoelectricity*. Piezoelectric Transducers Appl. doi:10.1007/978-3-540-77508-9_1

Arnold, M. S., Avouris, P., Pan, Z. W., & Wang, Z. L. (2003). Field-effect transistors based on single semiconducting oxide nanobelts. *The Journal of Physical Chemistry B*, *107*(3), 659–663. doi:10.1021/jp0271054

Arsalani, N., Tabrizi, A. G., & Ghadimi, L. S. (2018). Novel PANI/MnFe2O4 nanocomposite for low-cost supercapacitors with high rate capability. *Journal of Materials Science Materials in Electronics*, *29*(7), 6077–6085. doi:10.1007/s10854-018-8582-6

Aryana, A., Singh, S. M., Mugnai, G., de Asmundis, C., Kowalski, M., Pujara, D. K., Cohen, A. I., Singh, S. K., Fuenzalida, C. E., Prager, N., Bowers, M. R., O'Neill, P. G., Brugada, P., d'Avila, A., & Chierchia, G. B. (2016). Pulmonary vein reconnection following catheter ablation of atrial fibrillation using the second-generation cryoballoon versus open-irrigated radiofrequency: Results of a multicenter analysis. *Journal of Interventional Cardiac Electrophysiology*, *47*(3), 341–348. doi:10.1007/s10840-016-0172-z PMID:27475949

Asamitsu, A., Tomioka, Y., Kuwahara, H., & Tokura, Y. (1997). Current switching of resistive states in magnetoresistive manganites. *Nature*, *388*(6637), 50–52. doi:10.1038/40363

Asefi, G., Ma, T., & Wang, R. (2022). Parametric investigation of photovoltaic-thermal systems integrated with porous phase change material. *Applied Thermal Engineering*, *201*, 117727. doi:10.1016/j.applthermaleng.2021.117727

Ashton, D., Hilton, M., & Thomas, K. V. (2004). Investigating the environmental transport of human pharmaceuticals to streams in the United Kingdom. *The Science of the Total Environment*, *333*(1–3), 167–184. doi:10.1016/j.scitotenv.2004.04.062 PMID:15364527

Asif, M., & Muneer, T.ASIF. (2007). Energy supply, its demand and security issues for developed and emerging economies. *Renewable & Sustainable Energy Reviews*, *11*(7), 1388–1413. doi:10.1016/j.rser.2005.12.004

Assadi, M. K., Bakhoda, S., Saidur, R., & Hanaei, H. (2017). Recent progress in perovskite solar cells. *Renewable & Sustainable Energy Reviews*, *81*, 2812–2822. doi:10.1016/j.rser.2017.06.088

Assefi, M., Maroufi, S., Yamauchi, Y., & Sahajwalla, V. (2020). Pyrometallurgical recycling of Li-ion, Ni–Cd and Ni–MH batteries: A minireview. *Current Opinion in Green and Sustainable Chemistry*, *24*, 26–31. doi:10.1016/j.cogsc.2020.01.005

Assi, L. N., Carter, K., Deaver, E., & Ziehl, P. (2020). Review of availability of source materials for geopolymer/sustainable concrete. *Journal of Cleaner Production*, *263*, 121477. doi:10.1016/j.jclepro.2020.121477

Assi, L., Carter, K., Deaver, E. E., Anay, R., & Ziehl, P. (2018). Sustainable concrete: Building a greener future. *Journal of Cleaner Production*, *198*, 1641–1651. doi:10.1016/j.jclepro.2018.07.123

ASTM. (2006). Standard Test Method for Determination of Iodine Number of Activated Carbon 1. *ASTM International, 94*(Reapproved), 1–5. http://compass.astm.org.acces.bibl.ulaval.ca/download/D4607.6656.pdf

Ateş, Y., Yücesoy, C. A., Ünlü, M. A., Saygin, B., & Akkaş, N. (2000). The Mechanical Properties of Intact and Traumatized Epidural Catheters. *Anesthesia and Analgesia*, *90*(2), 393–399. doi:10.1213/00000539-200002000-00029 PMID:10648328

Auger, P. (1975). The Auger effect. *Surface Science*, *48*(1), 1–8. doi:10.1016/0039-6028(75)90306-4

Aydin, D., Casey, S. P., & Riffat, S. (2015, January). The latest advancements on thermochemical heat storage systems. *Renewable & Sustainable Energy Reviews*, *41*, 356–367. doi:10.1016/j.rser.2014.08.054

Ayornu, H., Anwar, S., & Thakur, Y. (2022). Fabrication and Characterization of Semi-Crystalline and Amorphous Dielectric Polymer Films for Energy Storage. *Materials Sciences and Applications*, *1*(11), 559–568. doi:10.4236/msa.2022.1311034

Azam, M. A., Safie, N. E., & Abdul Aziz, M. F. (2022). Characterization of Hierarchical Porous Materials. *Advanced Functional Porous Materials: From Macro to Nano Scale Lengths*, 407–429.

Aziz, I. H., Al Bakri Abdullah, M. M., Yong, H. C., Ming, L. Y., Hussin, K., Surleva, A., & Azimi, E. A. (2019). Manufacturing parameters influencing fire resistance of geopolymers: A review. *Proceedings of the Institution of Mechanical Engineers, Part L: Journal of Materials: Design and Applications, 233*(4), 721–733. 10.1177/1464420716668203

Baer, D. R., & Engelhard, M. H. (2010). XPS analysis of nanostructured materials and biological surfaces. *Journal of Electron Spectroscopy and Related Phenomena*, *178–179*, 415–432. doi:10.1016/j.elspec.2009.09.003

Baetens, R., Jelle, B. P., & Gustavsen, A. (2010). Phase change materials for building applications: A state-of-the-art review. *Energy and Building*, *42*(9), 1361–1368. doi:10.1016/j.enbuild.2010.03.026

Bai, C., & Colombo, P. (2018). Processing, properties and applications of highly porous geopolymers: A review. *Ceramics International*, *44*(14), 16103–16118. doi:10.1016/j.ceramint.2018.05.219

Baikie, T., Fang, Y., Kadro, J. M., Schreyer, M., Wei, F., Mhaisalkar, S. G., Graetzel, M., & White, T. J. (2013). Synthesis and crystal chemistry of the hybrid perovskite $(CH_3NH_3) PbI_3$ for solid-state sensitised solar cell applications. *Journal of Materials Chemistry. A, Materials for Energy and Sustainability, 1*(18), 5628–5641. doi:10.1039/c3ta10518k

Bais, A. R., Subhedhar, D. G., Joshi, N. C., & Panchal, S. (2022). Numerical investigation on thermal management system for lithium ion battery using phase change material. *Materials Today: Proceedings, 66*, 1726–1733. doi:10.1016/j.matpr.2022.05.269

Baker, S. N., & Baker, G. A. (2010). Luminescent carbon nanodots: Emergent nanolights. *Angewandte Chemie International Edition, 49*(38), 6726–6744. doi:10.1002/anie.200906623 PMID:20687055

Balasubramaniam, S., & Kangasharju, J. (2013). Realizing the Internet of Nano Things: Challenges, Solutions, and Applications. *Computer, 46*(2), 62–68. doi:10.1109/MC.2012.389

Ballou, B., Lagerholm, B. C., Ernst, L. A., Bruchez, M. P., & Waggoner, A. S. (2004). Noninvasive imaging of quantum dots in mice. *Bioconjugate Chemistry, 15*(1), 79–86. doi:10.1021/bc034153y PMID:14733586

Banerjee, M., & Kaur, H. (2022). A Comparison Among Lithium-Ion, Nickel-Cadmium & Nickel-Metal-Hydride Batteries for Charging and Discharging in Electric Vehicle by Bidirectional Dc-Dc Converter. In *2022 IEEE IAS Global Conference on Emerging Technologies (GlobConET)*. IEEE. 10.1109/GlobConET53749.2022.9872335

Barua, R., Das, S., Roy Chowdhury, A., & Datta, P. (2022). Experimental and simulation investigation of surgical needle insertion into soft tissue mimic biomaterial for minimally invasive surgery (MIS). *Proceedings of the Institution of Mechanical Engineers, Part H: Journal of Engineering in Medicine.* 10.1177/09544119221143860

Barua, R., Datta, S., Datta, P., & Chowdhury, A. R. (2020, September). Experimental analysis the tissue deformation of needle insertion process in tissue engineering. In *2020 IEEE 1st International conference for convergence in engineering (ICCE)* (pp. 83-85). IEEE. 10.1109/ICCE50343.2020.9290598

Barua, R., Datta, S., Chowdhury, A. R., & Datta, P. (2018). Advances in MEMS and Micro-Scale technologies for application in controlled Drug-Dosing systems: MEMS-Based drug delivery systems. In *Design and Development of Affordable Healthcare Technologies* (pp. 165–179). IGI Global. doi:10.4018/978-1-5225-4969-7.ch007

Barua, R., Datta, S., & Das, J. (2020). Application of Nanotechnology in Global Issues. In M. Merviö (Ed.), *Global Issues and Innovative Solutions in Healthcare, Culture, and the Environment* (pp. 292–300). IGI Global. doi:10.4018/978-1-7998-3576-9.ch015

Barua, R., Sarkar, A., & Datta, S. (2023). Modern Lab-on-Chip Biosensors Application on Infectious COVID-19 Detection. In R. Singh, R. Phanden, B. Sikarwar, & J. Davim (Eds.), *Advances in MEMS and Microfluidic Systems* (pp. 258–270). IGI Global. doi:10.4018/978-1-6684-6952-1.ch013

Beanland, R., & Thomas, P. A. (2014). Symmetry and defects in rhombohedral single-crystalline Na 0.5 Bi 0.5 TiO 3. *Physical Review B: Condensed Matter and Materials Physics, 89*(17), 174102. doi:10.1103/PhysRevB.89.174102

Behera, C., Patel, P., Pradhan, N., & Choudhary, R. N. P. (2022). Studies of structural, dielectric and electrical characteristics of nickel-modified barium titanate for device applications. *Journal of Materials Science Materials in Electronics, 33*(3), 1657–1669. doi:10.1007/s10854-022-07709-9

Benninghoven, A. (1973). New developments in the surface analysis of solids. *Applied Physics (Berlin), 1*(1), 3–16. doi:10.1007/BF00886799

Berardi, U., & Gallardo, A. A. (2019). Properties of concretes enhanced with phase change materials for building applications. *Energy and Building, 199*, 402–414. doi:10.1016/j.enbuild.2019.07.014

Berger, A. H., & Bhown, A. S. (2013). Optimizing solid sorbents for CO2 capture. *Energy Procedia, 37*, 25–32. doi:10.1016/j.egypro.2013.05.081

Bergveld, P. (1972). Development, operation, and application of the ion-sensitive field-effect transistor as a tool for electrophysiology. *IEEE Transactions on Biomedical Engineering, 19*(5), 342–351. doi:10.1109/TBME.1972.324137 PMID:5038390

Bernard, J. (1955). *U.S. Patent No. 2,708,244*. Washington, DC: U.S. Patent and Trademark Office.

Best Research-Cell Efficiency Chart. (n.d.). https://www.nrel.gov/pv/cell-efficiency.html

Best, R.-C. E. C. (n.d.). *The National Renewable Energy Laboratory*. https://www.nrel.gov/pv/cell-efficiency.html

Bharti, K., Kumar, A., Ahmed, G., Gupta, M., Bocchetta, P., Adalati, R., Chandra, R., & Kumar, Y. (2021). Theories and models of supercapacitors with recent advancements: Impact and interpretations. *Nano Express, 2*(2), 022004. doi:10.1088/2632-959X/abf8c2

Bhide, V. G., Hegde, M. S., & Deshmukh, K. G. (1968). Ferroelectric properties of lead titanate. *Journal of the American Ceramic Society, 51*(10), 565–568. doi:10.1111/j.1151-2916.1968.tb13323.x

Birks, T. A., Knight, J. C., & Russell, P. St. J. (1997). Endlessly single-mode photonic crystal fiber. *Optics Letters, 22*(13), 961. doi:10.1364/OL.22.000961 PMID:18185719

Bishop, H. E., & Rivière, J. C. (1970). Characteristic ionization losses observed in auger emission spectroscopy. *Applied Physics Letters, 16*(1), 21–23. doi:10.1063/1.1653016

Blanc, L., Delchambre, A., & Lambert, P. (2017). Flexible medical devices: Review of controllable stiffness solutions. *Actuators, 6*(3), 23. Advance online publication. doi:10.3390/act6030023

Bochenek, D., Skulski, R., Wawrzała, P., & Brzezińska, D. (2011). Dielectric and ferroelectric properties and electric conductivity of sol–gel derived PBZT ceramics. *Journal of Alloys and Compounds, 509*(17), 5356–5363. doi:10.1016/j.jallcom.2011.02.046

Boer, D. G., Langerak, J., & Pescarmona, P. P. (2023). Zeolites as Selective Adsorbents for CO2 Separation. *ACS Applied Energy Materials, 6*(5), 2634–2656. doi:10.1021/acsaem.2c03605

Bose, B. K. (2017). Artificial Intelligence Techniques in Smart Grid and Renewable Energy Systems—Some Example Applications. *Proceedings of the IEEE, 105*(11), 2262–2273. doi:10.1109/JPROC.2017.2756596

Boundzanga, H. M., Cagnon, B., Roulet, M., de Persis, S., Vautrin-Ul, C., & Bonnamy, S. (2022). Contributions of hemicellulose, cellulose, and lignin to the mass and the porous characteristics of activated carbons produced from biomass residues by phosphoric acid activation. *Biomass Conversion and Biorefinery, 12*(8), 3081–3096. doi:10.1007/s13399-020-00816-9

Boussaba, L., Foufa, A., Makhlouf, S., Lefebvre, G., & Royon, L. (2018). Elaboration and properties of a composite bio-based PCM for an application in building envelopes. *Construction & Building Materials, 185*, 156–165. doi:10.1016/j.conbuildmat.2018.07.098

Boussaba, L., Makhlouf, S., Foufa, A., Lefebvre, G., & Royon, L. (2019). vegetable fat: A low-cost bio-based phase change material for thermal energy storage in buildings. *Journal of Building Engineering, 21*, 222–229. doi:10.1016/j.jobe.2018.10.022

Bouzid, A., Bourim, E. M., Gabbay, M., & Fantozzi, G. (2005). PZT phase diagram determination by measurement of elastic moduli. *Journal of the European Ceramic Society, 25*(13), 3213–3221. doi:10.1016/j.jeurceramsoc.2004.07.018

Boyd, C. C., Cheacharoen, R., Leijtens, T., & McGehee, M. D. (2019). Understanding degradation mechanisms and improving stability of perovskite photovoltaics. *Chemical Reviews*, *119*(5), 119, 3418–3451. doi:10.1021/acs.chemrev.8b00336 PMID:30444609

Bo, Z., Wen, Z., Kim, H., Lu, G., Yu, K., & Chen, J. (2012). One-step fabrication and capacitive behavior of electrochemical double layer capacitor electrodes using vertically-oriented graphene directly grown on metal. *Carbon*, *50*(12), 4379–4387. doi:10.1016/j.carbon.2012.05.014

Bragg, W. H., & Bragg, W. L. (1913). The reflection of X-rays by crystals. *Proceedings of the Royal Society of London. Series A, Containing Papers of a Mathematical and Physical Character*, *88*(605), 428–438. doi:10.1098/rspa.1913.0040

Bryson, J. J. (2019). *Towards a New Enlightenment? A Transcendent Decade - The Past Decade and the Future of Globalization.* OpendMind BBVA.

Bucur, C. B., Gregory, T., Oliver, A. G., & Muldoon, J. (2015). Confession of a magnesium battery. *The Journal of Physical Chemistry Letters*, *6*(18), 3578–3591. doi:10.1021/acs.jpclett.5b01219 PMID:26722727

Bunaciu, A. A., Udriştioiu, E., & Aboul-Enein, H. Y. (2015). X-Ray Diffraction: Instrumentation and Applications. *Critical Reviews in Analytical Chemistry*, *45*(4), 289–299. doi:10.1080/10408347.2014.949616 PMID:25831472

Burschka, J., Pellet, N., Moon, S.-J., Humphry-Baker, R., Gao, P., Nazeeruddin, M. K., & Grätzel, M. (2013). Sequential deposition as a route to high-performance perovskite-sensitized solar cells. *Nature*, *499*(7458), 316–319. doi:10.1038/nature12340 PMID:23842493

Burton, T., Jenkins, N., Sharpe, D., & Bossanyi, E. (2011). *Wind Energy Handbook*. Wiley., doi:10.1002/9781119992714

Bzai, J., Alam, F., Dhafer, A., Bojović, M., Altowaijri, S. M., Niazi, I. K., & Mehmood, R. (2022). Machine Learning-Enabled Internet of Things (IoT): Data, Applications, and Industry Perspective. *Electronics (Basel)*, *11*(17), 2676. doi:10.3390/electronics11172676

Cabeza, L. F., Castellon, C., Nogues, M., Medrano, M., Leppers, R., & Zubillaga, O. (2007). Use of microencapsulated PCM in concrete walls for energy savings. *Energy and Building*, *39*(2), 113–119. doi:10.1016/j.enbuild.2006.03.030

Cady, W. G. (1963). Piezoelectricity and ultrasonics. *Sound: Its Uses and Control, 2*(1), 46-52.

Cai, L., Zhang, Q., Mwizerwa, J. P., Wan, H., Yang, X., Xu, X., & Yao, X. (2018). Highly crystalline layered VS2 nanosheets for all-solid-state lithium batteries with enhanced electrochemical performances. *ACS Applied Materials & Interfaces*, *10*(12), 10053–10063. doi:10.1021/acsami.7b18798 PMID:29498503

Calvert, J. G. (1990). Glossary of atmospheric chemistry terms. *Pure and Applied Chemistry*, *62*(11), 2167–2219. doi:10.1351/pac199062112167

Cao, L., Wang, X., Meziani, M. J., Lu, F., Wang, H., Luo, P. G., Lin, Y., Harruff, B. A., Veca, L. M., Murray, D., Xie, S.-Y., & Sun, Y. P. (2007). Carbon dots for multiphoton bioimaging. *Journal of the American Chemical Society*, *129*(37), 11318–11319. doi:10.1021/ja0735271 PMID:17722926

Cao, M., Gao, L., Lv, X., & Shen, Y. (2017). TiO2-B@ VS2 heterogeneous nanowire arrays as superior anodes for lithium-ion batteries. *Journal of Power Sources*, *350*, 87–93. doi:10.1016/j.jpowsour.2017.03.070

Cao, R., Wang, J., Zhao, S., Yang, W., Yuan, Z., Yin, Y., Du, X., Li, N.-W., Zhang, X., Li, X., Wang, Z. L., & Li, C. (2018). Self-powered nanofiber-based screen-print triboelectric sensors for respiratory monitoring. *Nano Research*, *11*(7), 3771–3779. doi:10.1007/s12274-017-1951-2

Carbonaro, C., Cascone, Y., Fantucci, S., Serra, V., Perino, M., & Dutto, M. (2015). Energy assessment of a PCM–embedded plaster: Embodied energy versus operational energy. *Energy Procedia*, *78*, 3210–3215. doi:10.1016/j.egypro.2015.11.782

Cárdenas, B., & León, N. (2013). High temperature latent heat thermal energy storage: Phase change materials, design considerations and performance enhancement techniques. *Renewable & Sustainable Energy Reviews*, *27*, 724–737. doi:10.1016/j.rser.2013.07.028

Carlson, T. (2013). *Photoelectron_and_Auger_Spectroscopy*. Springer US.

Carolin Mabel, M., & Fernandez, E. (2008). Analysis of wind power generation and prediction using ANN: A case study. *Renewable Energy*, *33*(5), 986–992. doi:10.1016/j.renene.2007.06.013

Castro-Gutiérrez, J., Celzard, A., & Fierro, V. (2020). Energy Storage in Supercapacitors: Focus on Tannin-Derived Carbon Electrodes. *Frontiers in Materials*, *7*(July), 1–25. doi:10.3389/fmats.2020.00217

Catalan, G., & Scott, J. F. (2009). Physics and applications of bismuth ferrite. *Advanced Materials*, *21*(24), 2463–2485. doi:10.1002/adma.200802849

Cavazos-Cuello, L. A., Ocampo-Pérez, R., Tututi-Ávila, S., Dávila-Guzmán, N. E., Soto-Regalado, E., & Salazar-Rábago, J. J. (2023). Comparative phenomenological modeling of the overall adsorption rate of drugs onto activated carbon. The case of the chlorpheniramine. *Chemical Engineering Research & Design*, *199*, 152–161. doi:10.1016/j.cherd.2023.09.043

Cellat, K., Beyhan, B., Güngör, C., Konuklu, Y., Karahan, O., Dündar, C., & Paksoy, H. (2015). Thermal enhancement of concrete by adding bio-based fatty acids as phase change materials. *Energy and Building*, *106*, 156–163. doi:10.1016/j.enbuild.2015.05.035

Chaffee, A. L., Knowles, G. P., Liang, Z., Zhang, J., Xiao, P., & Webley, P. A. (2007). CO2 capture by adsorption: Materials and process development. *International Journal of Greenhouse Gas Control*, *1*(1), 11–18. doi:10.1016/S1750-5836(07)00031-X

Chaffee, N., Ebelegi, A. N., & Wankasi, D. (2017). Modelling and interpretation of adsorption isotherms. *Journal of Chemistry*, •••, 2017.

Chaliha, C., & Maisnam, M. (2023). Observation of the phase evolution and associated dielectric properties of K0. 47Na0. 47Li0. 06NbO3+(x) Bi0. 5Na0. 5ZrO3 lead free piezoceramic under varying frequency. *Materials Today: Proceedings*. Advance online publication. doi:10.1016/j.matpr.2023.02.233

Chandel & Agarwal. (2017). Review of current state of research on energy storage, toxicity, health hazards and commercialization of phase changing materials. *Renewable and Sustainable Energy Reviews, 67*, 581–596. . doi:10.1016/j.rser.2016.09.070

Chandrasekhar, M., & Kumar, P. (2016). Processing and characterizations of BNT-KNN ceramics for actuator applications. *Processing and Application of Ceramics*, *10*(2), 73–77. doi:10.2298/PAC1602073C

Chandrasekhar, M., Negi, R. R., Nanda, D., Mallam, M., & Kumar, P. (2023). Microwave versus conventional sintering of NaNbO3 modified BNT–BT–BKT ceramics. *Journal of Materials Science Materials in Electronics*, *34*(6), 475. doi:10.1007/s10854-023-09918-2

Chang, C. C. (1971). Auger electron spectroscopy. *Surface Science*, *25*(1), 53–79. doi:10.1016/0039-6028(71)90210-X

Chang, Y., Yang, P., Ma, D., Liu, Z., & Wang, Z. (2008). Phase transitional behavior, microstructure, and electrical properties in Ta-modified [(K0.458Na0.542)0.96Li0.04]NbO3 lead-free piezoelectric ceramics. *Journal of Applied Physics*, *104*(2), 024109. doi:10.1063/1.2957591

Chatziasteriou, C. C., Kikkinides, E. S., & Georgiadis, M. C. (2022). Recent advances on the modeling and optimization of CO2 capture processes. *Computers & Chemical Engineering*, *165*, 107938. doi:10.1016/j.compchemeng.2022.107938

Chaudhary, P., Fatima, F., & Kumar, A. (2020). Relevance of Nanomaterials in Food Packaging and its Advanced Future Prospects. *Journal of Inorganic and Organometallic Polymers and Materials*, *30*(12), 5180–5192. doi:10.1007/s10904-020-01674-8 PMID:32837459

Chauhan, A. (2014). Powder XRD Technique and its Applications in Science and Technology. *Journal of Analytical & Bioanalytical Techniques*, *5*(6). Advance online publication. doi:10.4172/2155-9872.1000212

Chelnokov, A., Rowson, S., Lourtioz, J. M., Duvillaret, L., & Coutaz, J. L. (1997). Terahertzcharacterisation of mechanically machined 3D photonic crystal. *Electronics Letters*, *33*(23), 1981–1983. doi:10.1049/el:19971337

Chen, Chua, & Lee. (2020). The advances of metal sulfides and in situ characterization methods beyond Li ion batteries: sodium, potassium, and aluminum ion batteries. *Small Methods, 4*(1).

Chen, Shi, Zhang, Zhu, & Yan. (2006). *Size Dependence of Young's Modulus in ZnO Nanowires*. . doi:10.1103/PhysRevLett.96.075505

Chen, B., Niu, M., Peng, Z., Qiao, X., Liang, P., Yang, Z., & Chao, X. (2020). Dielectric properties of lead-free BNT-based ferroelectric ceramics near the morphotropic phase boundary. *Materials Chemistry and Physics*, *256*, 123639. doi:10.1016/j.matchemphys.2020.123639

Cheng & Verbrugge. (2010). Diffusion-induced stress, interfacial charge transfer, and criteria for avoiding crack initiation of electrode particles. *Journal of the Electrochemical Society, 157*(4).

Cheng, L. Q., Wang, K., Yao, F. Z., Zhu, F., & Li, J. F. (2013). Composition Inhomogeneity due to Alkaline Volatilization in Li-Modified (K, Na) NbO 3 Lead-Free Piezoceramics. *Journal of the American Ceramic Society*, *96*(9), 2693–2695. doi:10.1111/jace.12497

Cheng, Q., Tang, J., Shinya, N., & Qin, L. C. (2013). Polyaniline modified graphene and carbon nanotube composite electrode for asymmetric supercapacitors of high energy density. *Journal of Power Sources*, *241*, 423–428. doi:10.1016/j.jpowsour.2013.04.105

Cheng, X., Wu, J., Wang, X., Zhang, B., Lou, X., Wang, X., Xiao, D., & Zhu, J. (2013). Mediating the contradiction of d33 and TC in potassium– sodium niobate lead-free piezoceramics. *ACS Applied Materials & Interfaces*, *5*(21), 10409–10417. doi:10.1021/am403448r PMID:24219128

Chen, H., Dong, S., Zhang, Y., & He, P. (2022). Robust structure regulation of geopolymer as novel efficient amine support to prepare high-efficiency CO2 capture solid sorbent. *Chemical Engineering Journal*, *427*, 131577. doi:10.1016/j.cej.2021.131577

Chen, J., & Park, N. G. (2019). Causes and Solutions of Recombination in Perovskite Solar Cells. *Advanced Materials*, *31*(47), 1803019. doi:10.1002/adma.201803019 PMID:30230045

Chen, J., Tang, Z., Pan, Z., Shi, W., Wang, Y., Tian, Z. Q., & Shen, P. K. (2020). Template-free growth of spherical vanadium disulfide nanoflowers as efficient anodes for sodium/potassium ion batteries. *Materials & Design*, *192*, 108780. doi:10.1016/j.matdes.2020.108780

Chen, Q., Song, M., Zhang, X., Zhang, J., Hou, G., & Tang, Y. (2022). Ammonium ion pre-intercalation stabilized tunnel h-WO 3 for fast NH 4+ storage. *Journal of Materials Chemistry. A, Materials for Energy and Sustainability*, *10*(29), 15614–15622. doi:10.1039/D2TA03643F

Chen, Q., Zhou, H., Hong, Z., Luo, S., Duan, H. S., Wang, H. H., Liu, Y., Li, G., & Yang, Y. (2014). Planar heterojunction perovskite solar cells via vapor-assisted solution process. *Journal of the American Chemical Society*, *136*(2), 622–625. doi:10.1021/ja411509g PMID:24359486

Chen, S. M., Ramachandran, R., Mani, V., & Saraswathi, R. (2014). Recent advancements in electrode materials for the high-performance electrochemical supercapacitors: A review. *International Journal of Electrochemical Science*, *9*(8), 4072–4085. doi:10.1016/S1452-3981(23)08076-8

Chen, T., & Dai, L. (2013). Carbon nanomaterials for high-performance supercapacitors. *Materials Today*, *16*(7-8), 272–280. doi:10.1016/j.mattod.2013.07.002

Chen, T., Zhu, X., Chen, X., Zhang, Q., Li, Y., Peng, W., Zhang, F., & Fan, X. (2020). VS2 nanosheets vertically grown on graphene as high-performance cathodes for aqueous zinc-ion batteries. *Journal of Power Sources*, *477*, 228652. doi:10.1016/j.jpowsour.2020.228652

Chen, W., Wu, Y., Yue, Y., Liu, J., Zhang, W., Yang, X., Chen, H., Bi, E., Ashraful, I., Grätzel, M., & Han, L. (2015). Efficient and stable large-area perovskite solar cells with inorganic charge extraction layers. *Science*, *350*(6263), 944–948. doi:10.1126/science.aad1015 PMID:26516198

Chen, X., Wang, J., Yuan, G., Wu, D., Liu, J., Yin, J., & Liu, Z. (2012). Structure, ferroelectric and piezoelectric properties of multiferroic Bi0. 875Sm0. 125FeO3 ceramics. *Journal of Alloys and Compounds*, *541*, 173–176. doi:10.1016/j.jallcom.2012.06.094

Chen, Y., Tse, W. H., Chen, L., & Zhang, J. (2015). Ag nanoparticles-decorated ZnO nanorod array on a mechanical flexible substrate with enhanced optical and antimicrobial properties. *Nanoscale Research Letters*, *10*(1), 106. doi:10.1186/s11671-014-0712-3 PMID:25852401

Chianelli, R. R. (1976). Microscopic studies of transition metal chalcogenides. *Journal of Crystal Growth*, *34*(2), 239–244. doi:10.1016/0022-0248(76)90135-4

Ching-Jen Chen, Haik, Y., & Chatterjee, J. (2005). Development of nanotechnology for biomedical applications. Conference. *Emerging Information Technology*, *2005*, 9–12. Advance online publication. doi:10.1109/EITC.2005.1544329

Chiu. (2014). *A Review on Conduction Mechanisms in Dielectric Films*. Academic Press.

Chiu, F. C., Lee, C. Y., & Pan, T. M. (2009). Current conduction mechanisms in Pr2 O3 /oxynitride laminated gate dielectrics. *Journal of Applied Physics*, *105*(7), 074103. doi:10.1063/1.3103282

Choi, H., & Yoon, H. (2015). Nanostructured electrode materials for electrochemical capacitor applications. *Nanomaterials (Basel, Switzerland)*, *5*(2), 906–936. doi:10.3390/nano5020906 PMID:28347044

Choi, S. W., Shrout, T. R., Jang, S. J., & Bhalla, A. S. (1989). Morphotropic phase boundary in Pb(MgNb)O3-PbTiO3 system. *Materials Letters*, *8*(6-7), 253–255. doi:10.1016/0167-577X(89)90115-8

Cho, J., Li, X., Gu, Z., & Kurup, P. U. (2011). Recognition of explosive precursors using nanowire sensor array and decision tree learning. *IEEE Sensors Journal*, *12*(7), 2384–2391. doi:10.1109/JSEN.2011.2182042

Christians, J. A., Fung, R. C. M., & Kamat, P. V. (2014). An inorganic hole conductor for organo-lead halide perovskite solar cells. Improved hole conductivity with copper iodide. *Journal of the American Chemical Society*, *136*(2), 758–764. doi:10.1021/ja411014k PMID:24350620

Chu, K. H. (2020). Breakthrough curve analysis by simplistic models of fixed bed adsorption: In defense of the century-old Bohart-Adams model. *Chemical Engineering Journal*, *380*(June), 122513. doi:10.1016/j.cej.2019.122513

Chu, G., Zhao, J., Huang, Y., Zhou, D., Liu, Y., Wu, M., Peng, H., Zhao, Q., Pan, B., & Steinberg, C. E. W. (2018). Phosphoric acid pretreatment enhances the specific surface areas of biochars by generation of micropores. *Environmental Pollution*, *240*, 1–9. doi:10.1016/j.envpol.2018.04.003 PMID:29729563

Chu, S. H., & Kong, Y. K. (2021). Mathematical model for strength of alkali-activated materials. *Journal of Building Engineering*, *44*, 103189. doi:10.1016/j.jobe.2021.103189

Condeles, J. F., Lofrano, R. C. Z., Rosolen, J. M., & Mulato, M. (2006). Stoichiometry, surface and structural characterization of lead iodide thin films. *Brazilian Journal of Physics*, *36*(2a), 320–323. doi:10.1590/S0103-97332006000300023

Consonni, V., & Lord, A. M. (2021). Polarity in ZnO nanowires: A critical issue for piezotronic and piezoelectric devices. *Nano Energy*, *83*, 105789. doi:10.1016/j.nanoen.2021.105789

Coondoo, I., Panwar, N., & Kholkin, A. (2013). Lead-free piezoelectrics: Current status and perspectives. *Journal of Advanced Dielectrics*, *3*(02), 1330002. doi:10.1142/S2010135X13300028

Corker, D. L., Glazer, A. M., Whatmore, R. W., Stallard, A., & Fauth, F. (1998). A neutron diffraction investigation into the rhombohedral phases of the perovskite series. *Journal of Physics Condensed Matter*, *10*(28), 6251–6269. doi:10.1088/0953-8984/10/28/007

Correa-Baena, J. P., Saliba, M., Buonassisi, T., Grätzel, M., Abate, A., Tress, W., & Hagfeldt, A. (2017). Promises and Challenges of Perovskite Solar Cells. *Science*, *358*(6364), 739–744. doi:10.1126/science.aam6323 PMID:29123060

Cortenraad, R., Denier van der Gon, A. W., & Brongersma, H. H. (2000). Influence of analyser transmission and detection efficiency on the energy dependence of low-energy ion scattering signals. *Surface and Interface Analysis*, *29*(8), 524–534. doi:10.1002/1096-9918(200008)29:8<524::AID-SIA897>3.0.CO;2-J

Coursey, P. R., & Brand, K. G. (1946). *Dielectric constants of some titanates*. Academic Press.

Cox, D. E., Noheda, B., Shirane, G., Uesu, Y., Fujishiro, K., & Yamada, Y. (2001). Universal phase diagram for high-piezoelectric perovskite systems. *Applied Physics Letters*, *79*(3), 400–402. doi:10.1063/1.1384475

Cross, E. (2004). Lead-free at last. *Nature*, *432*(7013), 24–25. doi:10.1038/nature03142 PMID:15516920

Cucinotta, A., Selleri, S., Vincetti, L., & Zoboli, M. (2002). Holey fiber analysis through the finite-element method. *IEEE Photonics Technology Letters*, *14*(11), 1530–1532. doi:10.1109/LPT.2002.803375

Cui, Y., Zhong, Z., Wang, D., Wang, W. U., & Lieber, C. M. (2003). High performance silicon nanowire field effect transistors. *Nano Letters*, *3*(2), 149–152. doi:10.1021/nl0258751

Curie, J., & Curie, P. (1880). Développement par compression de l'électricité polaire dans les cristaux hémièdres à faces inclinées, Bull. *La Société Minéralogique Fr.*, *3*(4), 90–93. doi:10.3406/bulmi.1880.1564

D'Ambrogio, G., Zahhaf, O., Bordet, M., Le, M. Q., Della Schiava, N., Liang, R., Cottinet, P.-J., & Capsal, J.-F. (2021). Structuring BaTiO3 /PDMS Nanocomposite via Dielectrophoresis for Fractional Flow Reserve Measurement. *Advanced Engineering Materials*, *2100341*(10), 2100341. Advance online publication. doi:10.1002/adem.202100341

D'Ambrogio, G., Zahhaf, O., Hebrard, Y., Le, M. Q., Cottinet, P., & Capsal, J. (2020). Micro-Structuration of Piezoelectric Composites Using Dielectrophoresis: Toward Application in Condition Monitoring of Bearings. *Advanced Engineering Materials*, *2000773*. Advance online publication. doi:10.1002/adem.202000773

Damousis, I. G., & Dokopoulos, P. (2001). A fuzzy expert system for the forecasting of wind speed and power generation in wind farms. *IEEE Power Industry Computer Applications Conference*, 63–69. 10.1109/PICA.2001.932320

Das, C. K., Bass, O., Kothapalli, G., Mahmoud, T. S., & Habibi, D. (2018). Overview of energy storage systems in distribution networks: Placement, sizing, operation, and power quality. *Renewable & Sustainable Energy Reviews*, *91*, 1205–1230. doi:10.1016/j.rser.2018.03.068

Das, S., Datta, S., Barman, A., & Barua, R. (2023). Smart Biodegradable and Bio-Based Polymeric Biomaterials for Biomedical Applications. In A. Kumar, P. Kumar, A. Srivastava, & V. Goyat (Eds.), *Modeling, Characterization, and Processing of Smart Materials* (pp. 56–82). IGI Global. doi:10.4018/978-1-6684-9224-6.ch003

Datta, S., & Barua, R. (2023). Advanced Materials for Surgical Tools and Biomedical Implants. *Advanced Materials and Manufacturing Techniques for Biomedical Applications*, 25-35.

Datta, S., Barua, R., & Prasad, A. (2023). Additive Manufacturing for the Development of Artificial Organs. *Advanced Materials and Manufacturing Techniques for Biomedical Applications*, 411-427.

Datta, S., & Barua, R. (2023). Fluorescent Nanomaterials and Its Application in Biomedical Engineering. In A. Rakha, A. Munawar, V. Khanna, & S. Bansal (Eds.), *Modeling and Simulation of Functional Nanomaterials for Forensic Investigation* (pp. 164–186). IGI Global. doi:10.4018/978-1-6684-8325-1.ch009

Datta, S., Barua, R., Sarkar, R., Barui, A., Chowdhury, A. R., & Datta, P. (2018, September). Design and development of alginate: Poly-l-lysine scaffolds by 3D bio printing and studying their mechanical, structural and cell viability properties. *IOP Conference Series. Materials Science and Engineering*, *402*, 012113. doi:10.1088/1757-899X/402/1/012113

de Albuquerque Landi, F. F., Fabiani, C., & Pisello, A. L. (2020). Palm oil for seasonal thermal energy storage applications in buildings: The potential of multiple melting ranges in blends of bio-based fatty acids. *Journal of Energy Storage*, *29*, 101431. doi:10.1016/j.est.2020.101431

De Gracia, A., Rincón, L., Castell, A., Jiménez, M., Boer, D., Medrano, M., & Cabeza, L. F. (2010). Life Cycle Assessment of the inclusion of phase change materials (PCM) in experimental buildings. *Energy and Building*, *42*(9), 1517–1523. doi:10.1016/j.enbuild.2010.03.022

de Jong, E., Higson, A., Walsh, P., & Wellisch, M. (2012). Bio-based chemicals value added products from biorefineries. *IEA Bioenergy, Task42 Biorefinery, 34*, 1–33.

De Masi, R. F., Del Regno, N., Festa, V., Gigante, A., Ruggiero, S., & Vanoli, G. P. (2023). Innovative wall package made of macro-encapsulated phase change material and geopolymer concrete: In-field thermal analysis for a Mediterranean climate. *Architectural Science Review*, 1–21. doi:10.1080/00038628.2023.2243245

De Wolf, S., Holovsky, J., Moon, S.-J., Löper, P., Niesen, B., Ledinsky, M., Haug, F.-J., Yum, J.-H., & Ballif, C. (2014). Organometallic halide Perovskites: Sharp optical absorption edge and its relation to photovoltaic performance. *The Journal of Physical Chemistry Letters*, *5*(6), 1035–1039. doi:10.1021/jz500279b PMID:26270984

Degefu, D. M., Liao, Z., Berardi, U., & Doan, H. (2021). Salient parameters affecting the performance of foamed geopolymers as sustainable insulating materials. *Construction & Building Materials*, *313*, 125400. doi:10.1016/j.conbuildmat.2021.125400

Della Schiava, Thetpraphi, Le, Lermusiaux, Millon, Capsal, & Cottinet. (2018). Enhanced Figures of Merit for a High-Performing Actuator in Electrostrictive Materials. *Polym., 10*, 263. . doi:10.3390/polym10030263

Deng. (2018). Layered P2-type K0. 65Fe0. 5Mn0. 5O2 microspheres as superior cathode for high-energy potassium-ion batteries. *Advanced Functional Materials, 28*(28).

De, P., Halder, J., Priya, S., Srivastava, A. K., & Chandra, A. (2023). Two-Dimensional V2O5 Nanosheets as an Advanced Cathode Material for Realizing Low-Cost Aqueous Aluminum-Ion Batteries. *ACS Applied Energy Materials*, *6*(2), 753–762. doi:10.1021/acsaem.2c02979

Desai, A. V., & Haque, M. A. (2007). Mechanical properties of ZnO nanowires. *Sensors and Actuators. A, Physical*, *134*(1), 169–176. doi:10.1016/j.sna.2006.04.046

Dhyani, V., & Bhaskar, T. (2019). Pyrolysis of biomass. *Biomass, Biofuels, Biochemicals: Biofuels: Alternative Feedstocks and Conversion Processes for the Production of Liquid and Gaseous Biofuels*, 217–244. doi:10.1016/B978-0-12-816856-1.00009-9

Díaz-Blancas, V., Ocampo-Pérez, R., Leyva-Ramos, R., Alonso-Dávila, P. A., & Moral-Rodríguez, A. I. (2018). 3D modeling of the overall adsorption rate of metronidazole on granular activated carbon at low and high concentrations in aqueous solution. *Chemical Engineering Journal*, *349*(May), 82–91. doi:10.1016/j.cej.2018.05.076

Díaz, E. E. S., & Barrios, V. A. E. (2022). Development and use of geopolymers for energy conversion: An overview. *Construction & Building Materials*, *315*, 125774. doi:10.1016/j.conbuildmat.2021.125774

Dimitropoulos. (2022). Wrinkle-mediated CVD synthesis of wafer scale Graphene/h-BN heterostructures. *Nanotechnology, 34*(2).

Ding, Bonk, & Bauer. (2018). Corrosion behavior of metallic alloys in molten chloride salts for thermal energy storage in concentrated solar power plants: A review. *Frontiers of Chemical Science and Engineering, 12*(3), 564–576. . doi:10.1007/s11705-018-1720-0

Ding, Y., Wang, T. H., Yang, W. C., Lin, T. C., Tu, C. S., Yao, Y. D., & Wu, K. T. (2011). Magnetization, Magnetoelectric Effect, and Structure Transition in $\hbox {BiFeO} _ {3} $ and $(\hbox {Bi} _ {0.95}\hbox {La} _ {0.05})\hbox {FeO} _ {3} $ Multiferroic Ceramics. *IEEE Transactions on Magnetics*, *47*(3), 513–516. doi:10.1109/TMAG.2010.2103553

Dolcet, P., & Gross, S. (2021). Electronic and Chemical Properties: X-Ray Absorption and Photoemission. In O. Diwald & T. Berger (Eds.), *Metal Oxide Nanoparticles* (pp. 383–434). Wiley., doi:10.1002/9781119436782.ch11

Dong, H., Duan, A., Zhong, S., Zhang, Y., & Zhou, X. (2019). Chemical vapor deposition growth of broadband tunable light absorption and anti-reflection properties of VS2 nanostructure films. *Materials Letters*, *252*, 227–230. doi:10.1016/j.matlet.2019.06.002

Dong, Q., Fang, Y., Shao, Y., Mulligan, P., Qiu, J., Cao, L., & Huang, J. (2015). Electron-hole diffusion lengths > 175 μm in solution-grown $CH_3NH_3PbI_3$ single crystals. *Science*, *347*(6225), 967–970. doi:10.1126/science.aaa5760 PMID:25636799

Dong, Y. D., Zhang, H., Zhong, G. J., Yao, G., & Lai, B. (2021). Cellulose/carbon Composites and their Applications in Water Treatment – a Review. *Chemical Engineering Journal*, *405*, 126980. doi:10.1016/j.cej.2020.126980

Dong, Y., Liu, Y., Hu, Y., Ma, K., Jiang, H., & Li, C. (2020). Boosting reaction kinetics and reversibility in Mott-Schottky VS2/MoS2 heterojunctions for enhanced lithium storage. *Science Bulletin*, *65*(17), 1470–1478. doi:10.1016/j.scib.2020.05.007 PMID:36747404

Dong, Y., Wang, R., Li, G., Chen, C., Chi, Y., & Chen, G. (2012). Polyamine-functionalized carbon quantum dots as fluorescent probes for selective and sensitive detection of copper ions. *Analytical Chemistry*, *84*(14), 6220–6224. doi:10.1021/ac3012126 PMID:22686413

Donnelly, K. P., & Varlow, B. R. (2003). Non-linear dc and ac conductivity in electrically insulating composites. *IEEE Transactions on Dielectrics and Electrical Insulation*, *10*(4), 610–614. doi:10.1109/TDEI.2003.1219645

Donnelly, N. J., Shrout, T. R., & Randall, C. A. (2007). Addition of a Sr, K, Nb (SKN) combination to PZT (53/47) for high strain applications. *Journal of the American Ceramic Society*, *90*(2), 490–495. doi:10.1111/j.1551-2916.2006.01450.x

Donzel, L., Greuter, F., & Christen, T. (2011). Nonlinear resistive electric field grading part 2: Materials and applications. *IEEE Electrical Insulation Magazine*, *27*(2), 18–29. doi:10.1109/MEI.2011.5739419

Dorcet, V., Trolliard, G., & Boullay, P. (2008). Reinvestigation of phase transitions in Na0. 5Bi0. 5TiO3 by TEM. Part I: First order rhombohedral to orthorhombic phase transition. *Chemistry of Materials*, *20*(15), 5061–5073. doi:10.1021/cm8004634

Dressler, F., & Fischer, S. (2015). Connecting in-body nano communication with body area networks: Challenges and opportunities of the Internet of Nano Things. *Nano Communication Networks*, *6*(2), 29–38. doi:10.1016/j.nancom.2015.01.006

Du Pasquier, A., Plitz, I., Menocal, S., & Amatucci, G. (2003). A comparative study of Li-ion battery, supercapacitor and nonaqueous asymmetric hybrid devices for automotive applications. *Journal of Power Sources*, *115*(1), 171–178. doi:10.1016/S0378-7753(02)00718-8

Dubey, S. K., Kumar, A., Kumar, A., Pathak, A., & Srivastava, S. K. (2021). A study of highly sensitive d-shaped optical fiber surface plasmon resonance based refractive index sensor using grating structures of Ag-TiO2 and Ag-SnO2. *Optik (Stuttgart)*, *252*, 168527. doi:10.1016/j.ijleo.2021.168527

Du, C., Liang, D., Shang, M., Zhang, J., Mao, J., Liu, P., & Song, W. (2018). In situ engineering MoS2 NDs/VS2 lamellar heterostructure for enhanced electrocatalytic hydrogen evolution. *ACS Sustainable Chemistry & Engineering*, *6*(11), 15471–15479. doi:10.1021/acssuschemeng.8b03929

Du, C., Shang, A., Shang, M., Ma, X., & Song, W. (2018). Water-soluble VS2 quantum dots with unusual fluorescence for biosensing. *Sensors and Actuators. B, Chemical*, *255*, 926–934. doi:10.1016/j.snb.2017.08.070

Duck, F. (2022). Paul Langevin, U-boats, and ultrasonics. *Physics Today*, *75*(11), 42–48. doi:10.1063/PT.3.5122

Dufo-López, R., Bernal-Agustín, J. L., & Contreras, J. (2007). Optimization of control strategies for stand-alone renewable energy systems with hydrogen storage. *Renewable Energy*, *32*(7), 1102–1126. doi:10.1016/j.renene.2006.04.013

Duguet, E., Abboud, M., Morvan, F., Maheu, P., & Fontanille, M. (2000). PMMA encapsulation of alumina particles through aqueous suspension polymerisation processes. *Macromolecular Symposia*, *151*(1), 365–370. doi:10.1002/1521-3900(200002)151:1<365::AID-MASY365>3.0.CO;2-T

Du, H., Tang, F., Liu, D., Zhu, D., Zhou, W., & Qu, S. (2007). The microstructure and ferroelectric properties of (K0. 5Na0. 5) NbO3–LiNbO3 lead-free piezoelectric ceramics. *Materials Science and Engineering B*, *136*(2-3), 165–169. doi:10.1016/j.mseb.2006.09.031

Du, H., Wu, Z., Xu, Y., Liu, S., & Yang, H. (2020). Poly(3,4-ethylenedioxythiophene) based solid-state polymer supercapacitor with ionic liquid gel polymer electrolyte. *Polymers*, *12*(2), 297. Advance online publication. doi:10.3390/polym12020297 PMID:32024287

Du, H., Zhou, W., Luo, F., Zhu, D., Qu, S., Li, Y., & Pei, Z. (2008). Design and electrical properties' investigation of (K0.5Na0.5)NbO3-BiMeO3 lead-free piezoelectric ceramics. *Journal of Applied Physics*, *104*(3), 034104. doi:10.1063/1.2964100

Du, J., Yang, C., Li, Y., Yan, J., Qiu, L., Wang, L., & Zhu, K. (2023). Microstructural, dielectric, piezoelectric and ferroelectric properties of x PZN–PZT ternary ceramics. *Journal of Materials Science Materials in Electronics*, *34*(9), 780. doi:10.1007/s10854-023-10152-z

Dunn, B., Kamath, H., & Tarascon, J.-M. (2011). Electrical energy storage for the grid: A battery of choices. *Science*, *334*(6058), 928–935. doi:10.1126/science.1212741 PMID:22096188

Dwivedi, S., Pareek, T., & Kumar, S. (2018). Structure, dielectric, and piezoelectric properties of K 0.5 Na 0.5 NbO 3-based lead-free ceramics. *RSC Advances*, *8*(43), 24286–24296. doi:10.1039/C8RA04038A PMID:35539216

Dziejarski, B., Serafin, J., Andersson, K., & Krzyżyńska, R. (2023). CO2 capture materials: a review of current trends and future challenges. *Materials Today Sustainability*, 100483.

Eckel, S., Sushkov, A. O., & Lamoreaux, S. K. (2009). Magnetic susceptibility and magnetization fluctuation measurements of mixed gadolinium-yttrium iron garnets. *Physical Review B: Condensed Matter and Materials Physics*, *79*(1), 014422. doi:10.1103/PhysRevB.79.014422

Eda, K. (1989). Zinc Oxide Varistors. *IEEE Electrical Insulation Magazine*, *5*(6), 28–30. doi:10.1109/57.44606

Ediger, V. Ş. (2019). An integrated review and analysis of multi-energy transition from fossil fuels to renewables. *Energy Procedia*, *156*, 2–6. doi:10.1016/j.egypro.2018.11.073

Edsall, D. L. P. R. (1907). The nature of the general toxic reaction following exposure to the x-rays. *The American Journal of the Medical Sciences (1827-1924)*, *133*(3), 426.

Eduardo, S. R., Dávila-Guzmán, N. E., & Margarita, L. C. (2022). Agro-wastes as precursors of biochar, a cleaner adsorbent to remove pollutants from aqueous solutions. In *Biomass-Derived Materials for Environmental Applications* (pp. 349–376). Elsevier. doi:10.1016/B978-0-323-91914-2.00001-5

Eitel, R. E., Randall, C. A., Shrout, T. R., & Park, S.-E. (2014). Preparation and characterization of high temperature perovskite ferroelectrics, in the solid-solution (1-x)BiScO3–xPbTiO3. *Japanese Journal of Applied Physics*, *41*(4A), 2099–2104. doi:10.1143/JJAP.41.2099

Elahi, H., Munir, K., Eugeni, M., Abrar, M., Khan, A., Arshad, A., & Gaudenzi, P. (2020). A review on applications of piezoelectric materials in aerospace industry. *Integrated Ferroelectrics*, *211*(1), 25–44. doi:10.1080/10584587.2020.1803672

Elia, G. A., Kravchyk, K. V., Kovalenko, M. V., Chacón, J., Holland, A., & Wills, R. G. A. (2021). An overview and prospective on Al and Al-ion battery technologies. *Journal of Power Sources*, *481*, 228870. doi:10.1016/j.jpowsour.2020.228870

Ellenbogen, J. C. (2006). *Supercapacitors : A Brief Overview*. The MITRE Corporation. https://www.mitre.org/tech/nanotechSponsor

Espinal, L., Poster, D. L., Wong-Ng, W., Allen, A. J., & Green, M. L. (2013). Measurement, standards, and data needs for CO2 capture materials: A critical review. *Environmental Science & Technology*, *47*(21), 11960–11975. doi:10.1021/es402622q PMID:24060087

Esteban, M., & Leary, D. (2012). Current developments and future prospects of offshore wind and ocean energy. *Applied Energy*, *90*(1), 128–136. doi:10.1016/j.apenergy.2011.06.011

Etzioni, R., Urban, N., Ramsey, S., McIntosh, M., Schwartz, S., Reid, B., Radich, J., Anderson, G., & Hartwell, L. (2003). The case for early detection. *Nature Reviews. Cancer*, *3*(4), 243–252. doi:10.1038/nrc1041 PMID:12671663

Evliya, H. (2007). Energy Storage for Sustainable Future—A Solution to Global Warming. In *Thermal Energy Storage for Sustainable Energy Consumption: Fundamentals, Case Studies and Design* (pp. 87–99). Springer Netherlands. doi:10.1007/978-1-4020-5290-3_5

Ezz El-Din, H., & Manjaiah, D. H. (2017). Internet of Nano Things and Industrial Internet of Things. *Internet of Things: Novel Advances and Envisioned Applications*, 109–123. . doi:10.1007/978-3-319-53472-5_5

Fabini, D. H., Seshadri, R., & Kanatzidis, M. G. (2020). The underappreciated lone pair in halide perovskites underpins their unusual properties. *MRS Bulletin*, *45*(6), 467–477. doi:10.1557/mrs.2020.142

Falconi, C., Mantini, G., D'Amico, A., & Wang, Z. L. (2009). Studying piezoelectric nanowires and nanowalls for energy harvesting. *Sensors and Actuators. B, Chemical*, *139*(2), 511–519. doi:10.1016/j.snb.2009.02.071

Fan, F.-R., Tian, Z.-Q., & Wang, L. (2012). Z. L. *Nano Energy*, *1*, 328–334. doi:10.1016/j.nanoen.2012.01.004

Fang, F., & Yang, W. (2014). Phase fragility and mechatronic reliability for Pb (Mg 1/3 Nb 2/3) O 3–PbTiO 3 ferroelectric single crystals—A review. *Journal of Advanced Dielectrics*, *4*(01), 1430001. doi:10.1142/S2010135X14300011

Fang, W., Zhao, H., Xie, Y., Fang, J., Xu, J., & Chen, Z. (2015). Facile hydrothermal synthesis of VS2/graphene nanocomposites with superior high-rate capability as lithium-ion battery cathodes. *ACS Applied Materials & Interfaces*, *7*(23), 13044–13052. doi:10.1021/acsami.5b03124 PMID:26016687

Fang, Y., Ahmad, M. R., Lao, J.-C., Qian, L.-P., & Dai, J.-G. (2023). Development of artificial geopolymer aggregates with thermal energy storage capacity. *Cement and Concrete Composites*, *135*, 104834. doi:10.1016/j.cemconcomp.2022.104834

Fan, S., Xie, X., Zhang, F., Guo, X., Yang, S., & Zhang, L. (2016). Improved leakage and ferroelectric properties of Sr doped BiFe 0.95 Mn 0.05 O 3 thin films. *Journal of Materials Science Materials in Electronics*, *27*(7), 6854–6858. doi:10.1007/s10854-016-4638-7

Fan, X., Liu, B., Liu, J., Ding, J., Han, X., Deng, Y., Lv, X., Xie, Y., Chen, B., Hu, W., & Zhong, C. (2020). Battery technologies for grid-level large-scale electrical energy storage. *Transactions of Tianjin University*, *26*(2), 92–103. doi:10.1007/s12209-019-00231-w

Feng, J., Sun, X., Wu, C., Peng, L., Lin, C., Hu, S., Yang, J., & Xie, Y. (2011). Metallic few-layered VS2 ultrathin nanosheets: High two-dimensional conductivity for in-plane supercapacitors. *Journal of the American Chemical Society*, *133*(44), 17832–17838. doi:10.1021/ja207176c PMID:21951158

Feng, L., Zhang, Y., Zhou, H., Kang, Y., Zhang, S., Bao, L., Lei, J., Bian, L., & Wang, J. (2022). Quasi-monodispersed nanocapsules with form stability at high temperature and under shear force for thermal energy storage. *Chemical Engineering Journal*, *428*, 131088. doi:10.1016/j.cej.2021.131088

Fernandez-Benavides, D. A., Gutierrez-Perez, A. I., Benitez-Castro, A. M., Ayala-Ayala, M. T., Moreno-Murguia, B., & Muñoz-Saldaña, J. (2018). Comparative study of ferroelectric and piezoelectric properties of BNT-BKT-BT ceramics near the phase transition zone. *Materials (Basel)*, *11*(3), 361. doi:10.3390/ma11030361 PMID:29494486

Ferrando, A., Silvestre, E., Miret, J. J., Andres, P., & Andres, M. V. (1999). Full-vector analysis of a realistic photonic crystal fiber. *Optics Letters*, *24*(5), 276. doi:10.1364/OL.24.000276 PMID:18071478

Ferrer, G., Solé, A., Barreneche, C., Martorell, I., & Cabeza, L. F. (2015, April). Corrosion of metal containers for use in PCM energy storage. *Renewable Energy*, *76*, 465–469. doi:10.1016/j.renene.2014.11.036

Ferrer, P. R., Mace, A., Thomas, S. N., & Jeon, J. W. (2017). Nanostructured porous graphene and its composites for energy storage applications. *Nano Convergence*, *4*(1), 29. Advance online publication. doi:10.1186/s40580-017-0123-0 PMID:29152447

Ferrier, D. C., & Honeychurch, K. C. (2021). Carbon nanotube (CNT)-based biosensors. *Biosensors (Basel)*, *11*(12), 1–33. doi:10.3390/bios11120486 PMID:34940243

Festus, N. (2022). A review on selected durability parameters on performance of geopolymers containing industrial by-products, agro-wastes and natural pozzolan. *Journal of Sustainable Construction Materials and Technologies*, *7*(4), 375–400. doi:10.47481/jscmt.1190244

Fiala, L., Petříková, M., & Černý, R. (2019). Basic physical, electrical and thermal properties of geopolymers with enhanced electrical properties. *AIP Conference Proceedings*, *2133*(1), 020013. doi:10.1063/1.5120143

Fichtner, M. (2022). Recent research and progress in batteries for electric vehicles. *Batteries & Supercaps*, *5*(2).

Filip, M. R., Eperon, G. E., Snaith, H. J., & Giustino, F. (2014). Steric engineering of metal-halide perovskites with tunable optical band gaps. *Nature Communications*, *5*(1), 5757. doi:10.1038/ncomms6757 PMID:25502506

Fischer, H., Seidenstricker, S., Berger, T., & Holopainen, T. (2022). *Artificial intelligence in B2B sales: Impact on the sales process*. AHFE International. doi:10.54941/ahfe1001456

Fleming, J. G., & Lin, S. Y. (1999). Three-dimensional photonic crystal with a stop band from 1.35 to 1.95 μm. *Optics Letters*, *24*(1), 49–51. doi:10.1364/OL.24.000049 PMID:18071404

Fletcher, S., Black, V. J., & Kirkpatrick, I. (2014). A universal equivalent circuit for carbon-based supercapacitors. *Journal of Solid State Electrochemistry*, *18*(5), 1377–1387. doi:10.1007/s10008-013-2328-4

Foo, K. Y., & Hameed, B. H. (2010). Insights into the modeling of adsorption isotherm systems. *Chemical Engineering Journal*, *156*(1), 2–10. doi:10.1016/j.cej.2009.09.013

Forouzandeh, P., Kumaravel, V., & Pillai, S. C. (2020). Electrode materials for supercapacitors: A review of recent advances. *Catalysts*, *10*(9), 1–73. doi:10.3390/catal10090969

Foubert, A., Beloglazova, N. V., Rajkovic, A., Sas, B., Madder, A., Goryacheva, I. Y., & De Saeger, S. (2016). Bioconjugation of quantum dots: Review & impact on future application. *Trends in Analytical Chemistry*, *83*, 31–48. doi:10.1016/j.trac.2016.07.008

Frackowiak, E., Jurewicz, K., Delpeux, S., & Béguin, F. (2001). Nanotubular materials for supercapacitors. *Journal of Power Sources*, *97-98*, 822–825. doi:10.1016/S0378-7753(01)00736-4

Frackowiak, E., Khomenko, V., Jurewicz, K., Lota, K., & Béguin, F. (2006). Supercapacitors based on conducting polymers/nanotubes composites. *Journal of Power Sources*, *153*(2), 413–418. doi:10.1016/j.jpowsour.2005.05.030

Frank Ogletree, D., Bluhm, H., Hebenstreit, E. D., & Salmeron, M. (2009). Photoelectron spectroscopy under ambient pressure and temperature conditions. *Nuclear Instruments & Methods in Physics Research. Section A, Accelerators, Spectrometers, Detectors and Associated Equipment*, *601*(1–2), 151–160. doi:10.1016/j.nima.2008.12.155

Freer, E. M., Grachev, O., Duan, X., Martin, S., & Stumbo, D. P. (2010). High-yield self-limiting single-nanowire assembly with dielectrophoresis. *Nature Nanotechnology*, *5*(7), 525–530. doi:10.1038/nnano.2010.106 PMID:20526324

Freitas, R. A. Jr. (2006). Pharmacytes: An Ideal Vehicle for Targeted Drug Delivery. *Journal of Nanoscience and Nanotechnology*, *6*(9), 2769–2775. doi:10.1166/jnn.2006.413 PMID:17048481

Fröhlich, A. C., Ocampo-Pérez, R., Diaz-Blancas, V., Salau, N. P. G., & Dotto, G. L. (2018). Three-dimensional mass transfer modeling of ibuprofen adsorption on activated carbon prepared by sonication. *Chemical Engineering Journal*, *341*(January), 65–74. doi:10.1016/j.cej.2018.02.020

Frost, J. M., Butler, K. T., Brivio, F., Hendon, C. H., van Schilfgaarde, M., & Walsh, A. (2014). Atomistic origins of high-performance in hybrid halide perovskite solar cells. *Nano Letters*, *14*(5), 2584–2590. doi:10.1021/nl500390f PMID:24684284

Fu, M., Pan, D., Yang, Y., Shi, T., Zhang, Z., Zhao, J., Xu, H. Q., & Chen, Q. (2014). Electrical characteristics of field-effect transistors based on indium arsenide nanowire thinner than 10 nm. *Applied Physics Letters*, *105*(14), 143101. doi:10.1063/1.4897496

Furusawa, K., Nagashima, K., & Anzai, C. (1994). Synthetic process to control the total size and component distribution of multilayer magnetic composite particles. *Colloid & Polymer Science*, *272*(9), 1104–1110. doi:10.1007/BF00652379

Fu, Z., Yang, J., Lu, P., Zhang, L., Yao, H., Xu, F., & Li, Y. (2017). Influence of secondary phase on polymorphic phase transition in Li-doped KNN lead-free ceramics. *Ceramics International*, *43*(15), 12893–12897. doi:10.1016/j.ceramint.2017.06.185

Ganet, F., Le, M. Q., Capsal, J. F., Gérard, J. F., Pruvost, S., Duchet-Rumeau, J., Livi, S., Lermusiaux, P., Millon, A., & Cottinet, P. J. (2015). Haptic feedback using an all-organic electroactive polymer composite. *Sensors and Actuators. B, Chemical*, *220*, 1120–1130. doi:10.1016/j.snb.2015.06.071

Gan, G., Li, X., Fan, S., Wang, L., Qin, M., Yin, Z., & Chen, G. (2019). Carbon Aerogels for Environmental Clean-Up. *European Journal of Inorganic Chemistry*, *2019*(27), 3126–3141. doi:10.1002/ejic.201801512

Ganganboina, A. B., & Doong, R. A. (2019). Graphene quantum dots decorated gold-polyaniline nanowire for impedimetric detection of carcinoembryonic antigen. *Scientific Reports*, *9*(1), 7214. doi:10.1038/s41598-019-43740-3 PMID:31076624

Ganguly, S., Das, P., Banerjee, S., & Das, N. C. (2019). Advancement in science and technology of carbon dot-polymer hybrid composites: A review. *Functional Composites and Structures*, *1*(2), 022001. doi:10.1088/2631-6331/ab0c80

Ganne, J. P., Lebourgeois, R., Paté, M., Dubreuil, D., Pinier, L., & Pascard, H. (2007). The electromagnetic properties of Cu-substituted garnets with low sintering temperature. *Journal of the European Ceramic Society*, *27*(8-9), 2771–2777. doi:10.1016/j.jeurceramsoc.2006.11.054

Gan, Y. X., Jayatissa, A. H., Yu, Z., Chen, X., & Li, M. (2020). Hydrothermal synthesis of nanomaterials. *Journal of Nanomaterials*, *2020*, 1–3. doi:10.1155/2020/8917013

Gao, H., Liao, L., Liang, Y., Tang, X., Liu, H., Mei, L., Lv, G., & Wang, L. (2021). Improvement of durability of porous perlite geopolymer-based thermal insulation material under hot and humid environment. *Construction & Building Materials*, *313*, 125417. doi:10.1016/j.conbuildmat.2021.125417

Gao, H., Liu, H., Liao, L., Mei, L., Zhang, F., Zhang, L., Li, S., & Lv, G. (2020). A bifunctional hierarchical porous kaolinite geopolymer with good performance in thermal and sound insulation. *Construction & Building Materials*, *251*, 118888. doi:10.1016/j.conbuildmat.2020.118888

Gao, J., Xue, D., Liu, W., Zhou, C., & Ren, X. (2017, July). Recent progress on BaTiO3-based piezoelectric ceramics for actuator applications. *Actuators*, *6*(3), 24. doi:10.3390/act6030024

Gao, P. X., Song, J., Liu, J., & Wang, Z. L. (2007). Nanowire piezoelectric nanogenerators on plastic substrates as flexible power sources for nanodevices. *Advanced Materials*, *19*(1), 67–72. doi:10.1002/adma.200601162

Gao, P., Grätzel, M., & Nazeeruddin, M. K. (2014). Organohalide lead perovskites for photovoltaic applications. *Energy & Environmental Science*, *7*(8), 2448–2463. doi:10.1039/C4EE00942H

Gao, X., Cheng, Z., Chen, Z., Liu, Y., Meng, X., Zhang, X., Wang, J., Guo, Q., Li, B., Sun, H., Gu, Q., Hao, H., Shen, Q., Wu, J., Liao, X., Ringer, S. P., Liu, H., Zhang, L., Chen, W., ... Zhang, S. (2021). The mechanism for the enhanced piezoelectricity in multi-elements doped (K, Na) NbO3 ceramics. *Nature Communications*, *12*(1), 881. doi:10.1038/s41467-021-21202-7 PMID:33564001

García-Fernandez, A., Bermúdez-García, J. M., Castro-García, S., Llamas-Saiz, A. L., Artiaga, R., López-Beceiro, J., Hu, S., Ren, W., Stroppa, A., Sánchez-Andújar, M., & Señarís-Rodríguez, M. A. (2017). Phase transition, dielectric properties, and ionic transport in the $[(CH_3)_2NH_2]PbI_3$ organic-inorganic hybrid with 2H-hexagonal perovskite structure. *Inorganic Chemistry*, *56*(9), 4918–4927. doi:10.1021/acs.inorgchem.6b03095 PMID:28375611

Gayathri, Manimegalai, & Krishnakumar. (2021). Challenging Issues and Solutions on Battery Thermal Management for Electric Vehicles. *Electrical and Electronic Devices, Circuits, and Materials: Technological Challenges and Solutions,* 535-553.

Ge, C., Xue, Y. Z. B., Li, L., Tang, B., & Hu, H. (2020). Recent Progress in 2D/3D Multidimensional Metal Halide Perovskites Solar Cells. *Frontiers in Materials*, *7*, 601179. doi:10.3389/fmats.2020.601179

Gencel, O., Nodehi, M., Hekimoğlu, G., Ustaoğlu, A., Sarı, A., Kaplan, G., Bayraktar, O. Y., Sutcu, M., & Ozbakkaloglu, T. (2022). Foam concrete produced with recycled concrete powder and phase change materials. *Sustainability (Basel)*, *14*(12), 7458. doi:10.3390/su14127458

Genchi, G. (2020). The effects of cadmium toxicity. *International Journal of Environmental Research and Public Health*, *17*(11).

Georgiou, L., Konatzii, P., Morsink-Georgali, P.-Z., Klumbyte, E., Christou, P., & Fokaides, P. A. (2023). Numerical and environmental analysis of post constructive application of PCM coatings for the improvement of the energy performance of building structures. *Construction & Building Materials*, *364*, 129984. doi:10.1016/j.conbuildmat.2022.129984

Ge, P., & Jouaneh, M. (1996). Tracking control of a piezoceramic actuator. *IEEE Transactions on Control Systems Technology*, *4*(3), 209–216. doi:10.1109/87.491195

Ge, R., Xie, J., Yan, B., Liu, E., Tan, W., & Liu, J. (2018). Refractive index sensor with high sensitivity based on circular photonic crystal. *Journal of the Optical Society of America. A, Optics, Image Science, and Vision*, *35*(6), 992–997. doi:10.1364/JOSAA.35.000992 PMID:29877344

Ghder Soliman, W., Swathi, Ch., Yasasvi, T., Keerthi Priya, B., & Akhila Reddy, D. (2021). Review on poly(ethylene oxide)-based electrolyte and anode nanomaterials for the internet of things node-level lithium-ion batteries. *Materials Today: Proceedings*, *42*, 429–435. doi:10.1016/j.matpr.2020.09.806

Ghorui. (2022). Newly Designed One-Pot In-Situ Synthesis of VS2/rGO Nanocomposite to Explore Its Electrochemical Behavior towards Oxygen Electrode Reactions. *ChemElectroChem*, *9*(15).

Goel, P., Dobhal, D., & Sharma, R. C. (2020). Aluminum–air batteries: A viability review. *Journal of Energy Storage*, *28*, 101287. doi:10.1016/j.est.2020.101287

Goldman, A. (2006). *Modern ferrite technology*. Springer Science & Business Media.

Goldschmidt, V. M. (1927). Krystallbau und chemische Zusammensetzung. Ber. Dtsch. *Chem. Ges*, *60*, 1263–1296.

Gong, H., Zhang, Y., Quan, J., & Che, S. (2011). Preparation and properties of cement based piezoelectric composites modified by CNTs. *Current Applied Physics*, *11*(3), 653–656. doi:10.1016/j.cap.2010.10.021

Gong, J., Darling, S. B., & You, F. (2015). Perovskite photovoltaics: Life-cycle assessment of energy and environmental impacts. *Energy & Environmental Science*, *8*(7), 1953–1968. doi:10.1039/C5EE00615E

Gonzalez-Olmos, R., Gutierrez-Ortega, A., Sempere, J., & Nomen, R. (2022). Zeolite versus carbon adsorbents in carbon capture: A comparison from an operational and life cycle perspective. *Journal of CO2 Utilization*, *55*, 101791.

Goodman, G. (1953). Ferroelectric properties of lead metaniobate. *Journal of the American Ceramic Society*, *36*(11), 368–372. doi:10.1111/j.1151-2916.1953.tb12820.x

Goralski, M. A., & Tan, T. K. (2020). Artificial intelligence and sustainable development. *International Journal of Management Education*, *18*(1), 100330. doi:10.1016/j.ijme.2019.100330

Gould, T. (2023). *World Energy Investment 2023*. https://www.iea.org/reports/world-energy-investment-2023

Gou, Q., Xiao, D. Q., Wu, B., Xiao, M., Feng, S. S., Ma Zhao, D. D., Wu, J. G., & Zhu, J. G. (2015). New (1 _ xÞK0:5Na0:5NbO3–x (0.15Bi0:5Na0:5TiO3–0.85Bi0:5Na0:5ZrO3Þ ternary lead-free ceramics: Microstructure and electrical properties. *RSC Advances*, *5*, 30660. doi:10.1039/C4RA16780E

Govorov, A. O., Richardson, H. H., Soci, C., Zhang, A., Xiang, B., Dayeh, S. A., Aplin, D. P. R., Park, J., Bao, X. Y., Lo, Y. H., & Wang, D. (2007). ZnO nanowire UV photodetectors with high internal gain. *Nano Letters*, *7*(4), 30–38. doi:10.1021/nl070111x PMID:17358092

Green, M. A., Dunlop, E. D., Hohl-Ebinger, J., Yoshita, M., Kopidakis, N., & Ho-Baillie, A. W. Y. (2020). Solar cell efficiency tables (version 55). *Progress in Photovoltaics: Research and Applications*, *28*(1), 3–15. doi:10.1002/pip.3228

Green, M. A., Hishikawa, Y., Dunlop, E. D., Levi, D. H., Hohl-Ebinger, J., & Ho-Baillie, A. W. Y. (2018). Solar cell efficiency tables (version 52). *Progress in Photovoltaics: Research and Applications*, *26*(7), 427–436. doi:10.1002/pip.3040

Grinberg, D., Siddique, S., Le, M. Q., Liang, R., Capsal, J. F., & Cottinet, P. J. (2019). 4D Printing based piezoelectric composite for medical applications, J. Polym. Sci. Part B Polym. *Journal of Polymer Science. Part B, Polymer Physics*, *57*(2), 109–115. doi:10.1002/polb.24763

Guan, L., Pan, L., Peng, T., Qian, T., Huang, Y., Li, X., Gao, C., Li, Z., Hu, H., & Wu, M. (2019). Green and scalable synthesis of porous carbon nanosheet-assembled hierarchical architectures for robust capacitive energy harvesting. *Carbon*, *152*, 537–544. doi:10.1016/j.carbon.2019.06.059

Guerquin-Kern, J.-L., Wu, T.-D., Quintana, C., & Croisy, A. (2005). Progress in analytical imaging of the cell by dynamic secondary ion mass spectrometry (SIMS microscopy). Biochimica et Biophysica Acta (BBA) -. *Biochimica et Biophysica Acta. G, General Subjects*, *1724*(3), 228–238. doi:10.1016/j.bbagen.2005.05.013 PMID:15982822

Güngör, D., & Özen, S. (2021). Development and characterization of clinoptilolite-, mordenite-, and analcime-based geopolymers: A comparative study. *Case Studies in Construction Materials*, *15*, e00576. doi:10.1016/j.cscm.2021.e00576

Guo, Y., Luo, H., & Ling, D. (2003). The phase transition sequence and the location of the morphotropic phase boundary region in (1-x)[Pb(Mg1/3Nb2/3)O3]-xPbTiO3 single crystal. *J Phys: Conden Matter, 15*(2), L77–L82.

Guo, Q., Cao, G. Z., & Shen, I. Y. (2013). Measurements of piezoelectric coefficient d 33 of lead zirconate titanate thin films using a mini force hammer. *Journal of Vibration and Acoustics*, *135*(1), 011003. doi:10.1115/1.4006881

Guo, Y., Kakimoto, I., & Ohsato, H. (2004). Phase transitional behavior and piezoelectric properties of (Na0.5K0.5) NbO3-LiNbO3 ceramics. *Applied Physics Letters*, *85*(18), 4121–4123. doi:10.1063/1.1813636

Guo, Y., Kakimoto, I., & Ohsato, H. (2005). (Na0.5K0.5)NbO3-LiTaO3 lead-free piezoelectric ceramics. *Materials Letters*, *59*(2-3), 241–244. doi:10.1016/j.matlet.2004.07.057

Guo, Z., Yang, L., Wang, W., Cao, L., & Dong, B. (2018). Ultrathin VS 2 nanoplate with in-plane and out-of-plane defects for an electrochemical supercapacitor with ultrahigh specific capacitance. *Journal of Materials Chemistry. A, Materials for Energy and Sustainability*, *6*(30), 14681–14688. doi:10.1039/C8TA03812K

Gupta, A., & Garg, A. (2019). Adsorption and oxidation of ciprofloxacin in a fixed bed column using activated sludge derived activated carbon. *Journal of Environmental Management*, *250*(August), 109474. doi:10.1016/j.jenvman.2019.109474 PMID:31505384

Gür, T. M. (2022). Carbon dioxide emissions, capture, storage and utilization: Review of materials, processes and technologies. *Progress in Energy and Combustion Science*, *89*, 100965. doi:10.1016/j.pecs.2021.100965

Habib, M., Lee, M. H., Choi, H. I., Kim, M. H., Kim, W. J., & Song, T. K. (2020). Phase evolution and origin of the high piezoelectric properties in lead-free BiFeO3–BaTiO3 ceramics. *Ceramics International*, *46*(14), 22239–22252. doi:10.1016/j.ceramint.2020.05.301

Hadjipaschalis, I., Poullikkas, A., & Efthimiou, V. (2009). Overview of current and future energy storage technologies for electric power applications. *Renewable & Sustainable Energy Reviews*, *13*(6-7), 1513–1522. doi:10.1016/j.rser.2008.09.028

Haertling, G. I., & Buchanan, R. C. (1991). *Ceramic materials for electronics*. Buchanan.

Haily, E., Ousaleh, H. A., Zari, N., Faik, A., Bouhfid, R., & Qaiss, A. (2023). Use of a form-stable phase change material to improve thermal properties of phosphate sludge-based geopolymer mortar. *Construction & Building Materials*, *386*, 131570. doi:10.1016/j.conbuildmat.2023.131570

Hall, P. J., Mirzaeian, M., Fletcher, S. I., Sillars, F. B., Rennie, A. J. R., Shitta-Bey, G. O., Wilson, G., Cruden, A., & Carter, R. (2010). Energy storage in electrochemical capacitors: Designing functional materials to improve performance. *Energy & Environmental Science*, *3*(9), 1238–1251. doi:10.1039/c0ee00004c

Hammond, G., & Jones, C. (2011). *Embodied Carbon: The Inventory of Carbon and Energy (ICE), BSRIA Guide BG*. Services Research and Information Association. doi:10/2011. Building

Han, J., Yin, J., & Wu, J. (2020). BNT-based ferroelectric ceramics: Electrical properties modification by Ta2O5 oxide addition. *Journal of the American Ceramic Society*, *103*(1), 412–422. doi:10.1111/jace.16752

Hannan, M. A., Al-Shetwi, A. Q., Ker, P. J., Begum, R. A., Mansor, M., Rahman, S. A., Dong, Z. Y., Tiong, S. K., Mahlia, T. M. I., & Muttaqi, K. M. (2021). Impact of renewable energy utilization and artificial intelligence in achieving sustainable development goals. *Energy Reports*, *7*, 5359–5373. doi:10.1016/j.egyr.2021.08.172

Hansen, B. J., Liu, Y., Yang, R., & Wang, Z. L. (2010). Hybrid nanogenerator for concurrently harvesting biomechanical and biochemical energy. *ACS Nano*, *4*(7), 3647–3652. doi:10.1021/nn100845b PMID:20507155

Han, Y., & Ho, W. W. (2018). Recent advances in polymeric membranes for CO2 capture. *Chinese Journal of Chemical Engineering*, *26*(11), 2238–2254. doi:10.1016/j.cjche.2018.07.010

Han, Z., Li, S., Wu, Y., Yu, C., Cheng, S., & Xie, J. (2021). Challenges and key parameters in exploring the cyclability limitation of practical lithium–sulfur batteries. *Journal of Materials Chemistry. A, Materials for Energy and Sustainability*, *9*(43), 24215–24240. doi:10.1039/D1TA06499A

Hao, F., Zhao, X., Ouyang, W., Lin, C., Chen, S., Shan, Y., & Lai, X. (2013). Molecular structure of corncob-derived Biochars and the mechanism of Atrazine sorption. *Agronomy Journal*, *105*(3), 773–782. doi:10.2134/agronj2012.0311

Hao, X., Zhai, J., Kong, L. B., & Xu, Z. (2014). A comprehensive review on the progress of lead zirconate-based antiferroelectric materials. *Progress in Materials Science*, *63*, 1–57. doi:10.1016/j.pmatsci.2014.01.002

Hardouin Duparc, O. (2009). Pierre Auger – Lise Meitner: Comparative contributions to the Auger effect. *International Journal of Materials Research*, *100*(9), 1162–1166. doi:10.3139/146.110163

Hart, C. J., Cuisinier, M., Liang, X., Kundu, D., Garsuch, A., & Nazar, L. F. (2015). Rational design of sulphur host materials for Li–S batteries: Correlating lithium polysulphide adsorptivity and self-discharge capacity loss. *Chemical Communications*, *51*(12), 2308–2311. doi:10.1039/C4CC08980D PMID:25562067

Hassan, Q., Sameen, A. Z., Salman, H. M., Al-Jiboory, A. K., & Jaszczur, M. (2023). *The role of renewable energy and articial intelligence towards environmental sustainability and net zero.* Academic Press.

Hassan, F., Jamil, F., Hussain, A., Ali, H. M., Janjua, M. M., Khushnood, S., Farhan, M., Altaf, K., Said, Z., & Li, C. (2022, February). Recent advancements in latent heat phase change materials and their applications for thermal energy storage and buildings: A state of the art review. *Sustainable Energy Technologies and Assessments*, *49*, 101646. Advance online publication. doi:10.1016/j.seta.2021.101646

Haumont, R., Kreisel, J., Bouvier, P., & Hippert, F. (2006). Phonon anomalies and the ferroelectric phase transition in multiferroic BiFeO 3. *Physical Review B: Condensed Matter and Materials Physics*, *73*(13), 132101. doi:10.1103/PhysRevB.73.132101

Hayashi, S., Fujiki, K., & Tsubokawa, N. (2000). Grafting of hyperbranched polymers onto ultrafine silica: Postgraft polymerization of vinyl monomers initiated by pendant initiating groups of polymer chains grafted onto the surface. *Reactive & Functional Polymers*, *46*(2), 193–201. doi:10.1016/S1381-5148(00)00053-5

He. (2017). Layered VS2 nanosheet-based aqueous Zn ion battery cathode. *Advanced Energy Materials, 7*(11).

He, J. (2019). *Introduction of Varistor Ceramics, Met.* Oxide Varistors. doi:10.1002/9783527684038.ch1

He, J. H., Hsin, C. L., Liu, J., Chen, L. J., & Wang, Z. L. (2007). Piezoelectric Gated Diode of a Single ZnO Nanowire. *Advanced Materials*, *19*(6), 781–784. doi:10.1002/adma.200601908

Helseth, L. E. (2019). Modelling supercapacitors using a dynamic equivalent circuit with a distribution of relaxation times. *Journal of Energy Storage*, *25*(August), 100912. doi:10.1016/j.est.2019.100912

He, M.-R., Shi, Y., Zhou, W., Chen, J. W., Yan, Y. J., & Zhu, J. (2009). Diameter dependence of modulus in zinc oxide nanowires and the effect of loading mode: In situ experiments and universal core-shell approach. *Applied Physics Letters*, *95*(9), 091912. doi:10.1063/1.3205102

Heo, J. H., Im, S. H., Noh, J. H., Mandal, T. N., Lim, C.-S., Chang, J. A., Lee, Y. H., Kim, H., Sarkar, A., Nazeeruddin, M. K., Grätzel, M., & Seok, S. I. (2013). Efficient inorganicorganic hybrid heterojunction solar cells containing perovskite compound and polymeric hole conductors. *Nature Photonics*, *7*(6), 486–491. doi:10.1038/nphoton.2013.80

Hernández-Romero, I. M., Flores-Tlacuahuac, A., Nápoles-Rivera, F., Esquivel-Patiño, G. G., & García-Pardo, M. L. (2022). Multi objective optimization of the amines-CO2 capture absorption-desorption process by a non-equilibrium rate model. *Chemical Engineering Research & Design*, *187*, 93–104. doi:10.1016/j.cherd.2022.08.050

He, Z., Guo, W., & Zhang, P. (2022). Performance prediction, optimal design and operational control of thermal energy storage using artificial intelligence methods. *Renewable & Sustainable Energy Reviews*, *156*, 111977. doi:10.1016/j.rser.2021.111977

Hijazi, H., Desai, P., & Mariyappan, S. (2021). Non-Aqueous Electrolytes for Sodium-Ion Batteries: Challenges and Prospects Towards Commercialization. *Batteries & Supercaps*, *4*(6), 881–896. doi:10.1002/batt.202000277

Hirsh. (2020). Sodium-ion batteries paving the way for grid energy storage. *Advanced Energy Materials, 10*(32).

Ho, M. M. (1998). Sigalas, W. Zubrzycki, S. R. Kurtz, and J. Bur, "A three-dimensional photonic crystal operating at infrared wavelengths,". *Nature*, *394*(6690), 251–253. doi:10.1038/28343

Hong, G., Diao, S., Antaris, A. L., & Dai, H. (2015). Carbon nanomaterials for biological imaging and nanomedicinal therapy. *Chemical Reviews*, *115*(19), 10816–10906. doi:10.1021/acs.chemrev.5b00008 PMID:25997028

Hong, J., & Fang, D. (2008). Systematic study of the ferroelectric properties of Pb(Zr0.5Ti0.5)O3 nanowires. *Journal of Applied Physics*, *104*(6), 064118. doi:10.1063/1.2982090

Hong, X. L., Loy, D. J. J., Dananjaya, P. A., Tan, F., Ng, C. M., & Lew, W. S. (2018). Oxide-based RRAM materials for neuromorphic computing. *Journal of Materials Science*, *53*(12), 8720–8746. doi:10.1007/s10853-018-2134-6

Hossain, Qin, & Sen. (2021). Chemical Synthesis and Substrate Temperature Effect on Morphology of 2D Vanadium Disulfide. *Crystal Research and Technology, 56*(3).

Hossain. (2018). Chemical vapor deposition of 2D vanadium disulfide and diselenide and Raman characterization of the phase transitions. *Advanced Materials Interfaces, 5*(16).

Ho, Y. S., & McKay, G. (1998). Sorption of dye from aqueous solution by peat. *Chemical Engineering Journal*, *70*(2), 115–124. doi:10.1016/S0923-0467(98)00076-1

https://www.mordorintelligence.com/industry-reports/internet-of-nano-things-market

https://www.shiksha.com/online-courses/articles/applications-of-internet-of-things-iot-in-defence-and-military/

Hu. (2019). A portable and efficient solar-rechargeable battery with ultrafast photo-charge/discharge rate. *Advanced Energy Materials, 9*(28).

Hu. (2023). Recent Progresses on Vanadium Sulfide Cathodes for Aqueous Zinc-Ion Batteries. *Energies, 16*(2).

Huang, Mao, Feick, Yan, Wu, Kind, Weber, Russo, & Yang. (2001). Room-temperature ultraviolet nanowire nanolasers. *Science, 292*, 1897–1899. . doi:10.1126/science.1060367

Huang. (2020). Electrode design for lithium–sulfur batteries: problems and solutions. *Advanced Functional Materials, 30*(22).

Huang, B., Ren, R., Zhang, Z., & Zheng, S. (2013). The improvement of dispersibility of YIG precursor prepared via chemical coprecipitation. *Journal of Alloys and Compounds*, *558*, 56–61. doi:10.1016/j.jallcom.2013.01.037

Huang, J., Yuan, Y., Shao, Y., & Yan, Y. (2017). Understanding the physical properties of hybrid perovskites for photovoltaic applications. *Nature Reviews. Materials*, *2*(7), 17042. doi:10.1038/natrevmats.2017.42

Huang, K.-J., Liu, Y.-J., Shi, G.-W., Yang, X.-R., & Liu, Y.-M. (2014). Label-free aptamer sensor for 17β-estradiol based on vanadium disulfide nanoflowers and Au nanoparticles. *Sensors and Actuators. B, Chemical*, *201*, 579–585. doi:10.1016/j.snb.2014.05.055

Huang, L., Zhu, W., Zhang, W., Chen, K., Wang, J., Wang, R., Yang, Q., Hu, N., Suo, Y., & Wang, J. (2018). Layered vanadium (IV) disulfide nanosheets as a peroxidase-like nanozyme for colorimetric detection of glucose. *Mikrochimica Acta*, *185*(1), 1–8. doi:10.1007/s00604-017-2552-1 PMID:29594417

Huang, X., Chen, X., Li, A., Atinafu, D., Gao, H., Dong, W., & Wang, G. (2019). Shape-stabilized phase change materials based on porous supports for thermal energy storage applications. *Chemical Engineering Journal*, *356*, 641–661. doi:10.1016/j.cej.2018.09.013

Huang, X., Zhang, F., Zhu, L., Choi, K. Y., Guo, N., Guo, J., Tackett, K., Anilkumar, P., Liu, G., Quan, Q., Choi, H. S., Niu, G., Sun, Y.-P., Lee, S., & Chen, X. (2013). Effect of injection routes on the biodistribution, clearance, and tumor uptake of carbon dots. *ACS Nano*, *7*(7), 5684–5693. doi:10.1021/nn401911k PMID:23731122

Huang, Y., Duan, X., Wei, Q., & Lieber, C. M. (2001). Directed assembly of one-dimensional nanostructures into functional networks. *Science*, *291*(5504), 630–633. doi:10.1126/science.291.5504.630 PMID:11158671

Huang, Z., Han, X., Cui, X., He, C., Zhang, J., Wang, X., Lin, Z., & Yang, Y. (2020). Vertically aligned VS 2 on graphene as a 3D heteroarchitectured anode material with capacitance-dominated lithium storage. *Journal of Materials Chemistry. A, Materials for Energy and Sustainability*, *8*(12), 5882–5889. doi:10.1039/C9TA13835H

Hussain. (2023). Design of XS2 (X= W or Mo)-Decorated VS2 Hybrid Nano-Architectures with Abundant Active Edge Sites for High-Rate Asymmetric Supercapacitors and Hydrogen Evolution Reactions. *Small, 19*(8).

Hu, W., & Yu, X. (2012). Encapsulation of bio-based PCM with coaxial electrospun ultrafine fibers. *RSC Advances*, *2*(13), 5580–5584. doi:10.1039/c2ra20532g

Hu, W., & Yu, X. (2014). Thermal and mechanical properties of bio-based PCMs encapsulated with nanofibrous structure. *Renewable Energy*, *62*, 454–458. doi:10.1016/j.renene.2013.07.047

Hu, Y., & Fisher, T. S. (2018). Suggested standards for reporting power and energy density in supercapacitor research. *Bulletin of Materials Science*, *41*(5), 10–13. doi:10.1007/s12034-018-1641-z

Hu, Y., Xu, C., Zhang, Y., Lin, L., Snyder, R. L., & Wang, Z. L. (2011). A nanogenerator for energy harvesting from a rotating tire and its application as a self-powered pressure/speed sensor. *Advanced Materials*, *23*(35), 4068–4071. doi:10.1002/adma.201102067 PMID:21812038

Hu, Y., Zhang, Y., Xu, C., Lin, L., Snyder, R. L., & Wang, Z. L. (2011). Self-powered system with wireless data transmission. *Nano Letters*, *11*(6), 2572–2577. doi:10.1021/nl201505c PMID:21604749

Hwang, S. I., Franconi, N. G., Rothfuss, M. A., Bocan, K. N., Bian, L., White, D. L., Burkert, S. C., Euler, R. W., Sopher, B. J., Vinay, M. L., Sejdic, E., & Star, A. (2019). Tetrahydrocannabinol Detection Using Semiconductor-Enriched Single-Walled Carbon Nanotube Chemiresistors. *ACS Sensors*, *4*(8), 2084–2093. doi:10.1021/acssensors.9b00762 PMID:31321969

Ibarlucea, B., Fawzul Akbar, T., Kim, K., Rim, T., Baek, C. K., Ascoli, A., Tetzlaff, R., Baraban, L., & Cuniberti, G. (2018). Ultrasensitive detection of Ebola matrix protein in a memristor mode. *Nano Research*, *11*(2), 1057–1068. doi:10.1007/s12274-017-1720-2

IEA. (2022). *Renewables 2022 - December*. https://www.iea.org/reports/renewables-2022

IEA. (2023). *Renewable Energy Market Update - June*. https://www.iea.org/reports/renewable-energy-market-update-june-2023

Im, J.-H., Lee, C.-R., Lee, J.-W., Park, S.-W., & Park, N.-G. (2011). 6.5% efficient Perovskite quantum-dot-sensitized solar cell. *Nanoscale*, *3*(10), 4088–4093. doi:10.1039/c1nr10867k PMID:21897986

Imteyaz, S., Suresh, C. M., Kausar, T., & Ingole, P. P. (2023). Carbon dioxide capture and its electrochemical reduction study in deep eutectic solvent (DES) via experimental and molecular simulation approaches. *Journal of CO2 Utilization, 68*, 102349.

Imtiaz, L., Rehman, S. K. U., Ali Memon, S., Khizar Khan, M., & Faisal Javed, M. (2020). A review of recent developments and advances in eco-friendly geopolymer concrete. *Applied Sciences (Basel, Switzerland)*, *10*(21), 7838. doi:10.3390/app10217838

Inci, F., Tokel, O., Wang, S., Gurkan, U. A., Tasoglu, S., Kuritzkes, D. R., & Demirci, U. (2013). Nanoplasmonic quantitative detection of intact viruses from unprocessed whole blood. *ACS Nano*, *7*(6), 4733–4745. doi:10.1021/nn3036232 PMID:23688050

Ippili, S., Jella, V., Kim, J., Hong, S., & Yoon, S. G. (2018). Enhanced piezoelectric output performance via control of dielectrics in Fe_{2+}-incorporated $MAPbI_3$ perovskite thin films: Flexible piezoelectric generators. *Nano Energy*, *49*, 247–256. doi:10.1016/j.nanoen.2018.04.031

Ippili, S., Jella, V., Thomas, A. M., Yoon, C., Jung, J. S., & Yoon, S. G. (2021). ZnAl–LDH-induced electroactive β-phase and controlled dielectrics of PVDF for a high-performance triboelectric nanogenerator for humidity and pressure sensing applications. *Journal of Materials Chemistry. A, Materials for Energy and Sustainability*, *9*(29), 15993–16005. doi:10.1039/D1TA02966E

Iro, Z. S., Subramani, C., & Dash, S. S. (2016). A brief review on electrode materials for supercapacitor. *International Journal of Electrochemical Science*, *11*(12), 10628–10643. doi:10.20964/2016.12.50

Ishibe, Mori, & Kozako. (2014). *A New Concept Varistor With Epoxy / Microvaristor Composite*. Academic Press.

Ismagilov, Z. R., Shalagina, A. E., Podyacheva, O. Y., Ischenko, A. V., Kibis, L. S., Boronin, A. I., Chesalov, Y. A., Kochubey, D. I., Romanenko, A. I., Anikeeva, O. B., Buryakov, T. I., & Tkachev, E. N. (2009). Structure and electrical conductivity of nitrogen-doped carbon nanofibers. *Carbon*, *47*(8), 1922–1929. doi:10.1016/j.carbon.2009.02.034

Jaffe, B., Cook, W. R. Jr, & Jaffe, H. (1971). *Piezoelectric Ceramics*. Academic.

Jaffe, B., Roth, R. S., & Marzullo, S. (1954). Piezoelectric properties of lead zirconate-lead titanate solid-solution ceramics. *Journal of Applied Physics*, *25*(6), 809–810. doi:10.1063/1.1721741

Jaffe, H. (1958). Piezoelectric ceramics. *Journal of the American Ceramic Society*, *41*(11), 494–498. doi:10.1111/j.1151-2916.1958.tb12903.x

Jain, A., & KJ, PSharma, A. KJain, APN, R. (2015). Dielectric and piezoelectric properties of PVDF/PZT composites: A review. *Polymer Engineering and Science*, *55*(7), 1589–1616. doi:10.1002/pen.24088

Jakmuangpak, S., Prada, T., Mongkolthanaruk, W., Harnchana, V., & Pinitsoontorn, S. (2020). Engineering Bacterial Cellulose Films by Nanocomposite Approach and Surface Modification for Biocompatible Triboelectric Nanogenerator. *ACS Applied Electronic Materials*, *2*(8), 2498–2506. doi:10.1021/acsaelm.0c00421

Jamei, M., Ali, M., Karbasi, M., Xiang, Y., Ahmadianfar, I., & Yaseen, Z. M. (2022). Designing a Multi-Stage Expert System for daily ocean wave energy forecasting: A multivariate data decomposition-based approach. *Applied Energy*, *326*, 119925. doi:10.1016/j.apenergy.2022.119925

Janoschka, T., Martin, N., Martin, U., Friebe, C., Morgenstern, S., Hiller, H., Hager, M. D., & Schubert, U. S. (2015). An aqueous, polymer-based redox-flow battery using non-corrosive, safe, and low-cost materials. *Nature*, *527*(7576), 78–81. doi:10.1038/nature15746 PMID:26503039

Janssens, K. (2004). *X-ray based methods of analysis*. doi:10.1016/S0166-526X(04)80008-4

Jarmakiewicz, J., Parobczak, K., & Maslanka, K. (2016). On the Internet of Nano Things in healthcare network. *2016 International Conference on Military Communications and Information Systems (ICMCIS)*. 10.1109/ICMCIS.2016.7496572

Jayalakshmi, M., & Balasubramanian, K. (2008). Simple capacitors to supercapacitors - An overview. *International Journal of Electrochemical Science*, *3*(11), 1196–1217. doi:10.1016/S1452-3981(23)15517-9

Jayasimha. (n.d.). *Application of Scheffler Reflectors for Process Industry*. Academic Press.

Jella, V., Ippili, S., Eom, J.-H., Pammi, S. V. N., Jung, J.-S., Tran, V.-D., Nguyen, V. H., Kirakosyan, A., Yun, S., Kim, D., Sihn, M. R., Choi, J., Kim, Y.-J., Kim, H.-J., & Yoon, S.-G. (2019). A comprehensive review of flexible piezoelectric generators based on organic-inorganic metal halide perovskites. *Nano Energy*, *57*, 74–93. doi:10.1016/j.nanoen.2018.12.038

Jella, V., Ippili, S., & Yoon, S. G. (2020). Halide (Cl/Br)-incorporated organic–inorganic metal trihalide perovskite films: Study and investigation of dielectric properties and mechanical energy harvesting performance. *ACS Applied Electronic Materials*, *2*(8), 2579–2590. doi:10.1021/acsaelm.0c00473

Jena, A. K., Kulkarni, A., & Miyasaka, T. (2019). Halide perovskite photovoltaics: Background, status, and future prospects. *Chemical Reviews*, *119*(5), 3036–3103. doi:10.1021/acs.chemrev.8b00539 PMID:30821144

Jensen, M. O., & Brett, M. J. (2005). Square spiral 3D photonic bandgap crystals attelecommunications frequencies. *Optics Express*, *13*(9), 3348–3354. doi:10.1364/OPEX.13.003348 PMID:19495237

Jeong, D. S., Thomas, R., Katiyar, R. S., Scott, J. F., Kohlstedt, H., Petraru, A., & Hwang, C. S. (2012). Emerging memories: Resistive switching mechanisms and current status. *Reports on Progress in Physics*, *75*(7), 076502. doi:10.1088/0034-4885/75/7/076502 PMID:22790779

Jeong, S.-G., Chung, O., Yu, S., Kim, S., & Kim, S. (2013). Improvement of the thermal properties of Bio-based PCM using exfoliated graphite nanoplatelets. *Solar Energy Materials and Solar Cells*, *117*, 87–92. doi:10.1016/j.solmat.2013.05.038

Jeong, S.-G., Lee, J.-H., Seo, J., & Kim, S. (2014). Thermal performance evaluation of Bio-based shape stabilized PCM with boron nitride for energy saving. *International Journal of Heat and Mass Transfer*, *71*, 245–250. doi:10.1016/j.ijheatmasstransfer.2013.12.017

Jeon, I., Xiang, R., Shawky, A., Matsuo, Y., & Maruyama, S. (2019). Single-Walled Carbon Nanotubes in Emerging Solar Cells: Synthesis and Electrode Applications. *Advanced Energy Materials*, *9*(23), 1–27. doi:10.1002/aenm.201801312

Jeon, N. J., Noh, J. H., Kim, Y. C., Yang, W. S., Ryu, S. C., & Seok, S. I. (2014). Solvent engineering for high-performance inorganic–organic hybrid perovskite solar cells. *Nature Materials*, *13*(9), 897–903. doi:10.1038/nmat4014 PMID:24997740

Jerliu, B., Hüger, E., Dörrer, L., Seidlhofer, B.-K., Steitz, R., Oberst, V., Geckle, U., Bruns, M., & Schmidt, H. (2014). Volume expansion during lithiation of amorphous silicon thin film electrodes studied by in-operando neutron reflectometry. *The Journal of Physical Chemistry. C, Nanomaterials and Interfaces*, *118*(18), 9395–9399. doi:10.1021/jp502261t

Jeyaseelan, C. (2020). *Ni-Cd Batteries. Rechargeable Batteries: History*, Progress, and Applications.

Jha, R., Singh, A., Sharma, P. K., & Fuloria, N. K. (2020). Smart carbon nanotubes for drug delivery system: A comprehensive study. *Journal of Drug Delivery Science and Technology*, *58*(May), 101811. doi:10.1016/j.jddst.2020.101811

Jha, S., Bilalovic, J., Jha, A., Patel, N., & Zhang, H. (2017). Renewable energy: Present research and future scope of Artificial Intelligence. *Renewable & Sustainable Energy Reviews*, *77*, 297–317. doi:10.1016/j.rser.2017.04.018

Jia, D., He, P., Wang, M., Yan, S., Jia, D., He, P., Wang, M., & Yan, S. (2020). Geopolymers and Their Matrix Composites: A State-of-the-Art Review. *Geopolymer and Geopolymer Matrix Composites*, 7–34.

Jiang, L., Li, Y., Xie, L., Wu, J., Chen, Q., Xiao, D., & Zhu, J. (2017). Investigation of phase structure, microstructure, and electrical properties of LaAlO 3-modified alkali niobate lead-free perovskite. *Journal of Materials Science Materials in Electronics*, *28*(23), 17761–17771. doi:10.1007/s10854-017-7715-7

Jiang, L., Li, Y., Xing, J., Wu, J., Chen, Q., Liu, H., Xiao, D., & Zhu, J. (2017). Phase structure and enhanced piezoelectric properties in (1 _ xÞ(K0:48Na0:52Þ(Nb0:95Sb0:05ÞO3–x(Bi0:5Na0:42Li0:08Þ0:9- Sr0:1ZrO3 lead-free piezoelectric ceramics. *Ceramics International*, *43*, 2100. doi:10.1016/j.ceramint.2016.10.189

Jiang, L., Tan, Z., Xing, J., Wu, J., Chen, Q., Zhang, W., Xiao, D., & Zhu, J. (2016). New potassium–sodium niobate ternary system with large piezoelectric coefficient and high Curie temperature. *Journal of Materials Science Materials in Electronics*, *27*(9), 9812–9820. doi:10.1007/s10854-016-5047-7

Jiang, S., & Song, Z. (2022). A review on the state of health estimation methods of lead-acid batteries. *Journal of Power Sources*, *517*, 230710. doi:10.1016/j.jpowsour.2021.230710

Jiang, Z., Rivero, M. E. N., Liu, X., She, X., Xuan, Y., & Ding, Y. (2021). A novel composite phase change material for medium temperature thermal energy storage manufactured with a scalable continuous hot-melt extrusion method. *Applied Energy*, *303*, 117591. doi:10.1016/j.apenergy.2021.117591

Jiansirisomboon, S., Songsiri, K., Watcharapasorn, A., & Tunkasiri, T. (2006). Mechanical properties and crack growth behavior in poled ferroelectric PMN-PZT ceramics. *Current Applied Physics*, *6*(3), 299–302. doi:10.1016/j.cap.2005.11.004

Jiao, T., Yang, Q., Wu, S., Wang, Z., Chen, D., Shen, D., Liu, B., Cheng, J., Li, H., Ma, L., Zhi, C., & Zhang, W. (2019). Binder-free hierarchical VS 2 electrodes for high-performance aqueous Zn ion batteries towards commercial level mass loading. *Journal of Materials Chemistry. A, Materials for Energy and Sustainability*, *7*(27), 16330–16338. doi:10.1039/C9TA04798K

Jing, X., & Guo, Z. (2019). Fabrication of biocompatible super stable lubricant-immobilized slippery surfaces by grafting a polydimethylsiloxane brush: Excellent boiling water resistance, hot liquid repellency and long-term slippery stability. *Nanoscale*, *11*(18), 8870–8881. doi:10.1039/C9NR01556F PMID:31012900

Jin, Q., Long, X., & Liang, R. (2022). Numerical analysis on the thermal performance of PCM-integrated thermochromic glazing systems. *Energy and Building*, *257*, 111734. doi:10.1016/j.enbuild.2021.111734

Jin, X., Haider, M. Z., Ahn, J., Fang, G., & Hu, J. W. (2023). Development of nanomodified eco-friendly thermal energy storing cementitious composite using PCM microencapsulated in biosourced encapsulation shell. *Case Studies in Construction Materials*, *19*, e02447. doi:10.1016/j.cscm.2023.e02447

Ji, Q., Li, C., Wang, J., Niu, J., Gong, Y., Zhang, Z., Fang, Q., Zhang, Y., Shi, J., Liao, L., Wu, X., Gu, L., Liu, Z., & Zhang, Y. (2017). Metallic vanadium disulfide nanosheets as a platform material for multifunctional electrode applications. *Nano Letters*, *17*(8), 4908–4916. doi:10.1021/acs.nanolett.7b01914 PMID:28749686

Ji, R., Li, X., Lv, C., & Dong, Y. (2022). Energy performance optimization of phase change materials considering the building micro-environment. *International Journal of Energy Research*, *46*(13), 18067–18078. doi:10.1002/er.8299

Ji, X., Zhou, K., Zhao, Y., Sun, M., Dong, S., Zhang, H., Cao, H., Zheng, H., Wu, Q., & Zhang, Y. (2022). Crystal Structure, Magnetic, Dielectric and Ferromagnetic Resonance Properties of Pr-Zn-Zr Co-Doped Yttrium Iron Garnet. *Journal of Electronic Materials*, *51*(3), 1–9. doi:10.1007/s11664-021-09382-w

John, S. (1987). Strong localization of photons in certain disordered dielectric superlattices. *Physical Review Letters*, *58*(23), 2486–2489. doi:10.1103/PhysRevLett.58.2486 PMID:10034761

Jones, G. O., & Thomas, P. A. (2002). Investigation of the structure and phase transitions in the novel A-site substituted distorted perovskite compound Na0. 5Bi0. 5TiO3. *Acta Crystallographica. Section B, Structural Science*, *58*(2), 168–178. doi:10.1107/S0108768101020845 PMID:11910154

Jornet, J. M., & Akyildiz, I. F. (2012). The Internet of Multimedia Nano-Things. *Nano Communication Networks*, *3*(4), 242–251. doi:10.1016/j.nancom.2012.10.001

Joseph, A., & Aneesh, P. M. (2022). Efficient degradation of methylene blue: A comparative study using hydrothermally synthesised SnS2, WS2 and VS2 nanostructures. *Materials Research Bulletin*, *146*, 111623. doi:10.1016/j.materresbull.2021.111623

Joseph, J., Vimala, T. M., Sivasubramanian, V., & Murthy, V. R. K. (2000). Structural investigations on Pb (ZrxT1− x) O3 solid solutions using the X-ray Rietveld method. *Journal of Materials Science*, *35*(6), 1571–1575. doi:10.1023/A:1004778223721

Jost, K., Stenger, D., Perez, C. R., McDonough, J. K., Lian, K., Gogotsi, Y., & Dion, G. (2013). Knitted and screen printed carbon-fiber supercapacitors for applications in wearable electronics. *Energy & Environmental Science*, *6*(9), 2698–2705. doi:10.1039/c3ee40515j

Jo, W., Daniels, J. E., Jones, J. L., Tan, X., Thomas, P. A., Damjanovic, D., & Rödel, J. (2011). Evolving morphotropic phase boundary in lead-free (Bi1/2Na1/2) TiO3–BaTiO3 piezoceramics. *Journal of Applied Physics*, *109*(1), 014110. doi:10.1063/1.3530737

Jo, W., Dittmer, R., Acosta, M., Zang, J., Groh, C., Sapper, E., Wang, K., & Rödel, J. (2012). Giant electric-field-induced strains in lead-free ceramics for actuator applications–status and perspective. *Journal of Electroceramics*, *29*(1), 71–93. doi:10.1007/s10832-012-9742-3

Juarez-Perez, E. J., Ono, L. K., & Qi, Y. (2019). Thermal degradation of formamidinium based lead halide perovskites into *sym*-triazine and hydrogen cyanide observed by coupled thermogravimetry-mass spectrometry analysis. *Journal of Materials Chemistry. A, Materials for Energy and Sustainability*, *7*(28), 16912–16919. doi:10.1039/C9TA06058H

Jurewicz, K., Delpeux, S., Bertagna, V., Béguin, F., & Frackowiak, E. (2001). Supercapacitors from nanotubes/polypyrrole composites. *Chemical Physics Letters*, *347*(1-3), 36–40. doi:10.1016/S0009-2614(01)01037-5

Kagan, C., Mitzi, D., & Dimitrakopoulos, C. (1999). Organic-inorganic hybrid materials as semiconducting channels in thin-film field-effect transistors. *Science*, *286*(5441), 945–947. doi:10.1126/science.286.5441.945 PMID:10542146

Kalam, M. A., Hussam, A., Khaliquzzaman, M., Khan, A. H., Islam, M. M., Zaman, M. B., & Husain, M. (1978). Elemental analysis in solutions by radioisotope excited X-ray fluorescence. *Journal of Radioanalytical Chemistry*, *46*(2), 285–297. doi:10.1007/BF02519895

Kalem, V., Cam, I., & Timuçin, M. (2011). Dielectric and piezoelectric properties of PZT ceramics doped with strontium and lanthanum. *Ceramics International*, *37*(4), 1265–1275. doi:10.1016/j.ceramint.2010.12.003

Kanagaraj, B., Anand, N., Lukose, J., Andrushia, D., & Lubloy, E. (2023). Influence of elevated temperature exposure on the interfacial shear strength capacity of binary blended high strength self-compacting geopolymer concrete. *Case Studies in Construction Materials*, *18*, e01974. doi:10.1016/j.cscm.2023.e01974

Kang, L., Cui, M., Zhang, Z., & Jiang, F. (2020). Rechargeable Aqueous Zinc-Ion Batteries with Mild Electrolytes: A Comprehensive Review. *Batteries & Supercaps*, *3*(10), 966–1005. doi:10.1002/batt.202000060

Kang, Y., Jeong, S.-G., Wi, S., & Kim, S. (2015). Energy efficient Bio-based PCM with silica fume composites to apply in concrete for energy saving in buildings. *Solar Energy Materials and Solar Cells*, *143*, 430–434. doi:10.1016/j.solmat.2015.07.026

Kan, M., Wang, B., Lee, Y. H., & Sun, Q. (2015). A density functional theory study of the tunable structure, magnetism and metal-insulator phase transition in VS 2 monolayers induced by in-plane biaxial strain. *Nano Research*, *8*(4), 1348–1356. doi:10.1007/s12274-014-0626-5

Karaipekli, A., & Sarı, A. (2016). Development and thermal performance of pumice/organic PCM/gypsum composite plasters for thermal energy storage in buildings. *Solar Energy Materials and Solar Cells*, *149*, 19–28. doi:10.1016/j.solmat.2015.12.034

Karaki, T., Yan, K., & Adachi, M. (2007). Barium titanate piezoelectric ceramics manufactured by two-step sintering. *Japanese Journal of Applied Physics*, *46*(10S), 7035. doi:10.1143/JJAP.46.7035

Karthik, R., Vinoth Kumar, J., Chen, S.-M., Sundaresan, P., Mutharani, B., Chi Chen, Y., & Muthuraj, V. (2018). Simple sonochemical synthesis of novel grass-like vanadium disulfide: A viable non-enzymatic electrochemical sensor for the detection of hydrogen peroxide. *Ultrasonics Sonochemistry*, *48*, 473–481. doi:10.1016/j.ultsonch.2018.07.008 PMID:30080574

Katiyar, R., Patel, A. K., Nguyen, T.-B., Singhania, R. R., Chen, C.-W., & Dong, C.-D. (2021). Adsorption of Copper (II) in aqueous solution using biochars derived from Ascophyllum nodosum seaweed. *Bioresource Technology*, *328*(January), 124829. doi:10.1016/j.biortech.2021.124829 PMID:33618185

Kaushik, A., Kumar, R., Huey, E., Bhansali, S., Nair, N., & Nanir, M. (2014). Silica Nanowires: Growth, Integration, and Sensing Applications. *Mikrochimica Acta*, *181*(15-16), 1759–1780. doi:10.1007/s00604-014-1255-0 PMID:25382871

Kawai, H. (1969). The Piezoelectricity of Poly (vinylidene Fluoride). *Japanese Journal of Applied Physics*, *8*(7), 975–976. doi:10.1143/JJAP.8.975

Kayan, A. (2019). Inorganic-organic hybrid materials and their adsorbent properties. *Advanced Composites and Hybrid Materials*, *2*(1), 34–45. doi:10.1007/s42114-018-0073-y

Kazim, S., Nazeeruddin, M. K., Grätzel, M., & Ahmad, S. (2014). Perovskite as light harvester: A game changer in photovoltaics. *Angewandte Chemie International Edition*, *53*(11), 2812–2824. doi:10.1002/anie.201308719 PMID:24519832

Kethineni, P. (2017). Applications of internet of nano things: A survey. *2017 2nd International Conference for Convergence in Technology (I2CT)*. 10.1109/I2CT.2017.8226154

KhanbarehH. (2016). Expanding the Functionality of Piezo-Particulate Composites. doi:10.4233/UUID:AAB2497D-6EEC-4956-8C15-8F5FDEDC94F3

Kiamahalleh, M. V., Zein, S. H. S., Najafpour, G., Sata, S. A., & Buniran, S. (2012). Multiwalled carbon nanotubes based nanocomposites for supercapacitors: A review of electrode materials. *Nano*, *7*(2), 1–27. doi:10.1142/S1793292012300022

Kieslich, G., Sun, S., & Cheetham, A. K. (2014). Solid-state principles applied to organic–inorganic perovskites: New tricks for an old dog. *Chemical Science (Cambridge)*, *5*(12), 4712–4715. doi:10.1039/C4SC02211D

Kim & Schwartz. (2013). *An Inexpensive Scheffler Reflector for Developing Nations*. Academic Press.

Kim, D. M., & Jeong, Y. H. (Eds.). (2014). *Nanowire Field Effect Transistors: Principles and Applications* (Vol. 43). Springer. doi:10.1007/978-1-4614-8124-9

Kim, D. S., Kim, Y. H., Lee, C. E., & Kim, Y. T. (2006). Colossal electroresistance mechanism in a Au/Pr0.7Ca0.3MnO_3/Pt sandwich structure: Evidence for a Mott transition. *Physical Review B: Condensed Matter and Materials Physics*, *74*(17), 174430. doi:10.1103/PhysRevB.74.174430

Kim, H. D., Ohkita, H., Benten, H., & Ito, S. (2016). Photovoltaic performance of perovskite solar cells with different grain sizes. *Advanced Materials*, *28*(5), 917–922. doi:10.1002/adma.201504144 PMID:26639125

Kim, H. M., Park, J. H., & Lee, S. K. (2019). Fiber optic sensor based on ZnO nanowires decorated by Au nanoparticles for improved plasmonic biosensor. *Scientific Reports*, *9*(1), 15605. doi:10.1038/s41598-019-52056-1 PMID:31666617

Kim, H. S., Lee, J. W., Yantara, N., Boix, P. P., Kulkarni, S. A., Mhaisalkar, S., Gratzel, M., & Park, N. G. (2013). High efficiency solid-state sensitized solar cell-based on submicrometer rutile TiO_2 nanorod and $CH_3NH_3PbI_3$ perovskite sensitizer. *Nano Letters*, *13*(6), 2412–2417. doi:10.1021/nl400286w PMID:23672481

Kim, H.-S., Lee, C.-R., Im, J.-H., Lee, K.-B., Moehl, T., Marchioro, A., Moon, S.-J., Humphry-Baker, R., Yum, J.-H., Moser, J. E., Grätzel, M., & Park, N.-G. (2012). Lead iodide perovskite sensitized all-solid-state submicron thin film mesoscopic solar cell with efficiency exceeding 9%. *Scientific Reports*, *2*(1), 591. doi:10.1038/srep00591 PMID:22912919

Kim, K., Park, C., Kwon, D., Kim, D., Meyyappan, M., Jeon, S., & Lee, J. S. (2016). Silicon nanowire biosensors for detection of cardiac troponin I (cTnI) with high sensitivity. *Biosensors & Bioelectronics*, *77*, 695–701. doi:10.1016/j.bios.2015.10.008 PMID:26496224

Kim, Y. J., Dang, T. V., Choi, H. J., Park, B. J., Eom, J. H., Song, H. A., Seol, D., Kim, Y., Shin, S. H., Nah, H., & Yoon, S.-G. (2016). Piezoelectric properties of $CH_3NH_3PbI_3$ perovskite thin films and their applications in piezoelectric generators. *Journal of Materials Chemistry. A, Materials for Energy and Sustainability*, *4*(3), 756–763. doi:10.1039/C5TA09662F

Kitazawa, N., Watanabe, Y., & Nakamura, Y. (2002). Optical properties of $CH_3NH_3PbX_3$ (X = halogen) and their mixed-halide crystals. *Journal of Materials Science*, *37*(17), 3585–3587. doi:10.1023/A:1016584519829

Knight, J. C., Birks, T. A., Russel, P. St. J., & de Sandro, J. P. (1998). Properties of photonic crystal fiber and the effective index model. *Journal of the Optical Society of America. A, Optics, Image Science, and Vision*, *15*(3), 748. doi:10.1364/JOSAA.15.000748

Knutson, J. L., Martin, J. D., & Mitzi, D. B. (2005). Tuning the band gap in hybrid tin iodide perovskite semiconductors using structural templating. *Inorganic Chemistry*, *44*(13), 4699–4705. doi:10.1021/ic050244q PMID:15962978

Kojima, A., Teshima, K., Shirai, Y., & Miyasaka, T. (2009). Organometal Halide Perovskites as Visible-Light Sensitizers for Photovoltaic Cells. *Journal of the American Chemical Society*, *131*(17), 6050–6051. doi:10.1021/ja809598r PMID:19366264

Koketsu, T., Ma, J., Morgan, B. J., Body, M., Legein, C., Dachraoui, W., Giannini, M., Demortière, A., Salanne, M., Dardoize, F., Groult, H., Borkiewicz, O. J., Chapman, K. W., Strasser, P., & Dambournet, D. (2017). Reversible magnesium and aluminium ions insertion in cation-deficient anatase TiO2. *Nature Materials*, *16*(11), 1142–1148. doi:10.1038/nmat4976 PMID:28920941

Kong, W., Liu, J., Liu, R., Li, H., Liu, Y., Huang, H., Li, K., Liu, J., Lee, S.-T., & Kang, Z. (2014). Quantitative and real-time effects of carbon quantum dots on single living HeLa cell membrane permeability. *Nanoscale*, *6*(10), 5116–5120. doi:10.1039/c3nr06590a PMID:24722927

Koo, K. I., & Seo, J. M. (2015). A review of nanotechnology for highly sensitive photodetectors for vision sensors of insect-like robots. *Sensors and Materials*, *27*(6), 465–474.

Korniejenko, K., Figiela, B., Kozub, B., Azzopardi, B., & Łach, M. (2022). Environmental degradation of foamed geopolymers. *Continuum Mechanics and Thermodynamics*, 1–15.

Kothari, R., Buddhi, D., & Sawhney, R. L. (2008). Comparison of environmental and economic aspects of various hydrogen production methods. *Renewable & Sustainable Energy Reviews*, *12*(2), 553–563. doi:10.1016/j.rser.2006.07.012

Koutselas, I. B., Ducasse, L., & Papavassiliou, G. C. (1996). Electronic properties of three- and low- dimensional semiconducting materials with Pb halide and Sn halide units. *Journal of Physics Condensed Matter*, *8*(9), 1217–1227. doi:10.1088/0953-8984/8/9/012

Kowal, J., Avaroglu, E., Chamekh, F., Šenfelds, A., Thien, T., Wijaya, D., & Sauer, D. U. (2011). Detailed analysis of the self-discharge of supercapacitors. *Journal of Power Sources*, *196*(1), 573–579. doi:10.1016/j.jpowsour.2009.12.028

Kramar, P., Miklavčič, D., & Mir, L. M. (n.d.). Electroporation-based Technologies and Treatments. *Proceedings of the Electroporation-based Technologies and Treatments*. www.ebtt.org

Krishan, O., & Suhag, S. (2019). An updated review of energy storage systems: Classification and applications in distributed generation power systems incorporating renewable energy resources. *International Journal of Energy Research*, *43*(12), 6171–6210. doi:10.1002/er.4285

Krishna, R. S., Mishra, J., Zribi, M., Adeniyi, F., Saha, S., Baklouti, S., Shaikh, F. U. A., & Gökçe, H. S. (2021). A review on developments of environmentally friendly geopolymer technology. *Materialia*, *20*, 101212. doi:10.1016/j.mtla.2021.101212

Kulakowski, P., Solarczyk, K., & Wojcik, K. (2017). Routing in FRET-Based Nanonetworks. *IEEE Communications Magazine*, *55*(9), 218–224. doi:10.1109/MCOM.2017.1600760

Kulkarni, H., Zohaib, K., Khusru, A., & Aiyappa, K. S. (2018). Application of piezoelectric technology in automotive systems. *Materials Today: Proceedings*, *5*(10), 21299–21304. doi:10.1016/j.matpr.2018.06.532

Kulkarni, S. B., Patil, U. M., Salunkhe, R. R., Joshi, S. S., & Lokhande, C. D. (2011). Temperature impact on morphological evolution of ZnO and its consequent effect on physico-chemical properties. *Journal of Alloys and Compounds*, *509*(8), 3486–3492. doi:10.1016/j.jallcom.2010.12.036

Kumar, M. (2020). Social, Economic, and Environmental Impacts of Renewable Energy Resources. In Wind Solar Hybrid Renewable Energy System. IntechOpen. doi:10.5772/intechopen.89494

Kumar, A., & Kumar, A. (2020). Lead toxicity: Health hazards, influence on food chain, and sustainable remediation approaches. *International Journal of Environmental Research and Public Health*, *17*(7), 2179. doi:10.3390/ijerph17072179 PMID:32218253

Kumar, A., Prakash, O., & Kaviti, A. K. (2017, September). A comprehensive review of Scheffler solar collector. *Renewable & Sustainable Energy Reviews*, *77*, 890–898. doi:10.1016/j.rser.2017.03.044

Kumar, D., Alam, M., & Sanjayan, J. (2023). Experimental and numerical investigation of novel light weight concrete panels made with aerogel and phase change materials. *Energy and Building*, *283*, 112836. doi:10.1016/j.enbuild.2023.112836

Kumar, N., Kumar, D., Layek, A., & Yadav, S. (2022). Renewable energy and sustainable development. In *Artificial Intelligence for Renewable Energy Systems* (pp. 305–328). Elsevier. doi:10.1016/B978-0-323-90396-7.00011-0

Kumar, P., Pattanaik, M., & Sonia. (2013). Synthesis and characterizations of KNN ferroelectric ceramics near 50/50 MPB. *Ceramics International*, *39*(1), 65–69. doi:10.1016/j.ceramint.2012.05.093

Kumar, P., Singh, S., Juneja, J. K., Prakash, C., & Raina, K. K. (2009). Ferroelectric properties of substituted barium titanate ceramics. *Physica B, Condensed Matter*, *404*(12-13), 1752–1756. doi:10.1016/j.physb.2009.02.016

Kumar, V., Toffoli, G., & Rizzolio, F. (2013). Fluorescent carbon nanoparticles in medicine for cancer therapy. *ACS Medicinal Chemistry Letters*, *4*(11), 1012–1013. doi:10.1021/ml400394a PMID:24936239

Kumbhar, D., Palliyarayil, A., Reghu, D., Shrungar, D., Umapathy, S., & Sil, S. (2021). Rapid discrimination of porous bio-carbon derived from nitrogen rich biomass using Raman spectroscopy and artificial intelligence methods. *Carbon*, *178*, 792–802. doi:10.1016/j.carbon.2021.03.064

Kurkina, T., & Balasubramanian, K. (2012). Towards in vitro molecular diagnostics using nanostructures. *Cellular and molecular life sciences. Cellular and Molecular Life Sciences*, *69*(3), 373–388. doi:10.1007/s00018-011-0855-7 PMID:22009454

Kuscu, M., & Akan, O. B. (2016). The Internet of Molecular Things Based on FRET. *IEEE Internet of Things Journal*, *3*(1), 4–17. doi:10.1109/JIOT.2015.2439045

Kushwah, S., Mudgal, M., & Chouhan, R. K. (2021). The Process, Characterization and Mechanical properties of fly Solid form geopolymer via mechanical activation. *South African Journal of Chemical Engineering*, *38*(1), 104–114. doi:10.1016/j.sajce.2021.09.002

Kuwata, J., Uchino, K., & Nomura, S. (1981). Phase transitions in the Pb(Zn1/3Nb2/3)O3-PbTiO3 system. *Ferroelectrics*, *37*(1), 579–582. doi:10.1080/00150198108223490

Kuwata, J., Uchino, K., & Nomura, S. (2014). Dielectric and piezoelectric properties of 0.91Pb (Zn1/3Nb2/3)O3-0.09PbTiO3 single crystals. *Japanese Journal of Applied Physics*, *21*(21), 1298–1302.

Kylili, A., & Fokaides, P. A. (2016). Life cycle assessment (LCA) of phase change materials (PCMs) for building applications: A review. *Journal of Building Engineering*, *6*, 133–143. doi:10.1016/j.jobe.2016.02.008

Laasri, I. A., Es-sakali, N., Outzourhit, A., & Mghazli, M. O. (2022). Numerical building energy simulation with phase change materials including hysteresis effect for different square building cases in a semi-arid climate. *2022 13th International Renewable Energy Congress (IREC),* 1–4.

Łach, M., Pławecka, K., Bąk, A., Adamczyk, M., Bazan, P., Kozub, B., Korniejenko, K., & Lin, W.-T. (2021). Review of solutions for the use of phase change materials in geopolymers. *Materials (Basel)*, *14*(20), 6044. doi:10.3390/ma14206044 PMID:34683631

Laforgue, A., Simon, P., Fauvarque, J. F., Mastragostino, M., Soavi, F., Sarrau, J. F., Lailler, P., Conte, M., Rossi, E., & Saguatti, S. (2003). Activated Carbon/Conducting Polymer Hybrid Supercapacitors. *Journal of the Electrochemical Society*, *150*(5), A645. doi:10.1149/1.1566411

Lagou, A., Kylili, A., Šadauskienė, J., & Fokaides, P. A. (2019). Numerical investigation of phase change materials (PCM) optimal melting properties and position in building elements under diverse conditions. *Construction & Building Materials*, *225*, 452–464. doi:10.1016/j.conbuildmat.2019.07.199

Lakshmipathy, R., & Sarada, N. C. (2015). A fixed bed column study for the removal of Pb2+ ions by watermelon rind. *Environmental Science. Water Research & Technology*, *1*(2), 244–250. doi:10.1039/C4EW00027G

Law, M., Greene, L. E., Johnson, J. C., Saykally, R., & Yang, P. (2005). Nanowire dye-sensitized solar cells. *Nature Materials*, *4*(6), 455–459. doi:10.1038/nmat1387 PMID:15895100

Lazarus, M., & van Asselt, H. (2018). Fossil fuel supply and climate policy: Exploring the road less taken. *Climatic Change*, *150*(1-2), 1–13. doi:10.1007/s10584-018-2266-3

Lee, H., Cooper, R., Wang, K., & Liang, H. (2008). Nano-Scale Characterization of a Piezoelectric Polymer (Polyvinylidene Difluoride, PVDF). *Sensors (Basel)*, *8*(11), 7359–7368. doi:10.3390/s8117359 PMID:27873933

Lee, J. H., Chae, E. J., Park, S. J., & Choi, J. W. (2019). Label-free detection of γ-aminobutyric acid based on silicon nanowire biosensor. *Nano Convergence*, *6*(1), 1–6. doi:10.1186/s40580-019-0184-3 PMID:31041617

Lee, J. H., & Park, S. J. (2020). Recent advances in preparations and applications of carbon aerogels: A review. *Carbon*, *163*, 1–18. doi:10.1016/j.carbon.2020.02.073

Lee, K. Y., Kumar, B., Seo, J. S., Kim, K. H., Sohn, J. I., Cha, S. N., Choi, D., Wang, Z. L., & Kim, S. W. (2012). P-type polymer-hybridized high-performance piezoelectric nanogenerators. *Nano Letters*, *12*(4), 1959–1964. doi:10.1021/nl204440g PMID:22409420

Lee, M. H., Kim, D. J., Park, J. S., Kim, S. W., Song, T. K., Kim, M. H., Kim, W.-J., Do, D., & Jeong, I. K. (2015). High-performance lead-free piezoceramics with high curie temperatures. *Advanced Materials*, *27*(43), 6976–6982. doi:10.1002/adma.201502424 PMID:26444562

Lee, M. K., Yang, S. A., Park, J. J., & Lee, G. J. (2019). Proposal of a rhombohedral-tetragonal phase composition for maximizing piezoelectricity of (K, Na) NbO3 ceramics. *Scientific Reports, 9*(1), 4195. doi:10.1038/s41598-019-40943-6 PMID:30862966

Lee, M. M., Teuscher, J., Miyasaka, T., Murakami, T. N., & Snaith, H. J. (2012). Efficient Hybrid Solar Cells Based on Meso-Superstructured Organometal Halide Perovskites. *Science, 338*(6107), 643–647. doi:10.1126/science.1228604 PMID:23042296

Lee, S.-H., Park, Y. C., Chae, J., Kim, G., Kim, H. J., Choi, B. K., Lee, I. H., Chang, Y. J., Chun, S.-H., Jung, M., Seo, J., & Lee, S. (2022). Strong electron–phonon coupling driven charge density wave states in stoichiometric 1T-VS 2 crystals. *Journal of Materials Chemistry. C, Materials for Optical and Electronic Devices, 10*(44), 16657–16665. doi:10.1039/D2TC04186C

Lee, S., Hinchet, R., Lee, Y., Yang, Y., Lin, Z.-H., Ardila, G., Montès, L., Mouis, M., & Wang, Z. L. (2014). Ultrathin Nanogenerators as Self-Powered/Active Skin Sensors for Tracking Eye Ball Motion. *Advanced Functional Materials, 24*(8), 1163–1168. doi:10.1002/adfm.201301971

Lee, S.-W., Mao, C., Flynn, C. E., & Belcher, A. M. (2002). Ordering of Quantum Dots Using Genetically Engineered Viruses. *Science, 296*(5569), 892–895. doi:10.1126/science.1068054 PMID:11988570

Le, M. Q., Belhora, F., Cornogolub, A., Cottinet, P. J., Lebrun, L., & Hajjaji, A. (2014). Enhanced magnetoelectric effect for flexible current sensor applications. *Journal of Applied Physics, 115*(19), 194103. doi:10.1063/1.4876910

Le, M. Q., Capsal, J. F., Galineau, J., Ganet, F., Yin, X., Yang, M. D., Chateaux, J. F., Renaud, L., Malhaire, C., Cottinet, P. J., & Liang, R. (2015). All-organic electrostrictive polymer composites with low driving electrical voltages for micro-fluidic pump applications. *Scientific Reports, 5*(1), 1–13. doi:10.1038/srep11814 PMID:26139015

Leong, K. Y., Abdul Rahman, M. R., & Gurunathan, B. A. (2019, February). Nano-enhanced phase change materials: A review of thermo-physical properties, applications and challenges. *Journal of Energy Storage, 21*, 18–31. doi:10.1016/j.est.2018.11.008

Leontsev, S. O., & Eitel, R. E. (2010). Progress in engineering high strain lead-free piezoelectric ceramics. *Science and Technology of Advanced Materials, 11*(4), 044302. doi:10.1088/1468-6996/11/4/044302 PMID:27877343

Lewandowski, A., & Galinski, M. (2007). Practical and theoretical limits for electrochemical double-layer capacitors. *Journal of Power Sources, 173*(2), 822–828. doi:10.1016/j.jpowsour.2007.05.062

Leyva-ramos, R. (2007). Importancia y aplicaciones de la adsorción en fase líquida. Sólidos Porosos, Preparación. *Caracterización y Aplicaciones., 1*, 160–170.

Leyva-Ramos, R., Salazar-Rábago, J. J., & Ocampo-Pérez, R. (2021). A novel intraparticle mass transfer model for the biosorption rate of methylene blue on white pine (Pinus durangensis) sawdust. Diffusion-permeation. *Chemical Engineering Research & Design, 172*, 43–52. doi:10.1016/j.cherd.2021.05.029

Li. (2018). Synthesis of coin-like vanadium disulfide and its sodium storage performance. *Acta ChimicaSinica, 76*(4).

Li. (2019). Vanadium disulfide flakes with nanolayered titanium disulfide coating as cathode materials in lithium-ion batteries. *Nature Communications, 10*(1).

Li. (2021). Engineering textile electrode and bacterial cellulose nanofiber reinforced hydrogel electrolyte to enable high-performance flexible all-solid-state supercapacitors. *Advanced Energy Materials, 11*(12).

Liang, G., Mo, F., Ji, X., & Zhi, C. (2021). Non-metallic charge carriers for aqueous batteries. *Nature Reviews. Materials, 6*(2), 109–123. doi:10.1038/s41578-020-00241-4

Liang, H., Shi, H., Zhang, D., Ming, F., Wang, R., Zhuo, J., & Wang, Z. (2016). Solution growth of vertical VS2 nanoplate arrays for electrocatalytic hydrogen evolution. *Chemistry of Materials*, *28*(16), 5587–5591. doi:10.1021/acs.chemmater.6b01963

Liangliang, T., & Chunyang, J. (2010). Conducting polymers as electrode materials for supercapacitors. *Huaxue Jinzhan*, *22*(8), 1610–1618.

Liang, X., Chen, M., Zhu, H., Zhu, H., Cui, X., Yan, J., Chen, Q., Xia, X., & Liu, Q. (2020). Unveiling the solid-solution charge storage mechanism in 1T vanadium disulfide nanoarray cathodes. *Journal of Materials Chemistry. A, Materials for Energy and Sustainability*, *8*(18), 9068–9076. doi:10.1039/D0TA02922J

Liang, Y., Zhao, C.-Z., Yuan, H., Chen, Y., Zhang, W., Huang, J.-Q., Yu, D., Liu, Y., Titirici, M.-M., Chueh, Y.-L., Yu, H., & Zhang, Q. (2019). A review of rechargeable batteries for portable electronic devices. *InfoMat*, *1*(1), 6–32. doi:10.1002/inf2.12000

Liao, J.-Y., & Manthiram, A. (2015). High-performance Na2Ti2O5 nanowire arrays coated with VS2 nanosheets for sodium-ion storage. *Nano Energy*, *18*, 20–27. doi:10.1016/j.nanoen.2015.09.014

Li, C. P., Lee, C. S., Ma, X. L., Wang, N., Zhang, R. Q., & Lee, S. T. (2003). Growth direction and cross-sectional study of silicon nanowires. *Advanced Materials*, *15*(7-8), 607–609. doi:10.1002/adma.200304409

Lidjici, H., Lagoun, B., Berrahal, M., Rguitti, M., Hentatti, M. A., & Khemakhem, H. (2015). XRD, Raman and electrical studies on the (1− x)(Na0. 5Bi0. 5) TiO3− xBaTiO3 lead free ceramics. *Journal of Alloys and Compounds*, *618*, 643–648. doi:10.1016/j.jallcom.2014.08.161

Li, H. (2020). Investigation of the structural, magnetic, impedance properties in samarium-doped yttrium iron garnet. *Ceramics International*, *46*(10), 15408–15416. doi:10.1016/j.ceramint.2020.03.086

Li, H., Cheng, L., & Xia, Y. (2005). A hybrid electrochemical supercapacitor based on a 5 V Li-ion battery cathode and active carbon. *Electrochemical and Solid-State Letters*, *8*(9), A433. Advance online publication. doi:10.1149/1.1960007

Li, H., Kang, Z., Liu, Y., & Lee, S. T. (2012). Carbon nanodots: Synthesis, properties and applications. *Journal of Materials Chemistry*, *22*(46), 24230–24253. doi:10.1039/c2jm34690g

Li, J. F., Wang, K., Zhu, F. Y., Cheng, L. Q., & Yao, F. Z. (2013). (K, Na) NbO 3-Based Lead-Free Piezoceramics: Fundamental Aspects, Processing Technologies, and Remaining Challenges. *Journal of the American Ceramic Society*, *96*(12), 3677–3696. doi:10.1111/jace.12715

Li, J. W., Liu, Y. X., Thong, H. C., Du, Z., Li, Z., Zhu, Z. X., Nie, J.-K., Geng, J.-F., Gong, W., & Wang, K. (2020). Effect of ZnO doping on (K, Na) NbO3-based lead-free piezoceramics: Enhanced ferroelectric and piezoelectric performance. *Journal of Alloys and Compounds*, *847*, 155936. doi:10.1016/j.jallcom.2020.155936

Li, J., Li, Q., Qian, C., Wang, X., Lan, Y., Wang, B., & Yin, W. (2019). Volatile organic compounds analysis and characterization on activated biochar prepared from rice husk. *International Journal of Environmental Science and Technology*, *16*(12), 7653–7662. doi:10.1007/s13762-019-02219-4

Li, J., Xie, H., Li, Y., Liu, J., & Li, Z. (2011). Electrochemical properties of graphene nanosheets/polyaniline nanofibers composites as electrode for supercapacitors. *Journal of Power Sources*, *196*(24), 10775–10781. doi:10.1016/j.jpowsour.2011.08.105

Li, M., Sun, R., Chang, J., Dong, J., Tian, Q., Wang, H., Li, Z., Yang, P., Shi, H., Yang, C., Wu, Z., Li, R., Yang, Y., Wang, A., Zhang, S., Wang, F., Huang, W., & Qin, T. (2023). Orientated crystallization of FA-based perovskite via hydrogen-bonded polymer network for efficient and stable solar cells. *Nature Communications*, *14*(1), 573. doi:10.1038/s41467-023-36224-6 PMID:36732540

Lim, S. Y., Shen, W., & Gao, Z. (2015). Carbon quantum dots and their applications. *Chemical Society Reviews*, *44*(1), 362–381. doi:10.1039/C4CS00269E PMID:25316556

Lin, C. C., Lee, W. S., Sun, C. C., & Whu, W. H. (2008). A varistor-polymer composite with nonlinear electrical-thermal switching properties. *Ceramics International*, *34*(1), 131–136. doi:10.1016/j.ceramint.2006.09.018

Lin, D., & Kwok, K. W. (2010). Dielectric and piezoelectric properties of (Bi1− x− yNdxNa1− y) 0.5 BayTiO3 lead-free ceramics. *Current Applied Physics*, *10*(2), 422–427. doi:10.1016/j.cap.2009.06.044

Lin, D., Li, Z., Zhang, S., Xu, Z., & Yao, X. (2009). Dielectric/piezoelectric properties and temperature dependence of domain structure evolution in lead free (K0. 5Na0. 5) NbO3 single crystal. *Solid State Communications*, *149*(39-40), 1646–1649. doi:10.1016/j.ssc.2009.06.029

Lin, D., Li, Z., Zhang, S., Xu, Z., & Yao, X. (2010). Influence of MnO2 doping on the dielectric and piezoelectric properties and the domain structure in (K0. 5Na0. 5) NbO3 single crystals. *Journal of the American Ceramic Society*, *93*(4), 941–944. doi:10.1111/j.1551-2916.2009.03501.x

Lines, M., & Glass, A. (1979). *Principles and applications of ferroelectrics and related materials*. Clarendon Press.

Lingayat, A. B., & Suple, Y. R. (n.d.). *Review on Phase Change Material as Thermal Energy Storage Medium: Materials, Application*. Available: www.ijera.com

Lingayat, A. B., & Suple, Y. R. (n.d.). *Review On Phase Change Material As Thermal Energy Storage Medium: Materials, Application*. Available: www.ijera.com

Ling, T.-C., & Poon, C.-S. (2013). Use of phase change materials for thermal energy storage in concrete: An overview. *Construction & Building Materials*, *46*, 55–62. doi:10.1016/j.conbuildmat.2013.04.031

Ling, W., Wang, P., Chen, Z., Wang, H., Wang, J., Ji, Z., Fei, J., Ma, Z., He, N., & Huang, Y. (2020). Nanostructure design strategies for aqueous zinc-ion batteries. *ChemElectroChem*, *7*(14), 2957–2978. doi:10.1002/celc.202000372

Lin, K., Xing, J., Quan, L. N., de Arquer, F. P. G., Gong, X., Lu, J., Xie, L., Zhao, W., Zhang, D., Yan, C., Li, W., Liu, X., Lu, Y., Kirman, J., Sargent, E. H., Xiong, Q., & Wei, Z. (2018). Perovskite Light-Emitting Diodes with External Quantum Efficiency Exceeding 20 Per Cent. *Nature*, *56*(7726), 245–248. doi:10.1038/s41586-018-0575-3 PMID:30305741

Li, S., Deng, S., Zhao, L., Zhao, R., Lin, M., Du, Y., & Lian, Y. (2018). Mathematical modeling and numerical investigation of carbon capture by adsorption: Literature review and case study. *Applied Energy*, *221*, 437–449. doi:10.1016/j.apenergy.2018.03.093

Littlejohn, A. J., Li, Z., Lu, Z., Sun, X., Nawarat, P., Wang, Y., Li, Y., Wang, T., Chen, Y., Zhang, L., Li, H., Kisslinger, K., Shi, S., Shi, J., Raeliarijaona, A., Shi, W., Terrones, H., Lewis, K. M., Washington, M., ... Wang, G.-C. (2019). Large metallic vanadium disulfide ultrathin flakes for spintronic circuits and quantum computing devices. *ACS Applied Nano Materials*, *2*(6), 3684–3694. doi:10.1021/acsanm.9b00608

Liu, Liu, & Wu. (2023). Pre-intercalation Strategies to Layered Vanadium-based Electrode Materials for Aqueous Zinc Batteries. *Batteries & Supercaps, 6*(2).

Liu, A., Liang, X., Ren, X., Guan, W., & Ma, T. (2022). Recent progress in MXene-based materials for metal-sulfur and metal-air batteries: Potential high-performance electrodes. *Electrochemical Energy Reviews*, *5*(1), 1–33. doi:10.1007/s41918-021-00110-w

Liu, D., Xu, S., Cai, Y., Wang, Y., Guo, J., & Li, Y. (2022). A coupling technology of capacitive deionization and carbon-supported petal-like VS2 composite for effective and selective adsorption of lead (II) ions. *Journal of Electroanalytical Chemistry (Lausanne, Switzerland)*, *910*, 116152. doi:10.1016/j.jelechem.2022.116152

Liu, F., Peng, C., Porvali, A., Wang, Z., Wilson, B. P., & Lundström, M. (2019). Synergistic recovery of valuable metals from spent nickel–metal hydride batteries and lithium-ion batteries. *ACS Sustainable Chemistry & Engineering*, *7*(19), 16103–16111. doi:10.1021/acssuschemeng.9b02863

Liu, H. (2023). Local Strain engineering on Janus MoSSe Nanoribbons Induced Tunable Electronic Structures and Remarkable Magnetic Moments. *Journal of Physics. D, Applied Physics.*

Liu, H., Wang, L., & Lei, M. (2019). Positive impact of biochar amendment on thermal balance during swine manure composting at relatively low ambient temperature. *Bioresource Technology*, *273*, 25–33. doi:10.1016/j.biortech.2018.10.033 PMID:30399607

Liu, J., Fei, L. J., Chen, C. F., Han, X. L., & Luo, Y. (2011, December). Effect of preparation method on properties of BNT-BKT-BT piezoelectric ceramics. In *2011 Symposium on Piezoelectricity, Acoustic Waves and Device Applications (SPAWDA)* (pp. 279-281). IEEE. 10.1109/SPAWDA.2011.6167243

Liu, J., Peng, W., Li, Y., Zhang, F., & Fan, X. (2021). A VS 2@ N-doped carbon hybrid with strong interfacial interaction for high-performance rechargeable aqueous Zn-ion batteries. *Journal of Materials Chemistry. C, Materials for Optical and Electronic Devices*, *9*(19), 6308–6315. doi:10.1039/D1TC00531F

Liu, M., Johnston, M. B., & Snaith, H. J. (2013). Efficient planar heterojunction perovskite solar cells by vapour deposition. *Nature*, *501*(7467), 395–398. doi:10.1038/nature12509 PMID:24025775

Liu, M., Zhao, Z., Zhang, W., & Zheng, W. (2022). Perspective of Vanadium Disulfide: A Rising Star Finds Plenty of Room in Single and Multielectron Energy Storage. *Energy & Fuels*, *36*(23), 13931–13955. doi:10.1021/acs.energyfuels.2c02642

Liu, Q., Le, M. Q., Richard, C., Liang, R., Cottinet, P. J., & Capsal, J. F. (2019). Enhanced pseudo-piezoelectric dynamic force sensors based on inkjet-printed electrostrictive terpolymer. *Organic Electronics*, *67*, 259–271. doi:10.1016/j.orgel.2019.01.028

Liu, Q., Zhu, F. Y., Zhao, L., Wang, K., Li, L., & Li, J. F. (2016). Further enhancing piezoelectric properties by adding MnO2 in AgSbO3-modified (Li, K, Na)(Nb, Ta) O3 lead-free piezoceramics. *Journal of the American Ceramic Society*, *99*(11), 3670–3676. doi:10.1111/jace.14412

Liu, S., Hu, L., Xu, X., Al-Ghamdi, A. A., & Fang, X. (2015). Nickel Cobaltite Nanostructures for Photoelectric and Catalytic Applications. *Small*, *11*(34), 4267–4283. doi:10.1002/smll.201500315 PMID:26121217

Liu, X., Zhang, Y., Li, Z., Feng, R., & Zhang, Y. (2014). Characterization of corncob-derived biochar and pyrolysis kinetics in comparison with corn stalk and sawdust. *Bioresource Technology*, *170*, 76–82. doi:10.1016/j.biortech.2014.07.077 PMID:25125195

Liu, Y., & Wilcox, J. (2013). Molecular simulation studies of CO2 adsorption by carbon model compounds for carbon capture and sequestration applications. *Environmental Science & Technology*, *47*(1), 95–101. doi:10.1021/es3012029 PMID:22747244

Liu, Y.-Y., Xu, L., Guo, X.-T., Lv, T.-T., & Pang, H. (2020). Vanadium sulfide based materials: Synthesis, energy storage and conversion. *Journal of Materials Chemistry. A, Materials for Energy and Sustainability*, *8*(40), 20781–20802. doi:10.1039/D0TA07436E

Liu, Z., Sun, Y., Xing, C., Liu, J., He, Y., Zhou, Y., & Zhang, G. (2022). Artificial intelligence powered large-scale renewable integrations in multi-energy systems for carbon neutrality transition: Challenges and future perspectives. *Energy and AI*, *10*, 100195. doi:10.1016/j.egyai.2022.100195

Liu, Z., Yuan, Y., Luo, Z., Wan, H., Gao, P., Wu, H., Zhuang, J., Zhang, J., Zhang, N., Liu, H., Ren, W., & Ye, Z. G. (2020). Effects of antiferroelectric substitution on the structure and ferroelectric properties of a complex perovskite solid solution. *Journal of Materials Chemistry. C, Materials for Optical and Electronic Devices*, *8*(17), 5795–5806. doi:10.1039/C9TC05084A

Li, W., Hirbod, M. K. S., & Li, X. (2020). Emerging layered metallic vanadium disulfide for rechargeable metal-ion batteries: Progress and opportunities. *ChemSusChem*, *13*(6), 1172–1202. doi:10.1002/cssc.201903081 PMID:31777162

Li, W., Huang, J., Feng, L., Cao, L., Feng, Y., Wang, H., Li, J., & Yao, C. (2017). Facile in situ synthesis of crystalline VOOH-coated VS 2 microflowers with superior sodium storage performance. *Journal of Materials Chemistry. A, Materials for Energy and Sustainability*, *5*(38), 20217–20227. doi:10.1039/C7TA05205G

Li, X., Fu, J., Sun, Y., Sun, M., Cheng, S., Chen, K., Yang, X., Lou, Q., Xu, T., Shang, Y., Xu, J., Chen, Q., & Shan, C. (2019). Design and understanding of core/branch-structured VS 2 nanosheets@ CNTs as high-performance anode materials for lithium-ion batteries. *Nanoscale*, *11*(28), 13343–13353. doi:10.1039/C9NR03581H PMID:31271407

Li, X., & Wang, S. (2019). Energy management and operational control methods for grid battery energy storage systems. *CSEE Journal of Power and Energy Systems*, *7*(5), 1026–1040.

Li, X., Zhao, C., & Liu, X. (2015). A paper-based microfluidic biosensor integrating zinc oxide nanowires for electrochemical glucose detection. *Microsystems & Nanoengineering*, *1*(1), 1–7. doi:10.1038/micronano.2015.14

Li, Y., Çelik-Butler, Z., & Butler, D. P. (2016). An integrated piezoelectric zinc oxide nanowire micro-energy harvester. *Nano Energy*, *26*, 456–465. doi:10.1016/j.nanoen.2016.05.008

Li, Z., Yang, M., Park, J. S., Wei, S. H., Berry, J., & Zhu, K. (2016). Stabilizing perovskite structures by tuning tolerance factor: Formation of formamidinium and cesium lead iodide solid-state alloys. *Chemistry of Materials*, *28*(1), 284–292. doi:10.1021/acs.chemmater.5b04107

Li, Z., Yang, R., Yu, M., Bai, F., Li, C., & Wang, Z. L. (2008). Cellular level biocompatibility and biosafety of ZnO nanowires. *The Journal of Physical Chemistry. C, Nanomaterials and Interfaces*, *112*(51), 20114–20117. doi:10.1021/jp808878p

Li, Z., Zhu, G., Yang, R., Wang, A. C., & Wang, Z. L. (2010). Muscle-driven in vivo nanogenerator. *Advanced Materials*, *22*(23), 2534–2537. doi:10.1002/adma.200904355 PMID:20446305

Look, D. C. (2001). Recent advances in ZnO materials and devices. *Materials Science and Engineering B*, *80*(1-3), 383–387. doi:10.1016/S0921-5107(00)00604-8

Lopes, P. P., & Stamenkovic, V. R. (2020). Past, present, and future of lead–acid batteries. *Science*, *369*(6506), 923–924. doi:10.1126/science.abd3352 PMID:32820114

Loredo-Cancino, M., Soto-Regalado, E., Cerino-Córdova, F. J., García-Reyes, R. B., García-León, A. M., & Garza-González, M. T. (2013). Determining optimal conditions to produce activated carbon from barley husks using single or dual optimization. *Journal of Environmental Management*, *125*, 117–125. doi:10.1016/j.jenvman.2013.03.028 PMID:23651918

Loredo-Cancino, M., Soto-Regalado, E., García-Reyes, R. B., Cerino-Córdova, F. de J., Garza-González, M. T., Alcalá-Rodríguez, M. M., & Dávila-Guzmán, N. E. (2016). Adsorption and desorption of phenol onto barley husk-activated carbon in an airlift reactor. *Desalination and Water Treatment*, *57*(2), 845–860. doi:10.1080/19443994.2014.970579

Lourenssen, K., Williams, J., Ahmadpour, F., Clemmer, R., & Tasnim, S. (2019). Vanadium redox flow batteries: A comprehensive review. *Journal of Energy Storage*, *25*, 100844. doi:10.1016/j.est.2019.100844

Loya-González, D., Loredo-Cancino, M., Soto-Regalado, E., Rivas-García, P., Cerino-Córdova, F. de J., García-Reyes, R. B., Bustos-Martínez, D., & Estrada-Baltazar, A. (2019). Optimal activated carbon production from corn pericarp: A life cycle assessment approach. *Journal of Cleaner Production*, *219*, 316–325. doi:10.1016/j.jclepro.2019.02.068

Lueken, M., Mueller, L., Decker, M. G., Bollheimer, C., Leonhardt, S., & Ngo, C. (2020). Evaluation and Application of a Customizable Wireless Platform: A Body Sensor Network for Unobtrusive Gait Analysis in Everyday Life. *Sensors (Basel)*, *20*(24), 7325. doi:10.3390/s20247325 PMID:33419278

Lu, L., Ding, W., Liu, J., & Yang, B. (2020). Flexible PVDF based piezoelectric nanogenerators. *Nano Energy*, *78*, 105251. doi:10.1016/j.nanoen.2020.105251

Lu, X., Jin, D., Wei, S., Wang, Z., An, C., & Guo, W. (2015). Strategies to enhance CO 2 capture and separation based on engineering absorbent materials. *Journal of Materials Chemistry. A, Materials for Energy and Sustainability*, *3*(23), 12118–12132. doi:10.1039/C4TA06829G

Maculan, G., Sheikh, A. D., Abdelhady, A. L., Saidaminov, M. I., Haque, M. A., Murali, B., Alarousu, E., Mohammed, O. F., Wu, T., & Bakr, O. M. (2015). $CH_3NH_3PbCl_3$ single crystals: Inverse temperature crystallization and visible-blind UV-Photodetector. *The Journal of Physical Chemistry Letters*, *6*(19), 3781–3786. doi:10.1021/acs.jpclett.5b01666 PMID:26722870

Madeswaran, N., Desai, F. J., & Asmatulu, E. (2021). Life cycle inventory and performance analysis of phase change materials for thermal energy storages. *Emergent Materials*, *4*(6), 1697–1709. doi:10.1007/s42247-021-00235-0

Maeyoshi, Y., Yoshii, K., Shikano, M., & Sakaebe, H. (2021). Improving Cycling Stability of Vanadium Sulfide (VS4) as a Li Battery Cathode Material Using a Localized High-Concentration Carbonate-Based Electrolyte. *ACS Applied Energy Materials*, *4*(12), 13627–13635. doi:10.1021/acsaem.1c02312

Mahdavi, A., Sarau, G., Xavier, J., Paraïso, T. K., Christiansen, S., & Vollmer, F. (2016). Maximizing photoluminescence extraction in silicon photonic crystal slabs. *Scientific Reports*, *6*(1), 1–6. doi:10.1038/srep25135 PMID:27113674

Mahdavinejad, M. S., Rezvan, M., Barekatain, M., Adibi, P., Barnaghi, P., & Sheth, A. P. (2018). Machine learning for internet of things data analysis: A survey. *Digital Communications and Networks*, *4*(3), 161–175. doi:10.1016/j.dcan.2017.10.002

Majchrzak, A., & Nowak, W. (2017). Separation characteristics as a selection criteria of CO2 adsorbents. *Journal of CO2 Utilization, 17*, 69-79.

Malic, B., Bernard, J., Holc, J., Jenko, D., & Kosec, M. (2005). Alkaline-earth doping in (K, Na) NbO3 based piezoceramics. *Journal of the European Ceramic Society*, *25*(12), 2707–2711. doi:10.1016/j.jeurceramsoc.2005.03.127

Malik, R. A., Kang, J. K., Hussain, A., Ahn, C. W., Han, H. S., & Lee, J. S. (2014). High strain in lead-free Nb-doped Bi1/2 (Na0. 84K0. 16) 1/2TiO3–SrTiO3 incipient piezoelectric ceramics. *Applied Physics Express*, *7*(6), 061502. doi:10.7567/APEX.7.061502

Mallmann, E. J. J., Sombra, A. S. B., Goes, J. C., & Fechine, P. B. A. (2013). Yttrium iron garnet: Properties and applications review. *Diffusion and Defect Data, Solid State Data. Part B, Solid State Phenomena*, *202*, 65–96. doi:10.4028/www.scientific.net/SSP.202.65

Månberger, A. (2021). Reduced Use of Fossil Fuels can Reduce Supply of Critical Resources. *Biophysical Economics and Sustainability*, *6*(2), 1–15. doi:10.1007/s41247-021-00088-5

Mantler, M. (1986). X-ray fluorescence analysis of multiple-layer films. *Analytica Chimica Acta*, *188*, 25–35. doi:10.1016/S0003-2670(00)86026-2

Mao, Y., Zhu, Y., Zhao, T., Jia, C., Bian, M., Li, X., Liu, Y., & Liu, B. (2021). A portable and flexible self-powered multifunctional sensor for real-time monitoring in swimming. *Biosensors (Basel)*, *11*(5), 1–11. doi:10.3390/bios11050147 PMID:34066654

Marani, A., & Nehdi, M. L. (2019). Integrating phase change materials in construction materials: Critical review. *Construction & Building Materials*, *217*, 36–49. doi:10.1016/j.conbuildmat.2019.05.064

Mashiyama, H., & Kurihara, Y. (1998). Disordered cubic Perovskite structure of CH_3 NH_3PbX_3 (X = Cl, Br, I). *Journal of the Korean Physical Society*, *32*, 156–158.

Matschoss, P., Bayer, B., Thomas, H., & Marian, A. (2019). The German incentive regulation and its practical impact on the grid integration of renewable energy systems. *Renewable Energy*, *134*, 727–738. doi:10.1016/j.renene.2018.10.103

Matsimbe, J., Dinka, M., Olukanni, D., & Musonda, I. (2022). Geopolymer: A systematic review of methodologies. *Materials (Basel)*, *15*(19), 6852. doi:10.3390/ma15196852 PMID:36234194

Ma, X., Du, C., Shang, M., & Song, W. (2018). VS 2 quantum dot label-free fluorescent probe for sensitive and selective detection of ALP. *Analytical and Bioanalytical Chemistry*, *410*(5), 1417–1426. doi:10.1007/s00216-017-0778-8 PMID:29256077

Ma, Y., Chu, X., Li, W., & Tang, G. (2012). Preparation and characterization of poly (methyl methacrylate-co-divinylbenzene) microcapsules containing phase change temperature adjustable binary core materials. *Solar Energy*, *86*(7), 2056–2066. doi:10.1016/j.solener.2012.04.008

Mazhao, D., Xiao, D., Wu, J., & Zhu, J. (2015). Phase structure and electrical properties of 0.965(K0:45Na0:55Þ0:95Ag0:05(Nb1_xSbxÞ- O3–0.035Bi0:5(Na0:5Li0:5Þ0:5ZrO3 lead-free ceramics. *Journal of Materials Science Materials in Electronics*, *26*, 7309. doi:10.1007/s10854-015-3359-7

McDowall, W. (2012). Technology roadmaps for transition management: The case of hydrogen energy. *Technological Forecasting and Social Change*, *79*(3), 530–542. doi:10.1016/j.techfore.2011.10.002

Meadowcroft, W. H. (1896). The abc of x-rays.

Megaw. (1945). Crystal structure of barium titanate. *Nature*, *155*(3938), 484–485. doi:10.1038/155484b0

Mehrizi, A. A., Karimi-Maleh, H., Naddafi, M., & Karimi, F. (2023). Application of bio-based phase change materials for effective heat management. *Journal of Energy Storage*, *61*, 106859. doi:10.1016/j.est.2023.106859

Meijer, G. I., Staub, U., Janousch, M., Johnson, S. L., Delley, B., & Neisius, T. (2005). Valence states of Cr and the insulator-to-metal transition in Cr-doped $SrTiO_3$. *Physical Review B: Condensed Matter and Materials Physics*, *72*(15), 155102. doi:10.1103/PhysRevB.72.155102

Mellit, A., Kalogirou, S. A., Hontoria, L., & Shaari, S. (2009). Artificial intelligence techniques for sizing photovoltaic systems: A review. *Renewable & Sustainable Energy Reviews*, *13*(2), 406–419. doi:10.1016/j.rser.2008.01.006

Meng, L., You, J., & Yang, Y. (2018). Addressing the stability issue of perovskite solar cells for commercial applications. *Nature Communications*, *9*(1), 5265. doi:10.1038/s41467-018-07255-1 PMID:30532038

Meng, P., Wang, W., Shang, J., Liu, P., Xu, H., Wang, Q., Wang, S., Wang, F., & Wang, X. (2023). 2D VS2@ MXene Based Zinc Ion Batteries with SPANI-Contained Electrolyte Enables Dendrite-Free Anode for Stable Cycling. *Small Methods*, *7*(4), 2201471. doi:10.1002/smtd.202201471 PMID:36720008

Meng, T., Yang, X., Yu, Y., Yu, H., & Huang, M. (2021). Hybrid effect of pva fibre and carbon nanotube on the mechanical properties and microstructure of geopolymers. *Frontiers in Materials*, *8*, 773103. doi:10.3389/fmats.2021.773103

Meng, Y., Zhao, Y., Wang, D., Yang, D., Gao, Y., Lian, R., Chen, G., & Wei, Y. (2018). Fast Li+ diffusion in interlayer-expanded vanadium disulfide nanosheets for Li+/Mg 2+ hybrid-ion batteries. *Journal of Materials Chemistry. A, Materials for Energy and Sustainability*, *6*(14), 5782–5788. doi:10.1039/C8TA00418H

Men, T. L., Yao, F. Z., Zhu, Z. X., Wang, K., & Li, J. F. (2016). Piezoelectric properties of (K 0. 5 Na 0. 5) NbO3-BaTiO3 lead-free ceramics prepared by spark plasma sintering. *Journal of Advanced Dielectrics*, *6*(02), 1650013. doi:10.1142/S2010135X16500132

Menya, E., Olupot, P. W., Storz, H., Lubwama, M., Kiros, Y., & John, M. J. (2020). Optimization of pyrolysis conditions for char production from rice husks and its characterization as a precursor for production of activated carbon. *Biomass Conversion and Biorefinery*, *10*(1), 57–72. doi:10.1007/s13399-019-00399-0

Messenger, M. A., Troxler, C. J., Melendez, I., Freeman, T. B., Reed, N., Rodriguez, R. M., & Boetcher, S. K. S. (2023). Mechanical and Thermal Characterization of Phase-Change Material and High-Density Polyethylene Functional Composites for Thermal Energy Storage. *Journal of Solar Energy Engineering*, *145*(6), 61006. doi:10.1115/1.4062197

Messing, G. L., Trolier-McKinstry, S., Sabolsky, E. M., Duran, C., Kwon, S., Brahmaroutu, B., Park, P., Yilmaz, H., Rehrig, P. W., Eitel, K. B., Suvaci, E., Seabaugh, M., & Oh, K. S. (2004). Templated grain growth of textured piezoelectric ceramics. *Critical Reviews in Solid State and Material Sciences*, *29*(2), 45–96. doi:10.1080/10408430490490905

Meyer, E., Bede, A., Mutukwa, D., Taziwa, R., & Zingwe, N. (2020). Optimization, and analysis of carbon supported VS2 nanocomposites as potential electrodes in supercapacitors. *Journal of Energy Storage*, *27*, 101074. doi:10.1016/j.est.2019.101074

Miller & Kearnes. (2012). *Nanotechnology, Ubiquitous Computing and the Internet of Things: Challenges to Rights to Privacy and Data Protection*. Draft Report to the Council of Europe.

Miller, D. C., Agarwal, D., Bhattacharyya, D., Boverhof, J., Chen, Y., Eslick, J., ... & Zitney, S. E. (2017). Innovative computational tools and models for the design, optimization and control of carbon capture processes. *Process Systems and Materials for CO2 Capture: Modelling, Design, Control and Integration*, 311-342.

Minhong, J., Manjiao, D., Huaxin, L., Shi, W., & Xinyu, L. (2011). Piezoelectric and dielectric properties of K0.5Na0.5NbO3-LiSbO3-BiScO3 lead-free piezoceramics. *Materials Science and Engineering B*, *176*(2), 167–170. doi:10.1016/j.mseb.2010.10.007

Miraz, M. H., Ali, M., Excell, P. S., & Picking, R. (2015). *A review on Internet of Things (IoT), Internet of Everything (IoE) and Internet of Nano Things (IoNT). 2015 Internet Technologies and Applications*. ITA. doi:10.1109/ITechA.2015.7317398

Mirzaei, A., Bonyani, M., & Torkian, S. (2016). Effect of Nb doping on sintering and dielectric properties of PZT ceramics. *Processing and Application of Ceramics*, *10*(3), 175–182. doi:10.2298/PAC1603175M

Missoum, K., Belgacem, M. N., & Bras, J. (2013). Nanofibrillated Cellulose Surface Modification: A Review. *Materials (Basel)*, *6*(5), 1745–1766. doi:10.3390/ma6051745 PMID:28809240

Mitzi, D. B. (1999). *Synthesis, Structure, and roperties of Organic-Inorganic Perovskites and Related Materials*. John Wiley & Sons.

Mitzi, D. B. (2001). Templating and structural engineering in organic–inorganic perovskites. *Dalton Transactions (Cambridge, England)*, *1*, 1–12.

Mitzi, D. B., Chondroudis, K., & Kagan, C. R. (2001). Organic-inorganic electronics. *IBM Journal of Research and Development*, *45*(1), 29–45. doi:10.1147/rd.451.0029

Mitzi, D. B., Feild, C. A., Harrison, W. T. A., & Guloy, A. M. (1994). Conducting tin halides with a layered organic-based perovskite structure. *Nature*, *369*(6480), 467–469. doi:10.1038/369467a0

Mitzi, D. B., Prikas, M. T., & Chondroudis, K. (1999). Thin film deposition of organic-inorganic hybrid materials using a single source thermal ablation technique. *Chemistry of Materials*, *11*(3), 542–544. doi:10.1021/cm9811139

Mogilevtsev, D., Birks, T. A., & Russell, P. St. J. (2001). Localized function method for modeling defect modes in 2-D photonic crystals. *Journal of Lightwave Technology*, *17*(11), 2078–2081. doi:10.1109/50.802997

Mohan, P., Yang, J., Jena, A., & Suk Shin, H. (2015). VS2/rGO hybrid nanosheets prepared by annealing of VS4/rGO. *Journal of Solid State Chemistry*, *224*, 82–87. doi:10.1016/j.jssc.2014.06.031

Mohd Abdah, M. A. A., Azman, N. H. N., Kulandaivalu, S., & Sulaiman, Y. (2020). Review of the use of transition-metal-oxide and conducting polymer-based fibres for high-performance supercapacitors. *Materials & Design*, *186*, 108199. doi:10.1016/j.matdes.2019.108199

Mondal, P., Ghosal, K., Bhattacharyya, S. K., Das, M., Bera, A., Ganguly, D., Kumar, P., Dwivedi, J., Gupta, R. K., Martí, A. A., Gupta, B. K., & Maiti, S. (2014). Formation of a gold–carbon dot nanocomposite with superior catalytic ability for the reduction of aromatic nitro groups in water. *RSC Advances*, *4*(49), 25863–25866. doi:10.1039/C4RA02837F

Monika Aniz, I., & Maisnam, M. (2022). Microwave sintering of MnO2 added potassium sodium niobate ceramics and studies of their electrical properties. *Ferroelectrics*, *588*(1), 78–87. doi:10.1080/00150193.2022.2034441

Moore, R. W. (1919). A method of growing large perfect crystals from solution. *Journal of the American Chemical Society*, *41*(7), 1060–1066. doi:10.1021/ja02228a003

Moradbeigi, N., Bahari, A., & Ghasemi, S. (2023). Effect of molybdenum doping on the catalytic activity of VS 2/CNT for the oxygen reduction reaction in alkaline media. *New Journal of Chemistry*, *47*(3), 1291–1298. doi:10.1039/D2NJ04721G

Mordensky, S. P., Lipor, J. J., DeAngelo, J., Burns, E. R., & Lindsey, C. R. (2023). When less is more: How increasing the complexity of machine learning strategies for geothermal energy assessments may not lead toward better estimates. *Geothermics*, *110*, 102662. doi:10.1016/j.geothermics.2023.102662

Morkoç & Özgür. (2009). *General Properties of ZnO*. . doi:10.1002/9783527623945.ch1

Moubarak, E., Moosavi, S. M., Charalambous, C., Garcia, S., & Smit, B. (2023). A Robust Framework for Generating Adsorption Isotherms to Screen Materials for Carbon Capture. *Industrial & Engineering Chemistry Research*, *62*(26), 10252–10265. doi:10.1021/acs.iecr.3c01358 PMID:37425135

Moyassari. (2022). The role of silicon in silicon-graphite composite electrodes regarding specific capacity, cycle stability, and expansion. *Journal of The Electrochemical Society*, *169*(1).

Mukherjee. (2019). Electrode materials for high-performance sodium-ion batteries. *Materials*, *12*(12).

Muldoon, J., Bucur, C. B., & Gregory, T. (2017). Fervent hype behind magnesium batteries: An open call to synthetic chemists—electrolytes and cathodes needed. *Angewandte Chemie International Edition*, *56*(40), 12064–12084. doi:10.1002/anie.201700673 PMID:28295967

Munir, M. U., & Hussain, A. (2022). Thermal Stability Analysis of a PCM-Based Energy Storage System. *Engineering Proceedings*, *23*(1), 20.

Murayama, N., Nakamura, K., Obara, H., & Segawa, M. (1976). The strong piezoelectricity in polyvinylidene fluroide (PVDF). *Ultrasonics*, *14*(1), 15–24. doi:10.1016/0041-624X(76)90067-6

Murge, P., Dinda, S., & Roy, S. (2019). Zeolite-based sorbent for CO2 capture: Preparation and performance evaluation. *Langmuir*, *35*(46), 14751–14760. doi:10.1021/acs.langmuir.9b02259 PMID:31661624

Murugan, A. V. (2005). Entrapment of poly (3, 4-ethylenedioxythiophene) between VS 2 layers to form a new organic–inorganic intercalative nanocomposite. *Journal of Materials Chemistry*, *15*(8), 902–909. doi:10.1039/B414922J

Murutoğlu, M., Akça, E., & Yilmaz, H. (n.d.). *Effects of Antimony Doping and Sintering Atmosphere on the Properties of K4CuNb8O23 Modified K0. 5Na0. 5NbO3 (KNN-KCN) Lead-free Piezoceramics*. Academic Press.

Muzaffar, A., Ahamed, M. B., Deshmukh, K., & Thirumalai, J. (2019). A review on recent advances in hybrid supercapacitors: Design, fabrication and applications. *Renewable and Sustainable Energy Reviews, 101*(October), 123–145. doi:10.1016/j.rser.2018.10.026

Naderer, M., Kainz, T., Schütz, D., & Reichmann, K. (2014). The influence of Ti-nonstoichiometry in Bi0. 5Na0. 5TiO3. *Journal of the European Ceramic Society*, *34*(3), 663–667. doi:10.1016/j.jeurceramsoc.2013.10.010

Nagao, M., Suzuki, K., Imade, Y., Tateishi, M., Watanabe, R., Yokoi, T., Hirayama, M., Tatsumi, T., & Kanno, R. (2016). All-solid-state lithium–sulfur batteries with three-dimensional mesoporous electrode structures. *Journal of Power Sources*, *330*, 120–126. doi:10.1016/j.jpowsour.2016.09.009

Nagata, H., Shinya, T., Hiruma, Y., Takenaka, T., Sakaguchi, I., & Haneda, H. (2012). Piezoelectric properties of bismuth sodium titanate ceramics. *Developments in Dielectric Materials and Electronic Devices, 167*, 213-221.

Naik, V. M., Bhosale, S. V., & Kolekar, G. B. (2022). A brief review on the synthesis, characterisation and analytical applications of nitrogen doped carbon dots. *Analytical Methods*, *14*(9), 877–891. doi:10.1039/D1AY02105B PMID:35174374

Nakhleh, M. K., Amal, H., Jeries, R., Broza, Y. Y., Aboud, M., Gharra, A., Ivgi, H., Khatib, S., Badarneh, S., Har-Shai, L., Glass-Marmor, L., Lejbkowicz, I., Miller, A., Badarny, S., Winer, R., Finberg, J., Cohen-Kaminsky, S., Perros, F., Montani, D., ... Haick, H. (2017). Diagnosis and Classification of 17 Diseases from 1404 Subjects via Pattern Analysis of Exhaled Molecules. *ACS Nano*, *11*(1), 112–125. doi:10.1021/acsnano.6b04930 PMID:28000444

Nasrazadani, S., & Hassani, S. (2016). Modern analytical techniques in failure analysis of aerospace, chemical, and oil and gas industries. Handbook of Materials Failure Analysis with Case Studies from the Oil and Gas Industry, 39–54. doi:10.1016/B978-0-08-100117-2.00010-8

Nassef, A. M., Sayed, E. T., Rezk, H., Inayat, A., Yousef, B. A. A., Abdelkareem, M. A., & Olabi, A. G. (2020). Developing a fuzzy-model with particle swarm optimization-based for improving the conversion and gasification rate of palm kernel shell. *Renewable Energy*, *166*, 125–135. doi:10.1016/j.renene.2020.11.037

National Renewable Energy Laboratory. (n.d.). *Best Research-Cell Efficiencies*. Available online: https://www.nrel.gov/pv/assets/pdfs/best-research-cell efficiencies.20200925.pdf

Nature News Features. (2013). 365 days: Nature's 10. *Nature*, *504*(7480), 357–365. doi:10.1038/504357a PMID:24352276

Navarro, G., Torres, J., Blanco, M., Nájera, J., Santos-Herran, M., & Lafoz, M. (2021). Present and future of supercapacitor technology applied to powertrains, renewable generation and grid connection applications. *Energies*, *14*(11), 3060. Advance online publication. doi:10.3390/en14113060

Navarro, L., Solé, A., Martín, M., Barreneche, C., Olivieri, L., Tenorio, J. A., & Cabeza, L. F. (2019). Benchmarking of useful phase change materials for a building application. *Energy and Building*, *182*, 45–50. doi:10.1016/j.enbuild.2018.10.005

Nayak, P. K., Yang, L., Brehm, W., & Adelhelm, P. (2018). From lithium-ion to sodium-ion batteries: Advantages, challenges, and surprises. *Angewandte Chemie International Edition*, *57*(1), 102–120. doi:10.1002/anie.201703772 PMID:28627780

Nayak, S., Sahoo, B., Chaki, T. K., & Khastgir, D. (2014). Facile preparation of uniform barium titanate (BaTiO 3) multipods with high permittivity: Impedance and temperature dependent dielectric behavior. *RSC Advances*, *4*(3), 1212–1224. doi:10.1039/C3RA44815K

Nazari Sam, M., Caggiano, A., Mankel, C., & Koenders, E. (2020). A comparative study on the thermal energy storage performance of bio-based and paraffin-based PCMs using DSC procedures. *Materials (Basel)*, *13*(7), 1705. doi:10.3390/ma13071705 PMID:32260573

Nene, A. A., Ramachandran, S., & Suyambazhahan, S. (2019). Design and analysis of solar thermal energy storage system for Scheffler solar concentrator. *Comptes Rendus de l'Academie Bulgare des Sciences, 72*(10), 1321–1325. Available:https://inis.iaea.org/Search/search.aspx?orig_q=RN:51004309

Nenon, D. P., Christians, J. A., Wheeler, L. M., Blackburn, J. L., Sanehira, E. M., Dou, B., Olsen, M. L., Zhu, K., Berry, J. J., & Luther, J. M. (2016). Structural and chemical evolution of methylammonium lead halide perovskites during thermal processing from solution. *Energy & Environmental Science*, *9*(6), 2072–2082. doi:10.1039/C6EE01047D

Newnham, R. E. (1997). Molecular mechanisms in smart materials. *MRS Bulletin*, *22*(5), 20–34. doi:10.1557/S0883769400033170

Newnham, R. E., Skinner, D. P., & Cross, L. E. (1978). Connectivity and piezoelectric-pyroelectric composites. *Materials Research Bulletin*, *13*(5), 525–536. doi:10.1016/0025-5408(78)90161-7

Nikzad, L., Alibeigi, S., Vaezi, M. R., Yazdani, B., & Rahimipour, M. R. (2009). Synthesis of a graphite-polyaniline nanocomposite and evaluation of its electrochemical properties. *Chemical Engineering & Technology*, *32*(6), 861–866. doi:10.1002/ceat.200800470

Nitta, N., Wu, F., Lee, J. T., & Yushin, G. (2015). Li-ion battery materials: Present and future. *Materials Today*, *18*(5), 252–264. doi:10.1016/j.mattod.2014.10.040

Noda, S., Tomoda, K., Yamamoto, N., & Chutinan, A. (2000). Full three-dimensionalphotonic bandgap crystals at near-infrared wavelengths. *Science*, *289*(5479), 604–606. doi:10.1126/science.289.5479.604 PMID:10915619

Noheda, B. (2002). Structure and high-piezoelectricity in lead oxide solid solutions. *Current Opinion in Solid State and Materials Science*, *6*(1), 27–34. doi:10.1016/S1359-0286(02)00015-3

Noheda, B., Cox, D. E., Shirane, G., Gao, J., & Ye, Z.-G. (2002). Phase diagram of the ferroelectric relaxor (1-x) PbMg1/3Nb2/3O3-xPbTiO3. *Physical Review B: Condensed Matter*, *66*(5), 340–351. doi:10.1103/PhysRevB.66.054104

Noheda, B., Cox, D. E., Shirane, G., Gonzalo, J. A., Cross, L. E., & Park, S. E. (1999). A monoclinic ferroelectric phase in the Pb (Zr 1− x Ti x) O 3 solid solution. *Applied Physics Letters*, *74*(14), 2059–2061. doi:10.1063/1.123756

Noheda, B., Cox, D. E., Shirane, G., Guo, R., Jones, B., & Cross, L. E. (2000). Stability of the monoclinic phase in the ferroelectric perovskite PbZr 1− x Ti x O 3. *Physical Review B: Condensed Matter*, *63*(1), 014103. doi:10.1103/PhysRevB.63.014103

Noheda, B., Gonzalo, J. A., Cross, L. E., Guo, R., Park, S. E., Cox, D. E., & Shirane, G. (2000). Tetragonal-to-monoclinic phase transition in a ferroelectric perovskite: The structure of PbZr 0.52 Ti 0.48 O 3. *Physical Review B: Condensed Matter*, *61*(13), 8687–8695. doi:10.1103/PhysRevB.61.8687

Novais, R. M., Pullar, R. C., & Labrincha, J. A. (2020). Geopolymer foams: An overview of recent advancements. *Progress in Materials Science*, *109*, 100621. doi:10.1016/j.pmatsci.2019.100621

Ocampo-Perez, R., Aguilar-Madera, C. G., & Díaz-Blancas, V. (2017). 3D modeling of overall adsorption rate of acetaminophen on activated carbon pellets. *Chemical Engineering Journal*, *321*, 510–520. doi:10.1016/j.cej.2017.03.137

Ocampo-Pérez, R., Leyva-Ramos, R., Sanchez-Polo, M., & Rivera-Utrilla, J. (2013). Role of pore volume and surface diffusion in the adsorption of aromatic compounds on activated carbon. *Adsorption*, *19*(5), 945–957. doi:10.1007/s10450-013-9502-y

Ogino, M., Noguchi, Y., Kitanaka, Y., Miyayama, M., Moriyoshi, C., & Kuroiwa, Y. (2014). Polarization rotation and monoclinic distortion in ferroelectric (Bi0. 5Na0. 5) TiO3–BaTiO3 single crystals under electric fields. *Crystals*, *4*(3), 273–295. doi:10.3390/cryst4030273

Ohigashi, H. (2008). Electromechanical properties of polarized polyvinylidene fluoride films as studied by the piezoelectric resonance method. *Journal of Applied Physics*, *47*(3), 949–955. doi:10.1063/1.322685

Ojeda, J. J., Romero-González, M. E., Bachmann, R. T., Edyvean, R. G. J., & Banwart, S. A. (2008). Characterization of the Cell Surface and Cell Wall Chemistry of Drinking Water Bacteria by Combining XPS, FTIR Spectroscopy, Modeling, and Potentiometric Titrations. *Langmuir*, *24*(8), 4032–4040. doi:10.1021/la702284b PMID:18302422

Okpalugo, T. I. T., Papakonstantinou, P., Murphy, H., McLaughlin, J., & Brown, N. M. D. (2005). High resolution XPS characterization of chemical functionalised MWCNTs and SWCNTs. *Carbon*, *43*(1), 153–161. doi:10.1016/j.carbon.2004.08.033

Omitola, O. B., Abonyi, M. N., Akpomie, K. G., & Dawodu, F. A. (2022). Adams-Bohart, Yoon-Nelson, and Thomas modeling of the fix-bed continuous column adsorption of amoxicillin onto silver nanoparticle-maize leaf composite. *Applied Water Science*, *12*(5), 1–9. doi:10.1007/s13201-022-01624-4

Ong, P. J., Leow, Y., Soo, X. Y. D., Chua, M. H., Ni, X., Suwardi, A., Tan, C. K. I., Zheng, R., Wei, F., & Xu, J. (2023). Valorization of Spent coffee Grounds: A sustainable resource for Bio-based phase change materials for thermal energy storage. *Waste Management (New York, N.Y.)*, *157*, 339–347. doi:10.1016/j.wasman.2022.12.039 PMID:36603448

Orayech, B., Faik, A., López, G. A., Fabelo, O., & Igartua, J. M. (2015). Mode-crystallography analysis of the crystal structures and the low-and high-temperature phase transitions in Na0. 5K0. 5NbO3. *Journal of Applied Crystallography*, *48*(2), 318–333. doi:10.1107/S1600576715000941

Oschatz, M., & Antonietti, M. (2018). A search for selectivity to enable CO 2 capture with porous adsorbents. *Energy & Environmental Science*, *11*(1), 57–70. doi:10.1039/C7EE02110K

Ouhibi, A., Raouafi, A., Lorrain, N., Guendouz, M., Raouafi, N., & Moadhen, A. (2021). Functionalized SERS substrate based on silicon nanowires for rapid detection of prostate specific antigen. *Sensors and Actuators. B, Chemical*, *330*, 129352. doi:10.1016/j.snb.2020.129352

Ozeryanaya, I. N. (1985, March). Corrosion of metals by molten salts in heat-treatment processes. *Metal Science and Heat Treatment, 27*(3), 184–188. doi:10.1007/BF00699649

Özgür, Ü., Alivov, Y. I., Liu, C., Teke, A., Reshchikov, M. A., Doğan, S., Avrutin, V., Cho, S.-J., & Morkoç, H. (2005). A comprehensive review of ZnO materials and devices. *Journal of Applied Physics, 98*(4), 041301. doi:10.1063/1.1992666

Pal, B., Yang, S., Ramesh, S., Thangadurai, V., & Jose, R. (2019). Electrolyte selection for supercapacitive devices: A critical review. *Nanoscale Advances, 1*(10), 3807–3835. doi:10.1039/C9NA00374F PMID:36132093

Pan, H. (2016). Tension-enhanced hydrogen evolution reaction on vanadium disulfide monolayer. *Nanoscale Research Letters, 11*(1).

Panda. (n.d.). *Review: Environmental friendly lead-free piezoelectric materials.* . doi:10.1007/s10853-009-3643-0

Panda, P. K., & Sahoo, B. (2015). PZT to lead free piezo ceramics: A review. *Ferroelectrics, 474*(1), 128–143. doi:10.1080/00150193.2015.997146

Pandit, B., Rondiya, S. R., Cross, R. W., Dzade, N. Y., & Sankapal, B. R. (2022). Vanadium telluride nanoparticles on MWCNTs prepared by successive ionic layer adsorption and reaction for solid-state supercapacitor. *Chemical Engineering Journal, 429*, 132505. doi:10.1016/j.cej.2021.132505

Pandya, K. S., Veerraju, C., & Naik, N. K. (2011). Hybrid composites made of carbon and glass woven fabrics under quasi-static loading. *Materials & Design, 32*(7), 4094–4099. doi:10.1016/j.matdes.2011.03.003

Pang. (2020). Vanadium disulfide for ultrafast photonic application. *Nanotechnology, 32*(1).

Papadopoulos, A. I., & Seferlis, P. (Eds.). (2017). *Process systems and materials for CO2 capture: modelling, design, control and Integration.* John Wiley & Sons. doi:10.1002/9781119106418

Paper, C., Anodic, S., Wang, D., Li, F., Zhao, J., Ren, W., Chen, Z., Tan, J., Wu, Z., Gentle, I., Lu, G. Q., & Cheng, H. (2009). Fabrication of Graphene / Polyaniline. *ACS Nano, 3*(7), 1745–1752. PMID:19489559

Parekh, A., Chaturvedi, G., & Dutta, A. (2023). Sustainability analyses of CO2 sequestration and CO2 utilization as competing options for mitigating CO2 emissions. *Sustainable Energy Technologies and Assessments, 55*, 102942. doi:10.1016/j.seta.2022.102942

Parhizi, M., & Jain, A. (2019). The impact of thermal properties on performance of phase change based energy storage systems. *Applied Thermal Engineering, 162*, 114154. doi:10.1016/j.applthermaleng.2019.114154

Park, J. H., Berardi, U., Chang, S. J., Wi, S., Kang, Y., & Kim, S. (2021). Energy retrofit of PCM-applied apartment buildings considering building orientation and height. *Energy, 222*, 119877. doi:10.1016/j.energy.2021.119877

Park, J. H., Ok, Y. S., Kim, S. H., Cho, J. S., Heo, J. S., Delaune, R. D., & Seo, D. C. (2016). Competitive adsorption of heavy metals onto sesame straw biochar in aqueous solutions. *Chemosphere, 142*, 77–83. doi:10.1016/j.chemosphere.2015.05.093 PMID:26082184

Park, J., Rubiera Landa, H. O., Kawajiri, Y., Realff, M. J., Lively, R. P., & Sholl, D. S. (2019). How well do approximate models of adsorption-based CO2 capture processes predict results of detailed process models? *Industrial & Engineering Chemistry Research, 59*(15), 7097–7108. doi:10.1021/acs.iecr.9b05363

Park, N. G. (2015). Perovskite solar cells: An emerging photovoltaic technology. *Materials Today, 18*(2), 65–72. doi:10.1016/j.mattod.2014.07.007

Park, S. E., & Shrout, T. R. (1997). Ultrahigh strain and piezoelectric behavior in relaxor based ferroelectric single crystals. *Journal of Applied Physics, 82*(4), 1804–1811. doi:10.1063/1.365983

Park, W. I., & Yi, G.-C. (2004). Electroluminescence in n-ZnO Nanorod Arrays Vertically Grown on p-GaN. *Advanced Materials, 16*(1), 87–90. doi:10.1002/adma.200305729

Parsons & Alotaibi. (2023). The Application of Transition Metal Sulfide Nanomaterials and Their Composite Nanomaterials in the Electrocatalytic Reduction of CO2: A Review. *Applied Sciences, 13*(5).

Parwej, Akhtar, & Perwej. (2018). A Close-Up View About Spark in Big Data Jurisdiction. *International Journal of Engineering Research and Application, 8*(1), 26-41. 10.9790/9622-0801022641

Parwej, D. (2019). An empirical analysis of web of things (WOT). *International Journal of Advanced Research in Computer Science, 10*(3), 32–40. doi:10.26483/ijarcs.v10i3.6434

Patel & Patel. (2022). Scheffler reflector for cooking application: a review. *International Journal of Ambient Energy, 43*(1), 2606–2618. . doi:10.1080/01430750.2020.1758778

Patil, S. J., Pujari, R. B., Hou, T., Park, J., & Lee, D.-W. (2020). Supercapacitive performance of vanadium sulfide deposited on stainless steel mesh: Effect of etching. *Micro and Nano Systems Letters, 8*(1), 1–7. doi:10.1186/s40486-020-00110-3

Patolsky, F., Zheng, G., & Lieber, C. M. (2006). Nanowire sensors for medicine and the life sciences. *Nanomedicine (London), 1*(1), 51–65. doi:10.2217/17435889.1.1.51 PMID:17716209

Pattipaka, S., James, A. R., & Dobbidi, P. (2018). Enhanced dielectric and piezoelectric properties of BNT-KNNG piezoelectric ceramics. *Journal of Alloys and Compounds, 765*, 1195–1208. doi:10.1016/j.jallcom.2018.06.138

Peh, S. B., Farooq, S., & Zhao, D. (2023). Techno-economic analysis of MOF-based adsorption cycles for postcombustion CO2 capture from wet flue gas. *Chemical Engineering Science, 268*, 118390. doi:10.1016/j.ces.2022.118390

Pell, W. G., & Conway, B. E. (2004). Peculiarities and requirements of asymmetric capacitor devices based on combination of capacitor and battery-type electrodes. *Journal of Power Sources, 136*(2), 334–345. doi:10.1016/j.jpowsour.2004.03.021

Pell, W. G., Conway, B. E., Adams, W. A., & De Oliveira, J. (1999). Electrochemical efficiency in multiple discharge/recharge cycling of supercapacitors in hybrid EV applications. *Journal of Power Sources, 80*(1), 134–141. doi:10.1016/S0378-7753(98)00257-2

Perera, S. D., Ding, X., Bhargava, A., Hovden, R., Nelson, A., Kourkoutis, L. F., & Robinson, R. D. (2015). Enhanced Supercapacitor Performance for Equal Co-Mn Stoichiometry in Colloidal Co 3-x Mn x O 4 Nanoparticles, in Additive-Free Electrodes. *Chemistry of Materials, 27*(23), 7861–7873. doi:10.1021/acs.chemmater.5b02106

Perez, J. M., Josephson, L., & Weissleder, R. (2004). Use of magnetic nanoparticles as nanosensors to probe for molecular interactions. *ChemBioChem, 5*(3), 261–264. doi:10.1002/cbic.200300730 PMID:14997516

Perwej, Omer, Sheta, Harb, & Adrees. (2019). The Future of Internet of Things (IoT) and Its Empowering Technology. *International Journal of Engineering Science and Computing, 9*(3), 20192–20203.

Perwej, Y., Parwej, F., Mohamed Hassan, M. M., & Akhtar, N. (2019). The Internet-of-Things (IoT) Security : A Technological Perspective and Review. *International Journal of Scientific Research in Computer Science, Engineering and Information Technology*, 462–482. . doi:10.32628/CSEIT195193

Perwej, Y. (2016). A Literature Review of the Human Body as a Communication Medium using RedTacton. *Communications on Applied Electronics, 4*(9), 7–17. doi:10.5120/cae2016652161

Perwej, Y. (2017). The Next Generation of Wireless Communication Using Li-Fi (Light Fidelity) Technology. *Journal of Computer Networks, 4*(1), 20–29. doi:10.12691/jcn-4-1-3

Perwej, Y. (2017, March). An Experiential Study of the Big Data. *Science and Education Publishing*, *4*(1), 14–25. doi:10.12691/iteces-4-1-3

Perwej, Y. (2019). The Hadoop Security in Big Data: A Technological Viewpoint and Analysis. *International Journal of Scientific Research in Computer Science and Engineering*, *7*(3), 1–14. doi:10.26438/ijsrcse/v7i3.1014

Perwej, Y., Ahmed, M., Kerim, B., & Ali, H. (2019). An Extended Review on Internet of Things (IoT) and its Promising Applications. *Communications on Applied Electronics*, *7*(26), 8–22. doi:10.5120/cae2019652812

Perwej, Y., Haq, K., Parwej, F., & M, M. (2019). The Internet of Things (IoT) and its Application Domains. *International Journal of Computer Applications*, *182*(49), 36–49. doi:10.5120/ijca2019918763

Perwej, Y., Kerim, B., Adrees, M. S., & Sheta, O. E. (2017). An Empirical Exploration of the Yarn in Big Data. *International Journal of Applied Information Systems*, *12*(9), 19–29. doi:10.5120/ijais2017451730

Petrovic & Petrovic. (2021). Nickel–cadmium batteries. *Battery Technology Crash Course: A Concise Introduction,* 73-88.

Pham, K. N., Hussain, A., Ahn, C. W., Ill, W. K., Jeong, S. J., & Lee, J. S. (2010). Giant strain in Nb-doped Bi0. 5 (Na0. 82K0. 18) 0.5 TiO3 lead-free electromechanical ceramics. *Materials Letters*, *64*(20), 2219–2222. doi:10.1016/j.matlet.2010.07.048

Phillips, J. D. (2021). Energy Harvesting in Nanosystems: Powering the Next Generation of the Internet of Things. *Frontiers in Nanotechnology*, *3*, 633931. Advance online publication. doi:10.3389/fnano.2021.633931

Piccirillo, C., Binions, R., & Parkin, I. P. (2007). Synthesis and functional properties of vanadium oxides: V2O3, VO2, and V2O5 deposited on glass by aerosol-assisted CVD. *Chemical Vapor Deposition*, *13*(4), 145–151. doi:10.1002/cvde.200606540

Pierre, A. C. (2020). *Introduction to sol-gel processing*. Springer Nature. doi:10.1007/978-3-030-38144-8

Pijls & De Bruyne. (n.d.). *Coronary pressure measurement and fractional flow reserve.* Academic Press.

Pijls, N. H. J., de Bruyne, B., Peels, K., van der Voort, P. H., Bonnier, H. J. R. M., Bartunek, J., & Koolen, J. J. (2009). *Measurement of Fractional Flow Reserve to Assess the Functional Severity of Coronary-Artery Stenoses.* doi:10.1056/NEJM199606273342604

Pinho, R., Vilarinho, R., Moreira, J. A., Zorro, F., Ferreira, P., Ivanov, M., Tkach, A., Costa, M. E., & Vilarinho, P. M. (2023). Stress induced effects on piezoelectric polycrystalline potassium sodium niobate thin films. *Journal of Materials Chemistry. C, Materials for Optical and Electronic Devices*, *11*(23), 7758–7771. doi:10.1039/D2TC05538D

Podyacheva, O. Yu., & Ismagilov, Z. R. (2015). Nitrogen-doped carbon nanomaterials: To the mechanism of growth, electrical conductivity and application in catalysis. *Catalysis Today*, *249*, 12–22. doi:10.1016/j.cattod.2014.10.033

Polymer. (n.d.). https://www.britannica.com/science/polymer

Potyrailo, R. A. (2016). Multivariable Sensors for Ubiquitous Monitoring of Gases in the Era of Internet of Things and Industrial Internet. *Chemical Reviews*, *116*(19), 11877–11923. doi:10.1021/acs.chemrev.6b00187 PMID:27602947

Pouillot, C., Fournier, S., Glasenapp, J., Rambaud, G., Bougrini, K., Vi Fane, R., Geyer, C., & Adjedj, J. (2018). Pressure wire versus microcatheter for FFR measurement: A head-to-head comparison. *Archives of Cardiovascular Diseases. Supplements*, *10*(1), 25–26. doi:10.1016/j.acvdsp.2017.11.048 PMID:28804057

Pourkheirollah, H., Keskinen, J., Mäntysalo, M., & Lupo, D. (2023). Simplified exponential equivalent circuit models for prediction of printed supercapacitor's discharge behavior - Simulations and experiments. *Journal of Power Sources*, *567*(February), 232932. Advance online publication. doi:10.1016/j.jpowsour.2023.232932

Prakash, K., Ali, M., Siddique, M. N. I., Chand, A. A., Kumar, N. M., Dong, D., & Pota, H. R. (2022). A review of battery energy storage systems for ancillary services in distribution grids: Current status, challenges and future directions. *Frontiers in Energy Research*, *10*, 971704. doi:10.3389/fenrg.2022.971704

Prytulak, J. (2016). Stable vanadium isotopes as a redox proxy in magmatic systems. *Geochemical Perspectives Letters.*

Ptaszkowska-Koniarz, M., Goscianska, J., & Pietrzak, R. (2018a). Removal of rhodamine B from water by modified carbon xerogels. *Colloids and Surfaces. A, Physicochemical and Engineering Aspects*, *543*(January), 109–117. doi:10.1016/j.colsurfa.2018.01.057

Ptaszkowska-Koniarz, M., Goscianska, J., & Pietrzak, R. (2018b). Synthesis of carbon xerogels modified with amine groups and copper for efficient adsorption of caffeine. *Chemical Engineering Journal*, *345*(March), 13–21. doi:10.1016/j.cej.2018.03.132

Pullar, R. C., Zhang, Y., Chen, L., Yang, S., Evans, J. R., Salak, A. N., Kiselev, D. A., Kholkin, A. L., Ferreira, V. M., & Alford, N. M. (2009). Dielectric measurements on a novel Ba 1− x Ca x TiO 3 (BCT) bulk ceramic combinatorial library. *Journal of Electroceramics*, *22*(1-3), 245–251. doi:10.1007/s10832-008-9430-5

Pu, T., Chen, H., Xing, J., Luo, Y., Fan, S., Liu, H., Chen, Q., & Zhu, J. (2022). High piezoelectricity of low-temperature sintered Li2CO3-added PNN–PZT relaxor ferroelectrics. *Journal of Materials Science Materials in Electronics*, *33*(8), 4819–4830. doi:10.1007/s10854-021-07671-y

Putungan, D. B., Lin, S.-H., & Kuo, J.-L. (2016). Metallic VS2 monolayer polytypes as potential sodium-ion battery anode via ab initio random structure searching. *ACS Applied Materials & Interfaces*, *8*(29), 18754–18762. doi:10.1021/acsami.6b03499 PMID:27373121

Pu, X., Song, T., Tang, L., Tao, Y., Cao, T., Xu, Q., Liu, H., Wang, Y. G., & Xia, Y. (2019). Rose-like vanadium disulfide coated by hydrophilic hydroxyvanadium oxide with improved electrochemical performance as cathode material for aqueous zinc-ion batteries. *Journal of Power Sources*, *437*, 226917. doi:10.1016/j.jpowsour.2019.226917

Qian, L.-P., Xu, L.-Y., Alrefaei, Y., Wang, T., Ishida, T., & Dai, J.-G. (2022). Artificial alkali-activated aggregates developed from wastes and by-products: A state-of-the-art review. *Resources, Conservation and Recycling*, *177*, 105971. doi:10.1016/j.resconrec.2021.105971

Qie, L., & Manthiram, A. (2016). High-energy-density lithium–sulfur batteries based on blade-cast pure sulfur electrodes. *ACS Energy Letters*, *1*(1), 46–51. doi:10.1021/acsenergylett.6b00033

Qi, H., Wang, L., Zuo, T., Deng, S., Li, Q., Liu, Z.-H., Hu, P., & He, X. (2020). Hollow structure VS2@ reduced graphene oxide (RGO) architecture for enhanced sodium-ion battery performance. *ChemElectroChem*, *7*(1), 78–85. doi:10.1002/celc.201901626

Qin, P., Tanaka, S., Ito, S., Tetreault, N., Manabe, K., Nishino, H., Nazeeruddin, M. K., & Grätzel, M. (2014). Inorganic hole conductor-based lead halide perovskite solar cells with 12.4% conversion efficiency. *Nature Communications*, *5*(1), 3834. doi:10.1038/ncomms4834 PMID:24815001

Qin, Z., Wang, Z., & Zhao, J. (2022). Computational screening of single-atom catalysts supported by VS 2 monolayers for electrocatalytic oxygen reduction/evolution reactions. *Nanoscale*, *14*(18), 6902–6911. doi:10.1039/D2NR01671K PMID:35446333

Qiu, H., Lv, L., Pan, B. C., Zhang, Q. J., Zhang, W. M., & Zhang, Q. X. (2009). Critical review in adsorption kinetic models. *Journal of Zhejiang University. Science A*, *10*(5), 716–724. doi:10.1631/jzus.A0820524

Qi, X., Xu, H., Wang, X.-Q., Ma, W.-G., Qiu, C., An, M., Zhang, G., Wang, F., Zhang, X., & Bermak, A. (2020). Effective surface emissivity and heat dissipation among integrated bamboo-like super-black vertical carbon nanotube array electrodes in silicon via holes. *Carbon, 158*, 846–856. doi:10.1016/j.carbon.2019.11.068

Qi, Z., & Koenig, G. M. Jr. (2017). Review Article: Flow battery systems with solid electroactive materials. *Journal of Vacuum Science & Technology B. Journal of Vacuum Science and Technology. B, Nanotechnology & Microelectronics : Materials, Processing, Measurement, & Phenomena : JVST B, 35*(4), 040801. doi:10.1116/1.4983210

Qu, Y., Shao, M., Shao, Y., Yang, M., Xu, J., Kwok, C. T., Shi, X., Lu, Z., & Pan, H. (2017). Ultra-high electrocatalytic activity of VS 2 nanoflowers for efficient hydrogen evolution reaction. *Journal of Materials Chemistry. A, Materials for Energy and Sustainability, 5*(29), 15080–15086. doi:10.1039/C7TA03172F

Rached, A., Wederni, M. A., Khirouni, K., Alaya, S., Martín-Palma, R. J., & Dhahri, J. (2021). Structural, optical and electrical properties of barium titanate. *Materials Chemistry and Physics, 267*, 124600. doi:10.1016/j.matchemphys.2021.124600

Raganati, F., Chirone, R., & Ammendola, P. (2020). CO2 capture by temperature swing adsorption: Working capacity as affected by temperature and CO2 partial pressure. *Industrial & Engineering Chemistry Research, 59*(8), 3593–3605. doi:10.1021/acs.iecr.9b04901

Ragini, R., Ranjan, R., Mishra, S. K., & Pandey, D. (2002). Room temperature structure of Pb (Zr x Ti 1− x O 3) around the morphotropic phase boundary region: A Rietveld study. *Journal of Applied Physics, 92*(6), 3266–3274. doi:10.1063/1.1483921

Rahemipoor, S., Hasany, M., Mehrali, M., Almdal, K., Ranjbar, N., & Mehrali, M. (2023). Phase change materials incorporation into 3D printed geopolymer cement: A sustainable approach to enhance the comfort and energy efficiency of buildings. *Journal of Cleaner Production, 417*, 138005. doi:10.1016/j.jclepro.2023.138005

Rahjoo, M., Goracci, G., Gaitero, J. J., Martauz, P., Rojas, E., & Dolado, J. S. (2022). Thermal Energy Storage (TES) Prototype Based on Geopolymer Concrete for High-Temperature Applications. *Materials (Basel), 15*(20), 7086. doi:10.3390/ma15207086 PMID:36295154

Rahman, J. U., Hussain, A., Maqbool, A., Malik, R. A., Kim, M. S., & Kim, M. H. (2015). Effect of sintering temperature on the electromechanical properties of 0.945 Bi 0.5 Na 0.5 TiO 3-0.055 BaZrO 3 ceramics. *Journal of the Korean Physical Society, 66*(7), 1072–1076. doi:10.3938/jkps.66.1072

Rajagopalan. (2020). Advancements and challenges in potassium ion batteries: a comprehensive review. *Advanced Functional Materials, 30*(12).

Rakita, Y., Bar-Elli, O., Meirzadeh, E., Kaslasi, H., Peleg, Y., Hodes, G., Lubomirsky, I., Oron, D., Ehre, D., & Cahen, D. (2017). Tetragonal $CH_3NH_3PbI_3$ is ferroelectric. *Proceedings of the National Academy of Sciences of the United States of America, 114*(28), E5504–E5512. doi:10.1073/pnas.1702429114 PMID:28588141

Ramajo, L. A., Taub, J., & Castro, M. S. (2014). Effect of ZnO addition on the structure, microstructure and dielectric and piezoelectric properties of K0. 5Na0. 5NbO3 ceramics. *Materials Research, 17*(3), 728–733. doi:10.1590/S1516-14392014005000048

Ramam, K., & Lopez, M. (2007). Effect of strontium doping on dielectric, ferroelectric and piezoelectric properties of PLZTN ceramics. *Materials Science and Engineering B, 145*(1-3), 41–47. doi:10.1016/j.mseb.2007.09.085

Ramirez, F. D., Reddy, V. Y., Viswanathan, R., Hocini, M., & Jaïs, P. (2020). Emerging Technologies for Pulmonary Vein Isolation. *Circulation Research, 127*(1), 170–183. doi:10.1161/CIRCRESAHA.120.316402 PMID:32716722

Rao, B. N., Datta, R., Chandrashekaran, S. S., Mishra, D. K., Sathe, V., Senyshyn, A., & Ranjan, R. (2013). Local structural disorder and its influence on the average global structure and polar properties in Na 0.5 Bi 0.5 TiO 3. *Physical Review B: Condensed Matter and Materials Physics*, *88*(22), 224103. doi:10.1103/PhysRevB.88.224103

Rao, B. N., Fitch, A. N., & Ranjan, R. (2013). Ferroelectric-ferroelectric phase coexistence in Na 1/2 Bi 1/2 TiO 3. *Physical Review B: Condensed Matter and Materials Physics*, *87*(6), 060102. doi:10.1103/PhysRevB.87.060102

Rasta, I. M., & Suamir, I. N. (2019). Study on thermal properties of bio-PCM candidates in comparison with propylene glycol and salt based PCM for sub-zero energy storage applications. *IOP Conference Series. Materials Science and Engineering*, *494*(1), 12024. doi:10.1088/1757-899X/494/1/012024

Rathore, P. K. S., & Shukla, S. K. (2021). Enhanced thermophysical properties of organic PCM through shape stabilization for thermal energy storage in buildings: A state of the art review. *Energy and Building*, *236*, 110799. doi:10.1016/j.enbuild.2021.110799

Rathore, P. K. S., Shukla, S. K., & Gupta, N. K. (2020). Potential of microencapsulated PCM for energy savings in buildings: A critical review. *Sustainable Cities and Society*, *53*, 101884. doi:10.1016/j.scs.2019.101884

Raut, A. N., Murmu, A. L., & Alomayri, T. (2023). Physico-Mechanical and thermal behavior of prolong heat Cured geopolymer blocks. *Construction & Building Materials*, *370*, 130309. doi:10.1016/j.conbuildmat.2023.130309

Ravindra, S., Murali Mohan, Y., Narayana Reddy, N., & Mohana Raju, K. (2010). Fabrication of antibacterial cotton fibres loaded with silver nanoparticles via "Green Approach.". *Colloids and Surfaces. A, Physicochemical and Engineering Aspects*, *367*(1–3), 31–40. doi:10.1016/j.colsurfa.2010.06.013

Reddy, D. S., Khan, M. K., Alam, M. Z., & Rashid, H. (2018, March). Design charts for Scheffler reflector. *Solar Energy*, *163*, 104–112. doi:10.1016/j.solener.2018.01.081

Reddy, V. Y., Neuzil, P., Koruth, J. S., Petru, J., Funosako, M., Cochet, H., Sediva, L., Chovanec, M., Dukkipati, S. R., & Jais, P. (2019). Pulsed Field Ablation for Pulmonary Vein Isolation in Atrial Fibrillation. *Journal of the American College of Cardiology*, *74*(3), 315–326. doi:10.1016/j.jacc.2019.04.021 PMID:31085321

Redler, G., Jones, K. C., Templeton, A., Bernard, D., Turian, J., & Chu, J. C. H. (2018). Compton scatter imaging: A promising modality for image guidance in lung stereotactic body radiation therapy. *Medical Physics*, *45*(3), 1233–1240. doi:10.1002/mp.12755 PMID:29360151

Redondo, E., Fevre, L. W. L., Fields, R., Todd, R., Forsyth, A. J., & Dryfe, R. A. W. (2020). Enhancing supercapacitor energy density by mass-balancing of graphene composite electrodes. *Electrochimica Acta, 360.* doi:10.1016/j.electacta.2020.136957

Regin, A. F., Solanki, S. C., & Saini, J. S. (2008). Heat transfer characteristics of thermal energy storage system using PCM capsules: A review. *Renewable & Sustainable Energy Reviews*, *12*(9), 2438–2458. doi:10.1016/j.rser.2007.06.009

Rehman, A. U., Sheikh, S. R., Kausar, Z., Grimes, M., & McCormack, S. J. (2022). Experimental Thermal Response Study of Multilayered, Encapsulated, PCM-Integrated Building Construction Materials. *Energies*, *15*(17), 6356. doi:10.3390/en15176356

Reitz, W. (1997). Ceramic processing and sintering by MN Rahaman. *Materials and Manufacturing Processes*, *12*(3), 555–556. doi:10.1080/10426919708935164

Renault, S., Brandell, D., Gustafsson, T., & Edström, K. (2013). Improving the electrochemical performance of organic Li-ion battery electrodes. *Chemical Communications*, *49*(19), 1945–1947. doi:10.1039/c3cc39065a PMID:23370005

Ren, Y., Liu, T., Shen, Y., Lin, Y., & Nan, C. W. (2016). Chemical compatibility between garnet-like solid state electrolyte Li6. 75La3Zr1. 75Ta0. 25O12 and major commercial lithium battery cathode materials. *Journal of Materiomics*, *2*(3), 256–264. doi:10.1016/j.jmat.2016.04.003

Reuben, Whitten, Sarfati, & Kraiss. (2007). Increasing use of endovascular therapy in acute arterial injuries: Analysis of the National Trauma Data Bank. *J. Vasc. Surg.*, *46*, 1222-1226. doi:10.1016/j.jvs.2007.08.023

Reyez-Araiza, J. L., Pineda-Piñón, J., López-Romero, J. M., Gasca-Tirado, J. R., Arroyo Contreras, M., Jáuregui Correa, J. C., Apátiga-Castro, L. M., Rivera-Muñoz, E. M., Velazquez-Castillo, R. R., Pérez Bueno, J., & Manzano-Ramirez, A. (2021). Thermal energy storage by the encapsulation of phase change materials in building elements—A review. *Materials (Basel)*, *14*(6), 1420. doi:10.3390/ma14061420 PMID:33804151

Rheaume, P., Chen, J., & Casey, P. (2010). Open vs endovascular repair of blunt traumatic thoracic aortic injuries. *Journal of Vascular Surgery*, *51*(3), 763–769. doi:10.1016/j.jvs.2009.12.014 PMID:20206817

Riaz, M., Song, J., Nur, O., Wang, Z. L., & Willander, M. (2011). Study of the piezoelectric power generation of ZnO nanowire arrays grown by different methods. *Advanced Functional Materials*, *21*(4), 628–633. doi:10.1002/adfm.201001203

Richard, D., Martinez, J. M., Mizrahi, M., Andrini, L., & Rendtorff, N. M. (2022). Assessment of structural order indices in kaolinites: A multi-technique study including EXAFS. *Journal of Electron Spectroscopy and Related Phenomena*, *254*, 147128. doi:10.1016/j.elspec.2021.147128

Rietveld, H. M. (1969). A profile refinement method for nuclear and magnetic structures. *Journal of Applied Crystallography*, *2*(2), 65–71. doi:10.1107/S0021889869006558

Ringgaard, E., & Wurlitzer, T. (2005). Lead-free piezoceramics based on alkali niobates. *Journal of the European Ceramic Society*, *25*(12), 2701–2706. doi:10.1016/j.jeurceramsoc.2005.03.126

Robert, B., & Brown, E. B. (2006). Book Review: Wind Energy Explained: Theory, Design and Application. In Wind Engineering (Vol. 30, Issue 2). doi:10.1260/030952406778055054

Roberts, S. (1947). Dielectric and piezoelectric properties of barium titanate. *Physical Review*, *71*(12), 890–895. doi:10.1103/PhysRev.71.890

Rogers, J. A. (2009). Materials and mechanics for stretchable electronics - from electronic eye cameras to conformal brain monitors. *TRANSDUCERS 2009 - 2009 International Solid-State Sensors, Actuators and Microsystems Conference.* 10.1109/SENSOR.2009.5285765

Röhm, H., Leonhard, T., Hoffmann, M. J., & Colsmann, A. (2017). Ferroelectric domains in methylammonium lead iodide perovskite thin-films. *Energy & Environmental Science*, *10*(4), 950–955. doi:10.1039/C7EE00420F

Rojac, T., Bencan, A., Malic, B., Tutuncu, G., Jones, J. L., Daniels, J. E., & Damjanovic, D. (2014). BiFeO 3 ceramics: Processing, electrical, and electromechanical properties. *Journal of the American Ceramic Society*, *97*(7), 1993–2011. doi:10.1111/jace.12982

Romero-García, A. G., Ramírez-Corona, N., Sánchez-Ramírez, E., Alcocer-García, H., De Blasio, C., & Segovia-Hernández, J. G. (2022). Sustainability assessment in the CO2 capture process: Multi-objective optimization. *Chemical Engineering and Processing*, *182*, 109207. doi:10.1016/j.cep.2022.109207

Rong, Y., Hu, Y., Mei, A., Tan, H., Saidaminov, M. I., Seok, S. I., McGehee, M. D., Sargent, E. H., & Han, H. (2018). Challenges for commercializing perovskite solar cells. *Science*, *361*(6408), 8235. doi:10.1126/science.aat8235 PMID:30237326

Roy, A. K., Ahmad, Z., Prasad, A., & Prasad, K. (2012). Concentration dependent dielectric properties of Barium Titanate / Polyvenylidene Fluoride (PVDF) and. *Advanced Materials Research*, *1*(4), 285–297. doi:10.12989/amr.2012.1.4.285

Rubio-Marcos, F., Marchet, P., Merle-Méjean, T., & Fernandez, J. F. (2010). Role of sintering time, crystalline phases and symmetry in the piezoelectric properties of lead-free KNN-modified ceramics. *Materials Chemistry and Physics*, *123*(1), 91–97. doi:10.1016/j.matchemphys.2010.03.065

Russell, C., Welch, K., Jarvius, J., Cai, Y., Brucas, R., Nikolajeff, F., Svedlindh, P., & Nilsson, M. (2014). Gold nanowire based electrical DNA detection using rolling circle amplification. *ACS Nano*, *8*(2), 1147–1153. doi:10.1021/nn4058825 PMID:24433087

Rybyanets, A. N. (2011). Porous piezoceramics: Theory, technology, and properties. *IEEE Transactions on Ultrasonics, Ferroelectrics, and Frequency Control*, *58*(7), 1492–1507. doi:10.1109/TUFFC.2011.1968 PMID:21768033

Ryms, M., & Klugmann-Radziemska, E. (2019). Possibilities and benefits of a new method of modifying conventional building materials with phase-change materials (PCMs). *Construction & Building Materials*, *211*, 1013–1024. doi:10.1016/j.conbuildmat.2019.03.277

Saadeldin, M. M., Desouky, O. A., Ibrahim, M., Khalil, G. E., & Helali, M. Y. (2019). Investigation of structural and electrical properties of ZnO varistor samples doped with different additives. *NRIAG J. Astron. Geophys.*, *7*(2), 201–207. doi:10.1016/j.nrjag.2018.06.002

Sabbatini, A., Vidal, L., Pettinari, C., Sobrados, I., & Rossignol, S. (2017). Control of shaping and thermal resistance of metakaolin-based geopolymers. *Materials & Design*, *116*, 374–385. doi:10.1016/j.matdes.2016.12.039

Sadh, P. K., Duhan, S., & Duhan, J. S. (2018). Agro-industrial wastes and their utilization using solid state fermentation: a review. In Bioresources and Bioprocessing (Vol. 5, Issue 1). Springer Science and Business Media Deutschland GmbH. doi:10.1186/s40643-017-0187-z

Sahoo, B., & Panda, P. K. (2013). Effect of lanthanum, neodymium on piezoelectric, dielectric and ferroelectric properties of PZT. *Journal of Advanced Ceramics*, *2*(1), 37–41. doi:10.1007/s40145-013-0039-z

Sai Ramesh, A., Vigneshwar, S., Vickram, S., Manikandan, S., Subbaiya, R., Karmegam, N., & Kim, W. (2023). Artificial intelligence driven hydrogen and battery technologies – A review. *Fuel*, *337*, 126862. doi:10.1016/j.fuel.2022.126862

Saidaminov, M. I., Abdelhady, A. L., Murali, B., Alarousu, E., Burlakov, V. M., Peng, W., Dursun, I., Wang, L., He, Y., Maculan, G., Goriely, A., Wu, T., Mohammed, O. F., & Bakr, O. M. (2015). Highquality bulk hybrid perovskite single crystals within minutes by inverse temperature crystallization. *Nature Communications*, *6*(1), 7586. doi:10.1038/ncomms8586 PMID:26145157

Saikia, B. K., Benoy, S. M., Bora, M., Tamuly, J., Pandey, M., & Bhattacharya, D. (2020). A brief review on supercapacitor energy storage devices and utilization of natural carbon resources as their electrode materials. *Fuel*, *282*(July), 118796. doi:10.1016/j.fuel.2020.118796

Saito, Y., & Takao, H. (2006). High performance lead-free piezoelectric ceramics in the (K, Na) NbO3-LiTaO3 solid solution system. *Ferroelectrics*, *338*(1), 17–32. doi:10.1080/00150190600732512

Saito, Y., Takao, H., Tani, T., Nonoyama, T., Takatori, K., Homma, T., Nagaya, T., & Nakamura, M. (2004). Lead-free piezoceramics. *Nature*, *432*(7013), 84–87. doi:10.1038/nature03028 PMID:15516921

Salim, H. K., Padfield, R., Hansen, S. B., Mohamad, S. E., Yuzir, A., Syayuti, K., Tham, M. H., & Papargyropoulou, E. (2018). Global trends in environmental management system and ISO14001 research. *Journal of Cleaner Production*, *170*, 645–653. doi:10.1016/j.jclepro.2017.09.017

Salmeron, M., & Schlogl, R. (2008). Ambient pressure photoelectron spectroscopy: A new tool for surface science and nanotechnology. *Surface Science Reports*, *63*(4), 169–199. doi:10.1016/j.surfrep.2008.01.001

Salunkhe, P. B., & Jaya Krishna, D. (2017, August). Investigations on latent heat storage materials for solar water and space heating applications. *Journal of Energy Storage*, *12*, 243–260. doi:10.1016/j.est.2017.05.008

Sanchez-Silva, J. M., Collins-Martínez, V. H., Padilla-Ortega, E., Aguilar-Aguilar, A., Labrada-Delgado, G. J., Gonzalez-Ortega, O., Palestino-Escobedo, G., & Ocampo-Pérez, R. (2022). Characterization and transformation of nanche stone (Byrsonima crassifolia) in an activated hydrochar with high adsorption capacity towards metformin in aqueous solution. *Chemical Engineering Research & Design*, *183*, 580–594. doi:10.1016/j.cherd.2022.05.054

Santosh Junare, S., Vilas Zamre, S., & Madhukar Aware, M. (2017). *Scheffler Dish and its Applications.* Available: http://www.ijritcc.org

Sarkar, A., Ghosh, A. B., Saha, N., Bhadu, G. R., & Adhikary, B. (2018). Newly designed amperometric biosensor for hydrogen peroxide and glucose based on vanadium sulfide nanoparticles. *ACS Applied Nano Materials*, *1*(3), 1339–1347. doi:10.1021/acsanm.8b00076

Sarker, T. R., Pattnaik, F., Nanda, S., Dalai, A. K., Meda, V., & Naik, S. (2021). Hydrothermal pretreatment technologies for lignocellulosic biomass: A review of steam explosion and subcritical water hydrolysis. *Chemosphere*, *284*, 131372. doi:10.1016/j.chemosphere.2021.131372 PMID:34323806

Saski, M., Prochowicz, D., Marynowski, W., & Lewiński, J. (2019). Mechanosynthesis, optical, and morphological properties of MA, FA, Cs-SnX_3 (X = I, Br) and phase-pure mixed-halide $MASnI_xBr_{3-x}$ perovskites. *European Journal of Inorganic Chemistry*, *22*(22), 2680–2684. doi:10.1002/ejic.201801506

Sawadogo, M., Duquesne, M., Belarbi, R., Hamami, A. E. A., & Godin, A. (2021). Review on the integration of phase change materials in building envelopes for passive latent heat storage. *Applied Sciences (Basel, Switzerland)*, *11*(19), 9305. doi:10.3390/app11199305

Sawaguchi, E. (1953). Ferroelectricity versus antiferroelectricity in the solid solutions of PbZrO3 and PbTiO3. *Journal of the Physical Society of Japan*, *8*(5), 615–629. doi:10.1143/JPSJ.8.615

Sawaguchi, E., Shirane, G., & Takagi, Y. (1951). Phase transition in lead zirconate. *Journal of the Physical Society of Japan*, *6*(5), 333–339. doi:10.1143/JPSJ.6.333

Saxena, R., Rakshit, D., & Kaushik, S. C. (2019). Phase change material (PCM) incorporated bricks for energy conservation in composite climate: A sustainable building solution. *Solar Energy*, *183*, 276–284. doi:10.1016/j.solener.2019.03.035

Sayed, D. M., Taha, M. M., Ghanem, L. G., El-Deab, M. S., & Allam, N. K. (2020). Hybrid supercapacitors: A simple electrochemical approach to determine optimum potential window and charge balance. *Journal of Power Sources*, *480*(September), 229152. doi:10.1016/j.jpowsour.2020.229152

Scarmozzino, R., Gopinath, A., Pregla, R., & Helfert, S. (2000). Numerical techniques for modeling guided-wave photonic devices. *IEEE Journal of Selected Topics in Quantum Electronics*, *6*(1), 150–162. doi:10.1109/2944.826883

Scherrer, P. (1918). Bestimmung der Grosse und inneren Struktur von Kolloidteilchen mittels Rontgenstrahlen. *Nach Ges Wiss Gottingen*, *2*, 8–100.

Seah, M. P. (1984). A review of the analysis of surfaces and thin films by AES and XPS. *Vacuum*, *34*(3–4), 463–478. doi:10.1016/0042-207X(84)90084-8

Seibert, J. A. (1997). The AAPM/RSNA physics tutorial for residents. X-ray generators. *Radiographics*, *17*(6), 1533–1557. doi:10.1148/radiographics.17.6.9397462 PMID:9397462

Seibert, J. A., & Boone, J. M. (2005). X-ray imaging physics for nuclear medicine technologists. Part 2: X-ray interactions and image formation. *Journal of Nuclear Medicine Technology*, *33*(1), 3–18. PMID:15731015

Sekhar, B. C., Dhanalakshmi, B., Rao, B. S., Ramesh, S., Prasad, K. V., Rao, P. S., & Rao, B. P. (2021). Piezoelectricity and its applications. *Multifunct. Ferroelectr. Mater, 71.*

Sekijima, T., Itoh, H., Fujii, T., Wakino, K., & Okada, M. (2001). Influence of growth atmosphere on solubility limit of Ce3+ ions in Ce-substituted fibrous yttrium iron garnet single crystals. *Journal of Crystal Growth, 229*(1-4), 409–414. doi:10.1016/S0022-0248(01)01191-5

Sepúlveda-Cervantes, C. V., Soto-Regalado, E., Rivas-García, P., Loredo-Cancino, M., Cerino-Córdova, F. J., & García Reyes, R. B. (2018). Technical-environmental optimisation of the activated carbon production of an agroindustrial waste by means response surface and life cycle assessment. *Waste Management & Research, 36*(2), 121–130. doi:10.1177/0734242X17741680 PMID:29189111

Serafin, J., & Dziejarski, B. (2023). Application of isotherms models and error functions in activated carbon CO2 sorption processes. *Microporous and Mesoporous Materials, 354*, 112513. doi:10.1016/j.micromeso.2023.112513

Shafaatian, B., & Baharian, A. (2017). New zinc (II), copper (II), nickel (II), and vanadium (IV) complexes containing disulfide ligand: Synthesis, spectroscopic studies and calf thymus DNA interaction investigation. *Phosphorus, Sulfur, and Silicon and the Related Elements, 192*(10), 1102–1109. doi:10.1080/10426507.2017.1331232

Shang, W., Zhang, X., Zhang, M., Fan, Z., Sun, Y., Han, M., & Fan, L. (2014). The uptake mechanism and biocompatibility of graphene quantum dots with human neural stem cells. *Nanoscale, 6*(11), 5799–5806. doi:10.1039/C3NR06433F PMID:24740121

Shao, B., Qiu, J., Zhu, K., Pang, X., & Meng, Q. (2012). Tantalum influence on electrical properties of lead-free (K 0.4425 Na 0.52 Li 0.0375)(Nb 0.93− x Ta x Sb 0.07) O 3 piezoelectric ceramics. *Journal of Materials Science Materials in Electronics, 23*(4), 846–850. doi:10.1007/s10854-011-0504-9

Shao, Z., Wang, J., Jiang, Y., Zang, J., Wu, T., Ma, F., Qian, B., Wang, L., Hu, Y., & Ma, B. (2021). The performance of micropore-foamed geopolymers produced from industrial wastes. *Construction & Building Materials, 304*, 124636. doi:10.1016/j.conbuildmat.2021.124636

Shariati, M. (2018). The field effect transistor DNA biosensor based on ITO nanowires in label-free hepatitis B virus detecting compatible with CMOS technology. *Biosensors & Bioelectronics, 105*, 58–64. doi:10.1016/j.bios.2018.01.022 PMID:29355779

Sharker, S. M., & Do, M. (2021). Nanoscale carbon-polymer dots for theranostics and biomedical exploration. *Journal of Nanotheranostics, 2*(3), 118–130. doi:10.3390/jnt2030008

Sharma, Aroom, Naik, Gill, & Zhang. (n.d.). *Flexible Thin-Film PVDF-TrFE Based Pressure Sensor for Smart Catheter Applications.* doi:10.1007/s10439-012-0708-z

Sharma, A., Godara, S. K., Maji, P. K., & Srivastava, A. K. (2022). Influence of Temperature on Structural and Magnetic Properties of Gd3AlxFe5− xO12 (x= 2). *Crystal Research and Technology, 57*(2), 2100109. doi:10.1002/crat.202100109

Sharma, A., Godara, S. K., & Srivastava, A. K. (2022). Effect of Y3+, Bi3+, La3+ substitution on structural, optical and magnetic properties of gadolinium iron garnets. *Materials Today: Proceedings, 50*, 1867–1877. doi:10.1016/j.matpr.2021.09.234

Sharma, A., Godara, S. K., & Srivastava, A. K. (2022). Influence of composition variation on structural, magnetic and dielectric properties of Gd3Fe5O12 (x)/MgFe2O4 (1− x) composite. *Indian Journal of Physics, 96*(14), 4173–4184. doi:10.1007/s12648-022-02365-5

Sharma, A., Kapse, S., Verma, A., Bisoyi, S., Pradhan, G. K., Thapa, R., & Rout, C. S. (2022). All-Solid-State Supercapacitor Based on Advanced 2D Vanadium Disulfide/Black Phosphorus Hybrids for Wearable Electronics. *ACS Applied Energy Materials*, *5*(8), 10315–10327. doi:10.1021/acsaem.2c02116

Sharma, A., Mahajan, H., Mohammed, I., Sinha, S., Godara, S. K., & Srivastava, A. K. (2023). Effect of Sintering Temperature on the Structural, Dielectric, and Magnetic Properties of Garnet-Spinel Ferrite Composites for Use in L-Band Devices. *Materials Performance and Characterization*, *12*(1), 64–78. doi:10.1520/MPC20220084

Sharma, A., Mohammed, I., Godara, S. K., & Srivastava, A. K. (2022). Low dielectric losses and enhanced magnetic property of ErIG (x)/YIG (1− x)(x= 0.5) composite for antenna applications. *Bulletin of Materials Science*, *45*(4), 248. doi:10.1007/s12034-022-02810-5

Sharma, A., Tyagi, V. V., Chen, C. R., & Buddhi, D. (2009). Review on thermal energy storage with phase change materials and applications. *Renewable & Sustainable Energy Reviews*, *13*(2), 318–345. doi:10.1016/j.rser.2007.10.005

Sharma, P., & Bhatti, T. S. (2010). A review on electrochemical double-layer capacitors. *Energy Conversion and Management*, *51*(12), 2901–2912. doi:10.1016/j.enconman.2010.06.031

Sharma, R., Jang, J.-G., & Hu, J.-W. (2022). Phase-change materials in concrete: Opportunities and challenges for sustainable construction and building materials. *Materials (Basel)*, *15*(1), 335. doi:10.3390/ma15010335 PMID:35009481

Sharma, S., Panwar, A. K., & Tripathi, M. M. (2020). Storage technologies for electric vehicles. *Journal of Traffic and Transportation Engineering*, *7*(3), 340–361. doi:10.1016/j.jtte.2020.04.004

Sharshir, S. W., Joseph, A., Elsharkawy, M., Hamad, M. A., Kandeal, A. W., Elkadeem, M. R., Thakur, A. K., Ma, Y., Moustapha, M. E., & Rashad, M. (2023). Thermal energy storage using phase change materials in building applications: A review of the recent development. *Energy and Building*, *285*, 112908. doi:10.1016/j.enbuild.2023.112908

Shehada, N., Cancilla, J. C., Torrecilla, J. S., Pariente, E. S., Brönstrup, G., Christiansen, S., Johnson, D. W., Leja, M., Davies, M. P., Liran, O., Peled, N., & Haick, H. (2016). Silicon Nanowire Sensors Enable Diagnosis of Patients via Exhaled Breath. *ACS Nano*, *10*(7), 7047–7057. doi:10.1021/acsnano.6b03127 PMID:27383408

Shen. (2021). Advanced electrode materials in lithium batteries: retrospect and prospect. *Energy Material Advances*.

Shen, F., Wang, J., Xu, Z., Wu, Y., Chen, Q., Li, X., Jie, X., Li, L., Yao, M., Guo, X., & Zhu, T. (2012). Rapid flu diagnosis using silicon nanowire sensor. *Nano Letters*, *12*(7), 3722–3730. doi:10.1021/nl301516z PMID:22731392

Shen, P.-C., Lin, Y., Wang, H., Park, J.-H., Leong, W. S., Lu, A.-Y., Palacios, T., & Kong, J. (2018). CVD technology for 2-D materials. *IEEE Transactions on Electron Devices*, *65*(10), 4040–4052. doi:10.1109/TED.2018.2866390

Shen, Z. Y., & Li, J. F. (2010). Enhancement of piezoelectric constant d33 in BaTiO3 ceramics due to nano-domain structure. *Journal of the Ceramic Society of Japan*, *118*(1382), 940–943. doi:10.2109/jcersj2.118.940

Shi, D., Adinolfi, V., Comin, R., Yuan, M., Alarousu, E., Buin, A., Chen, Y., Hoogland, S., Rothenberger, A., Katsiev, K., Losovyj, Y., Zhang, X., Dowben, P. A., Mohammed, O. F., Sargent, E. H., & Bakr, O. M. (2015). Low trap-state density and long carrier diffusion in organolead trihalide perovskite single crystals. *Science*, *347*(6221), 519–522. doi:10.1126/science.aaa2725 PMID:25635092

Shinekumar, K., & Dutta, S. (2015). High-temperature piezoelectrics with large piezoelectric coefficients. *Journal of Electronic Materials*, *44*(2), 613–622. doi:10.1007/s11664-014-3534-2

Shin, J., Lee, J., Park, Y., & Choi, J. W. (2020). Aqueous zinc ion batteries: Focus on zinc metal anodes. *Chemical Science (Cambridge)*, *11*(8), 2028–2044. doi:10.1039/D0SC00022A PMID:32180925

Shirane, G., Suzuki, K., & Takeda, A. (1952). Phase transitions in solid solutions of PbZrO3 and PbTiO3 (II) X-ray study. *Journal of the Physical Society of Japan*, *7*(1), 12–18. doi:10.1143/JPSJ.7.12

Shirasaki, Yoshizawa, & Tateishi. (2001). Changes in the mechanical properties of silicone catheter in degrading process. *Nihon Reoroji Gakkaishi.*, *29*, 9–13. doi:10.1678/rheology.29.9

Shiri, H. M., Ehsani, A., & Shayeh, J. S. (2015). Synthesis and highly efficient supercapacitor behavior of a novel poly pyrrole/ceramic oxide nanocomposite film. *RSC Advances*, *5*(110), 91062–91068. doi:10.1039/C5RA19863A

Shrivastav, V., Sundriyal, S., Goel, P., Kaur, H., Tuteja, S. K., Vikrant, K., Kim, K.-H., Tiwari, U. K., & Deep, A. (2019). Metal-organic frameworks (MOFs) and their composites as electrodes for lithium battery applications: Novel means for alternative energy storage. *Coordination Chemistry Reviews*, *393*, 48–78. doi:10.1016/j.ccr.2019.05.006

Shuai. (2023). Recent Advances of Transition Metal Sulfides/Selenides Cathodes for Aqueous Zinc-Ion Batteries. *Advanced Energy Materials*, *13*(4).

Silver, D., Schrittwieser, J., Simonyan, K., Antonoglou, I., Huang, A., Guez, A., Hubert, T., Baker, L., Lai, M., Bolton, A., Chen, Y., Lillicrap, T., Hui, F., Sifre, L., van den Driessche, G., Graepel, T., & Hassabis, D. (2017). Mastering the game of Go without human knowledge. *Nature*, *550*(7676), 354–359. doi:10.1038/nature24270 PMID:29052630

Silverstein, M. S. (2017). Emulsion-templated polymers: Contemporary contemplations. *Polymer*, *126*, 261–282. doi:10.1016/j.polymer.2017.07.046

Simon, A. H. (2012). *Sputter Processing. Handbook of Thin Film Deposition: Techniques* (3rd ed.). Processes, and Technologies. doi:10.1016/B978-1-4377-7873-1.00004-8

Simonjan, J., Jornet, J. M., Akyildiz, I. F., & Rinner, B. (2018). Nano-cameras. *Proceedings of the 5th ACM International Conference on Nanoscale Computing and Communication*. 10.1145/3233188.3233209

Singh, A. K., & Pandey, D. (2001). Structure and the location of the morphotropic phase boundary region in (1-x) [Pb(Mg1/3Nb2/3)O3]-xPbTiO3. *Journal of Physics Condensed Matter*, *13*(48), 931–936. doi:10.1088/0953-8984/13/48/102

Singh, A. P., Tiwari, N. K., Karandikar, P. B., & Dubey, A. (2015). Effect of electrode shape on the parameters of supercapacitor. *2015 International Conference on Industrial Instrumentation and Control, ICIC 2015, Icic*, 669–673. 10.1109/IIC.2015.7150826

Singh, H. S., & Maisnam, M. (2023, December 10). Study of structural and electrical properties of oxides added K0. 47Na0. 47Li0. 06NbO3 ceramics. *Ferroelectrics*, *617*(1), 83–92. doi:10.1080/00150193.2023.2271138

Sitko, R. (2009). Quantitative X-ray fluorescence analysis of samples of less than 'infinite thickness': Difficulties and possibilities. *Spectrochimica Acta. Part B, Atomic Spectroscopy*, *64*(11–12), 1161–1172. doi:10.1016/j.sab.2009.09.005

Sitko, R., & Zawisz, B. (2012). Quantification in X-Ray Fluorescence Spectrometry. In S. K. Sharma (Ed.), *X-Ray Spectroscopy*. InTech. doi:10.5772/29367

Sizmur, T., Fresno, T., Akgül, G., Frost, H., & Moreno-Jiménez, E. (2017). Biochar modification to enhance sorption of inorganics from water. In *Bioresource Technology* (Vol. 246, pp. 34–47). Elsevier Ltd. doi:10.1016/j.biortech.2017.07.082

Skidmore, T. A., & Milne, S. J. (2007). Phase development during mixed-oxide processing of a [Na0. 5K0. 5NbO3] 1− x–[LiTaO3] x powder. *Journal of Materials Research*, *22*(8), 2265–2272. doi:10.1557/jmr.2007.0281

Smolenskii, G. A., Isupov, V. A., Agranovskaya, A. I., & Krainik, N. N. (1960). New materials of AIIBIVOVI type. *Sov Phys Solid State*, *2*, 2982–2985.

Snaith, H. J. (2010). Estimating the maximum attainable efficiency in dye-sensitized solar cells. *Advanced Functional Materials*, *20*(1), 13–19. doi:10.1002/adfm.200901476

Snaith, H., & Gratzel, M. (2006). Enhanced charge mobility in a molecular hole transporter via addition of redox inactive ionic dopant: Implication to dye-sensitized solar cells. *Applied Physics Letters*, *89*(26), 262114. doi:10.1063/1.2424552

Snyder, R. L. (1999). X-Ray Diffraction. In X-ray Characterization of Materials (pp. 1–103). Wiley. doi:10.1002/9783527613748.ch1

Som, T. (2021). Sustainability in Energy Economy and Environment: Role of AI Based Techniques. doi:10.1007/978-3-030-72929-5_31

Song, M. K., Cheng, S., Chen, H., Qin, W., Nam, K. W., Xu, S., Yang, X. Q., Bongiorno, A., Lee, J., Bai, J., Tyson, T. A., Cho, J., & Liu, M. (2012). Erratum: Anomalous pseudocapacitive behavior of a nanostructured, mixed-valent manganese oxide film for electrical energy storage. *Nano Letters, 12*(8), 4416. https://doi.org/ doi:10.1021/nl300984y

Song, P., Wang, Y. H., & Liu, X. (2017). Flexible physical sensors made from paper substrates integrated with zinc oxide nanostructures. *Flexible and Printed Electronics*, *2*(3), 034001. doi:10.1088/2058-8585/aa765d

Song, Z., McElvany, C. L., Phillips, A. B., Celik, I., Krantz, P. W., Watthage, S. C., Liyanage, G. K., Apul, D., & Heben, M. J. (2017). A technoeconomic analysis of perovskite solar module manufacturing with low-cost materials and techniques. *Energy & Environmental Science*, *10*(6), 1297–1305. doi:10.1039/C7EE00757D

Speirs, J., Contestabile, M., Houari, Y., & Gross, R. (2014). The future of lithium availability for electric vehicle batteries. *Renewable & Sustainable Energy Reviews*, *35*, 183–193. doi:10.1016/j.rser.2014.04.018

Stadler, K., Masignani, V., Eickmann, M., Becker, S., Abrignani, S., Klenk, H. D., & Rappuoli, R. (2003). SARS—Beginning to understand a new virus. *Nature Reviews. Microbiology*, *1*(3), 209–218. doi:10.1038/nrmicro775 PMID:15035025

Steirer, K. X., Schulz, P., Teeter, G., Stevanovic, V., Yang, M., Zhu, K., & Berry, J. J. (2016). Defect tolerance in methylammonium lead triiodide perovskite. *ACS Energy Letters*, *1*(2), 360–366. doi:10.1021/acsenergylett.6b00196

Stelzner, M., Dressler, F., & Fischer, S. (2016). Function Centric Networking. *Proceedings of the 3rd ACM International Conference on Nanoscale Computing and Communication*. 10.1145/2967446.2967479

Stoumpos, C. C., & Kanatzidis, M. G. (2015). The Renaissance of Halide Perovskites and Their Evolution as Emerging Semiconductors. *Accounts of Chemical Research*, *48*(10), 2791–2802. doi:10.1021/acs.accounts.5b00229 PMID:26350149

Stoumpos, C. C., Malliakas, C. D., & Kanatzidis, M. G. (2013). Semiconducting tin and lead iodide perovskites with organic cations: Phase transitions, high mobilities, and near-infrared photoluminescent properties. *Inorganic Chemistry*, *52*(15), 9019–9038. doi:10.1021/ic401215x PMID:23834108

Stranks, S. D., Eperon, G. E., Grancini, G., Menelaou, C., Alcocer, M. J. P., Leijtens, T., Herz, L. M., Petrozza, A., & Snaith, H. J. (2013). Electron-hole diffusion lengths exceeding 1 micrometer in an organometal trihalide perovskite absorber. *Science*, *342*(6156), 341–344. doi:10.1126/science.1243982 PMID:24136964

Stranks, S. D., & Snaith, H. J. (2015). Metal-halide perovskites for photovoltaic and light-emitting devices. *Nature Nanotechnology*, *10*(5), 391–402. doi:10.1038/nnano.2015.90 PMID:25947963

Streli, C., Wobrauschek, P., & Kregsamer, P. (1999). X-ray Fluorescence Spectroscopy, Applications. Encyclopedia of Spectroscopy and Spectrometry, 2478–2487. doi:10.1006/rwsp.2000.0337

Strüder, L., Lechner, P., & Leutenegger, P. (1998). Silicon drift detector – the key to new experiments. *Naturwissenschaften*, *85*(11), 539–543. doi:10.1007/s001140050545

Stukowski, A. (2012). Structure identification methods for atomistic simulations of crystalline materials. *Modelling and Simulation in Materials Science and Engineering, 20*(4).

Su. (2020). Sub-millimeter-scale monolayer p-type H-phase VS2. *Advanced Functional Materials, 30*(17).

Su, B., Holmes, J. E., Cheng, B. L., & Button, T. W. (2002). Processing effects on the microstructure and dielectric properties of barium strontium titanate (BST) ceramics. *Journal of Electroceramics, 9*(2), 111–116. doi:10.1023/A:1022850205284

Subbiah, A. S., Halder, A., Ghosh, S., Mahuli, N., Hodes, G., & Sarkar, S. K. (2014). Inorganic hole conducting layers for perovskite-based solar cells. *The Journal of Physical Chemistry Letters, 5*(10), 1748–1753. doi:10.1021/jz500645n PMID:26270378

Sun, H., Zhu, J., Baumann, D., Peng, L., Xu, Y., Shakir, I., Huang, Y., & Duan, X. (2019). Hierarchical 3D electrodes for electrochemical energy storage. *Nature Reviews. Materials, 4*(1), 45–60. doi:10.1038/s41578-018-0069-9

Sun, J., Lian, G., Jing, L., Wu, D., Cui, D., Wang, Q., Yu, H., Zhang, H., & Wong, C.-P. (2022). Assembly of flower-like VS2/N-doped porous carbon with expanded (001) plane on rGO for superior Na-ion and K-ion storage. *Nano Research, 15*(5), 4108–4116. doi:10.1007/s12274-021-4060-1

Sun, X., Zhang, Y., Xie, K., & Medina, M. A. (2022). A parametric study on the thermal response of a building wall with a phase change material (PCM) layer for passive space cooling. *Journal of Energy Storage, 47*, 103548. doi:10.1016/j.est.2021.103548

Sun, Z., Zhang, Y., Liu, Y., Hou, L., & Yuan, C. (2021). Recent Progress on In Situ/Operando Characterization of Rechargeable Alkali Ion Batteries. *ChemPlusChem, 86*(10), 1487–1496. doi:10.1002/cplu.202100393 PMID:34674379

Suppes, G. J., Goff, M. J., & Lopes, S. (2003). Latent heat characteristics of fatty acid derivatives pursuant phase change material applications. *Chemical Engineering Science, 58*(9), 1751–1763. doi:10.1016/S0009-2509(03)00006-X

Swanson, H. E. (1955). *Standard X-ray Diffraction Powder Patterns* (Vol. 5–10). U.S. Department of Commerce, National Bureau of Standards.

Sze, S. M., Li, Y., & Ng, K. K. (2021). *Physics of semiconductor devices*. John wiley & sons.

Tabassum, N. (2020). An empirical exploration of the nanotechnology. *International Journal of Advanced Research, 8*(7), 885–915. doi:10.21474/IJAR01/11352

Tahanian, H., Aliahmadi, M., & Faiz, J. (2020). Ferrite permanent magnets in electrical machines: Opportunities and challenges of a non-rare-earth alternative. *IEEE Transactions on Magnetics, 56*(3), 1–20. doi:10.1109/TMAG.2019.2957468

Takahashi, Y., Obara, R., Lin, Z.-Z., Takahashi, Y., Naito, T., Inabe, T., Ishibashi, S., & Terakura, K. (2011). Charge transport in tin-iodide perovskite: Origin of high conductivity. *Dalton Transactions (Cambridge, England), 40*(20), 5563–5568. doi:10.1039/c0dt01601b PMID:21494720

Takenaka. (2013). Lead-Free Piezoelectric Ceramics. Handbook of Advanced Ceramics.

Takenaka, T., & Nagata, H. (2005). Current status and prospects of lead-free piezoelectric ceramics. *Journal of the European Ceramic Society, 25*(12), 2693–2700. doi:10.1016/j.jeurceramsoc.2005.03.125

Tallury, P., Malhotra, A., Byrne, L. M., & Santra, S. (2010). Nanobioimaging and sensing of infectious diseases. *Advanced Drug Delivery Reviews, 62*(4–5), 424–437. doi:10.1016/j.addr.2009.11.014 PMID:19931579

Tang, N., Zhou, C., Xu, L., Jiang, Y., Qu, H., & Duan, X. (2019). A Fully Integrated Wireless Flexible Ammonia Sensor Fabricated by Soft Nano-Lithography. *ACS Sensors, 4*(3), 726–732. doi:10.1021/acssensors.8b01690 PMID:30793588

Tan, X., Li, Q., & Wang, H. (2013). Advances and trends of energy storage technology in microgrid. *International Journal of Electrical Power & Energy Systems*, *44*(1), 179–191. doi:10.1016/j.ijepes.2012.07.015

Tan, Z., Xing, J., Jiang, L., Zhu, J., & Wu, B. (2017). Ga 2 O 3 doping and vacancy effect in KNN—LT lead-free piezoceramics. *Frontiers of Materials Science*, *11*(4), 344–352. doi:10.1007/s11706-017-0403-y

Tao, F., Zhao, Y. Q., Zhang, G. Q., & Li, H. L. (2007). Electrochemical characterization on cobalt sulfide for electrochemical supercapacitors. *Electrochemistry Communications*, *9*(6), 1282–1287. doi:10.1016/j.elecom.2006.11.022

Tao, H., Wu, J., & Wang, H. (2016). Modification of strain and piezoelectricity in (K, Na) NbO3–(Bi, Na) HfO3 lead-free ceramics with high Curie temperature. *Journal of Alloys and Compounds*, *684*, 217–223. doi:10.1016/j.jallcom.2016.04.322

Tao, H., Wu, J., Zheng, T., Wang, X., & Lou, X. (2015). New (1 _ xÞ-K0:45Na0:55Nb0:96Sb0:04O3–xBi0:5Na0:5HfO3 lead-free ceramics: Phase boundary and their electrical properties. *Journal of Applied Physics*, *118*(4), 044102. doi:10.1063/1.4927281

Tao, H., Wu, W., & Wu, J. (2016). Electrical properties of holmium doped (K,Na)(Nb,Sb)O3–(Bi,Na)HfO3 ceramics with wide sintering and poling temperature range. *Journal of Alloys and Compounds*, *689*, 759–766. doi:10.1016/j.jallcom.2016.08.037

Tao, R., Parmar, M., Ardila, G., Oliveira, P., Marques, D., Montès, L., & Mouis, M. (2017). Performance of ZnO based piezo-generators under controlled compression. *Semiconductor Science and Technology*, *32*(6), 64003. doi:10.1088/1361-6641/aa691f

Tapia-Ruiz. (2021). 2021 roadmap for sodium-ion batteries. *Journal of Physics: Energy, 3*(3).

Tasker, P. W. (1979). The stability of ionic crystal surfaces. *Journal of Physics. C. Solid State Physics*, *12*(22), 4977–4984. doi:10.1088/0022-3719/12/22/036

Tatarenko, Gheevarughese, Srinivasan, Antonenkov, & Bichurin. (2007). Microwave magnetoelectric effects in ferrite—piezoelectric composites and dual electric and magnetic field tunable filters. *J. Electroceramics, 241*(24), 5–9. . doi:10.1007/s10832-007-9382-1

Tavakoli, S., Khojasteh, D., Haghani, M., & Hirdaris, S. (2023). A review on the progress and research directions of ocean engineering. *Ocean Engineering*, *272*, 113617. doi:10.1016/j.oceaneng.2023.113617

Taylor, G. W. (1985). Piezoelectricity. Academic Press.

Thangadurai, V., Narayanan, S., & Pinzaru, D. (2014). Garnet-type solid-state fast Li ion conductors for Li batteries: Critical review. *Chemical Society Reviews*, *43*(13), 4714–4727. doi:10.1039/c4cs00020j PMID:24681593

Thery, V., Bayart, A., Blach, J. F., Roussel, P., & Saitzek, S. (2015). Effective piezoelectric coefficient measurement of BaTiO3 thin films using the X-ray diffraction technique under electric field available in a standard laboratory. *Applied Surface Science*, *351*, 480–486. doi:10.1016/j.apsusc.2015.05.155

Thetpraphi, K., Le, M. Q., Houachtia, A., Cottinet, P. J., Petit, L., Audigier, D., Kuhn, J., Moretto, G., & Capsal, J. F. (2019). Surface Correction Control Based on Plasticized Multilayer P(VDF-TrFE-CFE) Actuator—Live Mirror. *Advanced Optical Materials*, *7*(13), 1900210. doi:10.1002/adom.201900210

Thierry, D., & Biegler, L. T. (2019). Dynamic real-time optimization for a CO2 capture process. *AIChE Journal. American Institute of Chemical Engineers*, *65*(7), e16511. doi:10.1002/aic.16511

Thirumalasetty, A. B., Krishnan, T., Kota Venkata, S. K., Adiraj, S., Xavier, D., Venkatachalam, S., Kesava Reddy, R. G., Dudekula, Z., & Wuppulluri, M. (2023). Strategic Design and Analysis of Energy Storage and ME Applications of Low Temperature Hybrid Microwave Sintered Electroceramic Composites. *ACS Applied Electronic Materials*, *5*(6), 3120–3129. doi:10.1021/acsaelm.3c00190

Thong, H. C., Zhao, C., Zhou, Z., Wu, C. F., Liu, Y. X., Du, Z. Z., Li, J.-F., Gong, W., & Wang, K. (2019). Technology transfer of lead-free (K, Na) NbO3-based piezoelectric ceramics. *Materials Today*, *29*, 37–48. doi:10.1016/j.mattod.2019.04.016

Tian, A., & Du, H. (2014). Enhanced ferroelectric property of manganese doped lead-free (K0. 5Na0. 5) NbO3 Ceramics. *Ferroelectrics*, *463*(1), 72–77. doi:10.1080/00150193.2014.891921

Tian, L., Li, Z., Wang, P., Zhai, X., Wang, X., & Li, T. (2021). Carbon quantum dots for advanced electrocatalysis. *Journal of Energy Chemistry*, *55*, 279–294. doi:10.1016/j.jechem.2020.06.057

Tobarameekul, P., & Worathanakul, P. (2021). Heat management options to reduce carbon footprint of green zeolite faujasite synthesis from rice husk ash. *Thai Environmental Engineering Journal*, *35*(3), 59–68.

Tolani, S. C., Golhar, A. R., & Rewatkar, K. G. (2019, May). A review of morphological, structural behaviour and technological applications of ferrites. In AIP Conference Proceedings (Vol. 2104, No. 1). AIP Publishing. doi:10.1063/1.5100459

Tomczyk, A., Kondracki, B., & Szewczuk-Karpisz, K. (2023). Chemical modification of biochars as a method to improve its surface properties and efficiency in removing xenobiotics from aqueous media. In *Chemosphere* (Vol. 312). Elsevier Ltd. doi:10.1016/j.chemosphere.2022.137238

Tran, V. A., Vo, G. N., Vo, T., Doan, V. D., Vo, V., & Le, V. T. (2023). Recent Applications and Prospects of Nanowire-Based Biosensors. *Processes (Basel, Switzerland)*, *11*(6), 1739. doi:10.3390/pr11061739

Trickett, C. A., Helal, A., Al-Maythalony, B. A., Yamani, Z. H., Cordova, K. E., & Yaghi, O. M. (2017). The chemistry of metal–organic frameworks for CO2 capture, regeneration and conversion. *Nature Reviews. Materials*, *2*(8), 1–16. doi:10.1038/natrevmats.2017.45

Trolliard, G., & Dorcet, V. (2008). Reinvestigation of phase transitions in Na0. 5Bi0. 5TiO3 by TEM. Part II: Second order orthorhombic to tetragonal phase transition. *Chemistry of Materials*, *20*(15), 5074–5082. doi:10.1021/cm800464d

Turk Sekulic, M., Boskovic, N., Slavkovic, A., Garunovic, J., Kolakovic, S., & Pap, S. (2019). Surface functionalised adsorbent for emerging pharmaceutical removal: Adsorption performance and mechanisms. *Process Safety and Environmental Protection*, *125*, 50–63. doi:10.1016/j.psep.2019.03.007

Tyagi, V. V., Chopra, K., Sharma, R. K., Pandey, A. K., Tyagi, S. K., Ahmad, M. S., Sarı, A., & Kothari, R. (2022). A comprehensive review on phase change materials for heat storage applications: Development, characterization, thermal and chemical stability. *Solar Energy Materials and Solar Cells*, *234*, 111392. doi:10.1016/j.solmat.2021.111392

Uchino, K. (2012). *Piezoelectric ceramics for transducers, Ultrason.* Transducers Mater. Des. Sensors, Actuators Med. Appl., doi:10.1533/9780857096302.1.70

Uchino, K. (2015). Glory of piezoelectric perovskites. *Science and Technology of Advanced Materials*, *16*(4), 046001. doi:10.1088/1468-6996/16/4/046001 PMID:27877827

Uchino, K., & Giniewicz, J. R. (2003). *Micromechatronics*. CRC. doi:10.4324/9780203912645

Valant, M., Axelsson, A. K., & Alford, N. (2007). Peculiarities of a solid-state synthesis of multiferroic polycrystalline BiFeO3. *Chemistry of Materials*, *19*(22), 5431–5436. doi:10.1021/cm071730+

Valasek, J. (1921). Piezo-Electric and Allied Phenomena in Rochelle Salt. *Physical Review*, *17*(4), 475–481. doi:10.1103/PhysRev.17.475

Valdez-garcía, G. D., & Leyva-ramos, R. (2023). Hindered diffusion of heavy metal cation in the adsorption rate on activated carbon fiber. *Chemical Engineering Research & Design*, *196*, 556–567. Advance online publication. doi:10.1016/j.cherd.2023.07.005

van der Heide, P. (2014). *Secondary Ion Mass Spectrometry An Introduction to Principles and Practices* (1st ed.). Wiley. doi:10.1002/9781118916780

van der Mei, H. C., de Vries, J., & Busscher, H. J. (2000). X-ray photoelectron spectroscopy for the study of microbial cell surfaces. *Surface Science Reports*, *39*(1), 1–24. doi:10.1016/S0167-5729(00)00003-0

van der Veer, E., Noheda, B., & Acuautla, M. (2021). Piezoelectric properties of PZT by an ethylene glycol-based chemical solution synthesis. *Journal of Sol-Gel Science and Technology*, *100*(3), 517–525. doi:10.1007/s10971-021-05651-6

van Driel, H. M. (2000). Large-scale synthesis of a silicon photonic crystal with a completethree-dimensional bandgap near 1.5 micrometres. *Nature*, *405*(6785), 437–440. doi:10.1038/35013024 PMID:10839534

Van Noorden, R. (2013). Sulphur back in vogue for batteries. *Nature*, *498*(7455), 416–417. doi:10.1038/498416a PMID:23803819

Vangari, M., Pryor, T., & Jiang, L. (2013). Supercapacitors: Review of Materials and Fabrication Methods. *Journal of Energy Engineering*, *139*(2), 72–79. doi:10.1061/(ASCE)EY.1943-7897.0000102

Vasiliu, F., Lucuta, P. G., & Constantinescu, F. (1983). Structure-sintering temperature relationship in PZT solid solutions at the morphotropic phase boundary. *Physica Status Solidi (A)*, *80*(2), 637-642.

Vicente, R., & Silva, T. (2014). Brick masonry walls with PCM macrocapsules: An experimental approach. *Applied Thermal Engineering*, *67*(1–2), 24–34. doi:10.1016/j.applthermaleng.2014.02.069

Vrakking, J. J., & Meyer, F. (1971). Auger electron spectroscopy made quantitative by ellipsometric calibration. *Applied Physics Letters*, *18*(6), 226–228. doi:10.1063/1.1653640

Vu, X. T., Eschermann, J. F., Stockmann, R., GhoshMoulick, R., Offenhäusser, A., & Ingebrandt, S. (2009). Top-down processed silicon nanowire transistor arrays for biosensing. *Physica Status Solidi (A)*, *206*(3), 426-434.

Vul, B. M. (1946). *High and ultrahigh dielectric constant materials.* Elektrichestvo. [inRussian]

Waidi, Y. O., Barua, R., & Datta, S. (2023). Metals, Polymers, Ceramics, Composites Biomaterials Used in Additive Manufacturing for Biomedical Applications. In A. Kumar, P. Kumar, A. Srivastava, & V. Goyat (Eds.), *Modeling, Characterization, and Processing of Smart Materials* (pp. 165–184). IGI Global. doi:10.4018/978-1-6684-9224-6.ch008

Wainer, E. (1946). High titania dielectrics. *Transactions of the Electrochemical Society*, *89*(1), 331. doi:10.1149/1.3071718

Walsh, A., Scanlon, D. O., Chen, S., Gong, X. G., & Wei, S.-H. (2015). Self-regulation mechanism for charged point defects in hybrid halide perovskites. *Angewandte Chemie International Edition*, *54*(6), 1791–1794. doi:10.1002/anie.201409740 PMID:25504875

Walter, M., Kravchyk, K. V., Ibáñez, M., & Kovalenko, M. V. (2015). Efficient and inexpensive sodium–magnesium hybrid battery. *Chemistry of Materials*, *27*(21), 7452–7458. doi:10.1021/acs.chemmater.5b03531

Walubita, L. F., Martinez-Arguelles, G., Polo-Mendoza, R., Ick-Lee, S., & Fuentes, L. (2022). Comparative environmental assessment of rigid, flexible, and perpetual pavements: A case study of Texas. *Sustainability (Basel)*, *14*(16), 9983. doi:10.3390/su14169983

Wang & Song. (2006). Piezoelectric nanogenerators based on zinc oxide nanowire arrays. *Science, 312*, 242–246. . doi:10.1126/science.1124005

Wang, Herbert, & Glass. (1988). The Applications of ferroelectric polymers. *Blackie, 160.*

Wang, J. B. N. J., Neaton, J. B., Zheng, H., Nagarajan, V., Ogale, S. B., Liu, B., ... Ramesh, R. (2003). Epitaxial BiFeO3 multiferroic thin film heterostructures. *Science, 299*(5613), 1719-1722.

Wang, X., Wu, J., Xiao, D., Zhu, J., Cheng, X., Zheng, T., & Wang, X. (2014). Giant piezoelectricity in potassium–sodium niobate lead-free ceramics. *Journal of the American Chemical Society, 136*(7), 2905-2910.

Wang, Z. (2011). *Nanogenerators for self-powered devices and systems.* http://smartech.gatech.edu/handle/1853/39262

Wang. (2004). *Zinc oxide nanostructures: growth, properties and applications.* . doi:10.1088/0953-8984/16/25/R01

Wang. (2009). ZnO nanowire and nanobelt platform for nanotechnology. *Mater. Sci. Eng. R Reports, 64*, 33–71. doi:10.1016/j.mser.2009.02.001

Wang. (2023). A customized strategy to design intercalation-type Li-free cathodes for all-solid-state batteries. *National Science Review.*

Wang, B., Ang, E. H., Yang, Y., Zhang, Y., Ye, M., Liu, Q., & Li, C. C. (2021). Post-lithium-ion battery era: Recent advances in rechargeable potassium-ion batteries. *Chemistry (Weinheim an der Bergstrasse, Germany), 27*(2), 512–536. doi:10.1002/chem.202001811 PMID:32510710

Wang, D., Hussain, F., Khesro, A., Feteira, A., Tian, Y., Zhao, Q., & Reaney, I. M. (2017). Composition and temperature dependence of structure and piezoelectricity in (1−x)(K1−yNay)NbO3-x (Bi1/2Na1/2)ZrO3 lead-free ceramics. *Journal of the American Ceramic Society, 100*(2), 627–637. doi:10.1111/jace.14589

Wang, D., Zhao, Y., Lian, R., Yang, D., Zhang, D., Meng, X., Liu, Y., Wei, Y., & Chen, G. (2018). Atomic insight into the structural transformation and anionic/cationic redox reactions of VS 2 nanosheets in sodium-ion batteries. *Journal of Materials Chemistry. A, Materials for Energy and Sustainability, 6*(33), 15985–15992. doi:10.1039/C8TA05396K

Wang, D., Zhu, K., Ji, H., & Qiu, J. (2009). Two-step sintering of the pure K0. 5Na0. 5NbO3 lead-free piezoceramics and its piezoelectric properties. *Ferroelectrics, 392*(1), 120–126. doi:10.1080/00150190903412622

Wang, F., Robert, R., Chernova, N. A., Pereira, N., Omenya, F., Badway, F., Hua, X., Ruotolo, M., Zhang, R., Wu, L., Volkov, V., Su, D., Key, B., Whittingham, M. S., Grey, C. P., Amatucci, G. G., Zhu, Y., & Graetz, J. (2011). Conversion reaction mechanisms in lithium ion batteries: Study of the binary metal fluoride electrodes. *Journal of the American Chemical Society, 133*(46), 18828–18836. doi:10.1021/ja206268a PMID:21894971

Wang, G., Liu, H., Wu, H., Li, X., Qiu, H., Yang, Y., Qu, B., Ren, T.-L., Han, X., Zhang, R., & Wang, H. (2016). Epitaxial yttrium iron garnet film for fabrication of high frequency on-chip inductors. *Applied Physics Letters, 109*(16), 162405. doi:10.1063/1.4964642

Wang, H., & Kim, D. H. (2017). Perovskite-based photodetectors: Materials and devices. *Chemical Society Reviews, 46*(17), 5204–5236. doi:10.1039/C6CS00896H PMID:28795697

Wang, H., Wang, Y., Xuan, Z., Chen, T., Zhang, J., Hao, X., Wu, L., Constantinou, I., & Zhao, D. (2021). Progress in perovskite solar cells towards commercialization—A review. *Materials (Basel), 14*(21), 6569. doi:10.3390/ma14216569 PMID:34772092

Wang, J., & Guo, X. (2020). Adsorption kinetic models: Physical meanings, applications, and solving methods. In *Journal of Hazardous Materials* (Vol. 390). Elsevier B.V. doi:10.1016/j.jhazmat.2020.122156

Wang, J., Huang, L., Yang, R., Zhang, Z., Wu, J., Gao, Y., Wang, Q., O'Hare, D., & Zhong, Z. (2014). Recent advances in solid sorbents for CO 2 capture and new development trends. *Energy & Environmental Science*, *7*(11), 3478–3518. doi:10.1039/C4EE01647E

Wang, J., & Hui, N. (2019). Electrochemical functionalization of polypyrrole nanowires for the development of ultrasensitive biosensors for detecting microRNA. *Sensors and Actuators. B, Chemical*, *281*, 478–485. doi:10.1016/j.snb.2018.10.131

Wang, J., Luo, N., Wu, J., Huang, S., Yu, L., & Wei, M. (2019). Hierarchical spheres constructed by ultrathin VS 2 nanosheets for sodium-ion batteries. *Journal of Materials Chemistry. A, Materials for Energy and Sustainability*, *7*(8), 3691–3696. doi:10.1039/C8TA11950C

Wang, K., Wang, S., Xiao, S., & Song, Q. (2018). Recent advances in perovskite micro- and Nanolasers. *Advanced Optical Materials*, *6*(18), 1800278. doi:10.1002/adom.201800278

Wang, R., Chung, C.-C., Liu, Y., Jones, J. L., & Augustyn, V. (2017). Electrochemical intercalation of Mg2+ into anhydrous and hydrated crystalline tungsten oxides. *Langmuir*, *33*(37), 9314–9323. doi:10.1021/acs.langmuir.7b00705 PMID:28732164

Wang, R., Mujahid, M., Duan, Y., Wang, Z.-K., Xue, J., & Yang, Y. (2019). A review of perovskites solar cell stability. *Advanced Functional Materials*, *29*(47), 1808843. doi:10.1002/adfm.201808843

Wang, R., Xie, J., Hanada, K., Matsusaki, K., Kawanaka, H., Bando, H., & Itoh, M. (2008). Enhanced piezoelectricity around the tetragonal/orthorhombic morphotropic phase boundary in (Na, K) NbO3-ATiO3 solid solutions. *Journal of Electroceramics*, *21*(1-4), 263–266. doi:10.1007/s10832-007-9136-0

WangS. (n.d.). Three-Dimensional Amorphous Carbon Rivet Vanadium Sulfide Enhanced Lithium Storage. *Available at* SSRN 4105060. doi:10.2139/ssrn.4105060

Wang, S. Y., Qiu, X., Gao, J., Feng, Y., Su, W. N., Zheng, J. X., Yu, D. S., & Li, D. J. (2011). Electrical reliability and leakage mechanisms in highly resistive multiferroic La0. 1Bi0. 9FeO3 ceramics. *Applied Physics Letters*, *98*(15), 152902. doi:10.1063/1.3580604

Wang, T., Lü, L., & Shi, L. (2020). Surface functionalization and shape tuning of carbon fiber monofilament via direct microplasma scanning for ultramicroelectrode application. *Applied Surface Science*, *531*(July), 147414. Advance online publication. doi:10.1016/j.apsusc.2020.147414

Wang, W., Sun, Z., Zhang, W., Fan, Q., Sun, Q., Cui, X., & Xiang, B. (2016). First-principles investigations of vanadium disulfide for lithium and sodium ion battery applications. *RSC Advances*, *6*(60), 54874–54879. doi:10.1039/C6RA07586J

Wang, X., Tan, G., Bai, Y., Wu, F., & Wu, C. (2021). Multi-electron reaction materials for high-energy-density secondary batteries: Current status and prospective. *Electrochemical Energy Reviews*, *4*(1), 35–66. doi:10.1007/s41918-020-00073-4

Wang, X., Wu, J., Cheng, X., Zhang, B., Xiao, D., Zhu, J., Wang, X., & Lou, X. (2013). Rhombohedral–tetragonal phase boundary and electrical properties of new K0:48Na0:52Nb0:98Sb0:02O3– Bi0:5Na0:5ZrO3 lead-free piezoceramics. *Journal of Physics. D, Applied Physics*, *46*(49), 495305. doi:10.1088/0022-3727/46/49/495305

Wang, X., Wu, J., Xiao, D., Zhu, J., Cheng, X., Zheng, T., Zhang, B., Lou, X., & Wang, X. (2014). Giant piezoelectricity in potassium–sodium niobate lead-free ceramics. *Journal of the American Chemical Society*, *136*(7), 2905–2910. doi:10.1021/ja500076h PMID:24499419

Wang, X., Wu, J., Zheng, T., Cheng, X., Zhang, B., Xiao, D., & Zhu, J. (2014). Phase structure, electrical properties, and stability of 0.96 (K0. 48Na0. 52) 1– xLixNbO3–0.04 Bi0. 5Na0. 5ZrO3 lead-free piezoceramics. *Current Applied Physics*, *14*(5), 809–813. doi:10.1016/j.cap.2014.03.016

Wang, X., Zhang, Y., Zheng, J., Liu, X., & Meng, C. (2019). Hydrothermal synthesis of VS4/CNTs composite with petal-shape structures performing a high specific capacity in a large potential range for high-performance symmetric supercapacitors. *Journal of Colloid and Interface Science*, *554*, 191–201. doi:10.1016/j.jcis.2019.06.105 PMID:31299547

Wang, X., & Zheng, J. P. (2004). The Optimal Energy Density of Electrochemical Capacitors Using Two Different Electrodes. *Journal of the Electrochemical Society*, *151*(10), A1683. doi:10.1149/1.1787841

Wang, X., Zhou, J., Song, J., Liu, J., Xu, N., & Wang, Z. L. (2006). Piezoelectric Field Effect Transistor and Nanoforce Sensor Based on a Single ZnO Nanowire. *Nano Letters*, *6*(12), 2768–2772. doi:10.1021/nl061802g

Wang, Y. P., Zhou, L., Zhang, M. F., Chen, X. Y., Liu, J. M., & Liu, Z. G. (2004). Room-temperature saturated ferroelectric polarization in BiFeO 3 ceramics synthesized by rapid liquid phase sintering. *Applied Physics Letters*, *84*(10), 1731–1733. doi:10.1063/1.1667612

Wang, Y., Anilkumar, P., Cao, L., Liu, J. H., Luo, P. G., Tackett, K. N. II, Sahu, S., Wang, P., Wang, X., & Sun, Y. P. (2011). Carbon dots of different composition and surface functionalization: Cytotoxicity issues relevant to fluorescence cell imaging. *Experimental Biology and Medicine*, *236*(11), 1231–1238. doi:10.1258/ebm.2011.011132 PMID:22036734

Wang, Y., Bao, L., Liu, Z., & Pang, D. W. (2011). Aptamer biosensor based on fluorescence resonance energy transfer from upconverting phosphors to carbon nanoparticles for thrombin detection in human plasma. *Analytical Chemistry*, *83*(21), 8130–8137. doi:10.1021/ac201631b PMID:21923110

Wang, Y., & Herron, N. (1991). Nanometer-sized semiconductor clusters: Materials synthesis, quantum size effects, and photophysical properties. *Journal of Physical Chemistry*, *95*(2), 525–532. doi:10.1021/j100155a009

Wang, Y., & Hu, A. (2014). Carbon quantum dots: Synthesis, properties and applications. *Journal of Materials Chemistry. C, Materials for Optical and Electronic Devices*, *2*(34), 6921–6939. doi:10.1039/C4TC00988F

Wang, Y., Mahmoudi, T., & Hahn, Y.-B. (2020). Highly stable and Efficient Perovskite Solar Cells Based on FAMA-Perovskite-Cu:NiO Composites with 20.7% Efficiency and 80.5% Fill Factor. *Advanced Energy Materials*, *10*(27), 2000967. doi:10.1002/aenm.202000967

Wang, Y., Wu, J., Xiao, D., Wu, W., Zhang, B., Wu, L., & Zhu, J. (2008). Microstructure and Electrical Properties of [(K0. 50Na0. 50) 0.95– xLi0. 05Agx](Nb0. 95Ta0. 05) O3 Lead-Free Ceramics. *Journal of the American Ceramic Society*, *91*(8), 2772–2775. doi:10.1111/j.1551-2916.2008.02529.x

Wang, Z. L. (2012). Progress in piezotronics and piezo-phototronics. *Advanced Materials*, *24*(34), 4632–4646. doi:10.1002/adma.201104365 PMID:22331639

Wang, Z. L., Chen, J., & Lin, L. (2015). Progress in triboelectric nanogenerators as a new energy technology and self-powered sensors. *Energy & Environmental Science*, *8*(8), 2250–2282. doi:10.1039/C5EE01532D

Wang, Z., Lee, S., Koo, K., & Kim, K. (2016). Nanowire-Based Sensors for Biological and Medical Applications. *IEEE Transactions on Nanobioscience*, *15*(3), 186–199. doi:10.1109/TNB.2016.2528258 PMID:26978831

Wang, Z., Su, H., Zhao, S., & Zhao, N. (2016). Influence of phase change material on mechanical and thermal properties of clay geopolymer mortar. *Construction & Building Materials*, *120*, 329–334. doi:10.1016/j.conbuildmat.2016.05.091

Wan, Q., Zhang, Y., & Zhang, R. (2021). The effect of pore behavior and gel structure on the mechanical property at different initial water content. *Construction & Building Materials*, *309*, 125146. doi:10.1016/j.conbuildmat.2021.125146

Weber, D. (1978). $CH_3NH_3SnBrxI_{3-x}$ (x = 0-3), a Sn(II)-system with the cubic perovskite structure. *Zeitschrift für Naturforschung Part B*, *33*, 1443–1445. doi:10.1515/znb-1978-1214

Webster. (1999). *The measurement, instrumentation, and sensors handbook.* Academic Press.

Weida, D., Richter, C., & Clemens, M. (2011). Design of ZnO microvaristor material stress-cone for cable accessories. *IEEE Transactions on Dielectrics and Electrical Insulation, 18*(4), 1262–1267. doi:10.1109/TDEI.2011.5976125

Wei, H., Wang, H., Xia, Y., Cui, D., Shi, Y., Dong, M., Liu, C., Ding, T., Zhang, J., Ma, Y., Wang, N., Wang, Z., Sun, Y., Wei, R., & Guo, Z. (2018). An overview of lead-free piezoelectric materials and devices. *Journal of Materials Chemistry. C, Materials for Optical and Electronic Devices, 6*(46), 12446–12467. doi:10.1039/C8TC04515A

Wei, Z., & Xing, J. (2019). The rise of perovskite light-emitting diodes. *The Journal of Physical Chemistry Letters, 10*(11), 3035–3042. doi:10.1021/acs.jpclett.9b00277 PMID:31117692

What It Is and How It Works. (n.d.). National Nanotechnology Initiative. https://www.nano.gov/nanotech-101/what

White, P., McPhedran, R. C., Botten, L. C., Smith, G. H., & de Sterke, C. M. (2001). Calculations of air-guided modes in photonic crystal fibers using the multipole method. *Optics Express, 9*(13), 721. doi:10.1364/OE.9.000721 PMID:19424312

Whitmore, A., Agarwal, A., & Da Xu, L. (2014). The Internet of Things—A survey of topics and trends. *Information Systems Frontiers, 17*(2), 261–274. doi:10.1007/s10796-014-9489-2

Wiegers, G. A., van der Meer, R., van Heiningen, H., Kloosterboer, H. J., & Alberink, A. J. A. (1974). The sodium intercalates of vanadium disulfide and their hydrolysis products. *Materials Research Bulletin, 9*(9), 1261–1265. doi:10.1016/0025-5408(74)90045-2

Wilson, J. S. (2005). *Sensor Technology Handbook.* Elsevier Inc. doi:10.1016/B978-0-7506-7729-5.X5040-X

Winter, M., & Brodd, R. J. (2004). What are batteries, fuel cells, and supercapacitors? *Chemical Reviews, 104*(10), 4245–4269. doi:10.1021/cr020730k PMID:15669155

Wołoszyn, J., Szopa, K., & Czerwiński, G. (2021). Enhanced heat transfer in a PCM shell-and-tube thermal energy storage system. *Applied Thermal Engineering, 196*, 117332. doi:10.1016/j.applthermaleng.2021.117332

Wong, K. L. (2021). Li-ion batteries state-of-charge estimation using deep lstm at various battery specifications and discharge cycles. *Proceedings of the Conference on Information Technology for Social Good.* 10.1145/3462203.3475878

Wong, S. I., Lin, H., Sunarso, J., Wong, B. T., & Jia, B. (2020). Optimization of ionic-liquid based electrolyte concentration for high-energy density graphene supercapacitors. *Applied Materials Today, 18*, 100522. doi:10.1016/j.apmt.2019.100522

Wooster, W. A. (1953). Physical properties and atomic arrangements in crystals. *Reports on Progress in Physics, 16*(1), 62–82. doi:10.1088/0034-4885/16/1/302

Worch, E. (2012). Adsorption Technology in Water Treatment. doi:10.1515/9783110240238

Worighi, I., Maach, A., Hafid, A., Hegazy, O., & Van Mierlo, J. (2019). Integrating renewable energy in smart grid system: Architecture, virtualization and analysis. *Sustainable Energy, Grids and Networks, 18*, 100226. doi:10.1016/j.segan.2019.100226

Wu, G., Wang, P., Cheng, J., Xiao, Q., & Zhu, G. (n.d.). *High-piezoelectricity (1-x)(K1-yNay)(Nb1-zSbz) O3 + xBi0.5(Na1-uKu)0.5ZrO3 lead-free ceramics and their preparation method.* Chinese Patent, CN103482977A

Wu, B., Wu, H., Wu, J., Xiao, D., Zhu, J., & Pennycook, J. (2016). Giant piezoelectricity and high curie temperature in nanostructured alkali niobate lead-free piezoceramics through phase coexistence. *Journal of the American Chemical Society, 138*(47), 15459–15464. doi:10.1021/jacs.6b09024 PMID:27933925

Wu, B., Wu, J., Xiao, D., & Zhu, J. (2015). Modification of both d33 and TC in a potassium–sodium niobate ternary system. *Dalton Transactions (Cambridge, England)*, *44*(48), 21141–21152. doi:10.1039/C5DT03680A PMID:26598931

Wu, C., Zhang, Q., Liu, Y., Luo, W., Guo, X., Huang, Z., Ting, H., Sun Zhong, X., Wei, S., & … . (2018). The dawn of lead-free perovskite solar cell: Highly stable double perovskite $Cs_2AgBiBr_6$ film. *Advancement of Science*, *5*(3), 1700759. doi:10.1002/advs.201700759 PMID:29593974

Wu, D., Wang, C., Wu, M., Chao, Y., He, P., & Ma, J. (2020). Porous bowl-shaped VS2 nanosheets/graphene composite for high-rate lithium-ion storage. *Journal of Energy Chemistry*, *43*, 24–32. doi:10.1016/j.jechem.2019.08.003

Wu, J. (2020). Perovskite lead-free piezoelectric ceramics. *Journal of Applied Physics*, *127*(19), 190901. doi:10.1063/5.0006261

Wu, J., Fan, Z., Xiao, D., Zhu, J., & Wang, J. (2016). Multiferroic bismuth ferrite-based materials for multifunctional applications: Ceramic bulks, thin films and nanostructures. *Progress in Materials Science*, *84*, 335–402. doi:10.1016/j.pmatsci.2016.09.001

Wu, J., Tao, H., Yuan, Y., Lv, X., Wang, X., & Lou, X. (2015). Role of antimony in the phase structure and electrical properties of potassium-sodium niobate lead-free ceramics. *RSC Advances*, *5*(19), 14575–14583. doi:10.1039/C4RA14271C

Wu, J., & Wang, Y. (2014). Two-step sintering of new potassium sodium niobate ceramics: A high d 33 and wide sintering temperature range. *Dalton Transactions (Cambridge, England)*, *43*(34), 12836–12841. doi:10.1039/C4DT01712A PMID:25051530

Wu, J., Wang, Y., Xiao, D., Zhu, J., & Pu, Z. (2007). Effects of Ag content on the phase structure and piezoelectric properties of (K0.44-xNa0.52Li0.04Agx)(Nb0.91Ta0.05Sb0.04)O3 lead-free ceramics. *Applied Physics Letters*, *91*(13), 132914. doi:10.1063/1.2793507

Wu, J., Xiao, D., Wang, Y., Zhu, J., Wu, L., & Jiang, Y. (2007). Effects of K/Na ratio on the phase structure and electrical properties of (KxNa0.96-xLi0.04)(Nb0.91Ta0.05Sb0.04)O3 lead-free ceramics. *Applied Physics Letters*, *91*(25), 252907. doi:10.1063/1.2827573

Wu, J., Xiao, D., & Zhu, J. (2015). Potassium–sodium niobate lead-free piezoelectric ceramics: Recent advances and perspectives. *Journal of Materials Science Materials in Electronics*, *26*(12), 9297–9308. doi:10.1007/s10854-015-3084-2

Wu, J., Xiao, D., & Zhu, J. (2015). Potassium–sodium niobate lead-free piezoelectric materials: Past, present, and future of phase boundaries. *Chemical Reviews*, *115*(7), 2559–2595. doi:10.1021/cr5006809 PMID:25792114

WuL. (2023). Interlayer Expanded Vs2 Nanosheet for Efficient Sodium Storage. *Available at* SSRN 4391411. doi:10.1016/j.matlet.2023.134551

Wu, L., Sun, R., Xiong, F., Pei, C., Han, K., Peng, C., Fan, Y., Yang, W., An, Q., & Mai, L. (2018). A rechargeable aluminum-ion battery based on a VS 2 nanosheet cathode. *Physical Chemistry Chemical Physics*, *20*(35), 22563–22568. doi:10.1039/C8CP04772C PMID:30159553

Wulfkuhle, J. D., Liotta, L. A., & Petricoin, E. F. (2003). Proteomic applications for the early detection of cancer. *Nature Reviews. Cancer*, *3*(4), 267–275. doi:10.1038/nrc1043 PMID:12671665

Wu, W., Xiao, D., Ma, J., Wu, J., Li, J., & Liang, W. (2014). Temperature stability, phase structure and electrical behavior of Li-modified 0.99 (K0. 48Na0. 52) NbO3–0.01 BiCoO3 piezoelectric ceramics. *Ceramics International*, *40*(1), 1133–1137. doi:10.1016/j.ceramint.2013.06.114

Wu, W., Xiao, D., Wu, J., Liang, W., Li, J., & Zhu, J. (2011). Polymorphic phase transition-induced electrical behavior of BiCoO3-modified (K0. 48Na0. 52) NbO3 lead-free piezoelectric ceramics. *Journal of Alloys and Compounds*, *509*(29), L284–L288. doi:10.1016/j.jallcom.2011.05.004

Wu, W., Xiao, D., Wu, J., Li, J., & Zhu, J. (2011). Piezoelectric properties of (K0. 474Na0. 474Li0. 052)(Nb0. 948Sb0. 052) O3–Co2O3 lead-free ceramics. *Journal of the Ceramic Society of Japan*, *119*(1392), 654–657. doi:10.2109/jcersj2.119.654

Wu, W., Xiao, D., Wu, J., Li, J., Zhu, J., & Zhang, B. (2012). Microstructure and electrical properties of relaxor (1− x) [(K0. 5Na0. 5) 0.95 Li0. 05](Nb0. 95Sb0. 05) O3–xBaTiO3 piezoelectric ceramics. *Ceramics International*, *38*(3), 2277–2282. doi:10.1016/j.ceramint.2011.10.079

Wu, Z., Wang, L., Huang, J., Zou, J., Chen, S., Cheng, H., Jiang, C., Gao, P., & Niu, X. (2019). Loofah-derived carbon as an anode material for potassium ion and lithium ion batteries. *Electrochimica Acta*, *306*, 446–453. doi:10.1016/j.electacta.2019.03.165

Xiao, M., Xiao, D., Wang, Z., Huang, T., Wu, J., Li, F., Wu, B., & Zhu, J. (2014). Microstructure and piezoelectric properties of lead-free 0.95 (Na 0.5 K 0.5) NbO 3–0.05 (Bi 0.5 K 0.5) Zr 1− x Ti x O 3 ceramics. *Journal of Materials Science Materials in Electronics*, *25*(4), 1938–1941. doi:10.1007/s10854-014-1824-3

Xiao, T., Shi, X., Gen, L., Dai, Y., Zhao, J., & Liu, C. (2023). *Synergistic Enhancement of Phase Change Materials through Three-Dimensional Macropore Lamellar Structured MOF/EG Composite for Solar Energy Storage and Beyond.* Eg Composite for Solar Energy Storage and Beyond. doi:10.1016/j.applthermaleng.2023.121378

Xiao, Z., Dong, Q., Bi, C., Shao, Y., Yuan, Y., & Huang, J. (2014). Solvent annealing of perovskite-induced crystal growth for photovoltaic-device efficiency enhancement. *Advanced Materials*, *26*(37), 6503–6509. doi:10.1002/adma.201401685 PMID:25158905

Xiao, Z., Yuan, Y., Wang, Q., Shao, Y., Bai, Y., Deng, Y., Dong, Q., Hu, M., Bi, C., & Huang, J. (2016). Thin-film semiconductor perspective of organometal trihalide perovskite materials for high-efficiency solar cells. *Materials Science and Engineering R Reports*, *101*, 1–38. doi:10.1016/j.mser.2015.12.002

Xie, X.-C., Shuai, H.-L., Wu, X., Huang, K.-J., Wang, L.-N., Wang, R.-M., & Chen, Y. (2020). Engineering ultra-enlarged interlayer carbon-containing vanadium disulfide composite for high-performance sodium and potassium ion storage. *Journal of Alloys and Compounds*, *847*, 156288. doi:10.1016/j.jallcom.2020.156288

Xie, Y., Xing, J., Tan, Z., Xie, L., Cheng, Y., Wu, X., Han, R., Chen, Q., & Zhu, J. (2022). High mechanical quality factor and piezoelectricity in potassium sodium niobate ceramics. *Ceramics International*, *48*(5), 6565–6573. doi:10.1016/j.ceramint.2021.11.204

Xing, G., Mathews, N., Sun, S., Lim, S. S., Lam, Y. M., Grätzel, M., Mhaisalkar, S., & Sum, T. C. (2013). Long-range balanced electron- and hole-transport lengths in organic-inorganic $CH_3NH_3PbI_3$. *Science*, *342*(6156), 344–347. doi:10.1126/science.1243167 PMID:24136965

Xing, J., Tan, Z., Jiang, L., Chen, Q., Wu, J., Zhang, W., Xiao, D., & Zhu, J. (2016). Phase structure and piezoelectric properties of (1 _ xÞ- K0:48Na0:52Nb0:95Sb0:05O3–x(Bi0:5Na0:5Þ0:9(Li0:5Ce0:5Þ0:1ZrO3 lead-free piezoelectric ceramics. *Journal of Applied Physics*, *119*, 034101. doi:10.1063/1.4940191

Xu, Z., Cai, J. G., & Pan, B. C. (2013). Mathematically modeling fixed-bed adsorption in aqueous systems. In Journal of Zhejiang University: Science A (Vol. 14, Issue 3, pp. 155–176). doi:10.1631/jzus.A1300029

Xu, C., Li, B., Du, H., & Kang, F. (2012). Energetic zinc ion chemistry: The rechargeable zinc ion battery. *Angewandte Chemie*, *124*(4), 957–959. doi:10.1002/ange.201106307 PMID:22170816

Xu, D., Wang, H., Qiu, R., Wang, Q., Mao, Z., Jiang, Y., Wang, R., He, B., Gong, Y., Li, D., & Hu, X. (2020). Coupling of bowl-like VS2 nanosheet arrays and carbon nanofiber enables ultrafast Na+-Storage and robust flexibility for sodium-ion hybrid capacitors. *Energy Storage Materials*, *28*, 91–100. doi:10.1016/j.ensm.2020.03.002

Xue. (2019). One-step synthesis of 2-ethylhexylamine pillared vanadium disulfide nanoflowers with ultralarge interlayer spacing for high-performance magnesium storage. *Advanced Energy Materials, 9*(22).

Xu, G., & Shi, X. (2018). Characteristics and applications of fly ash as a sustainable construction material: A state-of-the-art review. *Resources, Conservation and Recycling*, *136*, 95–109. doi:10.1016/j.resconrec.2018.04.010

Xu, H., & Shen, M. (2021). The control of lithium-ion batteries and supercapacitors in hybrid energy storage systems for electric vehicles: A review. *International Journal of Energy Research*, *45*(15), 20524–20544. doi:10.1002/er.7150

Xu, H., & Yang, H. (2007). Magnetic properties of Y3Fe5O12 nanoparticles doped Bi and Ce ions. *Materials and Manufacturing Processes*, *23*(1), 1–4. doi:10.1080/10426910701524212

Xu, H., Yang, H., Xu, W., & Feng, S. (2008). Magnetic properties of Ce, Gd-substituted yttrium iron garnet ferrite powders fabricated using a sol–gel method. *Journal of Materials Processing Technology*, *197*(1-3), 296–300. doi:10.1016/j.jmatprotec.2007.06.061

Xu, J., Wang, S., Huang, T., Chen, J., & Li, H. (2022). Three-dimensional amorphous carbon rivet vanadium sulfide enhanced lithium storage. *Journal of Alloys and Compounds*, *929*, 166966. doi:10.1016/j.jallcom.2022.166966

Xu, K., Li, J., Lv, X., Wu, J., Zhang, X., Xiao, D., & Zhu, J. (2016). Superior piezoelectric properties in potassium–sodium niobate lead-free ceramics. *Advanced Materials*, *28*(38), 8519–8523. doi:10.1002/adma.201601859 PMID:27441456

Xu, L.-Y., Huang, B.-T., & Dai, J.-G. (2021). Development of engineered cementitious composites (ECC) using artificial fine aggregates. *Construction & Building Materials*, *305*, 124742. doi:10.1016/j.conbuildmat.2021.124742

Xu, L.-Y., Huang, B.-T., Li, V. C., & Dai, J.-G. (2022). High-strength high-ductility Engineered/Strain-Hardening Cementitious Composites (ECC/SHCC) incorporating geopolymer fine aggregates. *Cement and Concrete Composites*, *125*, 104296. doi:10.1016/j.cemconcomp.2021.104296

Xu, L.-Y., Qian, L.-P., Huang, B.-T., & Dai, J.-G. (2021). Development of artificial one-part geopolymer lightweight aggregates by crushing technique. *Journal of Cleaner Production*, *315*, 128200. doi:10.1016/j.jclepro.2021.128200

Xu, S., Qin, Y., Xu, C., Wei, Y., Yang, R., & Wang, Z. L. (2010). Self-powered nanowire devices. *Nature Nanotechnology*, *5*(5), 366–373. doi:10.1038/nnano.2010.46 PMID:20348913

Xu, S., & Wang, Z. L. (2011). One-dimensional ZnO nanostructures: Solution growth and functional properties. *Nano Research*, *4*(11), 1013–1098. doi:10.1007/s12274-011-0160-7

Xu, X., Wang, Z., Yu, J., Li, L., & Yan, X. (2019). Phase engineering for highly efficient quasi-two-dimensional all-inorganic perovskite light-emitting diodes via adjusting the ratio of Cs cation. *Nanoscale Research Letters*, *14*(255), 1–8. doi:10.1186/s11671-019-3076-x PMID:31352526

Yablonovitch, E. (1987). Inhibited spontaneous emission in solid-state physics and electronics. *Physical Review Letters*, *58*(20), 2059.

Yadegarifard, A., Lee, H., Seok, H. J., Kim, I., Ju, B. K., Kim, H. K., & Lee, D. K. (2023). FA/Cs-based mixed Pb–Sn perovskite solar cells: A review of recent advances in stability and efficiency. *Nano Energy*, *112*, 108481. doi:10.1016/j.nanoen.2023.108481

Yamada, Y., Nakamura, T., Endo, M., Wakamiya, A., & Kanemitsu, Y. (2014). Photocarrier recombination dynamics in perovskite $CH_3NH_3PbI_3$ for solar cell applications. *Journal of the American Chemical Society*, *136*(33), 11610–11613. doi:10.1021/ja506624n PMID:25075458

Yang. (2021). Regulating intercalation of layered compounds for electrochemical energy storage and electrocatalysis. *Advanced Functional Materials, 31*(52).

Yang, D., Chen, L., Wang, L., Cui, Z., Wen, Z., Yang, B., He, H., Mei, S., Zhang, W., Xie, F., Ma, J., & Guo, R. (2021). One-step synthesis of high-quality vanadium disulfide quantum dots for long-term lysosome-targetable imaging. *Sensors and Actuators. B, Chemical*, *346*, 130544. doi:10.1016/j.snb.2021.130544

Yang, F., Jang, D., Dong, L., Qiu, S., Distler, A., Li, N., Brabec, C. J., & Egelhaaf, H. J. (2021). Upscaling Solution-Processed Perovskite Photovoltaics. *Advanced Energy Materials*, *11*(42), 2101973. doi:10.1002/aenm.202101973

Yang, H., Kannappan, S., Pandian, A. S., Jang, J. H., Lee, Y. S., & Lu, W. (2017). Graphene supercapacitor with both high power and energy density. *Nanotechnology*, *28*(44), 445401. Advance online publication. doi:10.1088/1361-6528/aa8948 PMID:28854156

Yang, J., Wang, J., Dong, X., Zhu, L., Hou, D., Zeng, W., & Wang, J. (2021). The potential application of VS2 as an electrode material for Mg ion battery: A DFT study. *Applied Surface Science*, *544*, 148775. doi:10.1016/j.apsusc.2020.148775

Yang, L., Zhang, Y., Chu, M., Deng, W., Tan, Y., Ma, M., Su, X., Xie, Q., & Yao, S. (2014). Facile fabrication of network film electrodes with ultrathin Au nanowires for nonenzymatic glucose sensing and glucose/O2 fuel cell. *Biosensors & Bioelectronics*, *52*, 105–110. doi:10.1016/j.bios.2013.08.038 PMID:24035853

Yang, Q., Wang, W., Xu, S., & Wang, Z. L. (2011). Enhancing Light Emission of ZnO Microwire-Based Diodes by Piezo-Phototronic Effect. *Nano Letters*, *11*(9), 4012–4017. doi:10.1021/nl202619d PMID:21823605

Yang, R., Qin, Y., Dai, L., & Wang, Z. L. (2009). Power generation with laterally packaged piezoelectric fine wires. *Nature Nanotechnology*, *4*(1), 34–39. doi:10.1038/nnano.2008.314 PMID:19119280

Yang, R., Qin, Y., Li, C., Zhu, G., & Wang, Z. L. (2009). Converting biomechanical energy into electricity by a muscle-movement- driven nanogenerator. *Nano Letters*, *9*(3), 1201–1205. doi:10.1021/nl803904b PMID:19203203

Yang, T. Y., Gregori, G., Pellet, N., Grätzel, M., & Maier, J. (2015). The significance of ion conduction in a hybrid organic-inorganic lead-iodide-based perovskite photosensitizer. *Angewandte Chemie International Edition*, *54*(27), 7905–7910. doi:10.1002/anie.201500014 PMID:25980541

Yang, W. G., Zhang, B. P., Ma, N., & Zhao, L. (2012). High piezoelectric properties of BaTiO3–xLiF ceramics sintered at low temperatures. *Journal of the European Ceramic Society*, *32*(4), 899–904. doi:10.1016/j.jeurceramsoc.2011.10.054

Yang, W. S., Noh, J. H., Jeon, N. J., Kim, Y. C., Ryu, S., Seo, J., & Seok, S. I. (2015). High-performance photovoltaic perovskite layers fabricated through intramolecular exchange. *Science*, *348*(6240), 1234–1237. doi:10.1126/science.aaa9272 PMID:25999372

Yang, Y., Hao, H., Zhang, L., Chen, C., Luo, Z., Liu, Z., Yao, Z., Cao, M., & Liu, H. (2018). Structure, electrical and dielectric properties of Ca substituted BaTiO3 ceramics. *Ceramics International*, *44*(10), 11109–11115. doi:10.1016/j.ceramint.2018.03.111

Yang, Y., Zhang, H., Chen, J., Jing, Q., Zhou, Y. S., Wen, X., & Wang, Z. L. (2013). Single-Electrode-Based Sliding Triboelectric Nanogenerator for Self-Powered Displacement Vector Sensor System. *ACS Nano*, *7*(8), 7342–7351. doi:10.1021/nn403021m PMID:23883397

Yang, Z., Surrente, A., Galkowski, K., Miyata, A., Portugall, O., Sutton, R. J., Haghighirad, A. A., Snaith, H. J., Maude, D. K., Plochocka, P., & Nicholas, R. J. (2017). Impact of the halide cage on the electronic properties of fully inorganic cesium lead halide perovskites. *ACS Energy Letters*, *2*(7), 1621–1627. doi:10.1021/acsenergylett.7b00416

Yang, Z., & Zu, J. (2016). Comparison of PZN-PT, PMN-PT single crystals and PZT ceramic for vibration energy harvesting. *Energy Conversion and Management*, *122*, 321–329. doi:10.1016/j.enconman.2016.05.085

Yan, K., Wang, F., Chen, X., Wu, D., & Zhu, K. (2019). Effects of A-site ionic size on the phase transition behavior of lead-free niobate ceramics. *Ceramics International*, *45*(16), 20323–20330. doi:10.1016/j.ceramint.2019.07.005

Yao, Y., Lei, J., Shi, Y., Ai, F., & Lu, Y.-C. (2021). Assessment methods and performance metrics for redox flow batteries. *Nature Energy*, *6*(6), 582–588. doi:10.1038/s41560-020-00772-8

Yassine, M., & Fabris, D. (2017). Performance of commercially available supercapacitors. *Energies*, *10*(9), 1340. Advance online publication. doi:10.3390/en10091340

Ye, Z. G., & Dong, M. (2000). Morphotropic domain structures and phase transitions in relaxor-based piezo-/ferroelectric (1−x)Pb(Mg1/3Nb2/3)O3−xPbTiO3 single crystals. *Journal of Applied Physics*, *87*(5), 2312–2319. doi:10.1063/1.372180

Yin, B.-S., Zhang, S.-W., Xiong, T., Shi, W., Ke, K., Lee, W. S. V., Xue, J., & Wang, Z.-B. (2020). Engineering sulphur vacancy in VS 2 as high performing zinc-ion batteries with high cyclic stability. *New Journal of Chemistry*, *44*(37), 15951–15957. doi:10.1039/D0NJ02917C

Yin, W.-J., Shi, T., & Yan, Y. (2014). Unusual defect physics in $CH_3NH_3PbI_3$ perovskite solar cell absorber. *Applied Physics Letters*, *104*(6), 063903. doi:10.1063/1.4864778

Yin, X., Chen, P., Que, M., Xing, Y., Que, W., Niu, C., & Shao, J. (2016). Highly efficient flexible perovskite solar cells using solution-derived NiOx hole contacts. *ACS Nano*, *10*(3), 3630–3636. doi:10.1021/acsnano.5b08135 PMID:26958704

Yokosuka, M. (1977). A new transparent ferroelectric ceramic, lanthanum-modified lead-barium metaniobate (PBLN). *Japanese Journal of Applied Physics*, *16*(2), 379–380. doi:10.1143/JJAP.16.379

Yongping, F. (2015). Solution growth of single crystal Methylammonium Lead halide Perovskite nanostructures for optoelectronic and photovoltaic applications. *Journal of the American Chemical Society*, *137*(17), 5810–5818. doi:10.1021/jacs.5b02651 PMID:25871732

Yonzon, C., Stuart, D., Zhang, X., McFarland, A., Haynes, C., & Vanduyne, R. (2005). Towards advanced chemical and biological nanosensors—An overview. *Talanta*, *67*(3), 438–448. doi:10.1016/j.talanta.2005.06.039 PMID:18970187

Yoo, J. J., Wieghold, S., Sponseller, M. C., Chua, M. R., Bertram, S. N., Hartono, N. T. P., Tresback, J. S., Hansen, E. C., Correa-Baena, J.-P., Bulović, V., Buonassisi, T., Shin, S. S., & Bawendi, M. G. (2019). An interface stabilized perovskite solar cell with high stabilized efficiency and low voltage loss. *Energy & Environmental Science*, *12*(7), 2192–2199. doi:10.1039/C9EE00751B

You, Z., Zhao, L., Zhao, K., Liao, H., Wen, S., Xiao, Y., Cheng, B., & Lei, S. (2023). Highly tunable three-dimensional porous carbon produced from tea seed meal crop by-products for high performance supercapacitors. *Applied Surface Science, 607*(September), 155080. doi:10.1016/j.apsusc.2022.155080

Yousefi, A., Tang, W., Khavarian, M., & Fang, C. (2021). Development of novel form-stable phase change material (PCM) composite using recycled expanded glass for thermal energy storage in cementitious composite. *Renewable Energy*, *175*, 14–28. doi:10.1016/j.renene.2021.04.123

Yu. (2021). Boosting Zn2+ and NH4+ Storage in Aqueous Media via In-Situ Electrochemical Induced VS2/VOx Heterostructures. *Advanced Functional Materials, 31*(11).

Yuan, J., Wu, J., Hardy, W. J., Loya, P., Lou, M., Yang, Y., Najmaei, S., Jiang, M., Qin, F., Keyshar, K., Ji, H., Gao, W., Bao, J., Kono, J., Natelson, D., Ajayan, P. M., & Lou, J. (2015). Facile synthesis of single crystal vanadium disulfide nanosheets by chemical vapor deposition for efficient hydrogen evolution reaction. *Advanced Materials*, *27*(37), 5605–5609. doi:10.1002/adma.201502075 PMID:26293810

Yuan, Y., Wu, J., Tao, H., Lv, X., Wang, X., & Lou, X. (2015). Composition design and electrical properties in (1 _ yÞ(K0:40- Na0:60Þ0:985Li0:015(Nb1_xSbxÞO3–yBi0:5Na0:5ZrO3 lead-free ceramics. *Journal of Applied Physics*, *117*, 084103. doi:10.1063/1.4913284

Yu, B., Li, M., Wang, J., Pei, L., Guo, D., & Zhao, X. (2008). Enhanced electrical properties in multiferroic BiFeO3 ceramics co-doped by La3+ and V5+. *Journal of Physics. D, Applied Physics*, *41*(18), 185401. doi:10.1088/0022-3727/41/18/185401

Yu, C. H., Huang, C. H., & Tan, C. S. (2012). A review of CO2 capture by absorption and adsorption. *Aerosol and Air Quality Research*, *12*(5), 745–769. doi:10.4209/aaqr.2012.05.0132

Yu, D., Pang, Q., Gao, Y., Wei, Y., Wang, C., Chen, G., & Du, F. (2018). Hierarchical flower-like VS2 nanosheets–A high rate-capacity and stable anode material for sodium-ion battery. *Energy Storage Materials*, *11*, 1–7. doi:10.1016/j.ensm.2017.09.002

Yu, H., Ryu, J., Lee, J. W., Roh, J., Lee, K., Yun, J., Lee, J., Kim, Y. K., Hwang, D., Kang, J., Kim, S. K., & Jang, J. (2017). Large grain-based hole-blocking layer free planar-type perovskite solar cell with best efficiency of 18.20%. *ACS Applied Materials & Interfaces*, *9*(9), 8113–8120. doi:10.1021/acsami.6b15710 PMID:28211274

Yu, L., & Chen, G. Z. (2020). Supercapatteries as High-Performance Electrochemical Energy Storage Devices. *Electrochemical Energy Reviews*, *3*(2), 271–285. doi:10.1007/s41918-020-00063-6

Yun. (2022). Review on Oxygen-Free Vanadium-Based Cathodes for Aqueous Zinc-Ion Batteries. *Journal of Electrochemistry, 28*(11).

Yun, S., Wang, X., & Xu, D. (2008). Influence of processing parameters on the structure and properties of barium strontium titanate ceramics. *Materials Research Bulletin*, *43*(8-9), 1989–1995. doi:10.1016/j.materresbull.2007.10.003

Yu, S. H., Tang, Z., Shao, Y., Dai, H., Wang, H. Y., Yan, J., Pan, H., & Chua, D. H. C. (2019). In situ hybridizing MoS2 microflowers on VS2 microflakes in a one-pot CVD process for electrolytic hydrogen evolution reaction. *ACS Applied Energy Materials*, *2*(8), 5799–5808. doi:10.1021/acsaem.9b00928

Yu, S., Jeong, S.-G., Chung, O., & Kim, S. (2014). Bio-based PCM/carbon nanomaterials composites with enhanced thermal conductivity. *Solar Energy Materials and Solar Cells*, *120*, 549–554. doi:10.1016/j.solmat.2013.09.037

Zaaba, N. I., Foo, K. L., Hashim, U., Tan, S. J., Liu, W.-W., & Voon, C. H. (2017). Synthesis of graphene oxide using modified hummers method: Solvent influence. *Procedia Engineering*, *184*, 469–477. doi:10.1016/j.proeng.2017.04.118

Zamarripa, M. A., Eslick, J. C., Matuszewski, M. S., & Miller, D. C. (2018). Multi-objective optimization of membrane-based CO2 capture. *Computer-Aided Chemical Engineering*, *44*, 1117–1122. doi:10.1016/B978-0-444-64241-7.50181-6

Zhang, L., Hu, X., Wang, Z., Sun, F., & Dorrell, D. G. (2018). A review of supercapacitor modeling, estimation, and applications: A control/management perspective. *Renewable and Sustainable Energy Reviews, 81*(February), 1868–1878. doi:10.1016/j.rser.2017.05.283

Zhang. (2020). Insights into the Storage Mechanism of Layered VS2 Cathode in Alkali Metal-Ion Batteries. *Advanced Energy Materials, 10*(22).

Zhang, B., Wu, J., Cheng, X., Wang, X., Xiao, D., Zhu, J., & Lou, X. (2013). Lead-free piezoelectrics based on potassium-sodium niobate with giant d33. *ACS Applied Materials & Interfaces*, *5*(16), 7718–7725. doi:10.1021/am402548x PMID:23937157

Zhang, G. J., Zhang, L., Huang, M. J., Luo, Z. H. H., Tay, G. K. I., Lim, E. J. A., Kang, T. G., & Chen, Y. (2010). Silicon nanowire biosensor for highly sensitive and rapid detection of Dengue virus. *Sensors and Actuators. B, Chemical*, *146*(1), 138–144. doi:10.1016/j.snb.2010.02.021

Zhang, H., Gao, H., Bernardo, E., Lei, S., & Wang, L. (2023). Thermal energy storage performance of hierarchical porous kaolinite geopolymer based shape-stabilized composite phase change materials. *Ceramics International*, *49*(18), 29808–29819. doi:10.1016/j.ceramint.2023.06.238

Zhang, J. L., Ji, P. F., Wu, Y. Q., Zhao, X., Tan, Y. Q., & Wang, C. L. (2014). Strong piezoelectricity exhibited by large-grained BaTiO3 ceramics. *Applied Physics Letters*, *104*(22), 222909. doi:10.1063/1.4881597

Zhang, J., Xu, Y., Liu, Z., Yang, W., & Liu, J. (2015). A highly conductive porous graphene electrode prepared via in situ reduction of graphene oxide using Cu nanoparticles for the fabrication of high performance supercapacitors. *RSC Advances*, *5*(67), 54275–54282. doi:10.1039/C5RA07857A

Zhang, L., Wei, Q., Sun, D., Li, N., Ju, H., Feng, J., Zhu, J., Mai, L., Cairns, E. J., & Guo, J. (2018). Conversion reaction of vanadium sulfide electrode in the lithium-ion cell: Reversible or not reversible? *Nano Energy*, *51*, 391–399. doi:10.1016/j.nanoen.2018.06.076

Zhang, M. Y., Miao, J. Y., Yan, X. H., Zhu, Y. H., Li, Y. L., Zhang, W. J., Zhu, W., Pan, J. M., Javed, M. S., & Hussain, S. (2022). Vanadium disulfide nanosheets loaded on carbon cloth as electrode for flexible quasi-solid-state asymmetric supercapacitors: Energy storage mechanism and electrochemical performance. *Journal of Materials Chemistry. C, Materials for Optical and Electronic Devices*, *10*(2), 640–648. doi:10.1039/D1TC03903B

Zhang, N., Glazer, A. M., Baker, D., & Thomas, P. A. (2009). Structures of K0. 05Na0. 95NbO3 (50–300 K) and K0. 30Na0. 70NbO3 (100–200 K). *Acta Crystallographica. Section B, Structural Science*, *65*(3), 291–299. doi:10.1107/S0108768109011057 PMID:19461138

Zhang, N., Zheng, T., & Wu, J. (2020). Lead-free (K, Na) NbO3-based materials: Preparation techniques and piezoelectricity. *ACS Omega*, *5*(7), 3099–3107. doi:10.1021/acsomega.9b03658 PMID:32118126

Zhang, R. R., Xu, Y. M., Harrison, D., Fyson, J., Qiu, F. L., & Southee, D. (2015). Flexible strip supercapacitors for future energy storage. *International Journal of Automation and Computing*, *12*(1), 43–49. doi:10.1007/s11633-014-0866-6

Zhang, S., Chang, P., Zhang, Y., Xu, X., Guan, L., & Tao, J. (2022). Effect of morphology and stacking on atomic interaction and magnetic characteristics in two-dimensional H-phase VS2 few layers. *Journal of Materials Science*, *57*(10), 5873–5884. doi:10.1007/s10853-022-06904-7

Zhang, W., Chen, X., Zhang, J., Tuo, C., Ji, L., Li, H., Zhang, X., & Yang, F. (2018). Exposure of active edge structure for electrochemical H2 evolution from VS2/MWCNTs hybrid catalysts. *International Journal of Hydrogen Energy*, *43*(51), 22949–22954. doi:10.1016/j.ijhydene.2018.10.168

Zhang, X., Bai, C., Qiao, Y., Wang, X., Jia, D., Li, H., & Colombo, P. (2021). Porous geopolymer composites: A review. *Composites. Part A, Applied Science and Manufacturing*, *150*, 106629. doi:10.1016/j.compositesa.2021.106629

Zhang, X., Le, M. Q., Zahhaf, O., Capsal, J. F., Cottinet, P. J., & Petit, L. (2020). Enhancing dielectric and piezoelectric properties of micro-ZnO/PDMS composite-based dielectrophoresis. *Materials & Design*, *192*, 108783. doi:10.1016/j.matdes.2020.108783

Zhang, X., Wang, J., Song, Z., & Zhou, T. (2021). Data-driven ionic liquid design for CO_2 capture: Molecular structure optimization and DFT verification. *Industrial & Engineering Chemistry Research*, *60*(27), 9992–10000. doi:10.1021/acs.iecr.1c01384

Zhang, Y., Hong, J., Liu, B., & Fang, D. (2010). Strain effect on ferroelectric behaviors of $BaTiO_3$ nanowires: A molecular dynamics study. *Nanotechnology*, *21*(1), 015701. Advance online publication. doi:10.1088/0957-4484/21/1/015701 PMID:19946153

Zhang, Y., Li, J., Ma, L., Li, H., Xu, X., Liu, X., Lu, T., & Pan, L. (2022). Insights into the storage mechanism of 3D nanoflower-like V_3S_4 anode in sodium-ion batteries. *Chemical Engineering Journal*, *427*, 130936. doi:10.1016/j.cej.2021.130936

Zhang, Y., Zhai, J., & Xue, S. (2020). Effect of three step sintering on piezoelectric properties of KNN-based lead free ceramics. *Chemical Physics Letters*, *758*, 137906. doi:10.1016/j.cplett.2020.137906

Zhang, Z., Huang, X., Wang, H., Teo, S. H., & Ma, T. (2019). Free-standing $NiCo_2S_4$@ VS_2 nanoneedle array composite electrode for high performance asymmetric supercapacitor application. *Journal of Alloys and Compounds*, *771*, 274–280. doi:10.1016/j.jallcom.2018.08.325

Zhao, Wang, & Mao. (2004). *Piezoelectric Characterization of Individual Zinc Oxide Nanobelt Probed by Piezoresponse Force Microscope*. doi:10.1021/nl035198a

Zhao, C., Wu, B., Thong, H. C., & Wu, J. (2018). Improved temperature stability and high piezoelectricity in lead-free barium titanate-based ceramics. *Journal of the European Ceramic Society*, *38*(16), 5411–5419. doi:10.1016/j.jeurceramsoc.2018.08.004

Zhao, M., Gao, Y., Sun, J., & Gao, F. (2015). Mediatorless glucose biosensor and direct electron transfer type glucose/air biofuel cell enabled with carbon nanodots. *Analytical Chemistry*, *87*(5), 2615–2622. doi:10.1021/acs.analchem.5b00012 PMID:25666266

Zhao, R., Deng, S., Wang, S., Zhao, L., Zhang, Y., Liu, B., Li, H., & Yu, Z. (2018). Thermodynamic research of adsorbent materials on energy efficiency of vacuum-pressure swing adsorption cycle for CO_2 capture. *Applied Thermal Engineering*, *128*, 818–829. doi:10.1016/j.applthermaleng.2017.09.074

Zhao, S. Y., & Bian, J. J. (2023). Effects of Li_2CO_3–CuO addition on the sintering behavior, dielectric and piezoelectric properties of PZT ceramics. *Journal of Materials Science Materials in Electronics*, *34*(15), 1202. doi:10.1007/s10854-023-10658-6

Zhao, X., Wang, H., Yuan, C., Xu, J., Cui, Y., & Ma, J. (2013). Effects of Co doping on microstructure and properties of $(K_{0.5}Na_{0.5})NbO_3$-$LiSbO_3$-$BiFe_{1-x}Co_xO_3$ lead-free piezoelectric ceramics. *Journal of Materials Science Materials in Electronics*, *24*(5), 1480–1484. doi:10.1007/s10854-012-0958-4

Zhao, Y., Tan, H., Yuan, H., Yang, Z., Fan, J. Z., Kim, J. H., Voznyy, O., Gong, X., Quan, L. N., Tan, C. S., Hofkens, J., Yu, D., Zhao, Q., & Sargent, E. H. (2018). Perovskite seeding growth of formamidinium-lead-iodide-based perovskites for efficient and stable solar cells. *Nature Communications*, *9*(1), 1607. doi:10.1038/s41467-018-04029-7 PMID:29686304

Zhao, Y., & Zhu, K. (2016). Organic–inorganic hybrid lead halide perovskites for optoelectronic and electronic applications. *Chemical Society Reviews*, *45*(3), 655–689. doi:10.1039/C4CS00458B PMID:26645733

Zhao, Z. H., Ge, R. F., & Dai, Y. (2019). Large electro-strain signal of the BNT–BT–KNN lead-free piezoelectric ceramics with CuO doping. *Journal of Advanced Dielectrics*, *9*(03), 1950022. doi:10.1142/S2010135X1950022X

Zheng, J. P. (2005). Theoretical Energy Density for Electrochemical Capacitors with Intercalation Electrodes. *Journal of the Electrochemical Society*, *152*(9), A1864. doi:10.1149/1.1997152

Zheng, T., Wu, H., Yuan, Y., Lv, X., Li, Q., Men, T., Zhao, C., Xiao, D., Wu, J., Wang, K., Li, J.-F., Gu, Y., Zhu, J., & Pennycook, S. J. (2017). The structural origin of enhanced piezoelectric performance and stability in lead free ceramics. *Energy & Environmental Science*, *10*(2), 528–537. doi:10.1039/C6EE03597C

Zheng, T., Wu, J., Cheng, X., Wang, X., Zhang, B., Xiao, D., Zhu, J., & Lou, X. (2014). Wide phase boundary zone, piezoelectric properties and stability in 0.97(K0:4Na0:6Þ(Nb1_xSbxÞO3–0.03Bi0:5Li0:5- ZrO3 lead-free ceramics. *Dalton Transactions (Cambridge, England)*, *43*(25), 9419. doi:10.1039/c4dt00768a PMID:24848676

Zheng, T., Wu, J., Xiao, D., Zhu, J., Wang, X., & Lou, X. (2015). Composition-driven phase boundary and piezoelectricity in potassium–sodium niobate-based ceramics. *ACS Applied Materials & Interfaces*, *7*(36), 20332–20341. doi:10.1021/acsami.5b06033 PMID:26302094

Zheng, T., Wu, J., Xiao, D., Zhu, J., Wang, X., & Lou, X. (2015). Potassium-sodium niobate lead-free ceramics: Modified strain as well as piezoelectricity. *Journal of Materials Chemistry. A, Materials for Energy and Sustainability*, *3*(5), 1868–1874. doi:10.1039/C4TA05423G

Zheng, X., Ahmad, T., & Chen, W. (2021). Challenges and strategies on Zn electrodeposition for stable Zn-ion batteries. *Energy Storage Materials*, *39*, 365–394. doi:10.1016/j.ensm.2021.04.027

Zhi, W., Guobin, R., Shuqin, L., & Shuisheng, J. (2003). Supercell lattice method for photonic crystal fibers. *Optics Express*, *11*(9), 980. doi:10.1364/OE.11.000980 PMID:19465960

Zhong, X., Tang, J., Wang, J., Shao, M., Chai, J., Wang, S., Yang, M., Yang, Y., Wang, N., Wang, S., Xu, B., & Pan, H. (2018). 3D heterostructured pure and N-Doped Ni3S2/VS2 nanosheets for high efficient overall water splitting. *Electrochimica Acta*, *269*, 55–61. doi:10.1016/j.electacta.2018.02.131

Zhou, C., & Liu, X. (2008). Dielectric and piezoelectric properties of bismuth-containing complex perovskite solid solution of Bi 1/2 Na 1/2 TiO 3– Bi (Mg 2/3 Nb 1/3) O 3. *Journal of Materials Science*, *43*(3), 1016–1019. doi:10.1007/s10853-007-2246-x

Zhou, H., Chen, Q., Li, G., Luo, S., Song, T. B., Duan, H. S., Hong, Z., Jingbi You, J., Liu, Y., & Yang, Y. (2014). Interface engineering of highly efficient perovskite solar cells. *Science*, *345*(6196), 542–546. doi:10.1126/science.1254050 PMID:25082698

Zhou, J. J., Cheng, L. Q., Wang, K., Zhang, X. W., Li, J. F., Liu, H., & Fang, J. Z. (2014). Low-temperature sintering of (K, Na) NbO3-based lead-free piezoceramics with addition of LiF. *Journal of the European Ceramic Society*, *34*(5), 1161–1167. doi:10.1016/j.jeurceramsoc.2013.11.029

Zhou, J., Xu, N., & Wang, Z. L. (2006). Dissolving behavior and stability of ZnO wires in biofluids: A study on biodegradability and biocompatibility of ZnO nanostructures. *Advanced Materials*, *18*(18), 2432–2435. doi:10.1002/adma.200600200

Zhou, X., Xue, G., Luo, H., Bowen, C. R., & Zhang, D. (2021). Phase structure and properties of sodium bismuth titanate lead-free piezoelectric ceramics. *Progress in Materials Science*, *122*, 100836. doi:10.1016/j.pmatsci.2021.100836

Zhou, Y., Sharma, S. K., Peng, Z., & Leblanc, R. M. (2017). Polymers in carbon dots: A review. *Polymers*, *9*(2), 67. doi:10.3390/polym9020067 PMID:30970747

Zhu, G., Chen, J., Liu, Y., Bai, P., Zhou, Y. S., Jing, Q., Pan, C., & Wang, Z. L. (2013). Linear-Grating Triboelectric Generator Based on Sliding Electrification. *Nano Letters*, *13*(5), 2282–2289. doi:10.1021/nl4008985 PMID:23577639

Zhu, J., Cai, L., Yin, X., Wang, Z., Zhang, L., Ma, H., Ke, Y., Du, Y., Xi, S., Wee, A. T. S., Chai, Y., & Zhang, W. (2020). Enhanced electrocatalytic hydrogen evolution activity in single-atom Pt-decorated VS2 nanosheets. *ACS Nano*, *14*(5), 5600–5608. doi:10.1021/acsnano.9b10048 PMID:32320212

Zhu, J., Jian, T., Wu, Y., Ma, W., Lu, Y., Sun, L., Meng, F., Wang, B., Cai, F., Gao, J., Li, G., Yang, L., Yan, X., & Xu, C. (2021). A highly stable aqueous Zn/VS2 battery based on an intercalation reaction. *Applied Surface Science*, *544*, 148882. doi:10.1016/j.apsusc.2020.148882

Zhuk, P. (2019). Lifecycle analysis of finishing products enhanced with phase changing materials. *IOP Conference Series. Earth and Environmental Science*, *323*(1), 12154. doi:10.1088/1755-1315/323/1/012154

Zhu, L. F., Zhang, B. P., Li, S., & Zhao, G. L. (2017). Large piezoelectric responses of Bi (Fe, Mg, Ti) O3-BaTiO3 lead-free piezoceramics near the morphotropic phase boundary. *Journal of Alloys and Compounds*, *727*, 382–389. doi:10.1016/j.jallcom.2017.08.014

Zhu, L. F., Zhang, B. P., Li, S., Zhao, L., Wang, N., & Shi, X. C. (2016). Enhanced piezoelectric properties of Bi (Mg1/2Ti1/2) O3 modified BiFeO3–BaTiO3 ceramics near the morphotropic phase boundary. *Journal of Alloys and Compounds*, *664*, 602–608. doi:10.1016/j.jallcom.2016.01.003

Zhu, Z., Cheng, F., & Chen, J. (2013). Investigation of effects of carbon coating on the electrochemical performance of Li 4 Ti 5 O 12/C nanocomposites. *Journal of Materials Chemistry. A, Materials for Energy and Sustainability*, *1*(33), 9484–9490. doi:10.1039/c3ta00114h

Zihan, Y. (2022). A Review of the Extremely Structural Properties of Carbon Nanotubes, Synthesis and Purification Methods. *American Journal of Materials Synthesis and Processing*, *7*(2), 29–35.

Zimele, Z., Sinka, M., Korjakins, A., Bajare, D., & Sahmenko, G. (2019). Life Cycle Assessment of Foam Concrete Production in Latvia. *Environmental & Climate Technologies, 23*(3).

Zubizarreta, L., Arenillas, A., Domínguez, A., Menéndez, J. A., & Pis, J. J. (2008). Development of microporous carbon xerogels by controlling synthesis conditions. *Journal of Non-Crystalline Solids*, *354*(10–11), 817–825. doi:10.1016/j.jnoncrysol.2007.08.015

Zuo, R., Fang, X., & Ye, C. (2007). Phase structures and electrical properties of new lead-free (Na0. 5K0. 5) NbO3–(Bi0. 5Na0. 5) TiO3 ceramics. *Applied Physics Letters*, *90*(9), 092904. doi:10.1063/1.2710768

Zuo, R., Fu, J., Lv, D., & Liu, Y. (2010). Antimony tuned rhombohedral-orthorhombic phase transition and enhanced piezoelectric properties in sodium potassium niobate. *Journal of the American Ceramic Society*, *93*(9), 2783–2787. doi:10.1111/j.1551-2916.2010.03804.x

Zuo, R., Wang, H., Ma, B., & Li, L. (2009). Effects of Nb 5+ doping on sintering and electrical properties of lead-free (Bi 0.5 Na 0.5) TiO 3 ceramics. *Journal of Materials Science Materials in Electronics*, *20*(11), 1140–1143. doi:10.1007/s10854-008-9840-9

About the Contributors

Shrikaant Kulkarni, Ph.D., is currently working as a Research and Technical Advisor, at Filantrofia Sarvada CleanTech Pvt. Ltd., Pune, India, and as an Adjunct Professor at Padm. Dr. V. B. Kolte College of Engineering, Malkapur, India. Dr. Kulkarni is an academician and researcher for 39 years Dr. Kulkarni has delivered invited lectures and conducted sessions at national and international conferences as well as faculty development programs. He has been teaching subjects such as engineering chemistry, green chemistry, nanotechnology, analytical chemistry, catalysis, chemical engineering materials, industrial organization, and management, to name a few. He has published over 100 research papers in national and international journals and conferences. He has authored 42 book chapters in CRC, Springer Nature, CRC, AAP Press, Elsevier, Wiley, and IET books. He has edited twelve books published by Apple Academic Press/CRC Press and Springer Nature. Another six books are in the offing. He authored four textbooks too and worked as a resource person for many national and international events. Dr. Kulkarni possesses M.Sc., M.Phil., and Ph.D. in Chemistry and Master's degrees in Economics, Business Management, and Political Science. He has expertise in Material Science, Green Chemistry & Engineering, Analytical Chemistry, Green Nanoscience & Nanotechnology.

Boris Kharisov, Dr Sci, is currently a professor and researcher at the Universidad Autónoma de Nuevo León (UANL), Monterrey, Mexico. He is a member of the Mexican Academy of Science, National Researchers System (Level III), and Materials Research Society. He is the coauthor of 16 books and 215 articles; he also has 2 patents. Kharisov is the coeditor of three invited special issues of international journals. He has over 5,000 citations and h+ index 34.

A. K. Haghi, PhD, is a retired professor and has written, co-written, edited or co-edited more than 1000 publications, including books, book chapters, and papers in refereed journals with over 3800 citations and h-index of 32, according to the Google Scholar database. He is currently a senior material engineering consultant in the UK. Professor Haghi has received several grants, consulted for several major corporations, and is a frequent speaker to national and international audiences. He is Founder and former Editor-in-Chief of the International Journal of Chemoinformatics and Chemical Engineering and Polymers Research Journal. Professor Haghi has acted as an editorial board member of many international journals. He has served as a member of the Canadian Research and Development Center of Sciences & Cultures and the Research Chemistry Centre, Coimbra, Portugal. He has supervised several PhD and MSc theses at the University of Guilan (UG) and co-supervised international doctoral projects. Professor Haghi holds a BSc in urban and environmental engineering from the University of North Carolina (USA) and holds two MSc, one in mechanical engineering from North Carolina State University (USA) and another one in applied mechanics, acoustics, and materials from the Université de Technologie de Compiègne (France). He was awarded a PhD in engineering sciences at Université de Franche-Comté (France). He is a persistent reviewer of leading international journals.

Vipul Srivastava is presently working as a Professor of Physics, Department of Physics, School of Chemical Engineering and Physical Sciences, Lovely Professional University, Phagwara, Punjab, India. He obtained his Ph.D. in Physics from Barkatullah University, Bhopal, India in 2005. He has a total of 21 years of research and teaching experience. His current areas of interest are materials science, condensed matter physics, electronic structure calculations, Density Functional Theory, optoelectronics, thermoelectrics, Heusler alloys, and Perovskite materials. He has guided seven Ph.D. scholars and many UG/PG students in their dissertation work. He received the Young Scientist award from the Government of India, Department of Science and Technology, New Delhi in 2005. He handled three research projects funded by the Government of India. He has published 100+ research articles, of which 90 articles are SCI indexed. His h-index is 23. He has attended more than 50 conferences in India and abroad and organized many workshops, seminars, and conferences. He is an active reviewer of 30+ indexed journals in Physics. In 2020 he has been awarded as research excellence award from Lovely Professional University, Punjab, India. Further, in 2021 he has been awarded a research appreciation award from his current University. Presently, he is contributing to the field of theoretical physics to discover novel materials for technological applications.

* * *

Abdul Aleem is a Professor having 17 years teaching experience and 19 years industrail experience.

Irom Monika Aniz, Ph. D. Scholar, Department of Physics, National Insitute of Technology Manipur. Special Interest in Material Science: Piezoceramics, processing and characterizations.

Ranjit Barua is a Senior Researchers at Centre for Healthcare Science and Technology, IIEST-SHIBPUR. Core research area Bio-Mechanical System, 3D Bioprinting, Tissue Engineering.

Jesús Botello-González is a full-time professor in the Chemical Engineering Department of Facultad de Ciencias Químicas, UANL, since 2013. Master of mathematics in the chemical engineering and environmental engineering careers. PhD since 2020. Researcher of mathematical models involved in adsorption processes. My main research indicators are h-Index =1 (Google Scholar,) and i10-Index =1

Luis Cavazos-Cuello was born on february 14th, 1997 in the city of Monterrey, Nuevo León, México, Luis Alfonso Cavazos is a Ph.D. student at Universidad Autónoma de Nuevo León. With a bachelors degree in Chemical engineering and a masters degree in sustainable industrial processes, he specializes in the study of the adsorption phenomenon using carbonaceous materials such as activated carbon and carbon xerogels among many others. He also studies the adsorption process from a phenomenological perspective, including batch and continuous mass transfer mathematical modelling of the adsorption process in carbonaceous materials.

Chiranjit Chaliha, Ph. D. Scholar (JRF), Department of Physics, National Insitute of Technology Manipur. Special Interest in Material Science: Piezoceramics, processing and characterizations.

Sudipto Datta completed my doctorate from IIEST, Shibpur, My research area are bioprinting, biomaterials and tissue engineering. I did B.tech in Electrical and Electronics Engineering from WBUT and M.Tech in Material Science and Engineering from IIEST, Shibpur. Associate Member of Institution of Engineers, Kolkata. Ex Research Associate at IIT Roorkee.

Nancy Elizabeth Dávila Guzmán received her PhD degree in Sustainable Processes from the Autonomous University of Nuevo León. She is part of the National Researchers System, and leads the research group of Chemical Engineering at UANL. As a founding member and president of the Mexican Adsorption Association, she's contributed significantly to its growth. With over 25 scientific articles and 4 book chapters, her research primarily focuses on adsorption-based separation processes. Her work delves into water treatment, air purification, biofuel purification, green synthesis of organometallic frameworks, and 3D-printed adsorbent synthesis. Her expertise earned her the prestigious UANL Research Award in 2022. Dr. Davila is also interested in educational innovation through mobile app development and augmented reality objects for STEM fields. Passionate about science communication, she initiated the YouTube channel "CienciasIQ UANL," offering fundamental science content. She remains committed to promoting women in science, evident in her role as former coordinator of the "Científicas Mexicanas" podcast, aimed at highlighting women's contributions in science.

Devaraj E. has earned his Ph.D. in Mechanical engineering from Bangalore University and a Master of Engineering in Machine Design from Bangalore University, UVCE. Dr. Devaraj has worked in teaching and industry for more than ten years. He has ten publications in Scopus-indexed international journals. In Product Life Cycle Management and Introduction to Business Management, he has written two books and has Australian patent on Development of Megawatt Wind Turbine for Optimal Management of Smart Agricultural Farms. Polymer Composites, Biomaterials, FEA, and Fracture Mechanics. are some of his research interests.

Lucy González is a professor at the Tecnológico de Monterrey since 2018, affiliated with the Research Group on Decarbonization, Climate Change, and Circular Economy at ITESM, and a member of the National System of Researchers, Level II. She earned her Bachelor's degree in Industrial Chemistry from the Faculty of Chemical Sciences at UANL, a Master's in Analytical Chemistry, and a Ph.D. in Environmental Analytical Chemistry from the same institution. With over 40 published scientific articles in high-impact international journals, she has spearheaded research projects to develop and implement cutting-edge analytical techniques to characterize environmental pollutants. Additionally, she has actively contributed to various investigations to advance new technologies for remediating contaminated sites. Finally, her research interests encompass the chemical characterization and monitoring of environmental pollutants, climate change, and advanced environmental remediation technologies.

Ilse María Hernández Romero is a researcher at the Institute of Advanced Materials for Sustainable Manufacturing at Tecnológico de Monterrey, Campus Monterrey. She obtained her Master's and Ph.D. in Chemical Engineering from the Universidad Michoacana de San Nicolás de Hidalgo in 2017 and 2021, respectively, with the latter receiving Honorable Mention. Additionally, she is a member of the National

System of Researchers, Level I. Her research interests are mathematical modeling, simulation, machine learning, optimization, and quantum computing. She has authored approximately 14 high-impact scientific articles on reducing CO_2 emissions, implementing technologies, developing optimal operational policies, decarbonization and circular economy. Furthermore, Dr. Hernández has conducted research visits at prominent foreign institutions such as the Technical University of Denmark, the University of Wisconsin-Madison, and Texas A&M University. She has actively participated in international and national conferences, including the American Institute of Chemical Engineers (AIChE), the International Federation of Automatic Control (IFAC), and the Mexican Academy of Energy. In addition to her research endeavors, she has led projects to secure funding at both international and national levels. Finally, Dr. Hernández plays a significant role in teaching and advising undergraduate and graduate students.

Karina Hernández Romero completed her undergraduate and graduate studies at the Universidad Michoacana de San Nicolas de Hidalgo, where she first obtained a Bachelor of Science in Mechanical Engineering (2018) and later the degree of Master of Science in Mechanical Engineering with an Honorable Mention (2021). He is currently pursuing her doctoral studies at the same institution. Her research interests include mechanical design, simulation, materials, computational modeling, mechanics, and biomechanics. He has published two scientific articles about hip prosthesis design and simulation. She has also participated in national and international congresses such as the International Symposium on Bioengineering, the International Congress of Academia Journals, and the Congress of Iberoamerican Researchers. In addition, she has given lectures in student chapters such as the Symposium "Innovation of Materials" at ITESI (Mex) and in the framework of International Women's Day with the lecture entitled "Women were also born with ingenuity" at the UMSNH.

Mukesh Kumar was born in Kangra, Himachal Pradesh, India in 1974, received the B.Sc. degree from the Himachal Pradesh University, Shimla in 1995. After completing his M.Sc. in Physics, he completed his Ph.D. degrees in Physics from Department of Physics, Guru Nanak Dev University, Amritsar, Punjab in 2005. During his Ph.D., he has studied the Talwandi Sabo area of the Bathinda district, Punjab for the possibility cause of cancer. At present he is working in the Department of Physics, Lovely Professional University, Phagwara, Punjab, India as Professor. He has been the part of the various committees related to the curriculum, admission, examination and other activities. His research interests include Radiation Physics, Natural radioactivity, Material Science (polymer composites and other materials for the energy storage as supercapacitors, gas sensing etc.). He has guided three research students and five students are pursuing the Ph.D. He has guided many PG students (M.Sc. and M.Phil.) in their dissertation/project work. He has published more than 40 research papers. He has reviewed many research papers from the journals of international repute. He has been the life time member of the Nuclear Track Society of India (NTSI) and Indian Science Congress Association.

Mamata Maisnam, Associate Professor, Department of Physics, National Institute of Technology Manipur. Special interest in Material Science: Electroceramics, processing and characterizations.

Arun Murugesan graduated B.E. – Civil Engineering in 2007, M.Tech. - Structural Engineering in 2009 and Completed Ph.D. in Civil Engineering in 2015.Joined Bannariamman Institute of Technology in 2009 as a lecturer, Currently working as Associate Professor, In total 14 years of Teaching Experience. Has published 28 research articles in reputed international journals (SCI/SCIE /WoS /CARE), 9 papers in National and 19 International conferences (which includes Kuwait and Turkey). Granted a patent on Preparation of Nano-Silica from waste Foundry Sand in 2021 (Patent no: 355945 - Indian Patent) Completed a DST – SERB funded project of worth Rs 18.3 lakhs (2019 – 2023). Has participated in consultancy works of worth around Rs 10,00,000/- (from 2012 – 2023) related to Testing on construction materials, Structural design and detailing, Field test on bearing capacity of soil and Site inspection on detailing of reinforcement, Rebar location, Strength assessment of reinforced concrete elements (Destructive) and structures (NDT) and Mix Design. Received "Young Faculty Achiever Award – 2018" from Engineering Professional Society. Has Guided 32 undergraduate, 14 Post Graduate projects and guiding Doctoral degree thesis. Supporting 10 Doctoral theses as Doctoral Committee member. Acted as a representative for foreign Examiner for Ph.D., Viva-Voce.

R. Nagarajan received his B.E. in Electrical and Electronics Engineering from Madurai Kamarajar University, Madurai, India, in 1997. He received his M.E. in Power Electronics and Drives from Anna University, Chennai, India, in 2008. He received his Ph.D in Electrical Engineering from Anna University, Chennai, India, in 2014. He has worked in the industry as an Electrical Engineer. He is currently working as Professor of Electrical and Electronics Engineering at Gnanamani College of Technology, Namakkal, and Tamilnadu, India. His current research interest includes Power Electronics, Power System, Network Security, Wireless Sensor Networks, Cloud Computing, Image Processing, Soft Computing Techniques and Renewable Energy Sources. He is published more than 120 research papers and more than 60 Books & Book Chapters in various referred international Publications.

Michael Pérez-Rodríguez is a researcher with a PhD in Analytical Chemistry and postdoctoral stays in Argentina (UNNE), Brazil (UNESP) and Mexico (UNAM) addressing issues related to the development, optimization and validation of new methods and procedures for analysis of pharmaceuticals, foodstuffs and biological matrices. I possess analytical and instrumental chemistry experience, focused on techniques of extraction/microextraction (SPE, SPME, HF-LPME, US-DLLME), molecular spectroscopy (UV-VIS, FT-IR, Raman), atomic spectroscopy (AAS, ICP-OES, ICP-MS, LIBS), microscopy (SEM), thermal analysis (DSC), chromatography (HPLC-UV, HPLC-MS, GC-ECD), and fabrication of microfluidic paper-based analytical devices (µPADs). I also possess experience in data chemometric analysis using supervised and unsupervised learning techniques for classification, authentication and/or certification purposes.

Sabyasachi Pramanik is a professional IEEE member. He obtained a PhD in Computer Science and Engineering from Sri Satya Sai University of Technology and Medical Sciences, Bhopal, India. Presently, he is an Associate Professor, Department of Computer Science and Engineering, Haldia Institute of Technology, India. He has many publications in various reputed international conferences, journals, and book chapters (Indexed by SCIE, Scopus, ESCI, etc). He is doing research in the fields of Artificial Intelligence, Data Privacy, Cybersecurity, Network Security, and Machine Learning. He also serves on the editorial boards of several international journals. He is a reviewer of journal articles from IEEE, Springer, Elsevier, Inderscience, IET and IGI Global. He has reviewed many conference papers, has been a keynote speaker, session chair, and technical program committee member at many international conferences. He has authored a book on Wireless Sensor Network. He has edited 8 books from IGI Global, CRC Press, Springer and Wiley Publications.

Sangeeta Prasher is an educator at Kanya Maha Vidyalaya, Jalandhar with an impressive record of 16 years of experience in academia. Alongside her extensive pedagogical expertise, she boast an research portfolio, encompassing 3 book chapters and 3 books in the subject of Mathematical Physics. Her extensive research contributions are exemplified by an impressive record of 50 publications in journals. Dr. Prasher has also actively contributed to the academic community by designing numerous experiments for students, facilitating hands-on learning experiences. Her dedication to professional growth is evident from the number of national and international conferences attended. Furthermore, she has taken on leadership roles, conducting impactful workshops and conferences, reviewing many papers, and serving as an editor for two esteemed journals. She has guided two students for their Ph.D. thesis, 7 students for M.Phil reports and more than 25 students for their masters research projects. Dr. Prasher has completed a research project of SHI irradiation of free standing polyaniline films at IUAC, New Delhi. She has been invited as a resource person in many international conferences. Recently she was a resource for a workshop "learning physics through life" conducted for Navodaya physics teachers by the Government of India. Currently along with teaching, she is acting as Deputy Controller of Examinations.

Jyoti Rajput received her Ph.D. degree from Dr. B.R. Ambedkar NIT Jalandhar, Punjab, India in 2019. She is currently an Associate professor of Physics at Lovely Professional University, Punjab, India with 14 years of teaching experience. Her research focused areas deal with laser induced electron acceleration in vacuum and plasma, harmonic generation and Terahertz radiation. She has published around 33 research articles in various international SCI journals and presented her research work at various international conferences/workshops. She has assisted 5 PG and 3 UG research projects. She is supervising 6 PhD scholars. She is also a member of different renowned associations/societies e.g., PSSI, ISCA, etc. She is also acting as a reviewer in many Scopus Indexed Journals in the relevant field.

Harikirshnan S., a highly motivated physicist, holds a Master's Honours in Condensed Matter Physics from Lovely Professional University, Punjab, following a BSc Honours in Physics from Sri Sathya Sai Institute of Higher Learning, Brindavan, Bengaluru. Specializing in material science during his Master's, he devoted his research efforts to developing an innovative electrode material for energy storage devices. Harikirshnan's academic journey reflects a deep commitment to advancing scientific understanding and contributing to the field of condensed matter physics, particularly in the realm of material science and energy storage technologies.

Jacob Salazar-Rábago is a professor Jacob Salazar has undergraduate (2009) and postgraduate studies in Chemical Engineering (2015). In 2017 he joined as a full professor at the Autonomous University of Nuevo León, in the Faculty of Chemical Sciences, where he develops the following lines of research: Evaluation and development of Surface Physicochemical Processes in the sanitation of impacted aqueous effluents. Synthesis and characterization of materials with environmental applications. Mathematical modeling with a phenomenological foundation of physicochemical processes.

Haobam Samananda Singh, Ph. D. Scholar (DST INSPIRE -JRF), Department of Physics, National Insitute of Technology Manipur. Special Interest in Material Science: Piezoceramics, processing and characterizations.

Preeti Singh Bahadur obtained her Ph.D Degree in Physics from Barkatullah University, Bhopal, M.P., India and is currently working as a Associate Professor in the Department of Applied Sciences, Amity University, Greater Noida, U.P., India. She has published many research papers in various reputed journals, indexed in SCI/SCOPUS and other indexing. She has presented more than 25 research papers in various international and national conferences.

Eduardo Soto-Regalado is a full-time professor in the Chemical Engineering Department. He teaches bachelor, master, and doctoral studies, and has directed several Thesis in master's and doctoral studies regarding to adsorption process as well as the production of biochar from agro-wastes. These adsorbent materials were studied in the adsorption process to remove pollutants, such as heavy metals, and dyes, among others. He is recognized as a national researcher by the National Research System (México) level 1. His main research indicators are h-Index =20 (Google Scholar,) h-index =16 (Scopus), and i10-Index =24.

Index

A

B

C

D

E

S

T

V

W

X

www.ingramcontent.com/pod-product-compliance
Lightning Source LLC
LaVergne TN
LVHW062109300926
849161LV00010B/2409